Lecture Notes in Mathematics

Volume 2316

This series reports on new developments in all areas of mathematics and their applications - quickly, informally and at a high level. Mathematical texts analysing new developments in modelling and numerical simulation are welcome. The type of material considered for publication includes:

1. Research monographs
2. Lectures on a new field or presentations of a new angle in a classical field
3. Summer schools and intensive courses on topics of current research.

Texts which are out of print but still in demand may also be considered if they fall within these categories. The timeliness of a manuscript is sometimes more important than its form, which may be preliminary or tentative.

Titles from this series are indexed by Scopus, Web of Science, Mathematical Reviews, and zbMATH.

David J. Grynkiewicz

The Characterization of Finite Elasticities

Factorization Theory in Krull Monoids via Convex Geometry

Springer

David J. Grynkiewicz
Department of Mathematical Sciences
University of Memphis
Memphis, TN, USA

ISSN 0075-8434 ISSN 1617-9692 (electronic)
Lecture Notes in Mathematics
ISBN 978-3-031-14868-2 ISBN 978-3-031-14869-9 (eBook)
https://doi.org/10.1007/978-3-031-14869-9

This Springer imprint is published by the registered company Springer Nature Switzerland AG
The registered company address is: Gewerbestrasse 11, 6330 Cham, Switzerland

For Sarah, who lovingly endured my many long absences to "mathematics land" while completing this project.

Preface

This work is a research monograph combining the areas of Convex Geometry, the Combinatorics of infinite subsets of lattice points in $\mathbb{R}^d$, and the arithmetic of Krull Domains and, more generally, of Transfer Krull Monoids. The content is principally new research, developed over the course of several years and begun during a visit at the University of Graz in June/July 2015. The main and original motivation concerns the study of factorization in the very general setting of Transfer Krull Monoids, a broad setting extending the notion of Krull Domains and encompassing many different areas of Algebra where abstract questions of factorization can be studied. Krull Domains are themselves one of the most ubiquitous classes of rings in Commutative Algebra, being the higher dimensional analog of Dedekind Domains. Each Krull Domain H has a class group G, and this work is primarily concerned in the setting when G is finitely generated.

While the content is principally new research, the monograph is generally self-contained and accessible with only a general mathematics background, with algebraic material and definitions needed for the applications in Factorization Theory and Transfer Krull Monoids presented from basic principles for the non-specialist. Chapter 2 summarizes the details of nearly all the prerequisite results used. More specialized algebraic background is only necessary for understanding the motivating examples behind the notion of Transfer Krull Monoids, but generally not for the actual line of inquiry undertaken in the text, which is combinatorial in nature. Due to the large number of new concepts and notation introduced throughout the manuscript, an index referencing the location of the relevant definitions has been provided.

The motivating results in Factorization Theory are accomplished by methodically developing an extended theory generalizing specialized areas of Convex Geometry, then linked to the algebraic study of factorization via intermediary Combinatorics involving infinite subsets of lattice points in $\mathbb{R}^d$. The bulk of the text is spent developing the convex geometric concepts and theory to the point where they can be of service for the target applications in Factorization Theory. The theory developed here is quite separate from current lines of inquiry in Convex Geometry, and so is presented in stages of increasing abstractness as a natural outgrowth from fundamental geometric concepts. At its core, it may be viewed as an attempt to quantify what it means for an infinite set of lattice points to nonetheless possess finite-like behavior.

Returning in more detail to the original algebraic motivation, factorization into irreducibles, called atoms, generally fails to be unique in most settings, but there are various measures of how badly this can fail. One of the most classical and important is the elasticity

$$\rho(H) = \lim_{k\to\infty} \rho_k(H)/k,$$

where $\rho_k(H)$ is the maximal number of atoms in any re-factorization of a product of k atoms in the monoid H. Having finite elasticity is a key indicator that factorization, while not unique, is not completely wild. In the setting of Transfer Krull Monoids, the elasticities, as well as many other arithmetic invariants, are the same as those of a combinatorial monoid $\mathcal{B}(G_0)$ of zero-sum sequences, where $G_0 \subseteq G$ is an associated subset of an abelian group G. In the case of Krull Domains, G is the class group with G_0 the classes containing height one primes.

The overarching goal is to characterize when finite elasticity holds for any Krull Domain with finitely generated class group G, and indeed, also for the more general class of Transfer Krull Monoids (over a subset G_0 of a finitely generated abelian group G). We will see there is a minimal $s \leq (d+1)m$, where d is the torsion-free rank and m is exponent of the torsion subgroup, such that $\rho_s(H) < \infty$ implies $\rho_k(H) < \infty$ for all $k \geq 1$. This ensures $\rho(H) < \infty$ if and only if $\rho_{(d+1)m}(H) < \infty$.

The characterization is in terms of a simple combinatorial obstruction to infinite elasticity: there existing a subset $G_0^\diamond \subseteq G_0$ and global bound N such that there are no nontrivial zero-sum sequences with terms from $G_0^\diamond$, and every minimal zero-sum sequence has at most N terms from $G_0 \setminus G_0^\diamond$. Like many other combinatorial characterizations, it is fairly straightforward to show such conditions guarantee finite elasticity, but much less apparent that this is the only way it can occur. We give an explicit description of $G_0^\diamond$ in terms of the Convex Geometry of G_0 modulo the torsion subgroup $G_T \leq G$, and show that finite elasticity is equivalent to there being no positive $\mathbb{R}$-linear combination of the elements of this explicitly defined subset $G_0^\diamond$ equal to 0 modulo G_T. Additionally, we use our results to show finite elasticity implies the set of distances $\Delta(H)$, the catenary degree $\mathsf{c}(H)$ (for Krull Monoids), and a weakened form of the tame degree (for Krull Monoids) are all also finite, and that the Structure Theorem for Unions holds—four of the most commonly used measurements of structured factorization, after the elasticity.

The results for factorization in Transfer Krull Monoids are accomplished by developing an extensive theory in Convex Geometry generalizing positive bases. The convex cone generated by $X \subseteq \mathbb{R}^d$ is

$$\mathsf{C}(X) = \{\sum_{i=1}^{n} \alpha_i x_i \,:\, n \geq 0,\ x_i \in X,\ \alpha_i \in \mathbb{R},\ \alpha_i \geq 0\}.$$

A positive basis for $\mathbb{R}^d$ is a minimal by inclusion subset $X \subseteq \mathbb{R}^d$ such that $\mathsf{C}(X) = \mathbb{R}^d$. Positive bases were first introduced and studied in the mid-twentieth century, and the structural work initiated by Reay led to a simplified proof and strengthening of Bonnice and Klee's celebrated generalization of Carathéordory's Theorem. They have since found increasing importance in areas of applied mathematics. We show that the structural result of Reay can be extended to special types of complete simplicial fans, which we term Reay systems. We extend these results to a general theory dealing with infinite sequences of Reay systems, as well as their limit structures. The latter, while more complex, avoid the introduction of linear dependencies into the limit structure that were not originally present, circumventing the general obstacle that a limit of linearly independent sets can degenerate into linear dependence. The resulting theory is used to study a broad family of infinite subsets G_0 of a lattice $\Lambda \subseteq \mathbb{R}^d$ that exhibit various types of finite-like behavior, and which generalize the class of subsets having finite elasticity.

Memphis, TN, USA
May 2022

David J. Grynkiewicz

Contents

Chapter 1
Introduction

This work comprises extensive developments combining the areas of Convex Geometry, the Combinatorics of infinite subsets of lattice points in $\mathbb{R}^d$, and the arithmetic of Krull Domains and, more generally, of Transfer Krull Monoids. Our main and original motivation concerns the study of factorization in the very general setting of Transfer Krull Monoids. This is accomplished by developing an extended theory significantly generalizing prior work in Convex Geometry, then linked to the algebraic study of factorization via intermediary Combinatorics involving infinite subsets of lattice points in $\mathbb{R}^d$.

That the geometry of $\mathbb{R}^d$ could be used to study similar algebraic questions has precedent in the study of Primitive Partition Identities [39], closely related to Toric Varieties [25, 120], with such geometric ideas employed in alternative combinatorial language in [14, 49, 67]. However, we will be concerned with more subtly defined algebraic questions, for whom a direct connection with Convex Geometry is much less evident. Nonetheless, we will show that such a connection does exist, though it will require considerable preparatory work in Convex Geometry.

We continue with a detailed overview of the structure and subject matter of the monograph, concluding with a section by section breakdown of all content. The motivation and connections with other parts of mathematics are described in detail suitable to experts in their respective fields, including describing several large classes of well-studied algebraic settings where our main results are applicable. The level of specialist familiarity assumed in parts of this chapter is only temporary, with fuller definitions and descriptions of all prerequisites carried out in Chap. 2.

1.1 Convex Geometry

Linear Algebra over $\mathbb{R}^d$ is concerned with subspaces and linear combinations. Imposing the seemingly benign condition that only non-negative scalars be con-

D. J. Grynkiewicz, *The Characterization of Finite Elasticities*, Lecture Notes in Mathematics 2316, https://doi.org/10.1007/978-3-031-14869-9_1

sidered leads to the subfield of Convex Geometry concerned with Convex Cones

$$\mathsf{C}(X) = \{\sum_{i=1}^{n} \alpha_i x_i \,:\, n \geq 0,\ x_i \in X,\ \alpha_i \in \mathbb{R},\ \alpha_i \geq 0\}$$

and positive linear combinations, where $X \subseteq \mathbb{R}^d$. This natural line of geometric inquiry was initiated and studied in the mid twentieth century by various authors [20, 21, 36, 87, 103, 105–107, 115]. Part of their motivation lay in the relationship with Linear and Integer Programming, Game Theory and classical combinatorial aspects of Convex Geometry [34, 106], particularly the widely studied subfield having the theorems of Helly, Rado, Steinitz and Carathéodory as core tenets (see the surveys [37, 41] and the hundreds of references there listed). In this setting, the natural analog of a linear basis is a positive basis for $\mathbb{R}^d$, which is a minimal by inclusion subset $X \subseteq \mathbb{R}^d$ such that $\mathsf{C}(X) = \mathbb{R}^d$. Unlike linear bases, the cardinality of a positive basis is not determined by the dimension, instead satisfying the basic bounds $d + 1 \leq |X| \leq 2d$ for $d \geq 1$. While their cardinality is not unique, Reay [105] gave a basic structural result for positive bases. The structural description of Reay closely tied positive bases to the combinatorial foundations of Convex Geometry by first giving a simplified proof [106] of Bonnice and Klee's common generalization of the theorems of Carathéordory and Steinitz [20, 41], and then significantly generalizing the Bonnice-Klee Theorem itself [21, 41]. Positive bases have since found renewed interest also in applied areas of mathematics, particularly in Derivative-Free Optimization [35, 110].

The first half of this work is devoted to an expansive extension of the basic structural theory initiated by Reay for positive bases. We begin by first extending the basic theory of positive bases to specialized complete simplicial fans. A fan is a finite collection of polyhedral cones $\mathsf{C}(X) \subseteq \mathbb{R}^d$, so $X \subseteq \mathbb{R}^d$ is finite, each containing no nontrivial subspace. The faces of $\mathsf{C}(X)$ are sub-cones $\mathsf{C}(Y)$ with $Y \subseteq X$ obtained by intersecting $\mathsf{C}(X)$ with a hyperplane defining a closed half-space that contains $\mathsf{C}(X)$, and it is required that each face of a cone from the fan be an element of the fan, and that the intersection of any two cones in the fan be a face of each. The fan is complete if the union of all its cones is the entire space, and it is simplicial if each cone $\mathsf{C}(X)$ has X linearly independent, in which case the faces of $\mathsf{C}(X)$ correspond to its subsets $Y \subseteq X$. From a purely geometric perspective, complete simplicial fans are in bijective correspondence with starshaped spheres (specialized polyhedral complexes defined on the unit sphere), and rational fans (those whose vertices X come from a lattice) are central to the definition of Toric Varieties [25], which constitute an entire subfield of Algebraic Geometry. The aforementioned structural partitioning result of Reay implicitly gives each positive basis the structure of a complete simplicial fan associated to the partitioning. In Chap. 4, we methodically extend all basic theory of positive bases, including the structural result of Reay and several additional properties crucial to the continued development of the theory, more generally to a type of complete simplicial fan which we term a Reay system.

One of our main later goals is to study infinite subsets G_0 of a lattice in $\mathbb{R}^d$. We wish to better understand the radial "directions" in which the set G_0 escapes to infinity. For a simple example like

$$G_0 = \{(-1, y) : y \in \mathbb{Z},\ y \geq 1\} \cup \{(x, 0) : x \in \mathbb{Z},\ x \geq 1\} \cup \{(0, -1)\},$$

it is intuitively clear that the positive x and y axes constitute the unbounded "directions" of the set G_0. However, more general subsets $G_0 \subseteq \mathbb{R}^d$ can exhibit much more complicated behaviour. Even for $d = 2$, we could replace the elements $(-1, y)$ in the previously defined $G_0 \subseteq \mathbb{Z}^2$ with the elements $(-f(y), y)$, for some sub-linear, monotonically increasing and unbounded function $f(y)$, and still obtain a set having the positive x and y axes as its unbounded "directions", yet with a marked 2nd order unbounded drift in the negative x-axis direction occurring as the set escapes to infinity along the positive y-axis. As even this simple example illustrates, to make such a vague notion of unbounded direction precise, we will need very careful definitions. The framework we adopt for making this precise is lain out in Chap. 3, where we define the notion of an asymptotically filtered sequence with limit $\vec{u}$. With this framework in place, we can then use asymptotically filtered sequences of terms from G_0, as well as their associated limits $\vec{u}$, as a means of studying the directions $\vec{u}$ for which the set G_0 is unbounded.

Our much more expansive extension of the structural work of Reay involves showing the fundamental properties of Reay systems can be extended to a general notion of convergent families of Reay systems, which we term a virtual Reay system, whose component Reay systems are defined by collections of asymptotically filtered sequences. The resulting theory is presented in Chap. 6. A key obstacle to accomplishing this is the fact that our convergent family of Reay systems will generally not converge to another Reay system. This is exemplary of the basic fact that a convergent family of linearly independent sets can have a linearly dependent set as limit. For instance, the linearly independent sets $X_i = \{(-1, 0), (1, 1/i)\}$ converge to the linearly dependent set $X = \{(-1, 0), (0, 1)\}$ as $i \to \infty$. We overcome this fundamental obstacle by extending the notion of Reay system to a specialized variation of polyhedral complex that has highly constrained boundary requirements for its faces despite allowing them to be neither fully open nor closed, as required for an ordinary polyhedral complex. We term this more general limit structure an Oriented Reay System, and show in Chap. 5 how all basic properties and associated theory for ordinary Reay systems carries over into the more complete setting of oriented Reay systems, which is then pre-requisite for the later work in the setting of virtual Reay systems.

In total, Chaps. 3–6 constitute the mostly self-contained extension of the basic theory of Reay and others from the original context of positive bases to the more expansive setting of virtual and oriented Reay systems, developed to the point where we are able to define and deal with the basic quantities essential to our later applications for Krull Domains and Transfer Krull Monoids.

1.2 Krull Domains, Transfer Krull Monoids and Factorization

Krull Domains are one of the most ubiquitous classes of rings in Commutative Algebraic, being the higher dimensional analog of Dedekind Domains. Each Krull Domain D has a class group G, and this work is primarily concerned with the case when G is finitely generated. Note this does *not* mean the monoid D is finitely generated. In this setting, it is well-known that unique factorization into primes corresponds to having a trivial class group, and there is unique factorization of (divisorial) ideals into (height one) prime ideals [22, 23, 47, 86]. When G is nontrivial, unique factorization fails, and there may be multiple ways to factor a non-unit $a = u_1 \cdot \ldots \cdot u_k$ into irreducibles u_i, called atoms, with k the length of the factorization. In such case, the degree of wildness of factorization is measured by various arithmetic invariants, the most common of which we introduce in more general context momentarily. The most fundamental question, then, is whether these invariants are finite, as this is a key indication that factorization is not completely wild.

Worth noting, most of the arithmetic invariants controlling factorization depend only on the distribution of height-one prime ideals in the class group, meaning the subset $G_0 \subseteq G$ consisting of all classes containing height-one prime ideals is of prime concern. If G_0 is finite or D is a tame domain, then sets of potential factorization lengths are well-structured and all invariants describing their structure are finite [61, 63]. It is easy to see that G_0 generates G as a semigroup, while realization theorems tell us this is essentially the only restriction for which subsets G_0 can occur [62, Theorems 2.5.4 and 3.7.8] and [40, 75]. To overview the most commonly studied arithmetic invariants for factorizations, we do so in the fairly broad class of unit-cancellative semigroups.

Let H be a multiplicatively written, commutative semigroup with identity element. Then H is said to be cancellative if $ab = ac$ implies $b = c$, whenever $a, b, c \in H$, and unit cancellative if $ua = a$ implies u is a unit, whenever $u, a \in H$. If D is a commutative domain, then the semigroup of nonzero elements of D is cancellative, and if D is noetherian, then the monoids of all nonzero ideals and of all invertible ideals are unit-cancellative (with usual ideal multiplication) but not cancellative in general. While it may not immediately come to mind, an important example of factorization occurs for semigroups of (isomorphism classes of) modules, in which case the semigroup operation is the direct-sum. In this setting, factorization of a module corresponds to a direct-sum decomposition $M = M_1 \oplus \ldots \oplus M_k$ into a finite number of irreducible submodules. Many classical unique decomposition theorems for modules are then equivalent to factorization in the associated semigroup of modules being unique. Here, unit-cancellativity means $M \cong M \oplus N$ implies $N = 0$. Thus unit-cancellativity states that all modules have to be directly finite or, in other words, Dedekind-finite. This is a frequent property (e.g., valid for all finitely generated modules over commutative rings), that is weaker than cancellativity [76, 90].

Among the oldest and most important arithmetic invariants of factorization are the elasticities. They were first studied (using alternative terminology) for rings of integers in algebraic fields (see [65] for references to the old literature), with the now standard term elasticity introduced by Valenza [121] in 1990. They have since been studied by numerous authors. To list a few examples (focussing on surveys and more recent papers), see [1–4, 6, 7, 9, 10, 13, 15, 18, 19, 26, 28–30, 33, 38, 48, 50, 52, 55–57, 62, 63, 68, 71, 77, 83–85, 89, 93–97, 104, 114, 122, 123]. It can be defined as $\rho(H) = \sup\{\sup \mathsf{L}(a)/\min \mathsf{L}(a) : a \in H \text{ a non-unit}\}$, where $\mathsf{L}(a) = \{k \in \mathbb{N} : a = u_1 \cdot \ldots \cdot u_k$ from some atoms $u_i \in H\}$ denotes the length set of a, that is, the set of all possible factorization lengths of a written as a product of atoms, or it can be defined equivalently [62, Proposition 1.4.2] as $\rho(H) = \lim_{k\to\infty} \rho_k(H)/k$, where the k-th elasticity $\rho_k(H)$ denotes the maximum number of atoms in any re-factorization of a product of k-atoms. The set of distances $\Delta(H)$, defined as the minimum difference of two consecutive elements of $\mathsf{L}(a)$ as we range over all non-units $a \in H$, and the catenary degree $\mathsf{c}(H)$ (see Chap. 2.3) are two other of the most oft studied arithmetic invariants [62]. Yet another measure of well-behaved factorization are structural results for $\mathcal{U}_k(H)$, which is the set of all ℓ for which there are atoms $u_1, \ldots, u_k, v_1, \ldots, v_\ell \in H$ with $u_1 \cdot \ldots \cdot u_k = v_1 \cdot \ldots \cdot v_\ell$, so the possible re-factorization lengths of some product of k atoms. For many semigroups, it is known that these sets must be highly structured in the following sense. A finite set $X \subseteq \mathbb{Z}$ is said to be an almost arithmetic progression with difference $d \geq 1$ and bound $N \geq 0$ if $X = P \setminus Y$, where P is an arithmetic progression with difference d and $Y \subseteq P$ is a subset contained in the union of the first N terms from P and the last N terms from P. If there exists a constant $N \geq 0$ and difference $d \geq 1$ such that $\mathcal{U}_k(H)$ is an almost arithmetic progression with difference d and bound N for all sufficiently large k, then H is said to satisfy the Structure Theorem for Unions. As shown in [54, Theorem 4.2], if $\Delta(H)$ is finite and there is a constant $M \geq 0$ such that $\rho_{k+1}(G_0) - \rho_k(G_0) \leq M$ for all $k \geq 1$, then the Structure Theorem for Unions holds for H, meaning this additional structure is implied by sufficiently strong finiteness results for the elasticities and set of distances.

The general class of semigroup treated in this paper are Transfer Krull Monoids. We remark that they include all Krull Domains, Krull Monoids, and many examples of natural semigroups of monoids, with a non-cancellative monoid of modules over Bass rings that is Transfer Krull studied in [12]. We defer the formal definitions to Chap. 2.3, but continue with a detailed list illustrating the broad extent of the class of semigroups covered by our results (see also [60, pp 972] and [70, Example 4.2]).

Commutative Domains. A Noetherian domain D is Krull if and only if it is integrally closed, and the integral closure of a Noetherian domain is Krull by the Mori-Nagata Theorem. If D is Krull and finitely generated over $\mathbb{Z}$, then its class group G is finitely generated [100, Chapter 2, Corollary 7.7]. Moreover, D is a Krull Domain if and only if its multiplicative monoid of nonzero elements is a Krull Monoid.

Submonoids of Commutative Domains. Regular congruence monoids defined in Krull Domains are Krull Monoids [62, Section 2.11]. Let D be a factorial domain with quotient field K. Then the ring of integer-valued polynomials $\mathsf{Int}(D) = \{f \in K[X] : f(D) \subseteq D\}$ is not Krull, but the divisor closed submonoid generated by f, denoted $[\![f]\!]$, is a Krull Monoid for any polynomial $f \in \mathsf{Int}(D)$ [51, 108]. Moreover, if $f \in D[X]$, then the class group of $[\![f]\!]$ is finitely generated by Reinhart [109, Theorem 4.1].

Finitely Generated Krull Monoids. A finitely generated monoid is a Krull Monoid if and only if it is root closed, and finitely generated Krull Monoids have finitely generated class groups [62, Theorem 2.7.14]. The rank of the class group is studied in geometric terms in [102, Corollary 1]. Two interesting special cases are the following.

Normal Affine Monoids. These are found in combinatorial Commutative Algebra, and are equivalent to reduced finitely generated Krull Monoids with torsion free quotient group. For the class group of a finite normal monoid algebra $D[M]$, where D is a Noetherian Krull Domain and M a normal affine monoid, its class group is the direct sum of the class group of D and that of M [24, Theorem 4.60], meaning it is finitely generated whenever the class group of D is finitely generated.

Diophantine Monoids. A Diophantine monoid is an additive monoid of non-negative solutions to a system of linear Diophantine equations, with the rank of the class group, in terms of the defining matrix, studied in [32].

Semigroups of Modules. A semigroup $\mathcal{V}$ of modules over a ring R (as described earlier) closed under finite direct sums, direct summands, and isomorphisms is a reduced commutative semigroup. If the endomorphism rings $\mathsf{End}_R(M)$ are semilocal for all modules M in the semigroup, then $\mathcal{V}$ will be a Krull Monoid [42, Theorem 3.4]. Lists of modules having this property may be found in [43], and examples when the class group is finitely generated are listed in [8]. Conversely, every reduced Krull Monoid is isomorphic to a monoid of modules [45, Theorem 2.1] with details available in the monograph [44].

Normalizing Krull Monoids. A semigroup H is normalizing if $aH = Ha$ for all $a \in H$. Normalizing Krull Monoids occur when studying Noetherian semigroup algebras [92] and are Transfer Krull by Geroldinger [59, Theorems 4.13 an 6.5].

Noncommutative rings. Any bounded hereditary Noetherian prime ring D for which every stably free left D-ideal is free is a Transfer Krull Domain [118, Theorem 4.4]. Results giving other noncommutative families of Transfer Krull Monoids may be found in [5, 11, 116, 117].

Commutative domains close to a Krull Domain. In many cases, if a domain R is "close" to a Krull Domain D, then R will be a Transfer Krull Domain. Results giving examples of this type may be found in [70, Section 5] and [72, Proposition 4.6 and Theorem 5.8].

1.3 Zero-Sum Sequences

The topics discussed above under the headings for Convex Geometry and for Transfer Krull Monoids may seem so extremely disparate as to be completely unrelated. It is one of our main goals to show that the opposite is true, with there in fact being an extremely close connection between these seemingly unrelated topics. In order to establish this connection, we will need an intermediary combinatorial structure involving subsets G_0 of a lattice $\Lambda \subseteq \mathbb{R}^d$.

Let G be an abelian group and let $G_0 \subseteq G$ be a subset. Following the tradition of Combinatorial Number Theory, a sequence over G_0 is a finite, unordered string of elements from G_0. The collection $\mathcal{B}(G_0)$ consisting of all sequences over G_0 whose terms sum to zero, equipped with the concatenation operation, makes $\mathcal{B}(G_0)$ into a monoid which is Krull. More importantly, every Krull Monoid M has a transfer homomorphism to a monoid $\mathcal{B}(G_0)$ of zero-sum sequences that directly translates nearly all factorization properties of the original monoid M into the corresponding ones for $\mathcal{B}(G_0)$. We go into more detail in Sect. 2.3. For our purposes, this means all factorization questions considered in this paper for a general Krull Monoid reduce to the study of the combinatorial object $\mathcal{B}(G_0)$ by established machinery, allowing us to focus solely on $\mathcal{B}(G_0)$. A similar statement is true for Transfer Krull Monoids, though the increased broadness of this class means the corresponding transfer homomorphism between M and $\mathcal{B}(G_0)$ is necessarily weaker. For our purposes, this means that, while all our main results regarding factorization will be valid in the highly general context of Transfer Krull Monoids, a small number of consequences will only be valid for Krull Monoids. Thus we work entirely with the combinatorial object $\mathcal{B}(G_0)$ in this work with the corresponding results transferred automatically to Transfer Krull Monoids or Krull Monoids by established machinery (referenced in Sect. 2.3) that we need only use in passing. Note, arithmetic invariants for $\mathcal{B}(G_0)$ are generally abbreviated as $\rho(G_0) := \rho(\mathcal{B}(G_0))$, etc.

While the connection between factorization in Transfer Krull Monoids and the combinatorial object $\mathcal{B}(G_0)$ is well-known, any further connections with Convex Geometry were much more limited. The behavior of factorization when G is not finitely generated is generally quite unrestricted. As such, the study of $\mathcal{B}(G_0)$ for subsets G_0 of a finitely generated abelian group G is the general framework for the foundational finiteness questions for the arithmetic invariants of Transfer Krull Monoids. Note we do not assume the *monoid* $\mathcal{B}(G_0)$ is finitely generated, as finite elasticity is guaranteed rather trivially in such case. In this regard, little in the way of characterization was known apart from the basic case when G_0 is finite or the case when G has rank one [4, 71]. Since all factorization invariants are finite when G is finite (by basic arguments), it is natural to expect the chief obstacles for characterizing their finiteness would already be present for torsion-free abelian groups. As it will turn out (though not initially clear), this is precisely the case, allowing us to focus almost exclusively on the torsion-free case, later adapting these argument applied to G/G_T, where G_T is the torsion subgroup. As such, we are reduced to considering infinite subsets $G_0 \subseteq \mathbb{Z}^d$. The abelian group $\mathbb{Z}^d$ is the prototypical lattice in $\mathbb{R}^d$, meaning it is a full rank discrete subgroup of $\mathbb{R}^d$. While any full rank lattice $\Lambda \subseteq \mathbb{R}^d$ is isomorphic to $\mathbb{Z}^d$, it will be convenient to expand consideration to subsets G_0 of a general full rank lattice $\Lambda \subseteq \mathbb{R}^d$, allowing the use of the geometry of $\mathbb{R}^d$ for studying G_0.

1.4 Overview of Main Results

Our driving goal is to characterize finite elasticity for $\mathcal{B}(G_0)$ (and thus for Transfer Krull Monoids in general) for any subset G_0 of a finitely generated abelian group G. One natural way to prevent infinite elasticity is if there is a subset $G_0^\diamond \subseteq G_0$ and global bound N such that there are no nontrivial zero-sum sequences with terms from $G_0^\diamond$, and every minimal zero-sum sequence has at most N terms from $G_0 \setminus G_0^\diamond$. Such a condition trivially implies the elasticity $\rho(G_0)$ is finite, in turn implying all refined elasticities $\rho_k(G_0)$ are also finite (see Proposition 8.3). In the spirit of results like Hall's Matching Theorem, which shows that the simple combinatorial obstruction to a perfect matching characterizes when they exist, we will show that this basic combinatorial obstruction characterizes finite elasticity. Indeed, it characterizes when $\rho_{(d+1)m}(G_0)$ is finite, where m is the exponent of the torsion subgroup of G and d is the torsion-free rank, meaning there is a minimal $s \leq (d+1)m$ such that $\rho_s(G_0) < \infty$ implies $\rho_k(G_0) < \infty$ for all k. Moreover, we give an explicit description of $G_0^\diamond$ in terms of the Convex Geometry of G_0 modulo the torsion subgroup $G_T \leq G$, and show that finite elasticity is equivalent to there being no positive $\mathbb{R}$-linear combination of the elements of this explicitly defined subset $G_0^\diamond$ equal to 0 modulo G_T. What this means is that the initial question of finite elasticity, involving equations of positive $\mathbb{Z}$-linear combinations of lattice points, is equivalent to one involving positive $\mathbb{R}$-linear combinations of lattice points. Additionally, we obtain a weak structural description of the atoms in $\mathcal{B}(G_0)$, assuming finite elasticities, and use this to show finite elasticity implies that the set of distances $\Delta(G_0)$, the catenary $\mathsf{c}(G_0)$ and a weakened form of the tame degree are all also finite, and that there are no arbitrarily large gaps in the sequence $\{\rho_k(G_0)\}_{k=1}^{\infty}$, which implies that the Structure Theorem for Unions also holds.

The key means of accomplishing our characterization and many of its consequences is by an in-depth study in Chap. 7 of an ample class of subsets $G_0 \subseteq \Lambda \subseteq \mathbb{R}^d$, defined using our generalized theory of Reay systems given in Chaps. 3–6, that possess several finite-like properties despite being (in general) infinite. We term such sets finitary, study their finite characteristics in detail, and show that finite elasticity implies G_0 is finitary (with the converse failing). The broader class of finitary sets, as it will turn out, shares most of the same structural properties as sets G_0 with finite elasticities, while at the same time behaving better with respect to inductive arguments using quotients.

Unfortunately, there is no quick bypass straight to the applications in Factorization Theory given in Chap. 8. Nearly every chapter either builds upon or extends the foundations of the preceding chapters. This means most arguments in Chap. 8 rely heavily on all the developed machinery in Convex Geometry from the preceding chapters. One of the rare exceptions here is that not all finiteness properties given in Sect. 7.3 are required. Specifically, a reader interested only in the Factorization Theory applications given in Chap. 8 can pass over the Finiteness from Above property of finitary sets given in Theorem 7.29, while the Finiteness from Below property of finitary sets, given in Theorem 7.32, serves as a gradual lead into the more refined finiteness properties surrounding the notion of minimal types later given in Proposition 7.33, and it is this stronger refinement which is later utilized in the arguments of Chap. 8. Additionally, while much of the material in Chap. 3 is presented for finite unions of polyhedral cones, as the arguments extend to this setting with little added effort, the case of a single polyhedral cone is all that is required.

We conclude the introduction by outlining the main results and content chapter by chapter. As a general remark, while somewhat subjective, we have labeled results as lemmas when they are more technical, often with highly restricted hypotheses, and generally needed as part of a larger proof, as propositions when they encode basic or fundamental properties of the concepts being explored, even when the proof may be quite involved, and as theorems when we wish to emphasize that it is one of our culminating results. Theorems that we wish to especially highlight will be referred to as main theorems.

Chapter 2. We introduce the basic notation and preliminaries for Convex Geometry, partially ordered sets (posets), lattices, zero-sum sequences, and Transfer Krull Monoids, as well as the requisite asymptotic notation. A more general notion of rational sequence over G_0 is introduced, where the multiplicities of terms are allowed to be non-negative rational numbers rather than integers, which will play a crucial role in later parts of Chap. 8. Precursors to this innovation may be found in [14].

Section 3.1. The concept of an asymptotically filtered sequence with limit $\vec{u}$ and related notation and definitions are introduced along with the basic properties of these definitions. This will be our main tool for measuring the directions in which a set escapes to infinity.

Section 3.2. The companion concepts of encasement and minimal encasement of $\vec{u}$, as well as several basic properties, are given in the context of finite unions of polyhedral cones.

Section 3.3. The notion of a set X being bound to another set Y, meaning every point in X is within some globally bounded distance of some point from Y, is introduced. The Chapter culminates with Theorem 3.14, which gives a characterization of X being bound to Y, assuming Y is a finite union of polyhedral cones, in terms of encasement of limits $\vec{u}$ of asymptotically filtered sequences. We will mostly need the results of Sects. 3.2 and 3.3 when Y is a polyhedral cone.

Section 4.1. We characterize some basic non-degeneracy assumptions for arithmetic properties of G_0 in terms of the geometry of $\mathbb{R}^d$.

Section 4.2. The concept of an elementary atom, which is a slight modification to the definition as presented in [14], is given. It corresponds to the notion of atom when factoring using rational sequences rather than ordinary sequences, so using multiplicities from $\mathbb{Q}$ rather than $\mathbb{Z}$. A version of Carathéordory's Theorem, formulated for rational sequences, is given in Theorem 4.7. Positive bases are introduced, with a lengthy list of equivalent defining conditions given in Proposition 4.8, including equivalent formulations involving elementary atoms.

Section 4.3. The notion of a Reay system and related concepts are introduced, and the basic theory of positive bases extended to this context. Proposition 4.14 corresponds to a refined statement of the structural result of Reay for positive bases, extended to Reay systems. The section concludes with Proposition 4.15, which contains the technical stability properties of complete simplicial fans, needed for handling some delicate arguments in Proposition 6.12.2 and Lemma 7.4.2, and thus in turn for Theorem 7.8 as well.

Chapter 5. The notion of Reay system is generalized to that of oriented Reay systems, with all related notation introduced. The theory and results for Reay systems presented in Sect. 4.3 are extended to this more general context.

Chapter 6. The notion of oriented Reay system is now extended to that of virtual Reay systems with all related notation introduced. The theory and results for ordinary and oriented Reay systems presented in Sects. 4.3 and 5 are now extended to this more general context.

Section 7.1. Three equivalent definitions for the subset $G_0^\diamond \subseteq G_0$, which plays the crucial role in the characterization of finite elasticities, are given in Proposition 7.2. The crucial notion of a finitary set is defined using the language of Chaps. 3–6. The remainder of the section is devoted to an in-depth study of finitary sets. Theorem 7.16 gives a condition on $G_0^\diamond$ that implies G_0 is finitary. In particular, $0 \notin \mathsf{C}^*(G_0^\diamond)$ implies G_0 is finitary. Combining this result with Proposition 8.3 gives us the means to use finitary sets, along with all their associated properties developed in Chap. 7, during our characterization of finite elasticities. Theorem 7.13 gives a 4th characterization of the set $G_0^\diamond$ in terms of the multiplicities of elements in elementary atoms, valid for finitary sets. This characterizes the elements of $G_0^\diamond$ in terms of positive $\mathbb{Q}$-linear primitive partition identities equal to zero. The section concludes with Theorem 7.8, giving geometric restrictions for the distribution of elements in a finitary set G_0. In particular, it ensures that a finitary set G_0 has a linearly independent subset $X \subseteq G_0^\diamond \subseteq G_0$ such that G_0 is bound to $-\mathsf{C}(X)$, meaning the set G_0 must be concentrated around the simplicial cone $-\mathsf{C}(X)$.

Section 7.2. The goal of this section and the next two is to derive more structural information regarding certain virtual Reay systems defined over a finitary set. To do this, the notion of a series decomposition of a purely virtual Reay system is introduced (inspired by the Jordan-Hölder Theorem and other similar results in Algebra), along with detailed notation and concepts.

Section 7.3. Series decompositions are used to study the structure of the sets $X(G_0)$ and $\mathfrak{X}(G_0)$ associated to a finitary set in Sect. 7.2. Four of the key finiteness properties of finitary sets are given in Theorems 7.24.1, 7.24.2, 7.29

and 7.32. Included is the striking property given in Proposition 7.31 that a finitary set G_0 has a finite subset $Y \subseteq G_0$ such that every zero-sum sequence with terms from G_0 must use at least one term from Y.

Section 7.4. The notions of lattice types and minimal types, as well as related notation and concepts, are introduced. Their crucial finiteness and interchangeability properties, needed for many results in Chap. 8, are given in the sole result of the section: Proposition 7.33.

Section 8.1. The core of this section is Theorem 8.6. Buried within the statement of Theorem 8.6 is a multi-dimensional generalization of a result of Lambert [99] and [71, Lemma 4.3]. The result of Lambert handles the case of dimension $d = 1$, where a novel variation for the argument giving the upper bound $\mathsf{D}(G) \leq |G|$ for the Davenport constant yields the result. Though, in the end, the proof of Theorem 8.6 was derived as a natural extension of the theory developed in Chaps. 3–7 rather than as a generalization of the argument of Lambert, this innocuous one-dimensional result nonetheless served as the inspiration for a good portion of the material contained in this work. Theorem 8.6 plays the central role in our characterization of finite elasticities, and its proof involves a detailed algorithm utilizing the theory developed for finitary sets in Chap. 7. The beginning of the section contains several basic results involving elasticity and $G_0^\diamond$. Corollary 8.8 contains a 5th characterization of the set $G_0^\diamond$ in terms of the multiplicities of elements in ordinary atoms, valid for finitary sets. This characterizes the elements of $G_0^\diamond$ in terms of positive $\mathbb{Z}$-linear primitive partition identities equal to zero. Theorem 8.9 is our characterization result in the torsion-free setting. Corollary 8.10 shows that finite elasticity is equivalent to a variation on elasticity defined using the more basic elementary atoms rather than ordinary atoms.

Section 8.2. This section contains all the additional consequences for factorization when G is torsion-free. Theorem 8.13 gives a weak structural characterization of the atoms over G_0 under the assumption of finite elasticities. Its principal use is to simulate globally bounded support for an atom. The number of distinct elements appearing in an atom over G_0 can generally be an arbitrarily large number. Omitting some details, Theorem 8.13 shows that there are a finite number of "types" of elements, with elements of the same type behaving in the same, well-controlled manner as given by Proposition 8.12, so that if one equates all elements of the same type in an atom (with some elements belonging to no type and left alone), then the total number of distinct elements/types in an atom is globally bounded. While $\mathcal{B}(G_0)$ may not be tame in general, we introduce a weaker notion of tameness in Theorem 8.14 which is finite (assuming the elasticities are finite). While this notion of tameness is weaker than what is generally studied in the literature, it is nonetheless sufficient, under an assumption of finite elasticities, to show that the set of distances and catenary degree are finite, as well as that the Structure Theorem for Unions holds. This is done in Theorem 8.15. At the end of Sect. 8.2, a more detailed summary of all the information derived for G_0 assuming finite elasticities is compiled.

Section 8.3. This final (comparatively short) section extends the results of Chap. 8 from the case G torsion-free to general finitely generated abelian groups, and sets results in the more general framework of (Transfer) Krull Monoids. It is set separate from other sections so that the main ideas, already quite involved, can be presented in the more natural torsion-free setting without the few additional technical issues that must be handled for the general case. Proposition 8.17 contains the basic observations that allow finiteness of arithmetic invariants to be transferred between G and G/G_T, where $G_T \leq G$ is the torsion subgroup. In Main Theorem 8.19, a technical modification to the torsion-free argument extends the validity of Theorem 8.14 from the torsion-free case to the general case (for Krull Monoids), establishing the finiteness of the weak tame degree. Main Theorem 8.21 likewise extends Theorem 8.15 to the general case, establishing the finiteness of the set of distances (for Transfer Krull Monoids), the finiteness of the Catenary degree (for Krull Monoids), and that the Structure Theorem for Unions holds (for Transfer Krull Monoids). Proposition 8.20 contains the basic argument showing that having finite elasticities implies a Krull Monoid is locally tame. Proposition 8.23 extends the two alternative characterizations of $G_0^\diamond$ given in Corollary 8.8 to the general case. Proposition 8.25 extends Proposition 7.31, showing that finite elasticities ensures there is finite subset $Y \subseteq G_0$ such that every zero-sum sequence with terms from G_0 must use at least one term from Y. Finally, Main Theorem 8.26 extends Theorem 8.9 to the general case (for Transfer Krull Monoids), giving our main characterization result for finite elasticities for general G. The section concludes with a more detailed summary of all the information derived for G_0 assuming finite elasticities.

Chapter 2
Preliminaries and General Notation

2.1 Convex Geometry

Variables introduced with inequalities, e.g., $x \geq 1$, are generally assumed to be integers unless otherwise stated. The letters $\alpha, \beta, \gamma, \delta, a, b$ and c generally indicate real/rational numbers. Intervals are also discrete unless otherwise stated, so $[x, y] = \{z \in \mathbb{Z} : y \leq x \leq z\}$ for $x, y \in \mathbb{R}$. We use $\subset$ to denote proper inclusion.

For $x \in \mathbb{R}^d$, we let $\|x\|$ denote the usual Euclidian L_2-norm. Then $\mathsf{d}(x, y) = \|x - y\|$ is the distance between two points $x, y \in \mathbb{R}^d$. Given two nonempty sets $X, Y \subseteq \mathbb{R}^d$, we let

$$\mathsf{d}(X, Y) = \inf\{\mathsf{d}(x, y) : x \in X, y \in Y\}$$

and define $X + Y = \{x + y : x \in X, y \in Y\}$, $-X = \{-x : x \in X\}$ and $X - Y = X + (-Y)$. For a real number $\epsilon > 0$, let $B_\epsilon(x)$ denote an open ball of radius ϵ centered at x.

Let $X \subseteq \mathbb{R}^d$ be a subset of the d-dimensional Euclidian space, where $d \geq 0$. Then

$$\mathbb{Z}\langle X\rangle = \{\sum_{i=1}^{n} \alpha_i x_i : n \geq 0,\ x_i \in X,\ \alpha_i \in \mathbb{Z}\}$$

$$\mathbb{Q}\langle X\rangle = \{\sum_{i=1}^{n} \alpha_i x_i : n \geq 0,\ x_i \in X,\ \alpha_i \in \mathbb{Q}\}, \quad \text{and}$$

$$\mathbb{R}\langle X\rangle = \{\sum_{i=1}^{n} \alpha_i x_i : n \geq 0,\ x_i \in X,\ \alpha_i \in \mathbb{R}\},$$

D. J. Grynkiewicz, *The Characterization of Finite Elasticities*, Lecture Notes in Mathematics 2316, https://doi.org/10.1007/978-3-031-14869-9_2

so $\mathbb{R}\langle X\rangle$ denotes the linear subspace spanned by X. Note $\mathbb{R}\langle\emptyset\rangle = \{0\}$. For $x_1,\ldots,x_n \in \mathbb{R}^d$, we let $\mathbb{R}\langle x_1,\ldots,x_n\rangle = \mathbb{R}\langle\{x_1,\ldots,x_n\}\rangle$, and likewise define $\mathbb{Q}\langle x_1,\ldots,x_n\rangle$ and $\mathbb{Z}\langle x_1,\ldots,x_n\rangle$. For a subspace $\mathcal{E} \subseteq \mathbb{R}^d$, we let $\dim\mathcal{E}$ denote the dimension of $\mathcal{E}$ and let $\mathcal{E}^\perp \subseteq \mathbb{R}^d$ denote the orthogonal complement to $\mathcal{E}$. A subset $X = \{x_1,\ldots,x_n\} \subseteq \mathbb{R}^d$ of size $n \geq 0$ is said to be **linearly independent modulo** $\mathcal{E}$ if $\sum_{i=1}^n \alpha_i x_i \in \mathcal{E}$ with all $\alpha_i \in \mathbb{R}$ implies every $\alpha_i = 0$. Equivalently, $\pi(X)$ is a linearly independent set of size $|X| = |\pi(X)|$ (so π is injective on X), where $\pi : \mathbb{R}^d \to \mathcal{E}^\perp$ denotes the orthogonal projection.

Let $\mathbb{Z}_+$, $\mathbb{Q}_+$ and $\mathbb{R}_+$ denote the set of *non-negative* integer, rational and real numbers, respectively. A **positive linear combination** of some $x_1,\ldots,x_n \in \mathbb{R}^n$ is an expression of the form $\sum_{i=1}^n \alpha_i x_i$ with all $\alpha_i \in \mathbb{R}_+$. A positive linear combination is **nontrivial** if some $\alpha_i > 0$, and it is **strictly positive** if every $\alpha_i > 0$. A set $X \subseteq \mathbb{R}^d$ is **convex** if $x, y \in X$ implies $\alpha x+(1-\alpha)y \in X$ for all real numbers $\alpha \in [0,1]$, that is, all points on the line segment between x and y lie in X. A set $C \subseteq \mathbb{R}^d$ is a **cone** if $x \in C$ implies $\alpha x \in C$ for all real numbers $\alpha > 0$, that is, the entire ray $\mathbb{R}_+x$ is contained in $C \cup \{0\}$. A **convex cone** is a set $C \subseteq \mathbb{R}^d$ that is both a cone and convex, i.e., C is closed under nontrivial positive linear combinations: $x, y \in C$ implies that $\alpha x + \beta y \in C \cup \{0\}$ for all $\alpha, \beta \in \mathbb{R}_+$. The convex cones spanned by a subset $X \subseteq \mathbb{R}^d$ are defined as

$$\mathsf{C}_{\mathbb{Z}}(X) = \{\sum_{i=1}^n \alpha_i x_i : n \geq 0,\ x_i \in X,\ \alpha_i \in \mathbb{Z}_+\},$$

$$\mathsf{C}_{\mathbb{Q}}(X) = \{\sum_{i=1}^n \alpha_i x_i : n \geq 0,\ x_i \in X,\ \alpha_i \in \mathbb{Q}_+\}, \qquad \text{and}$$

$$\mathsf{C}(X) = \mathsf{C}_{\mathbb{R}}(X) = \{\sum_{i=1}^n \alpha_i x_i : n \geq 0,\ x_i \in X,\ \alpha_i \in \mathbb{R}_+\}.$$

If X is finite, then $\mathsf{C}(X)$ is a closed convex cone. We likewise define

$$\mathsf{C}^*(X) = \{\sum_{i=1}^n \alpha_i x_i : n \geq 1,\ x_i \in X,\ \alpha_i \in \mathbb{R}_+,\ \alpha_i > 0\}.$$

Note $\mathsf{C}^*(X)$ differs from $\mathsf{C}(X)$ only in that $0 \in \mathsf{C}(X)$ is trivial while $0 \in \mathsf{C}^*(X)$ is not. As special cases, we have $\mathsf{C}(\emptyset) = \{0\}$ and $\mathsf{C}^*(\emptyset) = \emptyset$. For $x_1,\ldots,x_n \in \mathbb{R}^d$, we use the abbreviations $\mathsf{C}(x_1,\ldots,x_n) = \mathsf{C}(\{x_1,\ldots,x_n\})$, $\mathsf{C}_{\mathbb{Q}}(x_1,\ldots,x_n) = \mathsf{C}_{\mathbb{Q}}(\{x_1,\ldots,x_n\})$, $\mathsf{C}_{\mathbb{Z}}(x_1,\ldots,x_n) = \mathsf{C}_{\mathbb{Z}}(\{x_1,\ldots,x_n\})$ and also $\mathsf{C}^*(x_1,\ldots,x_n) = \mathsf{C}^*(\{x_1,\ldots,x_n\})$.

Given a subset $X \subseteq \mathbb{R}^d$, we let X° denote the **relative interior** of X, which is the interior of X relative to the topological subspace $\mathbb{R}\langle X - X\rangle + X$ (called the affine subspace spanned by X), let $\partial(X)$ denote the **relative boundary** of X, which is the boundary of X relative to the topological subspace $\mathbb{R}\langle X - X\rangle + X$, let $\overline{X}$

denote the **closure** of X (which is the same in both $\mathbb{R}\langle X - X\rangle + X$ and $\mathbb{R}^d$), and let $\mathsf{int}(X)$ denote the **interior** of X in $\mathbb{R}^d$. If $0 \in \overline{X}$ (e.g., if X is a convex cone), then $\mathbb{R}\langle X - X\rangle + X = \mathbb{R}\langle X\rangle$ is simply the linear subspace spanned by X. Note

$$\mathbb{R}^\circ_+ = \{\alpha \in \mathbb{R} : \alpha > 0\}$$

is the set of strictly positive real numbers. We set

$$\mathsf{C}^\circ(X) := \mathsf{C}(X)^\circ \quad \text{and} \quad \mathsf{C}^\circ(x_1, \ldots, x_n) := \mathsf{C}(x_1, \ldots, x_n)^\circ.$$

A cone $C \subseteq \mathbb{R}^d$ of the form $C = \mathsf{C}(X)$ for some finite subset $X \subseteq \mathbb{R}^d$ is called a finitely generated or **polyhedral cone**. If X is linearly independent, then $\mathsf{C}(X)$ is a **simplicial cone**, in which case $\mathsf{C}^\circ(X) = \{\sum_{i=1}^n \alpha_i x_i : \alpha_i > 0\}$, where $X = \{x_1, \ldots, x_n\}$ [119, Theorem 4.17]. A co-dimension one subspace $\mathcal{H} \subset \mathbb{R}^d$ naturally divides $\mathbb{R}^d$ into two closed half spaces $\mathcal{H}_+$ and $\mathcal{H}_-$ whose common boundary is $\mathcal{H}$.

The texts [34, 46, 74, 80, 81, 119] contains many of the basic properties of convexity that we will regularly use with little further reference. The following are several important highlights. We remark that the slightly more general version of Carathéodory's Theorem given below is rather difficult to find so stated despite encapsulating what is actually proved, particularly the portion regarding the representation of 0. The latter can be derived with ease from the argument used to prove Carathéodory's Theorem or from the basic theory of positive bases (see Chap. 4). It is also a special consequence of Theorem 4.7, whose brief proof we provide. Let $X \subseteq \mathbb{R}^d$.

- (Carathéodory's Theorem) If $x \in \mathsf{C}^*(X)$, then there exists a subset $Y \subseteq X$ with $|Y| \leq d + 1$ and $x \in \mathsf{C}^*(Y)$. Moreover, if $x \neq 0$, then Y is linearly independent, and if $x = 0$, then any proper subset of Y is linearly independent [119, Theorem 4.27].
- $\mathsf{C}(X) \cap -\mathsf{C}(X)$, called the **lineality space** of $\mathsf{C}(X)$, is the unique maximal linear subspace contained in $\mathsf{C}(X)$ [119, Theorem 4.15], and is nontrivial if and only if $0 \in \mathsf{C}^*(X \setminus \{0\})$.
- (Minkowski-Weyl Theorem) $\mathsf{C}(X)$ is a polyhedral cone if and only if $\mathsf{C}(X)$ is the intersection of a finite number of closed half-spaces [34, Theoerem 11.9].
- (Duality) A closed convex set $X \subset \mathbb{R}^d$ is either the empty set or the intersection of all closed half-spaces that contain X [74, Theorem 2.7].
- If $\mathsf{C}(X) \neq \mathbb{R}\langle X\rangle$, then $\mathsf{C}(X)$ lies in a closed half-space of $\mathbb{R}\langle X\rangle$ [46, Corollary 1].
- (Relative Interior and Closure) If $X \subseteq \mathbb{R}^d$ is convex, then $X^\circ = \overline{X}^\circ$ and $\overline{X} = \overline{X^\circ}$, with both these sets convex [119, Theorem 2.38] [119, Theorem 2.35] [119, Corollary 2.22].
- If $X \subseteq \mathbb{R}^d$ is convex and $X \neq \emptyset$, then $X^\circ \neq \emptyset$ [119, Corollary 2.18].

2.2 Lattices and Partially Ordered Sets

A **lattice** is a discrete subgroup $\Lambda \leq \mathbb{R}^d$, where $d \geq 0$, meaning *any bounded subset of* $\mathbb{R}^d$ *contains only finitely many lattice points* [27, 78, 79]. In particular, any convergent sequence of lattice points stabilizes, that is, if $\{x_i\}_{i=1}^{\infty}$ is a convergent sequence of lattice points $x_i \in \Lambda$ with $\lim_{i\to\infty} x_i = x$, then $x \in \Lambda$ and $x_i = x$ for all sufficiently large i. As is well-known [27, Theorem VI] and [79, Theorem 1], being a lattice is equivalent to there existing a linearly independent generating subset $X \subseteq \Lambda$, so $\mathbb{Z}\langle X\rangle = \Lambda$ with $X \subseteq \mathbb{R}^d$ linearly independent. Such a subset $X \subseteq \Lambda$ is called a **lattice basis**, and the parallelepiped generated by the vectors in a lattice basis is called its **fundamental parallelepiped** . The rank of the lattice Λ is $n = \dim \mathbb{R}\langle\Lambda\rangle = |X|$, and we say Λ has **full rank** in $\mathbb{R}^d$ if $n = d$, that is, $\mathbb{R}\langle\Lambda\rangle = \mathbb{R}^d$. In particular, $\Lambda \cong \mathbb{Z}^n$ for a rank n lattice Λ.

The following basic consequence of the Smith Normal Form [101, Theorem III.7.8] [79, Theorem 2] will be very important for us. We remark that the hypothesis that the kernel be generated by a subset of lattice points is quite necessary.

Proposition 2.1 *Let* $\Lambda \subseteq \mathbb{R}^d$ *be a full rank lattice and let* $\pi : \mathbb{R}^d \to \mathbb{R}^d$ *be a linear transformation with* $\ker \pi = \mathbb{R}\langle X\rangle$ *for some* $X \subseteq \Lambda$. *Then* $\pi(\Lambda)$ *is a lattice having full rank in the subspace* $\mathsf{im}\,\pi$.

Proof By passing to a subset of X, we can w.l.o.g. assume the elements $x_1, \ldots, x_s \in X \subseteq \Lambda$ are linearly independent lattice points, and thus form a linear basis for $\ker \pi = \mathbb{R}\langle X\rangle$ with $\mathbb{Z}\langle X\rangle \leq \Lambda$ a sublattice of rank s. Via the Smith Normal form, we can find a lattice basis $e_1, \ldots, e_d \in \Lambda$ and integers $m_1 \mid \ldots \mid m_d$, where $m_i > 0$ for $i \leq s$ and $m_i = 0$ for $i > s$, such that $y_i = m_i e_i$ for all $i \leq s$ with $\mathbb{Z}\langle y_1, \ldots, y_s\rangle = \mathbb{Z}\langle X\rangle = \mathbb{Z}\langle x_1, \ldots, x_s\rangle$. Since $\ker \pi = \mathbb{R}\langle x_1, \ldots, x_s\rangle = \mathbb{R}\langle y_1, \ldots, y_s\rangle = \mathbb{R}\langle m_1e_1, \ldots, m_se_s\rangle$, it is now clear that $\pi(e_i) = 0$ for $i \leq s$ while $\pi(e_{s+1}), \ldots, \pi(e_d) \in \pi(\Lambda)$ are distinct linearly independent elements which generate the subgroup $\pi(\Lambda)$. This shows $\pi(\Lambda) \leq \mathsf{im}\,\pi$ is a lattice, and since Λ has full rank in $\mathbb{R}^d$, it follows that $\pi(\Lambda)$ has full rank in $\mathsf{im}\,\pi$. □

A **partially ordered set (poset)** $(P, \preceq)$ is a set P together with a partial order $\preceq$, so $\preceq$ is a transitive, reflexive, anti-symmetric relation on P [112]. In such case, we use $x \prec y$ to indicate $x \preceq y$ but $x \neq y$. Any subset $X \subseteq P$ is also a partially ordered set using the partial order $\preceq$ inherited from P, and we implicitly consider subsets of posets to be posets using the inherited partial order. A maximal element $x \in P$ is one for which there is no $y \in P$ with $x \prec y$. Likewise, a minimal element $x \in P$ is one for which there is no $y \in P$ with $y \prec x$. We let

$$\mathsf{Min}(P) \subseteq P \quad \text{and} \quad \mathsf{Max}(P) \subseteq P$$

denote the set of minimal and maximal elements of P, respectively. Given a subset $X \subseteq P$, we let

$$\downarrow X = \{y \in P : y \preceq x \text{ for some } x \in X\} \quad \text{and}$$
$$\uparrow X = \{y \in P : x \preceq y \text{ for some } x \in X\}$$

denote the down-set and up-set generated by X, respectively. We likewise define $\downarrow x = \downarrow\{x\}$ and $\uparrow x = \uparrow\{x\}$ for $x \in X$. An **anti-chain** is a subset $X \subseteq P$ such that no two distinct elements of X are comparable. A **chain** is a totally ordered subset of P. An **ascending chain** is a sequence $x_1 \preceq x_2 \preceq \ldots \preceq x_n \preceq \ldots$ with $x_i \in P$. Likewise, a **descending chain** is a sequence $x_1 \succeq x_2 \succeq \ldots \succeq x_n \succeq \ldots$ with $x_i \in P$. In both cases, the chain can either be finite (stopping at n) or infinite, and the chain is **strict** if each $\preceq$ or $\succeq$ in the chain is always a strict inclusion $\prec$ or $\succ$.

If $(P, \preceq_1)$ and $(Q, \preceq_2)$ are both posets, then $P \times Q$ is also a poset using the product partial order: $(x, y) \preceq (x', y')$ when $x \preceq_1 x'$ and $y \preceq_2 y'$. An important example of a partially ordered set is $P = \mathbb{Z}_+^d$ equipped with the product partial order $(x_1, \ldots, x_d) \preceq (y_1, \ldots, y_d)$ when $x_i \leq y_i$ for all i. It is easily seen that the poset $\mathbb{Z}_+^d$ has no infinite strictly descending chain. Indeed, there are at most $x_1 + \ldots + x_d + 1$ elements in any strictly descending chain whose first element is $(x_1, \ldots, x_d)$. A well-known consequence of Hilbert's Basis Theorem [101, Theorem 4.1] is that the poset $\mathbb{Z}_+^d$ contains no infinite anti-chain: The points of $\mathbb{Z}_+^d$ correspond naturally to the monomials in $\mathbb{F}_2[x_1, \ldots, x_d]$, and if $X \subseteq \mathbb{Z}_+^d$ were an infinite anti-chain, then the monomial ideal generated by the monomials corresponding to the elements from X would not be finitely generated, contrary to Hilbert's Basis Theorem.

We will be interested in posets which neither contain an infinite anti-chain nor an infinite strictly descending chain. Such posets are called **well-quasi-orderings** [98]. We include the following basic propositions about such posets (mentioned in [98] without proof) for completeness, which together give an alternative proof (as opposed to Hilbert's Basis Theorem) that any subset $X \subseteq \mathbb{Z}_+^d$ has only a finite number of minimal points (a result known as Dickson's Lemma [62, Theorem 1.5.3]).

Proposition 2.2 *If $(P, \preceq)$ is a poset that contains neither infinite anti-chains nor infinite strictly descending chains, then $\mathsf{Min}(X)$ is finite and $\uparrow\mathsf{Min}(X) \cap X = X$ for any subset $X \subseteq P$.*

Proof The set of minimal points in a poset is an anti-chain and the hypothesis that P contains no infinite anti-chain inherits to X. Thus $\mathsf{Min}(X)$ is finite. If $Y = X \setminus \uparrow\mathsf{Min}(X)$ is nonempty, then $Y = \downarrow Y \cap X \subseteq X$ can contain no minimal point, and we can recursively select elements from Y to form an infinite descending chain in X, and thus also in P, contrary to hypothesis. □

Proposition 2.3 *If both $(P, \preceq)$ and $(Q, \preceq')$ are posets that contain neither infinite anti-chains nor infinite strictly descending chains, then the poset $P \times Q$, equipped* *with the product partial order, also contains no infinite anti-chains nor infinite strictly descending chains.*

Proof Assume to the contrary that $\{(x_i, y_i)\}_{i=1}^{\infty}$ is a strictly descending chain. Then, per definition of the product partial order, both $\{x_i\}_{i=1}^{\infty}$ and $\{y_i\}_{i=1}^{\infty}$ must be descending chains. Since neither P nor Q contains infinite strictly descending chains, it follows that there exists some index M such that $x_i = x_j$ for all $i, j \geq M$, and likewise some index N such that $y_i = y_j$ for all $i, j \geq N$. Hence $(x_i, y_i) = (x_j, y_j)$ for all $i, j \geq \max\{N, M\}$, contradicting that $\{(x_i, y_i)\}_{i=1}^{\infty}$ is a strictly descending chain. This shows that $P \times Q$ contains no infinite strictly descending chain.

Next assume to the contrary that $Z = \{(x_1, y_1), (x_2, y_2), \ldots, \}$ is an infinite anti-chain. Let $X = \{x_1, x_2, \ldots, \}$ and $Y = \{y_1, y_2, \ldots\}$. If X is finite, then the pigeonhole principle ensures that there is some $x \in X$ that occurs as the first coordinate of an infinite number of pairs from Z. Thus, by passing to a subset, we can w.l.o.g. assume $Z = \{(x, y_1), (x, y_2), \ldots, \}$, in which case Z can only be an infinite anti-chain, per definition of the product partial order, if $\{y_1, y_2, \ldots, \}$ is an infinite anti-chain in Q, which is assumed to not exist by hypothesis. Therefore instead assume that $X \subseteq P$ is infinite. Consequently, since any infinite poset contains either an infinite chain or an infinite anti-chain, and since any infinite chain contains either an infinite strictly descending or infinite strictly ascending chain [112, Theorem 1.14], our hypotheses ensure that $X \subseteq P$ contains an infinite strictly ascending chain. Hence, by passing to a subset of Z and re-indexing the elements of Z, we can w.l.o.g. assume $\{x_i\}_{i=1}^{\infty}$ is an infinite strictly ascending chain in P. Repeating the same argument using Y instead of X, we conclude that Y also contains an infinite strictly ascending chain, say $\{y_{i_j}\}_{j=1}^{\infty}$. However, since $\{x_i\}_{i=1}^{\infty}$ is an ascending chain but $Z = \{(x_1, y_1), (x_2, y_2), \ldots, \}$ is an anti-chain, the definition of the product partial order forces $i_j > i_{j+1}$ for all j. Thus $i_1 > i_2 > \ldots$ is an infinite strictly descending chain in $\mathbb{Z}_+$, which is not possible, completing the proof. □

2.3 Sequences and Rational Sequences

Let $G \cong \mathbb{Z}^d \oplus G_T$ be a finitely generated abelian group with torsion subgroup G_T. We let $\exp(G_T)$ denote the exponent of G_T, which is the minimal integer $m \geq 1$ such that $mg = 0$ for all $g \in G_T$. Let $G_0 \subseteq G$ be a subset. Regarding sequences and sequence subsums over G_0, we follow the standardized notation from Factorization Theory [53, 62, 66, 82]. The key parts are summarized here.

A **sequence** S of terms from G_0 is viewed formally as an element of the free abelian monoid with basis G_0, denoted $\mathcal{F}(G_0)$. Context will always distinguish between a sequence $S \in \mathcal{F}(G_0)$ and a sequence $\{x_i\}_{i=1}^{\infty}$ of terms $x_i \in G_0$. A sequence $S \in \mathcal{F}(G_0)$ is written as a finite multiplicative string of terms, using the boldsymbol dot operation $\boldsymbol{\cdot}$ to concatenate terms, and with the order irrelevant:

$$S = g_1 \boldsymbol{\cdot} \ldots \boldsymbol{\cdot} g_\ell = \prod^\bullet_{g \in G_0} g^{[\mathsf{v}_g(S)]}$$

with $g_i \in G_0$ the terms of S, with $\mathsf{v}_g(S) = |\{i \in [1, \ell] : g_i = g\}| \in \mathbb{Z}_+$ the multiplicity of the term $g \in G_0$, and with

$$|S| := \ell = \sum_{g \in G_0} \mathsf{v}_g(S) \geq 0 \quad \text{the } \textbf{length} \text{ of } S \quad \text{and} \quad \mathsf{v}_X(S) = \sum_{x \in X} \mathsf{v}_x(S),$$

where $X \subseteq G_0$. Here $g^{[n]} = \underbrace{g \boldsymbol{\cdot} \ldots \boldsymbol{\cdot} g}_{n}$ denotes the sequence consisting of the element g repeated n times, for $g \in G_0$ and $n \geq 0$. The notation is extended to sequences as well: $S^{[n]} = \underbrace{S \boldsymbol{\cdot} \ldots \boldsymbol{\cdot} S}_{n}$. If $S, T \in \mathcal{F}(G_0)$ are sequences, then $S \boldsymbol{\cdot} T \in \mathcal{F}(G_0)$ is the sequence obtained by concatenating the terms of T after those of S. We use $T \mid S$ to indicate that T is a subsequence of S and let $T^{[-1]} \boldsymbol{\cdot} S$ or $S \boldsymbol{\cdot} T^{[-1]}$ denote the sequence obtained by removing the terms of T from S. Then

$$\mathrm{Supp}(S) = \{g \in G_0 : \mathsf{v}_g(S) > 0\} \subseteq G_0 \quad \text{is the } \textbf{support} \text{ of } S, \quad \text{and}$$

$$\sigma(S) = \sum_{i=1}^{\ell} g_i = \sum_{g \in G_0} \mathsf{v}_g(S) g \in G \quad \text{is the sum of terms from } S.$$

Given two sequences $S, T \in \mathcal{F}(G_0)$, we let $\gcd(S, T) \in \mathcal{F}(G_0)$ denote the maximal length sequence diving both S and T.

Given a map $\varphi \colon G_0 \to G_0'$, we let $\varphi(S) = \varphi(g_1) \boldsymbol{\cdot} \ldots \boldsymbol{\cdot} \varphi(g_\ell) \in \mathcal{F}(G_0')$. The sequence S is called **zero-sum** if $\sigma(S) = 0$, and we let $\mathcal{B}(G_0) \subseteq \mathcal{F}(G_0)$ be the set of all zero-sum sequences over G_0. A *nontrivial* sequence $S \in \mathcal{F}(G_0)$ is called an **atom** or **minimal zero-sum sequence** if $\sigma(S) = 0$ but $\sigma(S') \neq 0$ for all proper, nontrivial subsequences $S' \mid S$. We let $\mathcal{A}(G_0) \subseteq \mathcal{B}(G_0)$ denote the set of all atoms over G_0. The **Davenport constant** for G_0 is

$$\mathsf{D}(G_0) = \sup\{|U| : U \in \mathcal{A}(G_0)\} \in \mathbb{Z}_+ \cup \{\infty\},$$

namely, the maximal length of an atom. Assuming G is finitely generated and every $g \in G_0$ is contained in some zero-sum sequence over G_0, it is known that $\mathsf{D}(G_0)$ is finite if and only if G_0 is finite [62, Theorem 3.4.2], and we have the general upper bound $\mathsf{D}(G) \leq |G|$ [62, Propositoin 5.1.4.4] and [82, Theorem 10.2].

A sequence $S \in \mathcal{F}(G_0)$ has the form $S = \prod^\bullet_{g \in G_0} g^{[\mathsf{v}_g(S)]}$ with $\mathsf{v}_g(S) \in \mathbb{Z}_+$ the multiplicity of g in S and $\mathsf{v}_g(S) > 0$ for only finitely many $g \in G_0$. Thus $\mathcal{F}(G_0)$ can viewed as tuples from the monoid $(\mathbb{Z}_+^{G_0}, +)$ with finite support, with the sequence $S = \prod^\bullet_{g \in G_0} g^{[\mathsf{v}_g(S)]}$ corresponding to the tuple $(\mathsf{v}_g(S))_{g \in G_0}$. It will be useful to sometimes allow more general exponents for terms in a sequence. This was done based on ideas from Matroid Theory in [14]. We continue with a natural framework utilizing Convex Geometry instead. The setup is as follows, which we present only in the context when G is torsion-free (though a similar framework holds for general abelian groups provided one restricts to rational multiplicities whose denominators are relatively prime to every $\mathrm{ord}(g) < \infty$ for $g \in G$). We let $\mathcal{F}_{\mathrm{rat}}(G_0)$ denote the multiplicative monoid whose elements have the form $S = \prod^\bullet_{g\in G_0} g^{[\mathsf{v}_g(S)]}$ with $\mathsf{v}_g(S) \in \mathbb{Q}_+$ the multiplicity of g in S and $\mathsf{v}_g(S) > 0$ for only finitely many $g \in G_0$. We refer to the elements $S \in \mathcal{F}_{\mathrm{rat}}(G_0)$ as **fractional** or **rational sequences**. Thus the rational sequences $\mathcal{F}_{\mathrm{rat}}(G_0)$ can be viewed as tuples from the monoid $(\mathbb{Q}_+^{G_0}, +)$ with finite support, with the sequence $S = \prod^\bullet_{g\in G_0} g^{[\mathsf{v}_g(S)]}$ corresponding to the tuple $(\mathsf{v}_g(S))_{g\in G_0}$. Any $S \in \mathcal{F}_{\mathrm{rat}}(G_0)$ has some positive integer $N > 0$ such that $S^{[N]} \in \mathcal{F}(G_0)$ (simply take the least common multiple of all denominators of the nonzero $\mathsf{v}_g(S)$). We extend all notation regarding $\mathcal{F}(G_0)$ to $\mathcal{F}_{\mathrm{rat}}(G_0)$. In particular, $\mathrm{Supp}(S) = \{g \in G_0 : \ \mathsf{v}_g(S) > 0\}$, which is always a finite set by definition, and $|S| = \sum_{g\in G_0} \mathsf{v}_g(S)$. We would also like to speak of the sum of a rational sequence $S \in \mathcal{F}(G_0)$. Since G is torsion-free, we have $G = \mathbb{Z}\langle X\rangle \cong \mathbb{Z}^{|X|}$ for some independent set X, and we can then embed G into $Q = \mathbb{Q}^{|X|}$ (thus we embed G into a divisible abelian group Q [91]). This means that the sum $\sigma(S) = \sum_{g\in G_0} \mathsf{v}_g(S)g \in Q$ of a rational sequence $S \in \mathcal{F}_{\mathrm{rat}}(G_0)$ is now well-defined. Let $\mathcal{B}_{\mathrm{rat}}(G_0)$ consist of all rational sequences $S \in \mathcal{F}_{\mathrm{rat}}(G_0)$ whose sum is zero $\sigma(S) = \sum_{g\in G_0} \mathsf{v}_g(S)g = 0$. This means $\mathcal{B}_{\mathrm{rat}}(G_0)$ is a rational generalized Krull Monoid [31, Proposition 2]. For $S \in \mathcal{F}_{\mathrm{rat}}(G_0)$, we define

$$\lfloor S\rfloor := \prod\nolimits^\bullet_{g\in G_0} g^{[\lfloor \mathsf{v}_g(S)\rfloor]} \quad \text{and} \quad \{S\} := \prod\nolimits^\bullet_{g\in G_0} g^{[\{\mathsf{v}_g(S)\}]},$$

where $\{\mathsf{v}_g(S)\} := \mathsf{v}_g(S) - \lfloor \mathsf{v}_g(S)\rfloor$ denotes the fractional part. Thus $S = \lfloor S\rfloor \cdot \{S\}$ with $\lfloor S\rfloor \mid S$ the unique maximal length subsequence with $\lfloor S\rfloor \in \mathcal{F}(G_0)$, and with $\mathsf{v}_g(\{S\}) < 1$ for all $g \in G_0$. It may be helpful to view the elements in $\{S\}$ as "damaged", and they will need to be avoided in some arguments as much as possible. Note $g \in \mathrm{Supp}(\{S\})$ if and only if $\mathsf{v}_g(S) \notin \mathbb{Z}$, while $g \in \mathrm{Supp}(\lfloor S\rfloor)$ if and only if $\mathsf{v}_g(S) \geq 1$. The following properties for $S, T \in \mathcal{F}_{\mathrm{rat}}(G_0)$ are immediate from the definitions but quite useful:

$$|\{S\}| \leq |\mathrm{Supp}(\{S\})|, \tag{2.1}$$

$$|\mathrm{Supp}(S)|,\ |S| \leq |\lfloor S\rfloor| + |\mathrm{Supp}(\{S\})|, \tag{2.2}$$

$$\lfloor S\rfloor \cdot \lfloor T\rfloor \mid \lfloor S\cdot T\rfloor \quad \text{and} \quad \mathrm{Supp}(\lfloor S\rfloor) \cup \mathrm{Supp}(\lfloor T\rfloor) \subseteq \mathrm{Supp}(\lfloor S\cdot T\rfloor), \tag{2.3}$$

$$\{S\cdot T\} \mid \{S\}\cdot\{T\} \quad \text{and} \quad \mathrm{Supp}(\{S\cdot T\}) \subseteq \mathrm{Supp}(\{S\}) \cup \mathrm{Supp}(\{T\}), \tag{2.4}$$

$$\mathrm{Supp}(\{S\cdot T^{[-1]}\}) \subseteq \mathrm{Supp}(\{S\}) \cup \mathrm{Supp}(\{T\}) \quad \text{for } T \mid S, \tag{2.5}$$

$$\lfloor T\rfloor \mid \lfloor S\rfloor \quad \text{and} \quad \lfloor S\cdot T^{[-1]}\rfloor \mid \lfloor S\rfloor \cdot \lfloor T\rfloor^{[-1]} \quad \text{for } T \mid S. \tag{2.6}$$

While generally the order of terms in a sequence will be irrelevant, there are times when we will need to view our sequences as indexed, so that two distinct terms of a sequence $S \in \mathcal{F}(G_0)$ that are equal as elements can still be viewed as distinct terms in the sequence S. For such occasions, we fix an indexing of the terms of S, say

$$S = g_1 \cdot \ldots \cdot g_\ell$$

with $g_i \in G_0$. Then for $I \subseteq [1, \ell]$, we let

$$S(I) = \prod^\bullet_{i \in I} g_i \in \mathcal{F}(G_0)$$

denote the subsequence of terms indexed by I.

2.4 Arithmetic Invariants for Transfer Krull Monoids

Let H be a unit-cancellative (associative and multiplicatively written) semigroup with identity 1_H. A semigroup homomorphism is map $\varphi : H \to H'$ between semigroups with identity such that $\varphi(1_H) = \varphi(1_{H'})$ and $\varphi(xy) = \varphi(x)\varphi(y)$ for $x, y \in H$. Let $H^\times \subseteq H$ denote the subgroup of units (invertible elements) in H. If a non-unit $a \in H \setminus H^\times$ has a factorization $a = u_1 \cdot \ldots \cdot u_k$ with each $u_i \in H$ a non-unit, then we call k the length of the factorization. We are generally interested in the case of factorizations into irreducible elements of H, called **atoms**, so $u_i \in \mathcal{A}(H)$ for all i with $\mathcal{A}(H)$ the set of all atoms (irreducible elements) in H. Let $\mathsf{L}(a)$ denote the set of all $k \geq 1$ for which a has a factorization *into atoms* of length k and let $\mathcal{L}(H) = \{\mathsf{L}(a) : a \in H \setminus H^\times\}$ denote the system of sets of lengths. Since, for purposes of factorization, elements of H equal up to units are essentially identical, it is often convenient to replace H (for H commutative) with the reduced quotient monoid $H_{\mathsf{red}} = H/H^\times$ where this has been formally implemented.

A **Krull Monoid** is a commutative, cancellative semigroup H with identity 1_H such that there is a **divisor homomorphism** $\varphi : H \to \mathcal{F}(P)$ into a free abelian monoid such that every $p \in P$ has some nonempty $X \subseteq H$ with $p = \gcd(\varphi(X))$. That φ is divisor homomorphism means φ is a semigroup homomorphism such that $x \mid y$ in H if and only if $\varphi(x) \mid \varphi(y)$ in F, where $x, y \in H$ [62, Theorem 2.4.8] and [70, Definition 4.]. The monoid $\mathcal{B}(G_0)$ of zero-sum sequences is always a Krull Monoid, for any subset G_0 of an abelian group G, and the reduced monoid $H_{\mathsf{red}} = H/H^\times$of a Krull Monoid H is also always a Krull Monoid. A **Transfer Krull Monoid** is a unit-cancellative semigroup H with identity 1_H such that there is a **(weak) transfer homomorphism** $\theta : H \to \mathcal{B}(G_0)$ for some subset G_0 of an abelian group G. The latter means θ is a surjective semigroup homomorphism such that $\theta(z)$ is the trivial sequence if and only if $z \in H^\times$ is a unit, for $z \in H$, and whenever $\theta(z) = U_1 \cdot \ldots \cdot U_k$ with $U_1, \ldots, U_k \in \mathcal{A}(G_0)$, for $z \in H$, then there is permutation $\tau : [1, k] \to [1, k]$ and factorization $z = x_1 \cdot \ldots \cdot x_k$ with $x_1, \ldots, x_k \in$

$\mathcal{A}(H)$ such that $\theta(x_i) = U_{\tau(i)}$ for all $i \in [1, k]$ (see [5, 11, 60, 64, 117, 118]). Then $z \in H$ is an atom if and only if $\theta(z) \in \mathcal{B}(G_0)$ is an atom, and the permutation τ may be taken to be the identity when H is commutative. Moreover, $\mathcal{L}(H) = \mathcal{L}(\mathcal{B}(G_0))$, meaning any arithmetic invariant regarding the factorization of elements from H that depends only on $\mathcal{L}(H)$ reduces to the study of the corresponding invariant for $\mathcal{B}(G_0)$.

We now briefly review the basic theory of Krull Monoids. To this end, consider a Krull Monoid H with divisor homomorphism $\varphi : H \to \mathcal{F}(P)$. The elements of $\mathcal{F}(P)$ are sequences $S = \prod^{\bullet}_{p \in P} p^{[\mathsf{v}_p(S)]}$ with $\mathsf{v}_p(S) \in \mathbb{Z}_+$. The set P does not live in an abelian group apriori, but this can be resolved by formally defining $\widetilde{G}_P$ to be a free abelian group with basis P, in which case $P \subseteq \widetilde{G}_P \cong \mathbb{Z}^P$. Now let $\widetilde{H}_P = \langle \sigma(\varphi(x)) : \ x \in H \rangle \leq \widetilde{G}_P$, let $G = \widetilde{G}_P / \widetilde{H}_P$, and let $[\cdot] : \widetilde{G}_P \to G = \widetilde{G}_P / \widetilde{H}_P$ be the natural homomorphism. For $S = \prod^{\bullet}_{p \in P} p^{[\mathsf{v}_p(S)]} \in \mathcal{F}(P)$, we have $[S] = \prod^{\bullet}_{p \in P} [p]^{[\mathsf{v}_p(S)]} \in \mathcal{F}(G_0)$, where $G_0 = \{[p] : \ p \in P\} \subseteq G$, referred to as the set of classes containing primes. We remark that our notation is slightly different from the presentation given in [62, 66]. There, the formal quotient group $q(\mathcal{F}(P))$ of $\mathcal{F}(P)$ is used in place of $\widetilde{G}_P$, with $q(\varphi(H))$ used in place of $\widetilde{H}_P$, at times leading to confusion since $[S]$, for $S \in \mathcal{F}(P)$, can both represent a sequence $[S] \in \mathcal{F}(G)$ as well as an element $[S] \in G$, the latter corresponding to the sum of terms in $[S]$ when considered as a sequence. The reformulation followed here avoids this potential confusion.

Any sequence $S = \prod^{\bullet}_{p \in P} p^{[\mathsf{v}_p(S)]} \in \varphi(H) \subseteq \mathcal{F}(P)$ will have $[S] \in \mathcal{B}(G_0)$ by definition of $[\cdot]$. This means the composition map $\theta : H \to \mathcal{B}(G_0)$, defined by $\theta(x) = [\varphi(x)]$, is a semigroup homomorphism taking each element of H to a zero-sum sequence over the subset G_0 of the abelian group G. Now further assume H is reduced, that is, assume we reduce our original Krull Monoid H by replacing it with the reduced Krull Monoid $H_{\mathsf{red}} = H/H^{\times}$. Under this assumption, the map θ is a transfer homomorphism, showing that the Krull Monoid H_{red} is a commutative Transfer Krull Monoid [66, Theorem 1.3.4]. Thus (up to units) every Krull Monoid is also a Transfer Krull Monoid. Moreover [62, Proposition 2.4.2.4, Corollary 2.4.3.2] (see also [66, Theorem 1.3.4.1]),

$$H_{\mathsf{red}} \cong \varphi(H) \quad \text{ and } \quad \varphi(H) = \{S \in \mathcal{F}(P) : [S] \in \mathcal{B}(G_0)\}. \tag{2.7}$$

We then say H is a Krull Monoid over the subset G_0 of the abelian group G.

We continue by describing some of the most commonly studied arithmetic invariants for factorizations that depend only on $\mathcal{L}(H)$. For an integer $k \geq 1$, we let $\mathcal{U}_k(H)$ denote the set of all $\ell \geq 1$ for which there are atoms $U_1, \ldots, U_k, V_1, \ldots, V_\ell \in \mathcal{A}(H)$ with

$$U_1 \cdot \ldots \cdot U_k = V_1 \cdot \ldots \cdot V_\ell. \tag{2.8}$$

Thus $\mathcal{U}_k(H)$ is the union of all sets from $\mathcal{L}(H)$ containing k. We then define

$$\rho_k(H) = \sup \mathcal{U}_k(H) \quad \text{for } k \geq 1, \text{ and}$$

$$\rho(H) = \sup\left\{\frac{\rho_k(H)}{k} : k \geq 1\right\} = \sup\left\{\frac{\sup \mathsf{L}(S)}{\min \mathsf{L}(S)} : S \in H\right\}.$$

The equality in the definition of $\rho(H)$ follows by a simple argument [62, Proposition 1.4.2.3]. Note $\ell \leq \rho_k(H)$ must hold for any re-factorization of k atoms as in (2.8). A basic argument [62, Proposition 1.4.2.1] shows

$$\rho_1(H) \leq \rho_2(H) \leq \ldots. \tag{2.9}$$

Moreover, the inequalities are strict when the $\rho_k(H)$ are finite. The constant $\rho(H)$ is the **elasticity** of H, and the $\rho_k(H)$ are its refinements, which we call the **elasticities** of H. Worth noting, if H is a Transfer Krull Monoid, then it is known that, for every $q \in \mathbb{Q}$ with $1 < q < \rho(H)$, there is some $L \in \mathcal{L}(H)$ with $q = \max L / \min L$ [69, Theorem 3.1]. When $H = \mathcal{B}(G_0)$, we abbreviate $\rho(G_0) = \rho(\mathcal{B}(G_0))$, and likewise for all other arithmetic invariants.

Given a set $X \subseteq \mathbb{Z}$, we let $\Delta(X) \subseteq \mathbb{Z}_+$ denote the set of successive distances in X, so $\Delta(X)$ consists of all elements of the form $k_2 - k_1$, where $k_2,\ k_2 \in X$ and $k_1 < k_2$, such that X contains no elements from $[k_1 + 1, k_2 - 1]$. Then the set of successive distances for H is

$$\Delta(H) = \bigcup_{L \in \mathcal{L}(H)} \Delta(L) = \bigcup_{a \in H \setminus H^\times} \Delta(\mathsf{L}(a)) \subseteq \mathbb{Z}_+.$$

A stronger measure of well-behaved factorization is indicated by structural results for $\mathcal{U}_k(H)$. A finite set $X \subseteq \mathbb{Z}$ is said to be an almost arithmetic progression with difference $d \geq 1$ and bound $N \geq 0$ if $X = P \setminus Y$, where P is an arithmetic progression with difference d and $Y \subseteq P$ is subset contained in the union of the first N terms from P and the last N terms from P. If there exists a constant $N \geq 0$ and difference $d \geq 1$ such that $\mathcal{U}_k(H)$ is an almost arithmetic progression with difference d and bound N for all sufficiently large k, then we say that H satisfies the **Structure Theorem for Unions**. As shown in [54, Theorem 4.2], if $\Delta(H)$ is finite and there is a constant $M \geq 0$ such that $\rho_{k+1}(H) - \rho_k(H) \leq M$ for all $k \geq 1$, then the Structure Theorem for Unions holds for H, providing additional motivation for the study of the elasticities.

The next invariant does not depend solely on $\mathcal{L}(H)$ but is known to nonetheless reduce to the study of $\mathcal{B}(G_0)$ for a large class of monoids, including Krull Monoids. Suppose we have two factorizations $S = U_1 \cdot \ldots \cdot U_k$ and $S = V_1 \cdot \ldots \cdot V_\ell$ of the same $S \in H$. Re-index the U_i and V_j such that $I \subseteq [1, \min\{k, \ell\}]$ is a maximal subset with $U_i = V_i$ for all $i \in I$, meaning $U_i \neq V_j$ for all $i \in [1, k] \setminus I$ and $j \in [1, \ell] \setminus I$. Then we say that the factorizations $V_1 \cdot \ldots \cdot V_\ell$ and $U_1 \cdot \ldots \cdot U_k$ **differ by** $\max\{\ell - |I|, k - |I|\}$ **factors**. The **catenary degree** $\mathsf{c}(H)$ is defined as the minimal integer $N \geq 0$ such that, given any two factorizations $U_1 \cdot \ldots \cdot U_k = S = U_1' \cdot \ldots \cdot U_{k'}'$ of the same $S \in H$ into *atoms* U_i, $U_j' \in \mathcal{A}(H)$, there is a sequence of factorizations $S = U_1^{(1)} \cdot \ldots \cdot U_{k_j}^{(j)}$ of S into atoms $U_i^{(j)} \in \mathcal{A}(H)$, for $j = 0, 1, \ldots, \ell$, such that

- $k_0 = k$ and $U_i^{(0)} = U_i$ for all $i \in [1, k]$,
- $k_\ell = k'$ and $U_i^{(\ell)} = U_i'$ for all $i \in [1, k']$, and
- each factorization $U_1^{(j)} \cdot \ldots \cdot U_{k_j}^{(j)}$ differs from the previous factorization $U_1^{(j-1)} \cdot \ldots \cdot U_{k_{j-1}}^{(j-1)}$, for $j \in [1, \ell]$, by at most $N = \mathsf{c}(H)$ factors.

If no such integer N exists, then $\mathsf{c}(H) = \infty$. Having finite catenary degree indicates that any factorization can be transformed into any other factorization of the same element by a sequence of small modifications, and we have $\mathsf{c}(G_0) \leq \mathsf{c}(H) \leq \max\{\mathsf{c}(G_0), 2\}$ when H is a Krull Monoid over the subset G_0 [62, Theorem 3.4.10.5].

Finally, we define one further invariant. Let $U \in \mathcal{A}(H)$. If U is prime, we set $\mathsf{t}(H, U) = 0$. Otherwise, we let $\mathsf{t}(H, U)$ be the minimal integer $N \geq 1$ such that, whenever $U_1, \ldots, U_k \in \mathcal{A}(H)$ with $U \mid U_1 \cdot \ldots \cdot U_k$, then there exists $I \subseteq [1, k]$ and $V_2, \ldots, V_r \in \mathcal{A}(H)$ with $\prod_{i \in I} U_i = U \cdot V_2 \cdot \ldots \cdot V_r$ and $\max\{|I|, r\} \leq N$. If no such N exists, we set $\mathsf{t}(H, U) = \infty$. The constant $\mathsf{t}(H, U)$ is called the **local tame degree**, and H is called **locally tame** if $\mathsf{t}(H, U) < \infty$ for all $U \in \mathcal{A}(H)$ [62].

2.5 Asymptotic Notation

Given sequences $\{a_i\}_{i=1}^{\infty}$ and $\{b_i\}_{i=1}^{\infty}$ of positive real numbers a_i, $b_i \in \mathbb{R}_+$, we use the following notation for gauging their comparative asymptotic growth: $a_i \in o(b_i)$ means $a_i/b_i \to 0$; $a_i \in O(b_i)$ means $\{a_i/b_i\}_{i=1}^{\infty}$ is a bounded sequence, and thus $a_i/b_i \to M$ for some real number $M \geq 0$ by passing to appropriate subsequences; $a_i \in \Theta(b_i)$ means there exist *positive* real numbers $\alpha, \beta \in \mathbb{R}_+$ such that $\alpha b_i \leq a_i \leq \beta b_i$ for all i, i.e., $a_i \in O(b_i)$ and $b_i \in O(a_i)$, in turn ensuring $a_i/b_i \to M$ for some positive real number $M > 0$ by passing to appropriate subsequences; and $a_i \sim b_i$ means $a_i/b_i \to 1$. We sometimes use $o(b_i)$ to represent some existent sequence $\{c_i\}_{i=1}^{\infty}$ with $|c_i| \in o(b_i)$, and likewise with the other asymptotic notation. In general, with only a few clear exceptions, all asymptotics will be with regard to the variable i in this work.

Chapter 3
Asymptotically Filtered Sequences, Encasement and Boundedness

3.1 Asymptotically Filtered Sequences

Given a subset $X \subseteq \mathbb{R}^d$, we will need a precise way to describe the directions in which X escapes to infinity, allowing us to characterize when X remains within bounded distance of another subset $Y \subseteq \mathbb{R}^d$. As a first approximation, we say that a sequence $\{x_i\}_{i=1}^{\infty}$ of nonzero points $x_i \in \mathbb{R}^d$ is **radially convergent** with **limit** u, where $u \in \mathbb{R}^d$ is a unit vector, if $\lim_{i\to\infty} \|x_i\| \in \mathbb{R}_+ \cup \{\infty\}$ exists and the sequence of unit vectors $\{x_i/\|x_i\|\}_{i=1}^{\infty}$ is convergent with limit $\lim_{i\to\infty} x_i/\|x_i\| = u$. Since the unit sphere is a compact metric space, any sequence of nonzero points contains a radially convergent subsequence.

Definition 3.1 For $X \subseteq \mathbb{R}^d$, let X^∞ denote all unit vectors which are a limit of an unbounded radially convergent sequence of terms from X.

The set X^∞ is a crude notion of the "directions" in which the set X escapes to infinity satisfying the following closure property.

Lemma 3.2 *Let $X \subseteq \mathbb{R}^d$, where $d \geq 0$. Then X^∞ is a closed subset of the unit sphere in $\mathbb{R}^d$.*

Proof For $d = 0$, X^∞ is empty, and there is nothing to show, so assume $d \geq 1$. To show X^∞ is closed, it suffices to show it contains all its limit points. To this end, let $\{u_i\}_{i=1}^{\infty}$ be a sequence of terms $u_i \in X^\infty$ with $u_i \to u$. Then u is another unit vector. By passing to a subsequence, we can w.l.o.g. assume $\mathsf{d}(u_j, u) < 1/2^j$ for all j. For each $u_j \in X^\infty$, there exists a radially convergent sequence $\{x_{ij}\}_{i=1}^{\infty}$ with $x_{ij} \in X$, $\|x_{ij}\| \to \infty$, and $\lim_{i\to\infty} x_{ij}/\|x_{ij}\| = u_j$. Thus we may take $y_j = x_{ij}$ for some fixed sufficiently large i such that $\|y_j\| > 2^j$ (possible as $\|x_{ij}\| \to \infty$) and $\mathsf{d}(y_j/\|y_j\|, u_j) < 1/2^j$. Now consider the sequence $\{y_j\}_{j=1}^{\infty}$ of terms $y_j \in X$. By construction, $\|y_j\| \to \infty$ while the triangle inequality ensures that $\mathsf{d}(y_j/\|y_j\|, u) \leq \mathsf{d}(y_j/\|y_j\|, u_j) + \mathsf{d}(u_j, u) < 1/2^j + 1/2^j = 1/2^{j-1}$, which tends to 0 as $j \to \infty$.

D. J. Grynkiewicz, *The Characterization of Finite Elasticities*, Lecture Notes in Mathematics 2316, https://doi.org/10.1007/978-3-031-14869-9_3

Thus $\{y_j\}_{j=1}^{\infty}$ is an unbounded radially convergent sequence of terms from X with limit u, showing that $u \in X^{\infty}$. As u was an arbitrary limit point, it follows that X^{∞} is closed. □

While the set X^{∞} provides a crude notion of the directions in which X escapes to infinity, it will turn out to be insufficient for our needs, leading us to the following more refined notion.

Definition 3.3 Let $\vec{u} = (u_1, \ldots, u_t)$ be a tuple of $t \geq 0$ orthonormal vectors in $\mathbb{R}^d$. A sequence $\{x_i\}_{i=1}^{\infty}$ of elements $x_i \in \mathbb{R}^d$ is called an **asymptotically filtered sequence** with **limit** $\vec{u}$ if

$$x_i = a_i^{(1)} u_1 + \ldots + a_i^{(t)} u_t + y_i \quad \text{for all } i \geq 1,$$

for some real numbers $a_i^{(j)} > 0$ and vectors $y_i \in \mathbb{R}\langle u_1, \ldots, u_t \rangle^{\perp}$ such that

- $\lim_{i \to \infty} a_i^{(j)} \in \mathbb{R}_+ \cup \{\infty\}$ exists for each $j \in [1, t]$,
- $\|y_i\| \in o(a_i^{(t)})$, and $a_i^{(j+1)} \in o(a_i^{(j)})$ for all $j \in [1, t-1]$.

The first bulleted condition above is added mostly for convenience and is not essential to the definition. We say that the limit $\vec{u} = (u_1, \ldots, u_t)$ is **fully unbounded** if $a_i^{(t)} \to \infty$ (and thus $a_i^{(j)} \to \infty$ for all $j \in [1, t]$). Note this requires $t \geq 1$. The empty tuple, corresponding to when $t = 0$, is referred to as the **trivial** tuple. We call the limit $\vec{u} = (u_1, \ldots, u_t)$ **anchored** if $t = 0$ or $\{a_i^{(t)}\}_{i=1}^{\infty}$ is instead a bounded sequence. The limit $\vec{u}$ is **complete** if $y_i = 0$ for all i, and $\vec{u}$ is a **complete fully unbounded** limit if $\vec{u}$ is fully unbounded but $\{y_i\}_{i=1}^{\infty}$ is bounded. Given any limit $\vec{u} = (u_1, \ldots, u_t)$ and $j \leq t$, we refer to $(u_1, \ldots, u_j)$ as a **truncation** of $\vec{u}$, which is **strict** when $j < t$. For $t \geq 1$, we let

$$\vec{u}^{\triangleleft} = (u_1, \ldots, u_{t-1})$$

denote the principal truncation of $\vec{u} = (u_1, \ldots, u_t)$. The choice of using tuples of orthonormal vectors to represent the limit of an asymptotically filtered sequence is purely a matter of canonical representation. Indeed, the tuple $(u_1, \ldots, u_t)$ really represents an ascending chain of half-spaces $\{0\} \subset \mathbb{R}_+^{\circ} u_1 \subset \mathbb{R}\langle u_1 \rangle + \mathbb{R}_+^{\circ} u_2 \subset \mathbb{R}\langle u_1, u_2 \rangle + \mathbb{R}_+^{\circ} u_3 \subset \ldots \subset \mathbb{R}\langle u_1, \ldots, u_{t-1} \rangle + \mathbb{R}_+^{\circ} u_t$ (or more compactly, the set $\bigcup_{j=0}^{t} (\mathbb{R}\langle u_1, \ldots, u_{j-1} \rangle + \mathbb{R}_+^{\circ} u_j)$, where $u_0 = 0$, from which this chain can be recovered), and any tuple $\vec{v} = (v_1, \ldots, v_t)$ of vectors $v_1, \ldots, v_t \in \mathbb{R}^d$ such that $\mathbb{R}\langle v_1 \ldots, v_{j-1} \rangle + \mathbb{R}_+^{\circ} v_j = \mathbb{R}\langle u_1 \ldots, u_{j-1} \rangle + \mathbb{R}_+^{\circ} u_j$ for all $j \in [1, t]$ is considered an **equivalent tuple**.

Proposition 3.4 *Let $\vec{u} = (u_1, \ldots, u_t)$ be a tuple of $t \geq 0$ orthonormal vectors in $\mathbb{R}^d$ and let $\{x_i\}_{i=1}^{\infty}$ be a sequence of elements $x_i \in \mathbb{R}^d$ with*

$$x_i = a_i^{(1)} u_1 + \ldots + a_i^{(t)} u_t + y_i \quad \textit{for all } i \geq 1,$$

for some real numbers $a_i^{(j)} > 0$ *and vectors* $y_i \in \mathbb{R}\langle u_1, \ldots, u_t\rangle^{\perp}$. *For* $j \in [0, t]$, *let* $\pi_j : \mathbb{R}^d \to \mathbb{R}\langle u_1, \ldots, u_j\rangle^{\perp}$ *denote the orthogonal projection. Then* $\{x_i\}_{i=1}^{\infty}$ *is an asymptotically filtered sequence with limit* $\vec{u}$ *if and only if* $\{\pi_{j-1}(x_i)\}_{i=1}^{\infty}$ *is a radially convergent sequence with limit* u_j *for all* $j \in [1, t]$. *Moreover, when this is the case, we have* $\lim_{i\to\infty} \|\pi_{j-1}(x_i)\| = \lim_{i\to\infty} a_i^{(j)}$ *for every* $j \in [1, t]$.

Proof For $j \in [0, t]$, let $y_i^{(j)} = a_i^{(j+1)} u_{j+1} + \ldots + a_i^{(t)} u_t + y_i$ and observe that

$$\pi_{j-1}(x_i) = a_i^{(j)} u_j + y_i^{(j)} = y_i^{(j-1)}$$

for $j \in [1, t]$. If $\{x_i\}_{i=1}^{\infty}$ is asymptotically filtered with limit $\vec{u}$, then $\|y_i^{(j)}\| \in o(a_i^{(j)})$ for $j \in [1, t]$. In view of the triangle inequality and $\|y_i^{(j)}\| \in o(a_i^{(j)})$, we have $\|a_i^{(j)} u_j + y_i^{(j)}\| = a_i^{(j)} + o(a_i^{(j)})$. Thus $\|y_i^{(j)}\| \in o(a_i^{(j)})$ further implies that

$$\|y_i^{(j)}\|/\|a_i^{(j)} u_j + y_i^{(j)}\| = \|y_i^{(j)}\|/(a_i^{(j)} + o(a_i^{(j)})) = (\|y_i^{(j)}\|/a_i^{(j)})/(1 + o(1)) \to 0,$$

ensuring that $y_i^{(j)}/\|a_i^{(j)} u_j + y_i^{(j)}\| \to 0$ and

$$\begin{aligned}\lim_{i\to\infty} \pi_{j-1}(x_i)/\|\pi_{j-1}(x_i)\| &= \lim_{i\to\infty} (a_i^{(j)} u_j + y_i^{(j)})/\|a_i^{(j)} u_j + y_i^{(j)}\| \\ &= \lim_{i\to\infty} a_i^{(j)} u_j/\|a_i^{(j)} u_j + y_i^{(j)}\| = u_j.\end{aligned}$$

Also, $\lim_{i\to\infty} \|\pi_{j-1}(x_i)\| = \lim_{i\to\infty} \|a_i^{(j)} u_j + y_i^{(j)}\| = \lim_{i\to\infty}(a_i^{(j)} + o(a_i^{(j)})) = \lim_{i\to\infty} a_i^{(j)} \in \mathbb{R}_+ \cup \{\infty\}$ exists, showing that $\{\pi_{j-1}(x_i)\}_{i=1}^{\infty}$ is radially convergent with limit u_j.

Next assume that $\{\pi_{j-1}(x_i)\}_{i=1}^{\infty}$ is radially convergent with limit u_j for each $j \in [1, t]$, so $(a_i^{(j)} u_j + y_i^{(j)})/\|a_i^{(j)} u_j + y_i^{(j)}\| \to u_j$ for all $j \in [1, t]$. In particular, since u_j and $y_i^{(j)}$ are linearly independent, $\|y_i^{(j)}\|/\|a_i^{(j)} u_j + y_i^{(j)}\| \to 0$ and also $\|a_i^{(j)} u_j + y_i^{(j)}\| \to a_i^{(j)}$. Thus

$$0 = \lim_{i\to\infty} \|y_i^{(j)}\|/\|a_i^{(j)} + y_i^{(j)}\| = \lim_{i\to\infty} \|y_i^{(j)}\|/a_i^{(j)},$$

ensuring that $\|y_i^{(j)}\| \in o(a_i^{(j)})$ for $j \in [1, t]$. In particular, $\|y_i\| = \|y_i^{(t)}\| \in o(a_i^{(t)})$, while

$$\|y_i^{(j)}\| = \|a_i^{(j+1)} u_{j+1} + y_i^{(j+1)}\| \in a_i^{(j+1)} + O(\|y_i^{(j+1)}\|) \subseteq a_i^{(j+1)} + o(a_i^{(j+1)})$$

for $j \in [1, t-1]$. Thus $\|y_i^{(j)}\| \in o(a_i^{(j)})$ for $j \in [1, t]$ ensures that $a_i^{(j+1)} \in o(a_i^{(j)})$ for $j \in [1, t-1]$. Furthermore, $\lim_{i\to\infty} a_i^{(j)} = \lim_{i\to\infty}(a_i^{(j)} + o(a_i^{(j)})) =$

$\lim_{i\to\infty}(\|a_i^{(j)}u_j + y_i^{(j)}\|) = \lim_{i\to\infty}\|\pi_{j-1}(x_i)\| \in \mathbb{R}_+ \cup \{\infty\}$ exists as $\{\pi_{j-1}(x_i)\}_{i=1}^{\infty}$ is radially convergent, completing the proof. □

Suppose $\{x_i\}_{i=1}^{\infty}$ is a sequence of elements $x_i \in \mathbb{R}^d$ that is not eventually the constant zero sequence. By passing to a subsequence, we can assume all x_i are nonzero with $\{x_i\}_{i=1}^{\infty}$ a radially convergent sequence with limit (say) u_1. Moreover, if $\{x_i\}_{i=1}^{\infty}$ is unbounded, then we can assume $\lim_{i\to\infty}\|x_i\| = \infty$. Write each $x_i = a_i^{(1)}u_1 + y_i^{(1)}$ with $y_i^{(1)} \in \mathbb{R}\langle u_1\rangle^{\perp}$. Since $x_i/\|x_i\| \to u_1$, we can assume $a_i^{(1)} > 0$ for all i by discarding the first few terms. Proposition 3.4 implies that the resulting subsequence is asymptotically filtered with limit u_1 and $\lim_{i\to\infty} a_i^{(1)} = \lim_{i\to\infty}\|x_i\|$. If $\{y_i^{(1)}\}_{i=1}^{\infty}$ is eventually zero, then discarding the first few terms allows us to assume $x_i = a_i^{(1)}u_1$ for all $i \geq 1$. Otherwise, we can repeat the above procedure and, passing to a yet more refined subsequence, conclude that $x_i = a_i^{(1)}u_1 + a_i^{(2)}u_2 + y_i^{(2)}$ is asymptotically filtered with limit (u_1, u_2). Continuing to iterate the procedure, we find that any sequence of terms from $\mathbb{R}^d$ contains an asymptotically filtered subsequence with complete limit. Likewise, truncating appropriately, any unbounded sequence in $\mathbb{R}^d$ must contain an asymptotically filtered subsequence with complete fully unbounded limit. Note Proposition 3.4 also ensures that if $\{x_i\}_{i=1}^{\infty}$ is an asymptotically filtered sequence both with limit $\vec{u}$ and with limit $\vec{v}$, then either $\vec{v}$ is a truncation of $\vec{u}$ or $\vec{u}$ is a truncation of $\vec{v}$.

The following proposition is routine but important and requires the following notation. Let $\vec{u} = (u_1, \ldots, u_t)$ be a tuple of orthonormal vectors $u_i \in \mathbb{R}^d$, let $\mathcal{E} \subseteq \mathbb{R}^d$ be a subspace, and let $\pi : \mathbb{R}^d \to \mathcal{E}^{\perp}$ be the orthogonal projection. Then

$$\pi(\vec{u}) := (\overline{u}_1, \ldots, \overline{u}_\ell),$$

where the $\overline{u}_i$ are defined as follows. Recursively define indices

$$0 = r_0 < r_1 < \ldots < r_\ell < r_{\ell+1} = t + 1 \tag{3.1}$$

by letting $r_j \in [r_{j-1} + 1, t]$ (for $j \in [1, \ell]$) be the minimal index such that $\pi_{j-1}(u_{r_j}) \neq 0$, where $\pi_{j-1} : \mathbb{R}^d \to (\mathcal{E} + \mathbb{R}\langle u_1, u_2 \ldots, u_{r_{j-1}}\rangle)^{\perp}$ is the orthogonal projection and $\ell \in [0, t]$ is the first index such that $\pi_\ell(u_i) = 0$ for all i. In particular, $\pi_0 = \pi$ and $r_1 \in [1, t]$ is the first index such that $\pi(u_{r_1}) \neq 0$ (unless $\ell = 0$, in which case no index with this property exists). Then $\overline{u}_j := \pi_{j-1}(u_{r_j})/\|\pi_{j-1}(u_{r_j})\|$, ensuring that $\pi(\vec{u})$ is a tuple of orthonormal vectors from $\mathcal{E}^{\perp}$. Equivalently, the indices r_j are those with $\mathbb{R}\langle\pi(u_1), \ldots, \pi(u_{r_j})\rangle \neq \mathbb{R}\langle\pi(u_1), \ldots, \pi(u_{r_j-1})\rangle$, and $\pi(\vec{u}) = (\overline{u}_1, \ldots, \overline{u}_\ell)$ is simply the canonical tuple of orthonormal vectors equivalent to the tuple $(\pi(u_{r_1}), \ldots, \pi(u_{r_\ell}))$. From this viewpoint, it is clear that, if $\pi' : \mathbb{R}^d \to (\mathcal{E}')^{\perp}$ is an orthogonal projection with $\mathcal{E} \subseteq \mathcal{E}'$, then $\pi'(\vec{u}) = \pi'(\pi(\vec{u}))$.

Proposition 3.5 *Let $X \subseteq \mathbb{R}^d$, where $d \geq 0$, let $\mathcal{E} \subseteq \mathbb{R}^d$ be a subspace, and let $\pi : \mathbb{R}^d \to \mathcal{E}^\perp$ be the orthogonal projection.*

1. *Let $\{x_i\}_{i=1}^\infty$ be an asymptotically filtered sequence of terms $x_i \in X$ with limit $\vec{u} = (u_1, \ldots, u_t)$, say $x_i = a_i^{(1)} u_1 + \ldots + a_i^{(t)} u_t + y_i$. Then the sufficiently large index terms in $\{\pi(x_i)\}_{i=1}^\infty$ form an asymptotically filtered sequence with limit $\pi(\vec{u}) = (\overline{u}_1, \ldots, \overline{u}_\ell)$ with*
$$\pi(x_i) = b_i^{(1)} \overline{u}_1 + \ldots + b_i^{(\ell)} \overline{u}_\ell + y_i',$$
$b_i^{(j)} \in \Theta(a_i^{(r_j)})$ for all $j \in [1, \ell]$, and $\|y_i'\| \in O(\|y_i\|)$, where the indices $r_1 < \ldots < r_\ell$ are those given by (3.1).
2. *If $\{y_i\}_{i=1}^\infty$ is an asymptotically filtered sequence of terms $y_i \in \pi(X)$ with limit $\vec{v}$, then there is an asymptotically filtered sequence $\{x_i\}_{i=1}^\infty$ of terms $x_i \in X$ with limit $\vec{u}$, such that, replacing $\{y_i\}_{i=1}^\infty$ with an appropriate subsequence, we have $\pi(x_i) = y_i$ for all i, $\pi(\vec{u}) = \vec{v}$, and $\pi(\vec{u}^\triangleleft) = \pi(\vec{u})^\triangleleft = \vec{v}^\triangleleft$ (if $\vec{v}$ is nontrivial).*

Proof

1. Let the r_j and π_j be as given in the definition of $\pi(\vec{u})$. For $j \in [1, \ell]$, we have $\pi_{j-1}(x_i) = a_i^{(1)} \pi_{j-1}(u_1) + \ldots + a_i^{(t)} \pi_{j-1}(u_t) + \pi_{j-1}(y_i)$. Since $a_i^{(j)} \in o(a_i^{(j-1)})$ for all $j \geq 2$ and $\|\pi_{j-1}(y_i)\| \in O(\|y_i\|) \subseteq o(a_i^{(t)})$ (the first inclusion follows as any linear transformation $\pi_{j-1} : \mathbb{R}^d \to \mathbb{R}^d$ is a bounded linear operator), it follows that $\|\pi_{j-1}(x_i)\| \sim a_i^{(r_j)} \|\pi_{j-1}(u_{r_j})\|$. Now $\pi(x_i) = b_i^{(1)} \overline{u}_1 + \ldots + b_i^{(\ell)} \overline{u}_\ell + y_i'$ with
$$b_i^{(j)} = a_i^{(r_j)} \|\pi_{j-1}(u_{r_j})\| \pm a_i^{(r_j+1)} \|\pi_j^\perp(u_{r_j+1})\| \pm \ldots \pm a_i^{(t)} \|\pi_j^\perp(u_t)\| \pm \|\pi_j^\perp(y_i)\|$$
and $y_i' = \pi_\ell(y_i)$, where $\pi_j^\perp : \mathbb{R}^d \to \mathbb{R}\overline{u}_j$ is the orthogonal projection onto $\mathbb{R}\overline{u}_j$. For each $j \in [1, \ell]$, we have $a_i^{(r_j)} \|\pi_{j-1}(u_{r_j})\| > 0$ for all i, $\|\pi_j^\perp(y_i)\| \in O(\|y_i\|) \subseteq o(a_i^{(r_j)})$ and $a_i^{(k)} \in o(a_i^{(r_j)})$ for all $k > r_j$. Thus $b_i^{(j)} \in \Theta(a_i^{(r_j)})$ with $b_i^{(j)} > 0$ for all sufficiently large i. Also, $\|y_i'\| = \|\pi_\ell(y_i)\| \in O(\|y_i\|) \subseteq o(a_i^{(r_\ell)}) = o(b_i^{(\ell)})$, and Item 1 now follows.
2. Since each $y_i \in \pi(G_0)$, there exists some $x_i \in G_0$ such that $\pi(x_i) = y_i$, for $i \geq 1$. By passing to a subsequence, we can assume $\{x_i\}_{i=1}^\infty$ is asymptotically filtered with complete limit $\vec{u}'$, in which case Item 1 implies that the sufficiently large index terms in the sequence $\{\pi(x_i)\}_{i=1}^\infty = \{y_i\}_{i=1}^\infty$ are asymptotically filtered with complete limit $\pi(\vec{u}')$. Since $\{y_i\}_{i=1}^\infty$ is also asymptotically filtered with limit $\vec{v}$, we conclude that $\vec{v}$ is a truncation of $\pi(\vec{u}')$ (cf. Proposition 3.4), whence $\pi(\vec{u}) = \vec{v}$ for an appropriate truncation $\vec{u}$ of $\vec{u}'$. Moreover, if $\vec{v}$ is nontrivial, then so is $\vec{u}$, and choosing a truncation $\vec{u} = (u_1, \ldots, u_t)$ with $t \geq 1$ minimal such that $\pi(\vec{u}) = \vec{v}$, we obtain that $\pi(\vec{u}^\triangleleft) = \pi(\vec{u})^\triangleleft$. We may consider $\{x_i\}_{i=1}^\infty$ as an asymptotically filtered sequence with the truncated limit $\vec{u}$, and Item 2 now follows.

□

Definition 3.6 Given a set $X \subseteq \mathbb{R}^d$, we let X^{lim} denote the set of all fully unbounded limits $\vec{u} = (u_1, \ldots, u_t)$ of an asymptotically filtered sequence of terms from X.

Each element of X^∞ occurs as a singleton tuple in X^{lim}, and we view the fully unbounded limits from X^{lim} as the more complete set of "directions" in which the set X escapes to infinity.

3.2 Encasement and Boundedness

The following definition associates a family of cones that "hug" the boundary of each potential direction from X^{lim}.

Definition 3.7 Let $\vec{u} = (u_1, \ldots, u_t)$ be a tuple of linearly independent vectors $u_1, \ldots, u_t \in \mathbb{R}^d$, where $t \in [0, d]$. A cone C **encases** $\vec{u}$ if, for each $j \in [1, t]$, there are $\alpha_{1,j}, \ldots, \alpha_{j,j} \in \mathbb{R}_+$ with

$$\alpha_{j,j} > 0, \quad z_j := \alpha_{j,j}u_j + \alpha_{j-1,j}u_{j-1} + \ldots + \alpha_{1,j}u_1 \quad \text{and} \quad \mathsf{C}(z_1, \ldots, z_t) \subseteq C.$$

In the context of convex cones, we will often also say a subset $X \subseteq \mathbb{R}^d$ **encases** $\vec{u}$ when $\mathsf{C}(X)$ does, and that X **minimally encases** $\vec{u}$ if, additionally, no proper subset of X encases $\vec{u}$.

When $t = 1$, we often speak of encasing the element u_1 rather than the tuple (u_1). Removing the requirement that $\alpha_{1,j}, \ldots, \alpha_{j-1,j} \in \mathbb{R}_+$ in the definition of encasement results in a seemingly weaker definition that is nonetheless equivalent. Indeed, if $\mathsf{C}(y_1, \ldots, y_t) \subseteq C$ with each $y_j = \beta_{j,j}u_j + \beta_{j-1,j}u_{j-1} + \ldots + \beta_{1,j}u_1$ for some $\beta_{i,j} \in \mathbb{R}$ with $\beta_{j,j} > 0$, then $y_j' = y_j + \max\{0, -\frac{\beta_{j-1,j}}{\beta_{j-1,j-1}}\}y_{j-1} \in \mathsf{C}(y_1, \ldots, y_t)$ is an element of the form $y_j' = \beta_{j,j}'u_j + \beta_{j-1,j}'u_{j-1} + \ldots + \beta_{1,j}'u_1$ for some $\beta_{i,j}' \in \mathbb{R}$ with $\beta_{j,j}' > 0$ and $\beta_{j-1,j}' \geq 0$. Repeating this process sequentially for $y_{j-1}, \ldots, y_1$ results in an element $z_j = \alpha_{j,j}u_j + \alpha_{j-1,j}u_{j-1} + \ldots + \alpha_{1,j}u_1 \in \mathsf{C}(y_1, \ldots, y_t) \subseteq C$, where $\alpha_{i,j} \in \mathbb{R}_+$ and $\alpha_{j,j} > 0$, showing that C encases $(u_1, \ldots, u_t)$. Thus C encasing $(u_1, \ldots, u_t)$ is equivalent to there existing a convex cone $C' \subseteq C$ that intersects $\mathbb{R}_+^\circ u_j + \mathbb{R}\langle u_1, \ldots, u_{j-1}\rangle$ for each $j \in [1, t]$. In particular, if C encases $\vec{u}$, then C encases all equivalent tuples to $\vec{u}$ too.

Lemma 3.8 *Let $(u_1, \ldots, u_t)$ be a tuple of linearly independent vectors $u_1, \ldots, u_t \in \mathbb{R}^d$, where $t \in [0, d]$. If the cones C, $C' \subseteq \mathbb{R}^d$ both encase $(u_1, \ldots, u_t)$, then $C \cap C'$ encases $(u_1, \ldots, u_t)$.*

Proof Since C, $C' \subseteq \mathbb{R}^d$ each encase $(u_1, \ldots, u_t)$, there are $X = \{x_1, \ldots, x_t\} \subseteq C$ and $Y = \{y_1, \ldots, y_t\} \subseteq C'$ with $\mathsf{C}(X) \subseteq C$ and $\mathsf{C}(Y) \subseteq C'$ such that, for each $j \in [1, t]$, $x_j = \alpha_{1,j}u_1 + \ldots + \alpha_{j,j}u_j \in \mathsf{C}(X)$ and $y_j = \beta_{1,j}u_1 + \ldots + \beta_{j,j}u_j \in \mathsf{C}(Y)$ for some $\alpha_{i,j}$, $\beta_{i,j} \geq 0$ with $\alpha_{j,j}$, $\beta_{j,j} > 0$. Let $A = (\alpha_{i,j})_{i,j}$ be the upper $t \times t$

triangular matrix given by the $\alpha_{i,j}$ with $i,\ j \in [1,t]$, so $\alpha_{i,j} = 0$ whenever $i > j$, and likewise define the upper $t \times t$ triangular matrix $B = (\beta_{i,j})_{i,j}$. For $j \in [1,t]$, we aim to construct $z_j = \gamma_{1,j}u_1 + \ldots + \gamma_{j,j}u_j \in \mathsf{C}(X) \cap \mathsf{C}(Y) \subseteq C \cap C'$ with $\gamma_{i,j} \geq 0$ for all $1 \leq i \leq j \leq t$ and each $\gamma_{j,j} > 0$.

Fix $j \in [1,t]$ arbitrary. Since the u_i are linearly independent, $z_j = \gamma_{1,j}u_1 + \ldots + \gamma_{j,j}u_j$ lies in $\mathsf{C}(X)$ precisely when there exists a vector $x = (r_1, \ldots, r_j, 0, \ldots, 0)$ with all entries non-negative such that $Ax = y$, where $y = (\gamma_{1,j}, \ldots, \gamma_{j,j}, 0, \ldots, 0)$. By well known back substitution formulas,

$$r_i = \frac{\gamma_{i,j} - \sum_{k=i+1}^{j} \alpha_{i,k} r_k}{\alpha_{i,i}} \quad \text{for } i \in [1,j]. \tag{3.2}$$

Likewise, $z_j = \gamma_{1,j}u_1 + \ldots + \gamma_{j,j}u_j$ lies in $\mathsf{C}(Y)$ precisely when there is an $x' = (r'_1, \ldots, r'_j, 0, \ldots, 0)$ with all entries non-negative such that $Ax' = y$. As before, we have

$$r'_i = \frac{\gamma_{i,j} - \sum_{k=i+1}^{j} \beta_{i,k} r'_k}{\beta_{i,i}} \quad \text{for } i \in [1,j]. \tag{3.3}$$

Taking $\gamma_{j,j} = 1$, we observe that both (3.2) and (3.3) ensure that $r_j > 0$ and $r'_j > 0$. But now we can recursively construct the $\gamma_{i,j} \geq 0$ for $i = j, j-1, \ldots, 1$ with $r_i, r'_i \geq 0$ by simply choosing $\gamma_{i,j} \geq 0$ sufficiently large. Indeed, taking $\gamma_{i,j} = \max\{\sum_{k=i+1}^{j} \alpha_{i,k} r_k,\ \sum_{k=i+1}^{j} \beta_{i,k} r'_k\}$ for $i \in [1, j-1]$ suffices, completing the proof. □

One of the main goals of this section is to give a local containment characterization of the more subtle notion of being bound to Y, which we will introduce in the next subsection. Finding a satisfying such characterization for a general set $Y \subseteq \mathbb{R}^d$ is more difficult, so we will instead restrict our attention to a suitably broad class of subsets of $\mathbb{R}^d$, namely, the class consisting of all *finite unions of polyhedral cones*. We formally allow the empty set to be considered a finite union of polyhedral cones, viewing it as an empty union. Since polyhedral cones are closed, a finite union of polyhedral cones is also closed. This class has several useful closure properties, summarized in the following lemma.

Lemma 3.9 *Suppose $X, X_1, \ldots, X_s \subseteq \mathbb{R}^d$ are finite unions of polyhedral cones.*

1. *$\bigcup_{i=1}^{s} X_i$ and $\bigcap_{i=1}^{s} X_i$ are both finite unions of polyhedral cones.*
2. *If $V \subseteq \mathbb{R}^d$ is a subspace containing $\mathbb{R}\langle X\rangle$, then $\overline{V \setminus X}$ is a finite union of polyhedral cones with $\partial(\overline{V \setminus X}) \subseteq \partial(X)$.*
3. *If $C_1, \ldots, C_s \subseteq \mathbb{R}^d$ are polyhedral cones with $\mathbb{R}\langle C_i\rangle = V$ for all $i \in [1,s]$, then $\partial(X) = \partial(\overline{V \setminus X})$, where $X = \bigcup_{j=1}^{s} C_j$.*

Proof

1. That $\bigcup_{i=1}^{s} X_i$ is a finite union of polyhedral cones is immediate, as this is the case for each X_i. That the intersection of a finite number of polyhedral cones is itself a polyhedral cone follows from the characterization of a polyhedral cone as the intersection of a finite number of half spaces. By hypothesis, each X_j is a finite union of polyhedral cones, say $X_j = \bigcup_{i=1}^{t_j} C_i^{(j)}$. Thus $\bigcap_{j=1}^{s} X_j = \bigcap_{j=1}^{s} \bigcup_{i=1}^{t_j} C_i^{(j)} = \bigcup_{i_1\in[1,t_1],\ldots,i_s\in[1,t_s]} \bigcap_{j=1}^{s} C_{i_j}^{(j)}$ with each $\bigcap_{j=1}^{s} C_{i_j}^{(j)}$ a polyhedral cone as previously remarked.
2. By hypothesis, $X \subseteq V \subseteq \mathbb{R}^d$ is a finite union of polyhedral cones, say $X = \bigcup_{j=1}^{s} C_j$ with each C_j a polyhedral cone, so each $C_j = \bigcap_{i=1}^{t_j} H_{i,j}$ for some closed half spaces $H_{i,j} \subseteq V$. We may w.l.o.g. assume $V = \mathbb{R}^d$. Then

$$V\setminus X = V\setminus(\bigcup_{j=1}^{s} C_j) = V\setminus(\bigcup_{j=1}^{s}\bigcap_{i=1}^{t_j} H_{i,j}) = \bigcap_{j=1}^{s}(V\setminus\bigcap_{i=1}^{t_j} H_{i,j}) = \bigcap_{j=1}^{s}\bigcup_{i=1}^{t_j} V\setminus H_{i,j}$$

$$= \bigcup_{i_1\in[1,t_1],\ldots,i_s\in[1,t_s]}\bigcap_{j=1}^{s} V\setminus H_{i_j,j} = \bigcup_{i_1\in[1,t_1],\ldots,i_s\in[1,t_s]}\bigcap_{j=1}^{s}\mathsf{int}(-H_{i_j,j}),$$

where each $\mathsf{int}(-H_{i_j,j})$ is an open half space in V. Thus

$$\overline{V\setminus X} = \overline{\bigcup_{i_1\in[1,t_1],\ldots,i_s\in[1,t_s]}\bigcap_{j=1}^{s}\mathsf{int}(-H_{i_j,j})} = \bigcup_{i_1\in[1,t_1],\ldots,i_s\in[1,t_s]}\overline{\bigcap_{j=1}^{s}\mathsf{int}(-H_{i_j,j})}$$

$$= \bigcup_{i_1\in[1,t_1],\ldots,i_s\in[1,t_s]}\overline{\mathsf{int}\Big(\bigcap_{j=1}^{s} -H_{i_j,j}\Big)}. \tag{3.4}$$

Each $\bigcap_{j=1}^{s} -H_{i_j,j}$ is a finite intersection of closed half spaces, and thus a polyhedral cone. In particular, $\bigcap_{j=1}^{s} -H_{i_j,j}$ is convex. If $\mathsf{int}\Big(\bigcap_{j=1}^{s} -H_{i_j,j}\Big) \neq \emptyset$, then $\mathsf{int}\Big(\bigcap_{j=1}^{s} -H_{i_j,j}\Big) = \Big(\bigcap_{j=1}^{s} -H_{i_j,j}\Big)^{\circ}$ (as any nonempty open set contains a ball of positive radius, which is a full dimensional subset), whence

$$\overline{\mathsf{int}\Big(\bigcap_{j=1}^{s} -H_{i_j,j}\Big)} = \overline{\Big(\bigcap_{j=1}^{s} -H_{i_j,j}\Big)^{\circ}} = \overline{\bigcap_{j=1}^{s} -H_{i_j,j}} = \bigcap_{j=1}^{s} -H_{i_j,j},$$

with the penultimate equality above in view of the convexity of $\bigcap_{j=1}^{s} -H_{i_j,j}$, and the final equality in view of $\bigcap_{j=1}^{s} -H_{i_j,j}$ being an intersection of closed subsets, and thus itself closed. Thus $\overline{\mathsf{int}\Big(\bigcap_{j=1}^{s} -H_{i_j,j}\Big)} = \bigcap_{j=1}^{s} -H_{i_j,j}$ is a polyhedral cone in such case. On the other hand, if $\mathsf{int}\Big(\bigcap_{j=1}^{s} -H_{i_j,j}\Big) = \emptyset$, then $\overline{\mathsf{int}\Big(\bigcap_{j=1}^{s} -H_{i_j,j}\Big)} = \emptyset$ too. As a result, (3.4) shows that $\overline{V \setminus X}$ is a finite union of polyhedral cones, as desired.

Let $\mathcal{E} = \mathbb{R}\langle X\rangle \subseteq V = \mathbb{R}^d$. If V properly contains $\mathcal{E}$, then $\overline{V \setminus X} = \overline{V \setminus \mathcal{E}} = V$ and $\partial(\overline{V \setminus X}) = \partial(V) = \emptyset \subseteq \partial(X)$. On the other hand, if $V = \mathcal{E}$, then

$$\partial(X) = \partial(V \setminus X) = (\overline{V \setminus X}) \setminus (V \setminus X)^{\circ} \quad \text{and}$$

$$\partial(\overline{V \setminus X}) = (\overline{\overline{V \setminus X}}) \setminus (\overline{V \setminus X})^{\circ} = (\overline{V \setminus X}) \setminus (\overline{V \setminus X})^{\circ},$$

where the equality $\partial(X) = \partial(V \setminus X)$ follows in view of X being closed (so that $V \setminus X$ lying in a proper subspace of V is only possible if $X = V$). Consequently, since $(V \setminus X)^{\circ} \subseteq (\overline{V \setminus X})^{\circ}$ (as both $V \setminus X$ and $\overline{V \setminus X}$ span the same subspace), it follows that $\partial(\overline{V \setminus X}) \subseteq \partial(X)$. This completes the proof of Item 2.

3. Each C_j has full dimension in V, and we can w.l.o.g. assume $V = \mathbb{R}^d$. Since X is closed and spans the subspace V, it follows that $V \setminus X$ is open in V and is thus either empty or has full dimension. If $V \setminus X = \emptyset$, then $\overline{V \setminus X} = \emptyset$ too, whence $\partial(\overline{V \setminus X}) = \partial(X) = \partial(V) = \emptyset$, as desired. So instead assume $V \setminus X$ has full dimension, whence $V \setminus X$ and X both span the entire space V. As a result, $\partial(X) = \partial(V \setminus X) = (\overline{V \setminus X}) \setminus (V \setminus X)^{\circ}$ and $\partial(\overline{V \setminus X}) = (\overline{\overline{V \setminus X}}) \setminus (\overline{V \setminus X})^{\circ} = (\overline{V \setminus X}) \setminus (\overline{V \setminus X})^{\circ}$. Letting $Y = V \setminus X$, it remains to show $\overline{Y}^{\circ} = Y^{\circ}$. The inclusion $Y^{\circ} \subseteq \overline{Y}^{\circ}$ is trivial (as Y and $\overline{Y}$ span the same subspace). Since $Y = V \setminus X$ is open in V, we have $Y^{\circ} = Y$. Thus we need to show $\overline{Y}^{\circ} \subseteq Y$. Let $x \in \overline{Y}^{\circ}$ be arbitrary. Then $x \in \overline{Y}$ and there exists an open ball $B_{\epsilon}(x) \subseteq \overline{Y}$ for some $\epsilon > 0$. Assume by contradiction that $x \notin Y = V \setminus X$, whence $x \in X = \bigcup_{j=1}^{s} C_j$, so w.l.o.g. $x \in C_1$, which is a polyhedral cone spanning V by hypothesis. Since C_1 is a convex set with $x \in C_1$, we may take a point $y \in C_1^{\circ}$ and consider the line segment between y and x. As C_1 is convex, all points on this line segment (apart from x) lie in C_1°, ensuring that $B_{\epsilon}(x) \cap C_1$ contains some point $x_0 \in C_1^{\circ}$. Thus $x_0 \in C_1^{\circ} \cap B_{\epsilon}(x)$. However, since $B_{\epsilon}(x) \subseteq \overline{Y} = \overline{V \setminus X}$, we also have $x_0 \in \overline{V \setminus X} \subseteq \overline{V \setminus C_1}$. Now C_1 is a closed set which spans the space V, which implies $V \setminus C_1$ is open in V, and thus also spans V (it is nonempty else its closure could not contain x_0). Since C_1 and $V \setminus C_1$ span the same subspace V, it follows that $C_1 \cap \overline{V \setminus C_1} = \partial(C_1) = \partial(V \setminus C_1)$. But now $x_0 \in \overline{V \setminus C_1} \cap C_1^{\circ} \subseteq \overline{V \setminus C_1} \cap C_1 = \partial(C_1) = C_1 \setminus C_1^{\circ}$, which contradicts that $x_0 \in C_1^{\circ}$, completing the proof.

□

Next, we need a notion of what it means for a set X to be bounded relative to another set Y, extending the notion of a bounded set (which is the case $Y = \{0\}$).

Definition 3.10 Let $X, Y \subseteq \mathbb{R}^d$ be subsets. We say that X is **bound** to Y if there is some real number $\epsilon > 0$ such that every $x \in X$ is distance less than ϵ from Y, i.e.,

$X \subseteq Y + B_\epsilon(0)$. A sequence $\{x_i\}_{i=1}^\infty$ is bound to Y if the set $X = \{x_i : i \geq 1\}$ is bound to Y.

We continue with a basic connection between boundedness and encasement.

Proposition 3.11 *Let $\{x_i\}_{i=1}^\infty$ be an asymptotically filtered sequence of terms $x_i \in \mathbb{R}^d$ with limit $\vec{u} = (u_1, \ldots, u_t)$ and let $\pi : \mathbb{R}^d \to \mathbb{R}\langle u_1, \ldots, u_t\rangle$ be the orthogonal projection. Suppose the cone $C \subseteq \mathbb{R}^d$ encases $\vec{u}$.*

1. *$\{\pi(x_i)\}_{i=1}^\infty$ is an asymptotically filtered sequence in $\mathbb{R}\langle u_1, \ldots, u_t\rangle \cong \mathbb{R}^t$ with limit $\vec{u}$.*
2. *$\pi(x_i) \in C$ for all sufficiently large i.*
3. *If $\vec{u}$ contains a complete fully unbounded limit of $\{x_i\}_{i=1}^\infty$ as a truncation, then the sequence $\{x_i\}_{i=1}^\infty$ is bound to C. Indeed, $\mathsf{d}(x_i, C) \leq \sup_i \|\pi_t(x_i)\| < \infty$ for all sufficiently large i, where $\pi_t : \mathbb{R}^d \to \mathbb{R}\langle u_1, \ldots, u_t\rangle^\perp$ is the orthogonal projection.*

Proof

1. For $j \in [0, t]$, let $\pi_j : \mathbb{R}^d \to \mathbb{R}\langle u_1, \ldots, u_j\rangle^\perp$ be the orthogonal projection. Since the cone C encases $(u_1, \ldots, u_t)$, there is some $Y = \{y_1, \ldots, y_t\} \subseteq C$ with $\mathsf{C}(Y) \subseteq C$ such that, for every $j \in [1, t]$, $y_j = \alpha_{1,j}u_1 + \alpha_{2,j}u_2 + \ldots + \alpha_{j,j}u_j$ for some $\alpha_{i,j} \geq 0$ with all $\alpha_{j,j} > 0$. Item 1 follows immediately from the definitions.
2. In view of Item 1, we may w.l.o.g. assume $x_i \in \mathbb{R}\langle u_1, \ldots, u_t\rangle$ for all i, so that each $\pi(x_i) = x_i = a_i^{(1)}u_1 + \ldots + a_i^{(t)}u_t$ for some $a_i^{(j)} > 0$ with $a_i^{(j)} \in o(a_i^{(j-1)})$ for $j \geq 2$. Let $A = (\alpha_{i,j})_{i,j}$ be the $t \times t$ upper triangular matrix given by the $\alpha_{i,j}$ with $i,\ j \in [1, t]$, so $\alpha_{i,j} = 0$ whenever $i > j$. Then $x_i \in \mathsf{C}(Y)$ precisely when there exists a vector $x = (r_i^{(1)}, \ldots, r_i^{(t)})$ with all $r_i^{(j)} \geq 0$ such that $Ax = z_i$, where $z_i = (a_i^{(1)}, \ldots, a_i^{(t)})$. As in the proof of Lemma 3.8, by well known back substitution formulas,

$$r_i^{(j)} = \frac{a_i^{(j)} - \sum_{k=j+1}^{t} \alpha_{j,k} r_i^{(k)}}{\alpha_{j,j}} \quad \text{for } j \in [1, t]. \tag{3.5}$$

   Consequently, since $a_i^{(k)} \in o(a_i^{(j)})$ for all $k > j$, we find that $r_i^{(j)} \in \Theta(a_i^{(j)})$ for all $j \in [1, t]$. In particular, $r_i^{(k)} \in o(a_i^{(j)})$ for $k \geq j + 1$, whence (3.5) ensures that $r_i^{(j)} > 0$ for all sufficiently large i. But this means $x_i \in \mathsf{C}(Y) \subseteq C$ for all sufficiently large i, completing Item 2.
3. By Item 2, $\pi(x_i) \in C$ for all sufficiently large i, say all $i \geq N$. Since $\{x_i\}_{i=1}^\infty$ is asymptotically filtered with limit $(u_1, \ldots, u_t)$ containing a complete fully unbounded limit of $\{x_i\}_{i=1}^\infty$, it follows that $\{\pi_t(x_i)\}_{i=1}^\infty$ is bounded, implying $\sup_i \|\pi_t(x_i)\| < \infty$. We have $x_i = \pi(x_i) + \pi_t(x_i)$, whence $\mathsf{d}(x_i, \pi(x_i)) = \|\pi_t(x_i)\| \leq \sup_i \|\pi_t(x_i)\|$ for all i. Thus, for $i \geq N$, we see $\pi(x_i) \in C$ is a point with distance at most $\sup_i \|\pi_t(x_i)\| < \infty$ from x_i, showing that the sequence $\{x_i\}_{i=1}^\infty$ is bound to C with the desired bounds.

□

For finite unions of polyhedral cones, the next proposition provides the key link between asymptotically filtered sequences and encasement, helping explain why we have restricted attention to this class of sets.

Proposition 3.12 *Let $Y \subseteq \mathbb{R}^d$ be a finite union of polyhedral cones. If $\{y_i\}_{i=1}^{\infty}$ is an asymptotically filtered sequence of terms $y_i \in Y$ with limit $\vec{u}$, then Y encases $\vec{u}$.*

Proof Let $\vec{u} = (u_1, \ldots, u_t)$ and extend $(u_1, \ldots, u_t)$ to an orthonormal basis $(u_1, \ldots, u_d)$ of $\mathbb{R}^d$. Since Y is a *finite* union of polyhedral cones, by replacing $\{y_i\}_{i=1}^{\infty}$ with an appropriate subsequence, we can w.l.o.g. assume $y_i \in C \subseteq Y$ for all i for some polyhedral cone C. Let $\pi : \mathbb{R}^d \to \mathbb{R}\langle u_1, \ldots, u_t\rangle$ be the orthogonal projection. Observing that $\mathsf{C}(u_1, \ldots, u_t)$ encases $\vec{u}$, we can apply Proposition 3.11.2 to conclude $\pi(y_i) \in \mathsf{C}(u_1, \ldots, u_t)$ for all sufficiently large i. Thus, by passing to a subsequence, we can w.l.o.g. assume all y_i lie in the polyhedral cone $C' = \mathsf{C}(u_1, \ldots, u_t, \pm u_{t+1}, \ldots, \pm u_d)$. As a result, since $C \cap C' \subseteq Y$ is also a polyhedral cone, with $y_i \in C \cap C'$ for all i, we see that we can replace C with $C \cap C'$ and thereby assume $C \subseteq \mathsf{C}(u_1, \ldots, u_t, \pm u_{t+1}, \ldots, \pm u_d)$.

Since C is a polyhedral cone, so too is $\pi_j(C)$ for each $j \in [0, t]$, where $\pi_j : \mathbb{R}^d \to \mathbb{R}\langle u_1, \ldots, u_j\rangle^{\perp}$ is the orthogonal projection. Since $\lim_{i\to\infty} \pi_{j-1}(x_i)/\|\pi_{j-1}(x_i)\| = u_j$, for $j \in [1, t]$, is a limit of points $\pi_{j-1}(x_i)/\|\pi_{j-1}(x_i)\|$ from the closed cone $\pi_{j-1}(C)$, it follows that $u_j \in \pi_{j-1}(C)$ for all $j \in [1, t]$. Consequently, since $\ker \pi_{j-1} = \mathbb{R}\langle u_1, \ldots, u_{j-1}\rangle$, there must be some $z_j \in C$ with $\pi_{j-1}(z_j) = u_j$, meaning $z_j = \alpha_{1,j}u_1 + \alpha_{2,j}u_2 + \ldots + \alpha_{j,j}u_j$ with $\alpha_{i,j} \in \mathbb{R}$ and $\alpha_{j,j} = 1$. However, since each $z_j \in C \subseteq \mathsf{C}(u_1, \ldots, u_t, \pm u_{t+1}, \ldots, \pm u_d)$, we must have $\alpha_{i,j} \geq 0$ for all i and j. This shows that C encases $\vec{u}$, and since $C \subseteq Y$, it follows that Y encases $\vec{u}$, as desired. □

Lemma 3.13 *Let $\{x_i\}_{i=1}^{\infty}$ be an asymptotically filtered sequence of points in $\mathbb{R}^d$ with fully unbounded limit $\vec{u}$ and let $Y \subseteq \mathbb{R}^d$. If $\{x_i\}_{i=1}^{\infty}$ is bound to Y, then there is an asymptotically filtered sequence $\{y_i\}_{i=1}^{\infty}$ of points $y_i \in Y$ having fully unbounded limit $\vec{u}$.*

Proof Let $\vec{u} = (u_1, \ldots, u_t)$ and let each $x_i = a_i^{(1)}u_1 + \ldots + a_i^{(t)}u_t + x_i'$ with $x_i' \in \mathbb{R}\langle u_1, \ldots, u_t\rangle^{\perp}$. Since $\vec{u}$ is fully unbounded, we have $a_i^{(j)} \to \infty$ for every $j \in [1, t]$. Since $\{x_i\}_{i=1}^{\infty}$ is bound to Y, there exists some finite $\epsilon > 0$ such that, for each term x_i, there is some $y_i \in Y$ with $\mathsf{d}(x_i, y_i) \leq \epsilon$. Consider the sequence $\{y_i\}_{i=1}^{\infty}$. For each $i \geq 1$, write $y_i = x_i + z_i$ and write $z_i = b_i^{(1)}u_1 + \ldots + b_i^{(t)}u_t + z_i'$ with $z_i' \in \mathbb{R}\langle u_1, \ldots, u_t\rangle^{\perp}$. Then $\{z_i\}_{i=1}^{\infty}$ is a bounded sequence in view of $\mathsf{d}(x_i, y_i) \leq \epsilon$, ensuring that $\|z_i\| \in O(1)$, and thus also $b_i^{(j)} \in O(1)$ for $j \in [1, t]$ and $z_i' \in O(1)$. Since $a_i^{(t)} \to \infty$, we have $O(1) \subseteq o(a_i^{(t)}) \subseteq o(a_i^{(j)})$ for all $j \in [1, t]$. In consequence, since $\|x_i'\| \in o(a_i^{(t)})$, it follows that $y_i = (a_i^{(1)} + b_i^{(1)})u_1 + \ldots + (a_i^{(t)} + b_i^{(t)})u_t + (x_i' + z_i')$ gives an asymptotically filtered representation of the y_i

having fully unbounded limit $\vec{u}$ (passing to sufficiently large index terms to ensure $a_i^{(j)} + b_i^{(j)} > 0$), as desired. □

We now come to the main result of this section, giving a local containment characterization for a set to be bound to a finite union of polyhedral cones. We remark that, if $Y \subseteq \mathbb{R}^d$ is only a cone, not a finite union of polyhedral cones, then the argument below shows 4. ⇒ 1. ⇒ 2. ⇒ 3. It is the implication 3. ⇒ 4. that requires Y to be a finite union of polyhedral cones.

Theorem 3.14 *Let $X, Y \subseteq \mathbb{R}^d$ be subsets with $Y \neq \emptyset$ a finite union of polyhedral cones. Then the following are equivalent.*

1. *X is bound to Y.*
2. *Every asymptotically filtered sequence $\{x_i\}_{i=1}^{\infty}$ of terms $x_i \in X$ with fully unbounded limit is bound to Y.*
3. *$X^{\mathsf{lim}} \subseteq Y^{\mathsf{lim}}$.*
4. *Y encases every $\vec{u} \in X^{\mathsf{lim}}$.*

Proof The implication 1. ⇒ 2. is trivial. The implication 2. ⇒ 3. follows from Lemma 3.13. As Y is a finite union of polyhedral cones, the implication 3. ⇒ 4. follows from Proposition 3.12. It remains to establish the implication 4. ⇒ 1. To this end, assume by contradiction that Y encases every $\vec{u} \in X^{\mathsf{lim}}$ but X is not bound to Y. Then there exists a sequence $\{x_i\}_{i=1}^{\infty}$ of points $x_i \in X$ with $\mathsf{d}(x_i, Y) \to \infty$. In particular, $\{\|x_i\|\}_{i=1}^{\infty}$ is unbounded (as $Y \neq \emptyset$). Thus, as discussed at the beginning of the section, there exists a subsequence which is asymptotically filtered with complete fully unbounded limit. Replacing $\{x_i\}_{i=1}^{\infty}$ with this subsequence, we can w.l.o.g. assume $\{x_i\}_{i=1}^{\infty}$ is itself an asymptotically filtered sequence of terms $x_i \in X$ with complete fully unbounded limit $\vec{u} = (u_1, \ldots, u_t)$ and $\mathsf{d}(x_i, Y) \to \infty$. Since Y encases $\vec{u} = (u_1, \ldots, u_t)$ by hypothesis, Proposition 3.11.3 implies that $\{x_i\}_{i=1}^{\infty}$ is bound to Y. However, this contradicts that $\mathsf{d}(x_i, Y) \to \infty$, completing the proof. □

Corollary 3.15 *Let $X, Y, Z \subseteq \mathbb{R}^d$ be subsets with X and Y each finite unions of polyhedral cones. If Z is bound to X as well as to Y, then Z is bound to $X \cap Y$.*

Proof By Lemma 3.9, $X \cap Y$ is also a finite union of polyhedral ones. Thus the corollary follows from Theorem 3.14.4 and Lemma 3.8. □

Chapter 4
Elementary Atoms, Positive Bases and Reay Systems

4.1 Basic Non-degeneracy Characterizations

We begin with a few basic properties regarding atoms and the representation of 0 as a positive linear combination.

Lemma 4.1 *Let $\Lambda \le \mathbb{R}^d$ be a full rank lattice, where $d \ge 0$, let $x_1, \ldots, x_r \in \Lambda$ and suppose $\alpha_1 x_1 + \ldots + \alpha_r x_r = 0$ for some $\alpha_i \in \mathbb{R}$. Then, for all $\epsilon > 0$, there exists $\alpha'_1, \ldots, \alpha'_r \in \mathbb{Q}$ with $\alpha'_1 x_1 + \ldots + \alpha'_r x_r = 0$ and $|\alpha_i - \alpha'_i| < \epsilon$ for all i.*

Proof Let $E = \{e_1, \ldots, e_d\} \subseteq \Lambda$ be a lattice basis for Λ. The lemma is vacuous for $r = 0$ and trivial if $\alpha_i = 0$ for all i, so assume $r \ge 1$ and that not all α_i are zero. Let x be the nonzero column vector whose i-th entry is α_i. Then, letting M be the $d \times r$ integer matrix whose j-th column is x_j expressed using the basis E, we see that x lies in the null space of M. Since the entries of M are integers, the null space of M is generated by integer vectors, say $y_1, \ldots, y_s \in \mathbb{Z}^r$ with $s \ge 1$. Thus $x = \beta_1 y_1 + \ldots + \beta_s y_s$ for some $\beta_i \in \mathbb{R}$. Since the rational numbers approximate the reals, we can find rational numbers $\beta'_1, \ldots, \beta'_s \in \mathbb{Q}$ with $|\beta_i - \beta'_i|$ sufficiently small so as to guarantee that, for every $i \in [1, r]$, the i-th coordinate α'_i of the vector $x' = \beta'_1 y_1 + \ldots + \beta'_s y_s$ satisfies $|\alpha'_i - \alpha_i| < \epsilon$. Moreover, $\alpha'_1 x_1 + \ldots + \alpha'_r x_r = 0$ (as x' lies in the null space of M) with $\alpha'_i \in \mathbb{Q}$ for all i (as each y_i is an integer vector and each $\beta'_i \in \mathbb{Q}$), as desired. □

Proposition 4.2 *Let $\Lambda \le \mathbb{R}^d$ be a full rank lattice, where $d \ge 0$, and let $G_0 \subseteq \Lambda$. Then*

$$\mathcal{A}(G_0) \ne \emptyset \quad \textit{if and only if} \quad 0 \in \mathsf{C}^*(G_0).$$

Moreover, if there exist $x_1, \ldots, x_r \in G_0$, where $r \ge 1$, and real numbers $\alpha_i > 0$ such that $\alpha_1 x_1 + \ldots + \alpha_r x_r = 0$, then there is a zero-sum subsequence $S \in \mathcal{B}(G_0)$ with $\mathrm{Supp}(S) = \{x_1, \ldots, x_r\}$.

D. J. Grynkiewicz, *The Characterization of Finite Elasticities*, Lecture Notes in Mathematics 2316, https://doi.org/10.1007/978-3-031-14869-9_4

Proof If $U \in \mathcal{A}(G_0)$ is an atom, then $\sum_{g\in G_0} \mathsf{v}_g(U)g = \sigma(U) = 0$ shows that $0 \in \mathsf{C}^*(G_0)$. On the other hand, if $0 \in \mathsf{C}^*(G_0)$, then there are $x_1, \dots, x_r \in G_0$, where $r \geq 1$, and real numbers $\alpha_i > 0$ such that $\alpha_1 x_1 + \dots + \alpha_r x_r = 0$. Applying Lemma 4.1 with $\epsilon = \min_i \alpha_i > 0$, we conclude that $\alpha'_1 x_1 + \dots + \alpha'_r x_r = 0$ for some $\alpha'_i \in \mathbb{Q}$ with $\alpha'_i > 0$ for all $i \in [1, r]$. By multiplying by a common denominator, we can w.l.o.g. assume the $\alpha'_i \in \mathbb{Z}$ for all i, and now $S = x_1^{[\alpha'_1]} \cdot \ldots \cdot x_r^{[\alpha'_r]} \in \mathcal{B}(G_0)$ is a nontrivial zero-sum sequence with $\mathrm{Supp}(S) = \{x_1, \dots, x_r\} \subseteq G_0$, showing that $\mathcal{A}(G_0) \neq \emptyset$. □

Proposition 4.3 *A subset $X \subseteq \mathbb{R}^d$, where $d \geq 0$, has $0 \notin \mathsf{C}^*(X)$ if and only if $0 \notin X$ and there exist a sequence of subspaces $\mathcal{H}^0 \subseteq \mathcal{H}^1 \subseteq \dots \subseteq \mathcal{H}^d$ with $\dim \mathcal{H}^i = i$ such that $X \cap \mathcal{H}^j$ is contained in the closed half space $\mathcal{H}^{j-1}_+ \subseteq \mathcal{H}^j$ for $j = 1, \dots, d$.*

Proof The case $d = 0$ is trivial, so we assume $d \geq 1$. A simple inductive argument shows that any $X \subseteq \mathbb{R}^d$ satisfying the stated conditions has $0 \notin \mathsf{C}^*(X)$. On the other hand, if $0 \notin \mathsf{C}^*(X)$, then $0 \notin X$ and $\mathsf{C}(X) \cap -\mathsf{C}(X) \neq \mathbb{R}^d$ (as $d \geq 1$). Thus $\mathsf{C}(X) \neq \mathbb{R}^d$, ensuring that $\mathsf{C}(X)$, and thus also $X \subseteq \mathsf{C}(X)$, must be contained in a closed half space of some subspace $\mathcal{H}^{d-1}$ with $\dim \mathcal{H}^{d-1} = d - 1$. Repeating this argument for $X \cap \mathcal{H}^{d-1}$ and iterating then yields the desired result. □

When studying zero-sum subsequences, it natural to focus on those subsets G_0 such that every $g \in G_0$ is contained in some atom. Otherwise, we could simply pass to a subset of G_0 having this property. The next proposition characterizes such sets $G_0 \subseteq \Lambda$.

Proposition 4.4 *Let $\Lambda \leq \mathbb{R}^d$ be a full rank lattice, where $d \geq 0$, and let $G_0 \subseteq \Lambda$. Then every $g \in G_0$ has some $U \in \mathcal{A}(G_0)$ with $g \in \mathrm{Supp}(U)$ if and only if $\mathsf{C}(G_0) = \mathbb{R}\langle G_0 \rangle$.*

Proof We may w.l.o.g. assume $\mathbb{R}\langle G_0 \rangle = \mathbb{R}^d$ (by passing to the lattice $\mathbb{Z}\langle G_0 \rangle \leq \mathbb{R}\langle G_0 \rangle$). If $d = 0$, then $G_0 \subseteq \{0\}$ and the result is clear. Therefore we may assume $d \geq 1$. Recall that G_0 is contained in a closed half-space if and only if $\mathsf{C}(G_0) \neq \mathbb{R}\langle G_0 \rangle = \mathbb{R}^d$. First assume every $g \in G_0$ has some $U \in \mathcal{A}(G_0)$ with $g \in \mathrm{Supp}(U)$. If G_0 were contained in a closed half space $\mathcal{H}_+$, then it is clear by reduction modulo $\mathcal{H}$ that no atom can contain a term outside of $\mathcal{H}$. Thus, as each $g_0 \in G_0$ lies in some atom by hypothesis, we must have $G_0 \subseteq \mathcal{H} \neq \mathbb{R}^d$, contradicting that $\mathbb{R}\langle G_0 \rangle = \mathbb{R}^d$. It remains to prove the other direction, for which we proceed by induction on d. The cases $d \leq 1$ are easily verified, so we assume $d \geq 2$.

Let $G'_0 \subseteq G_0$ be the subset of all $g \in G_0$ that are contained in the support of some atom. Assuming $G'_0 \neq G_0$, we need to show that G_0 is contained in a closed half space of $\mathbb{R}^d$. For this, we can w.l.o.g. assume $0 \in G_0$, and thus that G'_0 is nonempty. Let $\mathcal{E} = \mathbb{R}\langle G'_0 \rangle$. First suppose that $\dim \mathcal{E} = 0$, i.e., that $G'_0 = \{0\}$. Then $\mathcal{A}(G_0 \setminus \{0\}) = \mathcal{A}(G_0 \setminus G'_0) = \emptyset$, whence Propositions 4.2 and 4.3 imply that $G_0 \setminus \{0\}$ is contained in a closed half space of $\mathbb{R}^d$, and thus so too G_0, as desired. So we now assume $\dim \mathcal{E} \geq 1$.

Since every $g \in G_0'$ is contained in some atom over G_0 but no element in $G_0 \setminus G_0'$ is contained in an atom over G_0, we actually have that every $g \in G_0'$ is contained in some atom over G_0'. Thus by the already established direction of the proof, we conclude that $\mathsf{C}(G_0') = \mathcal{E}$. Consequently, any sequence $T \in \mathcal{F}(G_0)$ with $\sigma(T) \in \mathcal{E}$ can be completed to a zero-sum sequence $T' \in \mathcal{B}(G_0)$ with $\mathrm{Supp}(T) \subseteq \mathrm{Supp}(T')$ in view of Proposition 4.2. This means no element of $G_0 \setminus G_0' \neq \emptyset$ is contained in a zero-sum modulo $\mathcal{E}$. Thus, by the induction hypothesis, we conclude that $\pi(G_0)$ is contained in a closed half space $\mathcal{H}_+$ of $\mathcal{E}^\perp$, where $\pi : \mathbb{R}^d \to \mathcal{E}^\perp$ is the orthogonal projection, which implies G_0 is contained in the closed half space $\pi^{-1}(\mathcal{H}_+)$ of $\mathbb{R}^d$, as desired. □

As a corollary to Propositions 4.4 and 4.2, we have the following.

Corollary 4.5 *Let $\Lambda \leq \mathbb{R}^d$ be a full rank lattice, where $d \geq 0$, let $G_0 \subseteq \Lambda$, let $\mathcal{E} = \mathsf{C}(G_0) \cap -\mathsf{C}(G_0)$, and let $\widetilde{G}_0 = \{g \in G_0 : g \in \mathrm{Supp}(U)$ for some $U \in \mathcal{A}(G_0)\}$. Then $\widetilde{G}_0 = G_0 \cap \mathcal{E}$ and $\mathsf{C}(\widetilde{G}_0) = \mathcal{E}$.*

Proof By definition of $\widetilde{G}_0$, any $U \in \mathcal{A}(G_0)$ must have $\mathrm{Supp}(U) \subseteq \widetilde{G}_0$. Thus Proposition 4.4 applied to $\widetilde{G}_0$ (contained in the full rank lattice $\Lambda \cap \mathbb{R}\langle \widetilde{G}_0\rangle$ of $\mathbb{R}\langle \widetilde{G}_0\rangle$) implies that

$$\mathsf{C}(\widetilde{G}_0) = \mathbb{R}\langle \widetilde{G}_0\rangle.$$

If $g \in G_0 \cap \mathbb{R}\langle \widetilde{G}_0\rangle$, then $\mathsf{C}(\widetilde{G}_0 \cup \{g\}) = \mathsf{C}(\widetilde{G}_0) = \mathbb{R}\langle \widetilde{G}_0\rangle$, in which case Proposition 4.4 applied to $\widetilde{G}_0 \cup \{g\}$ (contained in the full rank lattice $\Lambda \cap \mathbb{R}\langle \widetilde{G}_0\rangle$ of $\mathbb{R}\langle \widetilde{G}_0\rangle$) implies that g is contained in the support of some atom $U \in \mathcal{A}(G_0)$, whence $g \in \widetilde{G}_0$ by definition of $\widetilde{G}_0$. This shows that

$$\widetilde{G}_0 = G_0 \cap \mathbb{R}\langle \widetilde{G}_0\rangle.$$

Let $\pi : \mathbb{R}^d \to \mathbb{R}\langle \widetilde{G}_0\rangle^\perp$ be the orthogonal projection, so $\widetilde{G}_0 = \{g \in G_0 : \pi(g) = 0\}$. Clearly, $\mathsf{C}(\widetilde{G}_0) = \mathbb{R}\langle \widetilde{G}_0\rangle$ is contained in the lineality space $\mathcal{E}$ for $\mathsf{C}(G_0)$. If $0 \in \mathsf{C}^*(\pi(G_0) \setminus \{0\}) = \mathsf{C}^*(\pi(G_0 \setminus \widetilde{G}_0))$, then there will be a nontrivial positive linear combination of elements from $G_0 \setminus \widetilde{G}_0$ contained in $\mathbb{R}\langle \widetilde{G}_0\rangle = \mathsf{C}(\widetilde{G}_0)$. But then Proposition 4.2 ensures that there is some zero-sum $U \in \mathcal{B}(G_0)$ whose support contains elements from $G_0 \setminus \widetilde{G}_0$, and thus an atom as well, contradicting the definition of $\widetilde{G}_0$. Therefore $0 \notin \mathsf{C}^*(\pi(G_0) \setminus \{0\})$, ensuring that $\mathsf{C}(\pi(G_0))$ has trivial lineality space, meaning $\mathbb{R}\langle \widetilde{G}_0\rangle = \ker \pi = \mathcal{E}$ is the lineality space for $\mathsf{C}(G_0)$. Hence $\mathsf{C}(\widetilde{G}_0) = \mathbb{R}\langle \widetilde{G}_0\rangle = \mathcal{E}$ and $\widetilde{G}_0 = G_0 \cap \mathbb{R}\langle \widetilde{G}_0\rangle = G_0 \cap \mathcal{E}$, as desired. □

4.2 Elementary Atoms and Positive Bases

The set of atoms $\mathcal{A}(G_0)$ is central to the study of factorizations over $\mathcal{B}(G_0)$, being the basic building block of all zero-sum sequences. However, some atoms are more elementary than others.

Definition 4.6 Let $G_0 \subseteq G$ be a subset of a torsion-free abelian group. An atom $U \in \mathcal{A}(G_0)$ is called **elementary** if $\mathcal{A}(X) = \emptyset$ for every proper subset $X \subset \mathrm{Supp}(U)$. We let $\mathcal{A}^{\mathsf{elm}}(G_0) \subseteq \mathcal{A}(G_0)$ denote the set of all elementary atoms.

The notion of an elementary atom was introduced to the context of factorization theory in [14] (with a variation in the definition), but has independent and much older origins in both Convex Geometry and Matroid Theory [111]. The approach taken in [14] followed the Matroid theoretic branch. However, the vein from Convex Geometry provides a natural framework for adapting the arguments, and one which we will generalize quite extensively in this work. We begin with the following theorem explaining why elementary atoms can, in some sense, be considered the basic building blocks for all atoms. Theorem 4.7 is essentially Carathéordory's Theorem translated into the language of zero-sum sequences. At the very least, the key idea used in the proof of Carathéodory's Theorem is the same used to prove Theorem 4.7. The details are given in [14, Theorem 3.7] albeit using an alternative definition for an elementary atom. To avoid any confusion, we include the short proof below.

Theorem 4.7 *Let $G_0 \subseteq G$ be a subset of a torsion-free abelian group G. Any zero-sum $S \in \mathcal{B}_{\mathsf{rat}}(G_0)$ can be written as a product of rational powers of elementary atoms, i.e., there are elementary atoms $U_1, \ldots, U_\ell \in \mathcal{A}^{\mathsf{elm}}(G_0)$ and positive rational numbers $\alpha_1, \ldots, \alpha_\ell \in \mathbb{Q}_+$ such that*

$$S = \prod\nolimits^{\bullet}_{i \in [1,\ell]} U_i^{[\alpha_i]} \quad \textit{with} \quad \ell \leq |\,\mathrm{Supp}(S)|.$$

Proof We show the theorem holds for any rational zero-sum $S \in \mathcal{B}_{\mathsf{rat}}(G_0)$ by induction on $|\,\mathrm{Supp}(S)|$. When $|\,\mathrm{Supp}(S)| = 1$, then S is itself a rational power of an elementary atom, and the theorem holds trivially. This completes the base of the induction. Given $S \in \mathcal{B}_{\mathsf{rat}}(G_0)$, there exists an elementary atom $U \in \mathcal{A}^{\mathsf{elm}}(G_0)$ with $\mathrm{Supp}(U) \subseteq \mathrm{Supp}(S)$. Indeed, simply consider an atom $U \in \mathcal{A}(G_0)$ with $\mathrm{Supp}(U) \subseteq \mathrm{Supp}(S) = \mathrm{Supp}(S^{[N]})$ and $\mathrm{Supp}(U)$ minimal subject to this constraint, where $S^{[N]} \in \mathcal{B}(G_0)$. Let $\alpha = \min\{\mathsf{v}_x(S)/\mathsf{v}_x(U) : \ x \in \mathrm{Supp}(U)\}$, which is a positive rational number as $\mathrm{Supp}(U) \subseteq \mathrm{Supp}(S)$ with $U, S \in \mathcal{B}_{\mathsf{rat}}(G_0)$. Then $\mathsf{v}_x(U^{[\alpha]}) \leq \mathsf{v}_x(S)$ for all $x \in \mathrm{Supp}(S)$ with equality holding for any x attaining the minimum in the definition of α. Thus $U^{[\alpha]} \mid S$ with $S \cdot U^{[-\alpha]} \in \mathcal{B}_{\mathsf{rat}}(G_0)$ a rational zero-sum having $|\,\mathrm{Supp}(S \cdot U^{[-\alpha]})| < |\,\mathrm{Supp}(S)|$. Applying the induction hypothesis to $S \cdot U^{[-\alpha]}$ now completes the proof. □

Given a convex cone $C \subseteq \mathbb{R}^d$, a subset $X \subseteq C$ such that $\mathsf{C}(X) = C$ but $\mathsf{C}(Y) \neq C$ for all proper subsets $Y \subset X$ is called a frame of C [36]. A frame for a *positive dimensional* subspace $\mathcal{E} \subseteq \mathbb{R}^d$ is called a **positive basis**. This is the natural extension of a linear basis to Convex Geometry. Unlike ordinary linear bases, positive bases can exhibit complex algebraic structure. Clearly, any positive basis X of an n-dimensional subspace $\mathcal{E}$ must have $|X| \geq n + 1$. If equality holds, X is called a **minimal positive basis** for the subspace $\mathcal{E}$, though the name is somewhat misleading. We do not allow positive bases of zero dimensional subspaces for technical reasons. While a general positive basis can be complex, minimal positive bases are easily described and closely related to elementary atoms, as the following proposition shows. In particular, Proposition 4.8 shows U is an elementary atom precisely when $\mathsf{Supp}(U)$ is a minimal positive basis or $\{0\}$.

Proposition 4.8 *For $X \subseteq \mathbb{R}^d$ with $\mathcal{E} = \mathbb{R}\langle X\rangle$ a* nontrivial *space, the following are equivalent.*

1. *X is a minimal positive basis for $\mathcal{E}$.*
2. *$0 \in \mathsf{C}^*(X)$ and any proper subset $Y \subset X$ is linearly independent.*
3. *$X \setminus \{x\}$ is linearly independent with $-x \in \mathsf{C}^\circ(X \setminus \{x\})$ for every $x \in X$.*
4. *$X \setminus \{x\}$ is linearly independent with $-x \in \mathsf{C}^\circ(X \setminus \{x\})$ for some $x \in X$.*
5. *If M is the $d \times |X|$ matrix whose columns are the elements from X, then the null space of M has dimension* 1 *and is generated by a vector having all coordinates strictly positive.*
6. *$0 \in \mathsf{C}^*(X)$ but $0 \notin \mathsf{C}^*(Y)$ for all proper subsets $Y \subset X$.*

If $X \subseteq \Lambda$ with $\Lambda \leq \mathbb{R}^d$ a full rank lattice, then the above are equivalent to each of the following.

7. *$X = \mathsf{Supp}(U)$ for some elementary atom U.*
8. *$|\mathcal{A}(X)| = 1$ and $\mathcal{A}(Y) = \emptyset$ for all proper $Y \subset X$.*

Proof 1. $\Rightarrow$ 2. Since X is a positive basis, $\mathsf{C}(X) = \mathcal{E}$, thus containing the subspace $\mathcal{E}$ of dimension $\dim \mathcal{E} \geq 1$, whence $0 \in \mathsf{C}^*(X)$, showing that X is linearly dependent. Since $\mathsf{C}(X) = \mathcal{E}$, it follows that $\mathbb{R}\langle X \setminus \{x\}\rangle = \mathcal{E}$ for all $x \in X$, for otherwise $\mathsf{C}(X)$ would be contained in the proper half space $\mathbb{R}\langle X \setminus \{x\}\rangle + \mathbb{R}_+ x \subset \mathcal{E}$, contradicting that $\mathsf{C}(X) = \mathcal{E}$. If a proper subset $Y \subset X$ were linearly dependent, then $X \setminus Y$ is nonempty and $\dim \mathcal{E} \leq (|Y| - 1) + |X \setminus Y| = |X| - 1 = \dim \mathcal{E}$, with the last equality following as X is a *minimal* positive basis. This forces $X \setminus Y$ to be a basis for $\mathcal{E}$ modulo the subspace $\mathcal{E}' = \mathbb{R}\langle Y\rangle$. In particular, $\mathbb{R}\langle X \setminus \{x\}\rangle \neq \mathcal{E}$ for any $x \in X \setminus Y$, contrary to what we established above. This shows each proper subset is linearly independent.

2. $\Rightarrow$ 3. For $x \in X$, we have $X \setminus \{x\}$ linearly independent by Item 2. Since $0 \in \mathsf{C}^*(X)$, there must be a strictly positive linear combination of *all* elements from X equal to zero, since any proper subset is linearly independent. Thus $-x$ can be written as a strictly positive linear combination of the linearly independent elements from $X \setminus \{x\}$, showing that $-x \in \mathsf{C}^\circ(X \setminus \{x\})$.

3. $\Rightarrow$ 4. This is immediate.

4. $\Rightarrow$ 5. The dimension of the null space of the $d \times |X|$ matrix M is equal to $|X| - \dim \mathbb{R}\langle X\rangle$, which equals 1 by Item 4, since $X \setminus \{x\}$ is linearly independent but $-x \in \mathsf{C}^\circ(X \setminus \{x\}) \subseteq \mathbb{R}\langle X \setminus \{x\}\rangle$. Since $X \setminus \{x\}$ is linearly independent, $\mathsf{C}^\circ(X \setminus \{x\})$ consists of all elements which are a strictly positive linear combination of *all* elements from $X \setminus \{x\}$, in which case $-x \in \mathsf{C}^\circ(X \setminus \{x\})$ ensures that the null space of M must contain a vector whose coordinates are all strictly positive, which must then be a generator.

5. $\Rightarrow$ 6. Let $x_1, \ldots, x_r \in X$ be the distinct elements of X. Let $z = (\alpha_1, \ldots, \alpha_r)$ be the generator for the null space of M. Then $\alpha_i > 0$ for all i by item 5 and $\alpha_1 x_1 + \ldots + \alpha_r x_r = 0$ showing that $0 \in \mathsf{C}^*(X)$. On the other hand, if $0 \in \mathsf{C}^*(Y)$ for some proper subset $Y \subset X$, then there would be $\beta_1, \ldots, \beta_r \in \mathbb{R}_+$ with $\beta_1 x_1 + \ldots + \beta_r x_r = 0$, not all β_i zero, and some $\beta_j = 0$. But then $(\beta_1, \ldots, \beta_r)$ would be a nonzero element of the null space of M and clearly not a multiple of z (as any non-zero multiple of z has all non-zero coordinates), contradicting Item 5.

6. $\Rightarrow$ 1. Since $0 \in \mathsf{C}^*(\{0\})$ and $\mathbb{R}\langle X\rangle$ is nontrivial, the hypotheses of Item 6 force $0 \notin X$. Let $x_1, \ldots, x_r \in X$ be the distinct elements of X. Observe that $\mathsf{C}(Y) = \mathbb{R}\langle Y\rangle$, for a nonempty subset $Y \subseteq X$, implies $\mathsf{C}(Y)$ contains the positive dimensional subspace $\mathbb{R}\langle Y\rangle$, yielding $0 \in \mathsf{C}^*(Y)$. By assumption of Item 6, the only nonempty subset $Y \subseteq X$ for which this can hold is $Y = X$. Thus to show X is a positive basis, we only need to show $\mathsf{C}(X) = \mathbb{R}\langle X\rangle$. Since $0 \in \mathsf{C}^*(X)$ but $0 \notin \mathsf{C}^*(Y)$ for all proper subsets $Y \subset X$, it follows that there is a strictly positive linear combination $\alpha_1 x_1 + \ldots + \alpha_r x_r = 0$ using all elements from X. Let $y \in \mathsf{C}(X)$ be nonzero. Then $y = \beta_1 x_1 + \ldots + \beta_r x_r$ for some $\beta_1, \ldots, \beta_r \in \mathbb{R}_+$. Let $\beta = \max_i \beta_i > 0$ and let $\alpha = \min_i \alpha_i > 0$. Then $-y = (-\beta_1 x_1 - \ldots - \beta_r x_r) + \frac{\beta}{\alpha}(\alpha_1 x_1 + \ldots + \alpha_r x_r) = \sum_{i=1}^{r}(-\beta_i + \frac{\beta}{\alpha}\alpha_i)x_i$, with $-\beta_i + \frac{\beta}{\alpha}\alpha_i \geq 0$ for all i, showing $-y \in \mathsf{C}(X)$. As $y \in \mathsf{C}(X)$ was an arbitrary non-zero element, this means the convex cone $\mathsf{C}(X)$ is in fact a subspace, which is only possible if $\mathsf{C}(X) = \mathbb{R}\langle X\rangle = \mathcal{E}$. This shows that X is a positive basis for $\mathcal{E}$.

Assuming X is not a *minimal* positive basis, we obtain $r = |X| \geq \dim \mathcal{E} + 2$. In such case, there must be a linear combination of a *proper* subset of X equal to 0. Thus, by re-indexing the x_i, we can w.l.o.g. assume $\gamma_1 x_1 + \ldots + \gamma_s x_s = 0$ for some $s \in [1, \dim \mathcal{E} + 1] \subseteq [1, r-1]$ and some $\gamma_i \in \mathbb{R}$ with not all $\gamma_i = 0$. If $\gamma_i \geq 0$ for all i, then this contradicts the hypothesis $0 \notin \mathsf{C}^*(Y)$ for $Y = \{x_1, \ldots, x_s\} \subset X$. Therefore we can define $\gamma = \min\{\alpha_i/|\gamma_i| : \gamma_i < 0,\ i \in [1, s]\} > 0$, which ensures that $\alpha_i + \gamma\gamma_i \geq 0$ for all $i \in [1, s]$, with equality holding for any i attaining the minimum in the definition of γ. But now $(\alpha_1 + \gamma\gamma_1)x_1 + \ldots + (\alpha_s + \gamma\gamma_s)x_s + \alpha_{s+1}x_{s+1} + \ldots + \alpha_r x_r = 0$ is a positive linear combination equal to zero. Moreover, the coefficient of x_r is $\alpha_r > 0$ as $r > s$, and the coefficient of some x_i with $i \in [1, s]$ is zero (namely, any $i \in [1, s]$ attaining the minimum in the definition of γ), contradicting that $0 \notin \mathsf{C}^*(Y)$ for all proper subsets $Y \subset X$. This completes the equivalence of the first 6 items.

Now assume $X \subseteq \Lambda$ with $\Lambda \leq \mathbb{R}^d$ a full rank lattice. Then, in view of Proposition 4.2, we see that Items 6 and 7 are equivalent reformulations of each other. The implication 8. $\Rightarrow$ 7. is also immediate. To complete the proof, we show that 5. $\Rightarrow$ 8. Since $X \subseteq \Lambda$, each vector $x \in X$ can be expressed as an integer linear combination of the elements from some lattice basis E for Λ, which is then also a linear basis for $\mathbb{R}^d$. If we use E to express the column vectors in M, it follows that M is an integer matrix, whence the generator z of the null space of M from Item 5 will have each of its i-th coordinates $\alpha_i > 0$ being a rational number. By clearing denominators, we can further assume $\alpha_i \in \mathbb{Z}$ with $\gcd(\alpha_1, \ldots, \alpha_r) = 1$ and $\alpha_i \geq 1$ for all i. It is then clear that any positive multiple of $z = (\alpha_1, \ldots, \alpha_r)$ having all of its coordinates integers must be an integer multiple of z. This means every zero-sum sequence $S \in \mathcal{B}(X)$ has the form $S = U^{[n]}$ for some integer $n \geq 1$, where $U = \prod^{\bullet}_{i \in [1,r]} x_i^{[\alpha_i]} \in \mathcal{B}(X)$, which implies that $U \in \mathcal{A}(X)$ is the unique atom with support contained in X. Hence $|\mathcal{A}(X)| = 1$, and since $\mathrm{Supp}(U) = X$, it follows that $\mathcal{A}(Y) = \emptyset$ for all proper $Y \subset X$, completing the proof. □

4.3 Reay Systems

Much of the material from this subsection can be extracted by a careful examination, variation and reformulation of key ideas from the proofs found in the early works [20, 36, 103, 105, 106, 115]. The material in the format and generality we require is not readily available, so this section will serve as the foundation for more extensive generalizations in later sections. The key definition is the following, which may be found buried in a proof of Reay in the specialized case when $X = X_1 \cup \ldots \cup X_s$ is a positive basis [105]. Since the authors of the aforementioned references were rather focussed on the study of positive bases, their notions were not developed beyond this context in the fuller generality needed for this paper.

Definition 4.9 Let $X_1, \ldots, X_s \subseteq \mathbb{R}^d$ be nonempty subsets, where $s \geq 0$ and $d \geq 0$. For $j \in [0, s]$, let $\pi_j : \mathbb{R}^d \to \mathbb{R}\langle X_1 \cup \ldots \cup X_j \rangle^{\perp}$ be the orthogonal projection. We say that $\mathcal{R} = (X_1, \ldots, X_s)$ is a **Reay (coordinate) system** for the subspace $\mathbb{R}\langle X_1 \cup \ldots \cup X_s \rangle \subseteq \mathbb{R}^d$ if

$$\pi_{j-1}(X_j) \text{ is a minimal positive basis of size } |\pi_{j-1}(X_j)| = |X_j|, \text{ for all } j \in [1, s].$$

We view the empty tuple as a Reay system for the trivial space. Let $\mathcal{R} = (X_1, \ldots, X_s)$ be a Reay system for the subspace $\mathcal{E} \subseteq \mathbb{R}^d$ and let $X = X_1 \cup \ldots \cup X_s$. We say that a subset $Y \subseteq \mathbb{R}^d$ **contains** the Reay system $\mathcal{R}$ if $X \subseteq Y$. If the Reay system $\mathcal{R}$ has $X = X_1 \cup \ldots \cup X_s$ being a positive basis, then we call $\mathcal{R}$ a **Reay basis**. We call $s \geq 0$ the **depth** of the Reay system, in which case a minimal positive basis is just a Reay system of depth 1, and an element $x \in X$ with $x \in X_j$ is said to be at depth j. The basic existence result for Reay systems is the following.

Proposition 4.10 *Let $X \subseteq \mathbb{R}^d$, where $d \geq 0$, and let $\mathcal{E} = \mathsf{C}(X) \cap -\mathsf{C}(X)$. Then X contains a Reay system for $\mathcal{E}$. Moreover, if $Y \subseteq X$ is a minimal positive basis, then X contains a Reay system $(X_1, \ldots, X_s)$ for $\mathcal{E}$ with $X_1 = Y$.*

Proof Recall that $\mathcal{E}$ is the maximal subspace contained in $\mathsf{C}(X)$. If $\dim \mathcal{E} = 0$, then X contains no positive basis, and the empty tuple gives the desired Reay system. Therefore assume $\dim \mathcal{E} \geq 1$ and proceed by induction on $\dim \mathcal{E}$, the base case having just been completed. Since $\mathsf{C}(X)$ contains the positive dimension subspace $\mathcal{E}$, we have $0 \in \mathsf{C}^*(X \setminus \{0\})$, and thus there must be a minimal (by inclusion) subset $X_1 \subseteq X \setminus \{0\}$ with $0 \in \mathsf{C}^*(X_1)$. In view of Proposition 4.8.6, such a subset $X_1 \subseteq X$ is a minimal positive basis contained in X. Let $\pi_1 : \mathbb{R}^d \to \mathbb{R}\langle X_1 \rangle^{\perp}$ be the orthogonal projection. Since there is nothing otherwise special about X_1, we can w.l.o.g. assume X_1 is equal to *any* minimal positive basis $Y \subseteq X$. In view of the maximality of $\mathcal{E}$, we have $\mathsf{C}(X_1) = \mathbb{R}\langle X_1\rangle \subseteq \mathcal{E}$. If $\mathbb{R}\langle X_1\rangle = \mathcal{E}$, we are done. So instead assume $\dim \mathbb{R}\langle X_1\rangle < \dim \mathcal{E}$. Then $\pi_1(\mathcal{E})$ will be the maximal subspace contained in $\pi_1(\mathsf{C}(X)) = \mathsf{C}(\pi_1(X))$, so by induction hypothesis $\pi_1(\mathcal{E})$ has a Reay system $(\pi_1(X_2), \ldots, \pi_1(X_s))$ with $X_i \subseteq X$ for all i, and by discarding elements with equal images under π_1, we can w.l.o.g. assume $|X_j| = |\pi_1(X_j)|$ for all j. It now follows that $(X_1, \ldots, X_s)$ will be a Reay system for $\mathcal{E}$. □

We continue with some basic observations regarding Reay systems.

Proposition 4.11 *Let $\mathcal{R} = (X_1, \ldots, X_s)$ be a Reay system for a subspace $\mathcal{E} \subseteq \mathbb{R}^d$ of dimension n and let $X = X_1 \cup \ldots \cup X_s$. For $j \in [0, s]$, let $\mathcal{E}_j = \mathbb{R}\langle X_1 \cup \ldots \cup X_j\rangle$ and let $\pi_j : \mathbb{R}^d \to \mathcal{E}_j^\perp$ be the orthogonal projection.*

1. *$\mathsf{C}(X) = \mathbb{R}\langle X\rangle$, and $\bigcup_{i=1}^s X_i \setminus \{x_i\}$ is a linear basis for $\mathcal{E}$ for any $x_i \in X_i$.*
2. *$X = X_1 \cup \ldots \cup X_s$ is a disjoint union with $|X| = \dim \mathcal{E} + s = n + s \leq 2n$.*
3. *$(X_1, \ldots, X_j)$ is a Reay system for $\mathcal{E}_j$ and $(\pi_{j-1}(X_j), \ldots, \pi_{j-1}(X_s))$ is a Reay system for $\pi_{j-1}(\mathcal{E})$, for any $j \in [1, s]$.*
4. *If X is a positive basis, then any Reay system $(Y_1, \ldots, Y_s)$ for $\mathcal{E}$ contained in X is a Reay basis and has $X = Y_1 \cup \ldots \cup Y_s$. Moreover, the Reay systems given in Item 3 are all Reay bases.*

Proof

1. A quick inductive argument on $j = 0, 1, \ldots, s$ shows that $\mathsf{C}(X_1 \cup \ldots \cup X_j) = \mathcal{E}_j$. Using Proposition 4.8.3 and an inductive argument on $j = 1, 2, \ldots, s$ shows that $\bigcup_{i=1}^j X_i \setminus \{x_i\}$ is a linear basis for $\mathbb{R}\langle X_1 \cup \ldots \cup X_j\rangle$ for any $x_i \in X_i$. The case $j = s$ yields Item 1.
2. That the elements in $X_1 \cup \ldots \cup X_j$ are distinct follows by a simple inductive argument on $j = 1, \ldots, s$ utilizing that π_{j-1} is injective on X_j with all elements in a minimal positive basis $\pi_{j-1}(X_j)$ nonzero. Hence $n = \dim \mathcal{E} = \sum_{i=1}^s (|X_i| - 1) = |X| - s$ by Item 1. Moreover, since $1 \leq \dim \mathcal{E}_1 < \dim \mathcal{E}_2 < \ldots < \dim \mathcal{E}_s = \dim \mathcal{E} = n$ (as each minimal positive basis $\pi_{j-1}(X_j)$ must span a *nontrivial* subspace), we have $s \leq n$, completing Item 2.
3. This follows immediately from the recursive definition of a Reay system.
4. That any Reay system $(Y_1, \ldots, Y_s)$ for $\mathcal{E}$ contained in the positive basis X must have $X = Y_1 \cup \ldots \cup Y_s$, and therefore be a Reay Basis, follows from Item 1, for otherwise $\mathsf{C}(Y) = \mathcal{E}$ for the proper subset $Y = Y_1 \cup \ldots \cup Y_s$, contradicting that X is a positive basis. If $Y \subseteq X_j \cup \ldots \cup X_s$ is a subset with $\mathsf{C}(\pi_{j-1}(Y)) = \pi_{j-1}(\mathcal{E})$, then $\mathsf{C}(X_1 \cup \ldots \cup X_{j-1} \cup Y) = \mathcal{E}$ follows in view of $\mathsf{C}(X_1 \cup \ldots \cup X_{j-1}) = \mathcal{E}_{j-1} = \ker \pi_{j-1}$ (which holds by Item 1 applied to the Reay system $(X_1, \ldots, X_{j-1})$). Likewise, if $Y \subseteq X_1 \cup \ldots \cup X_j$ with $\mathsf{C}(Y) = \mathcal{E}_j$, then $\mathsf{C}(Y \cup X_{j+1} \cup \ldots \cup X_s) = \mathcal{E}$ follows in view of $\mathsf{C}(\pi_j(X_{j+1}) \cup \ldots \pi_j(X_s)) = \pi_j(\mathcal{E})$ (which holds in view of Item 1 applied to the Reay system $(\pi_j(X_{j+1}), \ldots, \pi_j(X_s))$). Thus $\mathcal{R}$ being a Reay basis implies that $(X_1, \ldots, X_j)$ and $(\pi_{j-1}(X_j), \ldots, \pi_{j-1}(X_s))$ are Reay bases too, for any $j \in [1, s]$.

□

The key property of Reay systems is that they allow for a certain type of unique expression, a fact not highlighted in the original work of Reay.

Proposition 4.12 *Let $(X_1, \ldots, X_s)$ be a Reay system for the subspace $\mathcal{E} \subseteq \mathbb{R}^d$. Then every $z \in \mathcal{E}$ has a unique expression as*

$$z = \sum_{j=1}^{s} \sum_{x \in X_j} \alpha_x x$$

with all $\alpha_x \in \mathbb{R}_+$ but, for every $j \in [1, s]$, not all α_x with $x \in X_j$ are nonzero.

Proof If $\mathcal{E}$ is trivial, then $x = s = 0$ and the unique expression for x is the empty sum. Therefore assume $\dim \mathcal{E} \geq 1$. We proceed by induction on s. The first nontrivial case is $s = 1$, when we have only the minimal positive basis X_1. In view of the characterization given in Proposition 4.8.3, one can apply a linear transformation φ mapping the first $|X_1| - 1$ elements of X_1 to the standard basis vectors in $\mathbb{R}^d$, and then the remaining element of X_1 will map to an element with all coordinates strictly negative, a case for which the uniqueness of expression for $\varphi(z)$ is easily verified, and one which implies the same property for the original element z. This completes the base case. Now assume $s \geq 2$ and that we have unique expression for all smaller values of s. For $j \in [0, s]$, let $\mathcal{E}_j = \mathbb{R}\langle X_1 \cup \ldots \cup X_j \rangle$ and let $\pi_j : \mathbb{R}^d \to \mathcal{E}_j^\perp$ be the orthogonal projection, so $\mathcal{E}_s = \mathcal{E}$. By Proposition 4.11.3, $(X_1, \ldots, X_{s-1})$ is a Reay system for $\mathcal{E}_{s-1}$ and $\pi_{s-1}(X_s)$ is a minimal positive basis for $\pi_{s-1}(\mathcal{E})$. Since $\mathcal{E}_{s-1} \subseteq \mathcal{E}_s = \mathcal{E}$, every $z \in \mathcal{E}$ has a unique expression as $z = a + b$ with $a \in \ker \pi_{s-1} = \mathcal{E}_{s-1}$ and $b \in \mathcal{E}_{s-1}^\perp \cap \mathcal{E} = \pi_{s-1}(\mathcal{E})$. Applying the base case to the minimal positive basis $\pi_{s-1}(X_s)$ for $\pi_{s-1}(\mathcal{E})$, we find there is a unique expression $b = \sum_{x \in X_s} \alpha_x \pi_{s-1}(x) = \pi_{s-1}\left(\sum_{x \in X_s} \alpha_x x\right)$ with $\alpha_x \in \mathbb{R}_+$ for all $x \in X_s$ and not all α_x nonzero. Thus there is a unique expression $b = u_z + \sum_{x \in X_s} \alpha_x x$ with $u_z \in \ker \pi_{s-1} = \mathcal{E}_{s-1}$, with $\alpha_x \in \mathbb{R}_+$, and with not all α_x nonzero (since $b \in \pi_{s-1}(\mathcal{E})$ with π_{s-1} a projection, we have $\pi_{s-1}(b) = b$). Since b is uniquely determined by z, the element $u_z \in \mathcal{E}_{s-1}$ is uniquely determined by z. But now there is a unique expression $z = a' + \sum_{x \in X_s} \alpha_x x$ with $a' \in \mathcal{E}_{s-1}$, with $\alpha_x \in \mathbb{R}^+$, and with not all α_x nonzero (namely, $a' = a + u_z$). Applying the induction hypothesis to the element a' and Reay system $(X_1, \ldots, X_{s-1})$ for $\mathcal{E}_{s-1}$ now yields the desired unique expression for z. □

There is another way to view Proposition 4.12. If $\mathcal{R} = (X_1, \ldots, X_s)$ is a Reay system for the subspace $\mathcal{E} \subseteq \mathbb{R}^d$ with $X = X_1 \cup \ldots \cup X_s$, then define

$$\mathfrak{B} = \{\mathsf{C}(Y) : \; Y \subseteq X \text{ and } X_i \nsubseteq Y \text{ for every } i \in [1, s]\}.$$

For every choice of elements $x_i \in X_i$, for $i \in [1, s]$, the set $Y = \bigcup_{i=1}^{s} X_i \setminus \{x_i\}$ is a linear basis for $\mathcal{E}$ by Proposition 4.11.1. Thus Proposition 4.12 implies that $\mathcal{E} = \bigcup_{C \in \mathfrak{B}} C^\circ$ is the disjoint union of the relative interiors of the cones in $\mathfrak{B}$. Combining these observations, we see that if we intersect each cone from $\mathfrak{B}$ with the unit sphere in $\mathcal{E}$, we obtain an object homeomorphic to a *simplicial complex* of dimension $\dim \mathcal{E} - 1$ whose union is the unit sphere in $\mathcal{E}$. In the parlance of topologists, such an object is called a *simplicial sphere* or *triangulated sphere*. Indeed, owing to the method of construction, we obtain a more restricted class of simplicial sphere known as a *starshaped sphere* (somewhat surprisingly, not every simplicial sphere can be constructed this way). We direct the reader to [25] for a more comprehensive account, including more details of what follows below.

A **fan** is a *finite* collection $\mathfrak{B}$ of polyhedral cones $\mathsf{C}(Y) \subseteq \mathbb{R}^d$ each having trivial lineality space $\mathsf{C}(Y) \cap -C(Y) = \{0\}$. The **faces** of $\mathsf{C}(Y)$ are the sub-cones $\mathsf{C}(Z)$ with $Z \subseteq Y$ obtained by intersecting $\mathsf{C}(Y)$ with a hyperplane defining a closed half-space that contains $\mathsf{C}(Y)$, and it is required that each face of a cone from the fan be an element of the fan, and that the intersection of any two cones in the fan be a face of each. It is a **simplicial** fan if every $\mathsf{C}(Y) \in \mathfrak{B}$ is generated by a linearly independent set Y, in which case there is a face for each subset $Z \subseteq Y$, and it is a **complete** fan if the union of all cones in $\mathfrak{B}$ equals an entire subspace. Thus, the set $\mathfrak{B}$ defined above from a Reay system is a special type of **complete simplicial fan** for the subspace $\mathcal{E} \subseteq \mathbb{R}^d$. Complete simplicial fans are in bijective correspondence with starshaped spheres, and rational fans (those whose vertices come from a lattice) are central to the definition of Toric Varieties, though we will only need their more basic properties.

One easily notes that a Reay system (with depth $s \geq 2$) is not always stable even under small perturbations of its defining vectors. This forces us to work with the more general concept of a complete simplicial fan, which maintains many of the essential features of a Reay system while gaining the important property of being stable under small perturbations. Worth noting, any complete fan must be pure (that is, all maximal cones must have the same dimension). Indeed, if $\mathfrak{B}$ is a complete fan for $\mathbb{R}^d$, then the sub-collection of all d-dimensional cones from $\mathfrak{B}$, together with all their faces (sub-cones), would also be a fan. Their union must be all of $\mathbb{R}^d$, for if it were not, then its complement would be a d-dimensional subset of $\mathbb{R}^d$, which clearly cannot be written as a union of a finite number of lower dimensional objects. But once we know their union is all of $\mathbb{R}^d$, it then follows that any lower dimensional cone must lie in one of the d-dimensional cones, and thus only d-dimensional cones are maximal.

Let $\mathfrak{B}$ be a simplicial fan in $\mathbb{R}^d$. For an integer $k \in [0, d]$, we use $\mathfrak{B}_k$ to denote the subset of $\mathfrak{B}$ consisting of all k-dimensional cones in $\mathfrak{B}$ (generated by k elements). Thus the elements of $\mathfrak{B}_k$ when intersected with the unit sphere give rise (up to homeomorphism) to $(k-1)$-dimensional simplices. A vertex set V for $\mathfrak{B}$ is a collection of nonzero elements, one chosen from each $C \in \mathfrak{B}_1$. For instance, the set $V = \bigcup_{C \in \mathfrak{B}_1} \big(C \cap \partial(B_1(0))\big)$ consisting of the elements of the unit sphere contained in some $B \in \mathfrak{B}_1$ (i.e., the 0-dimensional vertices of the associated starshaped sphere) is one possible set of vertices for $\mathfrak{B}$. Note that $|V(\mathfrak{B})| = |\mathfrak{B}_1|$ is finite. Whenever dealing with a simplicial fan $\mathfrak{B}$, we will fix a vertex set $V(\mathfrak{B})$, and if none is explicitly mentioned, the unit sphere representatives are assumed to be the vertices. Every $x \in \bigcup_{C \in \mathfrak{B}} C$ has a unique cone $C \in \mathfrak{B}$ for which $x \in C^\circ$. The cone C is generated by a set of linearly independent vectors $B_C \subseteq V(\mathfrak{B})$, so

$C = \mathsf{C}(B_C)$ and x is a strictly positive linear combination of the elements of B_C (note: if $x = 0$, then $B_C = \emptyset$ and $C = \{0\}$). If $\mathfrak{B}$ is defined using a Reay system $(X_1, \ldots, X_s)$, then we take $V(\mathfrak{B}) = X_1 \cup \ldots \cup X_s$ for the vertices. The coefficients in the linear combination correspond to the baricentric coordinates for points inside the simplicial cone C. We define $\mathsf{Supp}_{\mathfrak{B}}(x) = B_C \subseteq V(\mathfrak{B})$ to be the **support** set of the element x with respect to the simplicial fan $\mathfrak{B}$, which is then the unique subset $\mathsf{Supp}_{\mathfrak{B}}(x) \subseteq V$ such that $x \in \mathsf{C}^{\circ}(\mathsf{Supp}_{\mathfrak{B}}(x))$. If $\mathfrak{B}$ arises from a Reay system $\mathcal{R}$, then we let

$$\mathsf{Supp}_{\mathcal{R}}(x) = \mathsf{Supp}_{\mathfrak{B}}(x)$$

and call this set the Reay support of the element x. By Proposition 4.11.1, $\mathsf{Supp}_{\mathcal{R}}(x)$ is always a linearly independent set. We carry on with some basic properties about Reay systems and positive bases.

Proposition 4.13 *Let $\mathcal{R} = (X_1, \ldots, X_s)$ be a Reay system for a subspace $\mathcal{E} \subseteq \mathbb{R}^d$ and let $\mathcal{E}_j = \mathbb{R}\langle X_1 \cup \ldots \cup X_j\rangle$ for $j \in [0, s]$. For every $k \in [1, s]$,*

$$\mathsf{C}(X_k) \cap \mathcal{E}_{k-1} = \mathbb{R}_+ u_k$$

for some (possibly zero) $u_k \in \mathcal{E}_{k-1}$. Moreover,

$$X_k' = \mathsf{Supp}_{\mathcal{R}}(-u_k) \cup X_k \subseteq X_1 \cup \ldots \cup X_k$$

is a minimal positive basis for some subspace $\mathcal{E}_k' \subseteq \mathcal{E}_k$.

Proof Let $k \in [1, s]$ be arbitrary. Since $X_k \setminus \{x_k\}$ is linearly independent for any $x_k \in X_k$ by Proposition 4.11.1, it follows that $\dim \mathbb{R}\langle X_k\rangle = |X_k|$ or $|X_k| - 1$, depending on whether the elements of X_k are linearly independent or linearly dependent. If they are linearly dependent, so $\dim \mathbb{R}\langle X_k\rangle = |X_k| - 1$, then, since $\mathcal{E}_k = \mathcal{E}_{k-1} + \mathbb{R}\langle X_k\rangle$, we have

$$\begin{aligned}\dim(\mathcal{E}_{k-1} \cap \mathbb{R}\langle X_k\rangle) &= \dim \mathcal{E}_{k-1} + \dim \mathbb{R}\langle X_k\rangle - \dim \mathcal{E}_k \\ &= \Big(\sum_{i=1}^{k-1}(|X_i| - 1)\Big) + |X_k| - 1 - \Big(\sum_{i=1}^{k}(|X_i| - 1)\Big) = 0,\end{aligned}$$

meaning $\mathbb{R}\langle X_k\rangle \cap \mathcal{E}_{k-1} = \{0\}$. In this case, $\mathsf{C}(X_k) \cap \mathcal{E}_{k-1} = \mathbb{R}_+ u_k$ for $u_k = 0$. On the other hand, if they are linearly independent, so $\dim \mathbb{R}\langle X_k\rangle = |X_k|$, then we instead have $\dim(\mathcal{E}_{k-1} \cap \mathbb{R}\langle X_k\rangle) = 1$, meaning $\mathbb{R}\langle X_k\rangle \cap \mathcal{E}_{k-1}$ is a one-dimensional subspace. Since $0 \in \mathsf{C}^*(\pi_{k-1}(X_k))$ (in view of Proposition 4.8.6), it follows that there is a nontrivial positive linear combination of elements from X_k that lies in $\mathcal{E}_{k-1}$. Since the elements of X_k are linearly independent, this linear combination must be a *nonzero* element of $\mathcal{E}_{k-1}$, say $u_k \in \mathcal{E}_{k-1} \cap \mathsf{C}^*(X_k)$. If $-u_k$ were also contained in $\mathsf{C}(X_k)$, then $0 \in \mathsf{C}^*(X_k)$, contradicting that the elements of X_k are linearly independent. Therefore we instead conclude that $\mathcal{E}_{k-1} \cap \mathsf{C}(X_k) = \mathbb{R}_+ u_k$ in this case as well.

If $u_k = 0$, then X_k is linearly dependent and $\mathrm{Supp}(-u_k) = \emptyset$, so $X'_k = X_k$. By Proposition 4.11.1, every proper subset of X_k is linearly independent, in which case $X'_k = X_k$ is a minimal positive basis by Proposition 4.8.2, as desired. Therefore we now assume $u_k \neq 0$ with X_k linearly independent.

Since $-u_k \in \mathcal{E}_{k-1}$, we have $\mathrm{Supp}_{\mathcal{R}}(-u_k) \subseteq X_1 \cup \ldots \cup X_{k-1}$. Since $u_k \in \mathsf{C}^*(X_k)$ and $-u_k \in \mathsf{C}(\mathrm{Supp}_{\mathcal{R}}(-u_k))$, we have $0 \in \mathsf{C}^*(X'_k)$. Consequently, to show X'_k is a minimal positive basis, it suffices by Proposition 4.8.6 to show $0 \notin \mathsf{C}^*(Y)$ for all proper subsets $Y \subset X'_k$. To this end, consider an arbitrary subset $Y \subseteq X'_k$ with $0 \in \mathsf{C}^*(Y)$. Since $\mathrm{Supp}_{\mathcal{R}}(-u_k)$ is a linearly independent subset, $Y \subseteq \mathrm{Supp}_{\mathcal{R}}(-u_k)$ is not possible, implying $Y \cap X_k \neq \emptyset$. However, since $0 \notin \mathsf{C}^*(\pi_{k-1}(Z))$ for any proper subset $Z \subseteq X_k$ (per Proposition 4.8.6), any strictly positive linear combination of a proper subset of terms from X_k lies outside the subspace $\mathcal{E}_{k-1}$, and thus cannot be combined with any linear combination of terms from $\mathcal{E}_{k-1}$ to yield 0. In consequence, we conclude that $X_k \subseteq Y$. Thus $0 \in \mathsf{C}^*(Y)$ ensures that $0 = a+b$ with $a \in \mathsf{C}^*(X_k)$ and $b \in \mathsf{C}(\mathrm{Supp}_{\mathcal{R}}(-u_k) \cap Y) \subseteq \mathcal{E}_{k-1}$. However, since $\mathcal{E}_{k-1} \cap \mathsf{C}(X_k) = \mathbb{R}_+ u_k$, we must have $a = \alpha u_k$ for some positive $\alpha > 0$, and by re-scaling we may w.l.o.g. assume $\alpha = 1$. Hence $-u_k = -a = b \in \mathsf{C}(\mathrm{Supp}_{\mathcal{R}}(-u_k) \cap Y)$. However, by definition of $\mathrm{Supp}_{\mathcal{R}}(-u_k)$, there is no proper subset of $\mathrm{Supp}_{\mathcal{R}}(-u_k)$ that contains $-u_k$ in its positive span, so we must have $\mathrm{Supp}_{\mathcal{R}}(-u_k) \subseteq Y$, which together with $X_k \subseteq Y$ implies that $Y = X_k$. As $Y \subseteq X'_k$ was an arbitrary subset with $0 \in \mathsf{C}^*(Y)$, we conclude that no proper subset $Y \subset X'_k$ has $0 \in \mathsf{C}^*(Y)$, completing the proof.

□

Let $\mathcal{R} = (X_1, \ldots, X_s)$ be a Reay system for $\mathbb{R}^d$ and let $X = X_1 \cup \ldots \cup X_s$. Then each $z \in \mathbb{R}^d$ corresponds via Proposition 4.12 uniquely to a tuple $\alpha(z) = (\alpha_x(z))_{x \in X} \in \mathbb{R}_+^{|X|}$ with $\sum_{x \in X} \alpha_x(z) x = z$ such that, for every $j \in [1, s]$, not all $\alpha_x(z)$ for $x \in X_j$ are non-zero. Unlike Euclidean coordinates, if $y, z \in \mathbb{R}^d$, we may not have $\alpha(y+z) = \alpha(y) + \alpha(z)$. The problem is that, in one (or more) of the s groupings of coordinates in $\alpha(y) + \alpha(z)$ corresponding to the X_j, all coordinates may be strictly positive, which is not allowed. Proposition 4.13 gives a means to quickly transform $\alpha(y) + \alpha(z)$ into $\alpha(y+z)$. For each $k \in [1, s]$, there is an expression $\sum_{x \in X_k} (\alpha_x^{(k)}) x = u_k = \sum_{i=1}^{k-1} \sum_{x \in X_i} \beta_x^{(k)} x$ with all $\alpha_x^{(k)} > 0$ and $\beta_x^{(k)} \geq 0$. We have $\beta_x^{(k)} > 0$ precisely when $x \in \mathrm{Supp}_{\mathcal{R}}(u_k)$. Let $\mathbf{a}_k = (\mathbf{a}_k(x))_{x \in X} \in \mathbb{R}^{|X|}$ be the vector with $\mathbf{a}_k(x) = -\alpha_x^{(k)}$ for $x \in X_k$, with $\mathbf{a}_k(x) = \beta_x^{(k)}$ for $x \in X_1 \cup \ldots \cup X_{k-1}$, and will all other coordinates zero. Then, if the k-th grouping in $\alpha(y) + \alpha(z)$ has all its coordinates positive, there will be a unique multiple of $\mathbf{a}_k$, namely $b_k \mathbf{a}_k$ with $b_k = \min_{x \in X_k} \frac{\alpha_x(y) + \alpha_x(y)}{|\mathbf{a}_k(x)|}$, such that $\alpha(y) + \alpha(z) + b_k \mathbf{a}_k$ has all coordinates non-negative but has at least one zero in the k-th grouping X_k. Since only coordinates in grouping k and lower change by adding $b_k \mathbf{a}_k$, we can sequentially apply the relations $b_k \mathbf{a}_k$ as needed for $k = s, s-1, \ldots, 1$ until we reduce $\alpha(y) + \alpha(z)$ to $\alpha(y+z)$ in at most $s \leq d$ steps.

The following proposition is a refined statement of one of the main goals in the original work of Reay [105].

Proposition 4.14 *Let $X \subseteq \mathbb{R}^d$ be a positive basis for $\mathbb{R}^d$ with $d \geq 1$. Then there exists a Reay basis $\mathcal{R} = (X_1, \ldots, X_s)$ for $\mathbb{R}^d$ with $X = X_1 \cup \ldots \cup X_s$ and*

$$|X_1| \geq |X_2'| \geq |X_2| \geq |X_3'| \geq |X_3| \geq \ldots \geq |X_s'| \geq |X_s| \geq 2,$$

where each $X'_j = (\mathrm{Supp}_{\mathcal{R}}(-u_j) \setminus \mathcal{E}_{j-2}) \cup X_j$ with $\mathcal{E}_{j-2} = \mathbb{R}\langle X_1 \cup \ldots \cup X_{j-2}\rangle$ and u_j as given in Proposition 4.13, for $j \in [2, s]$.

Proof If X is a minimal positive basis, then $X = X_1$ is itself a Reay basis of depth $s = 1$, in which case the proposition is trivial. Therefore we can assume otherwise. In particular, the proposition is true when $|X| \leq 2$, allowing us to proceed by induction on $|X|$. By Proposition 4.10, X contains a minimal positive basis X_1 for some subspace, so we may w.l.o.g. assume $X_1 \subseteq X$ is a *maximal cardinality* minimal positive basis contained in X. Let $\pi_1 : \mathbb{R}^d \to \mathbb{R}\langle X_1\rangle^\perp$ be the orthogonal projection. We can take any Reay system $(\pi_1(X_2), \ldots, \pi_1(X_s))$, where $X_i \subseteq X$ are subsets with $|\pi_1(X_i)| = |X_i|$, and then $(X_1, X_2, \ldots, X_s)$ will be a Reay system. By Proposition 4.10, there *is* some Reay system $(X_1, \ldots, X_s)$ with $X = X_1 \cup \ldots \cup X_s$. Now π_1 is injective on $X \setminus X_1$ (by definition of a Reay system) while Proposition 4.11.4 ensures that $\pi_1(X \setminus X_1)$ is a positive basis with $|\pi_1(X \setminus X_1)| = |X \setminus X_1| < |X|$. Apply the induction hypothesis to $\pi_1(X \setminus X_1)$ to find a Reay system $\mathcal{R}' = (\pi_1(X_2), \ldots, \pi_1(X_s))$ satisfying the conclusion of the proposition with $|\pi_1(X_i)| = |X_i|$ for all $i \geq 2$ and $X_2 \cup \ldots \cup X_s = X \setminus X_1$. Then $\mathcal{R} = (X_1, \ldots, X_s)$ is a Reay basis with $X = X_1 \cup \ldots \cup X_s$ as noted above. Let X'_j and u_j, for $j \in [2, s]$, be as defined by the proposition for the Reay Basis $\mathcal{R}$, and let $\pi_1(X_j)'$ and u'_j, for $j \in [3, s]$, be the corresponding quantities for the Reay Basis $\mathcal{R}'$. Then $\pi_1(u_j) = u'_j$ and $\pi_1(X'_j) = \pi_1(X_j)'$ for $j \geq 3$. Thus the induction hypothesis and injectivity of π_1 on $X \setminus X_1$ yield

$$|X_2| \geq |X_3'| \geq |X_3| \geq \ldots |X_s'| \geq |X_s| \geq 2.$$

By definition, $|X_2'| \geq |X_2|$, while Proposition 4.13 implies that X_2' is minimal positive basis, so that the maximality of X_1 ensures $|X_1| \geq |X_2'|$, completing the proof. □

Reay Bases can be used to help better understand positive bases. However, most of the useful properties of Reay Bases hold for the more general class of Reay systems or even complete simplicial fans. If one needs more refined structure for a positive basis, there is a geometric interpretation given by Shephard [115] involving the Gale diagram of a linear representation of the positive basis. We conclude the subsection with some important properties of complete simplicial fans.

Proposition 4.15 *Let $\mathfrak{B}$ be a complete simplicial fan for $\mathbb{R}^d$ with $d \geq 1$, and let $\{x_1, \ldots, x_s\} = V(\mathfrak{B})$ be the distinct vertices of $\mathfrak{B}$. Let $x \in \mathbb{R}^d$ and let $\mathfrak{B}_d(x) \subseteq \mathfrak{B}_d$ consist of all cones $\mathsf{C}(B) \in \mathfrak{B}_d$ with $\mathsf{Supp}_{\mathfrak{B}}(x) \subseteq B \subseteq V(\mathfrak{B})$.*

1. *If $\mathsf{C}(B) \in \mathfrak{B}$ with $\mathsf{Supp}_{\mathfrak{B}}(x) \nsubseteq B \subseteq V(\mathfrak{B})$, then $\mathsf{C}(B) \cap \mathbb{R}_+ x = \{0\}$.*
2. *$x \in \mathsf{Int}(\bigcup_{C \in \mathfrak{B}_d(x)} C)$ with $x \notin \mathsf{Int}(\bigcup_{C \in Y} C)$ for any proper subset $Y \subset \mathfrak{B}_d(x)$.*
3. *$\bigcap_{C \in \mathfrak{B}_d(x)} C = \mathsf{C}(\mathsf{Supp}_{\mathfrak{B}}(x))$.*
4. *There is a sufficiently small $\epsilon > 0$ (dependent on $\mathfrak{B}$) such that, for any $y_1, \ldots, y_s \in \mathbb{R}^d \setminus \{0\}$ with $\mathsf{d}(x_i/\|x_i\|, y_i/\|y_i\|) < \epsilon$ for all i, the map $\varphi : \mathfrak{B} \to \mathfrak{B}'$ given by $\varphi(C) = \mathsf{C}(\{y_i\}_{i \in I})$ for $C = \mathsf{C}(\{x_i\}_{i \in I}) \in \mathfrak{B}$, where $I \subseteq [1, s]$, is a simplicial isomorphism between $\mathfrak{B}$ and $\mathfrak{B}' := \{\varphi(C) : C \in \mathfrak{B}\}$. In particular, $\mathfrak{B}'$ is a complete simplicial fan for $\mathbb{R}^d$.*
5. *If $X = \{x_1, \ldots, x_s\}$ is a minimal positive basis for $\mathcal{E} \subseteq \mathbb{R}^d$, then there is an $\epsilon > 0$ such that any set $\{y_1, \ldots, y_{d+1}\} \subseteq \mathcal{E}$ with $\mathsf{d}(x_i/\|x_i\|, y_i/|y_i\|) < \epsilon$ for all i is also a minimal positive basis for $\mathcal{E}$.*
6. *Let $\mathfrak{B}'$ and φ be defined as in Item 4. Then, for all sufficiently small $\epsilon > 0$ (dependent on $\mathfrak{B}$ and x), $\varphi(\mathsf{Supp}_{\mathfrak{B}}(x)) \subseteq \mathsf{Supp}_{\mathfrak{B}'}(x)$.*

Proof

1. Per definition of a complete simplicial fan, we have a disjoint decomposition of $\mathbb{R}^d$ given by $\mathbb{R}^d = \biguplus_{C \in \mathfrak{B}} C^\circ$. Each polyhedral cone $C \in \mathfrak{B}$ corresponds uniquely to some linearly independent subset of vertices $B_C \subseteq V(\mathfrak{B})$ with $\mathsf{C}(B_C) = C$, with the faces of the cone C corresponding to the subsets of B_C. Thus $C' \subseteq C$ is a face when $C' = \mathsf{C}(B_{C'})$ with $B_{C'} \subseteq B_C$. By definition, the cone $\mathsf{C}(B_x) \in \mathfrak{B}$, where $B_x := \mathsf{Supp}_{\mathfrak{B}}(x)$, is the unique cone in $\mathfrak{B}$ that contains $\mathbb{R}_+^\circ x$ in its relative interior. If $\mathsf{C}(B) \in \mathfrak{B}$ is a cone that contains $\mathbb{R}_+ x$, then x will be contained in the relative interior of some face of $\mathsf{C}(B)$. Consequently, since all such faces of $\mathsf{C}(B)$ lie in $\mathfrak{B}$, the uniqueness of $\mathsf{C}(B_x)$ ensures that this face must be $\mathsf{C}(B_x)$, i.e., $B_x \subseteq B$.
2. If $x = 0$, then $\mathsf{Supp}_{\mathfrak{B}}(0) = \emptyset$ and $\mathfrak{B}_d(0) = \mathfrak{B}_d$, in which case the statement follows trivially in view of $\mathfrak{B}$ being complete. Otherwise, in view of Item 1, it follows that there is a small neighborhood around each cone $C \in \mathfrak{B}_d \setminus \mathfrak{B}_d(x)$ with the property that the closure of this neighborhood does not contain x. Thus this is also true for the finite union $\bigcup_{C \in \mathfrak{B}_d \setminus \mathfrak{B}_d(x)} C$, implying $x \in \mathsf{Int}(\bigcup_{C \in \mathfrak{B}_d(x)} C)$ in view of $\bigcup_{C \in \mathfrak{B}_d} C = \mathbb{R}^d$ (as $\mathfrak{B}$ is complete). Recall that the relative interiors of the cones from $\mathfrak{B}$ form a disjoint partition of $\mathbb{R}^d$ with x contained in the relative interior of the cone $C_x := \mathsf{C}(\mathsf{Supp}_{\mathfrak{B}}(x))$. Let $C' \in \mathfrak{B}_d$ be arbitrary. By definition of $\mathfrak{B}_d$, we have $C_x \subseteq C'$, so $x \in C_x \subseteq \overline{C'} = \overline{(C')^\circ}$, with the latter equality following since C' is a convex set. It follows that $B_\epsilon(x) \cap (C')^\circ \neq \emptyset$ for all $\epsilon > 0$, and since $(C')^\circ$ is disjoint from $\bigcup_{C \in \mathfrak{B}_d(x) \setminus \{C'\}} C$ (as the relative interiors of the cones in $\mathfrak{B}$ form a disjoint partition of $\mathbb{R}^d$), it follows that $B_\epsilon(x) \nsubseteq \bigcup_{C \in \mathfrak{B}_d(x) \setminus \{C'\}} C$ for all $\epsilon > 0$, showing that $x \notin \mathsf{Int}(\bigcup_{C \in \mathfrak{B}_d(x) \setminus \{C'\}} C)$. Since $C' \in \mathfrak{B}_d(x)$ was arbitrary, this implies $x \notin \mathsf{Int}(\bigcup_{C \in Y} C)$ for any proper subset $Y \subset \mathfrak{B}_d(x)$.

3. Let $C_x = \mathsf{C}(\mathrm{Supp}_{\mathfrak{B}}(x))$. By definition of $\mathfrak{B}_d(x)$, we have $C_x \subseteq \bigcap_{C \in \mathfrak{B}_d(x)} C$. If the reverse inclusion fails, then there must be some vertex $z \notin B_x := \mathrm{Supp}_{\mathfrak{B}}(x)$ contained in every $C \in \mathfrak{B}_d(x)$. In particular, $B_x \cup \{z\}$ is linearly independent (as the generating vertices of each cone are linearly independent) and linearly spans some subspace $\mathcal{E}$. For any $C = \mathsf{C}(B) \in \mathfrak{B}_d(x)$, we have $B_x \cup \{z\} \subseteq B$ with the vertices in $B \subseteq V(\mathfrak{B})$ linearly independent. Thus $C \cap \mathcal{E} = \mathsf{C}(B_x \cup \{z\})$. As this is true for every $C \in \mathfrak{B}_d(x)$, we conclude that $\left(\bigcup_{C \in \mathfrak{B}_d(x)} C\right) \cap \mathcal{E} = \mathsf{C}(B_x \cup \{z\})$. Consequently, since $B_x \cup \{z\}$ is linearly independent with $x \in \mathsf{C}^\circ(B_x)$, it follows that $\left(\bigcup_{C \in \mathfrak{B}_d(x)} C\right) \cap \mathsf{C}(-z, x) = \mathsf{C}(B_x \cup \{z\}) \cap \mathsf{C}(-z, x) = \mathbb{R}_+ x$, ensuring that x is *not* contained in the interior of $\bigcup_{C \in \mathfrak{B}_d(x)} C$ (else all points $x - \alpha z \in \mathsf{C}(-z, x)$ with $\alpha > 0$ sufficiently small would be contained in $\bigcup_{C \in \mathfrak{B}_d(x)} C$), contrary to Item 2.
4. For each vertex x_j, there are only a finite number of $\mathsf{C}(B) \in \mathfrak{B}$ with $x_j \in B \subseteq V(\mathfrak{B})$, defining a finite number of subspaces linearly spanned by the sets $B \setminus \{x_j\}$. Each such subspace $\mathcal{H}$ does not contain x_j, as the elements of B are linearly independent, so there is some finite positive distance (along the unit sphere) between $x_j/\|x_j\|$ and any such subspace $\mathcal{H}$ intersected with the unit sphere. Let $\epsilon' > 0$ be the minimum such distance, where the minimum runs over all possible vertices $x_j \in V(\mathfrak{B})$ and all possible subspace pairings. Given a subspace $\mathcal{H}$ generated by linearly independent unit vectors, if we perturb the generators of $\mathcal{H}$ by a small amount, replacing each with a new unit vector generator some sufficiently small distance $\epsilon > 0$ from the original vector, then the resulting set of perturbed generators will generate a perturbed subspace $\mathcal{H}'$ of equal dimension which, when intersected with the unit sphere, has all its points some small distance away from the original hyperplane $\mathcal{H}$ intersected with the sphere. Thus, by this continuity property, we may choose $\epsilon > 0$ sufficiently small with respect to $\epsilon' > 0$ to ensure that each perturbed point $y_j/\|y_j\|$ remains disjoint from each perturbed paired subspace $\mathcal{H}'$ (ensuring that each cone in $\mathfrak{B}'$ is still generated by linearly independent elements) and on the same side (in the case of the maximal co-dimension 1 subspaces) as the original vector $x_j/\|x_j\|$. By doing so, we ensure that the simplicial structure of $\mathfrak{B}$ is preserved in $\mathfrak{B}'$, and the result follows. See also [25, Section 5.2].
5. This is a special case of Item 4 in view of Proposition 4.8.
6. If $x = 0$, we have $\mathrm{Supp}_{\mathfrak{B}}(x) = \emptyset = \mathrm{Supp}_{\mathfrak{B}'}(x)$, and the result is clear. Therefore we may assume x is nonzero. Now, if x is a positive scaler multiple of a vertex of $\mathfrak{B}$, then we can assume by rescaling x that $x \in V(\mathfrak{B})$. On the other hand, if x is not a positive scaler multiple of any vertex of $\mathfrak{B}$, then we can perform a baricentric subdivision at x in $\mathfrak{B}$ to create a new complete simplicial fan $\mathfrak{C}$ having $V(\mathfrak{C}) = V(\mathfrak{B}) \cup \{x\}$ (so we remove each $\mathsf{C}(B)$ with $\mathrm{Supp}_{\mathfrak{B}}(x) \subseteq B \subseteq V(\mathfrak{B})$ and replace it with the collection of cones of the form $\mathsf{C}(B \setminus \{y\} \cup \{x\})$ for $y \in \mathrm{Supp}_{\mathfrak{B}}(x)$). In the former case (when $x \in V(\mathfrak{B})$), set $\mathfrak{C} = \mathfrak{B}$. By item 4 applied to $\mathfrak{C}$, for sufficiently small $\epsilon > 0$, replacing each vertex of $\mathfrak{C}$ with a new vertex at radial distance at most ϵ from the original vector results in a new complete simplicial fan $\mathfrak{C}'$ isomorphic to $\mathfrak{C}$. Let $\mathfrak{B}' \subseteq \mathfrak{C}'$ be the complete

simplicial fan associated to the image of the original vertex set $\varphi\big(V(\mathfrak{B})\big)$, in which case $\varphi(C) \in \mathfrak{B}'$ for $C \in \mathfrak{B}$. In view of Item 2 applied to $\mathfrak{B}$, we have $x \in \mathsf{Int}(\bigcup_{C \in \mathfrak{B}_d(x)} C)$, whence

$$x \in \mathsf{Int}(\bigcup_{C \in \mathfrak{B}_d(x)} \varphi(C)) \tag{4.1}$$

in view of our choice of $\epsilon > 0$. By Item 2 applied to $\mathfrak{B}'$, we know that $x \in \mathsf{Int}(\bigcup_{\varphi(C) \in \mathfrak{B}'_d(x)} \varphi(C))$ with this *failing* for any proper subset of $\mathfrak{B}'_d(x)$. Combining this with (4.1) and Item 1, we conclude that

$$\mathfrak{B}'_d(x) \subseteq \varphi(\mathfrak{B}_d(x)).$$

Consequently, since every element of $\mathfrak{B}_d(x)$ contains the set $\mathrm{Supp}_{\mathfrak{B}}(x)$, it follows that

$$\mathsf{C}\Big(\varphi(\mathrm{Supp}_{\mathfrak{B}}(x))\Big) \subseteq \bigcap_{\varphi(C) \in \mathfrak{B}'_d(x)} \varphi(C) = \mathsf{C}\Big(\mathrm{Supp}_{\mathfrak{B}'}(x)\Big), \tag{4.2}$$

with the final equality above in view of Item 3 applied to $\mathfrak{B}'$. However, (4.2) is equivalent to $\varphi(\mathrm{Supp}_{\mathfrak{B}}(x)) \subseteq \mathrm{Supp}_{\mathfrak{B}'}(x)$, which completes the proof.

□

4.4 $\mathcal{F}$-Filtered Sequences, Minimal Encasement and Reay Systems

Next, we extend the concept of an asymptotically filtered sequence, which we introduced in Chap. 3. Note, if we specialize below to the case when $w_i^{(j)} = 0$ for all $j \in [1, \ell]$ and $i \geq 1$ with $\mathcal{E}_j = \mathbb{R}\langle u_1, \ldots, u_j \rangle$ for $j \in [1, \ell]$, then we recover the notion of an asymptotically filtered sequence. We extend much of the terminology introduced in Chap. 3 from asymptotically filtered sequences to $\mathcal{F}$-filtered sequences.

Definition 4.16 Let $\vec{u} = (u_1, \ldots, u_\ell)$ be a tuple of $\ell \geq 0$ orthonormal vectors in $\mathbb{R}^d$ and let $\{0\} = \mathcal{E}_0 \subset \mathcal{E}_1 \subset \ldots \subset \mathcal{E}_\ell \subseteq \mathbb{R}^d$ be a chain of subspaces such that $u_j \in \mathcal{E}_j \cap \mathcal{E}_{j-1}^\perp$ for all $j \in [1, \ell]$. A sequence $\{x_i\}_{i=1}^\infty$ of terms $x_i \in \mathbb{R}^d$ is an $\mathcal{F}$-**filtered** sequence with filter $\mathcal{F} = (\mathcal{E}_1, \ldots, \mathcal{E}_\ell)$ and limit $\vec{u}$ if

$$x_i = (a_i^{(1)} u_1 + w_i^{(1)}) + \ldots + (a_i^{(\ell)} u_\ell + w_i^{(\ell)}) + y_i \quad \text{for all } i \geq 1,$$

for some real numbers $a_i^{(j)} > 0$, vectors u_j, $w_i^{(j)} \in \mathcal{E}_j \cap \mathcal{E}_{j-1}^{\perp}$, and $y_i \in \mathcal{E}_\ell^{\perp}$ such that

- $\lim_{i\to\infty} a_i^{(j)} \in \mathbb{R}_+ \cup \{\infty\}$ exists for each $j \in [1, \ell]$,
- $\|y_i\|$, $\|w_i^{(j)}\| \in o(a_i^{(j)})$ for all $j \in [1, \ell]$, and $a_i^{(j+1)} \in o(a_i^{(j)})$ for all $j \in [1, \ell - 1]$.

Let $\vec{u} = (u_1, \dots, u_t)$ be a tuple of orthonormal vectors $u_i \in \mathbb{R}^d$ and let $\mathcal{F} = (\mathcal{E}_1, \dots, \mathcal{E}_\ell)$ be a tuple of subspaces with

$$\{0\} = \mathcal{E}_0 \subset \mathcal{E}_1 \subset \dots \subset \mathcal{E}_\ell \quad \text{and} \quad \mathbb{R}\langle u_1, \dots, u_t\rangle \subseteq \mathcal{E}_\ell.$$

Then, for each $i \in [1, t]$, there is a unique $j \in [1, \ell]$ with $u_i \in \mathcal{E}_j \setminus \mathcal{E}_{j-1}$. Let $J \subseteq [1, \ell]$ consist of all indices j for which $\mathcal{E}_j \setminus \mathcal{E}_{j-1}$ contains some u_i with $i \in [1, t]$, and for each $j \in J$, let $r_j \in [1, t]$ be the minimal index with $u_{r_j} \in \mathcal{E}_j \setminus \mathcal{E}_{j-1}$. If $J = [1, \ell]$ and $r_i \le r_j$ holds whenever $i \le j$, for $i, j \in [1, \ell]$, then we say that $\mathcal{F}$ is a **compatible filter** for $\vec{u}$, define

$$\mathcal{F}(\vec{u}) = (\overline{u}_1, \dots, \overline{u}_\ell),$$

where $\overline{u}_j = \pi_{j-1}(u_{r_j})/\|\pi_{j-1}(u_{r_j})\|$ with $\pi_{j-1} : \mathbb{R}^d \to \mathcal{E}_{j-1}^{\perp}$ the orthogonal projection, and set $r_{\ell+1} = t + 1$, in which case

$$1 = r_1 < r_2 < \dots < r_\ell < r_{\ell+1} = t + 1,$$

which refer to as the **associated indices** for $\mathcal{F}(\vec{u})$. Note $\ell \ge 1$ except when $\vec{u}$ is the empty tuple, in which case $\mathcal{F}(\vec{u})$ is the empty tuple.

Proposition 4.17 *Let $\vec{u} = (u_1, \dots, u_t)$ be a tuple of $t \ge 0$ orthonormal vectors $u_i \in \mathbb{R}^d$ and let $\mathcal{F} = (\mathcal{E}_1, \dots, \mathcal{E}_\ell)$ be a compatible filter for $\vec{u}$ with $1 = r_1 < \dots < r_\ell < r_{\ell+1} = t + 1$ the associated indices. Then*

$$u_i \in \mathcal{E}_{j-1} \quad \textit{for all } i < r_j \textit{ and } j \le \ell + 1.$$

Moreover, if $\{x_i\}_{i=1}^{\infty}$ is an asymptotically filtered sequence with limit $\vec{u}$, say with $x_i = a_i^{(1)}u_1 + \dots + a_i^{(t)}u_t + y_i$, then $\{x_i\}_{i=1}^{\infty}$ is an $\mathcal{F}$-filtered sequence with limit $\mathcal{F}(\vec{u})$, say with

$$x_i = (\alpha_i^{(1)}\overline{u}_1 + w_i^{(1)}) + \dots + (\alpha_i^{(\ell)}\overline{u}_\ell + w_i^{(\ell)}) + y_i',$$

where $\alpha_i^{(j)} \in \Theta(a_i^{(r_j)})$ for $j \in [1, \ell]$ and $\|y_i'\| \in O(\|y_i\|)$.

Proof If $u_i \notin \mathcal{E}_{j-1}$ with $i < r_j$, then $u_i \in \mathcal{E}_k \setminus \mathcal{E}_{k-1}$ for some $k \ge j$ (as $\mathbb{R}\langle u_1, \dots, u_t\rangle \subseteq \mathcal{E}_\ell$), whence $r_k \le i < r_j$, contradicting that $\mathcal{F}$ is a compatible filter for $\vec{u}$ in view of $k \ge j$. Therefore we instead conclude that $u_i \in \mathcal{E}_{j-1}$ for

$i < r_j$. For $j \in [0, \ell]$, let $\pi_j : \mathbb{R}^d \to \mathcal{E}_j^{\perp}$ and $\pi_j^{\perp} : \mathbb{R}^d \to \mathcal{E}_j$ be the orthogonal projections, where $\mathcal{E}_0 = \{0\}$. We may assume $\ell \geq 1$, for $\ell = 0$ implies $t = 0$, in which case $x_i = y_i = y_i'$ is trivially an $\mathcal{F}$-filtered sequence. By definition of the $\overline{u}_j$, we have $\overline{u}_j \in \mathcal{E}_j \cap \mathcal{E}_{j-1}^{\perp}$ for $j \in [1, \ell]$. For $j \in [1, \ell]$, let $y_i^{(j)} = \pi_j^{\perp}(x_i) - \pi_{j-1}^{\perp}(x_i)$, and set $y_i' = \pi_\ell(x_i) = \pi_\ell(y_i) \in \mathcal{E}_\ell^{\perp}$ (since $\mathbb{R}\langle u_1, \ldots, u_t\rangle \subseteq \mathcal{E}_\ell = \ker \pi_\ell$). Then $x_i = y_i^{(1)} + \ldots + y_i^{(\ell)} + y_i'$ with each $y_i^{(j)} \in \mathcal{E}_j \cap \mathcal{E}_{j-1}^{\perp}$ and $y_i' \in \mathcal{E}_\ell^{\perp}$. Let $\alpha_i^{(j)} = a_i^{(r_j)} \|\pi_{j-1}(u_{r_j})\| > 0$ (since $u_{r_j} \notin \mathcal{E}_{j-1}$) and let

$$w_i^{(j)} = y_i^{(j)} - \alpha_i^{(j)}\overline{u}_j = y_i^{(j)} - \pi_{j-1}(a_i^{(r_j)} u_{r_j}) = \pi_{j-1}(y_i^{(j)} - a_i^{(r_j)} u_{r_j}) \in \mathcal{E}_j \cap \mathcal{E}_{j-1}^{\perp}$$

(since $y_i^{(j)} \in \mathcal{E}_j \cap \mathcal{E}_{j-1}^{\perp}$ and $u_{r_j} \in \mathcal{E}_j$ by definition of r_j). Thus

$$x_i = (\alpha_i^{(1)}\overline{u}_1 + w_i^{(1)}) + \ldots + (\alpha_i^{(\ell)}\overline{u}_\ell + w_i^{(\ell)}) + y_i',$$

with $\overline{u}_j, w_i^{(j)} \in \mathcal{E}_j \cap \mathcal{E}_{j-1}^{\perp}$ and $y_i' \in \mathcal{E}_\ell^{\perp}$. By definition of r_j, we are assured that $\|\pi_{j-1}(u_{r_j})\| > 0$. Thus $\alpha_i^{(j)} \in \Theta(a_i^{(r_j)})$ and $\|y_i'\| = \|\pi_\ell(y_i)\| \in O(\|y_i\|)$ (since linear operators between finite dimensional spaces are bounded). Consequently, since $x_i = a_i^{(1)} u_1 + \ldots + a_i^{(t)} u_t + y_i$ is asymptotically filtered, we have $\alpha_i^{(j)} \in o(a_i^{(r_j-1)}) \subseteq o(a_i^{(r_{j-1})}) = o(\alpha_i^{(j-1)})$ and $\|y_i'\| \in O(\|y_i\|) \subseteq o(a_i^{(t)}) \subseteq o(a_i^{(r_\ell)}) = o(\alpha_i^{(\ell)})$. Since $u_i \in \mathcal{E}_{j-1} \subset \mathcal{E}_j$ for $i < r_j$, and since $u_{r_j} \in \mathcal{E}_j \setminus \mathcal{E}_{j-1}$, it follows that

$$\begin{aligned} w_i^{(j)} &= \pi_{j-1}(y_i^{(j)} - a_i^{(r_j)} u_{r_j}) = \pi_{j-1}\pi_j^{\perp}(x_i - a_i^{(r_j)} u_{r_j}) \\ &= \pi_{j-1}\pi_j^{\perp}(a_i^{(r_j+1)} u_{r_j+1} + \ldots + a_i^{(t)} u_t + y_i). \end{aligned}$$

In consequence, if $r_j < t$, then $\|w_i^{(j)}\| \in O(a_i^{(r_j+1)}) \subseteq o(a_i^{(r_j)}) = o(\alpha_i^{(j)})$, and if $r_j = t$, then $j = \ell$ and $\|w_i^{(j)}\| \in O(\|y_i\|) \subseteq o(a_i^{(t)}) = o(a_i^{(r_j)}) = o(\alpha_i^{(j)})$, and the proof is complete in either case. □

Lemma 4.18 *Let $X \subseteq \mathbb{R}^d$ be a linearly independent subset, let $\{x_i\}_{i=1}^{\infty}$ be a sequence of terms $x_i \in \mathbb{R}\langle X\rangle$, and let $x_i = \sum_{x \in X} \alpha_i^{(x)} x$ with $\alpha_i^{(x)} \in \mathbb{R}$ for $i \geq 1$. Then $|\alpha_i^{(x)}| \in O(\|x_i\|)$ for all $x \in X$.*

Proof Let $|X| = s \leq d$ (as X is linearly independent). Let M' be the $d \times s$ matrix with column vectors the elements from X. Since X is linearly independent, the matrix M' has rank s, allowing us to add an additional $d - s$ columns to the right of M' to create an invertible $d \times d$ matrix M. Since $x_i \in \mathbb{R}\langle X\rangle$, for each i, there is a vector $y_i = (\alpha_{i,1}, \ldots, \alpha_{i,s}, 0, \ldots, 0)$ with $My_i = x_i$. Then $\|y_i\| = \|M^{-1}x_i\| \leq \|M^{-1}\| \, \|x_i\|$, where $\|M^{-1}\|$ is the matrix operator norm induced by the Euclidean

L_2-norm. This shows $\alpha_{i,j}^2 \leq \alpha_{i,1}^2 + \ldots + \alpha_{i,s}^2 = \|y_i\|^2 \leq C^2\|x_i\|^2$ for each $j \in [1, s]$, where $C = \|M^{-1}\| > 0$, implying $|\alpha_{i,j}| \leq C\|x_i\|$ for all i and $j \in [1, s]$, as desired. □

The next proposition links minimal encasement, Reay systems and $\mathcal{F}$-filtered sequences.

Proposition 4.19 *Let $X \subseteq \mathbb{R}^d$, and let $\vec{u} = (u_1, \ldots, u_t)$ be a tuple of $t \geq 0$ orthonormal vectors $u_i \in \mathbb{R}^d$, where $d \geq 0$. Then $-X$ minimally encases $\vec{u}$ if and only if*

1. *there exists a disjoint partition $X = \bigcup_{i=1}^{\ell} X_i$ such that $\mathcal{F} = (\mathcal{E}_1, \ldots, \mathcal{E}_\ell)$ is a compatible filter for $\vec{u}$, where $\mathcal{E}_j = \mathbb{R}\langle X_1 \cup \ldots \cup X_j\rangle$ for $j \in [1, \ell]$, and*
2. *$(X_1 \cup \{u_{r_1}\}, \ldots, X_\ell \cup \{u_{r_\ell}\})$ is a Reay system, where $1 = r_1 < \ldots < r_\ell < r_{\ell+1} = t + 1$ are the associated indices for $\mathcal{F}(\vec{u})$.*

Moreover, the Reay system $(X_1 \cup \{u_{r_1}\}, \ldots, X_\ell \cup \{u_{r_\ell}\})$ satisfying Items 1 and 2 is unique.

Proof If $t = 0$, then only the empty set $X = \emptyset$ minimally encases the empty tuple, and the empty partition with $\ell = 0$ satisfies the desired conditions. Likewise, if such a partition exists for a set X when $t = 0$, then $\ell = 0$ follows, implying that $X = \emptyset$. Thus we may assume $t \geq 1$.

Suppose Items 1 and 2 hold. Item 2 allows us to apply Proposition 4.11 to conclude that $-u_{r_j} \in \mathbb{R}\langle X_1 \cup \ldots \cup X_j\rangle = \mathsf{C}(X_1 \cup \{u_{r_1}\} \cup \ldots \cup X_j \cup \{u_{r_j}\})$ for every $j \in [1, \ell]$, whence $-u_{r_j} = z + a_1 u_{r_1} + \ldots + a_{j-1} u_{r_{j-1}}$ for some $z \in \mathsf{C}(X_1 \cup \ldots \cup X_j)$ and $a_i \geq 0$, implying that $u_{r_j} + a_1 u_{r_1} + \ldots + a_{j-1} u_{r_{j-1}} \in \mathsf{C}(-X_1 \cup \ldots \cup -X_j) \subseteq \mathsf{C}(-X)$. By item 1 and Proposition 4.17, we have $u_i \in \mathcal{E}_{j-1}$ whenever $i < r_j$ and $j \leq \ell + 1$. Thus, for any $i \in [1, t]$ with $r_j < i < r_{j+1}$ and $j \in [1, \ell]$, we have $u_i \in \mathbb{R}\langle X_1 \cup \ldots \cup X_j\rangle = \mathsf{C}(X_1 \cup \{u_{r_1}\} \ldots \cup X_j \cup \{u_{r_j}\})$, so that a similar argument yields $u_i + a_1 u_{r_1} + \ldots + a_j u_{r_j} \in \mathsf{C}(-X_1 \cup \ldots \cup -X_j) \subseteq \mathsf{C}(-X)$ for some $a_i \geq 0$. This shows that $-X$ encases $\vec{u} = (u_1, \ldots, u_t)$. It remains to establish the minimality of the encasement. To this end, it suffices to show, for an arbitrary $x \in X$, that $\mathsf{C}(-X \setminus \{-x\})$ does not encase $(u_1, \ldots, u_t)$. Let $j \in [1, \ell]$ be the index such that $x \in X_j$ and let $\pi_{j-1} : \mathbb{R}^d \to \mathcal{E}_{j-1}^{\perp}$ be the orthogonal projection. Then $\mathcal{R} = (\pi_{j-1}(X_j \cup \{u_{r_j}\}), \ldots, \pi_{j-1}(X_\ell \cup \{u_{r_\ell}\}))$ is a Reay system and $\mathsf{Supp}_{\mathcal{R}}(-\pi_{j-1}(u_{r_j})) = \pi_{j-1}(X_j)$. Assume by contradiction that $-X \setminus \{-x\}$ encases $(u_1, \ldots, u_t)$. Then, since $\pi_{j-1}(u_i) = 0$ for all $i < r_j$ (by Proposition 4.17), it follows that

$$-\pi_{j-1}(u_{r_j}) \in \mathsf{C}(\pi_{j-1}(X \setminus \{x\})) = \mathsf{C}(\pi_{j-1}(X_j \setminus \{x\}) \cup \ldots \cup \pi_{j-1}(X_\ell)),$$

implying that $\pi_{j-1}(x) \notin \mathsf{Supp}_{\mathcal{R}}(-\pi_{j-1}(u_{r_j}))$. However, this contradicts the fact that $\mathsf{Supp}_{\mathcal{R}}(-\pi_{j-1}(u_{r_j})) = \pi_{j-1}(X_j)$ with $x \in X_j$. So we conclude that $-X$ minimally encases $\vec{u}$, as desired.

Next suppose $-X$ minimally encases $\vec{u}$. Then X must be finite (by the minimality of $-X$), and there are vectors $v_1, \ldots, v_t \in \mathsf{C}(-X)$ with each

$$v_j = \alpha_{1,j} u_1 + \ldots + \alpha_{j,j} u_j \in \mathsf{C}(-X) \tag{4.3}$$

for some real numbers $\alpha_{i,j} \geq 0$ with $\alpha_{j,j} > 0$. In particular, $u_1 \in \mathsf{C}(-X)$, so there must be a subset $X_1 \subseteq X$ such that $X_1 \cup \{u_1\}$ is a minimal positive basis (in view of Proposition 4.8.4 and Carthéordory's Theorem). Let $r_1 = 1$, let $\mathcal{E}_0 = \{0\}$, and let $\mathcal{E}_1 = \mathbb{R}\langle X_1 \rangle$. We proceed to recursively construct, for $j = 1, 2, \ldots$, nonempty subsets $X_1, \ldots, X_j \subseteq X$, subspaces $\mathcal{E}_j = \mathbb{R}\langle X_1 \ldots \cup X_j \rangle$, and indices $1 = r_1 < r_2 < \ldots < r_j \leq t$ such that $r_j \in [1, t]$ is the minimal index with $u_{r_j} \notin \mathcal{E}_{j-1}$, and $(X_1 \cup \{u_{r_1}\}, \ldots, X_j \cup \{u_{r_j}\})$ is a Reay system. We have just shown this is possible for $j = 1$, so assume $j \geq 2$ and that the sets $X_1, \ldots, X_{j-1} \subseteq X$ and indices $1 = r_1 < \ldots < r_{j-1}$ have already been found such that $(X_1 \cup \{u_{r_1}\}, \ldots, X_{j-1} \cup \{u_{r_{j-1}}\})$ is a Reay system and each $r_i \in [1, t]$ is the minimal index with $u_{r_i} \notin \mathcal{E}_{i-1} = \mathbb{R}\langle X_1 \cup \ldots \cup X_{i-1} \rangle$ for $i \leq j - 1$. Let $r_j \in [1, t]$ be the minimal index such that $u_{r_j} \notin \mathcal{E}_{j-1} = \mathbb{R}\langle X_1 \cup \ldots \cup X_{j-1} \rangle$ (if such an index exists), and otherwise set $r_j = t + 1$. By Proposition 4.11.1, we have

$$\mathcal{E}_{j-1} = \mathbb{R}\langle X_1 \cup \ldots \cup X_{j-1} \rangle = \mathsf{C}(X_1 \cup \{u_{r_1}\} \cup \ldots \cup X_{j-1} \cup \{u_{r_{j-1}}\}).$$

Thus $r_j > r_{j-1}$. If $r_j = t + 1$, then $-u_i \in \mathcal{E}_{j-1} = \mathsf{C}(X_1 \cup \{u_{r_1}\} \cup \ldots \cup X_{j-1} \cup \{u_{r_{j-1}}\})$ for all $i \in [1, t]$. In such case, for each $i > r_{j-1}$, we have $u_i + \alpha_{i,1} u_{r_1} + \ldots \alpha_{i,j-1} u_{r_{j-1}} \in \mathsf{C}(-X_1 \cup \ldots \cup -X_{j-1})$ for some $\alpha_{i,r_1}, \ldots, \alpha_{i,r_{j-1}} \in \mathbb{R}_+$.

If no such index r_2 exists (so $r_2 = t + 1$), then $-u_i \in \mathcal{E}_1 = \mathsf{C}(X_1 \cup \{u_1\})$ for all $i \geq 1$, meaning each $-u_i = b_i + a_i u_1$ for some $a_i \geq 0$ and $b_i \in \mathsf{C}(X_1)$. But then $u_i + a_i u_1 \in \mathsf{C}(-X_1)$ for all $i \geq 2$, implying that $\mathsf{C}(-X_1)$ encases $(u_1, \ldots, u_t)$. In this case, the minimality of X ensures that $X = X_1$, and the desired partition of X follows with $\ell = 1$ and $\mathcal{F} = (\mathcal{E}_1)$ compatible with $\vec{u}$ (as $u_1 \in \mathcal{E}_1 \setminus \{0\}$ with $\mathbb{R}\langle u_1, \ldots, u_t \rangle \subseteq \mathcal{E}_1$). So we may now assume the index r_2 exists. Moreover, the previous argument ensures that $\mathsf{C}(-X_1)$ encases $(u_1, \ldots, u_{r_2 - 1})$. Now let $\pi_1 : \mathbb{R}^d \to \mathcal{E}_1^\perp$ be the orthogonal projection. Since $u_1, \ldots, u_{r_2-1} \in \mathcal{E}_1$, we have $-\pi_1(u_{r_2}) \in \mathsf{C}(\pi_1(X))$ by (4.3). Thus, as before, we can find a subset $X_2 \subseteq X \setminus X_1$ such that $|\pi_1(X_2)| = |X_2|$ and $\pi_1(X_2 \cup \{u_{r_2}\})$ is a minimal positive basis, in which case $(X_1 \cup \{u_{r_1}\}, X_2 \cup \{u_{r_2}\})$ is a Reay system. Since $\pi_1(X_2) \cup \{\pi_1(u_{r_2})\}$ is a minimal positive basis, we have $u_{r_2} + b \in \mathsf{C}^\circ(-X_2)$ for some $b \in \ker \pi_1 = \mathcal{E}_1 = \mathsf{C}(X_1 \cup \{u_{r_1}\})$, in turn implying $u_{r_2} + a_{r_2} u_{r_1} \in \mathsf{C}(-X_1 \cup -X_2)$ for some $a_{r_2} \geq 0$. Thus $-X_1 \cup -X_2$ encases $(u_1, \ldots, u_{r_2})$. Let $\mathcal{E}_2 = \mathbb{R}\langle X_1 \cup X_2 \rangle = \mathsf{C}(X_1 \cup X_2 \cup \{u_{r_1}, u_{r_2}\})$ (in view of Proposition 4.11.1) and let r_3 be the minimal index such that $u_{r_3} \notin \mathcal{E}_2$. If no such index r_3 exists, then $-u_i \in \mathcal{E}_2 = \mathsf{C}(X_1 \cup X_2 \cup \{u_{r_1}, u_{r_2}\})$ for all $i \geq 1$, meaning each $-u_i = c_i + b_i u_{r_2} + a_i u_{r_1}$ for some $a_i, b_i \geq 0$ and $c_i \in \mathsf{C}(X_1 \cup X_2)$. But then $u_i + a_i u_{r_1} + b_i u_{r_2} \in \mathsf{C}(-X_1 \cup -X_2)$ for all $i > r_2$, implying that $\mathsf{C}(-X_1 \cup -X_2)$ encases $(u_1, \ldots, u_t)$. As before, the minimality of X then ensures that $X = X_1 \cup X_2$, and the desired partition follows with $\ell = 2$. So we may now assume the index r_3 exists. Moreover, $\mathsf{C}(-X_1 \cup -X_2)$ encases $(u_1, \ldots, u_{r_3 - 1})$.

Continuing to iterate these arguments (as in Proposition 4.10) now leads to the desired partition of X after $\ell \le t$ steps.

Finally, it remains to show $\mathcal{R} = (X_1 \cup \{u_{r_1}\}, \ldots, X_\ell \cup \{u_{r_\ell}\})$ is unique, which we handle by induction on ℓ. To this end, suppose $\mathcal{R}' = (X_1' \cup \{u_{r_1'}\}, \ldots, X_{\ell'}' \cup \{u_{r_{\ell'}'}\})$ is another Reay system satisfying Items 1 and 2, so $1 = r_1' < \ldots < r_{\ell'}' < r_{\ell'+1}' = t + 1$. Thus $-u_1 = -u_{r_1} \in \mathsf{C}^\circ(X_1)$ and $-u_1 = -u_{r_1'} \in \mathsf{C}^\circ(X_1')$ by Proposition 4.8.3. Let $\pi_1 : \mathbb{R}^d \to \mathcal{E}_1^\perp$ be the orthogonal projection. Since $\mathcal{R}$ is a Reay system, it follows that $\pi_1(X_2) \cup \ldots \cup \pi_1(X_\ell)$ is a linearly independent set of size $|X_2| + \ldots + |X_\ell|$ (per definition of a Reay system and Proposition 4.11). Consequently, any linear combination of elements from X equal to $-u_1 \in \mathcal{E}_1 = \ker \pi_1$ can only involve terms from X_1. Thus $-u_1 = -u_{r_1'} \in \mathsf{C}^\circ(X_1')$ implies $X_1' \subseteq X_1$. Exchanging the roles of $\mathcal{R}$ and $\mathcal{R}'$ and repeating the argument, we find that $X_1 \subseteq X_1'$, whence $X_1 = X_1'$. As a result, $u_{r_2} = u_{r_2'}$ now follows from Item 1 (since $\mathbb{R}\langle X_1\rangle = \mathbb{R}\langle X_1'\rangle$). This completes the base case when $\ell = 1$, so we can assume $\ell \ge 2$. Now $(\pi_1(X_2) \cup \{\pi_1(u_{r_2})\}, \ldots, \pi_1(X_\ell) \cup \{\pi_1(u_{r_\ell})\})$ and $(\pi_1(X_2') \cup \{\pi_1(u_{r_2'})\}, \ldots, \pi_1(X_{\ell'}') \cup \{\pi_1(u_{r_{\ell'}'})\})$ are both Reay systems (per Proposition 4.11.3) showing that $-\pi_1(X \setminus X_1)$ minimally encases $\pi_1(\vec{u})$. Thus, by induction hypothesis, $\ell = \ell'$ and $\pi_1(X_j) = \pi_1(X_j')$ for all $j \in [2, \ell]$, implying $X_j = X_j'$ for all $j \in [2, \ell]$ as π_1 is injective on $X \setminus X_1$. This shows $\mathcal{F} = (\mathcal{E}_1, \ldots, \mathcal{E}_\ell)$ is uniquely defined, where $\mathcal{E}_j = \mathbb{R}\langle X_1 \cup \ldots \cup X_j\rangle$, in which case Item 1 ensures that the indices $1 = r_1 < \ldots < r_\ell < r_{\ell+1} = t + 1$ are also uniquely defined, i.e., $r_j = r_j'$ for all $j \in [1, \ell]$. □

If $-X$ minimally encases $\vec{u}$, we have a unique Reay system $\mathcal{R} = (X_1 \cup \{u_{r_1}\}, \ldots, X_\ell \cup \{u_{r_\ell}\})$ and indices $1 = r_1 < \ldots < r_\ell < r_{\ell+1} = t + 1$ given by Proposition 4.19, which we refer to as the Reay system and indices **associated** to the minimal encasement of $\vec{u}$ by $-X$.

Proposition 4.20 *Let $\vec{u} = (u_1, \ldots, u_t)$ be a tuple of orthonormal vectors $u_i \in \mathbb{R}^d$, where $t, d \ge 1$, let $X \subseteq \mathbb{R}^d$ be a subset minimally encasing $-\vec{u}$, and let $X = \bigcup_{i=1}^{\ell} X_i$ and $1 = r_1 < \ldots < r_\ell < r_{\ell+1} = t + 1$ be the Reay system and indices associated to the minimal encasement.*

If $\{x_i\}_{i=1}^\infty$ is an asymptotically filtered sequence with limit $\vec{u}$, say $x_i = a_i^{(1)} u_1 + \ldots + a_i^{(t)} u_t + y_i$, and $y_i \in \mathbb{R}\langle X\rangle$ for all i, then $X \cup \{x_i\}$ is a minimal positive basis for $\mathbb{R}\langle X\rangle$ for all sufficiently large i. Moreover, letting $-x_i = \sum_{x\in X} \alpha_i^{(x)} x$ be the unique positive linear combination with $\alpha_i^{(x)} > 0$ (for i sufficiently large), we have $\alpha_i^{(x)} \in \Theta(a_i^{(r_j)})$ for $x \in X_j$.

Proof For $j \in [1, \ell]$, let $\mathcal{E}_j = \mathbb{R}\langle X_1 \cup \ldots \cup X_j\rangle$ and let $\pi_j : \mathbb{R}^d \to \mathcal{E}_j^\perp$ be the orthogonal projection. Then $X = X_1 \cup \ldots \cup X_\ell$ is linearly independent by Proposition 4.11.1, so $X \cup \{x_i\}$ is a minimal positive basis if and only if $x_i \in -\mathsf{C}^\circ(X)$ (by Proposition 4.8.4). By Proposition 3.11, $\{x_i - y_i\}_{i=1}^\infty$ is asymptotically filtered with limit $\vec{u}$ and $x_i - y_i \in -\mathsf{C}(X)$ for all sufficiently large i.

Let us first show that, if the proposition holds for the sequence $\{x_i - y_i\}_{i=1}^{\infty}$, then it holds for the sequence $\{x_i\}_{i=1}^{\infty}$ as well. To this end, suppose $x_i - y_i = -\sum_{x\in X}\alpha_i^{(x)}x$ for all sufficiently large i, for some $\alpha_i^{(x)} > 0$ with $\alpha_i^{(x)} \in \Theta(a_i^{(r_j)})$ for all $x \in X_j$ and $j \in [1, \ell]$. Since $y_i \in \mathbb{R}\langle X\rangle$ with X linearly independent, Lemma 4.18 implies that $y_i = -\sum_{x\in X}\beta_i^{(x)}x$ for some $\beta_i^{(x)} \in \mathbb{R}$ with $|\beta_i^{(x)}| \in O(\|y_i\|) \subseteq o(a_i^{(t)}) \subseteq o(a_i^{(r)})$ for all $x \in X$ and $r \in [1, t]$. Thus $\alpha_i^{(x)} + \beta_i^{(x)} \in \Theta(a_i^{(r_j)})$ for all $x \in X_j$ and $j \in [1, \ell]$. Moreover, $\alpha_i^{(x)} + \beta_i^{(x)} > 0$ for all sufficiently large i. Thus, since $x_i = (x_i - y_i) + y_i = -\sum_{x\in X}(\alpha_i^{(x)} + \beta_i^{(x)})x$, it follows that $x_i \in -\mathsf{C}^\circ(X)$ for all sufficiently large i, ensuring that the proposition holds for $\{x_i\}_{i=1}^{\infty}$, as desired.

We proceed by induction on the depth ℓ to show the proposition holds when $y_i = 0$ for all i, which will complete the proof by what was just shown. If $\ell = 1$, then $x_i = a_i^{(1)}u_1$ with $u_1 \in -\mathsf{C}^\circ(X_1)$ (as $X_1 \cup \{u_{r_1}\}$ is a minimal positive basis and $r_1 = 1$). Then there is a unique strictly positive linear combination $-u_1 = \sum_{x\in X}\alpha_x x$, implying that $-x_i = a_i^{(1)}(-u_1) = \sum_{x\in X}(a_i^{(1)}\alpha_x)x$. Since $\alpha_x > 0$ and $\alpha_i^{(1)} > 0$, we have $a_i^{(1)}\alpha_x \in \Theta(a_i^{(1)}) = \Theta(a_i^{(r_1)})$ with $a_i^{(1)}\alpha_x > 0$, ensuring $-x_i \in \mathsf{C}^\circ(X_1) = \mathsf{C}^\circ(X)$, as desired. This completes the induction base, so now assume $\ell \geq 2$.

Let $z_i = a_i^{(r_{\ell-1}+1)}u_{r_{\ell-1}+1} + \ldots + a_i^{(t)}u_t$. Write $z_i = z_i^* + z_i^\perp$ with $z_i^* \in \mathcal{E}_{\ell-1}$ and $z_i^\perp \in \mathcal{E}_{\ell-1}^\perp$, so

$$x_i = a_i^{(1)}u_1 + \ldots + a_i^{(r_{\ell-1})}u_{r_{\ell-1}} + z_i^* + z_i^\perp.$$

Since $\pi_{\ell-1}(X_\ell) \cup \{\pi_{\ell-1}(u_{r_\ell})\}$ is a minimal positive basis, Proposition 3.5 allows us to apply the base case to $\{\pi_{\ell-1}(-x_i)\}_{i=1}^{\infty}$ yielding that $-z_i^\perp = \pi_{\ell-1}(-x_i) = \sum_{x\in X_\ell}\alpha_i^{(x)}\pi_{\ell-1}(x)$ for some $\alpha_i^{(x)} > 0$ with $\alpha_i^{(x)} \in \Theta(a_i^{(r_\ell)}) \subseteq o(a_i^{(r_{\ell-1})})$, for all $x \in X_\ell$ and all sufficiently large i. Thus

$$\sum_{x\in X_\ell}\alpha_i^{(x)}x = -z_i^\perp + \xi_i \tag{4.4}$$

for some $\xi_i \in \mathcal{E}_{\ell-1}$. Since z_i^* and $z_i^\perp$ are orthogonal, we have $\|z_i^*\|, \|z_i^\perp\| \leq \|z_i\|$, ensuring $\|z_i^*\|, \|z_i^\perp\| \in O(\|z_i\|) = O(a_i^{(r_{\ell-1}+1)}) \subseteq o(a_i^{(r_{\ell-1})})$. Thus $z_i^* + \xi_i \in \mathcal{E}_{\ell-1}$ with $\|z_i^* + \xi_i\| \in o(a_i^{(r_{\ell-1})})$, allowing us to apply the induction hypothesis to the sequence $\{x_i - z_i^\perp + \xi_i\}_{i=1}^{\infty}$ to conclude $-x_i + z_i^\perp - \xi_i = \sum_{x\in X\setminus X_\ell}\alpha_i^{(x)}x$ for some $\alpha_i^{(x)} > 0$ with $\alpha_i^{(x)} \in \Theta(a_i^{(r_j)})$, for all $x \in X_j$, all $j \in [1, \ell-1]$ and all sufficiently large i. Combined with (4.4), it follows that $-x_i = \sum_{x\in X}\alpha_i^{(x)}x$ with $\alpha_i^{(x)} > 0$ and $\alpha_i^{(x)} \in \Theta(a_i^{(r_j)})$, for $x \in X_j$ and sufficiently large i, showing that $x_i \in -\mathsf{C}^\circ(X)$, which completes the induction and the proof. □

Chapter 5
Oriented Reay Systems

If $\mathbf{x} \subseteq \mathbb{R}^d$ is a half-space inside the subspace $\mathbb{R}\langle \mathbf{x} \rangle$ with partial boundary (meaning $\mathbf{x}$ is obtained from a closed half-space in $\mathbb{R}\langle \mathbf{x} \rangle$ by removing elements from the boundary subspace), so $\mathbf{x}^\circ \subseteq \mathbf{x} \subseteq \overline{\mathbf{x}}$ with $\overline{\mathbf{x}} = \overline{\mathbf{x}^\circ}$ a closed half-space in $\mathbb{R}\langle \mathbf{x} \rangle$ and $\mathbf{x}^\circ$ an open half space, then we call $\mathbf{x}$ a **(relative) half-space**, though we will henceforth simply refer to such sets as half-spaces for brevity. With regards to Convex Geometry, there is little difference between a nonzero point $x \in \mathbb{R}^d$ and the one-dimensional ray $\mathbb{R}_+ x$ that it defines, which is a one-dimensional half-space. In this way, we informally view a higher dimensional half-space $\mathbf{x} \subseteq \mathbb{R}^d$ as a type of higher-dimensional element. We will later see that, in many ways, half-spaces share similar behavior with ordinary elements. An element $x \in \mathbf{x}^\circ$ is called a **representative** for the relative half-space $\mathbf{x}$. Thus $\overline{\mathbf{x}} = \partial(\mathbf{x}) + \mathbb{R}_+ x$ for any representative x. Recall that a simplicial cone is set of the form $\mathsf{C}(X)$ with $X \subseteq \mathbb{R}^d$ a linearly independent set. If $\pi : \mathbb{R}^d \to \mathbb{R}^d$ is a linear transformation with $\ker \pi \cap \mathbb{R}\langle \mathbf{x} \rangle \subseteq \partial(\mathbf{x})$, or equivalently, with $\mathbf{x} \nsubseteq \ker \pi + \partial(\mathbf{x})$ (as both are equivalent to $\pi(x) \neq 0$ for all $x \in \mathbf{x}^\circ$), then $\pi(\mathbf{x})$ will also be a relative half-space with

$$\partial(\pi(\mathbf{x})) = \pi(\partial(\mathbf{x})), \quad \pi(\mathbf{x}) \cap \partial(\pi(\mathbf{x})) = \pi(\mathbf{x} \cap \partial(\mathbf{x})) \quad \text{and} \quad \pi(\mathbf{x}^\circ) = \pi(\mathbf{x})^\circ. \tag{5.1}$$

We define a **blunted simplicial cone** to be a set of the form $\mathcal{E} + \mathsf{C}(X)$ with $\mathcal{E} \subseteq \mathbb{R}^d$ a subspace and $X \subseteq \mathbb{R}^d$ a subset for which $\pi(X)$ is a linearly independent subset of $|X|$ elements, where $\pi : \mathbb{R}^d \to \mathcal{E}^\perp$ is the orthogonal projection. In order to avoid excessive use of dummy variables, given a set $\mathcal{X}$ whose elements $\mathbf{x} \in \mathcal{X}$ are subsets $\mathbf{x} \subseteq \mathbb{R}^d$, we let

$$\mathbb{R}^\cup\langle \mathcal{X} \rangle = \mathbb{R}\langle \bigcup_{\mathbf{x} \in \mathcal{X}} \mathbf{x} \rangle, \quad \mathsf{C}^\cup(\mathcal{X}) = \mathsf{C}(\bigcup_{\mathbf{x} \in \mathcal{X}} \mathbf{x}), \quad \text{and} \quad \mathsf{C}^\cup(\mathcal{X})^\circ = \mathsf{C}^\circ(\bigcup_{\mathbf{x} \in \mathcal{X}} \mathbf{x}).$$

We adapt the convention that relative half-spaces will be denoted in boldface, e.g. $\mathbf{x}$, and that the corresponding non-boldface symbol denotes a fixed representative

D. J. Grynkiewicz, *The Characterization of Finite Elasticities*, Lecture Notes in Mathematics 2316, https://doi.org/10.1007/978-3-031-14869-9_5

for the half-space, e.g., $x \in \mathbf{x}^\circ$. Likewise, a collection of relative half-spaces will be denoted in calligraphic script, e.g. $\mathcal{X}$, with the corresponding non-calligraphic symbol denoting a set obtained by replacing each half-space $\mathbf{x} \in \mathcal{X}$ with a fixed representative $x \in \mathbf{x}^\circ$, e.g., X denotes a set of representatives for $\mathcal{X}$. Generally, we will use a representative set X only in contexts where it is irrelevant which representative is chosen for each $\mathbf{x} \in \mathcal{X}$, and the representative sets will be fixed.

Definition 5.1 For $j \in [1, s]$, let $\mathcal{X}_j \cup \{\mathbf{v}_j\}$ be a subset of relative half-spaces in $\mathbb{R}^d$ with distinguished element $\mathbf{v}_j \notin \mathcal{X}_j$, where $d \geq 0$ and $s \geq 0$. For $j \in [0, s]$, let $\mathcal{E}_j = \mathbb{R}^\cup\langle \mathcal{X}_1 \cup \{\mathbf{v}_1\} \cup \ldots \cup \mathcal{X}_j \cup \{\mathbf{v}_j\}\rangle$ and let $\pi_j : \mathbb{R}^d \to \mathcal{E}_j^\perp$ be the orthogonal projection. Suppose, for each $j \in [1, s]$, that the following hold.

(OR1) For every $\mathbf{x} \in \mathcal{X}_j \cup \{\mathbf{v}_j\}$, we have $\partial(\mathbf{x}) = \mathbb{R}^\cup\langle \mathcal{B}_\mathbf{x}\rangle$ and $\partial(\mathbf{x}) \cap \mathbf{x} = \mathsf{C}^\cup(\mathcal{B}_\mathbf{x})$ for some $\mathcal{B}_\mathbf{x} \subseteq \mathcal{X}_1 \cup \ldots \cup \mathcal{X}_{j-1}$ (which we will later denote by $\partial(\{\mathbf{x}\}) = \mathcal{B}_\mathbf{x}$).
(OR2) $\pi_{j-1}(X_j \cup \{v_j\})$ is a minimal positive basis with $|\pi_{j-1}(X_j \cup \{v_j\})| = |\mathcal{X}_j| + 1$.

Then we call $\mathcal{R} = (\mathcal{X}_1 \cup \{\mathbf{v}_1\}, \ldots, \mathcal{X}_s \cup \{\mathbf{v}_s\})$ an **orientated Reay system** for the subspace $\mathcal{E}_s$.

Note, in view of (OR1), that $\pi_{j-1}(\mathcal{X}_j \cup \{\mathbf{v}_j\})$ is a set of rays in $\mathcal{E}_{j-1}^\perp$, so (OR2) does not depend on the choice of representatives. Using (OR1) and (OR2) and a recursive argument for $j = 1, 2, \ldots, s$, it follows that $(X_1 \cup \{v_1\}, \ldots, X_j \cup \{v_j\})$ is a Reay system for $\mathcal{E}_j$ with $\mathcal{E}_j = \mathbb{R}\langle X_1 \cup \ldots \cup X_j\rangle = \mathbb{R}^\cup\langle \mathcal{X}_1 \cup \ldots \cup \mathcal{X}_j\rangle$ for all $j \in [0, s]$. Also, (OR2) ensures that $\pi_{j-1}(\mathbf{x}) \neq 0$ for every $\mathbf{x} \in \mathcal{X}_j \cup \{\mathbf{v}_j\}$, whence $\mathbf{x} \nsubseteq \mathcal{E}_{j-1} = \mathcal{E}_{j-1} + \partial(\mathbf{x})$, and thus $\mathbf{x} \nsubseteq \mathcal{E}_{i-1} + \partial(\mathbf{x})$ for any $i \leq j$ as well. Consequently, for any $j \in [1, s]$,

$$\pi_{j-1}(\mathcal{R}) := \big(\pi_{j-1}(\mathcal{X}_j) \cup \{\pi_{j-1}(\mathbf{v}_j)\}, \ldots, \pi_{j-1}(\mathcal{X}_s) \cup \{\pi_{j-1}(\mathbf{v}_s)\}\big)$$

is also an orientated Reay system in view of (5.1), while it is clear from the recursive nature of the definition that $(\mathcal{X}_1 \cup \{\mathbf{x}_1\}, \ldots, \mathcal{X}_j \cup \{\mathbf{x}_j\})$ is an orientated Reay system for any $j \in [0, s]$.

If $\mathcal{B} \subseteq \mathcal{X}_1 \cup \ldots \cup \mathcal{X}_s$ and $\mathbf{x} \subseteq \mathsf{C}^\cup(\mathcal{B})$ with $\mathbf{x} \in \mathcal{X}_j \cup \{\mathbf{v}_j\}$ for $j \in [1, s]$, then

$$\mathbf{x} \in \mathcal{X}_j \text{ and there is some } \mathbf{y} \in \mathcal{B} \text{ with } \mathbf{x} \subseteq \mathbf{y}, \tag{5.2}$$

which can be seen by the following short inductive proof on s using (OR1) and (OR2). When $j = s$, applying π_{s-1} and using (OR2) yields the desired result, which covers the case $s = 1$. Let $\mathcal{B}_s = \mathcal{B} \cap \mathcal{X}_s$. Observe that (OR2) ensures there is no nontrivial linear combination of elements from B_s lying in $\mathcal{E}_{s-1}$, whence $\mathcal{E}_{s-1} \cap \mathsf{C}^\cup(\mathcal{B}_s) = \mathsf{C}\big(\bigcup_{\mathbf{y} \in \mathcal{B}_s}(\mathcal{E}_{s-1} \cap \mathbf{y})\big) = \mathsf{C}\big(\bigcup_{\mathbf{y} \in \mathcal{B}_s}(\partial(\mathbf{y}) \cap \mathbf{y})\big)$, with the latter equality in view of (OR1) and (OR2). As a result, when $j < s$, we have $\mathbf{x} \subseteq \mathcal{E}_{s-1} \cap \mathsf{C}^\cup(\mathcal{B}) = \mathsf{C}^\cup(\mathcal{B} \setminus \mathcal{B}_s) + (\mathcal{E}_{s-1} \cap \mathsf{C}^\cup(\mathcal{B}_s)) = \mathsf{C}^\cup(\mathcal{B} \setminus \mathcal{B}_s) + \mathsf{C}\big(\bigcup_{\mathbf{y} \in \mathcal{B}_s}(\partial(\mathbf{y}) \cap \mathbf{y})\big)$. In view of (OR1), there is a subset $\mathcal{X}' \subseteq \mathcal{X}_1 \cup \ldots \cup \mathcal{X}_{s-1}$ with $\mathsf{C}^\cup(\mathcal{X}') = \mathsf{C}\big(\bigcup_{\mathbf{y} \in \mathcal{B}_s}(\partial(\mathbf{y}) \cap \mathbf{y})\big)$, namely $\mathcal{X}' = \bigcup_{\mathbf{y} \in \mathcal{B}_s} \mathcal{B}_\mathbf{y}$. Moreover, if $\mathbf{y} \in \mathcal{B}_s$ and $\mathbf{z} \in \mathcal{B}_\mathbf{y}$, then $\mathbf{z} \subseteq \mathsf{C}^\cup(\mathcal{B}_\mathbf{y}) =$

$\partial(\mathbf{y}) \cap \mathbf{y} \subseteq \mathbf{y}$, meaning every $\mathbf{z} \in \mathcal{X}'$ has some $\mathbf{y} \in \mathcal{B}$ with $\mathbf{z} \subseteq \mathbf{y}$. Applying the induction hypothesis to $\mathcal{B} \setminus \mathcal{B}_s \cup \mathcal{X}' \subseteq \mathcal{X}_1 \cup \ldots \cup \mathcal{X}_{s-1}$ now yields the desired result.

We define a partial order on the elements $\mathbf{x}, \mathbf{y} \in \mathcal{X}_1 \cup \{\mathbf{v}_1\} \cup \ldots \cup \mathcal{X}_s \cup \{\mathbf{v}_s\}$ by declaring $\mathbf{x} \preceq \mathbf{y}$ when $\mathbf{x} \subseteq \mathbf{y}$. If $\mathbf{x}, \mathbf{y} \in \mathcal{X}_j \cup \{\mathbf{v}_j\}$, where $j \in [1, s]$, then $\mathbf{x} \preceq \mathbf{y}$ is only possible if $\mathbf{x} = \mathbf{y}$ (which can be seen by applying the map π_{j-1} and using (OR2)). If $\mathbf{x} \in \mathcal{X}_{j'} \cup \{\mathbf{v}_{j'}\}$ and $\mathbf{y} \in \mathcal{X}_j \cup \{\mathbf{v}_j\}$ with $\mathbf{x} \prec \mathbf{y}$, then $j' < j$ (which can be seen by applying the map π_j to $\mathbf{x}$ and $\mathbf{y}$ to conclude $j' \leq j$ and then using the previous observation). In such case, we have $\mathbf{x} \subseteq \mathcal{E}_{j-1} \cap \mathbf{y} = \partial(\mathbf{y}) \cap \mathbf{y}$ (the equality follows in view of (OR1) and (OR2) as before), whence (OR1) and (5.2) ensure that $\mathbf{x} \in \mathcal{X}_{j'}$. Thus each $\mathbf{v}_j$ is a maximal element. If $\mathcal{B} \subseteq \mathcal{X}_1 \cup \{\mathbf{v}_1\} \cup \ldots \cup \mathcal{X}_s \cup \{\mathbf{v}_s\}$, we let

$$\downarrow\mathcal{B} = \{\mathbf{x} \in \mathcal{X}_1 \cup \{\mathbf{v}_1\} \cup \ldots \cup \mathcal{X}_s \cup \{\mathbf{v}_s\} : \ \mathbf{x} \preceq \mathbf{y} \text{ for some } \mathbf{y} \in \mathcal{B}\}$$

denote the down-set generated by $\mathcal{B}$. Likewise, we let $\downarrow B$ denote the set of representatives for $\downarrow\mathcal{B}$, where $B \subseteq X_1 \cup \{v_1\} \cup \ldots \cup X_s \cup \{v_s\}$ is the set of representatives for $\mathcal{B}$. Indeed, since (OR2) ensures there is a bijective correspondence between $\mathcal{X}_1 \cup \{\mathbf{v}_1\} \cup \ldots \cup \mathcal{X}_s \cup \{\mathbf{v}_s\}$ and $X_1 \cup \{v_1\} \cup \ldots \cup X_s \cup \{v_s\}$, the partial order defined above inherits to one on $X_1 \cup \{v_1\} \cup \ldots \cup X_s \cup \{v_s\}$. We let $\mathcal{B}^* \subseteq \mathcal{B}$ denote the subset of all maximal elements of $\mathcal{B}$, that is, all $\mathbf{x} \in \mathcal{B}$ for which there is no $\mathbf{y} \in \mathcal{B}$ with $\mathbf{x} \prec \mathbf{y}$. Clearly, $\downarrow(\mathcal{B}^*) = \downarrow\mathcal{B}$, $(\downarrow\mathcal{B})^* = \mathcal{B}^*$,

$$\mathsf{C}^{\cup}(\mathcal{B}^*) = \mathsf{C}^{\cup}(\mathcal{B}) = \mathsf{C}^{\cup}(\downarrow\mathcal{B}) \quad \text{and} \quad \mathbb{R}^{\cup}(\mathcal{B}^*) = \mathbb{R}^{\cup}(\mathcal{B}) = \mathbb{R}^{\cup}(\downarrow\mathcal{B}). \tag{5.3}$$

A short argument now shows there is a uniquely defined subset $\mathcal{B}_{\mathbf{x}}$ satisfying (OR1) with the additional property that $\mathcal{B}_{\mathbf{x}}^* = \mathcal{B}_{\mathbf{x}}$. Indeed, the existence of such a set follows in view of $\mathsf{C}^{\cup}(\mathcal{B}_{\mathbf{x}}^*) = \mathsf{C}^{\cup}(\mathcal{B}_{\mathbf{x}})$. On the other hand, if $\mathcal{C}_{\mathbf{x}} \subseteq \mathcal{X}_1 \cup \ldots \cup X_{j-1}$ is another set satisfying (OR1) with $\mathcal{C}_{\mathbf{x}}^* = \mathcal{C}_{\mathbf{x}}$, then we have $\mathsf{C}^{\cup}(\mathcal{C}_{\mathbf{x}}) = \partial(\mathbf{x}) \cap \mathbf{x} = \mathsf{C}^{\cup}(\mathcal{B}_x)$. Consequently, if $\mathbf{y} \in \mathcal{B}_{\mathbf{x}}$ is arbitrary, then $\mathbf{y} \subseteq \mathsf{C}^{\cup}(\mathcal{B}_{\mathbf{x}}) = \mathsf{C}^{\cup}(\mathcal{C}_{\mathbf{x}})$, whence (5.2) implies $\mathbf{y} \subseteq \mathbf{z}$ for some $\mathbf{z} \in \mathcal{C}_{\mathbf{x}}$. Thus $\mathcal{B}_{\mathbf{x}} \subseteq \downarrow\mathcal{C}_{\mathbf{x}}$, implying $\downarrow\mathcal{B}_{\mathbf{x}} \subseteq \downarrow\mathcal{C}_{\mathbf{x}}$. Swapping the roles of $\mathcal{B}_{\mathbf{x}}$ and $\mathcal{C}_{\mathbf{x}}$ and repeating this argument shows $\downarrow\mathcal{C}_{\mathbf{x}} \subseteq \downarrow\mathcal{B}_{\mathbf{x}}$. As a result, we find that $\downarrow\mathcal{B}_{\mathbf{x}} = \downarrow\mathcal{C}_{\mathbf{x}}$, in turn implying $\mathcal{B}_{\mathbf{x}} = \mathcal{B}_{\mathbf{x}}^* = (\downarrow\mathcal{B}_{\mathbf{x}})^* = (\downarrow\mathcal{C}_{\mathbf{x}})^* = \mathcal{C}_{\mathbf{x}}^* = \mathcal{C}_{\mathbf{x}}$, establishing the uniqueness of $\mathcal{B}_{\mathbf{x}}$. We now henceforth use

$$\partial(\{\mathbf{x}\}) := \mathcal{B}_{\mathbf{x}}$$

to denote the unique set satisfying (OR1) with $\partial(\{\mathbf{x}\})^* = \partial(\{\mathbf{x}\})$, and let $\partial(\{x\})$ denoting the set of representatives for $\partial(\{\mathbf{x}\})$. Note, if $\mathbf{x}, \mathbf{y} \in \mathcal{X}_1 \cup \{\mathbf{v}_1\} \cup \ldots \cup \mathcal{X}_s \cup \{\mathbf{v}_s\}$ with $\mathbf{y} \prec \mathbf{x}$ and $\mathbf{x} \in \mathcal{X}_j \cup \{\mathbf{v}_j\}$, then $\mathbf{y} \subseteq \mathcal{E}_{j-1} \cap \mathbf{x} = \partial(\mathbf{x}) \cap \mathbf{x} = \mathsf{C}^{\cup}(\partial(\{\mathbf{x}\}))$, whence $\mathbf{y} \in \downarrow\partial(\{\mathbf{x}\})$ by (5.2). In consequence, since $\mathsf{C}^{\cup}(\downarrow\partial(\{\mathbf{x}\})) = \mathsf{C}^{\cup}(\partial(\{\mathbf{x}\})) = \partial(\mathbf{x}) \cap \mathbf{x} \subseteq \mathbf{x}$, we find that

$$\downarrow\partial(\{\mathbf{x}\}) = \downarrow\mathbf{x} \setminus \{\mathbf{x}\}.$$

Also, if $\partial(\mathbf{x}) = \{0\}$, then $\partial(\{\mathbf{x}\}) = \emptyset$, which will be the case for any $\mathbf{x} \in \mathcal{X}_1 \cup \{\mathbf{v}_1\}$.

In view of (OR1), it follows that, for any subset $\mathcal{B} \subseteq \mathcal{X}_1 \cup \{\mathbf{v}_1\} \ldots \cup \mathcal{X}_j \cup \{\mathbf{v}_j\}$, where $j \in [1, s]$, there exists a subset $\partial(\mathcal{B}) \subseteq \mathcal{X}_1 \cup \ldots \cup \mathcal{X}_{j-1}$ with

$$\mathbb{R}\langle \bigcup_{\mathbf{x}\in\mathcal{B}} \partial(\mathbf{x})\rangle = \mathbb{R}^{\cup}\langle \partial(\mathcal{B})\rangle \quad \text{and} \quad \mathsf{C}\Big(\bigcup_{\mathbf{x}\in\mathcal{B}} (\partial(\mathbf{x}) \cap \mathbf{x})\Big) = \mathsf{C}^{\cup}(\partial(\mathcal{B})). \tag{5.4}$$

For instance, we could take $\partial(\mathcal{B}) = (\bigcup_{\mathbf{x}\in\mathcal{B}} \mathcal{B}_{\mathbf{x}})^*$. Moreover, as argued in the previous paragraph, if we set

$$\partial(\mathcal{B}) := \Big(\bigcup_{\mathbf{x}\in\mathcal{B}} \mathcal{B}_{\mathbf{x}}\Big)^* = \Big(\bigcup_{\mathbf{x}\in\mathcal{B}} \partial(\{\mathbf{x}\})\Big)^*,$$

then $\partial(\mathcal{B}) \subseteq \mathcal{X}_1 \cup \ldots \cup \mathcal{X}_{j-1}$ will be the unique set satisfying (5.4) with the additional property that $\partial(\mathcal{B})^* = \partial(\mathcal{B})$, which we henceforth assume is the case. We let $\partial(B) \subseteq X_1 \cup \ldots \cup X_{j-1}$ denote the set of representatives for $\partial(\mathcal{B})$. Note, from its definition, if $\mathcal{A} \subseteq \mathcal{B}$, then $\partial(\mathcal{A}) \subseteq \downarrow\partial(\mathcal{B})$. Indeed, $\downarrow\partial(\mathcal{B}) = \bigcup_{\mathbf{x}\in\mathcal{B}} \downarrow\partial(\{\mathbf{x}\})$, ensuring

$$\downarrow\partial(\mathcal{A} \cup \mathcal{B}) = \downarrow\partial(\mathcal{A}) \cup \downarrow\partial(\mathcal{B}) \quad \text{for } \mathcal{A}, \mathcal{B} \subseteq \mathcal{X}_1 \cup \{\mathbf{v}_1\} \cup \ldots \cup \mathcal{X}_s \cup \{\mathbf{v}_s\}. \tag{5.5}$$

We extend the partial order $\preceq$ to the subsets of $\mathcal{X}_1 \cup \{\mathbf{v}_1\} \cup \ldots \cup \mathcal{X}_s \cup \{\mathbf{v}_s\}$ as follows. Given any subset $\mathcal{B} \subseteq \mathcal{X}_1 \cup \{\mathbf{v}_1\} \cup \ldots \cup \mathcal{X}_s \cup \{\mathbf{v}_s\}$, we define the immediate predecessors to $\mathcal{B}$ to be the sets $\mathcal{B}' = \mathcal{B} \setminus \{\mathbf{x}\} \cup \partial(\{\mathbf{x}\})$ for $\mathbf{x} \in \mathcal{B}$. Since the tuple $(|\mathcal{B} \cap (\mathcal{X}_s \cup \{\mathbf{v}_s\})|, \ldots, |\mathcal{B} \cap (\mathcal{X}_1 \cup \{\mathbf{v}_1\})|)$ associated to $\mathcal{B}$ strictly decreases in the lexicographic order under this operation, extending this relation transitively then defines the partial order $\preceq$. From its definition and a short inductive argument on $(|\mathcal{B} \cap (\mathcal{X}_s \cup \{\mathbf{v}_s\})|, \ldots, |\mathcal{B} \cap (\mathcal{X}_1 \cup \{\mathbf{v}_1\})|)$, we find that

$$\mathcal{A} \preceq \mathcal{B} \quad \text{implies} \quad \mathcal{A} \subseteq \downarrow\mathcal{B}, \tag{5.6}$$

thus ensuring that the partial order on the subsets of $\mathcal{X}_1 \cup \{\mathbf{v}_1\} \cup \ldots \cup \mathcal{X}_s \cup \{\mathbf{v}_s\}$ is compatible with the pre-order induced from the partial order for the elements of $\mathcal{X}_1 \cup \{\mathbf{v}_1\} \cup \ldots \cup \mathcal{X}_s \cup \{\mathbf{v}_s\}$. As a result, (5.3) implies that, if $\mathcal{A} \preceq \mathcal{B}$, then $\mathsf{C}^{\cup}(\mathcal{A}) \subseteq \mathsf{C}^{\cup}(\mathcal{B})$. Also,

$$\mathcal{A} \subseteq \mathcal{B} \quad \text{implies} \quad \mathcal{A} \preceq \mathcal{B},$$

as the following argument shows. Note, it suffices to show $\mathcal{B} \setminus \{\mathbf{x}\} \preceq \mathcal{B}$ for $\mathbf{x} \in \mathcal{B}$, as this can then be iterated. To see this, we observe that we may simply replace each $\mathbf{x}$ with the half-spaces from $\partial(\{\mathbf{x}\})$, and then replace each $\mathbf{y} \in \partial(\{\mathbf{x}\}) \setminus \mathcal{B}$ with the half-spaces from $\partial(\{\mathbf{y}\})$, and so forth, until all such elements and their successors are replaced either using the empty set or an element already in $\mathcal{B} \setminus \{\mathbf{x}\}$.

Next,

$$\mathcal{A} \subseteq \mathcal{B} \text{ implies } (\mathcal{B} \setminus \mathcal{A}) \cup \partial(\mathcal{A}) \preceq \mathcal{B},$$

which can be seen as follows. Sequentially replacing each $\mathbf{x} \in \mathcal{A}$ with $\partial(\{\mathbf{x}\})$, always choosing the next $\mathbf{x} \in \mathcal{A}$ in the sequence to be an element minimal among the remaining elements of $\mathcal{A}$ with respect to $\preceq$ (for instance, we could first take all $\mathbf{x} \in \mathcal{A} \cap (\mathcal{X}_1 \cup \{\mathbf{v}_1\})$, then all $\mathbf{x} \in \mathcal{A} \cap (\mathcal{X}_2 \cup \{\mathbf{v}_2\})$, and so forth), shows $(\mathcal{B} \setminus \mathcal{A}) \cup \bigcup_{\mathbf{x}\in\mathcal{A}} \partial(\{\mathbf{x}\}) \preceq \mathcal{B}$. Then, since $(\mathcal{B} \setminus \mathcal{A}) \cup \partial(\mathcal{A}) = (\mathcal{B} \setminus \mathcal{A}) \cup \left(\bigcup_{\mathbf{x}\in\mathcal{A}} \partial(\{\mathbf{x}\})\right)^* \subseteq (\mathcal{B} \setminus \mathcal{A}) \cup \bigcup_{\mathbf{x}\in\mathcal{A}} \partial(\{\mathbf{x}\}) \preceq \mathcal{B}$, the claimed result $(\mathcal{B} \setminus \mathcal{A}) \cup \partial(\mathcal{A}) \preceq \mathcal{B}$ follows. In particular, $\partial(\mathcal{A}) \preceq \mathcal{B}$ when $\mathcal{A} \subseteq \mathcal{B}$.

We remark that

$$\{\mathbf{x} \in \mathcal{A} : \mathcal{A} \preceq \mathcal{B}\} = \downarrow\mathcal{B}, \tag{5.7}$$

as can be seen by an inductive argument on s. Indeed, the inclusion $\{\mathbf{x} \in \mathcal{A} : \mathcal{A} \preceq \mathcal{B}\} \subseteq \downarrow\mathcal{B}$ follows from (5.6), while $\mathbf{x} \in \downarrow\mathcal{B}$ implies $\mathbf{x} \preceq \mathbf{y}$ for some $\mathbf{y} \in \mathcal{B}$. If $\mathbf{x} = \mathbf{y} \in \mathcal{B} \preceq \mathcal{B}$, the reverse inclusion holds. Otherwise, $\mathbf{x} \subseteq \mathcal{E}_{j-1} \cap \mathbf{y} = \partial(\mathbf{y}) \cap \mathbf{y} = \mathsf{C}^{\cup}(\partial(\{\mathbf{y}\}))$, where $\mathbf{y} \in \mathcal{X}_j \cup \{\mathbf{v}_j\}$. In view of (5.2), we have $\mathbf{x} \preceq \mathbf{z}$ for some $\mathbf{z} \in \partial(\{\mathbf{y}\}) \preceq \mathcal{B}$, i.e., $\mathbf{x} \in \downarrow\partial(\{\mathbf{y}\})$, and now applying the induction hypothesis to $\partial(\{\mathbf{y}\}) \subseteq \mathcal{X}_1 \cup \ldots \cup \mathcal{X}_{s-1}$ yields the reverse inclusion.

Let $\partial^n(\mathcal{B}) = \underbrace{\partial(\partial(\ldots\partial}_{n}(\mathcal{B}))\ldots))$ for $n \geq 0$, so $\partial^0(\mathcal{B}) = \mathcal{B}$. A similar inductive argument on s yields

$$\downarrow\mathcal{B} = \bigcup_{n=0}^{s-1} \partial^n(\mathcal{B}). \tag{5.8}$$

Note $\partial^s(\mathcal{B}) = \emptyset$ for any $\mathcal{B} \subseteq \mathcal{X}_1 \cup \{\mathbf{v}_1\} \cup \ldots \cup \mathcal{X}_s \cup \{\mathbf{v}_s\}$ and that the case $s = 1$ is clear since $\partial(\{\mathbf{x}\}) = \emptyset$ for all $\mathbf{x} \in \mathcal{B}$ in this case. The inclusion $\bigcup_{n=0}^{s-1} \partial^n(\mathcal{B}) \subseteq \bigcup_{n=0}^{s-1} \downarrow\partial^n(\mathcal{B}) \subseteq \downarrow\mathcal{B}$ follows in view of $\partial^{s-1}(\mathcal{B}) \preceq \partial^{s-2}(\mathcal{B}) \preceq \ldots \preceq \partial^0(\mathcal{B}) = \mathcal{B}$ and (5.6). On the other hand, if $\mathbf{x} \in \downarrow\mathcal{B}$, then either $\mathbf{x} \in \mathcal{B} = \partial^0(\mathcal{B}) \subseteq \bigcup_{n=0}^{s-1} \partial^n(\mathcal{B})$, or else $\mathbf{x} \subseteq \mathcal{E}_{j-1} \cap \mathbf{y} = \partial(\mathbf{y}) \cap \mathbf{y} = \mathsf{C}^{\cup}(\partial(\{\mathbf{y}\}))$, for some $\mathbf{y} \in \mathcal{B}$ with $\mathbf{y} \in \mathcal{X}_j \cup \{\mathbf{v}_j\}$, in which case (5.2) yields $\mathbf{x} \in \downarrow\partial(\{\mathbf{y}\})$, and now applying the induction hypothesis to $\partial(\mathcal{B}) \subseteq \mathcal{X}_1 \cup \ldots \cup \mathcal{X}_{s-1}$ yields $\mathbf{x} \in \downarrow\partial(\{\mathbf{y}\}) \subseteq \bigcup_{n=1}^{s-1} \downarrow\partial^n(\mathcal{B}) = \downarrow\partial(\mathcal{B}) = \bigcup_{n=1}^{s-1} \partial^n(\mathcal{B}) \subseteq \bigcup_{n=0}^{s-1} \partial^n(\mathcal{B})$, establishing the reverse inclusion.

It may be helpful to view the half-spaces from $\mathcal{X}_1 \cup \{\mathbf{v}_1\} \cup \ldots \cup \mathcal{X}_s \cup \{\mathbf{v}_s\}$ as vertices in a directed graph with each half-space $\mathbf{x}$ connected to the half-spaces from $\partial(\{\mathbf{x}\})$ by a directed edge. Let $\mathcal{B} \subseteq \mathcal{X}_1 \cup \{\mathbf{v}_1\} \cup \ldots \cup \mathcal{X}_s \cup \{\mathbf{v}_s\}$ and let $\mathcal{Y}$ denote the subset of vertices reachable from some $\mathbf{x} \in \mathcal{B}$, so $\mathbf{y} \in \mathcal{Y}$ means there is a sequence $\mathbf{y}_0, \ldots, \mathbf{y}_r$ with $\mathbf{y}_i \in \partial(\{\mathbf{y}_{i-1}\})$ for $i \geq 1$, $\mathbf{y}_0 \in \mathcal{B}$ and $\mathbf{y}_r = \mathbf{y}$. Note (5.8) and (5.5) imply $\downarrow\partial(\downarrow\mathcal{A}) = \downarrow\partial(\bigcup_{n=0}^{s-1} \partial^n(\mathcal{A})) = \bigcup_{n=1}^{s-1} \downarrow\partial^n(\mathcal{A}) = \downarrow\downarrow\partial(\mathcal{A}) = \downarrow\partial(\mathcal{A})$ for any $\mathcal{A} \subseteq \mathcal{X}_1 \cup \{\mathbf{v}_1\} \cup \ldots \cup \mathcal{X}_s \cup \{\mathbf{v}_s\}$. This can be used in an inductive argument on $i = 0, 1, \ldots, r$ to show $\mathbf{y}_i \in \downarrow\partial^i(\mathcal{B})$ for $i \in [0, r]$. Indeed, $\mathbf{y}_0 \in \mathcal{B}$ ensures

$\mathbf{y}_0 \in \downarrow\mathcal{B} = \downarrow\partial^0(\mathcal{B})$, while $\mathbf{y}_i \in \partial(\{\mathbf{y}_{i-1}\})$ and $\mathbf{y}_{i-1} \in \downarrow\partial^{i-1}(\mathcal{B})$ then ensure $\mathbf{y}_i \in \downarrow\partial(\downarrow\partial^{(i-1)}(\mathcal{B})) = \downarrow\partial^i(\mathcal{B})$, completing the induction. In particular, $\mathbf{y} = \mathbf{y}_r \in \downarrow\partial^r(\mathcal{B})$, so that (5.8) yields $\mathcal{Y} \subseteq \downarrow\mathcal{B}$. The reverse inclusion follows more directly from (5.8), meaning $\downarrow\mathcal{B} = \mathcal{Y}$ consists of all vertices which can be reached via a directed path starting at some vertex from the subset $\mathcal{B}$. The grading condition that $\partial(\mathcal{B}) \subseteq \mathcal{X}_1 \cup \ldots \cup \mathcal{X}_{j-1}$ for $\mathcal{B} \subseteq \mathcal{X}_1 \cup \{\mathbf{v}_1\} \cup \ldots \cup \mathcal{X}_j \cup \{\mathbf{v}_j\}$ ensures there are no directed cycles.

As already remarked, $\mathcal{A} \preceq \mathcal{B}$ implies $\mathcal{A} \subseteq \downarrow\mathcal{B}$. The partial converse to this is

$$\mathcal{A} \subseteq \downarrow\mathcal{B} \quad \text{implies} \quad \mathcal{A}^* \preceq \mathcal{B}. \tag{5.9}$$

To see this, we modify the argument for showing $\mathcal{A} \subseteq \mathcal{B}$ implies $\mathcal{A} \preceq \mathcal{B}$. Order the elements of $\mathcal{B} \setminus \mathcal{A}$. Beginning with the first $\mathbf{x} \in \mathcal{B} \setminus \mathcal{A}$, replace $\mathbf{x}$ with the half-spaces from $\partial(\{\mathbf{x}\})$, and then replace each $\mathbf{y} \in \partial(\{\mathbf{x}\}) \setminus \mathcal{A}$ with the half-spaces from $\partial(\{\mathbf{y}\})$, and so forth, until all such elements and their successors are replaced either using the empty set or an element already in $\mathcal{A}$. Let $\mathcal{B}' \subseteq \mathcal{B} \cup \mathcal{A}$ be the resulting set. Take the next $\mathbf{x}' \in \mathcal{B}' \cap (\mathcal{B} \setminus \mathcal{A}) = \mathcal{B}' \setminus \mathcal{A}$ and repeat the procedure. Continue until all elements from $\mathcal{B} \setminus \mathcal{A}$ have been exhausted and let $C \subseteq \mathcal{A}$ be the resulting set. By construction, $C \preceq \mathcal{B}$. Since $\mathcal{A}^*$ is the set of maximal elements in $\mathcal{A}$, it follows from $\mathcal{A} \subseteq \downarrow\mathcal{B}$ and (5.8) that $\mathcal{A}^* \subseteq C$, whence $\mathcal{A}^* \preceq C \preceq \mathcal{B}$, as desired.

If $\mathcal{A} \subseteq \downarrow\mathcal{B}$, then $\partial(\mathcal{A}) \subseteq \downarrow\partial(\mathcal{B})$ can be seen as follows. Let $\mathbf{x} \in \partial(\mathcal{A})$. Then $\mathbf{x} \in \partial(\{\mathbf{x}'\})$ for some $\mathbf{x}' \in \mathcal{A} \subseteq \downarrow\mathcal{B}$, in which case there is some $\mathbf{y} \in \mathcal{B}$ with $\mathbf{x} \preceq \mathbf{x}' \preceq \mathbf{y}$. If $\mathbf{x}' = \mathbf{y}$, then $\mathbf{x} \in \partial(\{\mathbf{x}'\}) \subseteq \downarrow\partial(\mathcal{B})$. Otherwise, $\mathbf{x} \subseteq \mathcal{E}_{j-1} \cap \mathbf{y} = \partial(\mathbf{y}) \cap \mathbf{y} = \mathsf{C}^{\cup}(\partial(\{\mathbf{y}\}))$, where $\mathbf{y} \in \mathcal{X}_j \cup \{\mathbf{v}_j\}$, in which case (5.2) implies $\mathbf{x} \preceq \mathbf{y}'$ for some $\mathbf{y}' \in \partial(\{\mathbf{y}\})$, i.e., $\mathbf{x} \in \downarrow\partial(\{\mathbf{y}\}) \subseteq \downarrow\partial(\mathcal{B})$, as desired. Combining this with (5.9), we conclude that

$$\mathcal{A} \subseteq \downarrow\mathcal{B} \quad \text{implies} \quad \partial(\mathcal{A}) \preceq \partial(\mathcal{B}).$$

Definition 5.2 Let $\mathcal{R} = (\mathcal{X}_1 \cup \{\mathbf{v}_1\}, \ldots, \mathcal{X}_s\{\mathbf{v}_s\})$ be an oriented Reay system for the subspace $\mathcal{E}_s \subseteq \mathbb{R}^d$ and let $\mathcal{B} \subseteq \mathcal{X}_1 \cup \{\mathbf{v}_1\} \cup \ldots \cup \mathcal{X}_s \cup \{\mathbf{v}_s\}$. We say that $\mathcal{B}$ is a **support set** for $\mathcal{R}$ if

$$\mathcal{B}^* = \mathcal{B} \quad \text{and} \quad \mathcal{X}_i \cup \{\mathbf{v}_i\} \nsubseteq \downarrow\mathcal{B} \quad \text{for all } i \in [1, s].$$

We say that $\mathcal{B}$ is a **virtual independent** set if

$$\mathcal{B}^* = \mathcal{B} \quad \text{and} \quad \downarrow B \text{ is linearly independent.}$$

We remark that Proposition 5.3.3 ensures that the definition of a virtual independent set does not depend on the choice of representative set $\downarrow B$.

Proposition 5.3 *Let $\mathcal{R} = (\mathcal{X}_1 \cup \{\mathbf{v}_1\}, \ldots, \mathcal{X}_s \cup \{\mathbf{v}_s\})$ be a an orientated Reay system for the subspace $\mathcal{E}_s \subseteq \mathbb{R}^d$ and let $\mathcal{B} \subseteq \mathcal{X}_1 \cup \{\mathbf{v}_1\} \cup \ldots \cup \mathcal{X}_s \cup \{\mathbf{v}_s\}$. For $j \in [1, s]$, let $\mathcal{E}_{j-1} = \mathbb{R}\langle \mathcal{X}_1 \cup \ldots \cup \mathcal{X}_{j-1} \rangle$, let $\mathcal{E}'_{j-1} = \mathcal{E}_{j-1} + \mathbb{R}^{\cup}\langle \partial(\mathcal{B}) \rangle$, let $\pi'_{j-1} : \mathbb{R}^d \to (\mathcal{E}'_{j-1})^{\perp}$ be the orthogonal projection, and let $\mathcal{B}_j = \mathcal{B} \cap (\mathcal{X}_j \cup \{\mathbf{v}_j\})$.*

1. *$\mathbb{R}\langle \downarrow B \rangle = \mathbb{R}^{\cup}\langle \mathcal{B} \rangle$.*
2. *$\mathsf{C}^{\cup}(\mathcal{B}) = z_1 + \ldots + z_\ell$ is a convex cone containing 0, and $\overline{\mathsf{C}^{\cup}(\mathcal{B})} = \overline{z}_1 + \ldots + \overline{z}_\ell$ is a polyhedral cone, where $z_1, \ldots, z_\ell \in \mathcal{B}$ are the distinct half-spaces in $\mathcal{B}$.*
3. *$\mathcal{B}$ is a virtual independent set if and only if $\pi(B)$ is a linearly independent set of size $|\mathcal{B}|$, where $\pi : \mathbb{R}^d \to \mathbb{R}^{\cup}\langle \partial(\mathcal{B}) \rangle^{\perp}$ is the orthogonal projection.*
4. *If $\mathcal{B}$ is virtual independent, then $\overline{\mathsf{C}^{\cup}(\mathcal{B})} = \overline{z}_1 + \ldots + \overline{z}_\ell$ is a blunted simplicial cone with lineality space $\mathbb{R}^{\cup}\langle \partial(\mathcal{B}) \rangle$, $\mathsf{C}^{\cup}(\mathcal{B})$ has trivial lineality space, and $\mathsf{C}^{\cup}(\mathcal{B})^{\circ} = z_1^{\circ} + \ldots + z_\ell^{\circ}$, where $z_1, \ldots, z_\ell \in \mathcal{B}$ are the distinct half-spaces in $\mathcal{B}$.*
5. *If $\mathcal{B}$ is a support set for $\mathcal{R}$, then $\mathcal{B}$ is a virtual independent set.*
6. *If $\mathcal{B}$ is a support set for $\mathcal{R}$, then $\pi'_{j-1}\big(\bigcup_{i=j}^{s} B_i\big)$ is a linearly independent subset of $\sum_{i=j}^{s} |\mathcal{B}_i| \geq 0$ distinct elements, for any $j \in [1, s]$.*
7. *If $\mathcal{B}$ is virtual independent and $\mathcal{A} \subseteq \mathcal{B}$, then $C = ((\mathcal{B} \setminus \mathcal{A}) \cup \partial(\mathcal{A}))^*$ is virtual independent with $\mathcal{B} \setminus \mathcal{A} \subseteq C$. Moreover, if $\mathcal{B}$ is a support set, then so is C.*
8. *If $\mathcal{B} \subseteq \mathcal{X}_1 \cup \ldots \cup \mathcal{X}_s$, then $\mathcal{B}^*$ is a support set. In particular, $\partial(\mathcal{B})$ is always a support set for any $\mathcal{B} \subseteq \mathcal{X}_1 \cup \{\mathbf{v}_1\} \cup \ldots \cup \mathcal{X}_s \cup \{\mathbf{v}_s\}$.*
9. *If $\mathcal{B} \subseteq \mathcal{X}_1 \cup \ldots \cup \mathcal{X}_s$, $\mathbf{x} \in \mathcal{X}_1 \cup \ldots \cup \mathcal{X}_s$ and $\mathbf{x} \subseteq \mathbb{R}^{\cup}\langle \mathcal{B} \rangle$, then $\mathbf{x} \in \downarrow \mathcal{B}$.*

Proof

1. We proceed by induction on the depth s. In view of (5.8), we have

$$\downarrow \mathcal{B} = \mathcal{B}_s \cup \downarrow \partial(\mathcal{B}) \cup \downarrow (\mathcal{B} \setminus \mathcal{B}_s) \quad \text{and} \quad \downarrow B = B_s \cup \downarrow \partial(B) \cup \downarrow (B \setminus B_s). \tag{5.10}$$

By induction hypothesis, $\mathbb{R}\langle \downarrow \partial(B) \cup \downarrow (B \setminus B_s) \rangle = \mathbb{R}^{\cup}\langle \partial(\mathcal{B}) \cup (\mathcal{B} \setminus \mathcal{B}_s) \rangle$ (note $\downarrow \big(\partial(\mathcal{B}) \cup (\mathcal{B} \setminus \mathcal{B}_s)\big) = \emptyset$ for $s = 1$). But now (5.10) implies that

$$\mathbb{R}\langle \downarrow B \rangle = \mathbb{R}\langle B_s \rangle + \mathbb{R}\langle \downarrow \partial(B) \cup \downarrow (B \setminus B_s) \rangle = \mathbb{R}\langle B_s \rangle + \mathbb{R}^{\cup}\langle \partial(\mathcal{B}) \rangle + \mathbb{R}^{\cup}\langle \mathcal{B} \setminus \mathcal{B}_s \rangle = \mathbb{R}^{\cup}\langle \mathcal{B} \rangle,$$

as desired.

2. Suppose $\mathcal{B} \subseteq \mathcal{X}_1 \cup \{\mathbf{v}_1\} \cup \ldots \mathcal{X}_j \cup \{\mathbf{v}_j\}$, where $j \in [0, s]$, and let $\mathbf{z}_1, \ldots, \mathbf{z}_\ell \in \mathcal{B}$ be the distinct half-spaces in $\mathcal{B}$. For $j = 0$, we have $\mathcal{B} = \emptyset$ and $\mathsf{C}^{\cup}(\mathcal{B}) = \overline{\mathsf{C}^{\cup}(\mathcal{B})} = \{0\}$, which is the sum of an empty number of half-spaces. For $j = 1$, we have $\partial(\{\mathbf{z}_i\}) = \emptyset$, $\partial(\mathbf{z}_i) = \{0\}$ and $\mathbf{z}_i = \overline{\mathbf{z}}_i = \mathbb{R}_+ z_i$ for all i (by (OR2)), in which case $\mathsf{C}^{\cup}(\mathcal{B}) = \mathsf{C}(z_1, \ldots, z_r)$ with the result clear. Thus we assume $j \geq 2$ and proceed by induction on j. By induction hypothesis applied to $\partial(\{\mathbf{z}_i\})$, we have $0 \in \mathsf{C}^{\cup}(\partial(\{\mathbf{z}_i\})) = \mathbf{z}_i \cap \partial(\mathbf{z}_i)$, ensuring each $\mathbf{z}_i$ is a convex cone with $0 \in (\mathbf{z}_i \cap \partial(\mathbf{z}_i)) \subseteq \mathbf{z}_i$, for $i \in [1, \ell]$, in turn implying $0 \in \mathsf{C}^{\cup}(\mathcal{B}) = \mathbf{z}_1 + \ldots + \mathbf{z}_\ell$ with $\mathsf{C}^{\cup}(\mathcal{B})$ a convex cone.

   Note $\overline{\mathbf{z}}_i = \mathbb{R}_+ z_i + \partial(\mathbf{z}_i) = \mathsf{C}(Y_i \cup \{z_i\})$, where Y_i is any positive basis for the subspace $\partial(\mathbf{z}_i)$. Thus $\overline{\mathbf{z}}_1 + \ldots + \overline{\mathbf{z}}_\ell = \mathsf{C}\big(\bigcup_{i=1}^{\ell}(Y_i \cup \{z_i\})\big)$, which is polyhedral cone, and thus closed. Hence $\overline{\mathsf{C}^{\cup}(\mathcal{B})} = \overline{\mathbf{z}_1 + \ldots + \mathbf{z}_\ell} \subseteq \overline{\overline{\mathbf{z}}_1 + \ldots + \overline{\mathbf{z}}_\ell} \subseteq \overline{\mathbf{z}}_1 + \ldots + \overline{\mathbf{z}}_\ell$. On the other hand, since $0 \in \mathbf{z}_i$ for all i with $\mathsf{C}^{\cup}(\mathcal{B})$, and thus also $\overline{\mathsf{C}^{\cup}(\mathcal{B})}$, a convex cone, it follows that $\overline{\mathbf{z}}_1 + \ldots + \overline{\mathbf{z}}_\ell \subseteq \overline{\mathbf{z}_1 + \ldots + \mathbf{z}_\ell} = \overline{\mathsf{C}^{\cup}(\mathcal{B})}$, establishing the reverse inclusion, showing $\overline{\mathsf{C}^{\cup}(\mathcal{B})} = \overline{\mathbf{z}}_1 + \ldots + \overline{\mathbf{z}}_\ell = \mathsf{C}\big(\bigcup_{i=1}^{\ell}(Y_i \cup \{z_i\}\big)$ is a polyhedral cone.

3. Suppose $\mathcal{B}^* = \mathcal{B}$ and $\downarrow B$ is linearly independent. Since $\mathcal{B}^* = \mathcal{B}$, it follows that $\mathcal{B}$ is disjoint from $\downarrow\partial(\mathcal{B})$. By Item 1, $\mathbb{R}\langle\downarrow\partial(B)\rangle = \mathbb{R}^{\cup}\langle\partial(\mathcal{B})\rangle$. Thus $\ker\pi$ is generated by the linearly independent subset $\downarrow\partial(B) \subseteq \downarrow B$ (by Proposition 4.11.1), ensuring $\pi(B)$ is a linearly independent set of size $|\mathcal{B} \setminus \downarrow\partial(\mathcal{B})| = |\mathcal{B}|$, with the equality since $\mathcal{B}$ is disjoint from $\downarrow\partial(\mathcal{B})$. Next instead suppose $\pi(B)$ is linearly independent of size $|\mathcal{B}|$. Since $\pi(B)$ is linearly independent, no half-space in $\mathcal{B}$ is contained in $\mathbb{R}^{\cup}\langle\partial(\mathcal{B})\rangle$. If $\mathbf{x} \prec \mathbf{y}$ with $\mathbf{x},\ \mathbf{y} \in \mathcal{B}$, then we must have $\mathbf{x} \in \downarrow\partial(\{\mathbf{y}\})$, contradicting that $\mathbf{x} \nsubseteq \mathbb{R}^{\cup}\langle\partial(\mathcal{B})\rangle = \mathbb{R}^{\cup}\langle\downarrow\partial(\mathcal{B})\rangle$. Therefore we instead conclude that $\mathcal{B}^* = \mathcal{B}$. Since $\downarrow\partial(\mathcal{B}) \subseteq \mathcal{X}_1 \cup \ldots \cup \mathcal{X}_{s-1}$, Proposition 4.11.1 ensures that $\downarrow\partial(B)$ is always a linearly independent set. Consequently, since $\pi(B)$ is a linearly independent set of size $|\mathcal{B}|$ with $\ker\pi = \mathbb{R}\langle\downarrow\partial(B)\rangle$ (by Item 1), it follows that $\downarrow B = B \cup \downarrow\partial(B)$ is linearly independent, as desired.
4. In view of Items 2 and 3, $\overline{\mathsf{C}^{\cup}(\mathcal{B})} = \overline{\mathbf{z}}_1 + \ldots + \overline{\mathbf{z}}_\ell$ is a blunted simplicial cone with lineality space $\mathbb{R}^{\cup}\langle\partial(\mathcal{B})\rangle$, and $\mathsf{C}^{\cup}(\mathcal{B})^\circ = \mathbf{z}_1^\circ + \ldots + \mathbf{z}_\ell^\circ$. In particular, if $\partial(\mathcal{B}) = \emptyset$, then $\mathsf{C}^{\cup}(\mathcal{B})$ has trivial lineality space. Assuming Items 5 and 8 have been established, $\partial(\mathcal{B})$ will be virtual independent. We can then use an inductive argument on $|\downarrow\mathcal{B}|$ to show $\mathsf{C}^{\cup}(\mathcal{B})$ has trivial lineality space for a general virtual independent set $\mathcal{B}$. Indeed, in view of the established portions of Item 4, the lineality space of $\mathsf{C}^{\cup}(\mathcal{B})$ must be contained in $(\mathbf{z}_1 \cap \partial(\mathbf{z}_1)) + \ldots + (\mathbf{z}_\ell \cap \partial(\mathbf{z}_\ell)) = \mathsf{C}^{\cup}(\partial(\{\mathbf{z}_1\})) + \ldots + \mathsf{C}^{\cup}(\partial(\{\mathbf{z}_\ell\})) = \mathsf{C}^{\cup}(\partial(\mathcal{B}))$, and so applying the induction hypothesis to $\partial(\mathcal{B})$, completes the proof.
5. This follows from Proposition 4.11.1.
6. By definition, $\pi'_{j-1}(\mathbf{x})$ is either zero or a ray for any $\mathbf{x} \in \mathcal{B}$ and $j \in [1, s]$, so $\pi_{j-1}(x)$ is a representative for the half-space $\pi_{j-1}(\mathbf{x})$ whenever $\pi_{j-1}(\mathbf{x}) \neq \{0\}$. We proceed by induction on $j = s, s-1, \ldots, 1$. If Item 6 fails, then there must be a nontrivial linear combination of the elements of $\bigcup_{i=j}^{s} B_i$ equal to an element of $\mathcal{E}'_{j-1}$, say

$$\sum_{x \in B_{j+1} \cup \ldots \cup B_s} \alpha_x x + \sum_{y \in B_j} \beta_y y \in \mathcal{E}'_{j-1}, \tag{5.11}$$

where the $\alpha_x,\ \beta_y \in \mathbb{R}$ are not all zero. But then $\sum_{x \in B_{j+1} \cup \ldots \cup B_s} \alpha_x x \in \mathcal{E}'_{j-1} + \mathcal{E}_j = \mathcal{E}'_j = \ker\pi'_j$. By induction hypothesis, $\pi'_j(\bigcup_{i=j+1}^{s} B_i)$ is a linearly independent set of $\sum_{i=j+1}^{s} |B_i| \geq 0$ elements, meaning this is only possible if $\alpha_x = 0$ for all $x \in B_{j+1} \cup \ldots \cup B_s$ (note this is trivially true when $j = s$).

Hence

$$\sum_{y\in B_j} \beta_y y \in \mathcal{E}'_{j-1} = \mathcal{E}_{j-1} + \mathbb{R}^{\cup}\langle\partial(\mathcal{B})\rangle. \tag{5.12}$$

In view of Item 1, we have $\mathbb{R}^{\cup}\langle\partial(\mathcal{B})\rangle = \mathbb{R}\langle\downarrow\partial(B)\rangle$. Since $\mathcal{B}^* = \mathcal{B}$ for the support set $\mathcal{B}$, it follows that $\downarrow\partial(B)$ and B are disjoint subsets of $\downarrow B$, which is linearly independent by Item 5 as $\mathcal{B}$ is a support set. Also, all elements from $\downarrow\partial(B)\cap(X_1\cup\ldots\cup X_{j-1})$ are contained in $\mathcal{E}_{j-1}$. Thus (5.12) implies that there is a nontrivial linear combination

$$\sum_{y\in B_j} \beta_y y + \sum_{z\in Z} \gamma_z z \in \mathcal{E}_{j-1}, \tag{5.13}$$

for some $\gamma_z \in \mathbb{R}$, where $Z := \downarrow\partial(B)\cap(X_j\cup\ldots\cup X_s)$. Since $B_j\cup Z\subseteq\downarrow B$ is a disjoint union with $\mathcal{B}$ a support set, Propositions 4.11.3 and 4.11.1 imply that $\pi_{j-1}(B_j\cup Z)$ is a linearly independent set of size $|B_j\cup Z| = |B_j|+|Z|$. Thus, applying π_{j-1} to (5.13), it follows that $\beta_y = 0$ for all $y\in B_j$, and $\gamma_z = 0$ for all $z\in Z$, contradicting that the linear combination was nontrivial, which completes the proof of Item 6.

7. By definition of C, we have $C^* = C$. Since $\mathcal{A}\subseteq\mathcal{B}$ (implying $\partial(\mathcal{A})\preceq\mathcal{B}$ and then $\partial(\mathcal{A})\subseteq\downarrow\mathcal{B}$), we have $\downarrow C = \downarrow(\mathcal{B}\setminus\mathcal{A})\cup\downarrow\partial(\mathcal{A})\subseteq\downarrow\mathcal{B}$. If $\mathcal{B}$ is a support set, then $\mathcal{X}_j\cup\{\mathbf{v}_j\}\nsubseteq\downarrow\mathcal{B}$ for every $j\in[1,s]$. As a result, since $\downarrow C\subseteq\downarrow\mathcal{B}$, we also have $\mathcal{X}_j\cup\{\mathbf{v}_j\}\nsubseteq\downarrow C$ for every $j\in[1,s]$, which shows that C is a support set. If $\mathcal{B}$ is virtual independent, then $\downarrow B$ is linearly independent, whence $\downarrow C\subseteq\downarrow B$ is also linearly independent, showing that C is virtual independent.

   Suppose $\mathcal{B}\setminus\mathcal{A}\nsubseteq C$. Then, in view of the definition of C, there must be some $\mathbf{x}\in\mathcal{B}\setminus\mathcal{A}$ and $\mathbf{y}\in(\mathcal{B}\setminus\mathcal{A})\cup\partial(\mathcal{A})$ with $\mathbf{x}\prec\mathbf{y}$. If $\mathbf{y}\in\mathcal{B}\setminus\mathcal{A}$, then $\mathbf{x}\prec\mathbf{y}$ contradicts that $\mathcal{B}^* = \mathcal{B}$ for the virtual independent set $\mathcal{B}$. Therefore we must have $\mathbf{y}\in\partial(\mathcal{A})$, implying that $\mathbf{x}\prec\mathbf{y}\prec\mathbf{z}$ for some $\mathbf{z}\in\mathcal{A}\subseteq\mathcal{B}$, which again contradicts that $\mathcal{B}^* = \mathcal{B}$ for the virtual independent set $\mathcal{B}$. So we instead conclude that $\mathcal{B}\setminus\mathcal{A}\subseteq C$, completing the proof of Item 7.
8. Since $\mathcal{B}\subseteq\mathcal{X}_1\cup\ldots\cup\mathcal{X}_s$ implies $\downarrow\mathcal{B}\subseteq\mathcal{X}_1\cup\ldots\cup\mathcal{X}_s$, we have $\mathcal{X}_j\cup\{\mathbf{v}_j\}\nsubseteq\downarrow\mathcal{B} = \downarrow(\mathcal{B}^*)$ for all $j\in[1,s]$, while $\mathcal{B}^{**} = \mathcal{B}^*$, which shows that $\mathcal{B}^*$ is a support set.
9. In view of Item 1, we have $\mathbb{R}^{\cup}\langle\mathcal{B}\rangle = \mathbb{R}\langle\downarrow B\rangle$. Thus $x\in\mathbf{x}\subseteq\mathbb{R}^{\cup}\langle\mathcal{B}\rangle = \mathbb{R}\langle\downarrow B\rangle$. Since $\mathcal{B}\subseteq\mathcal{X}_1\cup\ldots\cup\mathcal{X}_s$, we also have $\downarrow B\subseteq X_1\cup\ldots\cup X_s$, while $x\in X_1\cup\ldots\cup X_s$ since $\mathbf{x}\in\mathcal{X}_1\cup\ldots\cup\mathcal{X}_s$. Proposition 4.11.1 implies that $X_1\cup\ldots\cup X_s$ is a linearly independent set, whence $x\in\mathbb{R}\langle\downarrow B\rangle$ with $x\in X_1\cup\ldots\cup X_s$ and $\downarrow B\subseteq X_1\cup\ldots\cup X_s$ is only possible if $x\in\downarrow B$, implying $\mathbf{x}\in\downarrow\mathcal{B}$.

□

Next, we show that we have a well-behaved quotient oriented Reay system defined modulo $\mathcal{R}^{\cup}\langle\mathcal{B}\rangle$, for any $\mathcal{B}\subseteq\mathcal{X}_1\cup\ldots\cup\mathcal{X}_s$. We will use the notation

$$C^{\pi} = \pi(C)\setminus\{\{0\}\}, \quad \pi(\mathcal{R}) \quad \text{and} \quad \pi^{-1}(\mathcal{D})$$

defined in Proposition 5.4 for the remainder of this work.

Proposition 5.4 *Let $\mathcal{R} = (\mathcal{X}_1 \cup \{\mathbf{v}_1\}, \ldots, \mathcal{X}_s \cup \{\mathbf{v}_s\})$ be an orientated Reay system for the subspace $\mathcal{E}_s \subseteq \mathbb{R}^d$, let $\mathcal{B} \subseteq \mathcal{X}_1 \cup \ldots \cup \mathcal{X}_s$ be a subset, let $\mathcal{E} = \mathbb{R}^{\cup}\langle \mathcal{B} \rangle$, and let $\pi : \mathbb{R}^d \to \mathcal{E}^{\perp}$ be the orthogonal projection. For a subset $C \subseteq \mathcal{X}_1 \cup \{\mathbf{v}_1\} \cup \ldots \cup \mathcal{X}_s \cup \{\mathbf{v}_s\}$, set*

$$C^{\pi} := \pi(C) \setminus \{\{0\}\}.$$

For $j \in [1, s]$, let $\mathcal{X}'_j \subseteq \mathcal{X}_j$ be all those $\mathbf{x} \in \mathcal{X}_j$ with $\pi(\mathbf{x}) \neq \{0\}$, and let $J \subseteq [1, s]$ be all those indices j with $\mathcal{X}^{\pi}_j \neq \emptyset$, say $J = \{j_1, \ldots, j_t\}$ with $1 \le j_1 < \ldots < j_t \le s$. For a subset $\mathcal{D} \subseteq \bigcup_{i \in J}(\mathcal{X}^{\pi}_i \cup \{\pi(\mathbf{v}_i)\})$, set

$$\pi^{-1}(\mathcal{D}) := \Big\{\mathbf{x} \in \mathcal{X}_1 \cup \ldots \cup \mathcal{X}_s \cup \{\mathbf{v}_i : i \in J\} : \pi(\mathbf{x}) \in \mathcal{D}\Big\} \subseteq \bigcup_{i \in J}(\mathcal{X}'_i \cup \{\mathbf{v}_i\}).$$

1.

$$\pi(\mathcal{R}) := \Big(\mathcal{X}^{\pi}_j \cup \{\pi(\mathbf{v}_j)\}\Big)_{j \in J} = (\mathcal{X}^{\pi}_{j_1} \cup \{\pi(\mathbf{v}_{j_1})\}, \ldots, \mathcal{X}^{\pi}_{j_t} \cup \{\pi(\mathbf{v}_{j_t})\})$$

is an oriented Reay system for $\pi(\mathcal{E}_s)$ with π injective on $\bigcup_{i \in J}(\mathcal{X}'_i \cup \{\mathbf{v}_i\})$ and $\pi(\mathbf{v}_i) \neq \{0\}$ for all $i \in J$.

2. If $C \subseteq \mathcal{X}_1 \cup \ldots \cup \mathcal{X}_s \cup \{\mathbf{v}_j : j \in J\}$, then

$$(C^{\pi})^* = (C^*)^{\pi}, \quad \downarrow(C^{\pi}) = (\downarrow C)^{\pi} \quad \text{and} \quad \partial(C^{\pi}) = \partial(C)^{\pi}.$$

3. (a) If $C_1, C_2 \subseteq \mathcal{X}_1 \cup \ldots \cup \mathcal{X}_s \cup \{\mathbf{v}_j : j \in J\}$ with $C_1 \prec C_2$, then $C_1^{\pi} \preceq C_2^{\pi}$. Moreover, if in a sequence of replacing elements $\mathbf{x}$ by $\partial(\{\mathbf{x}\})$ showing that $C_1 \prec C_2$ there is some $\mathbf{x}$ with $\pi(\mathbf{x}) \neq \{0\}$, then $C_1^{\pi} \prec C_2^{\pi}$.

(b) If $\mathcal{D}_1, \mathcal{D}_2 \subseteq \bigcup_{j \in J}(\mathcal{X}^{\pi}_j \cup \{\pi(\mathbf{v}_j)\})$ with $\mathcal{D}_1 \prec \mathcal{D}_2$, then $\pi^{-1}(\mathcal{D}_1) \prec \pi^{-1}(\mathcal{D}_2)$.

4. If $\mathcal{D} \subseteq \bigcup_{i \in J}(\mathcal{X}^{\pi}_i \cup \{\pi(\mathbf{v}_i)\})$, then

$$\pi\big((\pi^{-1}(\mathcal{D}))^*\big) = \mathcal{D}^*, \quad (\pi^{-1}(\mathcal{D}))^* \subseteq (\pi^{-1}(\mathcal{D}) \cup \mathcal{B})^* \quad \text{and}$$
$$\big((\pi^{-1}(\mathcal{D}) \cup \mathcal{B})^*\big)^{\pi} = \mathcal{D}^*.$$

In particular, $\mathcal{D}^* = \mathcal{D}$ if and only if $(\pi^{-1}(\mathcal{D}))^* = \pi^{-1}(\mathcal{D})$.

5. (a) If $\mathcal{D} \subseteq \bigcup_{i \in J}(\mathcal{X}^{\pi}_i \cup \{\pi(\mathbf{v}_i)\})$ is a virtual independent set, then $(\pi^{-1}(\mathcal{D}) \cup \mathcal{B})^*$ and $\pi^{-1}(\mathcal{D}) \subseteq (\pi^{-1}(\mathcal{D}) \cup \mathcal{B})^*$ are virtual independent sets.

(b) If $\mathcal{D} \subseteq \bigcup_{i \in J}(\mathcal{X}^{\pi}_i \cup \{\pi(\mathbf{v}_i)\})$ is a support set for $\pi(\mathcal{R})$, then $(\pi^{-1}(\mathcal{D}) \cup \mathcal{B})^*$ and $\pi^{-1}(\mathcal{D})$ are support sets for $\mathcal{R}$.

In either case, if $\mathcal{B}^* = \mathcal{B}$, then $(\pi^{-1}(\mathcal{D}) \cup \mathcal{B})^* = \pi^{-1}(\mathcal{D}) \cup \big(\mathcal{B} \setminus \downarrow\partial(\pi^{-1}(\mathcal{D}))\big)$.

6. (a) *If $C \subseteq \mathcal{X}_1 \cup \ldots \cup \mathcal{X}_s \cup \{v_j : j \in J\}$ is a virtual independent set with $\mathcal{B} \subseteq \downarrow C$, then C^π is a virtual independent set.*
   (b) *If $C \subseteq \mathcal{X}_1 \cup \{v_1\} \cup \ldots \cup \mathcal{X}_s \cup \{v_s\}$ is a support set with $\mathcal{B} \subseteq \downarrow C$, then C^π is a support set with $C \subseteq \mathcal{X}_1 \cup \ldots \cup \mathcal{X}_s \cup \{v_j : j \in J\}$.*
   (c) *If $C \subseteq \mathcal{X}_1 \cup \ldots \cup \mathcal{X}_s$ is a support set, then C^π is a support set.*

Proof

1. For $j \in [1, s+1]$, let $\mathcal{E}_{j-1} = \mathbb{R}^\cup\langle \mathcal{X}_1 \cup \ldots \cup X_{j-1}\rangle$ and let $\pi_{j-1} : \mathbb{R}^d \to \mathcal{E}_{j-1}^\perp$ and $\varpi_{j-1} : \mathbb{R}^d \to (\mathcal{E}_{j-1} + \mathcal{E})^\perp$ be the orthogonal projections. Note $\varpi_0 = \pi$. Since $\ker \pi_{j-1} \leq \ker \varpi_{j-1}$, we have $\varpi_{j-1} = \varpi_{j-1}\pi_{j-1}$ for all $j \in [1, s]$. In view of Proposition 5.3.1,

$$\mathbb{R}\langle \downarrow B\rangle = \mathbb{R}^\cup\langle \mathcal{B}\rangle = \mathcal{E}$$

with $\downarrow B \subseteq X_1 \cup \ldots \cup X_s$ (since $\mathcal{B} \subseteq \mathcal{X}_1 \cup \ldots \cup \mathcal{X}_s$). Thus $\mathcal{E}'_{j-1} = \mathcal{E}_{j-1} + \mathbb{R}\langle \downarrow B\rangle$, while $X_j \subseteq \mathcal{E}_j$ and $\mathcal{E}_{j-1} \subseteq \mathcal{E}_j$, implying $\mathcal{E}_{j-1} + \mathbb{R}\langle \downarrow B \cap X_j\rangle \subseteq \mathcal{E}'_{j-1} \cap \mathcal{E}_j$. Any element $x \in \mathcal{E}'_{j-1} \cap \mathcal{E}_j$ has $x = y + z$ with $y \in \mathcal{E}_{j-1}$, z a linear combination of elements from $\downarrow B$ and $x = y + z \in \mathcal{E}_j$. Then $\pi_j(z) = \pi_j(x) = 0$. Applying π_j to the linear combination representing z and using Propositions 4.11.1 and 4.11.3 shows that $x = y + z \in \mathcal{E}_{j-1} + \mathbb{R}\langle \downarrow B \cap X_j\rangle$. Hence

$$\mathcal{E}'_{j-1} \cap \mathcal{E}_j = \mathcal{E}_{j-1} + \mathbb{R}\langle \downarrow B \cap X_j\rangle. \tag{5.14}$$

Since $\varpi_{j-1} = \varpi_{j-1}\pi_{j-1}$, we have $\ker \varpi_{j-1} \cap \pi_{j-1}(\mathcal{E}_j) = \pi_{j-1}(\ker \varpi_{j-1}) \cap \pi_{j-1}(\mathcal{E}_j)$, meaning

$$\ker \varpi_{j-1} \cap \pi_{j-1}(\mathcal{E}_j) = \pi_{j-1}(\mathcal{E}'_{j-1}) \cap \pi_{j-1}(\mathcal{E}_j). \tag{5.15}$$

Any element $x \in \pi_{j-1}(\mathcal{E}'_{j-1}) \cap \pi_{j-1}(\mathcal{E}_j)$ is a linear combination of terms $\pi_{j-1}(b) \in \pi_{j-1}(\downarrow B)$ (with each $b \in \downarrow B$) that lies in $\pi_{j-1}(\mathcal{E}_j)$. Applying π_j to this linear combination and using Propositions 4.11.1 and 4.11.3, we find that only those $b \in \downarrow B \cap \mathcal{E}_j$ have non-zero coefficients, meaning $x \in \pi_{j-1}(\mathcal{E}'_{j-1} \cap \mathcal{E}_j)$. Thus $\pi_{j-1}(\mathcal{E}'_{j-1}) \cap \pi_{j-1}(\mathcal{E}_j) = \pi_{j-1}(\mathcal{E}'_{j-1} \cap \mathcal{E}_j)$ (the reverse inclusion is trivial), which combined with (5.14) and (5.15) yields

$$\ker \varpi_{j-1} \cap \pi_{j-1}(\mathcal{E}_j) = \mathbb{R}\langle \pi_{j-1}(\downarrow B \cap X_j)\rangle.$$

Thus the kernel of ϖ_{j-1} restricted to $\pi_{j-1}(\mathcal{E}_j)$ is generated by a subset of the linearly independent set $\pi_{j-1}(X_j)$. Consequently, since $\pi_{j-1}(X_j) \cup \{\pi_{j-1}(v_j)\}$ is a minimal positive basis for $\pi_{j-1}(\mathcal{E}_j)$ of size $|X_j| + 1$, it follows that $(X_j \cup \{v_j\})^{\varpi_{j-1}}$ is either empty or a minimal positive basis for $\varpi_{j-1}(\mathcal{E}_j)$ of size

$$|(X_j \cup \{v_j\})^{\varpi_{j-1}}| = |(X_j \cup \{v_j\}) \setminus (\mathcal{E}_{j-1} + \mathcal{E})| = |X_j| + 1 - |\downarrow B \cap X_j|. \tag{5.16}$$

Let $j \in J$. If $\varpi_{j-1}(x) = 0$ for some $x \in X_j \cup \{v_j\}$, so $x \in \mathcal{E}'_{j-1} \cap \mathcal{E}_j = \mathcal{E}_{j-1} + \mathbb{R}\langle \downarrow B \cap X_j \rangle$ (the equality follows from (5.14)), then $\pi_{j-1}(x) \in \mathbb{R}\langle \pi_{j-1}(\downarrow B \cap X_j) \rangle$ with $\pi_{j-1}(\downarrow B \cap X_j) \subseteq \pi_{j-1}(X_j)$. Thus, since $\pi_{j-1}(X_j) \cup \{\pi_{j-1}(v_j)\}$ is a minimal positive basis of size $|X_j| + 1$ and $x \in X_j \cup \{v_j\}$, we must either have $\downarrow B \cap X_j = X_j$ or $x \in \downarrow B$. In the former case, $\mathcal{X}_j \subseteq \downarrow \mathcal{B}$, ensuring that $\mathcal{X}_j^\pi = \emptyset$, which is contrary to the definition of $j \in J$. Therefore we must instead have the latter case, $x \in \downarrow B$, meaning $\pi(x) = 0$, $\mathbf{x} \in \downarrow \mathcal{B}$ and $\pi(\mathbf{x}) = \{0\}$. As a result, we find that

$$\varpi_{j-1}(\mathbf{x}) \neq \{0\} \quad \text{if and only if} \quad \pi(\mathbf{x}) \neq \{0\}, \qquad \text{for all } \mathbf{x} \in \mathcal{X}_j \cup \{\mathbf{v}_j\} \text{ and } j \in J. \tag{5.17}$$

Moreover, since each $\mathbf{v}_j \notin \downarrow \mathcal{B} \subseteq \mathcal{X}_1 \cup \ldots \cup \mathcal{X}_s$, the above work yields

$$\varpi_{j-1}(\mathbf{v}_j) \neq \{0\} \quad \text{for any } j \in J. \tag{5.18}$$

If $\pi(\mathbf{x}) \neq \{0\}$ for some $\mathbf{x} \in \mathcal{X}_j \cup \{\mathbf{v}_j\}$ with $j \in J$, then $\varpi_{j-1}(\mathbf{x}) \neq \{0\}$, meaning $\mathbf{x} \nsubseteq \mathcal{E}_{j-1} + \mathcal{E} = \ker \varpi_{j-1} + \partial(\mathbf{x})$ for $\mathbf{x} \in \mathcal{X}'_j \cup \{\mathbf{v}_j\}$ and $j \in J$, and thus also $\mathbf{x} \nsubseteq \ker \varpi_i + \partial(\mathbf{x})$ for any $i < j$. This ensures that, given any $\mathbf{x} \in \bigcup_{i \in J}(\mathcal{X}_i \cup \{\mathbf{v}_i\})$ and $j \in [1, s]$, we either have $\varpi_{j-1}(\mathbf{x}) = \{0\}$ or else $\varpi_{j-1}(\mathbf{x})$ remains a relative half-space. In particular, $\pi(\mathbf{x})$ is a relative half-space for all $\mathbf{x} \in \bigcup_{i \in J}(\mathcal{X}'_i \cup \{\mathbf{v}_i\})$.

Suppose $\mathbf{x}, \mathbf{y} \in \bigcup_{i \in J}(\mathcal{X}_i \cup \{\mathbf{v}_i\})$ and $j \in [1, s]$ with $\varpi_{j-1}(\mathbf{x}) = \varpi_{j-1}(\mathbf{y}) \neq \{0\}$. We aim to show $\mathbf{x} = \mathbf{y}$. Now w.l.o.g. $\mathbf{x} \in \mathcal{X}_i \cup \{\mathbf{v}_i\}$ and $\mathbf{y} \in \mathcal{X}_{i'} \cup \{\mathbf{v}_{i'}\}$ with $i, i' \in J$ and $i, i' \geq j$. Moreover, if $i > i'$, then $\varpi_{i-1}(\mathbf{x}) = \varpi_{i-1}\varpi_{j-1}(\mathbf{x}) = \varpi_{i-1}\varpi_{j-1}(\mathbf{y}) = \varpi_{i-1}(\mathbf{y}) = \{0\}$. Thus, since $i \in J$, it follows from (5.17) that $\varpi_{i-1}(\mathbf{x}) = \pi(\mathbf{x}) = 0$, implying $\varpi_{j-1}(\mathbf{x}) = \varpi_{j-1}\pi(\mathbf{x}) = \{0\}$, contrary to assumption. Hence $i \leq i'$, and likewise $i' \leq i$, whence $i = i'$. This argument also shows that $\varpi_{i-1}(\mathbf{x}) = \varpi_{i-1}(\mathbf{y}) \neq \{0\}$. Since $\partial(\mathbf{x}), \partial(\mathbf{y}) \subseteq \mathcal{E}_{i-1} \subseteq \ker \varpi_{i-1}$, it follows from $\varpi_{i-1}(\mathbf{x}) = \varpi_{i-1}(\mathbf{y}) \neq \{0\}$ that $\mathbb{R}\varpi_{i-1}(x) = \mathbb{R}\varpi_{i-1}(y) \neq \{0\}$. However, since $x, y \in X_i \cup \{v_i\}$, this contradicts that $(X_i \cup \{v_i\})^{\varpi_{i-1}}$ is a minimal positive basis of size $|(X_i \cup \{v_i\}) \setminus (\mathcal{E}_{i-1} + \mathcal{E})|$ (cf. (5.16)) unless $x = y$, in which case $\mathbf{x} = \mathbf{y}$. In summary, we have just shown that $\varpi_{j-1}(\mathbf{x}) = \varpi_{j-1}(\mathbf{y}) \neq \{0\}$, for $\mathbf{x}, \mathbf{y} \in \bigcup_{i \in J}(\mathcal{X}_i \cup \{\mathbf{v}_i\})$, implies $\mathbf{x} = \mathbf{y}$. In particular, $\pi = \varpi_0$ is injective on $\bigcup_{i \in J}(\mathcal{X}'_i \cup \{\mathbf{v}_i\})$.

From (5.17) and (5.18) and the injectivity of π just established, we conclude that

$$|(X_j \cup \{v_j\}) \setminus (\mathcal{E}_{j-1} + \mathcal{E})| = |(X_j \cup \{v_j\}) \setminus \mathcal{E}| = |X_j^\pi| + 1 \geq 2 \quad \text{for any } j \in J. \tag{5.19}$$

In view of (5.16), (5.18), and (5.19), it follows that $X_j^{\varpi_{j-1}} \cup \{\varpi_{j-1}(v_j)\}$ is a minimal positive basis for $\varpi_{j-1}(\mathcal{E}_j)$ of size $|X_j^\pi| + 1$, for $j \in J$. Hence $\pi(\mathcal{R})$ satisfies (OR2) in the definition of an orientated Reay system (note $\varpi_{j-1}\pi = \omega_{j-1}$ since $\ker \pi \leq \ker \varpi_{j-1}$). Also, if $\mathbf{x} \in \mathcal{X}'_j \cup \{\mathbf{v}_j\}$ and $j \in J$, then $\pi(\mathbf{x}) \neq \{0\}$ is a relative half-space with $\varpi_{j-1}(\mathbf{x}) \neq \{0\}$ as shown earlier, so $\mathbf{x} \nsubseteq \mathcal{E}_{j-1} + \ker \pi$, in which case (OR1) holding for $\mathcal{R}$ together with (5.1) implies (OR1) holds for $\pi(\mathcal{R})$ with

$$\mathcal{B}_{\pi(\mathbf{x})} = \mathcal{B}_{\mathbf{x}}^{\pi} = \partial(\{\mathbf{x}\})^{\pi}, \tag{5.20}$$

showing that $\pi(\mathcal{R})$ is an oriented Reay system for $\pi(\mathcal{E}_s)$. This establishes Item 1.

2. We begin by showing

$$(C^*)^{\pi} = (C^{\pi})^* \quad \text{for any } C \subseteq \mathcal{X}_1 \cup \ldots \cup \mathcal{X}_s. \tag{5.21}$$

Let $\pi(\mathbf{x}) \in (C^*)^{\pi}$ be arbitrary, so $\mathbf{x} \in C^*$ with $\pi(\mathbf{x}) \neq \{0\}$. If $\pi(\mathbf{x}) \notin (C^{\pi})^*$, then there must be some $\mathbf{y} \in C$ with $\{0\} \neq \pi(\mathbf{x}) \prec \pi(\mathbf{y})$. Thus $\mathbf{x} \subseteq \mathbf{y} + \mathcal{E} \subseteq \mathbb{R}^{\cup}\langle\{\mathbf{y}\} \cup \mathcal{B}\rangle$, in which case Proposition 5.3.9 ensures that $\mathbf{x} \in \downarrow\{\mathbf{y}\} \cup \downarrow\mathcal{B}$ (since $\mathbf{x}, \mathbf{y} \in C \cup \mathcal{B} \subseteq \mathcal{X}_1 \cup \ldots \cup \mathcal{X}_s$). Since $\pi(\mathbf{x}) \neq \{0\}$, we have $\mathbf{x} \notin \downarrow\mathcal{B}$, forcing $\mathbf{x} \in \downarrow\{\mathbf{y}\}$, i.e., $\mathbf{x} \preceq \mathbf{y}$. However, since $\mathbf{x} \in C^*$ and $\mathbf{y} \in C$, this is only possible if $\mathbf{x} = \mathbf{y}$. Hence $\pi(\mathbf{x}) = \pi(\mathbf{y})$, contradicting that $\pi(\mathbf{x}) \prec \pi(\mathbf{y})$. This shows that $(C^*)^{\pi} \subseteq (C^{\pi})^*$.

Next let $\pi(\mathbf{x}) \in (C^{\pi})^*$ be arbitrary, so $\mathbf{x} \in C$ and $\pi(\mathbf{x}) \neq \{0\}$. If $\mathbf{x} \notin C^*$, then there is some $\mathbf{y} \in C$ with $\mathbf{x} \prec \mathbf{y}$, implying $\{0\} \neq \pi(\mathbf{x}) \preceq \pi(\mathbf{y})$. Thus, since $\pi(\mathbf{x}) \in (C^{\pi})^*$, we must have $\{0\} \neq \pi(\mathbf{x}) = \pi(\mathbf{y})$, in which case the injectivity of π given in Item 1 implies $\mathbf{x} = \mathbf{y}$, contradicting that $\mathbf{x} \prec \mathbf{y}$. Therefore we instead conclude that $\mathbf{x} \in C^*$, whence $\pi(\mathbf{x}) \in (C^*)^{\pi}$, establishing the reverse inclusion $(C^{\pi})^* \subseteq (C^*)^{\pi}$, which establishes (5.21).

In view of (5.21), we have $(\partial(\{\mathbf{x}\})^{\pi})^* = (\partial(\{\mathbf{x}\})^*)^{\pi} = \partial(\{\mathbf{x}\})^{\pi}$ for any $\mathbf{x} \in \bigcup_{i \in j}(\mathcal{X}_i' \cup \{\mathbf{v}_i\})$, which combined with (5.20) and the definition of $\partial(\{\pi(\mathbf{x})\})$ implies

$$\partial(\{\pi(\mathbf{x})\}) = \partial(\{\mathbf{x}\})^{\pi} \quad \text{for any } \mathbf{x} \in \bigcup_{i \in j}(\mathcal{X}_i' \cup \{\mathbf{v}_i\}). \tag{5.22}$$

Let $C \subseteq \mathcal{X}_1 \cup \ldots \cup \mathcal{X}_s \cup \{\mathbf{v}_j : j \in J\}$. Then

$$\begin{aligned}
\partial(C^{\pi}) &= \Big(\bigcup_{\mathbf{x} \in C,\, \pi(\mathbf{x}) \neq \{0\}} \partial(\{\pi(\mathbf{x})\})\Big)^* = \Big(\bigcup_{\mathbf{x} \in C,\, \pi(\mathbf{x}) \neq \{0\}} \partial(\{\mathbf{x}\})^{\pi}\Big)^* \\
&= \Big(\Big(\bigcup_{\mathbf{x} \in C,\, \pi(\mathbf{x}) \neq \{0\}} \partial(\{\mathbf{x}\})\Big)^{\pi}\Big)^* = \Big(\Big(\bigcup_{\mathbf{x} \in C} \partial(\{\mathbf{x}\})\Big)^{\pi}\Big)^* \\
&= \Big(\Big(\bigcup_{\mathbf{x} \in C} \partial(\{\mathbf{x}\})\Big)^*\Big)^{\pi} = \partial(C)^{\pi},
\end{aligned} \tag{5.23}$$

where the first equality follows by definition of $\partial(C^{\pi})$, the second by (5.22) in view of $C \subseteq \mathcal{X}_1 \cup \ldots \cup \mathcal{X}_s \cup \{\mathbf{v}_j : j \in J\}$, the third is a trivial identity, the fourth follows since $\pi(\mathbf{x}) = \{0\}$ ensures $\pi(\mathbf{y}) = \{0\}$ for all $\mathbf{y} \in \downarrow\mathbf{x}$, and thus for all

$\mathbf{y} \in \partial(\{\mathbf{x}\}) \subseteq \downarrow\mathbf{x}$ as well, the fifth follows by (5.21) applied to $\left(\bigcup_{\mathbf{x}\in C} \partial(\{\mathbf{x}\})\right)$, and the sixth follows by definition of $\partial(C)$. The identity $(\downarrow C)^{\pi} = \downarrow(C^{\pi})$ now follows from repeated application of (5.23) to (5.8):

$$(\downarrow C)^{\pi} = \left(\bigcup_{n=0}^{s-1} \partial^n(C)\right)^{\pi} = \bigcup_{n=0}^{s-1} \left(\partial^n(C)\right)^{\pi} = \bigcup_{n=0}^{s-1} \partial^n(C^{\pi}) = \downarrow(C^{\pi}). \tag{5.24}$$

Finally, it remains to establish $(C^*)^{\pi} = (C^{\pi})^*$ in the case when $C \subseteq \mathcal{X}_1 \cup \ldots \cup \mathcal{X}_s \cup \{\mathbf{v}_j : j \in J\}$. We have $(C^{\pi})^* \subseteq (C^*)^{\pi}$ by the argument used to establish this inclusion for (5.21). To see the reverse inclusion $(C^*)^{\pi} \subseteq (C^{\pi})^*$, let $\pi(\mathbf{x}) \in (C^*)^{\pi}$ be arbitrary, so $\mathbf{x} \in C^*$ with $\pi(\mathbf{x}) \neq \{0\}$. If $\pi(\mathbf{x}) \notin (C^{\pi})^*$, then there must be some $\mathbf{y} \in C$ with $\{0\} \neq \pi(\mathbf{x}) \prec \pi(\mathbf{y})$. Thus (5.24) implies $\pi(\mathbf{x}) \in \downarrow(\{\mathbf{y}\}^{\pi}) = (\downarrow\{\mathbf{y}\})^{\pi}$. As a result, since $\pi(\mathbf{x})$ and $\pi(\mathbf{y})$ are both nonzero, it follows from the injectivity of π established in Item 1 that $\mathbf{x} \preceq \mathbf{y}$. However, since $\mathbf{x} \in C^*$ and $\mathbf{y} \in C$ by hypothesis, this is only possible if $\mathbf{x} = \mathbf{y}$. Thus $\pi(\mathbf{x}) = \pi(\mathbf{y})$, contradicting that $\pi(\mathbf{x}) \prec \pi(\mathbf{y})$. This shows that $(C^*)^{\pi} \subseteq (C^{\pi})^*$, whence $(C^*)^{\pi} = (C^{\pi})^*$, which completes the proof of Item 2.

3. Let $C \subseteq \mathcal{X}_1 \cup \ldots \cup \mathcal{X}_s \cup \{\mathbf{v}_j : j \in J\}$ and let $\mathbf{x} \in C$. If $\pi(\mathbf{x}) = \{0\}$, then we have $(C \setminus \{\mathbf{x}\} \cup \partial(\{\mathbf{x}\}))^{\pi} = C^{\pi}$. If $\pi(\mathbf{x}) \neq \{0\}$, then Item 2 and the injectivity of π from Item 1 imply that $(C \setminus \{\mathbf{x}\} \cup \partial(\{\mathbf{x}\}))^{\pi} = C^{\pi} \setminus \{\pi(\mathbf{x})\} \cup \partial(\{\pi(\mathbf{x})\})$. This shows $C_1 \prec C_2$ implies $C_1^{\pi} \preceq C_2^{\pi}$ when C_1 is the immediate predecessor of C_2, and iterating then yields the first part of Item 3. Moreover, if $\pi(\mathbf{x}) \neq \{0\}$, then $C_1^{\pi} \prec C_2^{\pi}$. Next let $\mathcal{D} \subseteq \bigcup_{j\in J}(\mathcal{X}_j^{\pi} \cup \{\pi(\mathbf{v}_j)\})$ and $\mathbf{x} \in \pi^{-1}(\mathcal{D})$. Then $\pi^{-1}\big(\mathcal{D} \setminus \{\pi(\mathbf{x})\} \cup \partial(\{\pi(\mathbf{x})\})\big) \subseteq \pi^{-1}(\mathcal{D}) \setminus \{\mathbf{x}\} \cup \partial(\{\mathbf{x}\}) \prec \pi^{-1}(\mathcal{D})$ in view of Item 2 and the injectivity of π given in Item 1, which ensures $\pi^{-1}\big(\mathcal{D} \setminus \{\pi(\mathbf{x})\} \cup \partial(\{\pi(\mathbf{x})\})\big) \prec \pi^{-1}(\mathcal{D})$. This shows that $\mathcal{D}_1 \prec \mathcal{D}_2$ implies $\pi^{-1}(\mathcal{D}_1) \prec \pi^{-1}(\mathcal{D}_2)$ when $\mathcal{D}_1$ is the immediate predecessor of $\mathcal{D}_2$, and iterating then yields the remaining part of Item 3.
4. Let $\mathcal{D} \subseteq \bigcup_{i\in J}(\mathcal{X}_i^{\pi} \cup \{\pi(\mathbf{v}_i)\})$ and note that $\pi^{-1}(\mathcal{D}) \subseteq \bigcup_{j\in J}(\mathcal{X}'_j \cup \{\mathbf{v}_j\})$ by definition and Item 1. Thus, Item 2 implies that

$$\pi\left((\pi^{-1}(\mathcal{D}))^*\right) = \left((\pi^{-1}(\mathcal{D}))^*\right)^{\pi} = \left((\pi^{-1}(\mathcal{D}))^{\pi}\right)^* = (\pi(\pi^{-1}(\mathcal{D})))^* = \mathcal{D}^*. \tag{5.25}$$

If $(\pi^{-1}(\mathcal{D}))^* = \pi^{-1}(\mathcal{D})$, then (5.25) yields $\mathcal{D}^* = \pi\left((\pi^{-1}(\mathcal{D}))^*\right) = \pi\left(\pi^{-1}(\mathcal{D})\right) = \mathcal{D}$. On the other hand, if $\mathcal{D}^* = \mathcal{D}$, then (5.25) combined with the injectivity of π established in Item 1 implies that $(\pi^{-1}(\mathcal{D}))^* = \pi^{-1}(\mathcal{D})$. This shows that $\mathcal{D}^* = \mathcal{D}$ if and only if $(\pi^{-1}(\mathcal{D}))^* = \pi^{-1}(\mathcal{D})$.

Let $\mathbf{x} \in (\pi^{-1}(\mathcal{D}))^*$ be arbitrary, so $\mathbf{x} \in \bigcup_{j\in J}(\mathcal{X}'_j \cup \{\mathbf{v}_j\})$ with $\pi(\mathbf{x}) \in \mathcal{D}$. If $\mathbf{x} \notin (\pi^{-1}(\mathcal{D}) \cup \mathcal{B})^*$, then there must be some $\mathbf{y} \in \pi^{-1}(\mathcal{D}) \cup \mathcal{B}$ with $\mathbf{x} \prec \mathbf{y}$. Since $\mathbf{x} \in (\pi^{-1}(\mathcal{D}))^*$, we cannot have $\mathbf{y} \in \pi^{-1}(\mathcal{D})$, forcing $\mathbf{x} \prec \mathbf{y} \in \mathcal{B}$. However, since $\ker \pi = \mathbb{R}^{\cup}\langle \mathcal{B} \rangle$, this implies $\pi(\mathbf{x}) = \pi(\mathbf{y}) = \{0\}$, contradicting that $\mathbf{x} \in$

$\bigcup_{j\in J}(\mathcal{X}'_j \cup \{\mathbf{v}_j\})$. This establishes that

$$(\pi^{-1}(\mathcal{D}))^* \subseteq (\pi^{-1}(\mathcal{D}) \cup \mathcal{B})^*. \tag{5.26}$$

By Item 2, $\big((\pi^{-1}(\mathcal{D}) \cup \mathcal{B})^*\big)^\pi = \big((\pi^{-1}(\mathcal{D}) \cup \mathcal{B})^\pi\big)^* = \big((\pi^{-1}(\mathcal{D}))^\pi \cup \mathcal{B}^\pi\big)^* = \mathcal{D}^*$, with the latter equality since $\mathcal{B}^\pi = \emptyset$, which completes Item 4.

5. Since $\mathcal{D}$ is virtual independent, we have $\mathcal{D}^* = \mathcal{D}$, whence Item 4 implies

$$\pi^{-1}(\mathcal{D}) = (\pi^{-1}(\mathcal{D}))^* \subseteq (\pi^{-1}(\mathcal{D}) \cup \mathcal{B})^*.$$

We have $\downarrow(\pi^{-1}(\mathcal{D}) \cup \mathcal{B})^* = \downarrow\pi^{-1}(\mathcal{D}) \cup \downarrow\mathcal{B}$. By Item 1 and Proposition 5.3.9, π is injective on $\downarrow\pi^{-1}(\mathcal{D}) \setminus \downarrow\mathcal{B}$, and thus on $\downarrow\pi^{-1}(D) \setminus \downarrow B$. By Item 2, $(\downarrow\pi^{-1}(\mathcal{D}))^\pi = \downarrow\mathcal{D}$. Consequently, if we have a nontrivial linear combination of terms from $\downarrow\pi^{-1}(D) \cup \downarrow B$ equal to zero, then applying π to this linear combination and using that $\mathcal{D}$ is virtual independent shows that only terms from $\downarrow B$ occur in the linear combination, contradicting that $\downarrow B$ is linearly independent in view of $\mathcal{B} \subseteq \mathcal{X}_1 \cup \ldots \cup \mathcal{X}_s$. This shows that $(\pi^{-1}(\mathcal{D}) \cup \mathcal{B})^*$ is virtual independent. As a result, any subset $C \subseteq \downarrow(\pi^{-1}(\mathcal{D}) \cup \mathcal{B})^*$ with $C^* = C$ is also virtual independent. In particular, $\pi^{-1}(\mathcal{D})$ is virtual independent.

Next suppose $\mathcal{D}$ is a support set for $\pi(\mathcal{R})$. To show $\pi^{-1}(\mathcal{D})$ and $(\pi^{-1}(\mathcal{D}) \cup \mathcal{B})^*$ are support sets, we need to show $\mathcal{X}_j \cup \{\mathbf{v}_j\} \nsubseteq \downarrow\mathcal{B} \cup \downarrow\pi^{-1}(\mathcal{D})$, for all $j \in [1, s]$. If $j \notin J$, then $\mathbf{v}_j \notin \downarrow\pi^{-1}(\mathcal{D}) \subseteq \mathcal{X}_1 \cup \ldots \cup \mathcal{X}_s \cup \{\mathbf{v}_i : i \in J\}$, while $\mathcal{B} \subseteq \mathcal{X}_1 \cup \ldots \cup \mathcal{X}_s$ ensures that $\mathbf{v}_j \notin \downarrow\mathcal{B}$, as desired. On the other hand, if $j \in J$, then, since $\mathcal{D}$ is a support set, we have

$$\mathcal{X}_j^\pi \cup \{\pi(\mathbf{v}_j)\} \nsubseteq \downarrow\mathcal{D} = \downarrow\pi(\pi^{-1}(\mathcal{D})) = (\downarrow\pi^{-1}(\mathcal{D}))^\pi,$$

with the last equality above in view of Item 2, which show that there is some $\mathbf{y} \in \mathcal{X}_j \cup \{\mathbf{v}_j\}$ with $\pi(\mathbf{y}) \neq \{0\}$ and $\mathbf{y} \notin \downarrow\pi^{-1}(\mathcal{D})$. Note that $\downarrow\mathcal{B} \subseteq \mathbb{R}^\cup\langle\downarrow\mathcal{B}\rangle = \mathbb{R}^\cup\langle\mathcal{B}\rangle = \mathcal{E}$, implying $\pi(\mathbf{x}) = \{0\}$ for every $\mathbf{x} \in \downarrow\mathcal{B}$. Thus $\mathbf{y} \notin \downarrow\mathcal{B}$ as well, whence $\mathcal{X}_j \cup \{\mathbf{v}_j\} \nsubseteq \downarrow\mathcal{B} \cup \downarrow\pi^{-1}(\mathcal{D})$ follows, as desired. This establishes that $\pi^{-1}(\mathcal{D})$ and $(\pi^{-1}(\mathcal{D}) \cup \mathcal{B})^*$ are support sets.

To complete Item 5, now suppose $\mathcal{B}^* = \mathcal{B}$. By Item 4, we have $\pi^{-1}(\mathcal{D}) = \pi^{-1}(\mathcal{D})^* \subseteq (\pi^{-1}(\mathcal{D}) \cup \mathcal{B})^*$, implying $(\pi^{-1}(\mathcal{D}) \cup \mathcal{B})^* = \pi^{-1}(\mathcal{D}) \cup \mathcal{B}'$ for some $\mathcal{B}' \subseteq \mathcal{B}$. If $\mathbf{x} \in \mathcal{B} \setminus \mathcal{B}'$, then we have $\mathbf{x} \prec \mathbf{y}$ for some $\mathbf{y} \in \pi^{-1}(\mathcal{D})$ (as $\mathcal{B}^* = \mathcal{B}$), implying $\mathbf{x} \subseteq \mathbb{R}^\cup\langle\partial(\pi^{-1}(\mathcal{D}))\rangle$, and then $\mathbf{x} \in \downarrow\partial(\pi^{-1}(\mathcal{D}))$ by Proposition 5.3.9. Hence $\mathcal{B} \setminus \mathcal{B}' \subseteq \downarrow\partial(\pi^{-1}(\mathcal{D}))$. On the other hand, if $\mathbf{y} \in \mathcal{B} \cap \downarrow\partial(\pi^{-1}(\mathcal{D}))$, then $\mathbf{y} \prec \mathbf{x}$ for some $\mathbf{x} \in \pi^{-1}(\mathcal{D})$, ensuring that $\mathbf{y} \notin \mathcal{B}'$ (since $(\pi^{-1}(\mathcal{D}) \cup \mathcal{B})^* = \pi^{-1}(\mathcal{D}) \cup \mathcal{B}'$). Thus $\mathcal{B}' \cap \downarrow\partial(\pi^{-1}(\mathcal{D})) = \emptyset$. Combining this with $\mathcal{B} \setminus \mathcal{B}^* \subseteq \downarrow\partial(\pi^{-1}(\mathcal{D}))$, we conclude that $\mathcal{B}' = \mathcal{B} \setminus \downarrow\partial(\pi^{-1}(\mathcal{D}))$, which completes Item 5.

6. Suppose $C \subseteq \mathcal{X}_1 \cup \{\mathbf{v}_1\} \cup \ldots \cup \mathcal{X}_s \cup \{\mathbf{v}_s\}$ is a support set with $\mathcal{B} \subseteq \downarrow C$. If $\mathbf{v}_j \in C$ for some $j \notin J$, then $\mathbf{x} \subseteq \ker\pi = \mathbb{R}^\cup\langle\mathcal{B}\rangle$ for all $\mathbf{x} \in \mathcal{X}_j$, whence Proposition 5.3.9 ensures that $\mathcal{X}_j \subseteq \downarrow\mathcal{B} \subseteq \downarrow C$, in which case $\mathcal{X}_j \cup \{\mathbf{v}_j\} \subseteq \downarrow C$, contradicting that C is a support set. Therefore we may assume otherwise, in

which case $(C^\pi)^* = (C^*)^\pi = C^\pi$ by Item 2 (as $C^* = C$ is a support set). But now, if $\mathcal{X}_j^\pi \cup \{\pi(\mathbf{v}_j)\} \subseteq {\downarrow}C^\pi = ({\downarrow}C)^\pi$ for some $j \in J$, with the equality form Item 2, then the injectivity of π established in Item 1 ensures that $\mathcal{X}'_j \cup \{\mathbf{v}_j\} \subseteq {\downarrow}C$. As before, since $\mathbf{x} \subseteq \ker\pi = \mathbb{R}^\cup\langle\mathcal{B}\rangle$ for all $\mathbf{x} \in \mathcal{X}_j \setminus \mathcal{X}'_j$ (by definition of $\mathcal{X}'_j$), Proposition 5.3.9 ensures that $\mathcal{X}_j \setminus \mathcal{X}'_j \subseteq {\downarrow}\mathcal{B} \subseteq {\downarrow}C$, in which case $\mathcal{X}_j \cup \{\mathbf{v}_j\} \subseteq {\downarrow}C$, contradicting that C is a support set. Therefore we instead find that $\mathcal{X}_j^\pi \cup \{\pi(\mathbf{v}_j)\} \nsubseteq C^\pi$ for $j \in J$, which combined with $(C^\pi)^* = C^\pi$ implies that C^π is a support set.

Suppose $C \subseteq \mathcal{X}_1 \cup \ldots \cup \mathcal{X}_s$ is a support set. Then $C^* = C$. Thus Item 2 implies that $(C^\pi)^* = (C^*)^\pi = C^\pi$ and ${\downarrow}C^\pi = ({\downarrow}C)^\pi \subseteq \bigcup_{j\in J} \mathcal{X}_j^\pi$, ensuring that C^π is a support set.

Finally, suppose $C \subseteq \mathcal{X}_1 \cup \ldots \cup \mathcal{X}_s \cup \{\mathbf{v}_j : j \in J\}$ is virtual independent with $\mathcal{B} \subseteq {\downarrow}C$. Then $C^* = C$, whence Item 2 implies $(C^\pi)^* = (C^*)^\pi = C^\pi$. Also, ${\downarrow}C$ is linearly independent. Thus, since $\mathcal{B} \subseteq {\downarrow}C$ ensures that ${\downarrow}B \subseteq {\downarrow}C$ with $\ker\pi = \mathbb{R}\langle{\downarrow}B\rangle$ by Proposition 5.3.1, it follows that $\pi({\downarrow}C) \setminus \{0\} = \pi({\downarrow}C \setminus {\downarrow}B)$ is linearly independent, while item 2 ensures ${\downarrow}(C^\pi) = ({\downarrow}C)^\pi = \pi({\downarrow}C) \setminus \{\{0\}\}$. Hence item 1 implies that ${\downarrow}(C^\pi) = \pi({\downarrow}C) \setminus \{0\}$ is linearly independent, implying C^π is virtual independent.

□

If $\mathcal{R} = (\mathcal{X}_1 \cup \{\mathbf{v}_1\}, \ldots, \mathcal{X}_s \cup \{\mathbf{v}_s\})$ is an oriented Reay system, $\mathcal{B} \subseteq \mathcal{X}_1 \cup \ldots \cup \mathcal{X}_s$, and $\pi : \mathbb{R}^d \to \mathbb{R}^\cup\langle\mathcal{B}\rangle^\perp$ is the orthogonal projection, Proposition 5.4.1 implies $\pi(\mathcal{R}) = (\mathcal{X}_j^\pi \cup \{\pi(\mathbf{v}_j)\})_{j\in J}$ is also an oriented Reay system. In particular, either $\pi(\mathbf{x}) = \{0\}$ or $\pi(\mathbf{x})$ is a relative half-space, for any $\mathbf{x} \in \mathcal{X}_1 \cup \ldots \cup \mathcal{X}_s \cup \{\mathbf{v}_j : j \in J\}$; thus $\pi(x) = 0$ if and only if $\pi(\mathbf{x}) = \{0\}$, for $\mathbf{x} \in \mathcal{X}_1 \cup \ldots \cup \mathcal{X}_s \cup \{\mathbf{v}_j : j \in J\}$. Most of the well-behaved properties of $\pi(\mathcal{R})$ that pull-back to $\mathcal{R}$ require $\mathbf{x} \in \mathcal{X}_1 \cup \ldots \cup \mathcal{X}_s \cup \{\mathbf{v}_j : j \in J\}$. This is a technical restriction that will stay with us throughout the remainder of this work.

Let $\mathcal{D} \subseteq \bigcup_{j\in J}(\mathcal{X}_j^\pi \cup \{\pi(\mathbf{v}_j)\})$. In view of Proposition 5.4.1, π is injective on all half-spaces of $\mathcal{X}_1 \cup \ldots \cup \mathcal{X}_1 \cup \{\mathbf{v}_j : j \in J\}$ not mapped to $\{0\}$, so $|\pi^{-1}(\mathcal{D})| = |\mathcal{D}|$. The set $\pi^{-1}(\mathcal{D}) \subseteq \bigcup_{i\in J}(\mathcal{X}'_i \cup \{\mathbf{v}_i\})$ is then the unique subset $C \subseteq \mathcal{X}_1 \cup \ldots \cup \mathcal{X}_s \cup \{\mathbf{v}_j : j \in J\}$ with $C^\pi = \pi(C) = \mathcal{D}$. We call $\pi^{-1}(\mathcal{D})$ the **pull-back** of $\mathcal{D}$ and $(\pi^{-1}(\mathcal{D}) \cup \mathcal{B})^*$ the **lift** of $\mathcal{D}$. These sets feature in Proposition 5.4 and will reoccur often. When $\mathcal{D}$ is a support/virtual independent set for $\pi(\mathcal{R})$, Propositions 5.4.5 and 5.4.4 ensure that both the pull-back and lift of $\mathcal{D}$ are support/virtual independent sets for $\mathcal{R}$ with

$$\pi^{-1}(\mathcal{D}) = (\pi^{-1}(\mathcal{D}))^* \subseteq (\pi^{-1}(\mathcal{D}) \cup \mathcal{B})^* \quad \text{and} \quad ((\pi^{-1}(\mathcal{D}) \cup \mathcal{B})^*)^\pi = \mathcal{D}^* = \mathcal{D}.$$

We continue with the generalization of Proposition 4.12 to oriented Reay systems.

Proposition 5.5 *Let $\mathcal{R} = (\mathcal{X}_1 \cup \{v_1\}, \ldots, \mathcal{X}_s \cup \{v_s\})$ be an orientated Reay system for the subspace $\mathcal{E}_s \subseteq \mathbb{R}^d$. Then $\mathcal{E}_s = \bigcup_{\mathcal{B}} \mathsf{C}^{\cup}(\mathcal{B})^\circ$ is a disjoint union, where the union runs over all support sets $\mathcal{B}$ for $\mathcal{R}$.*

Proof For $j \in [1, s]$, let $\mathcal{E}_{j-1} = \mathbb{R}^{\cup}\langle \mathcal{X}_1 \cup \ldots \cup \mathcal{X}_{j-1}\rangle$ and let $\pi_{j-1} : \mathbb{R}^d \to \mathcal{E}_{j-1}^\perp$ be the orthogonal projection. We proceed by induction on s. Let $x \in \mathcal{E}_s$ be arbitrary. We need to show there is a unique support set $\mathcal{B}$ with $x \in \mathsf{C}^{\cup}(\mathcal{B})^\circ$.

Since $\pi_{s-1}(X_s \cup \{v_s\})$ is a minimal positive basis of size $|X_s|+1$ (by definition of an oriented Reay system), it follows from Proposition 4.12 that there is a uniquely defined proper subset $\mathcal{B}_s \subset \mathcal{X}_s \cup \{\mathbf{v}_s\}$ with $\pi_{s-1}(x) \in \mathsf{C}^{\cup}(\pi_{s-1}(\mathcal{B}_s))^\circ$, which completes the proof in the base case when $s = 1$. Thus we may assume $s \geq 2$. Let $\mathbf{z}_1, \ldots, \mathbf{z}_\ell \in \mathcal{B}_s$ be the distinct elements of $\mathcal{B}_s$, let $\mathcal{E} = \mathbb{R}^{\cup}\langle \partial(\mathcal{B}_s)\rangle$, and let $\pi : \mathbb{R}^d \to \mathcal{E}^\perp$ be the orthogonal projection. Since $\partial(\mathcal{B}_s) \subseteq \mathcal{X}_1 \cup \ldots \cup \mathcal{X}_{s-1}$, we have $\ker \pi = \mathcal{E} \subseteq \mathcal{E}_{s-1} = \ker \pi_{s-1}$, ensuring $\pi_{s-1}\pi = \pi_{s-1}$. By Proposition 4.8, $\pi_{s-1}(B_s)$ is linearly independent, so there is a unique linear combination of the elements of $\pi_{s-1}(B_s)$ equal to $\pi_{s-1}(x)$, say $\sum_{i=1}^{\ell} \alpha_i \pi_{s-1}(z_i) = \pi_{s-1}(x)$ with $\alpha_i \in \mathbb{R}$, and this linear combination is strictly positive since $\pi_{s-1}(x) \in \mathsf{C}^{\cup}(\pi_{s-1}(\mathcal{B}_s))^\circ$, so $\alpha_i > 0$ for all i. Thus

$$\sum_{i=1}^{\ell} \alpha_i \pi(z_i) = \pi(x) - \xi = \pi(x - \xi) \quad \text{for some } \xi \in \mathcal{E}_{s-1} \cap \mathcal{E}^\perp. \tag{5.27}$$

If there were some other linear combination $\sum_{i=1}^{\ell} \alpha_i' \pi(z_i) = \pi(x) - \xi'$ for some $\xi' \in \mathcal{E}_{s-1} \cap \mathcal{E}^\perp$ and $\alpha_i' \in \mathbb{R}$, then, applying π_{s-1} to this linear combination and using that $\pi_{s-1}\pi = \pi_{s-1}$, we conclude from the uniqueness of the linear combination $\sum_{i=1}^{\ell} \alpha_i \pi_{s-1}(z_i) = \pi_{s-1}(x)$ that $\alpha_i' = \alpha_i$ for all i, whence $\xi = \xi'$ as well. In consequence, the α_i and $\xi \in \mathcal{E}_{s-1} \cap \mathcal{E}^\perp$ in (5.27) are uniquely defined.

Note $\mathcal{B}_s \subset \mathcal{X}_s \cup \{\mathbf{v}_s\}$ is a support set for $\mathcal{R}$ since $\mathbf{x} \prec \mathbf{y}$ is impossible for half-spaces from the same level $\mathcal{X}_s \cup \{\mathbf{v}_s\}$. Thus Proposition 5.3 ensures that $\mathsf{C}^{\cup}(\mathcal{B}_s)^\circ = \mathbf{z}_1^\circ + \ldots + \mathbf{z}_\ell^\circ$ with $\mathsf{C}^{\cup}(\mathcal{B}_s) = \overline{\mathbf{z}}_1 + \ldots + \overline{\mathbf{z}}_\ell$ a blunted simplicial cone having lineality space $\mathcal{E} = \partial(\mathbf{z}_1) + \ldots + \partial(\mathbf{z}_\ell) = \mathbb{R}^{\cup}\langle \partial(\mathcal{B}_s)\rangle \subseteq \mathcal{E}_{s-1}$. As a result, since $\alpha_i > 0$ for all i, (5.27) implies

$$x - \xi + \mathcal{E} \subseteq \mathsf{C}^{\cup}(\mathcal{B}_s)^\circ = \mathbf{z}_1^\circ + \ldots + \mathbf{z}_\ell^\circ. \tag{5.28}$$

Moreover, $\xi = \pi(\xi)$ is the unique element from $\mathcal{E}_{s-1} \cap \mathcal{E}^\perp$ such that $x - \xi \in \mathsf{C}^{\cup}(\mathcal{B}_s)^\circ$.

In view of Proposition 5.4.1,

$$\pi(\mathcal{R}) = \left(\mathcal{X}_j^\pi \cup \{\pi(\mathbf{v}_j)\}\right)_{j \in J}$$

is an orientated Reay system for $\pi(\mathcal{E}_s)$ for some $J \subseteq [1, s]$. Moreover, since $\mathcal{E} \subseteq \mathcal{E}_{s-1}$, we have $s \in J$. Since $\xi \in \mathcal{E}_{s-1} \cap \mathcal{E}^\perp = \pi(\mathcal{E}_{s-1})$, we can apply the induction hypothesis to the oriented Reay system $\left(\mathcal{X}_j^{\pi} \cup \{\pi(\mathbf{v}_j)\}\right)_{j\in J\setminus\{s\}}$ to conclude there exists a unique support set $\mathcal{D} \subseteq \bigcup_{j\in J\setminus\{s\}}(\mathcal{X}_j^{\pi} \cup \{\pi(\mathbf{v}_j)\})$ with

$$\xi \in \mathsf{C}^{\cup}(\mathcal{D})^{\circ} = \pi(\mathbf{y}_1)^{\circ} + \ldots + \pi(\mathbf{y}_r)^{\circ} = \pi(\mathbf{y}_1^{\circ}) + \ldots + \pi(\mathbf{y}_r^{\circ}), \tag{5.29}$$

where $\mathbf{y}_1, \ldots, \mathbf{y}_r \in \pi^{-1}(\mathcal{D})$ are the distinct half-spaces from $\pi^{-1}(\mathcal{D})$, with the first equality in (5.29) in view of Proposition 5.3.4 applied to the support set $\mathcal{D}$, and the second holding since each $\pi(\mathbf{y}_i)$ remains a relative half-space in $\pi(\mathcal{R})$.

Since $\mathcal{D}$ is a support set, Proposition 5.4.5 ensures that both $\pi^{-1}(\mathcal{D})$ and $(\pi^{-1}(\mathcal{D}) \cup \partial(\mathcal{B}_s))^*$ are both support sets with $\pi^{-1}(\mathcal{D}) \subseteq (\pi^{-1}(\mathcal{D}) \cup \partial(\mathcal{B}_s))^*$. Set

$$\mathcal{B} := \pi^{-1}(\mathcal{D}) \cup \mathcal{B}_s.$$

If $\mathcal{B}^* \neq \mathcal{B}$, then there are $\mathbf{y}_1, \mathbf{y}_2 \in \pi^{-1}(\mathcal{D}) \cup \mathcal{B}_s$ with $\mathbf{y}_1 \prec \mathbf{y}_2$. Since $\pi^{-1}(\mathcal{D})^* = \pi^{-1}(\mathcal{D}) \subseteq \mathcal{X}_1 \cup \{\mathbf{v}_1\} \cup \ldots \cup \mathcal{X}_{s-1} \cup \{\mathbf{v}_{s-1}\}$ (as $\pi^{-1}(\mathcal{D})$ is a support set) and $\mathcal{B}_s \subset \mathcal{X}_s \cup \{\mathbf{v}_s\}$, this is only possible if $\mathbf{y}_1 \in \pi^{-1}(\mathcal{D})$ and $\mathbf{y}_2 \in \mathcal{B}_s$, in which case $\mathbf{y}_1 \in \downarrow\partial(\mathcal{B}_s)$, whence $\pi(\mathbf{y}_1) = \{0\}$ as $\mathcal{E} = \mathbb{R}^{\cup}\langle\partial(\mathcal{B}_s)\rangle$, contradicting that $\mathbf{y}_1 \in \pi^{-1}(\mathcal{D})$ with $\mathcal{D}$ a set of non-zero half-spaces. Therefore we instead conclude that $\mathcal{B}^* = \mathcal{B}$. Since $\pi^{-1}(\mathcal{D}) \subseteq \mathcal{X}_1 \cup \{\mathbf{v}_1\} \cup \ldots \cup \mathcal{X}_{s-1} \cup \{\mathbf{v}_{s-1}\}$ and $\mathcal{B}_s \subset \mathcal{X}_s \cup \{\mathbf{v}_s\}$, we have $\mathcal{X}_s \cup \{\mathbf{v}_s\} \nsubseteq \downarrow\mathcal{B}$. Since $\mathcal{B}_s \subseteq \mathcal{X}_s \cup \{\mathbf{v}_s\}$, any $\mathbf{y} \in \mathcal{X}_j \cup \{\mathbf{v}_j\}$ with $j < s$ and $\mathbf{y} \in \downarrow\mathcal{B}$ must have $\mathbf{y} \in \downarrow\pi^{-1}(\mathcal{D})$ or $\mathbf{y} \in \downarrow\partial(\mathcal{B}_s)$. Thus $\mathcal{X}_j \cup \{\mathbf{v}_j\} \subseteq \downarrow\mathcal{B}$, for $j < s$, would imply $\mathcal{X}_j \cup \{\mathbf{v}_j\} \subseteq \downarrow(\pi^{-1}(\mathcal{D}) \cup \partial(\mathcal{B}_s))^*$, contradicting that $(\pi^{-1}(\mathcal{D}) \cup \partial(\mathcal{B}_s))^*$ is a support set. This shows that $\mathcal{B} = \mathcal{B}^*$ is itself a support set.

Since $\mathcal{B} = \pi^{-1}(\mathcal{D}) \cup \mathcal{B}_s$ is a support set, Proposition 5.3.4 implies

$$\mathsf{C}^{\cup}(\mathcal{B})^{\circ} = \mathsf{C}^{\cup}(\pi^{-1}(\mathcal{D}))^{\circ} + \mathsf{C}^{\cup}(\mathcal{B}_s)^{\circ}. \tag{5.30}$$

Since $\pi^{-1}(\mathcal{D})$ is a support set, Proposition 5.3.4 implies $\mathsf{C}^{\cup}(\pi^{-1}(\mathcal{D}))^{\circ} = \mathbf{y}_1^{\circ} + \ldots + \mathbf{y}_r^{\circ}$. Thus (5.29) implies that $\xi = \pi(\xi) \in \mathsf{C}^{\cup}(\pi^{-1}(\mathcal{D}))^{\circ} + \mathcal{E}$, which combined with (5.28) and (5.30) yields

$$x \in \mathsf{C}^{\cup}(\pi^{-1}(\mathcal{D}))^{\circ} + \mathsf{C}^{\cup}(\mathcal{B}_s)^{\circ} = \mathsf{C}^{\cup}(\mathcal{B})^{\circ}. \tag{5.31}$$

It remains to establish the uniqueness of $\mathcal{B}$.

Now suppose $\mathcal{B}'$ were any support set for $\mathcal{R}$ with $x \in \mathsf{C}^{\cup}(\mathcal{B}')^{\circ}$. We need to show it equals the support set $\mathcal{B}$ constructed above. Note that we have $\pi_{s-1}(x) \in \mathsf{C}^{\cup}\left(\pi_{s-1}\left(\mathcal{B}' \cap (\mathcal{X}_s \cup \{\mathbf{v}_s\})\right)\right)^{\circ}$ in view of Proposition 5.3.4, in which case we must have $\mathcal{B}_s = \mathcal{B}' \cap (\mathcal{X}_s \cup \{\mathbf{v}_s\})$ by the uniqueness property established with the existence of $\mathcal{B}_s$. Let $C = \mathcal{B}' \setminus \mathcal{B}_s$. We need to show $C = \pi^{-1}(\mathcal{D})$.

Since $\partial(\mathcal{B}_s) \subseteq \downarrow\mathcal{B}_s \subseteq \downarrow\mathcal{B}'$, Proposition 5.4.6 implies that $(\mathcal{B}')^{\pi} = (C \cup \mathcal{B}_s)^{\pi} = C^{\pi} \cup \mathcal{B}_s^{\pi}$ is a support set for $\pi(\mathcal{R})$ with $C \subseteq \mathcal{X}_1 \cup \ldots \cup \mathcal{X}_{s-1} \cup \{\mathbf{v}_j : j \in J \setminus \{s\}\}$. If there were some $\mathbf{y} \in C$ with $\pi(\mathbf{y}) = \{0\}$, then, since $\pi(\mathbf{v}_j) \neq \{0\}$ for all $j \in J$ by

Proposition 5.4.1, we must have $\mathbf{y} \in \mathcal{X}_1 \cup \ldots \cup \mathcal{X}_{s-1}$ and $\mathbf{y} \subseteq \ker \pi = \mathbb{R}^{\cup}\langle \partial(\mathcal{B}_s) \rangle$, whence Proposition 5.3.9 implies that $\mathbf{y} \in \downarrow\partial(\mathcal{B}_s) \subseteq \downarrow\mathcal{B}_s$. However this contradicts that $(C \cup \mathcal{B}_s)^* = (C \cup \mathcal{B}_s)$ for the support set $\mathcal{B}' = C \cup \mathcal{B}_s$. Therefore, we conclude that $C^\pi = \pi(C)$, ensuring $C = \pi^{-1}(\pi(C))$ in view of the injectivity of π given in Proposition 5.4.1. It remains to show $C^\pi = \pi(C) = \mathcal{D}$.

Since $\mathcal{B}' = C \cup \mathcal{B}_s$ and $(\mathcal{B}')^\pi = C^\pi \cup \mathcal{B}_s^\pi$ are support sets, C and C^π are also support sets. We have $x \in \mathsf{C}^{\cup}(C \cup \mathcal{B}_s)^\circ = \mathsf{C}^{\cup}(C)^\circ + \mathsf{C}^{\cup}(\mathcal{B}_s)^\circ$, ensuring $\pi(x) \in \mathsf{C}^{\cup}(C^\pi)^\circ + \mathsf{C}^{\cup}(\mathcal{B}_s^\pi)^\circ$, both in view of Proposition 5.3.4 (since $C \cup \mathcal{B}_s$ and $C^\pi \cup \mathcal{B}_s^\pi$ are both support sets). Thus $\pi(x) - \xi' \in \mathsf{C}^{\cup}(\mathcal{B}_s^\pi)$ for some $\xi' \in \mathsf{C}^{\cup}(C^\pi)^\circ \subseteq \pi(\mathcal{E}_{s-1}) = \mathcal{E}_{s-1} \cap \mathcal{E}^\perp$, in which case the uniqueness of ξ given in (5.27) ensures that $\xi' = \xi$. Hence $\xi = \xi' \in \mathsf{C}^{\cup}(C^\pi)^\circ$ with C^π a support set, in which case the uniqueness property established with the existence of $\mathcal{D}$ ensures that $\mathcal{D} = C^\pi = \pi(C)$, completing the proof. □

In view of Proposition 5.5, given any $x \in \mathcal{E}_s$, there is a uniquely defined support set $\mathcal{B}$ for $\mathcal{R}$ with $x \in \mathsf{C}^{\cup}(\mathcal{B})^\circ$, which we denote by $\mathcal{B} = \mathsf{Supp}_{\mathcal{R}}(x)$. Given a support set $\mathcal{B}$ for $\mathcal{R}$, we define

$$\mathrm{wt}(\mathcal{B}) = |\mathcal{B} \cap \{\mathbf{v}_1, \ldots, \mathbf{v}_s\}|. \tag{5.32}$$

Likewise, for $x \in \mathcal{E}_s$, we let $\mathrm{wt}(x) = \mathrm{wt}(\mathsf{Supp}_{\mathcal{R}}(x))$.

Definition 5.6 Let $\vec{u} = (u_1, \ldots, u_t)$ be a tuple of orthonormal vectors from $\mathbb{R}^d$, let $\mathcal{R} = (\mathcal{X}_1 \cup \{\mathbf{v}_1\}, \ldots, \mathcal{X}_s \cup \{\mathbf{v}_s\})$ be an oriented ray system, and let $\mathcal{B} \subseteq \mathcal{X}_1 \cup \{\mathbf{v}_1\} \cup \ldots \cup \mathcal{X}_s \cup \{\mathbf{v}_s\}$. We say that $\mathcal{B}$ **encases** $\vec{u}$ provided $\mathsf{C}^{\cup}(\mathcal{B})$ encases $\vec{u}$. We say that $\mathcal{B}$ **minimally encases** $\vec{u}$ if $\mathcal{B}$ encases $\vec{u}$ but no proper $\mathcal{B}' \prec \mathcal{B}$ encases $\vec{u}$.

Since $\mathsf{C}^{\cup}(\mathcal{B}^*) = \mathsf{C}^{\cup}(\mathcal{B})$ and $\mathcal{B}^* \preceq \mathcal{B}$, it is clear that $\mathcal{B}$ minimally encasing $\vec{u}$ is only possible if $\mathcal{B}^* = \mathcal{B}$. Thus any subset $\mathcal{B} \subseteq \mathcal{X}_1 \cup \ldots \cup \mathcal{X}_s$ which minimally encases $\vec{u}$ must be a support set. The following lemma deals with minimal encasement for $t = 1$.

Lemma 5.7 *Let $\mathcal{R} = (\mathcal{X}_1 \cup \{v_1\}, \ldots, \mathcal{X}_s \cup \{v_s\})$ be an oriented ray system for a subspace $\mathcal{E}_s \subseteq \mathbb{R}^d$, let $\mathcal{B} \subseteq \mathcal{X}_1 \cup \{v_1\} \cup \ldots \cup \mathcal{X}_s \cup \{v_s\}$, and let $u_1 \in \mathbb{R}^d$ be a unit vector.*

1. *If $\mathcal{B}$ minimally encases $-u_1$, then $-u_1 \in z_1^\circ + \ldots + z_\ell^\circ$, where $z_1, \ldots, z_\ell \in \mathcal{B}$ are the distinct half-spaces from $\mathcal{B}$.*
2. *If $\mathcal{B}$ is virtual independent, then $\mathcal{B}$ minimally encases $-u_1$ if and only if $-u_1 \in \mathsf{C}^{\cup}(\mathcal{B})^\circ$.*

Proof

1. Suppose $\mathcal{B}$ minimally encases $-u_1$. Then $-u_1 \in \mathsf{C}^{\cup}(\mathcal{B}) = \mathbf{z}_1 + \ldots + \mathbf{z}_\ell$ (by Proposition 5.3.2). If $-u_1 \in (\partial(\mathbf{z}_1) \cap \mathbf{z}_1) + \mathbf{z}_2 + \ldots + \mathbf{z}_\ell = \mathsf{C}(\partial(\{\mathbf{z}_1\})) + \mathbf{z}_2 + \ldots + \mathbf{z}_\ell$, then $-u_1 \in \mathsf{C}(\mathcal{B}')$, where $\mathcal{B}' = (\mathcal{B} \setminus \{\mathbf{z}_1\}) \cup \partial(\{\mathbf{z}_1\}) \prec \mathcal{B}$, contradicting the minimality of $\mathcal{B}$. A similar argument can be used for $\mathbf{z}_2, \ldots, \mathbf{z}_\ell$, and we conclude that $-u_1 \in \mathbf{z}_1^\circ + \ldots + \mathbf{z}_\ell^\circ$, as desired.

2. Since $\mathcal{B}$ is virtual independent, Proposition 5.3.4 implies that $\mathsf{C}^{\cup}(\mathcal{B})^{\circ} = \mathbf{z}_1^{\circ} + \ldots + \mathbf{z}_{\ell}^{\circ}$. Thus, if $\mathcal{B}$ minimally encases $-u_1$, then Item 1 implies $-u_1 \in \mathsf{C}^{\cup}(\mathcal{B})^{\circ}$. Next suppose that $-u_1 \in \mathsf{C}^{\cup}(\mathcal{B})^{\circ} = \mathbf{z}_1^{\circ} + \ldots + \mathbf{z}_{\ell}^{\circ}$. Let $\pi : \mathbb{R}^d \to \mathbb{R}^{\cup}\langle \partial(\mathcal{B})\rangle^{\perp}$ be the orthogonal projection. Since $-u_1 \in \mathbf{z}_1^{\circ} + \ldots + \mathbf{z}_{\ell}^{\circ}$, it follows that $-\pi(u_1)$ can be written as a strictly positive linear combination of the distinct linearly independent elements $\pi(z_1), \ldots, \pi(z_\ell) \in \mathbb{R}^{\cup}\langle \partial(\mathcal{B})\rangle^{\perp}$ (in view of Proposition 5.3.3). Since the $\pi(z_i)$ are distinct and linearly independent, this is the unique way to write $-\pi(u_1)$ as a linear combination of the elements $\pi(z_1), \ldots, \pi(z_\ell) \in \mathbb{R}^{\cup}\langle \partial(\mathcal{B})\rangle^{\perp}$. If $\mathcal{B}$ does not minimally encase $-u_1$, then w.l.o.g. $-u_1 \in \mathsf{C}^{\cup}\big(\mathcal{B} \setminus \{\mathbf{z}_1\} \cup \partial(\{\mathbf{z}_1\})\big) = \mathsf{C}^{\cup}(\partial(\{\mathbf{z}_1\})) + \mathbf{z}_2 + \ldots + \mathbf{z}_\ell = (\partial(\mathbf{z}_1) \cap \mathbf{z}_1) + \mathbf{z}_2 + \ldots + \mathbf{z}_\ell$. Thus $-\pi(u_1)$ can be also be written as a positive linear combination of the $\pi(z_2) \ldots, \pi(z_\ell) \in \mathbb{R}^{\cup}\langle \partial(\mathcal{B})\rangle^{\perp}$, contradicting that the unique way to express $-\pi(u_1)$ has the coefficient of $\pi(z_1)$ nonzero.

□

Any support set $\mathcal{B}$ is always virtual independent (by Proposition 5.3.5). Consequently, since $-u_1 \in \mathsf{C}(\mathsf{Supp}_{\mathcal{R}}(-u_1))^{\circ}$ by definition of $\mathsf{Supp}_{\mathcal{R}}$, we conclude via Lemma 5.7 that $\mathcal{B} = \mathsf{Supp}_{\mathcal{R}}(-u_1)$ is always a support set which minimally encases $-u_1 \in \mathbb{R}^{\cup}\langle \mathcal{X}_1 \cup \ldots \cup \mathcal{X}_s\rangle$, while Proposition 5.5 and Lemma 5.7 ensure that $\mathsf{Supp}_{\mathcal{R}}(-u_1)$ is the unique *support* set for $\mathcal{R}$ which minimally encases $-u_1$, though other non-support sets may also do the same. Indeed, if $\mathcal{B}$ is a virtual independent set and $\pi : \mathbb{R}^d \to \mathbb{R}^{\cup}\langle \partial(\mathcal{B})\rangle^{\perp}$ is the orthogonal projection, then Lemma 5.7 and Proposition 5.3 (Items 3–4) ensure that $\mathcal{B}$ minimally encasing the *nonzero* element $-u_1$ is equivalent to $\pi(B) \cup \{\pi(u_1)\}$ being a minimal positive basis.

As the above discussion shows, there is always a unique support set $\mathcal{B} = \mathsf{Supp}_{\mathcal{R}}(-u_1)$ which minimally encases the element $u_1 \in \mathbb{R}^{\cup}\langle \mathcal{X}_1 \cup \ldots \cup \mathcal{X}_s\rangle$. However, if $\vec{u} = (u_1, \ldots, u_t)$ is a tuple of orthonormal vectors $u_1, \ldots, u_t \in \mathbb{R}^{\cup}\langle \mathcal{X}_1 \cup \ldots \cup \mathcal{X}_s\rangle$, there is no guarantee that $-\vec{u}$ will be minimally encased by some support set from $\mathcal{R}$ when $t \geq 2$. We will later show that this problem does not occur if we impose additional conditions on $\mathcal{R}$. Until then, we will have need of the following definition.

Definition 5.8 Let $\mathcal{R} = (\mathcal{X}_1 \cup \{\mathbf{v}_1\}, \ldots, \mathcal{X}_s \cup \{\mathbf{v}_s\})$ be an oriented Reay system in $\mathbb{R}^d$, let $\mathcal{B} \subseteq \mathcal{X}_1 \cup \{\mathbf{v}_1\} \cup \ldots \cup \mathcal{X}_s \cup \{\mathbf{v}_s\}$, and let $\vec{u} = (u_1, \ldots, u_t)$ be a tuple of orthonormal vectors from $\mathbb{R}^d$. Suppose $\mathcal{B}$ minimally encases $\vec{u}$. If $\mathcal{B} \neq \emptyset$, then $t \geq 1$ and there will be a maximal index $t' \in [0, t-1]$ such that $\mathcal{B}$ does *not* minimally encase $(u_1, \ldots, u_{t'})$. Note $\mathcal{B}$ still encases $(u_1, \ldots, u_{t'})$, so there is some $\mathcal{A} \prec \mathcal{B}$ such that $\mathcal{A}$ minimally encases $(u_1, \ldots, u_{t'})$. If $\mathcal{B} = \emptyset$ or

$$\mathcal{B} \subseteq \mathcal{X}_1 \cup \ldots \cup \mathcal{X}_s \cup \{\mathbf{v}_j : j \in J\} \text{ is virtual independent} \quad \text{and}$$
$$\mathcal{A} \subseteq \mathcal{X}_1 \cup \ldots \cup \mathcal{X}_s,$$

where $\pi(\mathcal{A}) = (\mathcal{X}_i^{\pi} \cup \{\pi(\mathbf{v}_i)\})_{i \in J}$ and $\pi : \mathbb{R}^d \to \mathbb{R}^{\cup}\langle \mathcal{A}\rangle^{\perp}$ is the orthogonal projection, then we say that $\mathcal{B}$ minimally encases $\vec{u}$ **urbanely**.

When $\mathcal{B}$ is a support set, the condition $\mathcal{B} \subseteq \mathcal{X}_1 \cup \ldots \cup \mathcal{X}_s \cup \{\mathbf{v}_j : j \in J\}$ in the above definition holds automatically in view of Proposition 5.4.6, and so can be dropped. We will later see in Proposition 5.9.1 that the $\mathcal{A}$ occurring in the definition of urbane minimal encasement is uniquely defined. If $\mathcal{B} \subseteq \mathcal{X}_1 \cup \ldots \cup \mathcal{X}_s$ and $\mathcal{B}$ minimally encases $-\vec{u}$, then it must always do so urbanely and be a support set in view of $\mathcal{A} \subseteq \downarrow\mathcal{B} \subseteq \mathcal{X}_1 \cup \ldots \cup \mathcal{X}_s$. However, for more general subsets $\mathcal{B} \subseteq \mathcal{X}_1 \cup \{\mathbf{v}_1\} \cup \ldots \cup \mathcal{X}_s \cup \{\mathbf{v}_s\}$, it is possible for $\mathcal{B}$ to minimally encase $-\vec{u}$ non-urbanely. The following proposition contains the basic properties regarding urbane minimal encasement and is the analogue of Propositions 4.19 and 4.20 for oriented Reay systems.

Proposition 5.9 *Let $\mathcal{R} = (\mathcal{X}_1 \cup \{v_1\}, \ldots, \mathcal{X}_s \cup \{v_s\})$ be an oriented ray system in $\mathbb{R}^d$, let $\vec{u} = (u_1, \ldots, u_t)$ be a tuple of $t \geq 0$ orthonormal vectors in $\mathbb{R}^d$, and let $\mathcal{B} \subseteq \mathcal{X}_1 \cup \{v_1\} \cup \ldots \cup \mathcal{X}_s \cup \{v_s\}$.*

1. *If $\mathcal{B}$, $\mathcal{C} \subseteq \mathcal{X}_1 \cup \{v_1\} \cup \ldots \cup \mathcal{X}_s \cup \{v_s\}$ are support sets that both minimally encase $\vec{u}$ urbanely, then $\mathcal{B} = \mathcal{C}$. In particular, if $\mathcal{B}$, $\mathcal{C} \subseteq \mathcal{X}_1 \cup \ldots \cup \mathcal{X}_s$ with $\mathcal{B}$ minimally encasing $\vec{u}$ and $\mathcal{C}$ encasing $\vec{u}$, then $\mathcal{B} \preceq \mathcal{C}$.*
2. *$\mathcal{B}$ minimally encases $-\vec{u}$ urbanely if and only if there are indices $1 = r_1 < \ldots < r_\ell < r_{\ell+1} = t+1$ and virtual independent sets $\mathcal{C}_i$, for $i = 0, 1, \ldots, \ell$, satisfying*
$$\emptyset = \mathcal{C}_0 \prec \mathcal{C}_1 \prec \ldots \prec \mathcal{C}_{\ell-1} \subseteq \mathcal{X}_1 \cup \ldots \cup \mathcal{X}_s \quad \text{and} \quad \mathcal{C}_{\ell-1} \prec \mathcal{C}_\ell = \mathcal{B}$$
such that either $\ell = 0$ or else
   (a) *$\mathcal{C}_\ell^{\pi_{\ell-1}} = \mathcal{D}_\ell$ for some virtual independent set $\mathcal{D}_\ell$ from $\pi_{\ell-1}(\mathcal{R})$ that minimally encases $-\pi_{\ell-1}(u_{r_\ell})$ and $\mathcal{C}_\ell$ contains no v_i with $\mathcal{X}_i^{\pi_{\ell-1}} = \emptyset$ (the latter which necessarily holds when $\mathcal{C}_\ell$ is a support set),*
   (b) *$\mathcal{C}_j^{\pi_{j-1}} = \mathcal{D}_j := \mathrm{Supp}_{\pi_{j-1}(\mathcal{R})}(-\pi_{j-1}(u_{r_j}))$ for every $j \in [1, \ell-1]$, and*
   (c) *$u_i \in \mathbb{R}^{\cup}\langle \mathcal{C}_j \rangle$ for all $i < r_{j+1}$ and $j \in [1, \ell]$,*

   where $\pi_{j-1} : \mathbb{R}^d \to \mathbb{R}^{\cup}\langle \mathcal{C}_{j-1}\rangle^{\perp}$ is the orthogonal projection for $j \in [1, \ell]$.
   Moreover, $\mathcal{B}$ is a support set if and only if $\ell = 0$ or $\mathcal{D}_\ell = \mathrm{Supp}_{\pi_{\ell-1}(\mathcal{R})}(-\pi_{\ell-1}(u_{r_\ell}))$.

Now assume the conditions of Item 2 hold along with the relevant notation.

3. *$\mathcal{F} = (\mathbb{R}^{\cup}\langle \mathcal{C}_1\rangle, \ldots, \mathbb{R}^{\cup}\langle \mathcal{C}_\ell\rangle)$ is a compatible filter for $\vec{u}$ with associated indices $1 = r_1 < \ldots < r_\ell < r_{\ell+1} = t+1$, $\mathcal{F}(\vec{u}) = (\overline{u}_1, \ldots, \overline{u}_\ell)$, and for every $j \in [1, \ell]$, the following hold:*
   (a) *$\mathcal{C}_j = (\pi_{j-1}^{-1}(\mathcal{D}_j) \cup \mathcal{C}_{j-1})^*$ is the lift of $\mathcal{D}_j$,*
   (b) *$\mathsf{C}^{\cup}(\mathcal{C}_j) + \mathsf{C}(u_{r_1}, \ldots, u_{r_j}) = \mathsf{C}^{\cup}(\mathcal{C}_j) + \mathsf{C}(\overline{u}_1, \ldots, \overline{u}_j) = \mathbb{R}^{\cup}\langle \mathcal{C}_j\rangle$, and*
   (c) *$\mathcal{B}$ minimally encases $-\mathcal{F}(\vec{u})$ and $-(u_{r_1}, \ldots, u_{r_\ell})$ urbanely.*
4. *Let $\mathcal{A} \subseteq \mathcal{X}_1 \cup \ldots \cup \mathcal{X}_s$, let $\mathcal{E} = \mathbb{R}^{\cup}\langle \mathcal{A}\rangle$, let $\pi : \mathbb{R}^d \to \mathcal{E}^{\perp}$ be the orthogonal projection, let $J \subseteq [1, \ell]$ be all those indices $j \in [1, \ell]$ with $\mathcal{C}_{j-1}^{\pi} \prec \mathcal{C}_j^{\pi}$, and let $\mathcal{F}^{\pi} = (\mathbb{R}^{\cup}\langle \mathcal{C}_j^{\pi}\rangle)_{j \in J}$. For $j \in [1, \ell]$, let $\tau_{j-1} : \mathbb{R}^d \to (\mathcal{E} + \mathbb{R}^{\cup}\langle \mathcal{C}_{j-1}\rangle)^{\perp}$ be the*

orthogonal projection. Suppose

$$\mathcal{B} \subseteq \mathcal{X}_1 \cup \ldots \cup \mathcal{X}_s \quad \text{or} \quad \mathcal{A} \subseteq \downarrow\mathcal{B}, \quad \text{and either}$$

$$\mathcal{B} \text{ is a support set} \quad \text{or} \quad \mathcal{B} \text{ contains no } \mathbf{v}_i \text{ with } \mathcal{X}_i^{\tau_{\ell-1}} = \emptyset \text{ for } i \in [1, s].$$

Then $\mathcal{F}^\pi$ is a compatible filter for $\pi(\vec{u})$ with

$$\mathcal{F}^\pi(\pi(\vec{u})) = (\overline{u_i^*})_{i \in J},$$

where $\overline{u_i^*} = \tau_{i-1}(u_{r_i})/\|\tau_{i-1}(u_{r_i})\|$, and the virtual independent set $\mathcal{B}^\pi$ for $\pi(\mathcal{R})$ minimally encases $-\pi(\vec{u})$ urbanely with the sets C_j^π for $j \in J$ those satisfying Item 2 for $\pi(\vec{u})$.

Proof

1. We begin by proving Item 1 in the case when $\mathcal{B}, C \subseteq \mathcal{X}_1 \cup \ldots \cup \mathcal{X}_s$. For this, it suffices to consider subsets $\mathcal{B}, C \subseteq \mathcal{X}_1 \cup \ldots \cup \mathcal{X}_s$ which minimally encase $\vec{u}$ and show $\mathcal{B} = C$, in which case $\mathcal{B}$ and C are both support sets. Let $\mathcal{E} = \mathbb{R}^{\cup}\langle\mathcal{B}\rangle$, let $\pi : \mathbb{R}^d \to \mathcal{E}^\perp$ be the orthogonal projection, and let $C' \subseteq C$ be all those half-spaces $\mathbf{x} \in C$ with $\pi(\mathbf{x}) \neq \{0\}$. By Proposition 5.4.6, $C^\pi = \pi(C')$ is a support set for $\pi(\mathcal{R})$ with π injective on C' (by Proposition 5.4.1), while Proposition 5.4.2 implies that $\partial(C^\pi) = \partial(C)^\pi$. Thus Proposition 5.3 (Items 3 and 5) applied to the support set C^π for $\pi(\mathcal{R})$ implies that the elements of C' are linearly independent modulo $\mathcal{E} + \mathbb{R}\langle\partial(C)\rangle$. Consequently,
$$\mathsf{C}^{\cup}(C) \cap \mathcal{E} = (\mathsf{C}^{\cup}(C') \cap \mathcal{E}) + \mathsf{C}^{\cup}(C \setminus C') \subseteq \mathsf{C}^{\cup}(\partial(C')) + \mathsf{C}^{\cup}(C \setminus C'). \quad (5.33)$$
Since C encases $(u_1, \ldots, u_t)$ with $\mathbb{R}\langle u_1, \ldots, u_t\rangle \subseteq \mathbb{R}^{\cup}\langle\mathcal{B}\rangle = \mathcal{E}$ (as $\mathcal{B}$ encases $(u_1, \ldots, u_t)$), it follows that $\mathsf{C}^{\cup}(C) \cap \mathcal{E}$ also encases $(u_1, \ldots, u_t)$, whence (5.33) ensures that $(C \setminus C') \cup \partial(C') \preceq C$ encases $(u_1, \ldots, u_t)$. Consequently, since C *minimally* encases $(u_1, \ldots, u_t)$, we conclude that $(C \setminus C') \cup \partial(C') = C$, which is only possible if $C' = 0$, that is, if all half-spaces in $C \subseteq \mathcal{X}_1 \cup \ldots \cup \mathcal{X}_s$ are contained in $\mathcal{E} = \mathbb{R}^{\cup}\langle\mathcal{B}\rangle$. Hence Proposition 5.3.9 implies that $C \subseteq \downarrow\mathcal{B}$, further implying $C = C^* \preceq \mathcal{B}$ (the equality follows since C is a support set). But now, since $\mathcal{B}$ *minimally* encases $(u_1, \ldots, u_t)$, we must have $C = \mathcal{B}$, completing the proof of Item 1 in the case $\mathcal{B}, C \subseteq \mathcal{X}_1 \cup \ldots \cup \mathcal{X}_s$. We will complete the more general case for Item 1 later.

2–3. If $t = 0$, then Item 2 holds trivially with $\ell = 0$, since only $\mathcal{B} = \emptyset$ can minimally encase the empty tuple. If $\ell = 0$, then $1 = r_1 = r_{\ell+1} = t + 1$ implies $t = 0$. Therefore we may assume $t, \ell \geq 1$. Suppose the virtual independent sets C_i exist with the prescribed properties 2(a), 2(b) and 2(c). Since $C_j \subseteq \mathcal{X}_1 \cup \ldots \cup \mathcal{X}_s$ for $j < \ell$, it follows that each C_j with $j < \ell$ is not just a virtual independent set, but also a support set. We must show that $\mathcal{B}$ minimally encases $-(u_1, \ldots, u_t)$ and that the conditions in Item 3 all hold.

If there were some $\mathbf{v}_i \in C_j$ with $\mathcal{X}_i^{\pi_{j-1}} = \emptyset$, for some $i \in [1, s]$ and $j \in [1, \ell]$, then Proposition 5.3.9 implies $\mathcal{X}_i \subseteq \downarrow C_{j-1} \subseteq \downarrow C_j$, with the latter inclusion in view of $C_{j-1} \prec C_j$. Thus $\mathcal{X}_i \cup \{\mathbf{v}_i\} \subseteq \downarrow C_j$, ensuring that C_j is *not* a support set, which is only possible when $j = \ell$, in which case 2(a) gives us the contradiction $\mathbf{v}_i \notin C_\ell = C_j$ by hypothesis. Therefore we instead conclude there is no $\mathbf{v}_i \in C_j$ with $\mathcal{X}_i^{\pi_{j-1}} = \emptyset$ for any $i \in [1, s]$ and $j \in [1, \ell]$.

Suppose $\pi_{j-1}(u_{r_j}) = 0$ for some $j \in [1, \ell]$. Then, since $\mathcal{D}_j$ minimally encases $-\pi_{j-1}(u_{r_j}) = 0$, it follows that $\mathcal{D}_j = \emptyset$, whence 2(a) or 2(b) implies $C_j^{\pi_{j-1}} = \emptyset$, i.e., $\pi_{j-1}(\mathbf{x}) = \{0\}$ for all $\mathbf{x} \in C_j$. Combining this with the conclusion of the previous paragraph and Proposition 5.4.1, we find $C_j \subseteq \mathcal{X}_1 \cup \ldots \cup \mathcal{X}_s$, and now Proposition 5.3.9 implies $C_j \subseteq \downarrow C_{j-1}$. Thus $C_j = C_j^* \preceq C_{j-1}$ as C_j is virtual independent, which contradicts that $C_{j-1} \prec C_j$. So we instead conclude that $\pi_{j-1}(u_{r_j}) \neq 0$ for all $j \in [1, \ell]$, whence 2(c) ensures that $\mathcal{F} = (\mathbb{R}^{\cup}\langle C_1\rangle, \ldots, \mathbb{R}^{\cup}\langle C_\ell\rangle)$ is a compatible filter for $\vec{u}$ with $\mathcal{F}(\vec{u}) = (\overline{u}_1, \ldots, \overline{u}_\ell)$, where $\overline{u}_j = \pi_{j-1}(u_{r_j})/\|\pi_{j-1}(u_{r_j})\|$ for $j \in [1, \ell]$, as required for Item 3.

We proceed by induction on $j \in [0, \ell]$ to show that C_j minimally encases $-(u_1, \ldots, u_{r_{j+1}-1})$ as well as $-(u_{r_1}, \ldots, u_{r_j})$ and $-(\overline{u}_1, \ldots, \overline{u}_j)$, all urbanely, with $\mathsf{C}^{\cup}(C_j) + \mathsf{C}(u_{r_1}, \ldots, u_{r_j}) = \mathsf{C}^{\cup}(C_j) + \mathsf{C}(\overline{u}_1, \ldots, \overline{u}_j) = \mathbb{R}^{\cup}\langle C_j\rangle$. The case $j = \ell$ will then verify 3(b) and 3(c), and also show that $\mathcal{B}$ minimally encases $\vec{u}$ (as required for Item 2). During the course of the proof, we will also see that 3(a) holds. The base case is $j = 0$, in which case the empty set $C_0 = \emptyset$ minimally encases the empty tuple trivially with $\mathbb{R}^{\cup}\langle\emptyset\rangle = \{0\} = \mathsf{C}^{\cup}(\emptyset)$. Therefore we assume $j \geq 1$ and that this has been shown for $j - 1$. To lighten notation, let

$$\pi = \pi_{j-1}.$$

Since $C_{j-1} \prec C_j$ implies $C_{j-1} \subseteq \downarrow C_j$, we have $\mathsf{C}^{\cup}(C_{j-1}) \subseteq \mathsf{C}^{\cup}(C_j)$. As a result, since C_{j-1} minimally encases $-(u_1, \ldots, u_{r_j-1})$, $-(u_{r_1}, \ldots, u_{r_{j-1}})$ and $-(\overline{u}_1, \ldots, \overline{u}_{j-1})$ by induction hypothesis, it follows that C_j encases these tuples as well. By induction hypothesis, we also have $\ker \pi = \mathbb{R}^{\cup}\langle C_{j-1}\rangle = \mathsf{C}^{\cup}(C_{j-1}) + \mathsf{C}(u_{r_1}, \ldots, u_{r_{j-1}}) = \mathsf{C}^{\cup}(C_{j-1}) + \mathsf{C}(\overline{u}_1, \ldots, \overline{u}_{j-1})$. Consequently, in view of 2(c), to show C_j encases $-(u_1, \ldots, u_{r_j})$ and $-(u_{r_1}, \ldots, u_{r_j})$, we we just need to know $-\pi(u_{r_j}) \in \mathsf{C}^{\cup}(C_j^{\pi})$, for if this is the case, then $u_{r_j} + x \in -\mathsf{C}^{\cup}(C_j)$ for some $x \in \mathbb{R}^{\cup}\langle C_{j-1}\rangle = \mathsf{C}^{\cup}(C_{j-1}) + \mathsf{C}(u_{r_1}, \ldots, u_{r_{j-1}})$, whence $(-x + \mathsf{C}(u_{r_1}, \ldots, u_{r_{j-1}})) \cap -\mathsf{C}^{\cup}(C_j) \neq \emptyset$, and then the desired conclusion $((u_{r_j} + x) - x + \mathsf{C}(u_{r_1}, \ldots, u_{r_{j-1}})) \cap -\mathsf{C}^{\cup}(C_j) \neq \emptyset$ follows in view of $\mathsf{C}^{\cup}(C_j)$ being a convex cone. Likewise, to show C_j encases $-(\overline{u}_1, \ldots, \overline{u}_j)$, we just need to know $-\pi(\overline{u}_j) \in \mathsf{C}^{\cup}(C_j^{\pi})$, and since $\overline{u}_j = \pi(u_{r_j})/\|\pi(u_{r_j})\|$, this is equivalent to the previous condition $-\pi(u_{r_j}) \in \mathsf{C}^{\cup}(C_j^{\pi})$. However, that $-\pi(u_{r_j}) \in \mathsf{C}^{\cup}(C_j^{\pi})$ holds follows directly from 2(a) or 2(b) and the definition of $\mathcal{D}_j$. Indeed, Lemma 5.7

ensures that $C_j^\pi = \mathcal{D}_j$ is a virtual independent set which minimally encases $-\pi(u_{r_j})$. Thus we now know C_j encases $-(u_1,\ldots,u_{r_j})$, $-(u_{r_1},\ldots,u_{r_j})$ and $-(\overline{u}_1,\ldots,\overline{u}_j)$.

Since $\ker\pi = \mathbb{R}^{\cup}\langle C_{j-1}\rangle$ and C_j contains no $\mathbf{v}_i$ with $\mathcal{X}_i^\pi = \emptyset$, as shown earlier, it follows in view of $C_j^\pi = \mathcal{D}_j$ and Propositions 5.3.9 and 5.4.1 that $\pi^{-1}(\mathcal{D}_j) \subseteq C_j \subseteq \downarrow(\pi^{-1}(\mathcal{D}_j) \cup C_{j-1})$. Since $C_{j-1} \prec C_j$, we have $\downarrow C_{j-1} \subseteq \downarrow C_j$, whence $\downarrow C_j \subseteq \downarrow\pi^{-1}(\mathcal{D}_j) \cup \downarrow C_{j-1} \subseteq \downarrow C_j \cup \downarrow C_{j-1} \subseteq \downarrow C_j$, implying

$$C_j = C_j^* = (\downarrow\pi^{-1}(\mathcal{D}_j) \cup \downarrow C_{j-1})^* = (\pi^{-1}(\mathcal{D}_j) \cup C_{j-1})^*, \tag{5.34}$$

which shows that the virtual independent set C_j is the lift of the virtual independent set $\mathcal{D}_j$ as required for 3(a). Since C_{j-1} is a support set, Proposition 5.4.5 implies that

$$C_j \setminus \pi^{-1}(\mathcal{D}_j) = C_{j-1} \setminus \downarrow\partial(\pi^{-1}(\mathcal{D}_j)) = C_{j-1} \setminus \downarrow\pi^{-1}(\mathcal{D}_j), \tag{5.35}$$

where the second equality follows since C_{j-1} and $\pi^{-1}(\mathcal{D}_j)$ are disjoint (as $\pi(\mathbf{x}) = \{0\}$ for $\mathbf{x} \in C_{j-1}$ but $\pi(\mathbf{x}) \neq \{0\}$ for $\mathbf{x} \in \pi^{-1}(\mathcal{D}_j)$).

Now $\pi(\pi^{-1}(\mathcal{D}_j)) = \mathcal{D}_j$ is a virtual independent set with π injective on $\pi^{-1}(\mathcal{D}_j)$ (by Proposition 5.4.1) while $\partial(\mathcal{D}_j) = \partial(\pi(\pi^{-1}(\mathcal{D}_j))) = \partial(\pi^{-1}(\mathcal{D}_j))^\pi$ by Proposition 5.4.2. Thus Proposition 5.3.3 implies that $\pi^{-1}(D_j)$ is linearly independent modulo $\ker\pi + \mathbb{R}^{\cup}\langle\partial(\pi^{-1}(\mathcal{D}_j))\rangle = \mathbb{R}^{\cup}\langle C_{j-1}\rangle + \mathbb{R}^{\cup}\langle\partial(\pi^{-1}(\mathcal{D}_j))\rangle$, ensuring that

$$\mathsf{C}^{\cup}(\downarrow\pi^{-1}(\mathcal{D}_j)) \cap \mathbb{R}^{\cup}\langle C_{j-1}\rangle = \mathsf{C}^{\cup}(\pi^{-1}(\mathcal{D}_j)) \cap \mathbb{R}^{\cup}\langle C_{j-1}\rangle \subseteq \mathsf{C}^{\cup}\big(\partial(\pi^{-1}(\mathcal{D}_j))\big). \tag{5.36}$$

□

Claim A $C_j \setminus \pi^{-1}(\mathcal{D}_j) \cup \partial(\pi^{-1}(\mathcal{D}_j))$ encases the tuples $-(u_1,\ldots,u_{r_j-1})$, $-(u_{r_1},\ldots,u_{r_{j-1}})$ and $-(\overline{u}_1,\ldots,\overline{u}_{j-1})$, but $C'_{j-1} \cup \partial(\pi^{-1}(\mathcal{D}_j))$ encases none of these for any $C'_{j-1} \prec C_j \setminus \pi^{-1}(\mathcal{D}_j)$.

Proof We know C_j encases $-(u_1,\ldots,u_{r_j-1})$, $-(u_{r_1},\ldots,u_{r_{j-1}})$ and $-(\overline{u}_1,\ldots,\overline{u}_{j-1})$ with $u_i, \overline{u}_k \in \mathbb{R}^{\cup}\langle C_{j-1}\rangle$ for all $i < r_j$ and $k < j$. In view of (5.34) and (5.36), we have

$$\begin{aligned}\mathsf{C}^{\cup}(C_j) \cap \mathbb{R}^{\cup}\langle C_{j-1}\rangle &= \mathsf{C}^{\cup}(\downarrow\pi^{-1}(\mathcal{D}_j) \cup C_{j-1}) \cap \mathbb{R}^{\cup}\langle C_{j-1}\rangle \\ &\subseteq \mathsf{C}^{\cup}(\partial(\pi^{-1}(\mathcal{D}_j))) + \mathsf{C}^{\cup}(C_{j-1} \setminus \downarrow\pi^{-1}(\mathcal{D}_j)).\end{aligned}$$

Thus $C_j \setminus \pi^{-1}(\mathcal{D}_j) \cup \partial(\pi^{-1}(\mathcal{D}_j))$ encases each of the tuples $-(u_1,\ldots,u_{r_j-1})$, $-(u_{r_1},\ldots,u_{r_{j-1}})$ and $-(\overline{u}_1,\ldots,\overline{u}_{j-1})$ in view of (5.35). Suppose by contradiction that $C'_{j-1} \cup \partial(\pi^{-1}(\mathcal{D}_j))$ encases one of the tuples $-(u_1,\ldots,u_{r_j-1})$ or

$-(u_{r_1},\ldots,u_{r_{j-1}})$ or $-(\overline{u}_1,\ldots,\overline{u}_{j-1})$ for some

$$C'_{j-1} \prec C_j \setminus \pi^{-1}(\mathcal{D}_j) = C_{j-1} \setminus \downarrow\pi^{-1}(\mathcal{D}_j)$$

(in view of (5.35)). Note that $C_j \setminus \pi^{-1}(\mathcal{D}_j) \subseteq C_j$ is a virtual independent set as it is a subset of the virtual independent set C_j. As a result, $C'_{j-1} \prec C_{j-1} \setminus \downarrow\pi^{-1}(\mathcal{D}_j)$ implies that there is some

$$\mathbf{y} \in C_{j-1} \setminus \downarrow\pi^{-1}(\mathcal{D}_j) \quad \text{with} \quad \mathbf{y} \notin \downarrow C'_{j-1}. \tag{5.37}$$

Since both C_{j-1} and $C'_{j-1} \cup \partial(\pi^{-1}(\mathcal{D}_j))$ are subsets of $\mathcal{X}_1 \cup \ldots \cup \mathcal{X}_s$ that encase $-(u_1,\ldots,u_{r_j-1})$, $-(u_{r_1},\ldots,u_{r_{j-1}})$ or $-(\overline{u}_1,\ldots,\overline{u}_{j-1})$ with the encasement by C_{j-1} minimal by induction hypothesis, it follows from the already established case in Item 1 that $C_{j-1} \preceq C'_{j-1} \cup \partial(\pi^{-1}(\mathcal{D}_j))$, implying $\mathbf{y} \in C_{j-1} \subseteq \downarrow C'_{j-1} \cup \downarrow\partial(\pi^{-1}(\mathcal{D}_j)) \subseteq \downarrow C'_{j-1} \cup \downarrow\pi^{-1}(\mathcal{D}_j)$. However, this contradicts (5.37), and Claim A is established. □

We showed above that C_j encases the tuples $-(u_1,\ldots,u_{r_j})$, $-(u_{r_1},\ldots,u_{r_j})$ and $-(\overline{u}_1,\ldots,\overline{u}_j)$. Let us next show that is does so *minimally*. To this end, its suffices to show that the immediate predecessor $C'_j = C_j \setminus \{\mathbf{x}\} \cup \partial(\{\mathbf{x}\})$ encases neither $-(u_1,\ldots,u_{r_j})$ nor $-(u_{r_1},\ldots,u_{r_j})$ nor $-(\overline{u}_1,\ldots,\overline{u}_j)$ for any $\mathbf{x} \in C_j$. Suppose by contradiction that this fails for $\mathbf{x} \in C_j$. If $\pi(\mathbf{x}) \neq \{0\}$, then Proposition 5.4.3 implies that $(C'_j)^\pi \prec C_j^\pi$ with $-\pi(u_{r_j}) \in \mathsf{C}^\cup((C'_j)^\pi)$ or $-\pi(\overline{u}_{r_j}) \in \mathsf{C}^\cup((C'_j)^\pi)$. Noting that $\overline{u}_j = \pi(u_{r_j})/||\pi(u_{r_j})||$, we see that the latter case implies the former, and now both cases contradict that $C_j^\pi = \mathcal{D}_j$ minimally encases the element $-\pi(u_{r_j})$ by definition of $\mathcal{D}_j$. So instead suppose $\pi(\mathbf{x}) = \{0\}$, so $\mathbf{x} \in C_j \setminus \pi^{-1}(\mathcal{D}_j)$. In this case,

$$C'_j = \pi^{-1}(\mathcal{D}_j) \cup C'_{j-1}$$

with $C'_{j-1} \prec C_j \setminus \pi^{-1}(\mathcal{D}_j)$. Since C'_j encases $-(u_1,\ldots,u_{r_j})$, $-(u_{r_1},\ldots,u_{r_j})$ or $-(\overline{u}_1,\ldots,\overline{u}_j)$, it also encases $-(u_1,\ldots,u_{r_j-1})$, $-(u_{r_1},\ldots,u_{r_{j-1}})$ or $-(\overline{u}_1,\ldots,\overline{u}_{j-1})$ with $u_i, \overline{u}_{r_k} \in \mathbb{R}^\cup\langle C_{j-1}\rangle$ for all $i < r_j$ and $k < j$. As argued in Claim A, it follows in view of (5.36) that

$$\begin{aligned}\mathsf{C}^\cup(C'_j) \cap \mathbb{R}^\cup\langle C_{j-1}\rangle &= \mathsf{C}^\cup(\pi^{-1}(\mathcal{D}_j) \cup C'_{j-1}) \cap \mathbb{R}^\cup\langle C_{j-1}\rangle \\ &\subseteq \mathsf{C}^\cup(\partial(\pi^{-1}(\mathcal{D}_j))) + \mathsf{C}^\cup(C'_{j-1}).\end{aligned}$$

Thus $C'_{j-1} \cup \partial(\pi^{-1}(\mathcal{D}_j))$ also encases one of the tuples $-(u_1,\ldots,u_{r_j-1})$ or $-(u_{r_1},\ldots,u_{r_{j-1}})$ or $-(\overline{u}_1,\ldots,\overline{u}_{j-1})$, which is contrary to Claim A. This shows C_j minimally encases $-(u_1,\ldots,u_{r_j})$, $-(u_{r_1},\ldots,u_{r_j})$ and $-(\overline{u}_1,\ldots,\overline{u}_j)$.

Next, we show that $\mathbb{R}^\cup\langle C_j\rangle = \mathsf{C}^\cup(C_j) + \mathsf{C}(u_{r_1},\ldots,u_{r_j}) = \mathsf{C}^\cup(C_j) + \mathsf{C}(\overline{u}_1,\ldots,\overline{u}_j)$. By induction hypothesis, $\mathbb{R}^\cup\langle C_{j-1}\rangle = \mathsf{C}^\cup(C_{j-1}) + \mathsf{C}(u_{r_1},\ldots,u_{r_{j-1}}) \subseteq \mathsf{C}^\cup(C_j) + \mathsf{C}(u_{r_1},\ldots,u_{r_j})$ (the inclusion follows as $C_{j-1} \prec C_j$ implies $C_{j-1} \subseteq$

$\downarrow C_j$), and likewise $\mathbb{R}^{\cup}(C_{j-1}) \subseteq \mathsf{C}^{\cup}(C_j)+\mathsf{C}(\overline{u}_1,\ldots,\overline{u}_j)$. Consequently, it suffices to show $\mathsf{C}^{\cup}(C_j^{\pi}) + \mathsf{C}(\pi(u_{r_j})) = \mathbb{R}^{\cup}\langle C_j^{\pi}\rangle$ and $\mathsf{C}^{\cup}(C_j^{\pi}) + \mathsf{C}(\pi(\overline{u}_j)) = \mathbb{R}^{\cup}\langle C_j^{\pi}\rangle$. Recalling that $\overline{u}_j = \pi(u_{r_j})/\|\pi(u_{r_j})\|$, we find that the latter condition is equivalent to the former. As already remarked above, the virtual independent set $C_j^{\pi} = \mathcal{D}_j$ minimally encases the element $-\pi(u_{r_j})$, which is equivalent to $D_j \cup \{\pi(u_{r_j})\}$ being a minimal positive basis modulo $\mathbb{R}^{\cup}\langle\partial(\mathcal{D}_j)\rangle$ (as remarked after Lemma 5.7). Thus there is a strictly positive linear combination $\sum_{\mathbf{x}\in\mathcal{D}_j}\alpha_{\mathbf{x}}x + \beta\pi(u_{r_j}) = \xi \in \mathbb{R}^{\cup}\langle\partial(\mathcal{D}_j)\rangle$, so $\beta > 0$ and $\alpha_{\mathbf{x}} > 0$ for $\mathbf{x} \in \mathcal{D}_j$. But then $x + \partial(\mathbf{x}) \subseteq \mathbf{x}$ for each $\mathbf{x} \in \mathcal{D}_j$ ensures that $\mathbb{R}^{\cup}\langle\partial(\mathcal{D}_j)\rangle = \xi + \sum_{\mathbf{x}\in\mathcal{D}_j}\partial(\mathbf{x}) \subseteq \mathsf{C}^{\cup}(\mathcal{D}_j) + \mathsf{C}(\pi(u_{r_j}))$, and now $D_j \cup \{\pi(u_{r_j})\}$ being a minimal positive basis modulo $\mathbb{R}^{\cup}\langle\partial(\mathcal{D}_j)\rangle$ ensures that $\mathsf{C}^{\cup}(C_j^{\pi}) + \mathsf{C}(\pi(u_{r_j})) = \mathsf{C}^{\cup}(\mathcal{D}_j) + \mathsf{C}(\pi(u_{r_j})) = \mathbb{R}^{\cup}\langle\mathcal{D}_j\rangle = \mathbb{R}^{\cup}\langle C_j^{\pi}\rangle$. This establishes $\mathbb{R}^{\cup}\langle C_j\rangle = \mathsf{C}^{\cup}(C_j) + \mathsf{C}(u_{r_1},\ldots,u_{r_j}) = \mathsf{C}^{\cup}(C_j) + \mathsf{C}(\overline{u}_1,\ldots,\overline{u}_j)$.

By 2(c) and the conclusion of the previous paragraph, $-u_i \in \mathbb{R}^{\cup}\langle C_j\rangle = \mathsf{C}^{\cup}(C_j) + \mathsf{C}(u_{r_1},\ldots,u_{r_j})$ for $i < r_{j+1}$, meaning $(u_i + \mathsf{C}(u_{r_1},\ldots,u_{r_j})) \cap -\mathsf{C}^{\cup}(C_j) \neq \emptyset$. In particular, this is true for $i \in [r_j+1, r_{j+1}-1]$, which implies C_j encases not just $-(u_1,\ldots,u_{r_j})$ but also $-(u_1,\ldots,u_{r_{j+1}-1})$. As a result, since we already know C_j minimally encases $-(u_1,\ldots,u_{r_j})$, we conclude that C_j minimally encases $-(u_1,\ldots,u_{r_{j+1}-1})$. Since C_{j-1} minimally encases $-(u_1,\ldots,u_{r_j-1})$ by induction hypothesis with $C_{j-1} \prec C_j$, we cannot have C_j *minimally* encasing $-(u_1,\ldots,u_{r_j-1})$. Thus $t'_j = r_j - 1 \in [0, r_{j+1} - 1]$ is the maximal index such that C_j does not minimally encase $-(u_1,\ldots,u_{t'_j})$. Consequently, since $C_{j-1} \subseteq \mathcal{X}_1 \cup \ldots \cup \mathcal{X}_s$, we see that C_j minimally encases $-(u_1,\ldots,u_{r_{j+1}-1})$ urbanely. Likewise, the minimal encasement of $-(u_{r_1},\ldots,u_{r_j})$ and $-(\overline{u}_1,\ldots,\overline{u}_j)$ must also be urbane, and the induction is complete. As already noted, this completes one of the implications in Item 2 and all parts of Item 3

To prove the other implication in Item 2, now assume that $\mathcal{B}$ minimally encases $-(u_1,\ldots,u_t)$ urbanely, which ensures that $\mathcal{B}$ is a virtual independent set. Let $t' \in [0, t-1]$ be the maximal index such that $\mathcal{B}$ does not minimally encase $-(u_1,\ldots,u_{t'})$, and let $\mathcal{A} \prec \mathcal{B}$ be a subset which minimally encases $-(u_1,\ldots,u_{t'})$. Since $\mathcal{B}$ minimally encases $-\vec{u}$ urbanely, we have $\mathcal{A} \subseteq \mathcal{X}_1 \cup \ldots \cup \mathcal{X}_s$. In view of the already established portion of Item 1, it follows that $\mathcal{A}$ is the unique subset of $\mathcal{X}_1 \cup \ldots \cup \mathcal{X}_s$ which minimally encases $-(u_1,\ldots,u_{t'})$.

We first construct the support sets C_i and indices r_i satisfying 2(b) and 2(c) for the set $\mathcal{A}$ recursively. We will then show $C_{\ell-1} = \mathcal{A}$ and $t' = r_\ell - 1$ with $C_\ell = \mathcal{B}$ also satisfying 2(a) and 2(c) afterwards to complete the proof of Item 2. Suppose the sets $C_i \subseteq \mathcal{X}_1 \cup \ldots \cup \mathcal{X}_s$ have already been constructed for $i = 0, 1\ldots, j-1$, where $j \geq 1$ (we set $r_0 = 0$ and $C_0 = \emptyset$). If $C_{j-1} = \mathcal{A}$, we are done with the initial construction, so assume otherwise. Let $r_j \in [r_{j-1}+1, t']$ be the minimal index such that $u_{r_j} \notin \mathbb{R}^{\cup}\langle C_{j-1}\rangle$, or set $r_j = t'+1$ if no such index exists. In view of the already completed implication in Item 2, we see that C_{j-1} minimally encases $-(u_1,\ldots,u_{r_j-1})$. If $\mathcal{A}$ also minimally encases $-(u_1,\ldots,u_{r_j-1})$, then both C_{j-1} and $\mathcal{A}$ minimally encase $-(u_1,\ldots,u_{r_j-1})$, in which case the already completed portion of Item 1 implies that $C_{j-1} = \mathcal{A}$, contrary to assumption. Therefore we may assume $\mathcal{A}$ does not minimally encases $-(u_1, \ldots, u_{r_j-1})$. In particular, $r_j \le t'$. Thus, since $C_{j-1} \subseteq X_1 \cup \ldots \cup X_s$ minimally encases $-(u_1, \ldots, u_{r_j-1})$ and $\mathcal{A} \subseteq X_1 \cup \ldots \cup X_s$ encases $-(u_1, \ldots, u_{r_j-1})$, it again follows from the already completed portion of Item 1 that $C_{j-1} \prec \mathcal{A}$. As before, let

$$\pi = \pi_{j-1}$$

to simplify notation. By Proposition 5.4, $\pi(\mathcal{R})$ is an oriented Reay system with $\mathcal{A}^\pi \subseteq \bigcup_{i=1}^s X_i^\pi$ a support set. Since $\mathcal{A}$ encases $-(u_1, \ldots, u_{t'})$ but $u_{r_j} \notin \mathbb{R}^\cup\langle C_{j-1}\rangle = \ker\pi$ and $u_1, \ldots, u_{r_j-1} \in \ker\pi$ (by induction hypothesis), it follows that $\mathcal{A}^\pi$ encases $-\pi(u_{r_j})$. Thus there is some $\mathcal{D}_j \preceq \mathcal{A}^\pi$ that minimally encases $-\pi(u_{r_j})$, and by Proposition 5.4.3 we have $\pi^{-1}(\mathcal{D}_j) \preceq \pi^{-1}(\mathcal{A}^\pi) \subseteq \mathcal{A} \subseteq X_1 \cup \ldots \cup X_s$. Since $\mathcal{D}_j \preceq \mathcal{A}^\pi \subseteq \bigcup_{j \in J} X_j^\pi$ minimally encases $-\pi(u_{r_j})$, it follows by Lemma 5.7 that $\mathcal{D}_j$ is a support set with $-\pi(u_{r_j}) \in \mathsf{C}^\cup(\mathcal{D}_j)^\circ$, whence $\mathcal{D}_j = \mathsf{Supp}_{\pi(\mathcal{R})}(-\pi(u_{r_j}))$. By Proposition 5.4.5,

$$C_j := (\pi^{-1}(\mathcal{D}_j) \cup C_{j-1})^* = \pi^{-1}(\mathcal{D}_j) \cup C'_{j-1}$$

is a support set for $\mathcal{R}$, where $C'_{j-1} = C_{j-1} \setminus \downarrow\partial(\pi^{-1}(\mathcal{D}_j))$. Since $C_{j-1} \subseteq \downarrow(\pi^{-1}(\mathcal{D}_j) \cup C_{j-1}) = \downarrow C_j$, we have $C_{j-1} = C^*_{j-1} \preceq C_j$. Proposition 5.4.4 implies $C_j^{\pi_{j-1}} = C_j^\pi = \mathcal{D}_j$, whence 2(b) holds for C_j. Also, since $\mathcal{D}_j$ is a nonempty set of nonzero elements while $C_{j-1}^\pi = \emptyset$, we must have $C_{j-1} \prec C_j$. Letting $r_{j+1} \in [r_j + 1, \ldots, t']$ be the minimal index such that $u_{r_{j+1}} \notin \mathbb{R}^\cup\langle C_j\rangle$, or setting $r_{j+1} = t' + 1$ if no such index exists, we see that 2(c) also holds. This defines the support sets C_i and indices r_i for the set $\mathcal{A}$. Since we cannot have an infinite ascending chain $\emptyset = C_0 \prec C_1 \prec C_2 \prec \ldots$ of subsets from the finite set $X_1 \cup \ldots \cup X_s$, the process must eventually terminate with some index $r_\ell = t' + 1$ with $C_{\ell-1} = \mathcal{A}$. The remainder of the proof of Item 2 is similar to what we have just seen, with some important but subtle differences. We now set $j = \ell$, so

$$\pi = \pi_{\ell-1} \quad \text{and} \quad \ker\pi = \mathbb{R}^\cup\langle C_{\ell-1}\rangle = \mathbb{R}^\cup\langle\mathcal{A}\rangle.$$

By definition of t', $\mathcal{B}$ minimally encases $-(u_1, \ldots, u_{t'+1}) = -(u_1, \ldots, u_{r_\ell})$. Since $\mathcal{A} \prec \mathcal{B}$, this ensures that $\mathcal{A}$ does not minimally encase $-(u_1, \ldots, u_{r_\ell})$, and thus $\pi(u_{r_\ell}) \neq 0$. In consequence, r_ℓ is the minimal index such that $u_{r_\ell} \notin \mathbb{R}^\cup\langle C_{\ell-1}\rangle = \mathbb{R}^\cup\langle\mathcal{A}\rangle$. In view of Proposition 5.4.1, $\pi(\mathcal{R}) = (X_j^\pi \cup \{\pi(\mathbf{v}_j)\})_{j \in J}$ is an oriented Reay system. Since $\mathcal{B}$ is a virtual independent set with $C_{\ell-1} = \mathcal{A} \prec \mathcal{B}$, and since $\mathcal{B} \subseteq X_1 \cup \ldots \cup X_s \cup \{\mathbf{v}_j : j \in J\}$ per definition of urbane encasement, Proposition 5.4.6 implies that $\mathcal{B}^\pi$ is a virtual independent set.

Since $\mathcal{B}$ encases $-(u_1, \ldots, u_t)$ but $u_{r_\ell} \notin \mathbb{R}^\cup\langle C_{\ell-1}\rangle = \ker\pi$ and $u_1, \ldots, u_{r_\ell-1} \in \ker\pi$ (since $\mathcal{A} = C_{\ell-1}$ encases $-(u_1, \ldots, u_{r_\ell-1})$), it follows that $\mathcal{B}^\pi$ encases

$-\pi(u_{r_\ell})$. Thus there is some $\mathcal{D}_\ell \preceq \mathcal{B}^\pi$ which minimally encases $-\pi(u_{r_\ell})$, and by Proposition 5.4.3, we have

$$\pi^{-1}(\mathcal{D}_\ell) \preceq \pi^{-1}(\mathcal{B}^\pi) \subseteq \mathcal{B}.$$

Since $\mathcal{B}^\pi$ is a virtual independent set, and since $(\mathcal{D}_\ell)^* = \mathcal{D}_\ell$ holds by virtue of $\mathcal{D}_\ell$ minimally encasing $-\pi(u_{r_\ell})$, it follows that $\mathcal{D}_\ell \preceq \mathcal{B}^\pi$ is also a virtual independent set, and one which minimally encases the element $-\pi(u_{r_\ell})$. By Proposition 5.4.5,

$$\mathcal{C}_\ell := (\pi^{-1}(\mathcal{D}_\ell) \cup \mathcal{C}_{\ell-1})^* = \pi^{-1}(\mathcal{D}_\ell) \cup \mathcal{C}'_{\ell-1}$$

is a virtual independent set for $\mathcal{R}$, where $\mathcal{C}'_{\ell-1} = \mathcal{C}_{\ell-1} \setminus \downarrow\partial(\pi^{-1}(\mathcal{D}_\ell))$. In view of $\mathcal{C}_{\ell-1} \subseteq \downarrow(\pi^{-1}(\mathcal{D}_\ell) \cup \mathcal{C}_{\ell-1}) = \downarrow\mathcal{C}_\ell$, we have $\mathcal{C}_{\ell-1} = \mathcal{C}^*_{\ell-1} \preceq \mathcal{C}_\ell$. Proposition 5.4.4 implies $\mathcal{C}^\pi_\ell = \mathcal{D}_\ell$, whence 2(a) holds for $\mathcal{C}_\ell = (\pi^{-1}(\mathcal{D}_\ell) \cup \mathcal{C}_{\ell-1})^*$. Note $\pi^{-1}(\mathcal{D}_\ell)$ contains no $\mathbf{v}_i$ with $\pi(\mathbf{v}_i) = \{0\}$ by definition, while this is also the case for $\mathcal{C}_{\ell-1} \subseteq \mathcal{X}_1 \cup \ldots \cup \mathcal{X}_s$. Since $\mathcal{D}_\ell$ is a nonempty set of nonzero elements while $\mathcal{C}^\pi_{\ell-1} = \emptyset$, we must have $\mathcal{C}_{\ell-1} \prec \mathcal{C}_\ell$. Letting $r_{\ell+1} \in [r_\ell + 1, \ldots, t]$ be the minimal index such that $u_{r_{\ell+1}} \notin \mathbb{R}^\cup\langle \mathcal{C}_\ell \rangle$, or setting $r_{\ell+1} = t + 1$ if no such index exists, we see that 2(c) also holds. It remains to show $\mathcal{C}_\ell = \mathcal{B}$. Since $\pi^{-1}(\mathcal{D}_\ell) \preceq \pi^{-1}(\mathcal{B}^\pi) \preceq \mathcal{B}$ and $\mathcal{C}_{\ell-1} = \mathcal{A} \prec \mathcal{B}$, we have $\mathcal{C}_\ell \subseteq \downarrow\mathcal{B}$, which combined with $\mathcal{C}^*_\ell = \mathcal{C}_\ell$ ensures that $\mathcal{C}_\ell \preceq \mathcal{B}$. By the already completed direction in Item 2, $\mathcal{C}_\ell$ minimally encases $-(u_1, \ldots, u_{t'+1})$. Thus, since $\mathcal{C}_\ell \preceq \mathcal{B}$, and since $\mathcal{B}$ *minimally* encases $-(u_1, \ldots, u_{t'+1})$, it follows that $\mathcal{C}_\ell = \mathcal{B}$, completing the reverse implication in Item 2.

It remains only to prove the moreover part of Item 2. If $\mathcal{B} = \mathcal{C}_\ell$ is a support set, then Proposition 5.4.6 implies that $\mathcal{C}_\ell^{\pi_{\ell-1}} = \mathcal{D}_\ell$ is a support set, and one which minimally encases $-\pi_{\ell-1}(u_{r_\ell})$ (in view of 2(a)), forcing $\mathcal{D}_\ell = \mathsf{Supp}_{\pi_{\ell-1}(\mathcal{R})}(-\pi_{\ell-1}(u_{r_\ell}))$ by definition of $\mathsf{Supp}_{\pi_{\ell-1}(\mathcal{R})}$. Conversely, if $\mathcal{D}_\ell = \mathsf{Supp}_{\pi_{\ell-1}(\mathcal{R})}(-\pi_{\ell-1}(u_{r_\ell}))$, then $\mathcal{D}_\ell$ is a support set which minimally encases $-\pi_{\ell-1}(u_{r_\ell})$, in which case Proposition 5.4.5 implies that $(\pi^{-1}_{\ell-1}(\mathcal{D}_\ell) \cup \mathcal{C}_{\ell-1})^*$ is also a support set. However, $\mathcal{B} = \mathcal{C}_\ell = (\pi^{-1}_{\ell-1}(\mathcal{D}_\ell) \cup \mathcal{C}_{\ell-1})^*$ was shown to be the lift of $\mathcal{D}_\ell$ during the proof (cf. Item 3(a)), so we conclude that $\mathcal{B}$ is support set, completing the proof of Item 2.

1. Next, we establish Item 1 in the unrestricted case when $\mathcal{B}, \mathcal{C} \subseteq \mathcal{X}_1 \cup \{\mathbf{v}_1\} \cup \ldots \cup \mathcal{X}_s \cup \{\mathbf{v}_s\}$ are both *support sets* that minimally encase $\vec{u}$ urbanely. If $t = 0$, then $\mathcal{B} = \mathcal{C} = \emptyset$, as desired, so we may assume $t \geq 1$, ensuring that $\mathcal{B}$ and $\mathcal{C}$ are both nonempty. Let $\emptyset = \mathcal{C}_0 \prec \mathcal{C}_1 \prec \ldots \prec \mathcal{C}_{\ell_B - 1} \prec \mathcal{C}_{\ell_B} = \mathcal{B}$ be the support sets given by application of Item 2 to $\mathcal{B}$. Observe that $\mathcal{C}_1 = \mathsf{Supp}_{\mathcal{R}}(-u_1)$ depends only on $\vec{u}$ and not on $\mathcal{B}$, and thus by an iterative argument (using 2(a)–2(c) and 3(a)), none of the sets $\mathcal{C}_i$, for $i \in [0, \ell_B]$, depend on $\mathcal{B}$ at all (note the case $i = \ell_B$ requires the moreover statement in Item 1, which is available as $\mathcal{B}$ and $\mathcal{C}$ are support sets by hypothesis), meaning the only portion of Item 2 that is dependent on $\mathcal{B}$, and not $\vec{u}$, is the number of iterations ℓ_B that occur for $\mathcal{B}$. Applying Item 2 to $\mathcal{C}$, we arrive at the same conclusion. Thus, letting $\emptyset = \mathcal{C}'_0 \prec \mathcal{C}'_1 \prec \ldots \prec \mathcal{C}'_{\ell_C - 1} \prec \mathcal{C}'_{\ell_C} = \mathcal{C}$ be resulting support sets, and w.l.o.g. assuming $\ell_C \leq \ell_B$, we find that $\mathcal{C}'_i = \mathcal{C}_i$ for all

$i \in [0, \ell_C]$. If $\ell_B = \ell_C$, then $C = C'_{\ell_C} = C_{\ell_B} = \mathcal{B}$, as desired. Otherwise, $\ell_B > \ell_C$, in which case $C = C'_{\ell_C} = C_{\ell_C} \prec C_{\ell_B} = \mathcal{B}$. However, since both C and $\mathcal{B}$ *minimally* encase $-\vec{u}$, this is not possible, completing the proof of Item 1.

4. Since any virtual independent subset $\mathcal{B} \subseteq \mathcal{X}_1 \cup \ldots \cup \mathcal{X}_s$ must be a support set, the hypotheses of Item 4 imply that

$$\mathcal{B} \subseteq \mathcal{X}_1 \cup \ldots \cup \mathcal{X}_s \text{ is a support set} \quad \text{or} \quad \mathcal{A} \subseteq \downarrow\mathcal{B}. \tag{5.38}$$

They also imply that

$$\text{each } C_j, \text{ for } j \in [1, \ell], \text{ contains no } \mathbf{v}_i \text{ with } \mathcal{X}_i^{\tau_{\ell-1}} = \emptyset. \tag{5.39}$$

Note (5.39) is trivially true when $C_j \subseteq \mathcal{X}_1 \cup \ldots \cup \mathcal{X}_s$, and thus for $j < \ell$, while it holds directly by hypothesis for $C_\ell = \mathcal{B}$ except when $\mathcal{A} \subseteq \downarrow\mathcal{B}$ with $\mathcal{B}$ a support set. However, in this last remaining case in question, $\mathcal{X}_j^{\tau_{\ell-1}} = \emptyset$ implies $\mathcal{X}_j \subseteq \downarrow\mathcal{A} \cup \downarrow C_{\ell-1} \subseteq \downarrow C_\ell = \downarrow\mathcal{B}$ in view of Proposition 5.3.9 and $\ker \tau_{\ell-1} = \mathbb{R}^{\cup}\langle \mathcal{A} \cup C_{\ell-1}\rangle$ (note $\downarrow C_{\ell-1} \subseteq \downarrow C_\ell$ follows from $C_{\ell-1} \prec C_\ell$), so that the definition of support set instead ensures $\mathbf{v}_j \notin \downarrow\mathcal{B} = \downarrow C_\ell$. Thus (5.39) is established in all cases, which together with (5.38) allows us to apply Proposition 5.4.6 to each C_j, for $j \in [1, \ell]$, to conclude C_j^π is a support set for $j < \ell$, and thus virtual independent, and that $C_\ell^\pi = \mathcal{B}^\pi$ is also virtual independent (note $\ker \pi \leq \ker \tau_{\ell-1}$). Proposition 5.4.3 implies that $C_0^\pi \preceq C_1^\pi \preceq \ldots \preceq C_\ell^\pi = \mathcal{B}^\pi$.

If $i, j \in [0, \ell]$ with $i < j$ and $C_i^\pi \prec C_j^\pi$, then there must be some $\mathbf{y} \in C_j$ with $\pi(\mathbf{y}) \neq \{0\}$ and $\mathbf{y} \notin \downarrow C_i$, for otherwise $C_j^\pi \subseteq (\downarrow C_i)^\pi = \downarrow C_i^\pi$ (by Proposition 5.4.2), yielding the contradiction $C_j^\pi = (C_j^\pi)^* \preceq C_i^\pi$ (as C_j^π is virtual independent). Suppose $\mathbf{y} \subseteq \mathcal{E} + \mathbb{R}^{\cup}\langle C_i\rangle = \mathbb{R}^{\cup}\langle \mathcal{A} \cup C_i\rangle$. If $\mathcal{A} \subseteq \downarrow\mathcal{B}$, then $y \in \mathbf{y} \subseteq \mathbb{R}^{\cup}\langle \mathcal{A} \cup C_i\rangle = \mathbb{R}\langle \downarrow A \cup \downarrow C_i\rangle$ meaning y can be written as a linear combination of the elements from $\downarrow A \cup \downarrow C_i$. However, it follows in view of $\mathcal{A} \subseteq \downarrow\mathcal{B}$, $\mathbf{y} \in C_j$ and $C_i \prec C_j \preceq C_\ell = \mathcal{B}$ that $\{y\} \cup \downarrow A \cup \downarrow C_i \subseteq \downarrow B$ is a linearly independent subset (as $\mathcal{B}$ is a virtual independent set), meaning the only way y can be written as a linear combination of the elements from $\downarrow A \cup \downarrow C_i$ is if $y \in \downarrow A \cup \downarrow C_i$, i.e., if $\mathbf{y} \in \downarrow\mathcal{A} \cup \downarrow C_i$. However, in view of $\pi(\mathbf{y}) \neq \{0\}$ and $\mathbf{y} \notin \downarrow C_i$, neither of these is possible. On the other hand, if $\mathcal{A} \nsubseteq \downarrow\mathcal{B}$, then (5.38) ensures $\mathcal{B} \subseteq \mathcal{X}_1 \cup \ldots \cup \mathcal{X}_s$ is a support set and $\mathbf{y} \in C_j \subseteq \downarrow\mathcal{B} \subseteq \mathcal{X}_1 \cup \ldots \cup \mathcal{X}_s$. In this case, $\mathbf{y} \subseteq \mathbb{R}^{\cup}\langle \mathcal{A} \cup C_i\rangle$ combined with Proposition 5.3.9 implies that $\mathbf{y} \in \downarrow\mathcal{A} \cup \downarrow C_i$, and we obtain the same contradiction as before. So we instead conclude that any $\mathbf{y} \in C_j$ with $\pi(\mathbf{y}) \neq \{0\}$ and $\mathbf{y} \notin \downarrow C_i$ must satisfy $\mathbf{y} \nsubseteq \mathcal{E} + \mathbb{R}^{\cup}\langle C_i\rangle$.

By hypothesis, $\mathcal{F}^\pi = (\mathbb{R}^{\cup}\langle C_i^\pi\rangle)_{i \in J}$ with $J \subseteq [1, \ell]$ the subset of indices $j \in [1, \ell]$ with $C_{j-1}^\pi \prec C_j^\pi$. If $j_1, j_2 \in \{0\} \cup J$ are consecutive elements with $j_1 < j_2$, then $C_{j_1}^\pi = C_i^\pi$ for all $i \in [j_1, j_2 - 1]$ and $C_{j_1}^\pi \prec C_{j_2}^\pi$. Thus $\mathcal{E} + \mathbb{R}^{\cup}\langle C_{j_1}\rangle = \mathcal{E} + \mathbb{R}^{\cup}\langle C_i\rangle$ for $i \in [j_1, j_2 - 1]$. Since $C_{j_1}^\pi \prec C_{j_2}^\pi$, there must be some $\mathbf{y} \in C_{j_2}$ with $\pi(\mathbf{y}) \neq \{0\}$ and $\mathbf{y} \notin \downarrow C_{j_1}$, implying $\mathbf{y} \nsubseteq \mathcal{E} + \mathbb{R}^{\cup}\langle C_{j_1}\rangle$ as shown above. Thus $\mathcal{E} + \mathbb{R}^{\cup}\langle C_{j_1}\rangle \subset \mathcal{E} + \mathbb{R}^{\cup}\langle C_{j_2}\rangle$, and in particular, for any $j \in J$, there must be some $\mathbf{y} \in C_j$ with $\mathbf{y} \nsubseteq \mathcal{E} + \mathbb{R}^{\cup}\langle C_{j-1}\rangle$.

Let $j \in J$ be arbitrary, let $\mathbf{y} \in C_j$ with $\mathbf{y} \nsubseteq \mathcal{E} + \mathbb{R}^{\cup}\langle C_{j-1}\rangle$ be arbitrary (at least one such $\mathbf{y}$ exists as just noted), and let

$$\tau : \mathbb{R}^d \to \mathbb{R}^{\cup}\langle \mathcal{A} \cup C_{j-1} \cup \partial(C_j)\rangle^{\perp}$$

be the orthogonal projection. By Proposition 5.4.1, we know that $\tau(\mathcal{R})$ is an oriented Reay system. Moreover, C_j is a virtual independent set and either $C_j \subseteq \mathcal{X}_1 \cup \ldots \cup \mathcal{X}_s$ or else $j = \ell$ with $C_\ell = \mathcal{B}$ and $\mathcal{A} \subseteq \downarrow\mathcal{B}$. In the former case, Proposition 5.4.6 ensures that C_j^{τ} is a support set, and thus $\tau(C_j) \setminus \{0\}$ is a linearly independent set of size $|C_j \setminus \ker \tau|$. In the latter case, $C_j = C_\ell$, and Proposition 5.3.1 implies $\ker \tau = \mathbb{R}^{\cup}\langle \mathcal{A} \cup C_{\ell-1} \cup \partial(\mathcal{B})\rangle = \mathbb{R}\langle \downarrow A \cup \downarrow C_{\ell-1} \cup \downarrow\partial(B)\rangle$ with $\downarrow A \cup \downarrow C_{\ell-1} \cup \downarrow\partial(B) \subseteq \downarrow B$. Thus, since $\downarrow B$ is linearly independent (as $\mathcal{B}$ is virtual independent), it follows that $\tau(\downarrow B) \setminus \{0\}$ is a linearly independent set of size $|\downarrow B \setminus \ker \tau|$, and since $B \subseteq \downarrow B$, we then conclude that $\tau(B) \setminus \{0\} = \tau(C_\ell) \setminus \{0\}$ is a linearly independent set of size $|C_\ell \setminus \ker \tau|$. In both cases, $\tau(C_j) \setminus \{0\}$ is a linearly independent set of size $|C_j \setminus \ker \tau|$.

Now the virtual independent set $C_j^{\pi_{j-1}} = \mathcal{D}_j$ minimally encases $-\pi_{j-1}(u_{r_j})$ by 2(a) or 2(b). Thus, since we also have $\ker \pi_{j-1} \subseteq \mathbb{R}^{\cup}\langle C_{j-1} \cup \partial(C_j)\rangle \subseteq \ker \tau$, it follows from Lemma 5.7 that there exists a strictly positive linear combination of the elements from $C_j \setminus \ker \tau$ equal to $-u_{r_j} + \xi$ for some $\xi \in \ker \tau$.

Let $C'_j = C_j \setminus \{\mathbf{y}\} \cup \partial(\{\mathbf{y}\})$. Since $\mathbf{y} \nsubseteq \mathcal{E} + \mathbb{R}^{\cup}\langle C_{j-1}\rangle = \mathbb{R}^{\cup}\langle \mathcal{A} \cup C_{j-1}\rangle$, we have $\mathbf{y} \notin \downarrow\mathcal{A} \cup \downarrow C_{j-1}$. In particular, $\mathbf{y} \notin C_{j-1}$, ensuring that $C_{j-1} \subseteq \downarrow C'_j$ (as $C_{j-1} \prec C_j$ implies $C_{j-1} \subseteq \downarrow C_j$). Since $\mathbf{y} \in C_j = C_j^*$ (as C_j is a virtual independent set), we also have $\mathbf{y} \notin \downarrow C'_j$. In summary,

$$\mathbf{y} \notin \downarrow\mathcal{A} \cup \downarrow C_{j-1} \cup \downarrow C'_j \quad \text{and} \quad C_{j-1} \subseteq \downarrow C'_j. \tag{5.40}$$

Suppose $\mathbf{y} \subseteq \mathcal{E} + \mathbb{R}^{\cup}\langle C'_j\rangle = \mathbb{R}^{\cup}\langle \mathcal{A} \cup C'_j\rangle$. Consequently, if $C_j \subseteq \mathcal{X}_1 \cup \ldots \cup \mathcal{X}_s$, then Proposition 5.3.9 enures that $\mathbf{y} \in \downarrow\mathcal{A} \cup \downarrow C'_j$, contradicting (5.40). On the other hand, if $C_j \nsubseteq \mathcal{X}_1 \cup \ldots \cup \mathcal{X}_s$, then $j = \ell$, $C_j = \mathcal{B}$ and $\mathcal{A} \subseteq \downarrow\mathcal{B}$. In this case, $y \in \mathbf{y} \subseteq \mathbb{R}^{\cup}\langle \mathcal{A} \cup C'_j\rangle = \mathbb{R}\langle \downarrow A \cup \downarrow C'_j\rangle$ with $\{y\} \cup \downarrow A \cup \downarrow C'_j \subseteq \downarrow B$. However, since $\mathcal{B} = C_\ell = C_j$ is a virtual independent set, it follows that $\downarrow B$, and thus also $\{y\} \cup \downarrow A \cup \downarrow C'_j$, is a linearly independent set, contradicting that $y \in \mathbb{R}\langle \downarrow A \cup \downarrow C'_j\rangle$ with $y \notin \downarrow A \cup \downarrow C'_j$ (by (5.40)). So we instead conclude that

$$\mathbf{y} \nsubseteq \mathcal{E} + \mathbb{R}^{\cup}\langle C'_j\rangle. \tag{5.41}$$

Since $\ker \tau = \mathcal{E} + \mathbb{R}^{\cup}\langle C_{j-1} \cup \partial(C_j)\rangle \subseteq \mathcal{E} + \mathbb{R}^{\cup}\langle C_{j-1} \cup C'_j\rangle = \mathcal{E} + \mathbb{R}^{\cup}\langle C'_j\rangle$, with the final equality in view of (5.40), and since $\tau(C_j) \setminus \{0\}$ is a linearly independent set of size $|C_j \setminus \ker \tau|$, we conclude from (5.41) and $\mathbf{y} \in C_j$ that

$$\mathbb{R}^{\cup}\langle C_j\rangle \nsubseteq \mathcal{E} + \mathbb{R}^{\cup}\langle C_{j-1} \cup \partial(C_j)\rangle \quad \text{and} \quad \tau(C'_j) \setminus \{0\} \subset \tau(C_j) \setminus \{0\}. \tag{5.42}$$

Suppose $u_{r_j} \in \mathcal{E}+\mathbb{R}^{\cup}\langle C_{j-1}\cup\partial(C_j)\rangle = \ker\tau$. As remarked above, $\tau(C_j)\setminus\{0\}$ is a linearly independent set of size $|C_j\setminus\ker\tau|$ and there exists a strictly positive linear combination of the elements from $C_j\setminus\ker\tau$ equal to $-u_{r_j}+\xi$ for some $\xi\in\ker\tau$. However, since $u_{r_j}\in\ker\tau$, this contradicts that $\tau(C_j)\setminus\{0\}$ is a linearly independent set of size $|C_j\setminus\ker\tau|$ unless $C_j\subseteq\ker\tau$, i.e., $\mathbb{R}^{\cup}\langle C_j\rangle\subseteq\mathcal{E}+\mathbb{R}^{\cup}\langle C_{j-1}\cup\partial(C_j)\rangle$, which is contrary to (5.42). So we instead conclude that

$$u_{r_j}\notin\mathcal{E}+\mathbb{R}^{\cup}\langle C_{j-1}\cup\partial(C_j)\rangle \quad \text{for every } j\in J. \tag{5.43}$$

Thus $\tau(u_{r_j})\neq 0$ and, since $\tau(C_j)\setminus\{0\}$ is linearly independent with a strictly positive linear combination of the elements of $\tau(C_j)\setminus\{0\}$ equal to $-\tau(u_{r_j})$, we see that $\tau(C_j)\setminus\{0\}\cup\{\tau(u_{r_j})\}$ is minimal positive basis of size $|C_j\setminus\ker\tau|+1$.

By definition, we have $\pi(\vec{u})=(u_1^*,u_2^*,\ldots,u_{\ell_\pi}^*)$ with the u_i^* for $i\in[1,\ell_\pi]$ defined as follows. We recursively define the indices $0=s_0<s_1<\ldots<s_{\ell_\pi}<s_{\ell_\pi+1}=r_{\ell+1}=t+1$ by letting $s_i\in[s_{i-1}+1,t]$ be the minimal index such that $\pi_{i-1}^*(u_{s_i})\neq 0$, where

$$\pi_{i-1}^*:\mathbb{R}^d\to(\mathcal{E}+\mathbb{R}\langle u_1,u_2,\ldots,u_{s_{i-1}}\rangle)^{\perp}$$

is the orthogonal projection, and then

$$u_i^*=\pi_{i-1}^*(u_{s_i})/\|\pi_{i-1}^*(u_{s_i})\|.$$

Note this ensures

$$\mathcal{E}+\mathbb{R}\langle u_1,u_2,\ldots,u_{s_{i-1}}\rangle=\mathcal{E}+\mathbb{R}\langle u_1,u_2,\ldots,u_{s_i-1}\rangle.$$

Recall that $j\in J\subseteq[1,\ell]$ is arbitrary. By 2(c), we have $u_i\in\mathbb{R}^{\cup}\langle C_{j-1}\rangle\subseteq\mathcal{E}+\mathbb{R}^{\cup}\langle C_{j-1}\rangle$ for all $i<r_j$. Hence, if $s_{i-1}<r_j<s_i$ for some $i\in[1,\ell_\pi+1]$, then $\ker\pi_{i-1}^*\subseteq\mathcal{E}+\mathbb{R}^{\cup}\langle C_{j-1}\rangle$ and $\pi_{i-1}^*(u_{r_j})=0$, the latter in view of the minimality in the definition of s_i, which contradicts (5.43). Therefore, we instead conclude that, for each $j\in J$,

$$r_j=s_{j^*}\quad\text{for some } j^*\in[1,\ell_\pi].$$

For $1\leq i<j^*$, we have $1\leq s_i<s_{j^*}=r_j$, ensuring $u_{s_i}\in\mathbb{R}^{\cup}\langle C_{j-1}\rangle\subseteq\mathcal{E}+\mathbb{R}^{\cup}\langle C_{j-1}\rangle$ and $\mathcal{E}\subseteq\ker\pi_{i-1}^*=\mathcal{E}+\mathbb{R}\langle u_1,\ldots,u_{s_{i-1}}\rangle\subseteq\mathcal{E}+\mathbb{R}^{\cup}\langle C_{j-1}\rangle$ by 2(c) for $\vec{u}$. Thus

$$u_i^*\in\mathbb{R}^{\cup}\langle C_{j-1}^{\pi}\rangle\quad\text{for all } i<j^*, \tag{5.44}$$

where $j\in J$. Likewise, $u_k\in\mathbb{R}^{\cup}\langle C_{j-1}\rangle$ for $k\leq s_{j^*-1}<s_{j^*}=r_j$ and $u_{s_{j^*}}=u_{r_j}\in\mathbb{R}^{\cup}\langle C_j\rangle\subseteq\mathcal{E}+\mathbb{R}^{\cup}\langle C_j\rangle$, while (5.43) implies that $u_{s_{j^*}}=u_{r_j}\notin\mathcal{E}+\mathbb{R}^{\cup}\langle C_{j-1}\rangle$.

Thus

$$\mathcal{E} \subseteq \ker \pi^*_{j^*-1} = \mathcal{E} + \mathbb{R}\langle u_1, \ldots, u_{s_{j^*-1}}\rangle \subseteq \ker \tau_{j-1} = \mathcal{E} + \mathbb{R}^{\cup}\langle C_{j-1}\rangle \subseteq \mathcal{E} + \mathbb{R}^{\cup}\langle C_j\rangle \tag{5.45}$$

with

$$u^*_{j^*} \in \mathbb{R}^{\cup}\langle C^\pi_j\rangle \setminus \mathbb{R}^{\cup}\langle C^\pi_{j-1}\rangle \tag{5.46}$$

and $\tau_{j-1}\pi^*_{j^*-1}(u_{s_{j^*}}) = \tau_{j-1}(u_{r_j})$, for $j \in J$. As a result, (5.44) and (5.46) ensure $\mathcal{F}^\pi$ is compatible with $\pi(\vec{u})$ with $\mathcal{F}^\pi(\pi(\vec{u})) = (\overline{u^*_j})_{j\in J}$, where each

$$\begin{aligned}\overline{u^*_j} &= \tau_{j-1}(u^*_{j^*})/\|\tau_{j-1}(u^*_{j^*})\| = \tau_{j-1}\pi^*_{j^*-1}(u_{s_{j^*}})/\|\tau_{j-1}\pi^*_{j^*-1}(u_{s_{j^*}})\| \\ &= \tau_{j-1}(u_{r_j})/\|\tau_{j-1}(u_{r_j})\|.\end{aligned}$$

Let $j \in J$ be arbitrary and let j_+ be the next consecutive element of J after j, or set $j_+ = \ell + 1$ and $j^*_+ = (\ell+1)^* := \ell_\pi + 1$ if j is the final element of J. Since $j,\ j_+ \in J \cup \{\ell+1\}$ are consecutive, the definition of J implies $\mathcal{E} + \mathbb{R}^{\cup}\langle C_i\rangle = \mathcal{E} + \mathbb{R}^{\cup}\langle C_{j^+-1}\rangle$ for all $i \in [j, j^+ - 1]$. In particular,

$$\mathcal{E} + \mathbb{R}^{\cup}\langle C_j\rangle = \mathcal{E} + \mathbb{R}^{\cup}\langle C_{j_+-1}\rangle. \tag{5.47}$$

Let $r_j = s_{j^*}$ and $r_{j_+} = s_{j^*_+}$ with $j^*,\ j^*_+ \in [1, \ell_\pi + 1]$ as shown above. If $J = \emptyset$, then $\pi(\vec{u})$ is the trivial tuple, $\emptyset = C^\pi_0 = C^\pi_\ell = \mathcal{B}^\pi$, and all parts of Item 4 hold trivially. Therefore we can assume J is nonempty. We now proceed to show that 2(a), 2(b) and 2(c) hold for $\pi(\vec{u}) = (u^*_1, \ldots, u^*_{\ell_\pi})$ using the virtual independent sets C^π_j for $j \in \{0\} \cup J$, indices j^* for $j \in J \cup \{\ell+1\}$, and elements

$$u^*_{j^*} = \pi^*_{j^*-1}(u_{s_{j^*}})/\|\pi^*_{j^*-1}(u_{s_{j^*}})\| = \pi^*_{j^*-1}(u_{r_j})/\|\pi^*_{j^*-1}(u_{r_j})\| \quad \text{for } j \in J$$

in place of the virtual independent sets C_j for $j \in [0, \ell]$, indices r_j for $j \in [1, \ell+1]$, and elements u_{r_j} for $j \in [1, \ell]$, which will imply that $\mathcal{B}^\pi$ minimally encases $-\pi(\vec{u})$ urbanely since $\mathcal{B}^\pi = C^\pi_\ell = C^\pi_j$ for the final element $j \in J$ (and thereby complete the proof). Note the C^π_j for $j \in J$ are virtual independent sets for $\pi(\mathcal{R})$ in view of Proposition 5.4.6 (as remarked earlier).

If $j \in J$ is the final element, then $u_{s_i} \in \mathcal{E} + \mathbb{R}^{\cup}\langle C_j\rangle = \mathcal{E} + \mathbb{R}^{\cup}\langle C_\ell\rangle$ and $\mathcal{E} \subseteq \ker \pi^*_{i-1} \subseteq \mathcal{E} + \mathbb{R}^{\cup}\langle C_j\rangle = \mathcal{E} + \mathbb{R}^{\cup}\langle C_\ell\rangle$ for all $i < \ell_\pi + 1 = j^*_+$ follows by 2(c) for $\vec{u}$, in turn implying that $u^*_i \in \mathbb{R}^{\cup}\langle C^\pi_j\rangle$ for all $i < j^*_+ = \ell_\pi + 1$. Otherwise, (5.44) (applied with $j = j_+$) and (5.47) yield $u^*_i \in \mathbb{R}^{\cup}\langle C^\pi_{j_+-1}\rangle = \mathbb{R}^{\cup}\langle C^\pi_j\rangle$ for all $i < j^*_+$. As a result, 2(c) holds for $\pi(\vec{u})$ for all $j \in J$.

For each $j \in J$, $u^*_{j^*}$ is a positive multiple of $\pi^*_{j^*-1}(u_{s_{j^*}}) = \pi^*_{j^*-1}(u_{r_j})$ with $\ker \pi^*_{j^*-1} \subseteq \ker \tau_{j-1}$ by (5.45), and thus $\tau_{j-1}(u^*_{j^*})$ is a positive multiple of $\tau_{j-1}\pi^*_{j*-1}(u_{r_j}) = \tau_{j-1}(u_{r_j})$. By (5.39), C_j contains no $\mathbf{v}_i$ with $X_i^{\tau_{j-1}} = \emptyset$. Thus Proposition 5.4.6 and (5.38) ensure that $C_j^{\tau_{j-1}}$ is a virtual independent set, which will be a support set when $j \in J$ is not the final element (as this ensures $C_j \subseteq X_1 \cup \ldots \cup X_s$). Consequently, to establish 2(a) and 2(b) for $\pi(\vec{u})$, we just need to show $C_j^{\tau_{j-1}}$ minimally encases $-\tau_{j-1}(u^*_{j*})$ for each $j \in J$, and since $\tau_{j-1}(u^*_{j*})$ is a positive multiple of $\tau_{j-1}(u_{r_j})$ for $j \in J$ (as just shown above), this is equivalent to showing $C_j^{\tau_{j-1}}$ minimally encases $-\tau_{j-1}(u_{r_j})$.

Since $-\pi_{j-1}(u_{r_j}) \in \mathsf{C}^{\cup}(C_j^{\pi_{j-1}})$ by 2(a) or 2(b) for $\mathcal{B}$, and since $\ker \pi_{j-1} \leq \ker \tau_{j-1}$, we have $-\tau_{j-1}(u_{r_j}) \in \mathsf{C}^{\cup}(C_j^{\tau_{j-1}})$, showing that $C_j^{\tau_{j-1}}$ encases $-\tau_{j-1}(u_{r_j})$. To show the encasement is minimal, we need to show $(C'_j)^{\tau_{j-1}}$ does not encase $-\tau_{j-1}(u_{r_j})$ for any $C'_j = C_j \setminus \{\mathbf{y}\} \cup \partial(\{\mathbf{y}\})$ with $\mathbf{y} \in C_j$ and $\mathbf{y} \nsubseteq \mathcal{E} + \mathbb{R}^{\cup}\langle C_{j-1}\rangle = \ker \tau_{j-1}$ (cf. Proposition 5.4.3: note $\mathcal{D} \prec C_j^{\tau_{j-1}}$ implies $\tau_{j-1}^{-1}(\mathcal{D}) \prec \tau_{j-1}^{-1}(C_j^{\tau_{j-1}}) \subseteq C_j$). Recall that $\tau : \mathbb{R}^d \to (\mathcal{E} + \mathbb{R}^{\cup}\langle C_{j-1} \cup \partial(C_j)\rangle)^{\perp}$ is the orthogonal projection. Suppose by contradiction that $-\tau_{j-1}(u_{r_j}) \in \mathsf{C}^{\cup}((C'_j)^{\tau_{j-1}})$ for some $C'_j = C_j \setminus \{\mathbf{y}\} \cup \partial(\{\mathbf{y}\})$ with $\mathbf{y} \in C_j$ and $\mathbf{y} \nsubseteq \mathcal{E} + \mathbb{R}^{\cup}\langle C_{j-1}\rangle$. Consequently, since $\ker \tau_{j-1} \subseteq \ker \tau$, (5.42) ensures $\tau(C'_j) \setminus \{0\}$ is a proper subset of $\tau(C_j) \setminus \{0\}$ that encases $-\tau(u_{r_j})$. However, as noted immediately after (5.43), $\tau(C_j) \setminus \{0\} \cup \{\tau(u_{r_j})\}$ is minimal positive basis of size $|C_j \setminus \ker \tau| + 1$, meaning it is not possible for a proper subset $\tau(C'_j) \setminus \{0\}$ of $\tau(C_j) \setminus \{0\}$ to encase $-\tau(u_{r_j})$, and with this contradiction, we establish that $\mathcal{B}^{\pi}$ minimally encases $-\pi(\vec{u})$ as described in Item 4, completing the proof.

Proposition 5.10 *Let $\mathcal{R} = (X_1 \cup \{v_1\}, \ldots, X_s \cup \{v_s\})$ be an oriented Reay system in $\mathbb{R}^d$, let $\vec{u} = (u_1, \ldots, u_t)$ be a tuple of $t \geq 0$ orthonormal vectors from $\mathbb{R}^d$, and let $\mathcal{A}, \mathcal{B} \subseteq X_1 \cup \ldots \cup X_s$. Suppose $\mathcal{B}$ minimally encases $-\vec{u}$ and $u_1, \ldots, u_t \in \mathbb{R}^{\cup}\langle\mathcal{A}\rangle$. Then $\downarrow\mathcal{B} \subseteq \downarrow\mathcal{A}$.*

Proof Let $\pi : \mathbb{R}^d \to \mathbb{R}^{\cup}\langle\mathcal{A}\rangle^{\perp}$ be the orthogonal projection. Since $\mathcal{B} \subseteq X_1 \cup \ldots \cup X_s$, it follows that $\mathcal{B}$ minimally encases $-\vec{u}$ urbanely. By Proposition 5.9.4, $\mathcal{B}^{\pi}$ minimally encases $-\pi(\vec{u})$. However, since $u_1, \ldots, u_t \in \mathbb{R}^{\cup}\langle\mathcal{A}\rangle = \ker \pi$, we have $\pi(\vec{u})$ equal to the empty tuple, and now $\mathcal{B}^{\pi}$ minimally encasing the empty tuple forces $\mathcal{B}^{\pi} = \emptyset$. Hence $\mathbf{x} \in \ker \pi = \mathbb{R}^{\cup}\langle\mathcal{A}\rangle$ for all $\mathbf{x} \in \mathcal{B} \subseteq X_1 \cup \ldots \cup X_s$. Consequently, since $\mathcal{A} \subseteq X_1 \cup \ldots \cup X_s$, Proposition 5.3.9 implies $\mathbf{x} \in \downarrow\mathcal{A}$ for all $\mathbf{x} \in \mathcal{B}$. Thus $\mathcal{B} \subseteq \downarrow\mathcal{A}$, implying $\downarrow\mathcal{B} \subseteq \downarrow\mathcal{A}$, as desired. □

The next proposition shows that, if a half-space $\mathbf{x}$ occurs in two separate oriented Reay systems $\mathcal{R}$ and $\mathcal{R}'$, then the quantity $\partial(\{\mathbf{x}\})$ is the same for both, and is thus intrinsically defined by the half-space $\mathbf{x}$ itself.

The proof of Proposition 5.11 requires the translation invariant notion for the lineality subspace of a convex cone. Recall that, if $C \subseteq \mathbb{R}^d$ is convex cone, then

$C \cap -C$ is the lineality subspace of C, which is the maximal subspace contained in C. For a general set $X \subseteq \mathbb{R}^d$, we define

$$o(X) = \{x \in X : x + \mathbb{R}_+(y - x) \subseteq X \text{ for every } y \in X\}$$

to be the set of apex points of X (see [119]). If $C \subseteq \mathbb{R}^d$ is a convex cone containing 0, then

$$o(C) = C \cap -C, \tag{5.48}$$

as the following argument shows. If $x \in C \cap -C$ and $y \in C$, then $-x \in C$ and $y \in C$ imply via the convexity of C that $y - x \in C$, whence $x + \mathbb{R}_+(y - x) \subseteq C$ since $x \in C$ with C a convex cone. This shows that $C \cap -C \subseteq o(C)$. On the other hand, if $x \in o(C)$, then since $0 \in C$ by hypothesis, it follows from the definition of $o(C)$ that $x + \mathbb{R}_+(0-x) \subseteq C$. In particular, $-x = x - 2x \in C$, whence $x \in C \cap -C$ (as $o(C) \subseteq C$ by definition). This establishes the reverse inclusion for (5.48) It is also routine to verify that

$$o(z + C) = z + o(C)$$

for any $z \in \mathbb{R}^d$ and convex cone $C \subseteq \mathbb{R}^d$. Combined with (5.48), it follows that

$$o(z + C) - o(z + C) = C \cap -C \tag{5.49}$$

for any convex cone $C \subseteq \mathbb{R}^d$ containing 0 (recall that $C \cap -C$ is a subspace).

Proposition 5.11 *Let* $\mathcal{R} = (\mathcal{X}_1 \cup \{v_1\}, \ldots, \mathcal{X}_s \cup \{v_s\})$ *and* $\mathcal{R}' = (\mathcal{X}'_1 \cup \{v'_1\}, \ldots, \mathcal{X}'_{s'} \cup \{v'_{s'}\})$ *be oriented Reay systems in* $\mathbb{R}^d$. *Suppose* $\mathbf{x} \in \mathcal{X}_1 \cup \ldots \cup \mathcal{X}_s \cup \{v_i : i \in [1, s]\}$ *and* $\mathbf{x}' \in \mathcal{X}'_1 \cup \ldots \cup \mathcal{X}'_s \cup \{v'_i : i \in [1, s']\}$ *with* $\mathbf{x} = \mathbf{x}'$. *Then* $\partial_{\mathcal{R}}(\{\mathbf{x}\}) = \partial_{\mathcal{R}'}(\{\mathbf{x}'\})$, *where* $\partial_{\mathcal{R}}(\{\mathbf{x}\}) = \partial(\{\mathbf{x}\}) \subseteq \mathcal{X}_1 \cup \ldots \cup \mathcal{X}_s$ *and* $\partial_{\mathcal{R}'}(\{\mathbf{x}\}) = \partial(\{\mathbf{x}'\}) \subseteq \mathcal{X}'_1 \cup \ldots \cup \mathcal{X}'_s$ *are the respective sets for* $\mathbf{x} = \mathbf{x}'$ *when considered as a half-space for the oriented Reay system* $\mathcal{R}$, *and when considered as a half-space for the oriented Reay system* $\mathcal{R}'$.

Proof Let $\mathbf{x}_1, \ldots, \mathbf{x}_r \in \partial_{\mathcal{R}}(\{\mathbf{x}\})$ and $\mathbf{x}'_1, \ldots, \mathbf{x}'_{r'} \in \partial_{\mathcal{R}'}(\{\mathbf{x}'\})$ be the distinct half-spaces from $\partial_{\mathcal{R}}(\{\mathbf{x}\})$ and $\partial_{\mathcal{R}'}(\{\mathbf{x}'\})$. Observe that $\overline{\mathsf{C}^{\cup}(\partial_{\mathcal{R}}(\{\mathbf{x}\}))} = \overline{\mathbf{x}_1 + \ldots + \mathbf{x}_r} = \overline{\mathbf{x}}_1 + \ldots + \overline{\mathbf{x}}_r$, and likewise $\overline{\mathsf{C}^{\cup}(\partial_{\mathcal{R}'}(\{\mathbf{x}'\}))} = \overline{\mathbf{x}'_1 + \ldots + \mathbf{x}'_r} = \overline{\mathbf{x}}'_1 + \ldots + \overline{\mathbf{x}}'_{r'}$ (recall Proposition 5.3.2). Since $\mathbf{x} = \mathbf{x}'$, we have

$$\mathsf{C}^{\cup}(\partial_{\mathcal{R}}(\{\mathbf{x}\})) = \mathbf{x} \cap \partial(\mathbf{x}) = \mathbf{x}' \cap \partial(\mathbf{x}') = \mathsf{C}^{\cup}(\partial_{\mathcal{R}'}(\{\mathbf{x}'\})).$$

Let $\mathcal{E} \subseteq \overline{\mathsf{C}^{\cup}(\partial_{\mathcal{R}}(\{\mathbf{x}\}))} = \overline{\mathsf{C}^{\cup}(\partial_{\mathcal{R}'}(\{\mathbf{x}'\}))}$ be the lineality subspace. Since $\partial_{\mathcal{R}}(\{\mathbf{x}\})$ and $\partial_{\mathcal{R}'}(\{\mathbf{x}'\})$ are both support sets, and thus virtual independent, Proposition 5.3.4

implies that

$$\mathcal{E} = \mathbb{R}^{\cup}\langle \partial^2_{\mathcal{R}}(\{\mathbf{x}\})\rangle = \partial(\mathbf{x}_1) + \ldots + \partial(\mathbf{x}_r) = \partial(\mathbf{x}'_1) + \ldots + \partial(\mathbf{x}'_{r'}) = \mathbb{R}^{\cup}\langle \partial^2_{\mathcal{R}}(\{\mathbf{x}'\})\rangle.$$

Moreover, $\pi(x_1), \ldots, \pi(x_r)$ are distinct, linearly independent elements by Proposition 5.3.3, where $\pi : \mathbb{R}^d \to \mathcal{E}^{\perp}$ is the orthogonal projection. Likewise, we see $\pi(x'_1), \ldots, \pi(x'_{r'})$ are distinct, linearly independent elements. We have

$$\begin{aligned}\mathsf{C}(\pi(x_1), \ldots, \pi(x_r)) &= \pi(\mathbf{x}_1) + \ldots + \pi(\mathbf{x}_r) = \pi\Big(\mathsf{C}^{\cup}(\partial_{\mathcal{R}}(\{\mathbf{x}\}))\Big)\\ &= \pi\Big(\mathsf{C}^{\cup}(\partial_{\mathcal{R}'}(\{\mathbf{x}'\}))\Big) = \pi(\mathbf{x}'_1) + \ldots + \pi(\mathbf{x}'_{r'}) = \mathsf{C}(\pi(x'_1), \ldots, \pi(x'_{r'})).\end{aligned}$$

Thus, since $\{\pi(x_1), \ldots, \pi(x_r)\}$ and $\{\pi(x'_1), \ldots, \pi(x'_{r'})\}$ are both linearly independent subsets of distinct elements, we conclude that $r = r'$ with (after re-indexing appropriately)

$$\pi(\mathbf{x}_i) = \pi(\mathbf{x}'_i) \quad \text{for all } i \in [1, r].$$

It remains to show $\mathbf{x}_i = \mathbf{x}'_i$ for every $i \in [1, r]$. By replacing each x'_i with an appropriate positive scalar multiple, we can w.l.o.g. also assume $\pi(x_i) = \pi(x'_i)$ for all $i \in [1, r]$.

For $j \in [1, r]$, let $Z_j \subseteq \mathsf{C}^{\cup}(\partial_{\mathcal{R}}(\{\mathbf{x}\})) = \mathsf{C}^{\cup}(\partial_{\mathcal{R}'}(\{\mathbf{x}'\}))$ consist of all $z \in \mathsf{C}^{\cup}(\partial_{\mathcal{R}}(\{\mathbf{x}\})) = \mathsf{C}^{\cup}(\partial_{\mathcal{R}'}(\{\mathbf{x}'\}))$ with $\pi(z) = \pi(x_j) = \pi(x'_j)$. Since the set $\{\pi(x_1), \ldots, \pi(x_r)\} = \{\pi(x'_1), \ldots, \pi(x'_r)\}$ is a linearly independent subset of distinct elements and $\mathbf{x}_1 + \ldots + \mathbf{x}_r = \mathsf{C}^{\cup}(\partial_{\mathcal{R}}(\{\mathbf{x}\})) = \mathsf{C}^{\cup}(\partial_{\mathcal{R}'}(\{\mathbf{x}'\})) = \mathbf{x}'_1 + \ldots + \mathbf{x}'_r$, it follows that

$$Z_j = x_j + \partial(\mathbf{x}_j) + \mathsf{C}^{\cup}(\partial^2_{\mathcal{R}}(\{\mathbf{x}\})) = x'_j + \partial(\mathbf{x}'_j) + \mathsf{C}^{\cup}(\partial^2_{\mathcal{R}'}(\{\mathbf{x}'\})). \tag{5.50}$$

Now

$$C_j := Z_j - x_j = \partial(\mathbf{x}_j) + \mathsf{C}^{\cup}(\partial^2_{\mathcal{R}}(\{\mathbf{x}\})) \quad\text{and}\quad C'_j := Z_j - x'_j = \partial(\mathbf{x}'_j) + \mathsf{C}^{\cup}(\partial^2_{\mathcal{R}'}(\{\mathbf{x}'\}))$$

are both convex cones containing 0 by Proposition 5.3.2. Consequently, in view of (5.49), we have

$$o(Z_j) - o(Z_j) = C_j \cap -C_j = C'_j \cap -C'_j.$$

Since $\partial(\mathbf{x}_j)$ is a subspace, $\partial(\mathbf{x}_j) \subseteq C_j \cap -C_j$. If the inclusion is strict, then $\pi_j\Big(\mathsf{C}^{\cup}(\partial^2_{\mathcal{R}}(\{\mathbf{x}\}))\Big) = \pi_j(C_j)$ would have nontrivial lineality subspace, where $\pi_j : \mathbb{R}^d \to \partial(\mathbf{x}_j)^{\perp} = \mathbb{R}^{\cup}\langle \partial(\{\mathbf{x}_j\})\rangle^{\perp}$ is the orthogonal projection. Proposition 5.4 implies that $\pi_j(\mathcal{R})$ is an oriented Reay system with $\pi_j\Big(\mathsf{C}^{\cup}(\partial^2_{\mathcal{R}}(\{\mathbf{x}\}))\Big) =$

$\mathsf{C}^{\cup}(\partial^2_{\mathcal{R}}(\{\pi_j(\mathbf{x})\}))$, and Proposition 5.3.4 ensures that $\mathsf{C}^{\cup}(\partial^2_{\mathcal{R}}(\{\pi_j(\mathbf{x})\}))$ has trivial lineality subspace. As a result, the inclusion $\partial(\mathbf{x}_j) \subseteq C_j \cap -C_j$ cannot be strict, forcing $\partial(\mathbf{x}_j) = C_j \cap -C_j$. An analogous argument shows $\partial(\mathbf{x}'_j) = C'_j \cap -C'_j$. Thus

$$\partial(\mathbf{x}'_j) = C'_j \cap -C'_j = C_j \cap -C_j = \partial(\mathbf{x}_j), \quad \text{for every } j \in [1, r]. \tag{5.51}$$

Let us next show that

$$(\mathbf{x}_1 + \ldots + \mathbf{x}_r) \cap \partial(\mathbf{x}_j) = \mathbf{x}_j \cap \partial(\mathbf{x}_j), \quad \text{for each } j \in [1, r]. \tag{5.52}$$

The inclusion $\mathbf{x}_j \cap \partial(\mathbf{x}_j) \subseteq (\mathbf{x}_1 + \ldots + \mathbf{x}_r) \cap \partial(\mathbf{x}_j)$ is trivial. Let $y \in (\mathbf{x}_1 + \ldots + \mathbf{x}_r) \cap \partial(\mathbf{x}_j)$ be arbitrary. Since $y \in \mathbf{x}_1 + \ldots + \mathbf{x}_r$, it follows that $\{\mathbf{x}_1, \ldots, \mathbf{x}_r\} = \partial_{\mathcal{R}}(\{\mathbf{x}\})$ encases y, so there is some $\mathcal{Z} \preceq \partial_{\mathcal{R}}(\{\mathbf{x}\})$ with $\mathcal{Z} \subseteq \downarrow\{\mathbf{x}_1, \ldots, \mathbf{x}_r\} = \downarrow\partial_{\mathcal{R}}(\{\mathbf{x}\})$ that minimally encases y. In view of the minimality of $\mathcal{Z}$, there is a subset $Z \subseteq \mathbb{R}^d$ of representatives for the half-spaces from $\mathcal{Z}$ such that $y = \sum_{z \in Z} z$. Since $\mathcal{Z} \subseteq \downarrow\partial_{\mathcal{R}}(\{\mathbf{x}\})$, we can extend the representative set Z to a set of representatives $\downarrow\partial_{\mathcal{R}}(\{x\})$ for $\downarrow\partial_{\mathcal{R}}(\{\mathbf{x}\}) \subseteq \mathcal{X}$, which will then be linearly independent in view of Proposition 4.11.1. Thus, since $\sum_{z \in Z} z = y \in \partial(\mathbf{x}_j) = \mathbb{R}^{\cup}\langle\partial_{\mathcal{R}}(\{\mathbf{x}_j\})\rangle = \mathbb{R}\langle\downarrow\partial_{\mathcal{R}}(\{x_j\})\rangle$ (by Proposition 5.3.1) with $\downarrow\partial_{\mathcal{R}}(\{x_j\}) \subseteq \downarrow\partial_{\mathcal{R}}(\{x\})$, $Z \subseteq \downarrow\partial_{\mathcal{R}}(\{x\})$ and $\downarrow\partial_{\mathcal{R}}(\{x\})$ linearly independent, it follows that $Z \subseteq \downarrow\partial_{\mathcal{R}}(\{x_j\})$, whence $\mathcal{Z} \subseteq \downarrow\partial_{\mathcal{R}}(\{\mathbf{x}_j\})$. But this means $y = \sum_{z \in Z} z \in \mathsf{C}^{\cup}(\mathcal{Z}) \subseteq \mathsf{C}^{\cup}(\downarrow\partial_{\mathcal{R}}(\{\mathbf{x}_j\})) = \mathsf{C}^{\cup}(\partial_{\mathcal{R}}(\{\mathbf{x}_j\})) = \mathbf{x}_j \cap \partial(\mathbf{x}_j)$. Since $y \in (\mathbf{x}_1 + \ldots + \mathbf{x}_r) \cap \partial(\mathbf{x}_j)$ was arbitrary, this shows the nontrivial inclusion $(\mathbf{x}_1 + \ldots + \mathbf{x}_r) \cap \partial(\mathbf{x}_j) \subseteq \mathbf{x}_j \cap \partial(\mathbf{x}_j)$, and (5.52) is established.

By an analogous argument, we conclude that

$$(\mathbf{x}'_1 + \ldots + \mathbf{x}'_r) \cap \partial(\mathbf{x}'_j) = \mathbf{x}'_j \cap \partial(\mathbf{x}'_j), \quad \text{for each } j \in [1, r].$$

We have $\mathbf{x}_1 + \ldots + \mathbf{x}_r = \mathsf{C}^{\cup}(\partial_{\mathcal{R}}(\{\mathbf{x}\})) = \mathsf{C}^{\cup}(\partial_{\mathcal{R}'}(\{\mathbf{x}'\})) = \mathbf{x}'_1 + \ldots + \mathbf{x}'_r$, while (5.51) gives $\partial(\mathbf{x}_j) = \partial(\mathbf{x}'_j)$. In consequence, $\mathbf{x}'_j \cap \partial(\mathbf{x}'_j) = (\mathbf{x}'_1 + \ldots + \mathbf{x}'_r) \cap \partial(\mathbf{x}'_j) = (\mathbf{x}_1 + \ldots + \mathbf{x}_r) \cap \partial(\mathbf{x}_j) = \mathbf{x}_j \cap \partial(\mathbf{x}_j)$, so

$$\mathbf{x}_j \cap \partial(\mathbf{x}_j) = \mathbf{x}'_j \cap \partial(\mathbf{x}'_j), \quad \text{for every } j \in [1, r]. \tag{5.53}$$

As a result,

$$\begin{aligned}
\mathsf{C}^{\cup}(\partial^2_{\mathcal{R}}(\{\mathbf{x}\})) &= \mathsf{C}^{\cup}(\partial_{\mathcal{R}}(\{\mathbf{x}_1\})) + \ldots + \mathsf{C}^{\cup}(\partial_{\mathcal{R}}(\{\mathbf{x}_r\})) \\
&= (\mathbf{x}_1 \cap \partial(\mathbf{x}_1)) + \ldots + (\mathbf{x}_r \cap \partial(\mathbf{x}_r)) = (\mathbf{x}'_1 \cap \partial(\mathbf{x}'_1)) + \ldots + (\mathbf{x}'_r \cap \partial(\mathbf{x}'_r)) \\
&= \mathsf{C}^{\cup}(\partial_{\mathcal{R}'}(\{\mathbf{x}'_1\})) + \ldots + \mathsf{C}^{\cup}(\partial_{\mathcal{R}'}(\{\mathbf{x}'_r\})) = \mathsf{C}^{\cup}(\partial^2_{\mathcal{R}'}(\{\mathbf{x}'\})).
\end{aligned}$$

Combining the above equality with (5.51) and (5.50), we conclude that

$$x_j - x'_j + \partial(\mathbf{x}_j) + \mathsf{C}^{\cup}(\partial^2_{\mathcal{R}}(\{\mathbf{x}\})) = \partial(\mathbf{x}_j) + \mathsf{C}^{\cup}(\partial^2_{\mathcal{R}}(\{\mathbf{x}\})),$$

which readily implies that $x_j - x'_j$ is contained in the lineality subspace of $\partial(\mathbf{x}_j) + \mathsf{C}^{\cup}(\partial^2_{\mathcal{R}}(\{\mathbf{x}\})) = C_j$, which is equal to $\partial(\mathbf{x}_j)$ by (5.51). This shows $x_j - x'_j \in \partial(\mathbf{x}_j)$, which combined with (5.51) ensures that $\mathbf{x}_j$ and $\mathbf{x}'_j$ linearly span the same subspace, and then $x_j - x'_j \in \partial(\mathbf{x}_j)$ further ensures that x_j and x'_j lie on the same side of the codimension one subspace $\partial(\mathbf{x}_j) = \partial(\mathbf{x}'_j)$. Thus $\overline{\mathbf{x}}_j = \overline{\mathbf{x}}'_j$, which combined with (5.53) yields the desired conclusion $\mathbf{x}_j = \mathbf{x}'_j$, for all $j \in [1, r]$. □

Chapter 6
Virtual Reay Systems

When trying to better understand the geometric properties of infinite subsets $G_0 \subseteq \mathbb{R}^d$, one of the crucial problems encountered is that small perturbations of linearly dependent sets can result in linearly independent sets. This allows limits of linearly independent subsets to degenerate into linearly dependent ones. We aim to better understand G_0 by looking at the limiting behavior of sequences of terms from G_0. However, for this to be effective, we need to avoid introducing linear dependencies into the limiting structures that do not exist in the original sequences. This will be accomplished by a careful development of the following generalization of a Reay system. It may be helpful to view a virtual Reay system as a convergent family of ordinary Reay systems whose limiting structure is an Oriented Reay system. This will be made more formal later.

Definition 6.1 Let $G_0 \subseteq \mathbb{R}^d$ be a subset. If $\mathcal{R} = (\mathcal{X}_1 \cup \{\mathbf{v}_1\}, \ldots, \mathcal{X}_s \cup \{\mathbf{v}_s\})$ is an oriented Reay system in $\mathbb{R}^d$ such that

(V1) every $\mathbf{x} \in \mathcal{X}_1 \cup \{\mathbf{v}_1\} \cup \ldots \cup \mathcal{X}_1 \cup \{\mathbf{v}_s\}$ has an asymptotically filtered sequence $\{\mathbf{x}(i)\}_{i=1}^{\infty}$ of terms $\mathbf{x}(i) \in G_0$ with limit $\vec{u}_{\mathbf{x}} = (u_1^{\mathbf{x}}, \ldots, u_{t_{\mathbf{x}}}^{\mathbf{x}})$ and $t_{\mathbf{x}} \geq 1$ such that the truncation $-\vec{u}_{\mathbf{x}}^{\triangleleft} = -(u_1^{\mathbf{x}}, \ldots, u_{t_{\mathbf{x}}-1}^{\mathbf{x}})$ is minimally encased by $\partial(\{\mathbf{x}\})$ and $\overline{\mathbf{x}} = \mathbb{R}^{\cup}\langle \partial(\{\mathbf{x}\}) \rangle + \mathbb{R}_+ u_{t_{\mathbf{x}}}^{\mathbf{x}}$,

then we say that $\mathcal{R}$ is a **virtual Reay system** in (or over) G_0 for the subspace $\mathbb{R}^{\cup}\langle \mathcal{X}_1 \cup \ldots \cup \mathcal{X}_s \rangle$.

Definition 6.2 If $\mathcal{R}$ is a virtual Reay system in $G_0 \subseteq \mathbb{R}^d$ and

(V2) $\vec{u}_{\mathbf{x}}$ is anchored for every $\mathbf{x} \in \mathcal{X}_1 \cup \ldots \cup \mathcal{X}_s$,

then we call $\mathcal{R}$ an **anchored** virtual Reay system. If $\mathcal{R}$ is a virtual Reay system in $G_0 \subseteq \mathbb{R}^d$ and

(V3) $\vec{u}_{\mathbf{v}_j}$ is fully unbounded for every $j \in [1, s]$,

then we call $\mathcal{R}$ a **purely** virtual Reay system.

D. J. Grynkiewicz, *The Characterization of Finite Elasticities*, Lecture Notes in Mathematics 2316, https://doi.org/10.1007/978-3-031-14869-9_6

Let $\mathbf{x}(i) = a_i^{(1)}u_1 + \ldots + a_i^{(t)}u_t + z_i$ be the representation of $\{\mathbf{x}(i)\}_{i=1}^{\infty}$ as an asymptotically filtered sequence with limit $(u_1, \ldots, u_t)$, let $\pi : \mathbb{R}^d \to \mathbb{R}^{\cup}\langle\partial(\{\mathbf{x}\})\rangle^{\perp}$ and $\pi^{\perp} : \mathbb{R}^d \to \mathbb{R}^{\cup}\langle\partial(\{\mathbf{x}\})\rangle$ be the orthogonal projections, and let $\overline{u}_t = \pi(u_t)/\|\pi(u_t)\|$. Then $z_i = w_i' + a_i'\overline{u}_t + y_i$ for some $w_i' \in \mathbb{R}^{\cup}\langle\partial(\{\mathbf{x}\})\rangle$, $a_i' \in \mathbb{R}$ and $y_i \in \mathbb{R}\langle\mathbf{x}\rangle^{\perp} = (\mathbb{R}^{\cup}\langle\partial(\{\mathbf{x}\})\rangle + \mathbb{R}\overline{u}_t)^{\perp}$, and $a_i^{(t)}u_t = a_i^{(t)}\pi^{\perp}(u_t) + a_i^{(t)}\|\pi(u_t)\|\overline{u}_t$. Consequently, setting $b_i^{(t)} = a_i^{(t)}\|\pi(u_t)\| + a_i'$ and $w_i = w_i' + a_i^{(t)}\pi^{\perp}(u_t)$, we find that

$$\mathbf{x}(i) = (a_i^{(1)}u_1 + \ldots + a_i^{(t-1)}u_{t-1} + w_i) + b_i^{(t)}\overline{u}_t + y_i \tag{6.1}$$

with $w_i \in \mathbb{R}^{\cup}\langle\partial(\{\mathbf{x}\})\rangle$, $\|w_i\| \in o(a_i^{(t-1)})$ (since $\|w_i'\| = \|\pi^{\perp}(z_i)\| \in O(\|z_i\|) \subseteq o(a_i^{(t)})$ and $a_i^{(t)} \in o(a_i^{(t-1)})$) and $w_i = 0$ when $t = 1$, with $b_i^{(t)} \in \Theta(a_i^{(t)})$ and $b_i^{(t)} > 0$ for all sufficiently large i (since $a_i' = \|a_i'\overline{u}_t\| \leq \|z_i\| \in o(a_i^{(t)})$ with $\pi(u_t) \neq 0$ by (V1)), and with $y_i \in \mathbb{R}\langle\mathbf{x}\rangle^{\perp}$ and $\|y_i\| \in o(a_i^{(t)}) = o(b_i^{(t)})$ (since $\|y_i\| \leq \|z_i\| \in o(a_i^{(t)})$). We will now generally assume, by discarding the first few terms in $\{\mathbf{x}(i)\}_{i=1}^{\infty}$, that any representative sequence $\{\mathbf{x}(i)\}_{i=1}^{\infty}$ must satisfy $b_i^{(t)} > 0$ for all i in any virtual Reay system.

As a matter of notation, we let

$$\tilde{\mathbf{x}}(i) = (a_i^{(1)}u_1 + \ldots + a_i^{(t-1)}u_{t-1} + w_i) + b_i^{(t)}\overline{u}_t \quad \text{and}$$
$$\tilde{\mathbf{x}}^{(t-1)}(i) = (a_i^{(1)}u_1 + \ldots + a_i^{(t-1)}u_{t-1} + w_i).$$

Note that $\{\tilde{\mathbf{x}}^{(t-1)}(i)\}_{i=1}^{\infty}$ is an asymptotically filtered sequence of terms $\tilde{\mathbf{x}}^{(t-1)}(i) \in \mathbb{R}^{\cup}\langle\partial(\{\mathbf{x}\})\rangle$ with limit $(u_1, \ldots, u_{t-1})$ and that $\tilde{\mathbf{x}}(i)$ is always a representative for the half-space $\mathbf{x}$ for any $i \geq 1$ (assuming we have discarded terms to attain $b_i^{(t)} > 0$ for all i). If $\mathcal{B}, \mathcal{C} \subseteq \mathcal{X}_1 \cup \{\mathbf{v}_1\} \cup \ldots \cup \mathcal{X}_s \cup \{\mathbf{v}_s\}$ with $\mathcal{B} \subseteq \mathcal{C}$ and $k = (i_{\mathbf{x}})_{\mathbf{x} \in \mathcal{C}}$ is a tuple of indices $i_{\mathbf{x}} \geq 1$, then we let

$$B(k) = \{\mathbf{x}(i_{\mathbf{x}}) : \mathbf{x} \in \mathcal{B}\} \quad \text{and} \quad \tilde{B}(k) = \{\tilde{\mathbf{x}}(i_{\mathbf{x}}) : \mathbf{x} \in \mathcal{B}\} \quad \text{for } k = (i_{\mathbf{x}})_{\mathbf{x} \in \mathcal{C}}.$$

In view of Proposition 5.9.1 and (V1), the limit $\vec{u}_{\mathbf{x}}$ uniquely determines the half-space $\mathbf{x} \in \mathcal{X}_j \cup \{\mathbf{v}_j\}$ (assuming all $\mathbf{y} \in \mathcal{X}_1 \cup \{\mathbf{v}_1\} \cup \ldots \cup \mathcal{X}_{j-1} \cup \{\mathbf{v}_{j-1}\}$ have already been determined by their respective limits $\vec{u}_{\mathbf{y}}$). It is important to realize that the half-spaces $\mathbf{x}$ defined in (V1) depend only on the tuples $\vec{u}_{\mathbf{x}}$ and not on the representative sequences $\mathbf{x}(i)$. It is even possible for distinct half-spaces $\mathbf{x}, \mathbf{y} \in \mathcal{X}_1 \cup \{\mathbf{v}_1\} \cup \ldots \cup \mathcal{X}_s \cup \{\mathbf{v}_s\}$ to have their representative sequences $\{\mathbf{x}(i)\}_{i=1}^{\infty}$ and $\{\mathbf{y}(i)\}_{i=1}^{\infty}$ be the same, since the same sequence may be considered as an asymptotically filtered sequence with limit $(u_1, \ldots, u_t)$ and also one with limit $(u_1, \ldots, u_{t'})$ with $t < t'$. There are then many compatible sequences $\{\mathbf{x}(i)\}_{i=1}^{\infty}$ that can be used for the representative sequence associated to $\mathbf{x}$. Indeed, any asymptotically filtered sequence of terms from G_0 having limit $\vec{u}_{\mathbf{x}} = (u_1, \ldots, u_t)$

with $\vec{u}_{\mathbf{x}}^{\triangleleft}$ minimally encased by $\partial(\{\mathbf{x}\})$, and u_t a representative for $\mathbf{x}$, will do once the first few terms with $b_i^{(t)} \leq 0$ are removed. Of course, it is important at least one such asymptotically filtered sequence exist. Besides the existence of an asymptotically filtered sequence $\{\mathbf{x}(i)\}_{i=1}^{\infty}$ of terms $\mathbf{x}(i) \in G_0$ with limit $\vec{u}_{\mathbf{x}}$, it is relevant to the properties (V2) and (V3) whether $\{\mathbf{x}(i)\}_{i=1}^{\infty}$ can be chosen to have $\vec{u}_{\mathbf{x}}$ as a fully unbounded or anchored limit, as a different choice can change whether $\vec{u}_{\mathbf{x}}$ is considered as fully unbounded or anchored in $\mathcal{R}$.

We continue with the analogue of Proposition 5.4 for virtual Reay systems, showing we once again have a well-behaved notion of a quotient virtual Reay system modulo $\mathbb{R}^{\cup}\langle\mathcal{A}\rangle$, for any $\mathcal{A} \subseteq \mathcal{X}_1 \cup \ldots \cup \mathcal{X}_s$.

Proposition 6.3 *Let $G_0 \subseteq \mathbb{R}^d$ be a subset. Suppose $\mathcal{R} = (\mathcal{X}_1 \cup \{v_1\}, \ldots, \mathcal{X}_s \cup \{v_s\})$ is a virtual Reay system in G_0. Let $\mathcal{A} \subseteq \mathcal{X}_1 \cup \ldots \cup \mathcal{X}_s$ and let $\pi : \mathbb{R}^d \to \mathbb{R}^{\cup}\langle\mathcal{A}\rangle^{\perp}$ be the orthogonal projection. Then $\pi(\mathcal{R}) = \big(\mathcal{X}_j^{\pi} \cup \{\pi(v_j)\}\big)_{j\in J}$ is a virtual Reay system in $\pi(G_0)$ with*

$$\vec{u}_{\pi(x)} = \pi(\vec{u}_{x}), \quad \pi(\vec{u}_{x})^{\triangleleft} = \pi(\vec{u}_{x}^{\triangleleft}), \quad \text{and} \quad \partial(\{\pi(x)\}) = \partial(\{x\})^{\pi},$$

for each $x \in \bigcup_{i\in J}(\mathcal{X}_i \cup \{v_i\})$ with $\pi(x) \neq \{0\}$, having representative sequence $\{\pi(x)(i)\}_{i=1}^{\infty}$ the sufficiently large index terms in $\{\pi(x(i))\}_{i=1}^{\infty}$. Moreover, if $\mathcal{R}$ is purely virtual, then so is $\pi(\mathcal{R})$, and if $\mathcal{R}$ is anchored, then so is $\pi(\mathcal{R})$.

Proof It follows from Proposition 5.4 that $\pi(\mathcal{R}) = \big(\mathcal{X}_j^{\pi} \cup \{\pi(\mathbf{v}_j)\}\big)_{j\in J}$ is an oriented Reay system with $\partial(\{\pi(\mathbf{x})\}) = \partial(\{\mathbf{x}\})^{\pi}$ a support set for all half-spaces $\mathbf{x} \in \bigcup_{j\in J}(\mathcal{X}_j \cup \{\mathbf{v}_j\})$ with $\pi(\mathbf{x}) \neq \{0\}$. Let $\mathbf{x} \in \bigcup_{j\in J}(\mathcal{X}_j \cup \{\mathbf{v}_j\})$ with $\pi(\mathbf{x}) \neq \{0\}$ be arbitrary, let $\vec{u}_{\mathbf{x}} = (u_1, \ldots, u_t)$ and let $\mathcal{E} = \mathbb{R}^{\cup}\langle\mathcal{A}\rangle$. By Proposition 5.9.4 and (V1), the support set $\partial(\{\mathbf{x}\})^{\pi}$ minimally encases $-\pi(\vec{u}_{\mathbf{x}}^{\triangleleft})$. By Proposition 3.5, the sufficiently large index terms in $\{\pi(\mathbf{x}(i))\}_{i=1}^{\infty}$ are an asymptotically filtered sequence with limit $\pi(\vec{u}_{\mathbf{x}})$. If $\pi(\vec{u}_{\mathbf{x}}^{\triangleleft}) = \pi(\vec{u}_{\mathbf{x}})$, then this would mean $u_t \in \mathcal{E} + \mathbb{R}\langle u_1, \ldots, u_{t-1}\rangle \subseteq \mathcal{E} + \mathbb{R}^{\cup}\langle\partial(\{\mathbf{x}\})\rangle$, with inclusion following since $\partial(\{\mathbf{x}\})$ minimally encases $-\vec{u}_{\mathbf{x}}^{\triangleleft}$ by (V1). However, since $\pi(\mathbf{x}) \neq \{0\}$ is a relative half-space by Proposition 5.4.1, we must have $\mathbf{x} \not\subseteq \mathcal{E}+\partial(\mathbf{x}) = \mathcal{E}+\mathbb{R}^{\cup}\langle\partial(\{\mathbf{x}\})\rangle$, contrary to what was just shown in view of $u_t \in \mathbf{x}^{\circ}$ (since (V1) holds for $\mathcal{R}$). Therefore we instead conclude that $\pi(\vec{u}_{\mathbf{x}}^{\triangleleft}) \neq \pi(\vec{u}_{\mathbf{x}})$, whence $\pi(\vec{u}_{\mathbf{x}}^{\triangleleft}) = \pi(\vec{u}_{\mathbf{x}})^{\triangleleft}$ with the last coordinate in $\pi(\vec{u}_{\mathbf{x}})$ equal to $\overline{u}_t := \tau(u_t)/\|\tau(u_t)\|$, where $\tau : \mathbb{R}^d \to (\mathcal{E} + \mathbb{R}\langle u_1, \ldots, u_{t-1}\rangle)^{\perp}$ is the orthogonal projection. Thus (V1) for $\mathcal{R}$ and (5.1) give $\pi(\overline{\mathbf{x}}) = \mathbb{R}^{\cup}\langle\partial(\{\mathbf{x}\})^{\pi}\rangle + \mathbb{R}_+\overline{u}_t$ and $\partial(\pi(\mathbf{x})) \cap \pi(\mathbf{x}) = \pi(\partial(\mathbf{x}) \cap \mathbf{x}) = \mathsf{C}^{\cup}(\partial(\{\mathbf{x}\})^{\pi})$, which establishes (V1) for $\pi(\mathcal{R})$. Since the last coordinate of $\pi(\vec{u}_{\mathbf{x}})$ equals $\overline{u}_t = \tau(u_t)/\|\tau(u_t)\|$ (ensuring $r_{\ell} = t$ in Proposition 3.5 when applied to $\pi(\vec{u}_{\mathbf{x}})$), Proposition 3.5.1 ensures that the limit $\pi(\vec{u}_{\mathbf{x}})$ is anchored if and only if $\vec{u}_{\mathbf{x}}$ is anchored, ensuring that (V2) holds for $\pi(\mathcal{R})$ when it does for $\mathcal{R}$, and also that (V3) holds for $\pi(\mathcal{R})$ when it does for $\mathcal{R}$, which completes the proof. □

Let $\mathcal{R} = (\mathcal{X}_1 \cup \{\mathbf{v}_1\}, \ldots, \mathcal{X}_s \cup \{\mathbf{v}_s\})$ be a virtual Reay system in $G_0 \subseteq \mathbb{R}^d$. Now suppose $\mathcal{R}' = (\mathcal{Y}_1 \cup \{\mathbf{w}_1\}, \ldots, \mathcal{Y}_r \cup \{\mathbf{w}_r\})$ is another virtual Reay system in G_0 with $\mathcal{Y}_1 \cup \ldots \cup \mathcal{Y}_{r-1} \subseteq \bigcup_{i=1}^{s} \mathcal{X}_i$ and $\mathcal{Y}_r \subseteq \bigcup_{i=1}^{s}(\mathcal{X}_i \cup \{\mathbf{v}_i\})$. Then each $\mathbf{x} \in \mathcal{Y}_1 \cup \ldots \cup \mathcal{Y}_r$ has a boundary neighborhood $\partial_{\mathcal{R}}(\{\mathbf{x}\})$ when considered as a half-space from $\mathcal{R}$ and a boundary neighborhood $\partial_{\mathcal{R}'}(\{\mathbf{x}\})$ when considered as a half-space from $\mathcal{R}'$. Proposition 5.11 ensures both these neighborhoods are equal: $\partial_{\mathcal{R}}(\{\mathbf{x}\}) = \partial_{\mathcal{R}'}(\{\mathbf{x}\})$. Thus the partial order in $\mathcal{R}'$ agrees with that in $\mathcal{R}$ and means there is no need to distinguish whether $\mathbf{x}$ lies in $\mathcal{R}$ or $\mathcal{R}'$ when dealing with quantities like $\partial(\{\mathbf{x}\})$ or $\downarrow\mathbf{x}$.

By Proposition 5.9, urbane minimal encasement of a tuple $\vec{u}$ by a *support* set from $\mathcal{R}$ corresponds to an ascending chain of support sets from $\mathcal{R}$. Our next major goal is to show this chain can be completed to an entire virtual Reay system, allowing us to use our machinery for virtual Reay systems when dealing with the support sets associated to urbane minimal encasement by a support set. We begin with the following lemma. Informally, we consider a fixed virtual Reay system $\mathcal{R}$ together with a "sub-" virtual Reay system $\mathcal{R}_A$ and quotient virtual Reay system $\mathcal{R}_C$, both of $\mathcal{R}$. Lemma 6.4 then shows that, under certain conditions, these two virtual Reay systems can be combined to create a new sub- virtual Reay System $\mathcal{R}'$ of $\mathcal{R}$.

Lemma 6.4 *Let $\mathcal{R} = (\mathcal{X}_1 \cup \{\mathbf{v}_1\}, \ldots, \mathcal{X}_s \cup \{\mathbf{v}_s\})$ and $\mathcal{R}_A = (\mathcal{A}_1 \cup \{\mathbf{a}_1\}, \ldots, \mathcal{A}_t \cup \{\mathbf{a}_t\})$ be virtual Reay systems in $G_0 \subseteq \mathbb{R}^d$ with $\mathcal{A} = \bigcup_{i=1}^{t} \mathcal{A}_i \subseteq \bigcup_{i=1}^{s} \mathcal{X}_i$, and let $\pi : \mathbb{R}^d \to \mathbb{R}^{\cup}\langle\mathcal{A}\rangle^{\perp}$ be the orthogonal projection with $\pi(\mathcal{R}) = (\mathcal{X}_i^{\pi} \cup \{\pi(\mathbf{v}_i)\})_{i\in J}$. Suppose $\mathcal{R}_C = (\pi(\mathcal{B}_1) \cup \{\mathbf{c}_1\}, \ldots, \pi(\mathcal{B}_r) \cup \{\mathbf{c}_r\})$ is a virtual Reay system in $\pi(G_0)$, for some $\mathcal{B}_j \subseteq \bigcup_{i=1}^{s} \mathcal{X}_i$ for $j \in [1, r-1]$ and $\mathcal{B}_r \subseteq \mathcal{X}_1 \cup \ldots \mathcal{X}_s \cup \{\mathbf{v}_i : i \in J\}$ with $\vec{u}_{\pi(\mathbf{x})} = \pi(\mathbf{x})$ for each $\mathbf{x} \in \mathcal{B}_1 \cup \ldots \cup \mathcal{B}_r$, and that each $\mathbf{c}_j$ for $j \in [1, r]$ is defined by a limit $\pi(\vec{u}_j)$ with $\vec{u}_j$ the limit of an asymptotically filtered sequence of terms from G_0, $\pi(\vec{u}_j^{\triangleleft}) = \pi(\vec{u}_j)^{\triangleleft}$ and $-\vec{u}_j^{\triangleleft}$ encased by $\mathcal{A} \cup C_j$ for some $C_j \subseteq \mathcal{B}_1 \cup \ldots \cup \mathcal{B}_{j-1}$ with $\pi(C_j) = \partial(\{\mathbf{c}_j\})$. Then*

$$\mathcal{R}' = (\mathcal{A}_1 \cup \{\mathbf{a}_1\}, \ldots, \mathcal{A}_s \cup \{\mathbf{a}_s\}, \mathcal{B}_1 \cup \{\mathbf{b}_1\}, \ldots, \mathcal{B}_r \cup \{\mathbf{b}_r\})$$

is virtual Reay system in G_0 with each $\mathbf{b}_j$ for $j \in [1, r]$ defined by the limit $\vec{u}_j$, $\pi(\mathbf{b}_j) = \mathbf{c}_j$, $\partial(\{\mathbf{b}_j\}) \preceq \mathcal{A} \cup C_j$ and $C_j \subseteq \partial(\{\mathbf{b}_j\})$. Moreover, if $\mathcal{R}_A$ and $\mathcal{R}_C$ are purely virtual, then so is $\mathcal{R}'$, and if $\mathcal{R}_A$ and $\mathcal{R}_C$ are anchored, then so is $\mathcal{R}'$.

Proof By Proposition 6.3, $\pi(\mathcal{R})$ is a virtual Reay system in $\pi(G_0)$. Let $j \in [1, r]$ be arbitrary. Since $\mathcal{R}_C$ is an oriented Reay system, we have $\pi(\mathbf{x}) \neq \{0\}$ for every $\mathbf{x} \in \mathcal{B}_j$, and thus also for every $\mathbf{x} \in C_j$. Since $\mathcal{R}_A$ is an oriented Reay system, we must have $\downarrow\mathcal{A} = \mathcal{A}$.

Since $-\vec{u}_j^{\triangleleft}$ is encased by $\mathcal{A} \cup C_j$, there exists $C'_j \preceq \mathcal{A} \cup C_j$ which minimally encases $-\vec{u}_j^{\triangleleft}$, and since $\mathcal{A} \cup C_j \subseteq \mathcal{X}_1 \cup \ldots \cup \mathcal{X}_s$, it must do so urbanely and be a support set. If $C_j \nsubseteq C'_j$, then $(C'_j)^{\pi} \prec C_j^{\pi} = \partial(\{\mathbf{c}_j\})$ follows in view of Proposition 5.4.3 since $\pi(\mathbf{x}) \neq \{0\}$ for all $\mathbf{x} \in C_j$. By Proposition 5.9.4, $(C'_j)^{\pi}$ minimally encases $-\pi(\vec{u}_j^{\triangleleft}) = -\pi(\vec{u}_j)^{\triangleleft}$, with the equality holding by hypothesis, but then, in view of $(C'_j)^{\pi} \prec \partial(\{\mathbf{c}_j\})$, we contradict that $\partial(\{\mathbf{c}_j\})$ minimally encases $-\pi(\vec{u}_j)^{\triangleleft}$ by (V1). Therefore we conclude that $C_j \subseteq C'_j$. Hence, since $(C'_j)^* = C'_j$

(as C'_j is a support set), let $\mathcal{A}'_j = C'_j \setminus C_j \subseteq \downarrow\mathcal{A} = \mathcal{A}$. Then $C'_j = \mathcal{A}'_j \cup C_j \subseteq \mathcal{A} \cup \mathcal{B}_1 \cup \ldots \cup \mathcal{B}_{j-1} \subseteq \mathcal{X}_1 \cup \ldots \cup \mathcal{X}_s$ and $(C'_j)^\pi = C_j^\pi$.

By hypothesis, $\mathcal{R}_A = (\mathcal{A}_1 \cup \{\mathbf{a}_1\}, \ldots, \mathcal{A}_s \cup \{\mathbf{a}_s\})$ is a virtual Reay system. If $\mathbf{x} \in \mathcal{B}_j$ for $j \in [1, r]$, then we know (V1) and (OR1) hold for $\mathbf{x}$ with the set $\partial(\{\mathbf{x}\})$ when we consider $\mathbf{x}$ as part of the virtual Reay system $\mathcal{R}$. Thus, to show this is also the case when we consider $\mathbf{x}$ as part of $\mathcal{R}'$, we just need to know $\partial(\{\mathbf{x}\}) \subseteq \mathcal{A} \cup \mathcal{B}_1 \cup \ldots \cup \mathcal{B}_{j-1}$. Since $\pi(\mathbf{x}) \in \pi(\mathcal{B}_j)$, we have $\pi(\mathbf{x}) \neq \{0\}$ and $\partial(\{\mathbf{x}\})^\pi = \partial(\{\pi(\mathbf{x})\}) \subseteq \pi(\mathcal{B}_1) \cup \ldots \cup \pi(\mathcal{B}_{j-1})$, where the first equality follows from proposition 5.4.2. Thus, in view of the injectivity of π (Proposition 5.4.1) and Proposition 5.3.9, we have $\partial(\{\mathbf{x}\}) \subseteq \downarrow\mathcal{A} \cup \mathcal{B}_1 \cup \ldots \cup \mathcal{B}_{j-1} = \mathcal{A} \cup \mathcal{B}_1 \cup \ldots \cup \mathcal{B}_{j-1}$, as desired. By Proposition 6.3 and hypothesis, we have $\pi(\vec{u}_{\mathbf{x}}) = \vec{u}_{\pi(\mathbf{x})}$ (both in $\pi(\mathcal{R})$ and $\mathcal{R}_C$). Thus, if $\mathcal{R}_C$ is anchored, then $\pi(\vec{u}_{\mathbf{x}}) = \vec{u}_{\pi(\mathbf{x})}$ is anchored, which is only possible if $\vec{u}_{\mathbf{x}}$ is anchored (cf. Proposition 3.5.1).

Let $\vec{u}_j = (u_{j,1}, \ldots, u_{j,t_j})$ for $j \in [1, r]$. The hypothesis $\pi(\vec{u}_j^{\triangleleft}) = \pi(\vec{u}_j)^{\triangleleft}$ simply means $u_{j,t_j} \notin \mathbb{R}^{\cup}\langle\mathcal{A}\rangle + \mathbb{R}\langle u_{j,1}, \ldots, u_{j,t_j-1}\rangle$. Thus, since $\mathcal{R}_C$ is a virtual Reay system with $(C'_j)^\pi = \partial(\{\mathbf{c}_j\})$ and $\mathbf{c}_j$ defined by the limit $\pi(\vec{u}_j)$, it follows that $u_{j,1}, \ldots, u_{j,t_j-1} \in \mathbb{R}^{\cup}\langle\mathcal{A}\rangle + \mathbb{R}^{\cup}\langle C'_j\rangle$ with $u_{j,t_j} \notin \mathbb{R}^{\cup}\langle\mathcal{A}\rangle + \mathbb{R}^{\cup}\langle C'_j\rangle$. Consequently, since $C'_j \subseteq \mathcal{A} \cup \mathcal{B}_1 \cup \ldots \cup \mathcal{B}_{j-1}$ minimally encases $-\vec{u}_j^{\triangleleft}$ with $\vec{u}_j$ the limit of an asymptotically filtered sequence of terms from G_0, it follows that we can define a half-space $\mathbf{b}_j$ by the limit $\vec{u}_j$ and it will satisfy the needed requirements in (V1) and (OR1) with $\partial(\{\mathbf{b}_j\}) = C'_j$. Since $\mathbf{c}_j$ is defined by the limit $\pi(\vec{u}_j)$ with $\pi(\vec{u}_j^{\triangleleft}) = -\pi(\vec{u}_j)^{\triangleleft}$, we have $\bar{\mathbf{c}}_j = \partial(\mathbf{c}_j) + \mathbb{R}_+\pi(u_{j,t_j}) = \mathbb{R}^{\cup}\langle\partial(\{\mathbf{c}_j\})\rangle + \mathbb{R}_+\pi(u_{j,t_j})$ and $\mathbf{c}_j \cap \partial(\mathbf{c}_j) = \mathsf{C}^{\cup}(\partial(\{\mathbf{c}_j\}))$. Thus, since

$$\partial(\{\mathbf{b}_j\})^\pi = (C'_j)^\pi = C_j^\pi = \pi(C_j) = \partial(\{\mathbf{c}_j\})$$

with $\mathbf{b}_j$ defined by the limit $\vec{u}_j$, we have $\pi(\mathbf{b}_j) \cap \partial(\pi(\mathbf{b}_j)) = \pi(\mathbf{b}_j \cap \partial(\mathbf{b}_j)) = \mathsf{C}^{\cup}(\partial(\{\mathbf{b}_j\})^\pi) = \mathsf{C}^{\cup}(\partial(\{\mathbf{c}_j\})) = \mathbf{c}_j \cap \partial(\mathbf{c}_j)$ and $\pi(\bar{\mathbf{b}}_j) = \pi(\partial(\mathbf{b}_j)) + \mathbb{R}_+\pi(u_{j,t_j}) = \mathbb{R}^{\cup}\langle\partial(\{\mathbf{b}_j\})^\pi\rangle + \mathbb{R}_+\pi(u_{j,t_j}) = \mathbb{R}^{\cup}\langle\partial(\{\mathbf{c}_j\})\rangle + \mathbb{R}_+\pi(u_{j,t_j}) = \bar{\mathbf{c}}_j$, whence $\pi(\mathbf{b}_j) = \mathbf{c}_j$. Moreover, if $\mathcal{R}_C$ is purely virtual, then $\pi(\vec{u}_j)$ will be fully unbounded, implying $\vec{u}_j$ is fully unbounded in view of $\pi(\vec{u}_j^{\triangleleft}) = -\pi(\vec{u}_j)^{\triangleleft}$ and Proposition 3.5.1. It remains to show (OR2) holds to complete the proof.

For $j \in [1, r]$, let $\pi_{j-1} : \mathbb{R}^d \to (\mathbb{R}^{\cup}\langle\mathcal{A}\rangle + \mathbb{R}^{\cup}\langle\mathcal{B}_1 \cup \ldots \cup \mathcal{B}_{j-1}\rangle)^\perp$ be the orthogonal projection. Then $\pi_{j-1}(\mathcal{B}_j \cup \{\mathbf{b}_j\}) = \pi_{j-1}\pi(\mathcal{B}_j \cup \{\mathbf{b}_j\}) = \pi_{j-1}(\pi(\mathcal{B}_j)) \cup \{\pi_{j-1}(\mathbf{c}_j)\}$. Hence it follows from the injectivity of π (Proposition 5.4.1) and (OR2) holding for $\mathcal{R}_C$ that (OR2) holds for $\mathcal{R}'$. □

The proof of Proposition 6.5 gives an algorithm by which $\mathcal{R}'$ can be constructed. Also worth noting, if $\vec{u}$ is fully unbounded and the strict truncation of any limit defining a half-space from $\downarrow\mathcal{B}$ in $\mathcal{R}$ is either trivial or fully unbounded, then $\mathcal{R}'$ will be purely virtual, and if every $\mathbf{x} \in \downarrow\mathcal{B}$ has $\vec{u}_{\mathbf{x}}$ anchored (e.g., if $\mathcal{B} \subseteq \mathcal{X}_1 \cup \ldots, \cup \mathcal{X}_s$ with $\mathcal{R}$ anchored), then $\mathcal{R}'$ will be anchored.

Proposition 6.5 *Let $G_0 \subseteq \mathbb{R}^d$ be a subset. Suppose $\mathcal{R} = (\mathcal{X}_1 \cup \{\mathbf{v}_1\}, \ldots, \mathcal{X}_s \cup \{\mathbf{v}_s\})$ is a virtual Reay system in G_0 and $\{x_i\}_{i=1}^{\infty}$ is an asymptotically filtered sequence of terms $x_i \in G_0$ with limit $\vec{u} = (u_1, \ldots, u_t)$ such that $-\vec{u}$ is minimally encased urbanely by a support set $\mathcal{B} \subseteq \mathcal{X}_1 \cup \{\mathbf{v}_1\} \cup \ldots \cup \mathcal{X}_s \cup \{\mathbf{v}_s\}$. Let*

$$\emptyset = \mathcal{C}_0 \prec \mathcal{C}_1 \prec \ldots \prec \mathcal{C}_{\ell-1} \subseteq \mathcal{X}_1 \cup \ldots \cup \mathcal{X}_s \quad \text{and} \quad \mathcal{C}_{\ell-1} \prec \mathcal{C}_\ell = \mathcal{B}$$

be the support sets and let $1 = r_1 < \ldots < r_\ell < r_{\ell+1} = t+1$ be the indices given by Proposition 5.9 for $\mathcal{B}$. Then there exists a virtual Reay system

$$\mathcal{R}' = (\mathcal{C}_1^{(1)} \cup \{\mathbf{v}_1^{(1)}\}, \ldots, \mathcal{C}_{s_1}^{(1)} \cup \{\mathbf{v}_{s_1}^{(1)}\}, \mathcal{C}_1^{(2)} \cup \{\mathbf{v}_1^{(2)}\}, \ldots, \mathcal{C}_{s_2}^{(2)} \cup \{\mathbf{v}_{s_2}^{(2)}\}, \\ \ldots, \mathcal{C}_1^{(\ell)} \cup \{\mathbf{v}_1^{(\ell)}\}, \ldots, \mathcal{C}_{s_\ell}^{(\ell)} \cup \{\mathbf{v}_{s_\ell}^{(\ell)}\})$$

in G_0 for the subspace $\mathbb{R}^{\cup}\langle\mathcal{B}\rangle$ such that the following hold for each $j \in [1, \ell]$.

(a) $\mathcal{C}_{s_j}^{(j)} = \mathcal{C}_j$ and $\bigcup_{k\in[1,s_j]} \mathcal{C}_k^{(j)} = \downarrow\mathcal{C}_j \setminus \downarrow\mathcal{C}_{j-1}$, whence

$$\downarrow\mathcal{B} = \bigcup_{\alpha\in[1,\ell]} \bigcup_{\beta\in[1,s_\alpha]} \mathcal{C}_\beta^{(\alpha)}.$$

In particular, $\mathcal{C}_\beta^\alpha \subseteq \mathcal{X}_1 \cup \ldots \cup \mathcal{X}_{s-1}$ for all β and α except possibly when $\alpha = \ell$ and $\beta = s_\ell$.

(b) $\mathbf{v}_{s_j}^{(j)}$ is the relative half-space defined by the limit $(u_1, \ldots, u_{r_j})$ taking $\mathbf{v}_{s_j}^{(j)}(i) = x_i$ (for all sufficiently large i), with $-(u_1, \ldots, u_{r_{j+1}-1})$ minimally encased by $\mathcal{C}_{s_j}^{(j)} = \mathcal{C}_j$ and $\partial(\{\mathbf{v}_{s_j}^{(j)}\}) = \mathcal{C}_{j-1}$.

(c) Each $\mathbf{v}_k^{(j)}$ with $k < s_j$ is defined by a strict truncation of a limit associated to some $\mathbf{y} \in \downarrow\mathcal{C}_j$ taking $\mathbf{v}_k^{(j)}(i) = \mathbf{y}(i)$ (for all sufficiently large i).

Proof Since $\mathcal{B}$ is a support set, implying $\mathcal{B}^* = \mathcal{B}$, we have $\downarrow\mathcal{B}\setminus\mathcal{B} = \downarrow\partial(\mathcal{B}) \subseteq \mathcal{X}_1 \cup \ldots \cup \mathcal{X}_{s-1}$. Thus, the in particular statement in (a) follows in view of $\mathcal{C}_{s_\ell}^{\ell} = \mathcal{C}_\ell = \mathcal{B}$ and $\downarrow\mathcal{B} = \bigcup_{\alpha\in[1,\ell]} \bigcup_{\beta\in[1,s_\alpha]} \mathcal{C}_\beta^{(\alpha)}$.

We define a directed graph with vertices $\mathcal{X}_1 \cup \{\mathbf{v}_1\} \cup \ldots \cup \mathcal{X}_s \cup \{\mathbf{v}_s\}$ as follows. Each vertex $\mathbf{y}$ is defined by an associated limit $\vec{u}_{\mathbf{y}} = (u_1^{\mathbf{y}}, \ldots, u_{t_{\mathbf{y}}}^{\mathbf{y}})$ such that $\partial(\{\mathbf{y}\})$ minimally encases $-\vec{u}_{\mathbf{y}}^{\triangleleft}$, and thus via Proposition 5.9 there is a uniquely defined associated sequence of support sets $\emptyset = \mathcal{C}_0^{\mathbf{y}} \prec \mathcal{C}_1^{\mathbf{y}} \prec \ldots \prec \mathcal{C}_{\ell_{\mathbf{y}}}^{\mathbf{y}} = \partial(\{\mathbf{y}\}) \subset \downarrow\{\mathbf{y}\}$ and indices $1 = r_1^{\mathbf{y}} < \ldots < r_{\ell_{\mathbf{y}}}^{\mathbf{y}} < r_{\ell_{\mathbf{y}}+1}^{\mathbf{y}} = t_{\mathbf{y}}+1$. We define a directed edge between $\mathbf{y}$ and each half-space from $\bigcup_{j=1}^{\ell_{\mathbf{y}}} \mathcal{C}_j^{\mathbf{y}}$. Moreover, each $\mathbf{z} \in \bigcup_{j=1}^{\ell_{\mathbf{y}}} \mathcal{C}_j^{\mathbf{y}}$ lies in some $\mathcal{C}_j^{\mathbf{y}}$, and associating to $\mathbf{z}$ the minimal index j for which this is true gives a way to group the neighbors of $\mathbf{y}$ in way that places all vertices from $\mathcal{C}_1^{\mathbf{y}}$ before those from $\mathcal{C}_2^{\mathbf{y}} \setminus \mathcal{C}_1^{\mathbf{y}}$, which then come before those from $\mathcal{C}_3^{\mathbf{y}} \setminus (\mathcal{C}_2^{\mathbf{y}} \cup \mathcal{C}_1^{\mathbf{y}})$, and so forth. Since $\mathcal{B}$ is a support set by hypothesis, each $\mathcal{C}_j$ for $j \in [1, \ell]$ is also a support set. If we start with the vertices from $\mathcal{B}$ and include all their neighbors, followed by all the neighbors of their neighbors, and continue in this fashion, we obtain an induced subgraph on $\downarrow\mathcal{B}$. We can add a vertex $\mathbf{y}_0$ to this subgraph and connect it to all vertices from $\mathcal{B}$ to obtain a graph rooted at $\mathbf{y}_0$. Set $\vec{u}_{\mathbf{y}_0} = (u_1, \ldots, u_t)$, $\mathbf{y}_0(i) = x_i$ for $i \geq 1$, and $C_j^{\mathbf{y}_0} = C_j$ for $j \in [1, \ell]$, which minimally encases $-(u_1, \ldots, u_{r_{j+1}-1})$ by Proposition 5.9.2. We now describe how the sets $C_k^{(j)} \cup \{\mathbf{v}_k^{(j)}\}$ can be constructed via a depth-first search argument rooted at the vertex $\mathbf{y}_0$ which respects the ordering of neighbors described above.

Starting at $\mathbf{y}_0$, consider a directed path, say with vertex sequence $\mathbf{y}_0, \mathbf{y}_1, \ldots, \mathbf{y}_{\alpha+1}$, such that the vertex $\mathbf{y}_j$ is always chosen from among the minimal available neighbors of $\mathbf{y}_{j-1}$, that is, $\mathbf{y}_j \in C_1^{\mathbf{y}_{j-1}}$. Moreover, when possible, choose $\mathbf{y}_j \in C_1^{\mathbf{y}_{j-1}}$ with $\partial(\{\mathbf{y}_j\}) \neq \emptyset$. Choosing vertices this way ensures that $\mathbf{y}_{\alpha+1} \prec \mathbf{y}_\alpha \prec \mathbf{y}_{\alpha-1} \prec \ldots \prec \mathbf{y}_1$. Suppose some $\mathbf{y}_j$ were a neighbor of some $\mathbf{y}_i$ with $i < j-1$. Then $\mathbf{y}_j \prec \mathbf{y}_{i+1} \in C_1^{\mathbf{y}_i}$ and $\mathbf{y}_j \in C_k^{\mathbf{y}_i}$ for some $k \in [1, \ell_{\mathbf{y}_i}]$. Since $C_1^{\mathbf{y}_i} \preceq C_k^{\mathbf{y}_i}$ implies $C_1^{\mathbf{y}_i} \subseteq \downarrow C_k^{\mathbf{y}_i}$, we see this would mean there is some $\mathbf{z} \in C_k^{\mathbf{y}_i}$ with $\mathbf{y}_j \prec \mathbf{y}_{i+1} \preceq \mathbf{z}$. Hence $(C_k^{\mathbf{y}_i})^* \neq C_k^{\mathbf{y}_i}$. However, this contradicts that $C_k^{\mathbf{y}_i}$ is a support set. So we instead conclude that each $\mathbf{y}_j$ is not in the neighborhood of any $\mathbf{y}_i$ with $i < j-1$. If we consider such a directed path with non-extendable length, then (in view of our preference for choosing $\mathbf{y}_j$ with $\partial(\{\mathbf{y}_j\}) \neq \emptyset$) we must have $\partial(\{\mathbf{z}\}) = \emptyset$ for every $\mathbf{z} \in C_1^{\mathbf{y}_\alpha}$. By definition, $C_1^{\mathbf{y}_\alpha}$ minimally encases $-u_1^{\mathbf{y}_\alpha}$, with $u_1^{\mathbf{y}_\alpha}$ a truncation of the limit $(u_1^{\mathbf{y}_\alpha}, \ldots, u_{t_{\mathbf{y}_\alpha}}^{\mathbf{y}_\alpha})$ which defines $\mathbf{y}_\alpha$. For $\alpha > 0$, we have $\mathbf{y}_{\alpha+1} \in C_1^{\mathbf{y}_\alpha} \preceq \partial(\{\mathbf{y}_\alpha\})$, thus ensuring $t_{\mathbf{y}_\alpha} > 1$ (if $t_{\mathbf{y}_\alpha} = 1$, then (V1) implies that $\partial(\{\mathbf{y}_\alpha\}) = \emptyset$, contradicting that $\mathbf{y}_{\alpha+1} \in C_1^{\mathbf{y}_\alpha} \preceq \partial(\{\mathbf{y}_\alpha\})$). Hence $u_1^{\mathbf{y}_\alpha}$ is a *strict* truncation for $\alpha > 0$. Define the half-space $\mathbf{v}_1^{(1)} = \mathbb{R}_+ u_1^{\mathbf{y}_\alpha}$ via this truncated limit $u_1^{\mathbf{y}_\alpha}$ and set $\mathbf{v}_1^{(1)}(i) = \mathbf{y}_\alpha(i)$ for all $i \geq 1$. Note this ensures that $(C_1^{(1)} \cup \{\mathbf{v}_1^{(1)}\})$ is a virtual Reay system in G_0, where $C_1^{(1)} = C_1^{\mathbf{y}_\alpha}$. If $\alpha = 0$, then $\mathbf{y}_\alpha = \mathbf{y}_0$, and we have $C_1^{(1)} = C_1$ and $u_1^{\mathbf{y}_\alpha} = u_1 = u_{r_1}$. Thus, for $\ell = 1$ and $\alpha = 0$ (meaning $\partial(\{\mathbf{z}\}) = \emptyset$ for all $\mathbf{z} \in \mathcal{B} = C_\ell = C_1 = C_1^{\mathbf{y}_\alpha}$), the proof is complete. So we can proceed by induction on $|\downarrow\mathcal{B}|$ with the base case when $|\downarrow\mathcal{B}| = 1$ complete.

Note that the only way $C_1^{(1)} = C_1^{\mathbf{y}_\alpha} \subseteq \mathcal{X}_1 \cup \ldots \cup \mathcal{X}_s$ can fail is when $\alpha = 0$ and $\ell = 1$, which was the case covered in the base of the induction. Therefore we may assume $C_1^{(1)} \subseteq \mathcal{X}_1 \cup \ldots \cup \mathcal{X}_s$. Let $\mathcal{E} = \mathbb{R}^\cup \langle C_1^{(1)} \rangle$ and let $\pi : \mathbb{R}^d \to \mathcal{E}^\perp$ be the orthogonal projection. In view of Proposition 6.3, $\pi(\mathcal{R}) = \big(\mathcal{X}_i^\pi \cup \{\pi(\mathbf{v}_i)\}\big)_{i \in J}$ is a virtual Reay system in $\pi(G_0)$. By proposition 5.4.6, $\downarrow\mathcal{B} \subseteq \mathcal{X}_1 \cup \ldots \cup \mathcal{X}_s \cup \{\mathbf{v}_j : j \in J\}$ with $\mathcal{B}^\pi$ a support set.

Let $\mathcal{A} \subseteq \downarrow\mathcal{B}$ be arbitrary. Since $\mathcal{B} \subseteq \mathcal{X}_1 \cup \ldots \cup \mathcal{X}_s \cup \{\mathbf{v}_j : j \in J\}$ and $\partial(\{\mathbf{z}\}) = \emptyset$ for every $\mathbf{z} \in C_1^{\mathbf{y}_\alpha} = C_1^{(1)}$, Propositions 5.3.9 and 5.4.1 ensure that $\mathcal{A}^\pi = \pi(\mathcal{A} \setminus C_1^{(1)})$, while Proposition 5.4.2 implies that $\downarrow(\mathcal{A}^\pi) = (\downarrow\mathcal{A})^\pi$. Since $\mathcal{B}$ is a support set, it follows that $\mathcal{A} \subseteq \downarrow\mathcal{B}$ is a support set if and only if $\mathcal{A}^* = \mathcal{A}$.

When this is the case, Proposition 5.4.2 implies that $(\mathcal{A}^\pi)^* = \mathcal{A}^\pi$, and thus $\mathcal{A}^\pi = \pi(\mathcal{A}\setminus C_1^{(1)})$ is also a support set since $\mathcal{B}^\pi$ is a support set with $\mathcal{A}^\pi \subseteq (\downarrow\mathcal{B})^\pi = \downarrow\mathcal{B}^\pi$.

By Proposition 3.5.1, the sufficiently large index terms in $\{\pi(x_i)\}_{i=1}^\infty$ form an asymptotically filtered sequence with limit $\pi(\vec{u})$, and by discarding the first few terms, we can w.l.o.g. assume $\{\pi(x_i)\}_{i=1}^\infty$ is an asymptotically filtered sequence with limit $\pi(\vec{u})$.

Case 1: $\alpha = 0$. In this case, $\ell > 1$ and $t > 1$ (else we fall in the already completed base of the induction), $C_1^{(1)} = C_1$ and $u_1, \ldots, u_{r_2-1} \in \mathcal{E} = \mathbb{R}^\cup\langle C_1\rangle$ (as $-(u_1, \ldots, u_{r_2-1})$ is minimally encased by C_1 by Proposition 5.9). If $C_j \subseteq \mathcal{X}_1 \cup \ldots \cup \mathcal{X}_s$ with $j \geq 2$, then Proposition 5.3.9 together with $C_{j-1} \prec C_j$ ensures that there is some $\mathbf{y} \in C_j$ with $\mathbf{y} \nsubseteq \mathbb{R}^\cup\langle C_{j-1}\rangle$ (else $C_j \subseteq \downarrow C_{j-1}$, yielding the contradiction $C_j = C_j^* \preceq C_{j-1}$), and thus with $\pi(\mathbf{y}) \neq \{0\}$ as well, so that $C_{j-1}^\pi \prec C_j^\pi$ (by Proposition 5.4.3). On the other hand, if $C_j \nsubseteq \mathcal{X}_1 \cup \ldots \cup \mathcal{X}_s$, then $j = \ell$ with $C_\ell = \mathcal{B}$ and $C_{\ell-1} \subseteq \mathcal{X}_1 \cup \ldots \cup \mathcal{X}_s$ (as $\mathcal{B}$ minimally encases $-\vec{u}$ urbanely). In this case, C_ℓ contains some $\mathbf{v}_i$ with $\mathbf{v}_i \notin C_{\ell-1} \subseteq \mathcal{X}_1 \cup \ldots \cup \mathcal{X}_s$. Since $\mathbf{v}_i \in \downarrow\mathcal{B} \subseteq \mathcal{X}_1 \cup \ldots \cup \mathcal{X}_s \cup \{\mathbf{v}_j : j \in J\}$, Proposition 5.4.1 implies $\pi(\mathbf{v}_i) \neq 0$, and thus $C_{\ell-1}^\pi \prec C_\ell^\pi$ holds in this case as well (in view of Proposition 5.4.3). Therefore, we find that $\emptyset = C_1^\pi \prec C_2^\pi \prec \ldots \prec C_\ell^\pi$. In such case, Proposition 5.9.4 implies that $\mathcal{B}^\pi$ minimally encases $-\pi(\vec{u})$ with $\mathcal{F}^\pi(\pi(\vec{u})) = (\overline{u}_{r_2}, \ldots, \overline{u}_{r_\ell})$, where $\mathcal{F}(\vec{u}) = (\overline{u}_{r_1}, \ldots, \overline{u}_{r_\ell})$ and $\mathcal{F} = (\mathbb{R}^\cup(C_1), \ldots, \mathbb{R}^\cup(C_\ell))$. Moreover, $C_2^\pi \prec \ldots \prec C_\ell^\pi$ are the support sets given by Proposition 5.9.1 for $\pi(\vec{u})$, and $r_2 < \ldots < r_\ell < r_{\ell+1} = t+1$ are the indices given by Proposition 5.9.1, ensuring that the encasement is urbane as it was for $\mathcal{B} = C_\ell$. By induction hypothesis (and Proposition 5.4), there exists a virtual Reay system (found by depth first search)

$$\mathcal{R}'' = (C_{1,2}^\pi \cup \{\mathbf{w}_1^{(2)}\}, \ldots, C_{s_2,2}^\pi \cup \{\mathbf{w}_{s_2}^{(2)}\}, \ldots, C_{1,\ell}^\pi \cup \{\mathbf{w}_1^{(\ell)}\}, \ldots, C_{s_\ell,\ell}^\pi \cup \{\mathbf{w}_{s_\ell}^{(\ell)}\})$$

in $\pi(G_0)$ for the subspace $\mathbb{R}^\cup\langle\mathcal{B}^\pi\rangle$ satisfying (a)–(c), with each $C_{i,j}$ disjoint from $C_1^{(1)}$ (so $C_{i,j}^\pi = \pi(C_{i,j})$), with each $\mathbf{w}_k^{(j)}$ with $k < s_j$ defined by a strict truncation of the limit associated to some $\pi(\mathbf{y})$ with $\mathbf{y} \in \downarrow C_j \setminus C_1^{(1)}$, and with each $\mathbf{w}_{s_j}^{(j)}$ defined by the limit $\pi((u_1, \ldots, u_{r_j}))$ with $\mathcal{F}^\pi(\pi((u_1, \ldots, u_{r_j}))) = (\overline{u}_{r_2}, \ldots, \overline{u}_{r_j})$, and with $C_{i,j} \subseteq \mathcal{X}_1 \cup \ldots \cup \mathcal{X}_s$ for all i and j except possibly when $i = s_\ell$ and $j = \ell$. We proceed to show we can apply Lemma 6.4.

For $\mathbf{w}_k^{(j)}$ with $k < s_j$, we have the half-space $\mathbf{w}_k^{(j)}$ defined by a strict truncated limit associated to some $\pi(\mathbf{y})$ with $\mathbf{y} \in \downarrow C_j \setminus C_1^{(1)}$. If $\vec{u}_\mathbf{y}$ is the limit which defines $\mathbf{y}$, then Proposition 3.5.1 ensures that $\vec{u}_\mathbf{y}$ has a strict truncation $\vec{v}_\mathbf{y}$ with $\pi(\vec{v}_\mathbf{y})$ the limit defining $\mathbf{w}_k^{(j)}$, and by choosing such a truncation of minimal length, we can assume $\pi(\vec{v}_\mathbf{y})^\triangleleft = \pi(\vec{v}_\mathbf{y}^\triangleleft)$. Since $\partial(\{\mathbf{y}\})$ encases $-\vec{u}_\mathbf{y}^\triangleleft$, it also encases the truncation $-\vec{v}_\mathbf{y}^\triangleleft$, so there must be some $\mathcal{B}_{k,j} \preceq \partial(\{\mathbf{y}\}) \subseteq \downarrow C_j$ which minimally encases $-\vec{v}_\mathbf{y}^\triangleleft$, and since $\mathcal{B}_{k,j} \preceq \partial(\{\mathbf{y}\}) \subseteq \mathcal{X}_1 \cup \ldots \cup \mathcal{X}_{s-1}$, it does so urbanely. Thus Proposition 5.9.4 ensures that $\mathcal{B}_{k,j}^\pi$ minimally encases $-\pi(\vec{v}_\mathbf{y}^\triangleleft) = -\pi(\vec{v}_\mathbf{y})^\triangleleft$. Since $\mathcal{B}_{k,j} \subseteq \downarrow C_j$, it follows from (a) holding for $\mathcal{R}''$ and Proposition 5.4.2 that $\mathcal{B}_{k,j}^\pi \subseteq C_{1,2}^\pi \cup \ldots \cup C_{s_j,j}^\pi$, in which case $\mathcal{B}^{\pi}_{k,j}$ minimally encases $-\pi(\vec{v}^{\triangleleft}_{\mathbf{y}}) = -\pi(\vec{v}_{\mathbf{y}})^{\triangleleft}$ urbanely in $\mathcal{R}''$. As a result, since the support set $\partial(\{\mathbf{w}^{(j)}_k\})$ in $\mathcal{R}''$ also minimally encases $-\pi(\vec{v}_{\mathbf{y}})^{\triangleleft}$ (urbanely) by (V1), it follows from Proposition 5.9.1 that $\mathcal{B}^{\pi}_{k,j} = \partial(\{\mathbf{w}^{(j)}_k\})$, and now we must have $\mathcal{B}^{\pi}_{k,j} = \partial(\{\mathbf{w}^{(j)}_k\}) \subseteq C^{\pi}_{1,2} \cup \ldots \cup C^{\pi}_{k,j}$ (with $\mathcal{B}^{\pi}_{k,j}$ disjoint from $C^{\pi}_{k,j}$), and thus $\mathcal{B}_{k,j} \subseteq C^{(1)}_1 \cup C_{1,2} \cup \ldots \cup C_{k,j}$ (with $\mathcal{B}_{k,j}$ disjoint from $C_{k,j}$).

Next consider a half-space $\mathbf{w}^{(j)}_{s_j}$ with $j \in [2, \ell]$, which is defined by the limit $\pi((u_1, \ldots, u_{r_j}))$ with $\pi((u_1, \ldots, u_{r_j}))^{\triangleleft} = \pi((u_1, \ldots, u_{r_j-1})) = \pi((u_1, \ldots, u_{r_j})^{\triangleleft})$ (since r_j occurs as the index of the last coordinate of $\mathcal{F}^{\pi}(\pi((u_1, \ldots, u_{r_j}))) = (\overline{u}_{r_2}, \ldots, \overline{u}_{r_j}))$. Since $\mathcal{B} = C_\ell$ minimally encases $-\vec{u}$, it also encases the truncation $-(u_1, \ldots, u_{r_j-1})$, so there must be some $\mathcal{B}_{s_j,j} \preceq \mathcal{B} \subseteq \downarrow C_\ell$ which minimally encases $-(u_1, \ldots, u_{r_j-1})$. Indeed, $\mathcal{B}_{s_j,j} = C_{j-1} \subseteq \mathcal{X}_1 \cup \ldots \cup \mathcal{X}_s$ by Proposition 5.9.2, ensuring the encasement is urbane. Thus Proposition 5.9.4 ensures that $\mathcal{B}^{\pi}_{s_j,j}$ minimally encases $-\pi((u_1, \ldots, u_{r_j-1}))$. Since $\mathcal{B}_{s_j,j} \subseteq \downarrow C_\ell$, it follows from (a) holding for $\mathcal{R}''$ and Proposition 5.4.2 that $\mathcal{B}^{\pi}_{s_j,j} \subseteq C^{\pi}_{1,2} \cup \ldots \cup C^{\pi}_{s_\ell,\ell}$, in which case $\mathcal{B}^{\pi}_{s_j,j}$ minimally encases $-\pi((u_1, \ldots, u_{r_j-1}))$ urbanely. As a result, since the support set $\partial(\{\mathbf{w}^{(j)}_{s_j}\})$ also minimally encases $-\pi(u_1, \ldots, u_{r_j-1})$ (urbanely) by (V1), it follows from Proposition 5.9.1 that $\mathcal{B}^{\pi}_{s_j,j} = \partial(\{\mathbf{w}^{(j)}_{s_j}\})$, and now we must have $\mathcal{B}^{\pi}_{s_j,j} = \partial(\{\mathbf{w}^{(j)}_{s_j}\}) \subseteq C^{\pi}_{1,2} \cup \ldots \cup C^{\pi}_{s_j,j}$ (with $\mathcal{B}^{\pi}_{s_j,j}$ disjoint from $C^{\pi}_{s_j,j}$), and thus $\mathcal{B}_{s_j,j} \subseteq C^{(1)}_1 \cup C_{1,2} \cup \ldots \cup C_{s_j,j}$ (with $\mathcal{B}_{s_j,j}$ disjoint from $C_{s_j,j}$). Additionally, Proposition 6.3 ensures that $\vec{u}_{\pi(\mathbf{x})} = \pi(\vec{u}_{\mathbf{x}})$ for $\mathbf{x} \in C_{1,2} \cup \ldots \cup C_{s_\ell,\ell}$.

The work of the above two paragraphs allows us to invoke Lemma 6.4 with $\mathcal{R}_{\mathcal{A}} = (C^{(1)}_1 \cup \{\mathbf{v}^{(1)}_1\})$ and $\mathcal{R}_C = \mathcal{R}''$, where $\mathbf{v}^{(1)}_1 = \mathbb{R}_+ u_1$: here the $\mathcal{B}_j$ for $j \in [1, r]$ are taken to be the $C_{k,j}$ with $j \in [2, \ell]$ and $k \in [1, s_j]$, the $\mathbf{c}_j$ and C_j for $j \in [1, r]$ are taken to be the $\mathbf{w}^{(j)}_k$ and $\mathcal{B}_{k,j}$ with $j \in [2, \ell]$ and $k \in [1, s_j]$, and the $\vec{u}_j$ for $j \in [1, r]$ are taken to be the $\vec{v}_{\mathbf{y}}$ for $\mathbf{y} \in C_{k,j}$ when $k < s_j$ as well as the $(u_1, \ldots, u_{r_j})$ when $k = s_j$, for $j \in [2, \ell]$ (note $\mathcal{B}_{k,j}$ minimally encasing $-\vec{v}^{\triangleleft}_{\mathbf{y}}$ ensures that $C^{(1)}_1 \cup \mathcal{B}_{k,j}$ encases $-\vec{v}^{\triangleleft}_{\mathbf{y}}$). As a result, we find that

$$\mathcal{R}' = (C^{(1)}_1 \cup \{\mathbf{v}^{(1)}_1\}, C_{1,2} \cup \{\mathbf{v}^{(2)}_1\}, \ldots, C_{s_2,2} \cup \{\mathbf{v}^{(2)}_{s_2}\}, \ldots, C_{1,\ell} \cup \{\mathbf{v}^{(\ell)}_1\}, \ldots, C_{s_\ell,\ell} \cup \{\mathbf{v}^{(\ell)}_{s_\ell}\})$$

is a virtual Reay system in G_0, where each $\pi(\mathbf{v}^{(j)}_k) = \mathbf{w}^{(j)}_k$ with $\mathbf{v}^{(j)}_k$ for $k < s_j$ defined by the limit $\vec{v}_{\mathbf{y}}$ (described above) and each $\mathbf{v}^{(j)}_{s_j}$ defined by the limit $(u_1, \ldots, u_{r_j})$. Since (a) held for $\mathcal{R}''$, it follows by construction that (a) holds for $\mathcal{R}'$. Since both $\partial(\{\mathbf{v}^{(j)}_{s_j}\})$ and C_{j-1} minimally encase $-(u_1, \ldots, u_{r_j-1})$ urbanely (the first from (V1) for $\mathcal{R}'$ and the second from Proposition 5.9.2), Proposition 5.9.1 ensures that $\partial(\{\mathbf{v}^{(j)}_{s_j}\}) = C_{j-1}$, and the remaining parts of (b)–(c) follow directly by construction, completing the case.

Case 2: $\alpha > 0$. In this case, $\mathbf{y}_\alpha \in \downarrow\mathcal{B}$. To lighten the notation, we use the following abbreviations for $\beta \in [0,\alpha]$: $C_j^{\mathbf{y}_\beta} = C_j^\beta$, $\vec{u}\mathbf{y}_\beta = \vec{u}_\beta$, $u_j^{\mathbf{y}_\beta} = u_j^\beta$, $\ell\mathbf{y}_\beta = \ell_\beta$, $t\mathbf{y}_\beta = t_\beta$, $r_j^{\mathbf{y}_\beta} = r_j^\beta$ and $u_{r_j^{\mathbf{y}_\beta}}^{\mathbf{y}_\beta} = u_{r_j}^\beta$, etc., for $j \in [1,\ell_\beta]$. Since C_{j-1}^β minimally encases the tuple $-(u_1^\beta,\ldots,u_{r_j-1}^\beta)$ (by Proposition 5.9.2), we have $u_1^\beta,\ldots,u_{r_j-1}^\beta \in \mathbb{R}^\cup\langle C_{j-1}^\beta\rangle$ for any $\beta \in [0,\alpha]$. We have $C_1^\beta \preceq C_{\ell_\beta}^\beta = \partial(\{\mathbf{y}_\beta\}) \prec \{\mathbf{y}_\beta\} \subseteq C_1^{\beta-1}$ for $\beta \in [1,\alpha]$. Thus $C_1^{(1)} = C_1^\alpha \prec C_1^{\alpha-1} \prec \ldots \prec C_1^0 = C_1$, ensuring $\mathcal{E} = \mathbb{R}^\cup\langle C_1^\alpha\rangle \subseteq \mathbb{R}^\cup\langle C_1^\beta\rangle$ for $\beta \in [0,\alpha]$. Observe that C_1^β minimally encases $-u_1^\beta$ by Proposition 5.9.2, implying that $C_1^\beta \cup \{u_1^\beta\}$ is a minimal positive basis modulo $\mathbb{R}^\cup\langle\partial(C_1^\beta)\rangle$ (see the comments after Lemma 5.7). For $\beta \in [0,\alpha-1]$, we have $C_1^\alpha \preceq \partial(\{\mathbf{y}_\alpha\}) \prec C_1^{\alpha-1} \preceq \partial(\{\mathbf{y}_{\alpha-1}\}) \prec \ldots \preceq \partial(\{\mathbf{y}_{\beta+1}\})$ with $\mathbf{y}_{\beta+1} \in C_1^\beta$, ensuring that $C_1^\alpha \subseteq \downarrow\partial(C_1^\beta)$ and $\mathcal{E} = \mathbb{R}^\cup\langle C_1^\alpha\rangle \subseteq \mathbb{R}^\cup\langle\partial(C_1^\beta)\rangle$. Thus, since $C_1^\beta \cup \{u_1^\beta\}$ is a minimal positive basis modulo $\mathbb{R}^\cup\langle\partial(C_1^\beta)\rangle$, we must have $u_1^\beta \notin \mathcal{E}$ with $(C_1^\beta)^\pi \neq \emptyset$. Likewise, for $j \geq 2$ and $\beta \in [0,\alpha]$, we have $\mathcal{E} \subseteq \mathbb{R}^\cup\langle C_1^\beta\rangle \subseteq \mathbb{R}^\cup\langle C_{j-1}^\beta\rangle$, while $C_j^\beta \cup \{u_{r_j}^\beta\}$ is a minimal positive basis modulo $\mathbb{R}^\cup\langle C_{j-1}^\beta \cup \partial(C_j^\beta)\rangle$ (cf. Proposition 5.9 and the comments after Lemma 5.7). In particular, $u_{r_j}^\beta \notin \mathcal{E} + \mathbb{R}^\cup\langle C_{j-1}^\beta\rangle = \mathbb{R}^\cup\langle C_{j-1}^\beta\rangle$. If C_j^β contains some $\mathbf{v}_i$, then $\mathbf{v}_i \in C_j^\beta \subseteq \downarrow\mathcal{B} \subseteq \mathcal{X}_1 \cup \ldots \cup \mathcal{X}_s \cup \{\mathbf{v}_j : j \in J\}$ ensures that $\{0\} \neq \pi(\mathbf{v}_i) \in C_j^\beta$ with $\mathbf{v}_i \notin C_{j-1}^\beta \subseteq \mathcal{X}_1 \cup \ldots \cup \mathcal{X}_s$. Otherwise, if every $\mathbf{y} \in C_j^\beta$ has $\mathbf{y} \subseteq \mathcal{E} + \mathbb{R}^\cup\langle C_{j-1}^\beta\rangle = \mathbb{R}^\cup\langle C_{j-1}^\beta\rangle$, then Proposition 5.3.9 would imply $C_j^\beta = (C_j^\beta)^* \preceq C_{j-1}^\beta$, contradicting that $C_{j-1}^\beta \prec C_j^\beta$. In either case, Proposition 5.4.3 now ensures that $(C_{j-1}^\beta)^\pi \prec (C_j^\beta)^\pi$. In summary, the above works shows that, for every $j \geq 1$ and $\beta \in [0,\alpha-1]$, we have $u_{r_j}^\beta \notin \mathcal{E} + \mathbb{R}\langle u_1^\beta,\ldots,u_{r_j-1}^\beta\rangle$ and $(C_{j-1}^\beta)^\pi \neq (C_j^\beta)^\pi$.

As a result, Proposition 5.9.4 implies that $(C_{\ell_\beta}^\beta)^\pi$ minimally encases $-\pi(\vec{u}_\beta)$ urbanely with $\mathcal{F}^\pi(\pi(\vec{u}_\beta)) = (\overline{u}_{r_1}^\beta,\ldots,\overline{u}_{r_{\ell_\beta}}^\beta)$, for each $\beta \in [0,\alpha-1]$. Moreover, $(C_1^\beta)^\pi \prec \ldots \prec (C_\ell^\beta)^\pi$ are the support sets given by Proposition 5.9.1 for $\pi(\vec{u}_\beta)$, and $r_1^\beta < \ldots < r_{\ell_\beta}^\beta < r_{\ell_\beta+1}^\beta = t_\beta + 1$ are the indices given by Proposition 5.9.1, ensuring that the encasement is urbane as it was for C_ℓ^β. Additionally, since (as remarked at the start of the proof) no element of C_1^α is in the neighborhood of any $\mathbf{y}_\beta$ with $\beta < \alpha$ (as any element of C_1^α may be taken as $\mathbf{y}_{\alpha+1}$), it follows that each C_j^β is disjoint from $C_1^\alpha = C_1^{(1)}$ for $\beta \in [0,\alpha-1]$, and thus $\pi(C_j^\beta) = (C_j^\beta)^\pi$.

The case $\beta = 0$ above tells us $C_\ell^\pi = \mathcal{B}^\pi$ minimally encases $-\pi(\vec{u})$ urbanely. By induction hypothesis (and Proposition 5.4), there exists a virtual Reay system (found by depth first search)

$$\mathcal{R}'' = (C_{2,1}^\pi \cup \{\mathbf{w}_2^{(1)}\},\ldots,C_{s_1,1}^\pi \cup \{\mathbf{w}_{s_1}^{(1)}\},\ldots,C_{1,\ell}^\pi \cup \{\mathbf{w}_1^{(\ell)}\},\ldots,C_{s_\ell,\ell}^\pi \cup \{\mathbf{w}_{s_\ell}^{(\ell)}\})$$

in $\pi(G_0)$ for the subspace $\mathbb{R}^{\cup}\langle\mathcal{B}^\pi\rangle$ satisfying (a)–(c), with each $C_{i,j}$ disjoint from $C_1^{(1)}$ (so $C_{i,j}^\pi = \pi(C_{i,j})$), with each $\mathbf{w}_k^{(j)}$ with $k < s_j$ defined by a strict truncation of the limit associated to some $\pi(\mathbf{y})$ with $\mathbf{y} \in \downarrow C_j \setminus C_1^{(1)}$, with each $\mathbf{w}_{s_j}^{(j)}$ defined by the limit $\pi((u_1, \ldots, u_{r_j}))$ with $\mathcal{F}^\pi(\pi((u_1, \ldots, u_{r_j}))) = (\overline{u}_{r_1}, \ldots, \overline{u}_{r_j})$, and with $C_{i,j} \subseteq \mathcal{X}_1 \cup \ldots \cup \mathcal{X}_s$ for all i and j except possibly when $i = s_\ell$ and $j = \ell$. In view of Proposition 5.4.2 and the work above (which holds for all $\beta < \alpha$), the path $\mathbf{y}_0, \mathbf{y}_1, \ldots, \mathbf{y}_\alpha$ remains a path modulo $\mathcal{E}$ with the ordering of vertices in each neighborhood preserved. If $\ell_\alpha > 1$ (so $C_1^{(1)} = C_1^\alpha \neq C_{\ell_\alpha}^\alpha$), then $\partial(\{\pi(\mathbf{y}_\alpha)\})$ remains nonempty, and we can assume we used some continuation of the path $\mathbf{y}_0, \mathbf{y}_1, \ldots, \mathbf{y}_\alpha$ when constructing $\mathcal{R}''$ via a depth-first search. If $\ell_\alpha = 1$, then $\partial(\{\pi(\mathbf{y}_\alpha)\}) = \emptyset$, and we may need to use some other $\mathbf{y}'_\alpha \in C_1^{\alpha-1}$ with $\partial(\{\pi(\mathbf{y}'_\alpha)\}) \neq \emptyset$ (assuming such $\mathbf{y}'_\alpha$ exists) in place of $\mathbf{y}_\alpha$ when constructing $\mathcal{R}''$. Nonetheless, in either case, we can assume we used some continuation of the path $\mathbf{y}_0, \mathbf{y}_1, \ldots, \mathbf{y}_{\alpha-1}, \mathbf{y}'_\alpha$, where $\mathbf{y}'_\alpha \in C_1^{\alpha-1}$, when constructing $\mathcal{R}''$ via a depth-first search.

We can apply Lemma 6.4, taking $\mathcal{R}_{\mathcal{A}} = (C_1^{(1)} \cup \{\mathbf{v}_1^{(1)}\})$ and $\mathcal{R}_C = \mathcal{R}''$, where $\mathbf{v}_1^{(1)} = \mathbb{R}_+ u_1^\alpha$, as can be seen by a near identical argument to that used in Case 1, the only difference being that the filtered limits begin with $\overline{u}_{r_1}$ instead of $\overline{u}_{r_2}$, and the $\mathbf{c}_j$ for $j \in [1, r]$ correspond to the $\mathbf{w}_k^{(j)}$ for $(j, k) \in \big([1, \ell] \times [1, s_j]\big) \setminus \{(1, 1)\}$. Thus

$$\mathcal{R}' = (C_1^{(1)} \cup \{\mathbf{v}_1^{(1)}\}, C_{2,1} \cup \{\mathbf{v}_2^{(1)}\}, \ldots, C_{s_1,1} \cup \{\mathbf{v}_{s_1}^{(1)}\},$$
$$\ldots, C_{1,\ell} \cup \{\mathbf{v}_1^{(\ell)}\}, \ldots, C_{s_\ell,\ell} \cup \{\mathbf{v}_{s_\ell}^{(\ell)}\})$$

is a virtual Reay system in G_0, where each $\pi(\mathbf{v}_k^{(j)}) = \mathbf{w}_k^{(j)}$ with $\mathbf{v}_k^{(j)}$ for $k < s_j$ defined by the limit $\vec{v}_{\mathbf{y}}$ (described above) and each $\mathbf{v}_{s_j}^{(j)}$ defined by the limit $(u_1, \ldots, u_{r_j})$. Recall that $\mathbf{v}_1^{(1)}$ is defined by $u_1^{\mathbf{y}_\alpha}$, which is a strict truncation of $\vec{u}_{\mathbf{y}_\alpha}$ when $\alpha > 0$ (as remarked at the start of the proof) with $\mathbf{y}_\alpha \in C_1^{\alpha-1} \subseteq \downarrow C_1$. As a result, since (a) held for $\mathcal{R}''$, it follows by construction (and Proposition 5.4.2) that (a) holds for $\mathcal{R}'$, and (b)–(c) follow directly by construction as argued in Case 1, completing the case and proof. □

Lemma 6.6 *Let $\vec{u} = (u_1, \ldots, u_s)$ a tuple of $s \geq 0$ orthonormal vectors in $\mathbb{R}^d$, let $t \in [0, s]$ be an index, let $X \subseteq \mathbb{R}^d$ be a subset which minimally encases $-(u_1, \ldots, u_t)$, and let $\pi : \mathbb{R}^d \to \mathbb{R}\langle X\rangle^\perp$ be the orthogonal projection.*

1. *$\mathbb{R}\langle X\rangle = \mathsf{C}(X \cup \{u_1, \ldots, u_t\})$.*
2. *If $Y \subseteq \mathbb{R}^d$ is subset such that $\pi(Y)$ minimally encases $-\pi(\vec{u})$ with $|\pi(Y)| = |Y|$, then $X \cup Y$ minimally encases $-(u_1, \ldots, u_s)$.*

Proof

1. Since X minimally encases $-(u_1, \ldots, u_t)$, Proposition 4.19 implies there is a Reay system $\mathcal{R} = (X_1 \cup \{u_{r_1}\}, \ldots, X_\ell \cup \{u_{r_\ell}\})$ with $X = \bigcup_{i=1}^\ell X_i$ such that $\mathcal{F} = (\mathcal{E}_1, \ldots, \mathcal{E}_\ell)$ is a compatible filter for $-(u_1, \ldots, u_t)$ having associated

indices $1 = r_1 < \ldots < r_\ell$, where $\mathcal{E}_j = \mathbb{R}\langle X_1 \cup \ldots \cup X_j \rangle$ for $j \in [0, \ell]$. Since $\mathcal{F}$ is a compatible filter, we have $u_i \in \mathcal{E}_\ell = \mathbb{R}\langle X \rangle$ for all i, and now applying Proposition 4.11.1 to $\mathcal{R}$ yields $\mathsf{C}(X \cup \{u_{r_1}, \ldots, u_{r_\ell}\}) = \mathbb{R}\langle X \rangle = \mathsf{C}(X \cup \{u_1, \ldots, u_t\})$.

2. Since X minimally encases $-(u_1, \ldots, u_t)$, it follows by Propositions 4.19 and 4.11.1 that X is linearly independent. Since $\pi(Y)$ minimally encases $-\pi(\vec{u})$ with $|\pi(Y)| = |Y|$, it likewise follows that $\pi(Y)$ and $X \cup Y$ are both linearly independent.

Let us first show that it suffices to know $X \cup Y$ simply encases $-\vec{u}$ to complete the proof. Assuming this is the case, there will be a subset $Z \subseteq X \cup Y$ that *minimally* encases $-\vec{u} = -(u_1, \ldots, u_s)$. We need to show $Z = X \cup Y$. Since Z encases $-(u_1, \ldots, u_s)$, it also encases the truncation $-(u_1, \ldots, u_t)$. Hence, since the elements of $Z \subseteq X \cup Y$ are linearly independent and $u_i \in \mathbb{R}\langle X \rangle$ for $i \leq t$ (as X encases $-(u_1, \ldots, u_t)$), it follows that $Z \cap X \subseteq X$ encases $-(u_1, \ldots, u_t)$. But since X *minimally* encases $-(u_1, \ldots, u_t)$ by hypothesis, this is only possible if $Z \cap X = X$, that is, $X \subseteq Z$. Next, since Z encases $-\vec{u}$, and since $X \cup Y$ is linearly independent with $\ker \pi = \mathbb{R}\langle X \rangle$, it follows that $\pi(Z) \setminus \{0\} = \pi(Z \setminus X) \subseteq \pi(Y)$ encases $-\pi(\vec{u})$ (cf. the equivalent version of encasement mentioned immediately after its definition and the commentary regarding equivalent tuples found there). But since $\pi(Y)$ *minimally* encases $-\pi(\vec{u})$ by hypothesis, this is only possible if $\pi(Y) = \pi(Z \setminus X)$, in turn implying $Z \setminus X = Y$ as π is injective on Y by hypothesis. Combined with $X \subseteq Z$, it follows that $Z = X \cup Y$, as desired. Thus it remains to show $X \cup Y$ simply encases $-\vec{u}$ to complete the proof, as the above argument shows any such encasement will necessarily be minimal.

If $t = s$, then $\pi(\vec{u})$ is the empty tuple, in which case we must have $Y = \emptyset$ (as $\pi(Y)$ minimally encases $-\pi(\vec{u})$), and now Item 2 is trivial.

If $t = s - 1$, then X minimally encases $-\vec{u}^{\triangleleft} = -(u_1, \ldots, u_{s-1})$, while $\pi(Y)$ encases $-\pi(\vec{u}) = -\pi(u_s)/\|\pi(u_s)\|$, and thus also $-\pi(u_s)$, meaning $-\pi(u_s) \in \mathsf{C}(\pi(Y))$ (a fact trivially true if $\pi(u_s) = 0$). Hence $u_s + a \in -\mathsf{C}(Y)$ for some $a \in \ker \pi = \mathbb{R}\langle X \rangle = \mathsf{C}(X \cup \{u_1, \ldots, u_{s-1}\})$, with the final equality by Item 1, then implying that $(u_s + \mathsf{C}(u_1, \ldots, u_{s-1})) \cap -\mathsf{C}(X \cup Y) \neq \emptyset$. Thus, since $X \subseteq X \cup Y$ encases $-(u_1, \ldots, u_{s-1})$, it now follows that $X \cup Y$ encases $-(u_1, \ldots, u_s)$, completing the proof as noted above. So we may assume $t \leq s - 2$ and proceed by induction on $s - t$.

Let $Y' \subseteq Y$ be a subset such that $\pi(Y')$ minimally encases $-\pi((u_1, \ldots, u_{s-1}))$. Then $X \cup Y'$ minimally encases $-(u_1, \ldots, u_{s-1})$ by induction hypothesis. Letting $\tau : \mathbb{R}^d \to \mathbb{R}\langle X \cup Y' \rangle^\perp$ be the orthogonal projection, we find that $\tau(Y) \setminus \{0\} = \tau(Y \setminus Y')$ is a linearly independent set of size $|Y \setminus Y'|$ (since $X \cup Y$ is linearly independent with $X \cup Y' \subseteq X \cup Y$). Since $\pi(Y)$ encases $-\pi(\vec{u})$, it follows that $\tau(\pi(Y)) \setminus \{0\} = \tau(Y \setminus Y')$ encases $-\tau(\pi(\vec{u})) = -\tau(\vec{u})$ (cf. the equivalent version of encasement mentioned immediately after its definition and the commentary regarding equivalent tuples found there). Thus there must be some subset $Z \subseteq Y \setminus Y'$ such that $\tau(Z)$ *minimally* encases $-\tau(\vec{u})$. Since $\tau(Y \setminus Y')$ is a linearly independent set of size $|Y \setminus Y'|$, it follows that τ is injective on $Y \setminus Y'$, and thus also on $Z \subseteq Y \setminus Y'$. As a result, we can apply the base of the induction using $X \cup Y'$, Z and τ (in place of X, Y and π, with $t = s - 1$) to conclude that $X \cup Y' \cup Z \subseteq X \cup Y$ minimally encases $-\vec{u}$, implying that $X \cup Y$ encases $-\vec{u}$, which completes the proof as shown above. □

We now make precise the idea that the underlying oriented Reay system to a virtual Reay system is a limit structure. Suppose for each tuple $k = (i_1, \ldots, i_t) \in \mathbb{Z}^t$ of indices $i_j \geq 1$ we have a set $X_k \subseteq \mathbb{R}^d$. Then we write $\lim_{k\to\infty} X_k = X$ if $X_k \subseteq X$ holds once all i_j are sufficiently large and, for every $x \in X$, there exists some bound $N_x > 0$ such that $x \in X_k$ for all tuples k having all coordinates $i_j \geq N_x$ for $j \in [1, t]$. We say that $\lim_{k\to\infty} X_k = X$ is an **order uniform limit** if (additionally) there is a global constant $N > 0$ such that, for every integer $m \geq N$, there is a relative constant $N_m \geq N$ such that, for every tuple $k = (i_1, \ldots, i_t)$ of indices with all coordinates $i_j \geq N$ and $i_\alpha = m$ for some $\alpha \in [1, t]$ and every tuple $k' = (i'_1, \ldots, i'_t)$ of indices with $i'_j = i_j$ for $j \in [1, t] \setminus \{\alpha\}$ and $i'_\alpha \geq N_m$, we have $X_k \subseteq X_{k'}$. (The terms global and relative are introduced above to later make referencing these constants easier.) When this is the case, we can find an increasing subsequences of indices $N \leq \iota_1 < \iota_2 < \ldots$ so that, given any tuples $k = (\iota_{\alpha_1}, \ldots, \iota_{\alpha_t})$ and $k' = (\iota_{\beta_1}, \ldots, \iota_{\beta_t})$ having all coordinates equal except one which is larger in k', we have $X_k \subseteq X_{k'}$. Indeed, we could take $\iota_1 = N$, $\iota_2 = \max\{N_N, \iota_1 + 1\}$, $\iota_3 = \max\{N_{(N_N)}, \iota_2 + 1\}$, and so forth. However, it is easily derived from this property that, if $k = (\iota_{\alpha_1}, \ldots, \iota_{\alpha_t})$ and $k' = (\iota_{\beta_1}, \ldots, \iota_{\beta_t})$ are any pair of tuples with $\iota_{\beta_j} \geq \iota_{\alpha_j}$ for all $j \in [1, t]$, then $X_k \subseteq X_{k'}$. In such case, the natural partial order on the tuples $k = (\iota_{\alpha_1}, \ldots, \iota_{\alpha_t})$ corresponds to set theoretic inclusion among the sets indexed by k. (In other words, these additional properties are obtained by passing to appropriate subsequences.) Additionally, if $k = (i_1, \ldots, i_t)$ is any tuple with $i_j \geq N$ for all j, then there exists a constant N_k such that $X_k \subseteq X_{k'}$ for any tuple $k' = (i'_1, \ldots, i'_t)$ with $i'_j \geq N_k$ for all j. Indeed, we may simply take $N_k = \max\{N_{i_1}, \ldots, N_{i_t}\}$, and this will then follow from the definition of order uniform limit.

Proposition 6.7 *Let $G_0 \subseteq \mathbb{R}^d$ be a subset and let $\mathcal{R} = (\mathcal{X}_1 \cup \{\mathbf{v}_1\}, \ldots, \mathcal{X}_s \cup \{\mathbf{v}_s\})$ be a virtual Reay system in G_0. There is a bound $N > 0$ such that the following hold for all $\mathbf{x} \in \mathcal{X}_1 \cup \{\mathbf{v}_1\} \cup \ldots \cup \mathcal{X}_1 \cup \{\mathbf{v}_s\}$.*

1. *$\widetilde{\downarrow\partial(\{x\})}(k)$ minimally encases $-\vec{u}_{\mathbf{x}}^{\triangleleft}$ for any tuple $k = (i_{\mathbf{z}})_{\mathbf{z}\in\downarrow\partial(\{\mathbf{x}\})}$ with all $i_{\mathbf{z}} \geq N$.*
2. *$\lim_{k\to\infty} \mathsf{C}\big((\downarrow\tilde{\mathbf{x}})(k)\big) = \mathbf{x}$, $\lim_{k\to\infty} \mathsf{C}^\circ\big((\downarrow\tilde{\mathbf{x}})(k)\big) = \mathbf{x}^\circ$ and*

$$\lim_{k\to\infty} \Big(\mathsf{C}\big((\downarrow\tilde{\mathbf{x}})(k)\big) \cap \partial(\mathbf{x})\Big) = \lim_{k\to\infty} \mathsf{C}(\widetilde{\downarrow\partial(\{x\})}(k)) = \partial(\mathbf{x}) \cap \mathbf{x},$$

with all limits holding order uniformly.

3. *$\widetilde{\downarrow\partial(\{x\})}(k) \cup \{\tilde{\mathbf{x}}^{(t-1)}(i)\}$ is a minimal positive basis when $t > 1$, where $\vec{u}_{\mathbf{x}} = (u_1, \ldots, u_t)$, for any tuple $k = (i_{\mathbf{z}})_{\mathbf{z}\in\downarrow\partial(\{\mathbf{x}\})}$ with all $i_{\mathbf{z}} \geq N$ and any $i \geq N$.*

Proof Let $\mathbf{x} \in \mathcal{X}_1 \cup \{\mathbf{v}_1\} \cup \ldots \cup \mathcal{X}_1 \cup \{\mathbf{v}_s\}$ be arbitrary, let $\vec{u}_{\mathbf{x}} = (u_1, \ldots, u_t)$, let $\mathcal{B}_{\mathbf{x}} = \partial(\{\mathbf{x}\})$ and let $\emptyset = \mathcal{C}_0 \prec \mathcal{C}_1 \prec \ldots \prec \mathcal{C}_\ell = \mathcal{B}_{\mathbf{x}}$ and $1 = r_1 < \ldots < r_\ell < r_{\ell+1} = t$ be the support sets and indices given by Proposition 5.9 applied to urbane minimal encasement of $-\vec{u}_{\mathbf{x}}^{\triangleleft}$ by $\mathcal{B}_{\mathbf{x}} = \partial(\{\mathbf{x}\})$. For $\mathbf{y} \in \mathcal{B}_{\mathbf{x}} = \partial(\{\mathbf{x}\})$, let $\mathcal{B}_{\mathbf{y}} = \partial(\{\mathbf{y}\})$ and $\vec{u}_{\mathbf{y}} = (u_1^{\mathbf{y}}, \ldots, u_{t_{\mathbf{y}}}^{\mathbf{y}})$. Let $\mathcal{B}'_{\mathbf{x}} \subseteq \mathcal{B}_{\mathbf{x}}$ be the subset of all half-spaces $\mathbf{y} \in \mathcal{B}_{\mathbf{x}}$ with $\partial(\{\mathbf{y}\}) \neq \emptyset$. Let $\hat{k} = (i_{\mathbf{y}})_{\mathbf{y} \in \downarrow \mathbf{x}}$ be a tuple of sufficiently large indices $i_{\mathbf{y}} \geq N_{\mathbf{x}}$ (the constant $N_{\mathbf{x}} > 0$ will be determined during the course of the proof) and let $k = (i_{\mathbf{y}})_{\mathbf{y} \in \downarrow \mathcal{B}_{\mathbf{x}}}$ be the sub-tuple consisting of all indices $i_{\mathbf{y}}$ with $\mathbf{y} \in \downarrow \mathcal{B}_{\mathbf{x}} = \downarrow \mathbf{x} \setminus \{\mathbf{x}\}$. For $\mathbf{y} \in \mathcal{B}_{\mathbf{x}}$, let $k_{\mathbf{y}} = (i_{\mathbf{z}})_{\mathbf{z} \in \downarrow \mathcal{B}_{\mathbf{y}}}$ be the sub-tuple of k consisting of coordinates indexed by $\downarrow \mathcal{B}_{\mathbf{y}}$, and let $\hat{k}_{\mathbf{y}} = (i_{\mathbf{z}})_{\mathbf{z} \in \downarrow \mathbf{y}}$ be the sub-tuple of k consisting of coordinates indexed by $\downarrow \mathbf{y} = \downarrow \mathcal{B}_{\mathbf{y}} \cup \{\mathbf{y}\}$. Since each C_j is a support set and replacing each half-space in an oriented Reay system with a representative yields an ordinary Reay system, Propositions 4.11.1 implies that $\downarrow \tilde{C}_j(k')$ is a basis for $\mathbb{R}^{\cup}\langle C_j \rangle = \mathbb{R}\langle \downarrow \tilde{C}_j(k') \rangle$ for any tuple k' indexed by $\downarrow C_j$ and $j \in [0, \ell]$. Likewise, $(\downarrow \tilde{\mathbf{x}})(\hat{k})$ is a linearly independent set. Since there are only a finite number of $\mathbf{x} \in \mathcal{X}_1 \cup \{\mathbf{v}_1\} \cup \ldots \mathcal{X}_s \cup \{\mathbf{v}_s\}$, if we can show Items 1–3 hold for our arbitrary $\mathbf{x}$ using some bound $N_{\mathbf{x}}$, with $N_{\mathbf{x}}$ also being the global constant required in the definition of order uniform limits, then the proposition will follow by taking $N = \max_{\mathbf{x}} N_{\mathbf{x}}$.

We first handle the case when $\partial(\{\mathbf{z}\}) = \emptyset$ for all $\mathbf{z} \in \mathcal{X}_1 \cup \{\mathbf{v}_1\} \cup \ldots \cup \mathcal{X}_1 \cup \{\mathbf{v}_s\}$. In this case, $t = 1$, $\mathcal{B}_{\mathbf{x}} = \emptyset$ and $\mathsf{C}\big((\downarrow \tilde{\mathbf{x}})(k)\big) = \mathbf{x}$ for all tuples k, in which case Items 1–3 hold trivially with $N = 1$. In particular, this completes the case $s = 1$, and thus also the case $\dim \mathbb{R}^{\cup}\langle \mathcal{X}_1 \cup \ldots \cup \mathcal{X}_s \rangle = \dim \mathbb{R}^{\cup}\langle \mathcal{X}_1 \rangle = 1$. We proceed by induction on $\dim \mathbb{R}^{\cup}\langle \mathcal{X}_1 \cup \ldots \cup \mathcal{X}_s \rangle$. Since $\dim \mathbb{R}^{\cup}\langle \mathcal{X}_1 \cup \ldots \cup \mathcal{X}_{s-1} \rangle < \dim \mathbb{R}^{\cup}\langle \mathcal{X}_1 \cup \ldots \cup \mathcal{X}_s \rangle$, it suffices by induction hypothesis and the base case to consider $\mathbf{x} \in \mathcal{X}_s \cup \{\mathbf{v}_s\}$ with $s \geq 2$. If $t = 1$, then $\partial(\{\mathbf{x}\}) = \emptyset$, and Items 1–3 hold trivially as before, so we can assume $t \geq 2$, and thus also $\ell \geq 1$.

1. We first show Item 1 holds, which we do in two cases based on whether $\ell > 1$ or $\ell = 1$.

   Case 1: Suppose that $\ell > 1$. Let $\mathcal{E} = \mathbb{R}^{\cup}\langle C_{\ell-1} \rangle$ and let $\pi : \mathbb{R}^d \to \mathcal{E}^{\perp}$ be the orthogonal projection. By proposition 6.3, $\pi(\mathcal{R})$ is a virtual Reay system with $\vec{u}_{\pi(\mathbf{x})} = \pi(\vec{u}_{\mathbf{x}})$, $\pi(\vec{u}_{\mathbf{x}})^{\triangleleft} = \pi(\vec{u}_{\mathbf{x}}^{\triangleleft})$ and $\partial(\{\pi(\mathbf{x})\}) = \partial(\{\mathbf{x}\})^{\pi} = \mathcal{B}_{\mathbf{x}}^{\pi}$. By proposition 6.5, there exists a virtual Reay system $\mathcal{R}' = (\mathcal{Y}_1 \cup \{\mathbf{w}_1\}, \ldots, \mathcal{Y}_{s'} \cup \{\mathbf{w}_{s'}\})$ in G_0 for the subspace $\mathbb{R}^{\cup}\langle \mathcal{B}_{\mathbf{x}} \rangle = \mathbb{R}^{\cup}\langle \partial(\{\mathbf{x}\}) \rangle$ such that $\mathcal{Y}_{s'} = \mathcal{B}_{\mathbf{x}}$, $\downarrow \mathcal{B}_{\mathbf{x}} = \bigcup_{i=1}^{s'} \mathcal{Y}'_i$, $\partial(\{\mathbf{w}_{s'}\}) = C_{\ell-1}$ and $\mathbf{w}_{s'}$ is defined by the limit $(u_1, \ldots, u_{r_\ell})$. By the induction hypothesis applied to $\mathcal{R}'$, $\downarrow \tilde{C}_{\ell-1}(k)$ minimally encases $-(u_1, \ldots, u_{r_\ell - 1})$ once all indices in the tuple k are sufficiently large. By Propositions 5.9.4 and 5.4.2, $\partial(\{\pi(\mathbf{x})\}) = \mathcal{B}_{\mathbf{x}}^{\pi} = C_{\ell}^{\pi}$ minimally encases $-\pi(\vec{u}_{\mathbf{x}}^{\triangleleft}) = -\pi(\vec{u}_{\mathbf{x}})^{\triangleleft}$. Thus, by the induction hypothesis applied to $\pi(\mathcal{R})$, $\pi(\downarrow \tilde{C}_{\ell}(k)) \setminus \{0\}$ minimally encases $-\pi(\vec{u}_{\mathbf{x}}^{\triangleleft})$ once all indices in the tuple k are sufficiently large. Since $C_\ell = \mathcal{B}_{\mathbf{x}} = \partial(\{\mathbf{x}\}) \subseteq \mathcal{X}_1 \cup \ldots \cup \mathcal{X}_s$, Propositions 5.3.9 and 5.4.2 ensure that $\downarrow C_{\ell}^{\pi} = (\downarrow C_\ell)^{\pi} = \pi(\downarrow C_\ell \setminus \downarrow C_{\ell-1})$. As noted above, $\mathcal{E} = \mathbb{R}^{\cup}\langle C_{\ell-1} \rangle = \mathbb{R}\langle \downarrow \tilde{C}_{\ell-1}(k) \rangle$ with $\downarrow \tilde{C}_\ell(k)$ a set of $|\downarrow C_\ell|$ linearly independent elements, and we have $\downarrow C_{\ell-1} \subsetneq \downarrow C_\ell$ in view of $C_{\ell-1} \prec C_\ell$. Consequently, Lemma 6.6 (applied with $X = \downarrow \tilde{C}_{\ell-1}(k')$ and $Y = (\downarrow \tilde{C}_\ell \setminus \downarrow \tilde{C}_{\ell-1})(k'')$, where k' and k'' are the appropriate sub-tuples of k indexed by $\downarrow C_{\ell-1}$ and $\downarrow C_\ell \setminus \downarrow C_{\ell-1}$, respectively, and with $\vec{u} = \vec{u}^{\triangleleft}_{\mathbf{x}}$ and $t = r_\ell - 1$) implies that $\downarrow\tilde{\mathcal{B}}_{\mathbf{x}}(k) = \downarrow\tilde{C}_\ell(k)$ minimally encases $-\vec{u}^{\triangleleft}_{\mathbf{x}}$, as desired. (Note $N_{\mathbf{x}}$ must be at least the values obtained inductively for $\mathcal{R}'$ and $\pi(\mathcal{R})$.)

Case 2: Next suppose that $\ell = 1$. Then $\mathcal{B}_{\mathbf{x}} = C_1$ minimally encases $-u_1$ with $u_i \in \mathbb{R}^{\cup}\langle\mathcal{B}_{\mathbf{x}}\rangle$ for all $i < t$. In consequence, since $\mathcal{B}_{\mathbf{x}}$ is a support set, Lemma 5.7 and Proposition 5.3.4 imply $-u_1 \in \mathsf{C}^{\cup}(\mathcal{B}_{\mathbf{x}})^\circ = \sum_{\mathbf{y}\in\mathcal{B}_{\mathbf{x}}} \mathbf{y}^\circ$. Thus we must have a representation of the form $\sum_{\mathbf{y}\in\mathcal{B}_{\mathbf{x}}} b_{\mathbf{y}} = -u_1$ with each $b_{\mathbf{y}} \in \mathbf{y}^\circ$. Applying Item 2 of the induction hypothesis to each $\mathbf{y} \in \mathcal{B}_{\mathbf{x}}$, we have $\mathbf{y}^\circ = \lim_{\hat{k}_{\mathbf{y}}\to\infty} \mathsf{C}^\circ((\downarrow\tilde{\mathbf{y}})(\hat{k}_{\mathbf{y}}))$, allowing us to assume $b_{\mathbf{y}} \in \mathsf{C}^\circ((\downarrow\tilde{\mathbf{y}})(\hat{k}_{\mathbf{y}}))$ so long as all indices from the tuple $\hat{k}_{\mathbf{y}}$ are sufficiently large. (Note $N_{\mathbf{x}}$ must be at least the value of the constant $N_{b_{\mathbf{y}}}$ in the definition of $b_{\mathbf{y}} \in \lim_{\hat{k}_{\mathbf{y}}\to\infty} \mathsf{C}^\circ((\downarrow\tilde{\mathbf{y}})(\hat{k}_{\mathbf{y}})))$, for each $\mathbf{y} \in \mathcal{B}_{\mathbf{x}}$, and we do not require order uniformity.) Consequently, since $\downarrow\mathbf{y} = \mathcal{B}_{\mathbf{y}} \cup \{\mathbf{y}\} \subseteq \downarrow\mathcal{B}_{\mathbf{x}}$, we see that $\downarrow\tilde{\mathcal{B}}_{\mathbf{x}}(k)$ encases $-u_1$. If it does not do so minimally, then there must be some proper subset $\mathcal{B} \subset \downarrow\mathcal{B}_{\mathbf{x}}$ such that $\tilde{B}(k)$ encases $-u_1$. Let $\mathcal{E} = \mathbb{R}^{\cup}\langle\partial(\mathcal{B}_{\mathbf{x}})\rangle = \mathbb{R}^{\cup}\langle\partial^2(\{\mathbf{x}\})\rangle$, and let $\pi : \mathbb{R}^d \to \mathcal{E}^\perp$ and $\pi^\perp : \mathbb{R}^d \to \mathcal{E}$ be the orthogonal projections. By Proposition 5.3 (Items 3 and 5), $\pi(\tilde{\mathcal{B}}_{\mathbf{x}}(k))$ is a linearly independent set of size $|\mathcal{B}_{\mathbf{x}}|$ with $-\pi(u_1) \in \sum_{\mathbf{y}\in\mathcal{B}_{\mathbf{x}}} \pi(b_{\mathbf{y}}) \in \mathsf{C}^{\cup}(\pi(\tilde{\mathcal{B}}_{\mathbf{x}}(k)))^\circ$. Thus Proposition 4.8.4 implies that $\pi(\tilde{\mathcal{B}}_{\mathbf{x}}(k)) \cup \{\pi(u_1)\}$ is a minimal positive basis of size $|\tilde{\mathcal{B}}_{\mathbf{x}}(k)|+1$, meaning $\pi(\tilde{\mathcal{B}}_{\mathbf{x}}(k))$ minimally encases $-\pi(u_1)$. Consequently, we must have $\mathcal{B}_{\mathbf{x}} \subseteq \mathcal{B}$, as otherwise $\pi(\tilde{B}(k)) \setminus \{0\}$ would be a proper subset of $\pi(\tilde{\mathcal{B}}_{\mathbf{x}}(k))$ which encases $-\pi(u_1)$, contradicting the minimality of $\pi(\tilde{\mathcal{B}}_{\mathbf{x}}(k))$. Therefore, there must be some element from $\downarrow\mathcal{B}_{\mathbf{x}} \setminus \mathcal{B}_{\mathbf{x}}$ missing from the proper subset $\mathcal{B} \subset \downarrow\mathcal{B}_{\mathbf{x}}$, which ensures that $\mathcal{B}'_{\mathbf{x}} \neq \emptyset$ (since $\mathcal{B}'_{\mathbf{x}} = \emptyset$ implies $\downarrow\mathcal{B}_{\mathbf{x}} = \mathcal{B}_{\mathbf{x}}$), and thus that $\mathcal{E}$ is nontrivial.

For $\mathbf{y} \in \mathcal{B}_{\mathbf{x}}$, we have $\tilde{\mathbf{y}}(i_{\mathbf{y}}) = (a^{\mathbf{y}}_{i_{\mathbf{y}},1}u^{\mathbf{y}}_1 + \ldots + a^{\mathbf{y}}_{i_{\mathbf{y}},t_{\mathbf{y}}-1}u^{\mathbf{y}}_{t_{\mathbf{y}}-1} + w^{\mathbf{y}}_{i_{\mathbf{y}}}) + b^{\mathbf{y}}_{i_{\mathbf{y}}}\overline{u}^{\mathbf{y}}_{t_{\mathbf{y}}}$ with $t_{\mathbf{y}} \geq 1, u^{\mathbf{y}}_1, \ldots, u^{\mathbf{y}}_{t_{\mathbf{y}}-1}, w^{\mathbf{y}}_{i_{\mathbf{y}}} \in \mathbb{R}^{\cup}\langle\partial(\{\mathbf{y}\})\rangle \subseteq \mathcal{E}$ and $b^{\mathbf{y}}_{i_{\mathbf{y}}}, \|w^{\mathbf{y}}_{i_{\mathbf{y}}}\| \in o(a^{\mathbf{y}}_{i_{\mathbf{y}},t_{\mathbf{y}}-1})$ (as given in (6.1)). Thus

$$\pi(\tilde{\mathbf{y}}(i_{\mathbf{y}})) = b^{\mathbf{y}}_{i_{\mathbf{y}}}\pi(\overline{u}^{\mathbf{y}}_{t_{\mathbf{y}}}) \neq 0, \tag{6.2}$$

in view of Proposition 5.3.3, and

$$\begin{aligned}\pi^\perp(\tilde{\mathbf{y}}(i_{\mathbf{y}})) &= (a^{\mathbf{y}}_{i_{\mathbf{y}},1}u^{\mathbf{y}}_1 + \ldots + a^{\mathbf{y}}_{i_{\mathbf{y}},t_{\mathbf{y}}-1}u^{\mathbf{y}}_{t_{\mathbf{y}}-1} + w^{\mathbf{y}}_{i_{\mathbf{y}}}) + b^{\mathbf{y}}_{i_{\mathbf{y}}}\pi^\perp(\overline{u}^{\mathbf{y}}_{t_{\mathbf{y}}}) \\ &= \tilde{\mathbf{y}}^{(t_{\mathbf{y}}-1)}(i_{\mathbf{y}}) + b^{\mathbf{y}}_{i_{\mathbf{y}}}\pi^\perp(\overline{u}^{\mathbf{y}}_{t_{\mathbf{y}}}).\end{aligned} \tag{6.3}$$

Note $t_{\mathbf{y}} \geq 2$ holds precisely for those $\mathbf{y} \in \mathcal{B}'_{\mathbf{x}}$, while $\tilde{\mathbf{y}}^{(t_{\mathbf{y}}-1)}(i_{\mathbf{y}}) = 0$ for all $i_{\mathbf{y}}$ and $\mathbf{y} \in \mathcal{B}_{\mathbf{x}} \setminus \mathcal{B}'_{\mathbf{x}}$.

Since $\pi(\tilde{B}_{\mathbf{x}}(k)) \cup \{\pi(u_1)\}$ is a minimal positive basis of size $|B_{\mathbf{x}}(k)| + 1$, so too is the re-scaled set $\{\pi(\overline{u}^{\mathbf{y}}_{t_{\mathbf{y}}}) : \mathbf{y} \in \mathcal{B}_{\mathbf{x}}\} \cup \{\pi(u_1)\}$. As a result, there is a unique linear combination

$$\sum_{\mathbf{y}\in\mathcal{B}_{\mathbf{x}}} \alpha_{\mathbf{y}}\pi(\overline{u}^{\mathbf{y}}_{t_{\mathbf{y}}}) = -\pi(u_1), \tag{6.4}$$

and this linear combination has $\alpha_{\mathbf{y}} > 0$ for all $\mathbf{y} \in \mathcal{B}_{\mathbf{x}}$. Consequently, in view of (6.3), (6.4) and (6.2), we have

$$\sum_{\mathbf{y}\in\mathcal{B}_{\mathbf{x}}} \frac{\alpha_{\mathbf{y}}}{b^{\mathbf{y}}_{i_{\mathbf{y}}}}\tilde{\mathbf{y}}(i_{\mathbf{y}}) = -\pi(u_1) + \sum_{\mathbf{y}\in\mathcal{B}_{\mathbf{x}}} \Big(\frac{\alpha_{\mathbf{y}}}{b^{\mathbf{y}}_{i_{\mathbf{y}}}}\tilde{\mathbf{y}}^{(t_{\mathbf{y}}-1)}(i_{\mathbf{y}}) + \alpha_{\mathbf{y}}\pi^{\perp}(\overline{u}^{\mathbf{y}}_{t_{\mathbf{y}}})\Big). \tag{6.5}$$

Since $\tilde{B}(k)$ encases $-u_1$, there is a positive linear combination of the elements from $\tilde{B}(k)$ equal to $-u_1$. Moreover, considering this linear combination modulo $\mathcal{E}$, we see (in view of the uniqueness of the linear combination in (6.4)) that the coefficient of each $\mathbf{y} \in \mathcal{B}_{\mathbf{x}} \subseteq \mathcal{B}$ must be $\frac{\alpha_{\mathbf{y}}}{b^{\mathbf{y}}_{i_{\mathbf{y}}}}$. Thus, in view of (6.5), we find that we have a positive linear combination of the elements from $(\tilde{B} \setminus \tilde{B}_{\mathbf{x}})(k)$ equal to

$$-z_k := -\sum_{\mathbf{y}\in\mathcal{B}'_{\mathbf{x}}} \frac{\alpha_{\mathbf{y}}}{b^{\mathbf{y}}_{i_{\mathbf{y}}}}\tilde{\mathbf{y}}^{(t_{\mathbf{y}}-1)}(i_{\mathbf{y}}) - \xi, \tag{6.6}$$

where $\xi = \pi^{\perp}(u_1) + \sum_{\mathbf{y}\in\mathcal{B}_{\mathbf{x}}} \alpha_{\mathbf{y}}\pi^{\perp}(\overline{u}^{\mathbf{y}}_{t_{\mathbf{y}}}) \in \mathcal{E}$ is a fixed element (independent of $i_{\mathbf{y}}$), i.e.,

$$-z_k \in \mathsf{C}((\tilde{B} \setminus \tilde{B}_{\mathbf{x}})(k)). \tag{6.7}$$

Applying Item 1 of the induction hypothesis to each $\mathbf{y} \in \mathcal{B}_{\mathbf{x}}$, we find that $\downarrow\tilde{B}_{\mathbf{y}}(k_{\mathbf{y}})$ minimally encases $-(u^{\mathbf{y}}_1, \ldots, u^{\mathbf{y}}_{t_{\mathbf{y}}-1})$ for each $\mathbf{y} \in \mathcal{B}_{\mathbf{x}}$ so long as all coordinates in the tuple $k_{\mathbf{y}}$ are sufficiently large (recall that $\mathcal{B}_{\mathbf{y}} = \partial(\{\mathbf{y}\})$). (Note $N_{\mathbf{x}}$ must be at least the value obtained inductively for each $\mathbf{y} \in \mathcal{B}_{\mathbf{x}} \subseteq \mathcal{X}_1 \cup \ldots \cup \mathcal{X}_{s-1}$.) For each $\mathbf{y} \in \mathcal{B}_{\mathbf{x}}$, fix one particular tuple $\kappa_{\mathbf{y}} = (\iota_{\mathbf{z}})_{\mathbf{z}\in\downarrow\mathcal{B}_{\mathbf{y}}}$ such that $\downarrow\tilde{B}_{\mathbf{y}}(\kappa_{\mathbf{y}})$ minimally encases $-(u^{\mathbf{y}}_1, \ldots, u^{\mathbf{y}}_{t_{\mathbf{y}}-1})$. For convenience, we can assume that the coordinate $\iota_{\mathbf{z}}$ is the same among all tuples $\kappa_{\mathbf{y}}$ that contain a coordinate indexed by $\mathbf{z} \in \downarrow\mathcal{B}_{\mathbf{y}}$. By Item 2 of the induction hypothesis applied to each $\mathbf{y} \in \mathcal{B}_{\mathbf{x}}$, we have $\lim_{k_{\mathbf{y}}\to\infty} \mathsf{C}(\downarrow\tilde{B}_{\mathbf{y}}(k_{\mathbf{y}})) = \partial(\mathbf{y}) \cap \mathbf{y}$ order uniformly. Consequently, so long as we choose the coordinates in the tuples $\kappa_{\mathbf{y}}$ sufficiently large, we can be assured that $\mathsf{C}(\downarrow\tilde{B}_{\mathbf{y}}(\kappa_{\mathbf{y}})) \subseteq \mathsf{C}(\downarrow\tilde{B}_{\mathbf{y}}(k_{\mathbf{y}}))$ for any tuple $k_{\mathbf{y}}$ with all coordinates sufficiently large. (Note each coordinate $\iota_{\mathbf{z}}$ in $\kappa_{\mathbf{y}}$, for $\mathbf{y} \in \mathcal{B}_{\mathbf{x}}$ and $\mathbf{z} \in \downarrow\mathcal{B}_{\mathbf{y}}$, must be at least the value of the global constant from the order uniform limit $\lim_{k_{\mathbf{y}}\to\infty} \mathsf{C}(\downarrow\tilde{B}_{\mathbf{y}}(k_{\mathbf{y}}))$ given by the inductive hypothesis applied to $\mathbf{y} \in \mathcal{B}_{\mathbf{x}}$, and then $N_{\mathbf{x}}$ must be at least $\max_{\mathbf{y}\in\mathcal{B}_{\mathbf{x}},\,\mathbf{z}\in\downarrow\mathcal{B}_{\mathbf{y}}} N_{\iota_{\mathbf{z}}}$, where $N_{\iota_{\mathbf{z}}}$ is the relative constant given in the definition of the order uniform limit $\lim_{k_{\mathbf{y}}\to\infty} \mathsf{C}(\downarrow\tilde{B}_{\mathbf{y}}(k_{\mathbf{y}}))$.)

Observe that

$$\bigcup_{\mathbf{y}\in\mathcal{B}'_{\mathbf{x}}} \downarrow\mathcal{B}_{\mathbf{y}} = \bigcup_{\mathbf{y}\in\mathcal{B}_{\mathbf{x}}} \downarrow\mathcal{B}_{\mathbf{y}} = \downarrow\partial(\mathcal{B}_{\mathbf{x}}) = \downarrow\mathcal{B}_{\mathbf{x}} \setminus \mathcal{B}_{\mathbf{x}},$$

with the final equality since $\mathcal{B}^*_{\mathbf{x}} = \mathcal{B}_{\mathbf{x}} = \partial(\{\mathbf{x}\})$ is a support set. Thus

$$\mathcal{B}\setminus\mathcal{B}_{\mathbf{x}} \subset \bigcup_{\mathbf{y}\in\mathcal{B}'_{\mathbf{x}}} \downarrow\mathcal{B}_{\mathbf{y}} \quad \text{and} \quad \mathcal{E} = \sum_{\mathbf{y}\in\mathcal{B}'_{\mathbf{x}}} \mathbb{R}^{\cup}\langle\mathcal{B}_{\mathbf{y}}\rangle = \sum_{\mathbf{y}\in\mathcal{B}'_{\mathbf{x}}} \mathbb{R}\langle\downarrow\tilde{B}_{\mathbf{y}}(k_{\mathbf{y}})\rangle. \tag{6.8}$$

(Recall that $\mathcal{B}_{\mathbf{y}} = \partial(\{\mathbf{y}\}) = \emptyset$ for $\mathbf{y} \in \mathcal{B}_{\mathbf{x}} \setminus \mathcal{B}'_{\mathbf{x}}$.) Hence, since $\xi \in \mathcal{E}$, we may write $\xi = \sum_{\mathbf{y}\in\mathcal{B}'_{\mathbf{x}}} \xi_{\mathbf{y}}$ with each $\xi_{\mathbf{y}} \in \mathbb{R}^{\cup}\langle\mathcal{B}_{\mathbf{y}}\rangle = \mathbb{R}\langle\downarrow\tilde{B}_{\mathbf{y}}(k_{\mathbf{y}})\rangle$.

Since $b^{\mathbf{y}}_{i_{\mathbf{y}}} \in o(a^{\mathbf{y}}_{i_{\mathbf{y}},j})$, we have $a^{\mathbf{y}}_{i_{\mathbf{y}},j}/b^{\mathbf{y}}_{i_{\mathbf{y}}} \to \infty$, for all $\mathbf{y} \in \mathcal{B}'_{\mathbf{x}}$ and $j \in [1, t_{\mathbf{y}} - 1]$. Hence, in view of (6.3), we see that $\{\frac{\alpha_{\mathbf{y}}}{b^{\mathbf{y}}_{i_{\mathbf{y}}}}\tilde{\mathbf{y}}^{(t_{\mathbf{y}}-1)}(i_{\mathbf{y}}) + \xi_{\mathbf{y}}\}_{i_{\mathbf{y}}=1}^{\infty}$ is an asymptotically filtered sequence of terms from $\mathbb{R}^{\cup}\langle\mathcal{B}_{\mathbf{y}}\rangle = \mathbb{R}^{\cup}\langle\downarrow\tilde{B}_{\mathbf{y}}(k_{\mathbf{y}})\rangle$ with limit $(u^{\mathbf{y}}_1, \ldots, u^{\mathbf{y}}_{t_{\mathbf{y}}-1})$ (once $i_{\mathbf{y}}$ is sufficiently large), for each $\mathbf{y} \in \mathcal{B}'_{\mathbf{x}}$. Consequently, since $\downarrow\tilde{B}_{\mathbf{y}}(\kappa_{\mathbf{y}})$ minimally encases $-(u^{\mathbf{y}}_1, \ldots, u^{\mathbf{y}}_{t_{\mathbf{y}}-1})$ (as noted above), it follows from Proposition 4.20 that $-\frac{\alpha_{\mathbf{y}}}{b^{\mathbf{y}}_{i_{\mathbf{y}}}}\tilde{\mathbf{y}}^{(t_{\mathbf{y}}-1)}(i_{\mathbf{y}}) - \xi_{\mathbf{y}} \in \mathsf{C}^{\circ}(\downarrow\tilde{B}_{\mathbf{y}}(\kappa_{\mathbf{y}})) \subseteq \mathsf{C}^{\circ}(\downarrow\tilde{B}_{\mathbf{y}}(k_{\mathbf{y}}))$ when $i_{\mathbf{y}}$ is sufficiently large. (Note this requires $N_{\mathbf{x}}$ to be at least the value resulting from the application of Proposition 4.20 to the *fixed* set $\downarrow\tilde{B}_{\mathbf{y}}(\kappa_{\mathbf{y}})$, for $\mathbf{y} \in \mathcal{B}_{\mathbf{x}}$, and thus does not depend on the infinite possible values for $k_{\mathbf{y}}$, which is a subtle but important point.) Hence (6.6) yields $-z_k \in \sum_{\mathbf{y}\in\mathcal{B}'_{\mathbf{x}}} \mathsf{C}^{\circ}(\downarrow\tilde{B}_{\mathbf{y}}(k_{\mathbf{y}}))$. As a result, since the elements of $\bigcup_{\mathbf{y}\in\mathcal{B}'_{\mathbf{x}}} \downarrow\tilde{B}_{\mathbf{y}}(k_{\mathbf{y}}) \subseteq \downarrow\tilde{C}_{\ell}(k)$ are linearly independent for any tuple k, it follows that $-z_k \in \mathsf{C}^{\circ}(\bigcup_{\mathbf{y}\in\mathcal{B}'_{\mathbf{x}}} \downarrow\tilde{B}_{\mathbf{y}}(k_{\mathbf{y}}))$, implying that $\bigcup_{\mathbf{y}\in\mathcal{B}'_{\mathbf{x}}} \downarrow\tilde{B}_{\mathbf{y}}(k_{\mathbf{y}})$ *minimally* encases $-z_k$. However, since $(\tilde{B} \setminus \tilde{B}_{\mathbf{x}})(k) \subset \bigcup_{\mathbf{y}\in\mathcal{B}'_{\mathbf{x}}} \downarrow\tilde{B}_{\mathbf{y}}(k_{\mathbf{y}})$ is a proper subset (in view of (6.8)), this contradicts (6.7). So we have now established that $\downarrow\tilde{B}_{\mathbf{x}}(k)$ minimally encases $-u_1$ when all coordinates in the tuple k are sufficiently large. Thus, since $u_i \in \mathbb{R}^{\cup}\langle\mathcal{B}_{\mathbf{x}}\rangle = \mathbb{R}\langle\downarrow\tilde{B}_{\mathbf{x}}(k)\rangle = \mathsf{C}(\downarrow\tilde{B}_{\mathbf{x}}(k) \cup \{u_1\})$ for all $i < t$ (the final equality follows by Lemma 6.6.1 while the initial inclusion was remarked at the start of Case 2), it follows that $\downarrow\tilde{B}_{\mathbf{x}}(k)$ must also minimally encase $-(u_1, \ldots, u_{t-1})$, and Item 1 is established for Case 2 as well.

2. We next show that Item 2 holds, for which we will implicitly make use of the fact that $(\downarrow\tilde{\mathbf{x}})(\hat{k})$ is always a linearly independent subset (noted at the beginning of the proof).

Since $\downarrow\tilde{B}_{\mathbf{x}}(k)$ encases $-(u_1,\ldots,u_{t-1})$, it follows that $u_1,\ldots,u_{t-1} \in \mathbb{R}\langle\downarrow\tilde{B}_{\mathbf{x}}(k)\rangle = \mathbb{R}^{\cup}\langle\mathcal{B}_{\mathbf{x}}\rangle = \partial(\mathbf{x})$. Thus $\mathsf{C}\big((\downarrow\tilde{\mathbf{x}})(\hat{k})\big)$ is contained in the closed half space $\partial(\mathbf{x}) + \mathbb{R}_+ u_t = \bar{\mathbf{x}}$, and likewise $\mathsf{C}^{\circ}\big((\downarrow\tilde{\mathbf{x}})(\hat{k})\big) \subseteq \mathbf{x}^{\circ}$. Moreover,

$$\mathsf{C}\big((\downarrow\tilde{\mathbf{x}})(\hat{k})\big) \cap \partial(\mathbf{x}) = \mathsf{C}(\downarrow\tilde{B}_{\mathbf{x}}(k)) = \mathsf{C}\Big(\bigcup_{\mathbf{y}\in\mathcal{B}_{\mathbf{x}}} (\downarrow\tilde{\mathbf{y}})(\hat{k}_{\mathbf{y}})\Big) = \sum_{\mathbf{y}\in\mathcal{B}_{\mathbf{x}}} \mathsf{C}\big((\downarrow\tilde{\mathbf{y}})(\hat{k}_{\mathbf{y}})\big). \tag{6.9}$$

Applying the induction hypothesis to each $\mathbf{y} \in \mathcal{B}_{\mathbf{x}}$, we find $\mathbf{y} = \lim_{\hat{k}_{\mathbf{y}}\to\infty} \mathsf{C}\big((\downarrow\tilde{\mathbf{y}})(\hat{k}_{\mathbf{y}})\big)$ order uniformly. Consequently, since $\partial(\mathbf{x}) \cap \mathbf{x} = \mathsf{C}^{\cup}(\partial(\{\mathbf{x}\})) = \mathsf{C}^{\cup}(\mathcal{B}_{\mathbf{x}}) = \sum_{\mathbf{y}\in\mathcal{B}_{\mathbf{x}}} \mathbf{y}$, it now follows in view of (6.9) that

$$\lim_{\hat{k}\to\infty} \Big(\mathsf{C}\big((\downarrow\tilde{\mathbf{x}})(\hat{k})\big) \cap \partial(\mathbf{x})\Big) = \lim_{k\to\infty} \mathsf{C}(\downarrow\tilde{B}_{\mathbf{x}}(k)) = \partial(\mathbf{x}) \cap \mathbf{x} \tag{6.10}$$

also order uniformly. It remains to show $\lim_{\hat{k}\to\infty} \mathsf{C}^{\circ}\big((\downarrow\tilde{\mathbf{x}})(\hat{k})\big) = \mathbf{x}^{\circ}$ order uniformly, as this together with (6.10) implies that $\lim_{\hat{k}\to\infty} \mathsf{C}\big((\downarrow\tilde{\mathbf{x}})(\hat{k})\big) = \mathbf{x}$ order uniformly as well, which will complete the proof. (Note this requires the global constant $N_{\mathbf{x}}$ in the order uniform limits to be at least the value for the global constant obtained inductively for each $\mathbf{y} \in \mathcal{B}_{\mathbf{x}} \subseteq \mathcal{X}_1 \cup \ldots \mathcal{X}_{s-1}$.) Let

$$\tilde{\mathbf{x}}(i_{\mathbf{x}}) = (a^{(1)}_{i_{\mathbf{x}}} u_1 + \ldots + a^{(t-1)}_{i_{\mathbf{x}}} u_{t-1} + w_{i_{\mathbf{x}}}) + b_{i_{\mathbf{x}}} \bar{u}_t = \tilde{\mathbf{x}}^{(t-1)}(i_{\mathbf{x}}) + b_{i_{\mathbf{x}}} \bar{u}_t$$

for $i_{\mathbf{x}} \geq 1$ be as given by (6.1). In particular, $w_{i_{\mathbf{x}}} \in \mathbb{R}^{\cup}\langle\mathcal{B}_{\mathbf{x}}\rangle$ with $b_{i_{\mathbf{x}}}, \|w_{i_{\mathbf{x}}}\| \in o(a^{(t-1)}_{i_{\mathbf{x}}})$.

Since $\downarrow\tilde{B}_{\mathbf{x}}(k)$ minimally encases $-(u_1,\ldots,u_{t-1})$ so long as all coordinates in k are sufficiently large, we can fix one particular tuple $\kappa = (\iota_{\mathbf{y}})_{\mathbf{y}\in\downarrow\mathcal{B}_{\mathbf{x}}}$ such that $\downarrow\tilde{B}_{\mathbf{x}}(\kappa)$ minimally encases $-(u_1,\ldots,u_{t-1})$. As shown above, $\lim_{k\to\infty} \mathsf{C}(\downarrow\tilde{B}_{\mathbf{x}}(k)) = \partial(\mathbf{x}) \cap \mathbf{x}$ order uniformly. Consequently, so long as we choose the coordinates in the tuple κ sufficiently large, we can be assured that $\mathsf{C}(\downarrow\tilde{B}_{\mathbf{x}}(\kappa)) \subseteq \mathsf{C}(\downarrow\tilde{B}_{\mathbf{x}}(k))$ for any tuple k with all coordinates sufficiently large. (Note each coordinate $\iota_{\mathbf{z}}$ in κ, for $\mathbf{z} \in \downarrow\mathcal{B}_{\mathbf{x}}$, must be at least the value of the global constant from the order uniform limit $\lim_{k\to\infty} \mathsf{C}(\downarrow\tilde{B}_{\mathbf{x}}(k))$ obtained above inductively, and then $N_{\mathbf{x}}$ must be at least $\max_{\mathbf{z}\in\downarrow\mathcal{B}_{\mathbf{x}}} N_{\iota_{\mathbf{z}}}$, where $N_{\iota_{\mathbf{z}}}$ is the relative constant given in the definition of the order uniform limit $\lim_{k\to\infty} \mathsf{C}(\downarrow\tilde{B}_{\mathbf{x}}(k))$.)

To show $\lim_{\hat{k}\to\infty} \mathsf{C}^{\circ}\big((\downarrow\tilde{\mathbf{x}})(\hat{k})\big) = \mathbf{x}^{\circ}$, let $z = \alpha\bar{u}_t + w \in \mathbf{x}^{\circ} = (\partial(\mathbf{x}) + \mathbb{R}_+\bar{u}_t)^{\circ}$ be an arbitrary element with $\alpha > 0$ and $w \in \partial(\mathbf{x}) = \mathbb{R}\langle\downarrow\tilde{B}_{\mathbf{x}}(k)\rangle$. Now

$$(\alpha/b_{i_{\mathbf{x}}})\tilde{\mathbf{x}}(i_{\mathbf{x}}) = \alpha\bar{u}_t + (\alpha/b_{i_{\mathbf{x}}})\tilde{\mathbf{x}}^{(t-1)}(i_{\mathbf{x}}) = z - w + (\alpha/b_{i_{\mathbf{x}}})\tilde{\mathbf{x}}^{(t-1)}(i_{\mathbf{x}}), \tag{6.11}$$

and $\{w - (\alpha/b_{i_{\mathbf{x}}})\tilde{\mathbf{x}}^{(t-1)}(i_{\mathbf{x}})\}_{i_{\mathbf{x}}=1}^{\infty}$ is an asymptotically filtered sequence of terms (once $i_{\mathbf{x}}$ is sufficiently large) from $\partial(\mathbf{x}) = \mathbb{R}\langle\downarrow\tilde{B}_{\mathbf{x}}(k)\rangle$ with limit $-(u_1, \ldots, u_{t-1})$ minimally encased by $(\downarrow\tilde{B}_{\mathbf{x}})(\kappa)$, in view of $b_{i_{\mathbf{x}}} \in o(a_{i_{\mathbf{x}}}^{(j)})$ for all $j < t$. Thus, by Proposition 4.20, $w - (\alpha/b_{i_{\mathbf{x}}})\tilde{\mathbf{x}}^{(t-1)}(i_{\mathbf{x}}) \in \mathsf{C}^{\circ}((\downarrow\tilde{B}_{\mathbf{x}})(\kappa)) \subseteq \mathsf{C}^{\circ}((\downarrow\tilde{B}_{\mathbf{x}})(k))$ for all sufficiently large $i_{\mathbf{x}}$ (independent of k since $\mathsf{C}^{\circ}((\downarrow\tilde{B}_{\mathbf{x}})(\kappa))$ is a fixed set), which combined with (6.11) ensures that $z \in \mathsf{C}^{\circ}((\downarrow\tilde{\mathbf{x}})(\hat{k}))$ for any tuple $\hat{k}$ with all coordinates sufficiently large. This shows that $\lim_{\hat{k}\to\infty} \mathsf{C}^{\circ}\big((\downarrow\tilde{\mathbf{x}})(\hat{k})\big) = \mathbf{x}^{\circ}$. It remains to show the limit holds order uniformly. For this, it suffices, in view of the order uniform limit in (6.10), to show that, for each sufficiently large integer $m \geq N_{\mathbf{x}}$, there is a bound N_m such that $\mathsf{C}^{\circ}(\downarrow\tilde{\mathbf{x}}(\hat{k})) \subseteq \mathsf{C}^{\circ}(\downarrow\tilde{\mathbf{x}}(\hat{k}'))$ whenever $\hat{k} = (i_{\mathbf{z}})_{i_{\mathbf{z}}\in\downarrow\mathbf{x}}$ and $\hat{k}' = (i'_{\mathbf{z}})_{i'_{\mathbf{z}}\in\downarrow\mathbf{x}}$ are tuples of sufficiently large indices $i_{\mathbf{z}}, i'_{\mathbf{z}} \geq N_{\mathbf{x}}$ with

$$i_{\mathbf{x}} = m, \quad i'_{\mathbf{x}} \geq N_m, \quad \text{and} \quad i_{\mathbf{z}} = i'_{\mathbf{z}} \quad \text{for } \mathbf{z} \in \downarrow\mathcal{B}_{\mathbf{x}}.$$

(Note this reduction requires the global bound $N_{\mathbf{x}}$ be at least the global bound for the order uniform limit in (6.10).) In particular, we have $k = k'$ as only the entry indexed by $\mathbf{x}$ differs between $\hat{k}$ and $\hat{k}'$.

Let $z + \alpha\overline{u}_t \in \mathbf{x}^{\circ} = (\partial(\mathbf{x}) + \mathbb{R}_+\overline{u}_t)^{\circ}$ be an arbitrary element, where $\alpha > 0$ and $z \in \partial(\mathbf{x})$. Since $\overline{u}_t \notin \partial(\mathbf{x})$, any linear combination of elements from $\downarrow\tilde{\mathbf{x}}(\hat{k})$ equal to $z + \alpha\overline{u}_t$ must have the coefficient of $\tilde{\mathbf{x}}(i_{\mathbf{x}})$ being $\alpha/b_{i_{\mathbf{x}}}$. In consequence, $z+\alpha\overline{u}_t \in \mathsf{C}^{\circ}(\downarrow\tilde{\mathbf{x}}(\hat{k}))$ is equivalent to $z+\alpha\overline{u}_t - (\alpha/b_{i_{\mathbf{x}}})\tilde{\mathbf{x}}(i_{\mathbf{x}}) = z - (\alpha/b_{i_{\mathbf{x}}})\tilde{\mathbf{x}}^{(t-1)}(i_{\mathbf{x}}) \in \mathsf{C}^{\circ}(\downarrow\tilde{B}_{\mathbf{x}}(k))$. This means the elements $z+\alpha\overline{u}_t \in \mathsf{C}^{\circ}(\downarrow\tilde{\mathbf{x}}(\hat{k}))$ are those with

$$z - (\alpha/b_{i_{\mathbf{x}}})\tilde{\mathbf{x}}^{(t-1)}(i_{\mathbf{x}}) \in \mathsf{C}^{\circ}(\downarrow\tilde{B}_{\mathbf{x}}(k)), \tag{6.12}$$

and we simply need to know

$$z - (\alpha/b_{i'_{\mathbf{x}}})\tilde{\mathbf{x}}^{(t-1)}(i'_{\mathbf{x}}) \in \mathsf{C}^{\circ}(\downarrow\tilde{B}_{\mathbf{x}}(k))$$

also holds for all sufficiently large $i'_{\mathbf{x}}$ (independent of $k = k'$) to show $z + \alpha u_t \in \mathsf{C}^{\circ}(\downarrow\tilde{\mathbf{x}}(\hat{k}'))$ (since $i_{\mathbf{z}} = i'_{\mathbf{z}}$ for all $\mathbf{z} \in \downarrow\mathcal{B}_{\mathbf{x}}$). To achieve this, in view of (6.12) and the fact that $\mathsf{C}(\downarrow\tilde{B}_{\mathbf{x}}(k))$ is a convex cone, it suffices to know

$$z_{i'_{\mathbf{x}}} := (1/b_{i_{\mathbf{x}}})\tilde{\mathbf{x}}^{(t-1)}(i_{\mathbf{x}}) - (1/b_{i'_{\mathbf{x}}})\tilde{\mathbf{x}}^{(t-1)}(i'_{\mathbf{x}}) \in \mathsf{C}(\downarrow\tilde{B}_{\mathbf{x}}(\kappa)) \subseteq \mathsf{C}(\downarrow\tilde{B}_{\mathbf{x}}(k)) \tag{6.13}$$

for all sufficiently large $i'_{\mathbf{x}}$ (independent of k and $z+\alpha\vec{u}_t$ but dependent on $m = i_{\mathbf{x}}$). However, each

$$z_{i'_{\mathbf{x}}} = \sum_{j=1}^{t-1}\Big(a^{(j)}_{i_{\mathbf{x}}}/b_{i_{\mathbf{x}}} - a^{(j)}_{i'_{\mathbf{x}}}/b_{i'_{\mathbf{x}}}\Big)u_j + (1/b_{i_{\mathbf{x}}})w_{i_{\mathbf{x}}} - (1/b_{i'_{\mathbf{x}}})w_{i'_{\mathbf{x}}}$$

$$= -\sum_{j=1}^{t-1}(a^{(j)}_{i'_{\mathbf{x}}}/b_{i'_{\mathbf{x}}})u_j + \Big(\sum_{j=1}^{t-1}(a^{(j)}_m/b_m)u_j + (1/b_m)w_m - (1/b_{i'_{\mathbf{x}}})w_{i'_{\mathbf{x}}}\Big)$$

with $a^{(j)}_{i'_{\mathbf{x}}}/b_{i'_{\mathbf{x}}} \to \infty$ in view of $b_{i'_{\mathbf{x}}} \in o(a^{(j)}_{i'_{\mathbf{x}}})$ and $j < t$ and with $\|(1/b_{i'_{\mathbf{x}}})w_{i'_{\mathbf{x}}}\| \in o(a^{(t-1)}_{i'_{\mathbf{x}}}/b_{i'_{\mathbf{x}}})$ (recall that $i_{\mathbf{x}} = m$ is fixed). Thus the sequence $\{z_{i'_{\mathbf{x}}}\}_{i'_{\mathbf{x}}=N_{\mathbf{x}}}^{\infty}$ is an asymptotically filtered sequence of terms from $\mathbb{R}^{\cup}\langle\mathcal{B}_{\mathbf{x}}\rangle$ with limit $-(u_1,\ldots,u_{t-1})$ (once $i'_{\mathbf{x}}$ is sufficiently large). As a result, since $\downarrow\mathcal{B}_{\mathbf{x}}(\kappa)$ minimally encases $-(u_1,\ldots,u_{t-1})$, it follows from Proposition 4.20 that $z_{i'_{\mathbf{x}}} \in \mathsf{C}(\downarrow\tilde{B}_{\mathbf{x}}(\kappa)) \subseteq \mathsf{C}(\downarrow\tilde{B}_{\mathbf{x}}(k))$ for all sufficiently large $i'_{\mathbf{x}} \geq N_m$ (independent of k and $z+\alpha\vec{u}_t$ but dependent on $i_{\mathbf{x}} = m$), as desired. This establishes (6.13) and completes the proof. (Note this final step for Item 2 requires the relative constant N_m be at least the value resulting from the application of Proposition 4.20 to the *fixed* set $\downarrow\tilde{B}_{\mathbf{x}}(\kappa)$).

3. In view of the main part in Item 1, let $\kappa = (\iota_{\mathbf{z}})_{\mathbf{z}\in\downarrow\mathcal{B}_{\mathbf{x}}}$ be a *fixed* tuple with all $\iota_{\mathbf{z}}$ sufficiently large that $\downarrow\tilde{B}_{\mathbf{x}}(\kappa)$ minimally encases $-(u_1,\ldots,u_{t-1})$. Moreover, choose the $\iota_{\mathbf{z}}$ to each be at least the global constant from the order uniform limits given in Item 2. Then, for any tuple $k = (i_{\mathbf{z}})_{\mathbf{z}\in\downarrow\mathcal{B}_{\mathbf{x}}}$ with all $i_{\mathbf{z}}$ sufficiently large, we have $\mathsf{C}^{\circ}(\downarrow\tilde{B}_{\mathbf{x}}(\kappa)) = \sum_{\mathbf{y}\in\mathcal{B}_{\mathbf{x}}}\mathsf{C}^{\circ}(\downarrow\tilde{\mathbf{y}}(\kappa)) \subseteq \sum_{\mathbf{y}\in\mathcal{B}_{\mathbf{x}}}\mathsf{C}^{\circ}(\downarrow\tilde{\mathbf{y}}(k)) = \mathsf{C}^{\circ}(\downarrow\tilde{B}_{\mathbf{x}}(k))$ by Item 2 applied to each $\mathbf{y}\in\mathcal{B}_{\mathbf{x}}$. By Proposition 4.20, $\downarrow\tilde{B}_k(\kappa)\cup\{\tilde{x}^{(t-1)}(i_{\mathbf{x}})\}$ is a minimal positive basis for all sufficiently large $i_{\mathbf{x}}$, implying $-\tilde{x}^{(t-1)}(i_{\mathbf{x}}) \in \mathsf{C}^{\circ}(\downarrow\tilde{B}_{\mathbf{x}}(\kappa)) \subseteq \mathsf{C}^{\circ}(\downarrow\tilde{B}_{\mathbf{x}}(k))$, and thus ensuring that $\downarrow\tilde{B}_k(k)\cup\{\tilde{x}^{(t-1)}(i_{\mathbf{x}})\}$ is a minimal positive basis for all sufficiently large $i_{\mathbf{x}}$ (dependent only on the *fixed* set $\downarrow\tilde{B}_k(\kappa)$), which completes the proof. (Note each coordinate ι_z for κ must be at least the constant $N_{\mathbf{x}}$ needed to obtain the main part of Item 1 for $\mathbf{x}$ as well as the global constant from the order uniform limits $\lim_{k\to\infty}\mathsf{C}^{\circ}(\downarrow\tilde{\mathbf{y}}(k)) = \mathbf{y}^{\circ}$ for $\mathbf{y}\in\mathcal{B}_{\mathbf{x}}$, and then $N_{\mathbf{x}}$ must be at least $\max_{\mathbf{y}\in\mathcal{B}_{\mathbf{x}},\,\mathbf{z}\in\downarrow\mathbf{y}} N_{\iota_{\mathbf{z}}}$, where $N_{\iota_{\mathbf{z}}}$ is the relative constant given in the definition of the order uniform limits $\lim_{k\to\infty}\mathsf{C}^{\circ}(\downarrow\tilde{\mathbf{y}}(k)) = \mathbf{y}^{\circ}$ for $\mathbf{y}\in\mathcal{B}_{\mathbf{x}}$, as well as at least the value given by Proposition 4.20 applied to the fixed set $\downarrow\tilde{B}_{\mathbf{x}}(\kappa)$).

□

For a half-space $\mathbf{x}$ from a general virtual Reay system, we only have $\tilde{\mathbf{x}}(i)$ as a representative for $\mathbf{x}$, which is a truncated approximation of the actual element $\mathbf{x}(i)\in G_0$. The following proposition shows that, when G_0 is a subset of a lattice, there is no need to truncate when each $\mathbf{y}\in\downarrow\mathbf{x}$ has $\vec{u}_{\mathbf{y}}$ anchored. We remark that the hypothesis in Proposition 6.8 that $\vec{u}_{\mathbf{y}}$ be anchored for every $\mathbf{y} \in \downarrow\mathbf{x}$ holds automatically when $\mathcal{R}$ is anchored and $\mathbf{x} \in \mathcal{X}_1 \cup \ldots \cup \mathcal{X}_s$.

Proposition 6.8 *Let $\Lambda \subseteq \mathbb{R}^d$ be a full rank lattice, let $G_0 \subseteq \Lambda$ be a subset, and let $\mathcal{R} = (\mathcal{X}_1 \cup \{v_1\}, \ldots, \mathcal{X}_s \cup \{v_s\})$ be a virtual Reay system in G_0. Suppose $\mathbf{x} \in \mathcal{X}_1 \cup \{v_1\} \cup \ldots \cup \mathcal{X}_s \cup \{v_s\}$ with $\vec{u}_{\mathbf{y}}$ anchored for every $\mathbf{y} \in \downarrow\mathbf{x}$. Then $\mathbf{x}(i) = \tilde{\mathbf{x}}(i)$ for all sufficiently large i and $\vec{u}^{\triangleleft}_{\mathbf{x}}$ is either trivial or fully unbounded. Moreover, if $\vec{u}_{\mathbf{x}} = (u_1, \ldots, u_t)$, then $\mathbf{x}(i) - \tilde{\mathbf{x}}^{(t-1)}(i) = \xi \neq 0$ is constant for all sufficiently large i. In particular, if $t = 1$, $\mathbf{x}(i) = \xi$ for all sufficiently large i.*

Proof Let $\vec{u}_{\mathbf{x}} = (u_1, \ldots, u_t)$, where $t \geq 1$, be the limit associated to $\mathbf{x}$, let $\mathcal{B}_{\mathbf{x}} = \partial(\{\mathbf{x}\})$ and let $\pi : \mathbb{R}^d \to \mathbb{R}^{\cup}\langle\mathcal{B}_{\mathbf{x}}\rangle^{\perp}$ and $\pi^{\perp} : \mathbb{R}^d \to \mathbb{R}^{\cup}\langle\mathcal{B}_{\mathbf{x}}\rangle$ be the orthogonal projections. Let

$$\mathbf{x}(i) = (a_i^{(1)} u_1 + \ldots + a_i^{(t-1)} u_{t-1} + w_i) + b_i \overline{u}_t + y_i$$

be the asymptotically filtered sequence of lattice points $\mathbf{x}(i) \in G_0 \subseteq \Lambda$ with limit $(u_1, \ldots, u_t)$ in the form given by (6.1), where $\overline{u}_t = \pi(u_t)/\|\pi(u_t)\|$. In particular, $\pi(\overline{u}_t) = \overline{u}_t$ is a positive multiple of $\pi(u_t)$, $y_i \in \mathbb{R}\langle\mathbf{x}\rangle^{\perp}$, $\|y_i\| \in o(b_i)$, $w_i \in \mathbb{R}^{\cup}\langle\partial(\{\mathbf{x}\})\rangle$, and b_i, $\|w_i\| \in o(a_i^{(t-1)})$. We proceed inductively on s. In particular, since $\vec{u}_{\mathbf{y}}$ is anchored for every $\mathbf{y} \in \downarrow\mathcal{B}_{\mathbf{x}} = \downarrow\partial(\{\mathbf{x}\}) \subseteq \downarrow\mathbf{x}$, and since $\downarrow\mathcal{B}_{\mathbf{x}} \subseteq \mathcal{X}_1 \cup \ldots \cup \mathcal{X}_{j-1}$ when $\mathbf{x} \in \mathcal{X}_j$, it follows from the induction hypothesis that $\mathbb{R}^{\cup}\langle\mathcal{B}_{\mathbf{x}}\rangle = \mathbb{R}\langle\downarrow\tilde{B}_{\mathbf{x}}(k)\rangle = \mathbb{R}\langle\downarrow B_{\mathbf{x}}(k)\rangle$ is a subspace generated by the lattice points $\downarrow B_{\mathbf{x}}(k) \subseteq G_0 \subseteq \Lambda$ (this is trivially true when $s = 1$, in which case π is the identity map and $\downarrow\mathcal{B}_{\mathbf{x}} = \emptyset$) for any tuple k with all coordinates sufficiently large. Thus, by Proposition 2.1, $\pi(G_0) \subseteq \pi(\Lambda)$ is a subset of the full rank lattice $\pi(\Lambda) \subseteq \mathbb{R}^{\cup}\langle\mathcal{B}_{\mathbf{x}}\rangle^{\perp}$. Hence $\{\pi(\mathbf{x}(i))\}_{i=1}^{\infty}$ is a bounded sequence of lattice points from $\mathbb{R}^{\cup}\langle\mathcal{B}_{\mathbf{x}}\rangle^{\perp}$ (in view of $\vec{u}_{\mathbf{x}}$ being *anchored* by hypothesis), meaning there are only a *finite* number of possibilities for the values of

$$\pi(\mathbf{x}(i)) = \pi(\tilde{\mathbf{x}}(i)) + \pi(y_i) = b_i \pi(\overline{u}_t) + \pi(y_i) = b_i \overline{u}_t + \pi(y_i).$$

Since b_i is bounded (as $\vec{u}_{\mathbf{x}}$ is anchored) and $\|y_i\| \in o(b_i)$, it follows that $\pi(y_i) \to 0$ and

$$\lim_{i\to\infty} \pi(\mathbf{x}(i)) = (\lim_{i\to\infty} b_i)\overline{u}_t = \xi, \tag{6.14}$$

for some $\xi \in \mathbb{R}^{\cup}\langle\mathcal{B}_{\mathbf{x}}\rangle^{\perp}$ (recall that $b_i \in \Theta(a_i^{(t)})$ is a convergent sequence by definition of an asymptotically filtered sequence $\mathbf{x}(i)$). Consequently, since any convergent sequence of lattice points must eventually stabilize, it follows, for all sufficiently large i, that

$$b_i \overline{u}_i + \pi(y_i) = \pi(\mathbf{x}(i)) = \xi \in \mathbb{R}^{\cup}\langle\mathcal{B}_{\mathbf{x}}\rangle^{\perp} \cap \pi(\Lambda).$$

If $\xi = 0$, then $b_i\|\overline{u}_t\| - \|\pi(y_i)\| \le \|\pi(\mathbf{x}(i))\| = 0$ for all sufficiently large i (by the triangle inequality), implying $1 = \|\overline{u}_t\| \le \|\pi(y_i)\|/b_i$ for all sufficiently large i. However, $\|\pi(y_i)\|/b_i \to 0$ in view of $\|y_i\| \in o(b_i)$, making this impossible. Therefore we conclude that $\xi \neq 0$, in which case the limit of b_i must be nonzero. Thus, since $b_i \in o(a_i^{(t-1)})$ when $t \ge 2$, we must have $a_i^{(t-1)}$ being unbounded for $t \ge 2$, meaning $\vec{u}_{\mathbf{x}}^{\triangleleft} = (u_1, \ldots, u_{t-1})$ is either trivial or fully unbounded. Also, (6.14) now ensures that ξ is a positive multiple of $\overline{u}_t$, so $\xi = \|\xi\|\overline{u}_t$ and $\mathbb{R}\langle\mathbf{x}\rangle = \mathbb{R}^{\cup}\langle\mathcal{B}_{\mathbf{x}}\rangle + \mathbb{R}\xi$.

Since $\xi \neq 0$, it follows that the space $\mathbb{R}\overline{u}_t = \mathbb{R}\xi$ is linearly spanned by the lattice point $\xi \in \pi(\Lambda)$, in which case Proposition 2.1 ensures $\pi'(\Lambda)$ is a lattice, where $\pi' : \mathbb{R}^d \to (\mathbb{R}^{\cup}\langle\mathcal{B}_{\mathbf{x}}\rangle + \mathbb{R}\xi)^{\perp}$ is the orthogonal projection. Now $\pi'(\mathbf{x}(i)) = \pi'(y_i) = y_i$, with the latter equality in view of $y_i \in \mathbb{R}\langle\mathbf{x}\rangle^{\perp} = (\mathbb{R}^{\cup}\langle\mathcal{B}_{\mathbf{x}}\rangle + \mathbb{R}\xi)^{\perp}$, and we also have $\pi'(y_i) \to 0$ (since $\pi(y_i) \to 0$). Hence, since any convergent sequence of lattice points must stabilize, we conclude $y_i = \pi'(y_i) = \pi'(\mathbf{x}(i)) = 0$ for all sufficiently large i, in turn implying $\tilde{\mathbf{x}}(i) = \mathbf{x}(i)$, and all parts of the proposition follow. □

When dealing with a virtual Reay system over a subset of lattice points G_0, we will now always assume, by removing the first few terms, that the representative sequences $\{\mathbf{x}(i)\}_{i=1}^{\infty}$ for $\mathbf{x} \in \mathcal{X}_1 \cup \{\mathbf{v}_1\} \cup \ldots \cup \mathcal{X}_s \cup \{\mathbf{v}_s\}$ when $\vec{u}_{\mathbf{y}}$ is anchored for all $\mathbf{y} \in \downarrow\mathbf{x}$ (in particular, for all $\mathbf{x} \in \mathcal{X}_1 \cup \ldots \cup \mathcal{X}_s$ when $\mathcal{R}$ is anchored) satisfy the conclusion of Proposition 6.8 for all $i \ge 1$. Then, combining Propositions 6.8 and 6.7, we can remove all $\sim$'s from the statement of Proposition 6.7 for such $\mathbf{x}$. Likewise, if $\mathcal{B} \subseteq \mathcal{X}_1 \cup \ldots \cup \mathcal{X}_s$ and $\mathcal{R}$ is anchored, then Proposition 6.8 implies that $\vec{u}_{\mathbf{x}}^{\triangleleft}$ is trivial or fully unbounded for every half-space $\mathbf{x}$ from $\downarrow\mathcal{B}$, in which case Proposition 6.5 outputs a virtual Reay system $\mathcal{R}'$ which is anchored, and also purely virtual provided the limit $\vec{u}$ is fully unbounded. Thus we gain important simplifications when restricting to virtual Reay systems over a subset of lattice points $G_0 \subseteq \Lambda$.

Our next goal is to provide the analog of Proposition 4.13 for virtual Reay systems (and thus for oriented Reay systems as well), which will be done with some effort in Proposition 6.12. However, we first need some lemmas (Lemma 6.11 will be needed in the next chapter).

Lemma 6.9 *Let $G_0 \subseteq \mathbb{R}^d$ be a subset, where $d \ge 1$, let $u \in \mathbb{R}^d$ be a unit vector, and let $\mathcal{R} = (\mathcal{X}_1 \cup \{v_1\}, \ldots, \mathcal{X}_s \cup \{v_s\})$ be a virtual Reay system in G_0. Suppose there is some virtual independent set $\mathcal{A} \subseteq \mathcal{X}_1 \cup \{v_1\} \cup \ldots \cup \mathcal{X}_s \cup \{v_s\}$ which minimally encases $-u$. Then, for any tuple $k = (i_{\mathbf{y}})_{\mathbf{y} \in \downarrow\mathcal{A}}$ with all $i_{\mathbf{y}}$ sufficiently large, $\downarrow\tilde{A}(k)$ minimally encases $-u$ and $\downarrow\tilde{A}(k) \cup \{u\}$ is a minimal positive basis.*

Proof Since $\mathcal{A}$ is a virtual independent set, $\downarrow\tilde{A}(k)$ is linearly independent for any tuple k. Let $\mathbf{z}_1, \ldots, \mathbf{z}_\ell \in \mathcal{A}$ be the distinct half-spaces in $\mathcal{A}$. Since $\mathcal{A}$ minimally encases $-u$, Lemma 5.7 and Proposition 5.3.4 imply $-u \in \mathbf{z}_1^{\circ} + \ldots + \mathbf{z}_\ell^{\circ} = \mathsf{C}^{\cup}(\mathcal{A})^{\circ}$. Hence, by Proposition 6.7.2 applied to each $\mathbf{z}_j^{\circ}$, we find that $-u \in \mathsf{C}^{\circ}(\downarrow\tilde{A}(k))$ for any tuple k with all $i_{\mathbf{y}}$ sufficiently large, and since $\downarrow\tilde{A}(k)$ is linearly independent, it follows that $\downarrow\tilde{A}(k)$ minimally encases $-u$ with $\downarrow\tilde{A}(k)\cup\{u\}$ a minimal positive basis, completing the proof. □

Lemma 6.10 *Let $\mathcal{R}=(\mathcal{X}_1\cup\{\mathbf{v}_1\},\ldots,\mathcal{X}_s\cup\{\mathbf{v}_s\})$ be an oriented Reay system in $\mathbb{R}^d$, let $\vec{u}=(u_1,\ldots,u_t)$ be a tuple of $t\geq 1$ orthonormal vectors in $\mathbb{R}^d$, let $\mathcal{B}\subseteq\mathcal{X}_1\cup\ldots\cup\mathcal{X}_s$ be a subset minimally encasing $-\vec{u}^{\triangleleft}$, let $\pi:\mathbb{R}^d\to\mathbb{R}^{\cup}\langle\mathcal{B}\rangle^{\perp}$ be the orthogonal projection, and let $\pi(\mathcal{R})=(\mathcal{X}_j^{\pi}\cup\{\pi(\mathbf{v}_j)\})_{j\in J}$. Suppose $\mathcal{D}\subseteq\bigcup_{j\in J}(\mathcal{X}_j^{\pi}\cup\{\mathbf{v}_j\})$ is a virtual independent set that minimally encases $-\pi(\vec{u})$. Then $\mathcal{C}=(\pi^{-1}(\mathcal{D})\cup\mathcal{B})^*$ is virtual independent and minimally encases $-\vec{u}$ urbanely, $\mathcal{C}^{\pi}=\mathcal{D}$ and $\pi^{-1}(\mathcal{D})\subseteq\mathcal{C}$.*

Proof By Proposition 5.4.1, $\pi(\mathcal{R})$ is an oriented Reay system. By Proposition 5.4.5, $\mathcal{C}$ is a virtual independent set with $\pi^{-1}(\mathcal{D})\subseteq\mathcal{C}$, and $\mathcal{C}^{\pi}=\mathcal{D}^*=\mathcal{D}$ follows by Proposition 5.4.4 and the fact that $\mathcal{D}$ is virtual independent. Since $\mathcal{B}\subseteq\downarrow\mathcal{C}$, it follows that $\mathcal{B}=\mathcal{B}^*\preceq\mathcal{C}$ (with the first equality in view of $\mathcal{B}$ minimally encasing $-\vec{u}^{\triangleleft}$). If $\mathcal{B}=\mathcal{C}$, then $\mathcal{D}=\mathcal{C}^{\pi}=\mathcal{B}^{\pi}=\emptyset$, whence $\pi(\vec{u})$ must be the empty tuple. In this case, $\mathcal{B}=\mathcal{C}$ minimally encases $-\vec{u}$ (cf. Proposition 5.9), with the encasement urbane since $\mathcal{B}\subseteq\mathcal{X}_1\cup\ldots\cup\mathcal{X}_s$, as desired. Therefore, we may assume $\mathcal{B}\prec\mathcal{C}$. Thus, since $\mathcal{B}\subseteq\mathcal{X}_1\cup\ldots\cup\mathcal{X}_s$ minimally encases $-\vec{u}^{\triangleleft}$, since $\mathcal{C}^{\pi}=\mathcal{D}$ is a virtual independent set which minimally encases $-\pi(\vec{u})=-\pi(u_t)/\|\pi(u_t)\|$, and since $\mathcal{C}\subseteq\mathcal{X}_1\cup\ldots\cup\mathcal{X}_s\cup\{\mathbf{v}_j:\ j\in J\}$ per definition of $\mathcal{C}$ and $\pi^{-1}(\mathcal{D})$, it follows from Proposition 5.9.2 that $\mathcal{C}$ minimally encases $-\vec{u}$ urbanely, completing the proof. □

Lemma 6.11 *Let $G_0\subseteq\mathbb{R}^d$, where $d\geq 1$, let $\mathcal{R}=(\mathcal{X}_1\cup\{\mathbf{v}_1\},\ldots,\mathcal{X}_s\cup\{\mathbf{v}_s\})$ be a purely virtual Reay system over G_0, let $\vec{u}=(u_1,\ldots,u_t)$ be a tuple of orthonormal vectors $u_1,\ldots,u_t\in\mathbb{R}^d$, where $t\geq 1$, and let $\mathcal{A}\subseteq\mathcal{X}_1\cup\{\mathbf{v}_1\}\cup\ldots\cup\mathcal{X}_s\cup\{\mathbf{v}_s\}$ be a support set. Suppose $\mathcal{A}$ minimally encases $-\vec{u}$ and there is some $\mathbf{x}\in\mathcal{A}$ with $\vec{u}_{\mathbf{x}}$ fully unbounded. Then there exists some $t_0\in[1,t]$ and $\mathcal{A}'\preceq\mathcal{A}$ such that $\mathcal{A}'$ minimally encases $-(u_1,\ldots,u_{t_0})$* urbanely *and has some $\mathbf{y}\in\mathcal{A}'$ with $\vec{u}_{\mathbf{y}}$ fully unbounded.*

Proof If $t=1$, then the *support* set $\mathcal{A}$ minimally encases $-u_1$ and $\mathcal{A}$ does not *minimally* encase the empty tuple, while the empty set $\emptyset\subseteq\mathcal{X}_1\cup\ldots\cup\mathcal{X}_s$ does. Thus $\mathcal{A}$ minimally encases $-u_1$ urbanely (as the regularity condition required for urbane encasement holds automatically for support sets, as remarked after the definition of urbane encasement), and the lemma follow taking $\mathcal{A}'=\mathcal{A}$, $t_0=t=1$ and $\mathbf{y}=\mathbf{x}$. Therefore we may assume $t\geq 2$ and proceed by induction on t. Let $t'\in[0,t-1]$ be the minimal index such that $\mathcal{A}$ does not minimally encase $-(u_1,\ldots,u_{t'})$ and let $\mathcal{A}'\prec\mathcal{A}$ be such that $\mathcal{A}'$ minimally encases $-(u_1,\ldots,u_{t'})$. If $\mathcal{A}'\subseteq\mathcal{X}_1\cup\ldots\cup\mathcal{X}_s$, then the *support* set $\mathcal{A}$ minimally encases $-\vec{u}$ urbanely, and the lemma follows taking $\mathcal{A}'=\mathcal{A}$, $t_0=t$ and $\mathbf{y}=\mathbf{x}$. Otherwise, $t'\geq 1$ and there must be some $\mathbf{v}_j\in\mathcal{A}'$, and since $\mathcal{R}$ is purely virtual, $\vec{u}_{\mathbf{v}_j}$ is fully unbounded. Since $\mathcal{A}'$ minimally encases $-(u_1,\ldots,u_{t'})$, we must have $(\mathcal{A}')^*=\mathcal{A}'$, while $\downarrow\mathcal{A}'\subseteq\downarrow\mathcal{A}$ ensures $\mathcal{X}_j\subseteq\{\mathbf{v}_j\}\nsubseteq\downarrow\mathcal{A}'$ for all $j\in[1,s]$ as $\mathcal{X}_j\subseteq\{\mathbf{v}_j\}\nsubseteq\downarrow\mathcal{A}$ for the support set $\mathcal{A}$. In consequence, $\mathcal{A}'$ is also a support set. Thus we can apply the induction hypothesis to $\mathcal{A}'$ to find some $\mathcal{A}''\preceq\mathcal{A}'\prec\mathcal{A}$ and some $t_0\in[1,t']\subseteq[1,t]$ such that $\mathcal{A}''$

minimally encases $-(u_1, \ldots, u_{t_0})$ urbanely and contains some $\mathbf{y} \in \mathcal{A}''$ with $\vec{u}_{\mathbf{y}}$ fully unbounded, as desired. □

Proposition 6.12.2(c) ensures that $\mathcal{A}_{\mathbf{x}}$ is the lift modulo $\partial(\mathbf{x})$ of the unique support set minimally encasing $-\vec{u}_{\mathbf{x}}$ modulo $\partial(\mathbf{x})$, and $\mathcal{A}_j$ is simply the pull-back of this same set union $\mathbf{x}$ with it not mattering which half-space $\mathbf{x} \in \mathcal{X}_j \cup \{\mathbf{v}_j\}$ is used to perform this construction by Proposition 6.12.1. Furthermore, the other half-spaces $\mathbf{y} \in \mathcal{A}_j$ behave symmetrically in this regard, with the exception that we do *not* have $\mathcal{A}_{\mathbf{y}}$ and $\mathcal{A}_{\mathbf{y}}^{\pi_{\mathbf{y}}}$ being support sets; instead, they are simply virtual independent sets, which is only slightly weaker.

Proposition 6.12 *Let $G_0 \subseteq \mathbb{R}^d$ be a subset, where $d \geq 1$, and let $\mathcal{R} = (\mathcal{X}_1 \cup \{\mathbf{v}_1\}, \ldots, \mathcal{X}_s \cup \{\mathbf{v}_s\})$ be a virtual Reay system in G_0. For $\mathbf{x} \in \mathcal{X}_1 \cup \{\mathbf{v}_1\} \cup \ldots \cup \mathcal{X}_s \cup \{\mathbf{v}_s\}$, let $\pi_{\mathbf{x}} : \mathbb{R}^d \to \mathbb{R}^{\cup}\langle\partial(\{\mathbf{x}\})\rangle^{\perp}$ denote the orthogonal projection and $\mathcal{D}_{\mathbf{x}} = \mathsf{Supp}_{\pi_{\mathbf{x}}(\mathcal{R})}(-\pi_{\mathbf{x}}(\vec{u}_{\mathbf{x}}))$. Let $j \in [1, s]$.*

1. *We have*

$$\mathcal{A}_j := \pi_{\mathbf{x}}^{-1}(\mathcal{D}_{\mathbf{x}}) \cup \{\mathbf{x}\} = \pi_{\mathbf{z}}^{-1}(\mathcal{D}_{\mathbf{z}}) \cup \{\mathbf{z}\} \quad \text{for every } \mathbf{x}, \mathbf{z} \in \mathcal{X}_j \cup \{\mathbf{v}_j\}.$$

2. *For every $\mathbf{x} \in \mathcal{X}_j \cup \{\mathbf{v}_j\}$ and every $\mathbf{y} \in \mathcal{A}_j$, the following hold.*
   (a) *$\mathcal{X}_j \cup \{\mathbf{v}_j\} \subseteq \mathcal{A}_j \subseteq \mathcal{X}_1 \cup \{\mathbf{v}_1\} \cup \ldots \cup \mathcal{X}_j \cup \{\mathbf{v}_j\}$ and $\mathcal{A}_j^* = \mathcal{A}_j$.*
   (b) *$\mathcal{A}_{\mathbf{y}} := \big(\mathcal{A}_j \setminus \{\mathbf{y}\} \cup \partial(\{\mathbf{y}\})\big)^*$ and $\mathcal{A}_{\mathbf{y}}^{\pi_{\mathbf{y}}} = \pi_{\mathbf{y}}(\mathcal{A}_j \setminus \{\mathbf{y}\})$ are virtual independent sets minimally encasing $-\vec{u}_{\mathbf{y}}$ and $-\pi_{\mathbf{y}}(\vec{u}_{\mathbf{y}})$, respectively, with $\mathcal{A}_j \setminus \{\mathbf{y}\} \subseteq \mathcal{A}_{\mathbf{y}}$.*
   (c) *$\mathcal{A}_{\mathbf{x}} = \big(\pi_{\mathbf{x}}^{-1}(\mathcal{D}_{\mathbf{x}}) \cup \partial(\{\mathbf{x}\})\big)^*$ and $\mathcal{A}_{\mathbf{x}}^{\pi_{\mathbf{x}}} = \mathcal{D}_{\mathbf{x}}$ are support sets.*
   (d) *$\mathcal{A}_j = \pi_{\mathbf{y}}^{-1}(\mathcal{A}_{\mathbf{y}}^{\pi_{\mathbf{y}}}) \cup \{\mathbf{y}\}$ and $\pi_{\mathbf{y}}^{-1}(\mathcal{A}_{\mathbf{y}}^{\pi_{\mathbf{y}}}) = \mathcal{A}_j \setminus \{\mathbf{y}\}$.*
   (e) *$\mathcal{A}_j \subseteq \mathcal{X}_1 \cup \ldots \cup \mathcal{X}_s \cup \{\mathbf{v}_i : i \in J\}$ and A_j^{π} is a minimal positive basis of size $|\mathcal{A}_j|$, where $\pi : \mathbb{R}^d \to \mathbb{R}^{\cup}\langle\partial(\mathcal{A}_j)\rangle^{\perp}$ is the orthogonal projection and $\pi(\mathcal{R}) = (\mathcal{X}_i^{\pi} \cup \{\pi(\mathbf{v}_i)\})_{i \in J}$.*
3. *If $\mathsf{C}(G_0) = \mathbb{R}^d$, then there exists a finite set $Y \subseteq G_0$ such that, for any $\mathbf{x} \in \mathcal{A}_j$ and any tuple $k = (i_{\mathbf{z}})_{\mathbf{z} \in \downarrow \mathcal{A}_{\mathbf{x}}}$ with all $i_{\mathbf{z}}$ sufficiently large, there is a subset $Y_k \subseteq Y$ such that*

$$(A_j \setminus \{x\})(k) \cup \downarrow\widetilde{\partial(A_j)}(k) \cup Y_k$$

minimally encases $-\vec{u}_{\mathbf{x}}$. Moreover, if $(A_j \setminus \{x\})(k) \subseteq \mathbb{R}^{\cup}\langle\mathcal{A}_j\rangle$, then $Y_k = \emptyset$.

Proof If $\mathbf{x} \in \mathcal{X}_j \cup \{\mathbf{v}_j\}$ with $\vec{u}_{\mathbf{x}} = (u_1, \ldots, u_t)$, then $\partial(\{\mathbf{x}\})$ encases $\vec{u}_{\mathbf{x}}^{\triangleleft} = (u_1, \ldots, u_{t-1})$, ensuring that $\pi_{\mathbf{x}}(\vec{u}_{\mathbf{x}}) = \pi_{\mathbf{x}}(u_t)/\|\pi_{\mathbf{x}}(u_t)\|$, both in view of (V1). Thus (V1) further ensures that $\pi_{\mathbf{x}}(\vec{u}_{\mathbf{x}})$ is a unit vector contained in the subspace $\mathbb{R}^{\cup}\langle\mathcal{X}_1^{\pi_{\mathbf{x}}} \cup \ldots \cup \mathcal{X}_j^{\pi_{\mathbf{x}}}\rangle$, in which case Lemma 5.7 (and the comments after its statement) imply that $\mathcal{D}_{\mathbf{x}} \subseteq \bigcup_{i \in J_{\mathbf{x}}, i \leq j} (\mathcal{X}_i^{\pi_{\mathbf{x}}} \cup \{\pi_{\mathbf{x}}(\mathbf{v}_i)\})$, where $\pi_{\mathbf{x}}(\mathcal{R}) = (\mathcal{X}_i^{\pi_{\mathbf{x}}} \cup \{\pi_{\mathbf{x}}(\mathbf{v}_i)\})_{i \in J_{\mathbf{x}}}$ is the virtual Reay system given by Proposition 6.3. As a result, $\pi_{\mathbf{x}}^{-1}(\mathcal{D}_{\mathbf{x}}) \cup \{\mathbf{x}\}$, $\mathcal{A}_{\mathbf{y}} \subseteq \mathcal{X}_1 \cup \{\mathbf{v}_1\} \cup \ldots \cup \mathcal{X}_j \cup \{\mathbf{v}_j\}$ for every $\mathbf{x} \in \mathcal{X}_j \cup \{\mathbf{v}_j\}$

and $\mathbf{y} \in \mathcal{A}_j$. It is now clear that all sets and half-spaces occurring in the proposition are found in $\mathcal{X}_1 \cup \{\mathbf{v}_1\} \cup \ldots \cup \mathcal{X}_j \cup \{\mathbf{v}_j\}$, so we may w.l.o.g. assume $j = s$, freeing the variable j for other use later.

1. and 2. Let $\mathbf{x} \in \mathcal{X}_s \cup \{\mathbf{v}_s\}$ be arbitrary and let $\vec{u}_{\mathbf{x}} = (u_1, \ldots, u_t)$. By (V1), we know that the support set $\partial(\{\mathbf{x}\}) \subseteq \mathcal{X}_1 \cup \ldots \cup \mathcal{X}_{s-1}$ minimally encases $-\vec{u}_{\mathbf{x}}^{\triangleleft} = -(u_1, \ldots, u_{t-1})$. By proposition 6.3, $\pi_{\mathbf{x}}(\mathcal{R}) = \big(\mathcal{X}_i^{\pi_{\mathbf{x}}} \cup \{\pi_{\mathbf{x}}(\mathbf{v}_i)\}\big)_{i \in J_{\mathbf{x}}}$ is a virtual Reay system with $\vec{u}_{\pi_{\mathbf{x}}(\mathbf{x})} = \pi_{\mathbf{x}}(\vec{u}_{\mathbf{x}})$. Note $\pi_{\mathbf{x}}(u_i) = 0$ for $i < t$ as $\partial(\{\mathbf{x}\})$ encases $\vec{u}_{\mathbf{x}}^{\triangleleft}$, but $\pi_{\mathbf{x}}(u_t) \neq 0$ and

$$\pi_{\mathbf{x}}(\mathbf{x}) = \mathbb{R}_+ \pi_{\mathbf{x}}(u_t)$$

in view of (V1). By definition, $\mathcal{D}_{\mathbf{x}}$ is the unique support set minimally encasing $-\pi_{\mathbf{x}}(u_t)$, and $\pi_{\mathbf{x}}^{-1}(\mathcal{D}_{\mathbf{x}})$ is a support set by Proposition 5.4.5. In view of $\pi_{\mathbf{x}}(\mathbf{x}) = \mathbb{R}_+ \pi_{\mathbf{x}}(u_t)$, we cannot have $\pi_{\mathbf{x}}(\mathbf{x}) \in \mathcal{D}_{\mathbf{x}}$ (by definition of $\mathcal{D}_{\mathbf{x}}$), whence $\mathbf{x} \notin \pi_{\mathbf{x}}^{-1}(\mathcal{D}_{\mathbf{x}})$. Let

$$\mathcal{A}_{\mathbf{x}} := \big(\pi_{\mathbf{x}}^{-1}(\mathcal{D}_{\mathbf{x}}) \cup \partial(\{\mathbf{x}\})\big)^*. \tag{6.15}$$

Thus Proposition 5.4.5 and Proposition 5.4.4 imply that $\mathcal{A}_{\mathbf{x}}$ is a support set with

$$\mathcal{A}_{\mathbf{x}}^{\pi_{\mathbf{x}}} = \mathcal{D}_{\mathbf{x}} \quad \text{and} \quad \pi_{\mathbf{x}}^{-1}(\mathcal{D}_{\mathbf{x}}) \subseteq \mathcal{A}_{\mathbf{x}}. \tag{6.16}$$

By Lemma 6.10 (applied to $\mathcal{B} = \partial(\{\mathbf{x}\})$ and $\vec{u} = \vec{u}_{\mathbf{x}}$), $\mathcal{A}_{\mathbf{x}}$ minimally encases $-\vec{u}_{\mathbf{x}}$ urbanely.

Let $\pi_{s-1} : \mathbb{R}^d \to \mathbb{R}^{\cup}\langle \mathcal{X}_1 \cup \ldots \cup \mathcal{X}_{s-1}\rangle^{\perp}$ be the orthogonal projection. Note that $\ker \pi_{\mathbf{x}} \leq \ker \pi_{s-1}$. Since $-\pi_{\mathbf{x}}(u_t) \in \mathsf{C}^{\cup}(\mathcal{D}_{\mathbf{x}})$ (by definition of $\mathcal{D}_{\mathbf{x}}$) and $\ker \pi_{\mathbf{x}} \subseteq \ker \pi_{s-1}$, it follows that

$$-\pi_{s-1}(u_t) \in \mathsf{C}^{\cup}(\pi_{s-1}(\mathcal{D}_{\mathbf{x}})) = \mathsf{C}\big(\pi_{s-1}(D_{\mathbf{x}}) \cap \big(\pi_{s-1}(X_s) \cup \{\pi_{s-1}(v_s)\}\big)\big). \tag{6.17}$$

By (OR2), $\pi_{s-1}(X_s) \cup \{\pi_{s-1}(v_s)\}$ is a minimal positive basis of size $|\mathcal{X}_s| + 1$ which contains (a positive multiple of) $\pi_{s-1}(u_t)$ with $\mathbb{R}_+ \pi_{s-1}(u_t) = \pi_{s-1}(\mathbf{x}) \neq \{0\}$ (recall that $\mathbb{R}_+ \pi_{\mathbf{x}}(u_t) = \pi_{\mathbf{x}}(\mathbf{x})$). Thus, since $\pi_{\mathbf{x}}(\mathbf{x}) \notin \mathcal{D}_{\mathbf{x}}$, we conclude from (6.16) and (6.17) that

$$(\mathcal{X}_s \cup \{\mathbf{v}_s\}) \setminus \{\mathbf{x}\} \subseteq \pi_{\mathbf{x}}^{-1}(\mathcal{D}_{\mathbf{x}}) \subseteq \mathcal{A}_{\mathbf{x}}. \tag{6.18}$$

Now fix $\mathbf{x} \in \mathcal{X}_s \cup \{\mathbf{v}_s\}$ and let

$$\mathcal{A}_s = (\mathcal{A}_{\mathbf{x}} \setminus \partial(\{\mathbf{x}\})) \cup \{\mathbf{x}\} = \pi_{\mathbf{x}}^{-1}(\mathcal{D}_{\mathbf{x}}) \cup \{\mathbf{x}\},$$

where the second equality follows from (6.15) and (6.16) and the fact that all half-spaces from $\pi_{\mathbf{x}}^{-1}(\mathcal{D}_{\mathbf{x}})$ are nonzero modulo $\ker \pi_{\mathbf{x}} = \mathbb{R}^{\cup}\langle \partial(\{\mathbf{x}\})\rangle$. In view of (6.18), we see that

$$\mathcal{X}_s \cup \{\mathbf{v}_s\} \subseteq \mathcal{A}_s. \tag{6.19}$$

Since $\mathcal{A}^*_{\mathbf{x}} = \mathcal{A}_{\mathbf{x}}$ (as $\mathcal{A}_{\mathbf{x}}$ is a support set) and there is no half-space $\mathbf{z} \in \mathcal{A}_s$ with $\mathbf{x} \prec \mathbf{z}$ (as this would require $\mathbf{z} \in \mathcal{X}_j \cup \{\mathbf{v}_j\}$ for some $j \geq s+1$) nor any half-space $\mathbf{z} \in \mathcal{A}_{\mathbf{x}} \setminus \partial(\{\mathbf{x}\}) = \pi_{\mathbf{x}}^{-1}(\mathcal{D}_{\mathbf{x}})$ (the equality follows from (6.15) and (6.16)) with $\mathbf{z} \prec \mathbf{x}$ (as this would imply $\pi_{\mathbf{x}}(\mathbf{z}) = 0$, contrary to the definition of $\pi_{\mathbf{x}}^{-1}(\mathcal{D}_{\mathbf{x}})$), we conclude that $\mathcal{A}^*_s = \mathcal{A}_s$. Thus 2(a) holds. By definition and (6.18), $\mathcal{A}_s \setminus \{\mathbf{x}\} \subseteq \mathcal{A}_{\mathbf{x}}$. Also,

$$(\mathcal{A}_s \setminus \{\mathbf{x}\} \cup \partial(\{\mathbf{x}\}))^* = (\mathcal{A}_{\mathbf{x}} \cup \partial(\{\mathbf{x}\}))^* = \mathcal{A}^*_{\mathbf{x}} = \mathcal{A}_{\mathbf{x}},$$

where the second equality follows in view of $\partial(\{\mathbf{x}\}) \subseteq \downarrow(\pi_{\mathbf{x}}^{-1}(\mathcal{D}_{\mathbf{x}}) \cup \partial(\{\mathbf{x}\})) = \downarrow\mathcal{A}_{\mathbf{x}}$. Let $\mathbf{y} \in \mathcal{A}_s$ be arbitrary, let

$$\mathcal{A}_{\mathbf{y}} = (\mathcal{A}_s \setminus \{\mathbf{y}\} \cup \partial(\{\mathbf{y}\}))^*,$$

and let $\vec{u}_{\mathbf{y}} = (u'_1, \ldots, u'_{t'})$. Note the case $\mathbf{y} = \mathbf{x}$ agrees with the previous definition for $\mathcal{A}_{\mathbf{x}}$ as just shown. Since $\mathbf{y} \in \mathcal{A}_s = \mathcal{A}^*_s$, the definition of $\mathcal{A}_{\mathbf{y}}$ implies

$$\downarrow\mathcal{A}_s = \downarrow\mathcal{A}_{\mathbf{y}} \cup \{\mathbf{y}\}, \tag{6.20}$$

with the union disjoint. Let $\pi(\mathcal{R}) = (\mathcal{X}^\pi_i \cup \{\pi(\mathbf{v}_i)\})_{i \in J}$, where $\pi : \mathbb{R}^d \to \mathbb{R}^\cup\langle\partial(\mathcal{A}_s)\rangle^\perp$ is the orthogonal projection. □

Claim A $\mathcal{A}_s \subseteq \mathcal{X}_1 \cup \ldots \cup \mathcal{X}_s \cup \{\mathbf{v}_j : j \in J\}$ with $\pi(\mathbf{y}) \neq \{0\}$ for all $\mathbf{y} \in \mathcal{A}_s$.

Proof Consider an arbitrary index $j \in [1, s] \setminus J$, so $\mathcal{X}^\pi_j = \emptyset$. Then, since $\partial(\mathcal{A}_s) \subseteq \mathcal{X}_1 \cup \ldots \cup \mathcal{X}_{s-1}$, it follows from (OR2) that $j < s$, and it follows from Proposition 5.3.9 that $\mathcal{X}_j \subseteq \downarrow\partial(\mathcal{A}_s) \subseteq \downarrow\mathcal{A}_{\mathbf{x}}$, with the latter inclusion in view of (6.20) (using $\mathbf{y} = \mathbf{x}$) and $\mathcal{A}^*_s = \mathcal{A}_s$. Thus, since $\mathcal{A}_{\mathbf{x}}$ is a support set, we must have $\mathbf{v}_j \notin \downarrow\mathcal{A}_{\mathbf{x}}$. Since $j < s$ and $\mathbf{x} \in \mathcal{X}_s \cup \{\mathbf{v}_s\}$, we also have $\mathbf{v}_j \neq \mathbf{x}$, in which case $\mathbf{v}_j \notin \downarrow\mathcal{A}_{\mathbf{x}} \cup \{\mathbf{x}\} = \downarrow\mathcal{A}_s$, with the equality in view of (6.20) (used in the case $\mathbf{y} = \mathbf{x}$), which shows that $\mathcal{A}_s \subseteq \mathcal{X}_1 \cup \ldots \cup \mathcal{X}_s \cup \{\mathbf{v}_j : j \in J\}$.

Next suppose that $\pi(\mathbf{y}) = \{0\}$ for some $\mathbf{y} \in \mathcal{A}_s$. In consequence, if $\mathbf{y} \in \mathcal{X}_1 \cup \ldots \cup \mathcal{X}_s$, then Proposition 5.3.9 implies that $\mathbf{y} \in \downarrow\partial(\mathcal{A}_s)$, implying that $\mathbf{y} \prec \mathbf{z}$ with $\mathbf{z}, \mathbf{y} \in \mathcal{A}_s$, which contradicts that $\mathcal{A}^*_s = \mathcal{A}_s$. Therefore we must instead have $\mathbf{y} = \mathbf{v}_j \in \mathcal{A}_s$ for some $j \in [1, s]$, which as just shown implies $j \in J$. However Proposition 5.4.1 implies that $\pi(\mathbf{v}_j) \neq \{0\}$ for all $j \in J$, contrary to assumption, which completes the claim. □

From Claim A and Proposition 5.4.1, we conclude that $\pi_{\mathbf{y}}$ is injective on $\mathcal{A}_s$ with $\pi_{\mathbf{y}}(\mathcal{A}_s) = \mathcal{A}_s^{\pi_{\mathbf{y}}}$ for all $\mathbf{y} \in \mathcal{A}_s$ (since $\ker \pi_{\mathbf{y}} \subseteq \ker \pi$). Since $\mathcal{A}^*_s = \mathcal{A}_s$ and $\mathbf{y} \in \mathcal{A}_s$, ensuring that $\downarrow\partial(\{\mathbf{y}\})$ is disjoint from $\mathcal{A}_s$, we have

$$\mathcal{A}_s \setminus \{\mathbf{y}\} \subseteq \mathcal{A}_{\mathbf{y}} = (\mathcal{A}_s \setminus \{\mathbf{y}\} \cup \partial(\{\mathbf{y}\}))^* \subseteq \mathcal{A}_s \setminus \{\mathbf{y}\} \cup \partial(\{\mathbf{y}\}). \tag{6.21}$$

Thus

$$\mathcal{A}_{\mathbf{y}}^{\pi\mathbf{y}} = (\mathcal{A}_s \setminus \{\mathbf{y}\})^{\pi\mathbf{y}} = \pi\mathbf{y}(\mathcal{A}_s \setminus \{\mathbf{y}\}) \quad \text{and} \quad \pi_{\mathbf{y}}^{-1}(\mathcal{A}_{\mathbf{y}}^{\pi\mathbf{y}}) = \mathcal{A}_s \setminus \{\mathbf{y}\}, \tag{6.22}$$

with the second and third equalities in view of Claim A and the injectivity of $\pi\mathbf{y}$ on $\mathcal{A}_s$, establishing 2(d). Consequently, (6.22), Proposition 5.4.2 and $\ker \pi\mathbf{y} = \mathbb{R}^{\cup}\langle\partial(\{\mathbf{y}\})\rangle$ imply

$$\partial(\mathcal{A}_{\mathbf{y}}^{\pi\mathbf{y}}) = \partial((\mathcal{A}_s \setminus \{\mathbf{y}\})^{\pi\mathbf{y}}) = \partial(\mathcal{A}_s \setminus \{\mathbf{y}\})^{\pi\mathbf{y}} = \partial(\mathcal{A}_s)^{\pi\mathbf{y}}. \tag{6.23}$$

In view of Claim A and Proposition 5.4.1, we see that π is injective on $\mathcal{A}_s$ with $\mathcal{A}_s^{\pi} = \pi(\mathcal{A}_s)$. Thus, since $\ker \pi\mathbf{x} \subseteq \ker \pi$, (6.22) (applied with $\mathbf{y} = \mathbf{x}$) yields $\pi(A_{\mathbf{x}}^{\pi\mathbf{x}}) = \pi\pi\mathbf{x}(A_s \setminus \{x\}) = \pi(A_s \setminus \{x\})$. Since $\mathcal{A}_{\mathbf{x}}^{\pi\mathbf{x}} = \mathcal{D}_{\mathbf{x}}$ is a support set that minimally encases $-\pi\mathbf{x}(u_t)$, Proposition 5.3 (Items 3 and 5) applied to $\mathcal{A}_{\mathbf{x}}^{\pi\mathbf{x}}$, combined with (6.23) (used with $\mathbf{y} = \mathbf{x}$) and Lemma 5.7, ensures that

$$\pi(A_{\mathbf{x}}^{\pi\mathbf{x}}) = \pi(A_s \setminus \{x\})$$

is a linearly independent set that minimally encases $-\pi\pi\mathbf{x}(u_t) = -\pi(u_t)$ and $\mathbb{R}_+\pi(u_t) = \pi(\mathbf{x}) \neq \{0\}$. Consequently, $A_s^{\pi} = \pi(A_s) = \pi(A_s \setminus \{x\}) \cup \{\pi(x)\} = \pi(A_{\mathbf{x}}^{\pi x}) \cup \{\pi(x)\}$ is a minimal positive basis of size $|\mathcal{A}_s|$, showing 2(e) holds.

Since $\mathbf{y} \notin \downarrow\mathcal{A}_{\mathbf{y}}$, two applications of (6.20) (once taking $\mathbf{y} = \mathbf{y}$ and once taking $\mathbf{y} = \mathbf{x}$) gives $\downarrow A_{\mathbf{y}} = \downarrow A_s \setminus \{y\} = (\downarrow A_{\mathbf{x}} \cup \{x\}) \setminus \{y\}$. Since $\mathcal{A}_{\mathbf{x}}$ is a support set, and thus also a virtual independent set, $\downarrow A_{\mathbf{x}}$ is linearly independent. Thus, if $\downarrow A_{\mathbf{y}}$ were not linearly independent, we could write the representative x for $\mathbf{x}$ as a linear combination of the elements from $\downarrow A_s \setminus \{x, y\}$, which when considered modulo $\mathbb{R}^{\cup}\langle\downarrow\partial(A_s)\rangle$ would contradict that $\pi(A_s)$ is a minimal positive basis of size $|\mathcal{A}_s|$ with $x, y \in A_s$. Therefore $\downarrow A_{\mathbf{y}}$ is linearly independent, which combined with $\mathcal{A}_{\mathbf{y}}^* = \mathcal{A}_{\mathbf{y}}$ implies that $\mathcal{A}_{\mathbf{y}}$ is a virtual independent set. Thus, in view of Claim A, Proposition 5.4.6 and $\partial(\{\mathbf{y}\}) \subseteq \downarrow(\mathcal{A}_s \setminus \{\mathbf{y}\} \cup \partial(\{\mathbf{y}\}))^* = \downarrow\mathcal{A}_{\mathbf{y}}$, it follows that $\mathcal{A}_{\mathbf{y}}^{\pi\mathbf{y}}$ is also a virtual independent set.

Let us show that $\mathcal{A}_{\mathbf{y}}^{\pi\mathbf{y}} = \pi\mathbf{y}(\mathcal{A}_s \setminus \{\mathbf{y}\})$ (the equality follow from (6.22)) minimally encases $-\pi\mathbf{y}(\vec{u}_{\mathbf{y}}) = -\pi\mathbf{y}(u'_{t'})/\|\pi\mathbf{y}(u'_{t'})\|$. Note $\mathbb{R}_+\pi\mathbf{y}(u'_{t'}) = \pi\mathbf{y}(\mathbf{y})$ by (V1). As just seen, $A_s^{\pi} = \pi(A_s)$ is a minimal positive basis of size $|\mathcal{A}_s|$ with π is injective on $\mathcal{A}_s$. Thus π is also injective on $\pi\mathbf{y}(\mathcal{A}_s \setminus \{\mathbf{y}\}) = \mathcal{A}_{\mathbf{y}}^{\pi\mathbf{y}}$ (as $\pi\pi\mathbf{y} = \pi$) and

$$-\pi(u'_{t'}) \in \mathsf{C}^{\circ}(\pi(\mathcal{A}_s \setminus \{\mathbf{y}\})) = \mathsf{C}^{\circ}(\pi(\mathcal{A}_{\mathbf{y}}^{\pi\mathbf{y}})). \tag{6.24}$$

In view (6.24) and (6.23), we find that $-\pi_{\mathbf{y}}(u'_{t'}) + a \in C^\circ(\mathcal{A}_{\mathbf{y}}^{\pi_{\mathbf{y}}})$ for some $a \in \mathbb{R}^\cup \langle \partial (\mathcal{A}_s)^{\pi_{\mathbf{y}}} \rangle = \mathbb{R}^\cup \langle \partial (\mathcal{A}_{\mathbf{y}}^{\pi_{\mathbf{y}}}) \rangle$, which implies $-\pi_{\mathbf{y}}(u'_{t'}) \in C^\circ(\mathcal{A}_{\mathbf{y}}^{\pi_{\mathbf{y}}})$ by Proposition 5.3.4. Thus the virtual independent set $\mathcal{A}_{\mathbf{y}}^{\pi_{\mathbf{y}}}$ minimally encases $-\pi_{\mathbf{y}}(u'_{t'}) = -\pi_{\mathbf{y}}(\vec{u}_{\mathbf{y}})$ by Lemma 5.7, while $\partial(\{\mathbf{y}\})$ minimally encases $-\vec{u}_{\mathbf{y}}^{\triangleleft}$ by (V1). As a result, Lemma 6.10, Claim A and (6.22) imply that $(\pi_{\mathbf{y}}^{-1}(\mathcal{A}_{\mathbf{y}}^{\pi_{\mathbf{y}}}) \cup \partial(\{\mathbf{y}\}))^* = (\mathcal{A}_s \setminus \{\mathbf{y}\} \cup \partial(\{\mathbf{y}\}))^* = \mathcal{A}_{\mathbf{y}}$ minimally encases $-\vec{u}_{\mathbf{y}}$. Consequently, since we have already shown that $\mathcal{A}_{\mathbf{y}}$ and $\mathcal{A}_{\mathbf{y}}^{\pi_{\mathbf{y}}}$ are virtual independent sets, 2(b) now follows in view of (6.21).

Let $\mathbf{z} \in \mathcal{X}_s \cup \{\mathbf{v}_s\}$ be arbitrary. By definition of $\mathcal{A}_{\mathbf{z}}$ and $\mathcal{A}_s$, we have

$$\mathcal{A}_{\mathbf{z}} = (\mathcal{A}_s \setminus \{\mathbf{z}\} \cup \partial(\{\mathbf{z}\}))^* = ((\pi_{\mathbf{x}}^{-1}(\mathcal{D}_{\mathbf{x}}) \cup \{\mathbf{x}\}) \setminus \{\mathbf{z}\} \cup \partial(\{\mathbf{z}\}))^*. \qquad (6.25)$$

Since $\mathcal{D}_{\mathbf{x}}$ is a support set, Proposition 5.4.5 implies that $(\pi_{\mathbf{x}}^{-1}(\mathcal{D}_{\mathbf{x}}) \cup \partial(\{\mathbf{x}\}))^*$ is a support set. We have already seen that $\mathcal{A}_{\mathbf{x}}$ is a support set. If $\mathbf{z} \neq \mathbf{x}$, then $\mathbf{z} \in \pi_{\mathbf{x}}^{-1}(\mathcal{D}_{\mathbf{x}})$ by (6.18), whence (6.25) implies $\downarrow\mathcal{A}_{\mathbf{z}} \subseteq \downarrow(\pi_{\mathbf{x}}^{-1}(\mathcal{D}_{\mathbf{x}}) \cup \partial(\{\mathbf{x}\}))^* \cup \{\mathbf{x}\}$. Consequently, since $(\pi_{\mathbf{x}}^{-1}(\mathcal{D}_{\mathbf{x}}) \cup \partial(\{\mathbf{x}\}))^*$ is a support set and $\mathbf{x} \in \mathcal{X}_s \cup \{\mathbf{v}_s\}$, the only way $\mathcal{A}_{\mathbf{z}} = \mathcal{A}_{\mathbf{z}}^*$ can fail to be a support set is if $\mathcal{X}_s \cup \{\mathbf{v}_s\} \subseteq \downarrow\mathcal{A}_{\mathbf{z}}$. However this contradicts that $\mathbf{z} \in \mathcal{X}_s \cup \{\mathbf{v}_s\}$ but $\mathbf{z} \notin \downarrow\mathcal{A}_{\mathbf{z}} = \downarrow(\mathcal{A}_s \setminus \{\mathbf{z}\} \cup \partial(\{\mathbf{z}\}))$ (as (6.20) using is a disjoint union, using $\mathbf{y} = \mathbf{z}$). In summary, we now conclude that $\mathcal{A}_{\mathbf{z}}$ is a support set for every $\mathbf{z} \in \mathcal{X}_s \cup \{\mathbf{v}_s\}$, in which case Proposition 5.4.6 implies that $\mathcal{A}_{\mathbf{z}}^{\pi_{\mathbf{z}}}$ is also a support set. By the established Item 2(b), the support set $\mathcal{A}_{\mathbf{z}}^{\pi_{\mathbf{z}}}$ minimally encases $-\pi_{\mathbf{z}}(\vec{u}_{\mathbf{z}})$, and must then equal $\mathcal{D}_{\mathbf{z}} = \text{Supp}_{\pi_{\mathbf{z}}(\mathcal{R})}(-\pi_{\mathbf{z}}(\vec{u}_{\mathbf{z}}))$, so

$$\mathcal{A}_{\mathbf{z}}^{\pi_{\mathbf{z}}} = \mathcal{D}_{\mathbf{z}} \quad \text{for } \mathbf{z} \in \mathcal{X}_s \cup \{\mathbf{v}_s\}. \qquad (6.26)$$

Since (6.22) and (6.26) imply that

$$\mathcal{A}_{\mathbf{z}} = (\mathcal{A}_s \setminus \{\mathbf{z}\} \cup \partial(\{\mathbf{z}\}))^* = (\pi_{\mathbf{z}}^{-1}(\mathcal{A}_{\mathbf{z}}^{\pi_{\mathbf{z}}}) \cup \partial(\{\mathbf{z}\}))^* = (\pi_{\mathbf{z}}^{-1}(\mathcal{D}_{\mathbf{z}}) \cup \partial(\{\mathbf{z}\}))^*$$

for $\mathbf{z} \in \mathcal{X}_s \cup \{\mathbf{v}_s\}$, all parts of 2(c) are now established, completing Item 2. Additionally, Item 2(a), (6.22) (applied with $\mathbf{y} = \mathbf{z}$) and (6.26) imply $\mathcal{A}_s = \pi_{\mathbf{z}}^{-1}(\mathcal{A}_{\mathbf{z}}^{\pi_{\mathbf{z}}}) \cup \{\mathbf{z}\} = \pi_{\mathbf{z}}^{-1}(\mathcal{D}_{\mathbf{z}}) \cup \{\mathbf{z}\}$ for $\mathbf{z} \in \mathcal{X}_s \cup \{\mathbf{v}_s\}$, establishing Item 1.

3. Let $\mathbf{x} \in \mathcal{A}_s$ be arbitrary and let $k = (i_{\mathbf{z}})_{\mathbf{z} \in \downarrow\mathcal{A}_{\mathbf{x}}}$ be an arbitrary tuple of indices with all $i_{\mathbf{z}}$ sufficiently large (as will be determined during the course of the proof). We maintain the general notation used in Items 1 and 2 and let

$$\mathcal{B}_{\mathbf{x}} = \partial(\{\mathbf{x}\}).$$

By Proposition 5.3.1, we have

$$\mathbb{R}^\cup \langle \mathcal{A}_s \rangle = \mathbb{R}\langle \downarrow\tilde{A}_s(k) \rangle \quad \text{and} \quad \mathbb{R}^\cup \langle \mathcal{B}_{\mathbf{x}} \rangle = \mathbb{R}\langle \downarrow\tilde{B}_{\mathbf{x}}(k) \rangle$$

for any tuple k. Let

$$\varpi : \mathbb{R}^d \to \mathbb{R}^{\cup}\langle \mathcal{A}_s \rangle^{\perp}$$

be the orthogonal projection. Since $\mathsf{C}(G_0) = \mathbb{R}^d$, we also have $\mathsf{C}(\varpi(G_0)) = \mathbb{R}^{\cup}\langle \mathcal{A}_s\rangle^{\perp}$. Thus, by Proposition 4.10, we can find a finite subset $Y = Y_1 \cup \ldots \cup Y_\ell \subseteq G_0$ such that $(\varpi(Y_1), \ldots, \varpi(Y_\ell))$ is a Reay system for $\mathbb{R}^{\cup}\langle \mathcal{A}_s\rangle^{\perp}$ with ϖ injective on Y. We will show Item 3 holds for an arbitrary such Y.

In view of Proposition 6.7.1, once all $i_{\mathbf{z}}$ are sufficiently large, we can assume $\downarrow\tilde{B}_{\mathbf{X}}(k)$ minimally encases $-\vec{u}_{\mathbf{X}}^{\triangleleft}$. In view of Item 2(b)(e), Proposition 5.4.1 and Proposition 5.3.9, it follows that the half-spaces from $\downarrow\mathcal{A}_{\mathbf{X}}$ mapped to $\{0\}$ by $\pi_{\mathbf{X}}$ are precisely those in $\downarrow\mathcal{B}_{\mathbf{X}}$. Consequently, in view of Lemma 6.6.2 (applied with $X = \downarrow\tilde{B}_{\mathbf{X}}(k)$ and $Y = (A_s \setminus \{x\})(k) \cup \big(\widetilde{\downarrow\partial(A_s)} \setminus \downarrow\tilde{B}_{\mathbf{X}}\big)(k) \cup Y_k$), in order to complete the proof, it suffices to show

$$\pi_{\mathbf{X}}(A_s \setminus \{x\})(k) \cup \pi_{\mathbf{X}}\big(\widetilde{\downarrow\partial(A_s)} \setminus \downarrow\tilde{B}_{\mathbf{X}}\big)(k) \cup \pi_{\mathbf{X}}(Y_k)$$

minimally encases $-\pi_{\mathbf{X}}(u_t)$ with $\pi_{\mathbf{X}}$ injective on $(A_s \setminus \{x\})(k) \cup \big(\widetilde{\downarrow\partial(A_s)} \setminus \downarrow\tilde{B}_{\mathbf{X}}\big)(k) \cup Y_k$, for some $Y_k \subseteq Y$ (as well as the additional moreover statement). We begin with the injectivity of $\pi_{\mathbf{X}}$.

Claim B $\pi_{\mathbf{X}}$ is injective on $(A_s \setminus \{x\})(k) \cup \big(\widetilde{\downarrow\partial(A_s)} \setminus \downarrow\tilde{B}_{\mathbf{X}}\big)(k) \cup Y$ for any tuple $k = (i_{\mathbf{z}})_{\mathbf{z} \in \downarrow\mathcal{A}_s \setminus \{\mathbf{x}\}}$ with all $i_{\mathbf{z}}$ sufficiently large.

Proof Item 2(b) implies that

$$\mathcal{A}_{\mathbf{X}}^{\pi_{\mathbf{X}}} = \pi_{\mathbf{X}}(\mathcal{A}_s \setminus \{\mathbf{x}\}) \tag{6.27}$$

minimally encases $-\pi_{\mathbf{X}}(u_t)$ with $\mathcal{A}_{\mathbf{X}}^{\pi_{\mathbf{X}}}$ a virtual independent set. Hence, by Lemma 6.9 (applied to $\mathcal{A}_{\mathbf{X}}^{\pi_{\mathbf{X}}}$) and Proposition 5.4.2, $\downarrow\tilde{A}_{\mathbf{X}}^{\pi_{\mathbf{X}}}(k) = (\downarrow\tilde{A}_{\mathbf{X}})^{\pi_{\mathbf{X}}}(k)$ minimally encases $-\pi_{\mathbf{X}}(u_t)$ (once all $i_{\mathbf{z}}$ are sufficiently large). In view of Item 2(e), Proposition 5.3.9 and Proposition 5.4.1, it follows that $(\downarrow\tilde{A}_{\mathbf{X}})^{\pi_{\mathbf{X}}}(k) = \pi_{\mathbf{X}}\big(\downarrow\tilde{A}_{\mathbf{X}} \setminus \downarrow\tilde{B}_{\mathbf{X}}\big)(k)$ with $\pi_{\mathbf{X}}$ injective on $\downarrow\mathcal{A}_{\mathbf{X}} \setminus \downarrow\mathcal{B}_{\mathbf{X}}$ and $(\downarrow\tilde{A}_{\mathbf{X}} \setminus \downarrow\tilde{B}_{\mathbf{X}})(k)$. As a result, $\pi_{\mathbf{X}}(\downarrow\tilde{A}_{\mathbf{X}} \setminus \downarrow\tilde{B}_{\mathbf{X}})(k)$ minimally encases $-\pi_{\mathbf{X}}(u_t)$ and it follows that

$$\pi_{\mathbf{X}}(\downarrow\tilde{A}_{\mathbf{X}} \setminus \downarrow\tilde{B}_{\mathbf{X}})(k) \cup \{\pi_{\mathbf{X}}(u_t)\} \text{ is a minimal positive basis of size } |\downarrow\mathcal{A}_{\mathbf{X}} \setminus \downarrow\mathcal{B}_{\mathbf{X}}| + 1$$

for the fixed (independent of k) subspace $\mathbb{R}^{\cup}\langle \mathcal{A}_{\mathbf{X}}^{\pi_{\mathbf{X}}} \cup \{\pi_{\mathbf{X}}(u_t)\}\rangle$ (by Proposition 5.3.1). Recall that $\mathcal{A}_{\mathbf{X}}^{\pi_{\mathbf{X}}} = \pi_{\mathbf{X}}(\mathcal{A}_s \setminus \{\mathbf{x}\})$ and $\mathbb{R}_+\pi_{\mathbf{X}}(\mathbf{x}) = \mathbb{R}_+\pi_{\mathbf{X}}(u_t)$. As a result,

$$\mathbb{R}^{\cup}\langle \mathcal{A}_{\mathbf{X}}^{\pi_{\mathbf{X}}} \cup \{\pi_{\mathbf{X}}(u_t)\}\rangle = \mathbb{R}^{\cup}\langle \pi_{\mathbf{X}}(\mathcal{A}_s)\rangle.$$

Thus, since $\ker \pi_{\mathbf{X}} \le \ker \varpi$, Proposition 4.10 implies that, for all k with every $i_{\mathbf{z}}$ sufficiently large,

$$\mathcal{R}_k = (\pi_{\mathbf{X}}(\downarrow \tilde{A}_{\mathbf{X}} \setminus \downarrow \tilde{B}_{\mathbf{X}})(k) \cup \{\pi_{\mathbf{X}}(u_t)\}, \pi_{\mathbf{X}}(Y_1), \ldots, \pi_{\mathbf{X}}(Y_\ell))$$

is an ordinary Reay system for $\mathbb{R}^{\cup}\langle \mathcal{B}_{\mathbf{X}}\rangle^{\perp}$. In view of $\ker \pi_{\mathbf{X}} \le \ker \varpi$, $\pi_{\mathbf{X}}$ must be injective on Y as ϖ is injective on Y. Thus, since $\mathcal{R}_k$ is a Reay system, $\pi_{\mathbf{X}}$ is injective on

$$(\downarrow \tilde{A}_{\mathbf{X}} \setminus \downarrow \tilde{B}_{\mathbf{X}})(k) \cup Y = (\tilde{A}_s \setminus \{\tilde{x}\})(k) \cup \big(\downarrow \widetilde{\partial(A_s)} \setminus \downarrow \tilde{B}_{\mathbf{X}}\big)(k) \cup Y$$

(as it is injective on each component set in $\mathcal{R}_k$ as already noted). We still must show that $\pi_{\mathbf{X}}$ is injective on $(A_s \setminus \{x\})(k) \cup \big(\downarrow \widetilde{\partial(A_s)} \setminus \downarrow \tilde{B}_{\mathbf{X}}\big)(k) \cup Y$ for k with all $i_{\mathbf{z}}$ sufficiently large.

Observe by Item 2(b) that

$$\mathbb{R}^{\cup}\langle \partial(\mathcal{A}_s)\rangle = \mathbb{R}^{\cup}\langle \mathcal{B}_{\mathbf{X}} \cup \partial(\mathcal{A}_s \setminus \{\mathbf{x}\})\rangle = \mathbb{R}^{\cup}\langle \mathcal{B}_{\mathbf{X}} \cup \partial(\mathcal{A}_{\mathbf{X}})\rangle.$$

Thus $\pi : \mathbb{R}^d \to \mathbb{R}^{\cup}\langle \partial(\mathcal{A}_s)\rangle^{\perp} = \mathbb{R}^{\cup}\langle \mathcal{B}_{\mathbf{X}} \cup \partial(\mathcal{A}_{\mathbf{X}})\rangle^{\perp}$ is the orthogonal projection with $\pi\pi_{\mathbf{X}} = \pi$ (as $\ker \pi_{\mathbf{X}} \le \pi$). Moreover, $\partial(\mathcal{A}_{\mathbf{X}})^{\pi_{\mathbf{X}}} = \partial(\mathcal{A}_{\mathbf{X}}^{\pi_{\mathbf{X}}}) \subseteq \downarrow \mathcal{A}_{\mathbf{X}}^{\pi_{\mathbf{X}}} = (\downarrow \mathcal{A}_{\mathbf{X}})^{\pi_{\mathbf{X}}} = \pi_{\mathbf{X}}(\downarrow \mathcal{A}_{\mathbf{X}} \setminus \downarrow \mathcal{B}_{\mathbf{X}})$ in view of Item 2(b)(e), Proposition 5.4.2 and Proposition 5.3.9. As a result, since an ordinary Reay system may be considered as an oriented Reay system (for which all half-spaces have trivial boundary) by replacing each element with the ray it defines, Propositions 5.4.1 and 5.3.1 ensure that $\pi(\mathcal{R}_k)$ is a Reay system for $\mathbb{R}^{\cup}\langle \partial(\mathcal{A}_s)\rangle^{\perp}$. Now $\mathcal{A}_{\mathbf{X}}^{\pi_{\mathbf{X}}}$ is a virtual independent set which minimally encases $-\pi_{\mathbf{X}}(u_t)$ (by Item 2(b)). Thus Lemma 5.7.1 ensures that $-\pi_{\mathbf{X}}(u_t)$ can be written as a sum of representatives from all the half-spaces in $\mathcal{A}_{\mathbf{X}}^{\pi_{\mathbf{X}}}$. Since $\pi(\mathcal{A}_{\mathbf{X}}^{\pi_{\mathbf{X}}}) = \pi(\mathcal{A}_s \setminus \{\mathbf{x}\})$ by (6.27), applying π to this sum shows that $-\pi(u_t) = -\pi\pi_{\mathbf{X}}(u_t)$ is a sum of representatives of all the half-spaces from $\pi(\mathcal{A}_{\mathbf{X}}^{\pi_{\mathbf{X}}}) = \pi(\mathcal{A}_s \setminus \{\mathbf{x}\})$, with these representatives being linearly independent in view of Item 2(e). In particular, since each half-space from $\pi(\mathcal{A}_s)$ has trivial boundary, it follows that $\pi(\mathcal{A}_{\mathbf{X}}^{\pi_{\mathbf{X}}}) = \pi(\mathcal{A}_s \setminus \{\mathbf{x}\})$ is a virtual independent set in $\pi(\mathcal{R}_k)$ (identifying this set of 1-dimensional rays with the corresponding set of representatives in $\pi(\mathcal{R}_k)$), and now Lemma 5.7.2 applied to the virtual independent set $\pi(\mathcal{A}_s \setminus \{\mathbf{x}\})$ shows that $\pi(\mathcal{A}_{\mathbf{X}}^{\pi_{\mathbf{X}}}) = \pi(\mathcal{A}_s \setminus \{\mathbf{x}\})$ minimally encases $-\pi(u_t)$. Note π is injective on $\mathcal{A}_s \setminus \{\mathbf{x}\}$ in view of Item 2(e). However, by definition of π, we have $\mathbb{R}_+\pi(\tilde{\mathbf{y}}(i_{\mathbf{y}})) = \pi(\mathbf{y})$ for all $\mathbf{y} \in \mathcal{A}_s \setminus \{\mathbf{x}\}$. Thus $\pi(\tilde{A}_s \setminus \{\tilde{x}\})(k) \cup \{\pi(u_t)\}$ is a minimal positive basis of size $|\mathcal{A}_s|$, and $\pi(\mathcal{R}_k) = (\pi(\tilde{A}_s \setminus \{\tilde{x}\})(k) \cup \{\pi(u_t)\}, \pi(Y_1), \ldots, \pi(Y_\ell))$ is a Reay system with π injective on Y in view of $\ker \pi \le \ker \varpi$. For each $\mathbf{y} \in \mathcal{A}_s \setminus \{\mathbf{x}\}$, we have $\pi(\mathbf{y}(i_{\mathbf{y}}))/\|\pi(\mathbf{y}(i_{\mathbf{y}}))\| \to v_{\mathbf{y}}$, where $v_{\mathbf{y}} = \pi(\tilde{\mathbf{y}}(i_{\mathbf{y}}))/\|\pi(\tilde{\mathbf{y}}(i_{\mathbf{y}}))\|$ is the constant unit vector pointing in the direction given by the 1-dimensional ray $\pi(\mathbf{y})$ (in view of $\mathbb{R}_+\pi(\tilde{\mathbf{y}}(i_{\mathbf{y}})) = \pi(\mathbf{y})$). Let $\mathfrak{V}_k^{\pi}$ be the associated complete simplicial fan with $V(\mathfrak{V}_k^{\pi}) = \pi(\tilde{A}_s \setminus \{\tilde{x}\})(k) \cup \{\pi(u_t)\} \cup \pi(Y)$. Note $\mathfrak{V}_k^{\pi}$ does not depend on k apart from the choice of vertices used to represent the one-dimensional rays from $\mathfrak{V}_k^{\pi}$

(since $\mathbb{R}_+\pi(\tilde{\mathbf{y}}(i_\mathbf{y})) = \pi(\mathbf{y})$). Then, by Proposition 4.15.4, for any tuple k with all $i_\mathbf{z}$ sufficiently large, there is a simplicial isomorphism between $\mathfrak{V}_k^\pi$ and the simplicial fan $(\mathfrak{V}'_k)^\pi$ with vertices $\pi(A_s \setminus \{x\})(k) \cup \{\pi(u_t)\} \cup \pi(Y)$. In particular, we have $|\pi(A_s \setminus \{x\})(k) \cup \{\pi(u_t)\} \cup \pi(Y)| = |\pi(\tilde{A}_s \setminus \{\tilde{x}\})(k) \cup \{\pi(u_t)\} \cup \pi(Y)| = |\mathcal{A}_s| + |Y|$, which shows that π, and thus also $\pi_\mathbf{X}$ (in view of $\ker \pi_\mathbf{X} \leq \ker \pi$), is injective on

$$(A_s \setminus \{x\})(k) \cup Y,$$

mapping all such elements to distinct non-zero elements. Thus, since all half-spaces from $\downarrow\partial(\mathcal{A}_s)$ are mapped to zero by π, it follows that

$$\pi_\mathbf{X}(A_s \setminus \{x\})(k) \cap \pi_\mathbf{X}(\widetilde{\downarrow\partial(A_s)} \setminus \downarrow\tilde{B}_\mathbf{X})(k) = \emptyset.$$

Since $\pi_\mathbf{X}$ is injective on $(\tilde{A}_s \setminus \{\tilde{x}\})(k) \cup \big(\widetilde{\downarrow\partial(A_s)} \setminus \downarrow\tilde{B}_\mathbf{X}\big)(k) \cup Y$, it is, in particular, injective on $\big(\widetilde{\downarrow\partial(A_s)} \setminus \downarrow\tilde{B}_\mathbf{X}\big)(k) \cup Y$. Combining the last three conclusions, it follows that $\pi_\mathbf{X}$ is injective on $(A_s \setminus \{x\})(k) \cup \big(\widetilde{\downarrow\partial(A_s)} \setminus \downarrow\tilde{B}_\mathbf{X}\big)(k) \cup Y$, completing the claim. □

In view of claim B, it remains to show $\pi_\mathbf{X}(A_s \setminus \{x\})(k) \cup \pi_\mathbf{X}\big(\widetilde{\downarrow\partial(A_s)} \setminus \downarrow\tilde{B}_\mathbf{X}\big)(k) \cup \pi_\mathbf{X}(Y_k)$ minimally encases $-\pi_\mathbf{X}(u_t)$ for some $Y_k \subseteq Y$ (as noted earlier), as well as the additional moreover statement. By Proposition 6.3, $\pi_\mathbf{X}(\mathcal{R})$ is a virtual Reay system.

Claim C $\pi_\mathbf{X}(\mathcal{A}_s)$ is the set given by Proposition 6.12.1 for $\mathcal{X}_s^{\pi_\mathbf{X}} \cup \{\pi_\mathbf{X}(\mathbf{v}_s)\}$. In particular, $\mathcal{A}_\mathbf{X}^{\pi_\mathbf{X}} = \mathcal{A}_{\pi_\mathbf{X}(\mathbf{X})}$.

Proof Let $\mathbf{y} \in \mathcal{X}_s$ and $\mathcal{B}_\mathbf{y} = \partial(\{\mathbf{y}\})$. Let $\pi_\mathbf{xy} : \mathbb{R}^d \to \mathbb{R}^\cup\langle \mathcal{B}_\mathbf{X} \cup \mathcal{B}_\mathbf{y}\rangle^\perp$ be the orthogonal projection. Now $\mathcal{A}_\mathbf{y}^{\pi_\mathbf{y}} = \mathcal{D}_\mathbf{y} = \mathsf{Supp}_{\pi_\mathbf{y}(\mathcal{R})}(-\pi_\mathbf{y}(\vec{u}_\mathbf{y}))$ is the unique support set minimally encasing $-\pi_\mathbf{y}(\vec{u}_\mathbf{y})$ by Item 2(c), and we have

$$\downarrow\mathcal{B}_\mathbf{X}^{\pi_\mathbf{y}} = (\downarrow\mathcal{B}_\mathbf{X})^{\pi_\mathbf{y}} \subseteq \downarrow\partial(\mathcal{A}_s)^{\pi_\mathbf{y}} = (\downarrow\partial(\mathcal{A}_s))^{\pi_\mathbf{y}} \subseteq (\downarrow\mathcal{A}_\mathbf{y})^{\pi_\mathbf{y}} = \downarrow\mathcal{A}_\mathbf{y}^{\pi_\mathbf{y}} \quad (6.28)$$

in view of Item 2(b)(e) and Proposition 5.4.2, in which case Proposition 5.4.6 ensures that $\mathcal{A}_\mathbf{y}^{\pi_\mathbf{xy}} = (\mathcal{A}_\mathbf{y}^{\pi_\mathbf{y}})^{\pi_\mathbf{xy}}$ is a support set. Since $-\pi_\mathbf{y}(\vec{u}_\mathbf{y})$ consists of a single element, its minimal encasement by the support set $\mathcal{A}_\mathbf{y}^{\pi_\mathbf{y}}$ is urbane. Thus Proposition 5.9.4 and (6.28) ensure that the support set $\mathcal{A}_\mathbf{y}^{\pi_\mathbf{xy}} = (\mathcal{A}_\mathbf{y}^{\pi_\mathbf{y}})^{\pi_\mathbf{xy}}$ minimally encases $-\pi_\mathbf{xy}(\vec{u}_\mathbf{y})$, meaning $\mathsf{Supp}_{\pi_\mathbf{xy}(\mathcal{R})}(-\pi_\mathbf{xy}(\vec{u}_\mathbf{y})) = \mathcal{A}_\mathbf{y}^{\pi_\mathbf{xy}}$. In view of Item 2(b), we have $\mathcal{A}_\mathbf{y}^{\pi_\mathbf{y}} = \pi_\mathbf{y}(\mathcal{A}_s \setminus \{\mathbf{y}\})$, while Item 2(e) ensures that π, and thus also $\pi_\mathbf{xy}$, is injective on $\mathcal{A}_s \setminus \{\mathbf{y}\}$ mapping no half-space to 0, whence $\pi_\mathbf{xy}^{-1}(\mathcal{A}_\mathbf{y}^{\pi_\mathbf{xy}}) = \mathcal{A}_s \setminus \{\mathbf{y}\}$, which implies that $(\mathcal{A}_s \setminus \{\mathbf{y}\})^{\pi_\mathbf{X}}$ is the pull-back of $\mathsf{Supp}_{\pi_\mathbf{xy}(\mathcal{R})}(-\pi_\mathbf{xy}(\vec{u}_\mathbf{y})) = \mathcal{A}_\mathbf{y}^{\pi_\mathbf{xy}}$ to $\pi_\mathbf{X}(\mathcal{R})$. Thus, in view of Item 1, we find

$(\mathcal{A}_s \setminus \{\mathbf{y}\})^{\pi\mathbf{x}} \cup \{\pi_{\mathbf{x}}(\mathbf{y})\}$ is the set given by Proposition 6.12.1 for $\mathcal{X}_s^{\pi\mathbf{x}} \cup \{\pi_{\mathbf{x}}(\mathbf{v}_s)\}$. Since $\pi_{\mathbf{x}}(\mathcal{A}_s) = \pi_{\mathbf{x}}(\mathcal{A}_s \setminus \{\mathbf{y}\}) \cup \{\pi_{\mathbf{x}}(\mathbf{y})\} = (\mathcal{A}_s \setminus \{\mathbf{y}\})^{\pi\mathbf{x}} \cup \{\pi_{\mathbf{x}}(\mathbf{y})\}$ by Item 2(e), the main part of the claim is complete. To establish the in particular statement, note the main part together with Item 2(e) and Proposition 5.4.2 yields $\mathcal{A}_{\pi_{\mathbf{x}}(\mathbf{x})} = \big(\pi_{\mathbf{x}}(\mathcal{A}_s) \setminus \{\pi_{\mathbf{x}}(\mathbf{x})\} \cup \partial(\{\pi_{\mathbf{x}}(\mathbf{x})\})\big)^* = \big((\mathcal{A}_s \setminus \{\mathbf{x}\})^{\pi\mathbf{x}} \cup \partial(\{\mathbf{x}\})^{\pi\mathbf{x}}\big)^* = \big(((\mathcal{A}_s \setminus \{\mathbf{x}\}) \cup \partial(\{\mathbf{x}\}))^*\big)^{\pi\mathbf{x}} = \mathcal{A}_{\mathbf{x}}^{\pi\mathbf{x}}$. □

If we knew Item 3 held for any $\mathbf{x} \in \mathcal{A}_s$ whenever $\mathcal{B}_{\mathbf{x}} = \emptyset$, that is, when $\pi_{\mathbf{x}} : \mathbb{R}^d \to \mathbb{R}^d$ is the identity map, $t = 1$ and $\mathcal{A}_{\mathbf{x}} = \mathcal{A}_s \setminus \{\mathbf{x}\}$, then applying this case to $\pi_{\mathbf{x}}(\mathbf{x}) \in \pi_{\mathbf{x}}(\mathcal{A}_s)$ in the virtual Reay system $\pi_{\mathbf{x}}(\mathcal{R})$ would yield (in view of Claim C) that

$$\pi_{\mathbf{x}}(A_s \setminus \{x\})(k) \cup \downarrow\partial(\widetilde{\pi_{\mathbf{x}}(A_s)})(k) \cup \pi_{\mathbf{x}}(Y_k)$$

minimally encases $-\pi_{\mathbf{x}}(u_t) = -\vec{u}_{\pi(\mathbf{x})}$ (by Proposition 6.3) for some $Y_k \subseteq Y$, whenever all $i_{\mathbf{z}}$ are sufficiently large. Moreover, if $(\mathcal{A}_s \setminus \{x\})(k) \subseteq \mathbb{R}^{\cup}\langle \mathcal{A}_s \rangle$, then $\pi_{\mathbf{x}}(\mathcal{A}_s \setminus \{x\})(k) \subseteq \mathbb{R}^{\cup}\langle \pi_{\mathbf{x}}(\mathcal{A}_s)\rangle$, and Item 3 would further give $\pi_{\mathbf{x}}(Y_k) = \emptyset$, forcing $Y_k = \emptyset$. However, Item 2(e), Propositions 5.4.2 and 5.3.9 imply $\downarrow\partial(\widetilde{\pi_{\mathbf{x}}(A_s)})(k) = \pi_{\mathbf{x}}(\downarrow\partial(\widetilde{\mathcal{A}_s}) \setminus \downarrow\tilde{B}_{\mathbf{x}})(k)$, and thus we obtain the needed conclusion that $\pi_{\mathbf{x}}(A_s \setminus \{x\})(k) \cup \pi_{\mathbf{x}}(\downarrow\partial(\widetilde{\mathcal{A}_s}) \setminus \downarrow\tilde{B}_{\mathbf{x}})(k) \cup \pi_{\mathbf{x}}(Y_k)$ minimally encases $-\pi_{\mathbf{x}}(u_t)$, which would complete the proof as already remarked. Thus it suffices to handle this case, so we now assume that $\partial(\{\mathbf{x}\}) = \mathcal{B}_{\mathbf{x}} = \emptyset$, so that $\pi_{\mathbf{x}} : \mathbb{R}^d \to \mathbb{R}^d$ is the identity map, $t = 1$,

$$\mathcal{A}_{\mathbf{x}} = \mathcal{A}_s \setminus \{\mathbf{x}\} \quad \text{and} \quad \partial(\mathcal{A}_{\mathbf{x}}) = \partial(\mathcal{A}_s). \tag{6.29}$$

This will simplify notation. In particular, we now have

$$\mathcal{R}_k = (\downarrow\tilde{A}_{\mathbf{x}}(k) \cup \{u_1\}, Y_1, \ldots, Y_\ell), \tag{6.30}$$

which is an ordinary Reay system for $\mathbb{R}^d$, for any tuple k with all $i_{\mathbf{z}}$ sufficiently large (as argued in Claim B). Note that

$$\mathsf{Supp}_{\mathcal{R}_k}(-u_1) = \downarrow\tilde{A}_{\mathbf{x}}(k). \tag{6.31}$$

We also have $-u_1 \in \mathsf{C}^{\cup}(\mathcal{A}_{\mathbf{x}})^{\circ}$ (cf. Lemma 5.7) as the virtual independent set $\mathcal{A}_{\mathbf{x}} = \mathcal{A}_s \setminus \{\mathbf{x}\} = \mathcal{A}_{\mathbf{x}}^{\pi\mathbf{x}}$ minimally encases $-u_1 = -\pi_{\mathbf{x}}(u_t)$.

Now let $\kappa = (\iota_{\mathbf{z}})_{\mathbf{z} \in \downarrow\mathcal{A}_{\mathbf{x}}}$ be a *fixed* tuple with all $\iota_{\mathbf{z}}$ sufficiently large that all statements derived above (for the arbitrary tuple k with all $i_{\mathbf{z}}$ sufficiently large) are applicable for κ and such that all coordinates $\iota_{\mathbf{z}}$ are also greater than the global constant for the order uniform limit $\lim_{k\to\infty} \mathsf{C}((\downarrow\tilde{\mathbf{y}})(k)) = \mathbf{y}$ given by Proposition 6.7.2 for each $\mathbf{y} \in \mathcal{A}_{\mathbf{x}}$. In view of Proposition 6.7.2 and the definition of order uniform limits, by increasing how large each $i_{\mathbf{z}}$ must be (relative to κ), we find that

$$\mathsf{C}((\downarrow\tilde{\mathbf{y}})(\kappa)) \subseteq \mathsf{C}((\downarrow\tilde{\mathbf{y}})(k)) \tag{6.32}$$

for each $\mathbf{y} \in \mathcal{A}_{\mathbf{x}}$ and any tuple k with all $i_{\mathbf{z}}$ sufficiently large.

Let $\mathbf{y} \in \mathcal{A}_{\mathbf{x}}$ be arbitrary and let $\overline{v}_{\mathbf{y}} \in \mathcal{E}_{\mathbf{y}}^{\perp} := \mathbb{R}^{\cup}\langle\partial(\{\mathbf{y}\})\rangle^{\perp}$ be the unit vector such that $\overline{\mathbf{y}} = \mathcal{E}_{\mathbf{y}} + \mathbb{R}_{+}\overline{v}_{\mathbf{y}}$. By (V1) and (6.1), we know that

$$\mathbf{y}(i_{\mathbf{y}}) = -x_{i_{\mathbf{y}}}^{\mathbf{y}} + \alpha_{i_{\mathbf{y}}}^{\mathbf{y}}\overline{v}_{\mathbf{y}} + \epsilon_{i_{\mathbf{y}}}^{\mathbf{y}} \quad \text{and} \quad \tilde{\mathbf{y}}(i_{\mathbf{y}}) = -x_{i_{\mathbf{y}}}^{\mathbf{y}} + \alpha_{i_{\mathbf{y}}}^{\mathbf{y}}\overline{v}_{\mathbf{y}}$$

for some $x_{i_{\mathbf{y}}}^{\mathbf{y}} \in \mathbb{R}^{\cup}(\partial(\{\mathbf{y}\})) = \mathcal{E}_{\mathbf{y}}$, $\alpha_{i_{\mathbf{y}}}^{\mathbf{y}} > 0$ and $\epsilon_{i_{\mathbf{y}}}^{\mathbf{y}} \in \mathbb{R}\langle\mathbf{y}\rangle^{\perp} = (\mathcal{E}_{\mathbf{y}} + \mathbb{R}\overline{v}_{\mathbf{y}})^{\perp}$, with

$$\alpha_{i_{\mathbf{y}}}^{\mathbf{y}} \in o(\|x_{i_{\mathbf{y}}}^{\mathbf{y}}\|) \quad \text{if } \partial(\{\mathbf{y}\}) \neq \emptyset \quad \text{and} \quad \|\epsilon_{i_{\mathbf{y}}}^{\mathbf{y}}\| \in o(\alpha_{i_{\mathbf{y}}}^{\mathbf{y}}). \tag{6.33}$$

In particular,

$$\tilde{\mathbf{y}}(\iota_{\mathbf{y}}) = -x_{\iota_{\mathbf{y}}}^{\mathbf{y}} + \alpha_{\iota_{\mathbf{y}}}^{\mathbf{y}}\overline{v}_{\mathbf{y}} = -\chi_{\mathbf{y}} + a_{\mathbf{y}}\overline{v}_{\mathbf{y}},$$

where $\chi_{\mathbf{y}} := x_{\iota_{\mathbf{y}}}^{\mathbf{y}} \in \mathcal{E}_{\mathbf{y}}$ and $a_{\mathbf{y}} := \alpha_{\iota_{\mathbf{y}}}^{\mathbf{y}} > 0$, is a fixed, nonzero vector.

In view of (6.32), there is a positive linear combination of the elements from $\downarrow\tilde{\mathbf{y}}(k)$ equal to $\tilde{\mathbf{y}}(\iota_{\mathbf{y}}) = -\chi_{\mathbf{y}} + a_{\mathbf{y}}\overline{v}_{\mathbf{y}} \in \mathsf{C}((\downarrow\tilde{\mathbf{y}})(\kappa)) \subseteq \mathsf{C}((\downarrow\tilde{\mathbf{y}})(k))$ (the element inclusion holds as $\tilde{\mathbf{y}}(\iota_{\mathbf{y}}) \in \downarrow\tilde{\mathbf{y}}(\kappa)$). Since $\tilde{\mathbf{y}}(i_{\mathbf{y}}) \in \downarrow\tilde{\mathbf{y}}(k)$ is the only representative for a half-space not contained in $\mathcal{E}_{\mathbf{y}}$, the coefficient of $\mathbf{y}(i_{\mathbf{y}})$ in this linear combination must be $a_{\mathbf{y}}/\alpha_{i_{\mathbf{y}}}^{\mathbf{y}} > 0$. Replacing $\tilde{\mathbf{y}}(i_{\mathbf{y}})$ with $\mathbf{y}(i_{\mathbf{y}})$ in this linear combination, we find that

$$z_{i_{\mathbf{y}}}^{\mathbf{y}} := \tilde{\mathbf{y}}(\iota_{\mathbf{y}}) + (a_{\mathbf{y}}/\alpha_{i_{\mathbf{y}}}^{\mathbf{y}})\epsilon_{i_{\mathbf{y}}}^{\mathbf{y}} = -\chi_{\mathbf{y}} + a_{\mathbf{y}}\overline{v}_{\mathbf{y}} + (a_{\mathbf{y}}/\alpha_{i_{\mathbf{y}}}^{\mathbf{y}})\epsilon_{i_{\mathbf{y}}}^{\mathbf{y}} \in \mathbb{R}_{+}^{\circ}\mathbf{y}(i_{\mathbf{y}}) + \mathsf{C}\Big(\widetilde{\downarrow\partial(\{\mathbf{y}\})(k)}\Big). \tag{6.34}$$

Since $\|\epsilon_{i_{\mathbf{y}}}^{\mathbf{y}}\| \in o(\alpha_{i_{\mathbf{y}}}^{\mathbf{y}})$ (by (6.33)) with $a_{\mathbf{y}} > 0$, we have $\|(a_{\mathbf{y}}/\alpha_{i_{\mathbf{y}}}^{\mathbf{y}})\epsilon_{i_{\mathbf{y}}}^{\mathbf{y}}\| \to 0$, so $\|(a_{\mathbf{y}}/\alpha_{i_{\mathbf{y}}}^{\mathbf{y}})\epsilon_{i_{\mathbf{y}}}^{\mathbf{y}}\| \in o(\|\tilde{\mathbf{y}}(\iota_{\mathbf{y}})\|) = o(1)$ (as $\tilde{\mathbf{y}}(\iota_{\mathbf{y}}) \neq 0$). In consequence

$$\{z_{i_{\mathbf{y}}}^{\mathbf{y}}\}_{i_{\mathbf{y}}=1}^{\infty} = \{\tilde{\mathbf{y}}(\iota_{\mathbf{y}}) + (a_{\mathbf{y}}/\alpha_{i_{\mathbf{y}}}^{\mathbf{y}})\epsilon_{i_{\mathbf{y}}}^{\mathbf{y}}\}_{i_{\mathbf{y}}=1}^{\infty}$$

is a radially convergent sequence of terms $z_{i_{\mathbf{y}}}^{\mathbf{y}} \in \mathbb{R}_{+}^{\circ}\mathbf{y}(i_{\mathbf{y}}) + \mathsf{C}\Big(\widetilde{\downarrow\partial(\{\mathbf{y}\})(k)}\Big)$ that has limit $\tilde{\mathbf{y}}(\iota_{\mathbf{y}})/\|\tilde{\mathbf{y}}(\iota_{\mathbf{y}})\|$. Moreover, if $(A_s \setminus \{x\})(k) \subseteq \mathbb{R}^{\cup}\langle\mathcal{A}_s\rangle$, then $\mathbf{y}(i_{\mathbf{y}}) \in \mathbb{R}^{\cup}\langle\mathcal{A}_s\rangle$, whence $\epsilon_{i_{\mathbf{y}}}^{\mathbf{y}} \in \mathbb{R}^{\cup}\langle\mathcal{A}_s\rangle$ and $z_{i_{\mathbf{y}}}^{\mathbf{y}} \in \mathbb{R}^{\cup}\langle\mathcal{A}_s\rangle$.

Let $\mathfrak{F}$ be the complete simplicial fan associated to the Reay system $\mathcal{R}_\mathcal{K}$ with vertex set $V(\mathfrak{F}) = \{\tilde{\mathbf{y}}(\iota_\mathbf{y}) : \mathbf{y} \in \mathcal{A}_\mathbf{x}\} \cup \{\tilde{\mathbf{y}}(\iota_\mathbf{y}) : \mathbf{y} \in \downarrow\partial(\mathcal{A}_\mathbf{x})\} \cup Y \cup \{u_1\}$. Since each $\{z^\mathbf{y}_{i_\mathbf{y}}\}_{i_\mathbf{y}=1}^\infty$ is a radially convergent sequence with limit $\tilde{\mathbf{y}}(\iota_\mathbf{y})/\|\tilde{\mathbf{y}}(\iota_\mathbf{y})\|$, Proposition 4.15.4 implies that, for any tuple k with all $i_\mathbf{z}$ sufficiently large, the map $\varphi_k : \mathfrak{F} \to \mathfrak{F}^{(k)}$, which replaces each vertex $\tilde{\mathbf{y}}(\iota_\mathbf{y})$ for $\mathbf{y} \in \mathcal{A}_\mathbf{x}$ with the slightly perturbed vertex $z^\mathbf{y}_{i_\mathbf{y}}$ (and leaves all other vertices fixed), is a simplicial isomorphism of $\mathfrak{F}$ with $\mathfrak{F}^{(k)}$, where $\mathfrak{F}^{(k)}$ denotes the resulting complete simplicial fan. Moreover, further assume all coordinates $i_\mathbf{y}$ are sufficiently large that Proposition 4.15.6 can be applied to $\mathfrak{F}^{(k)}$ with $x = -u_1$ (possible as the sequences $\{z^\mathbf{y}_{i_\mathbf{y}}\}_{i_\mathbf{y}=1}^\infty$ are radially convergent). Then, by Proposition 4.15.6 and (6.31), we have

$$\{z^\mathbf{y}_{i_\mathbf{y}} : \mathbf{y} \in \mathcal{A}_\mathbf{x}\} \cup \{\tilde{\mathbf{y}}(i_\mathbf{y}) : \mathbf{y} \in \downarrow\partial(\mathcal{A}_\mathbf{x})\} = \varphi_k\big(\downarrow\tilde{A}_\mathbf{x}(k)\big) \subseteq \mathsf{Supp}_{\mathfrak{F}^{(k)}}(-u_1) \tag{6.35}$$

for any tuple k with all $i_\mathbf{z}$ sufficiently large. Consequently,

$$\mathsf{Supp}_{\mathfrak{F}^{(k)}}(-u_1) = \{z^\mathbf{y}_{i_\mathbf{y}} : \mathbf{y} \in \mathcal{A}_\mathbf{x}\} \cup \{\tilde{\mathbf{y}}(i_\mathbf{y}) : \mathbf{y} \in \downarrow\partial(\mathcal{A}_\mathbf{x})\} \cup Y_k$$

for some subset $Y_k \subseteq Y$ (note u_1 can never be in the support set $\mathsf{Supp}(-u_1)$ in view of the definition of Supp). Since $\mathfrak{F}$ is a complete simplicial fan associated to the Reay system $\mathcal{R}_\mathcal{K}$, any support set for $\mathfrak{F}$ cannot contain all elements from Y_j, for any $j \in [1, \ell]$. Thus, since the simplicial isomorphism φ_k gives a correspondence between support sets in $\mathfrak{F}$ and support sets in $\mathfrak{F}^{(k)}$, we have

$$Y_j \nsubseteq Y_k \quad \text{for all } j \in [1, \ell]. \tag{6.36}$$

If $(A_s \setminus \{x\})(k) \subseteq \mathbb{R}^\cup\langle\mathcal{A}_s\rangle$, then we can perform all the above using the complete simplicial fan $\mathfrak{F}'$ with $V(\mathfrak{F}') = \{\tilde{\mathbf{y}}(\iota_\mathbf{y}) : \mathbf{y} \in \mathcal{A}_\mathbf{x}\} \cup \{\tilde{\mathbf{y}}(\iota_\mathbf{y}) : \mathbf{y} \in \downarrow\partial(\mathcal{A}_\mathbf{x})\} \cup \{u_1\}$ in place of $\mathfrak{F}$ (since all $z^\mathbf{y}_{i_\mathbf{y}} \in \mathbb{R}^\cup\langle\mathcal{A}_s\rangle$ in this case) and thereby conclude that $Y_k = \emptyset$. Now by definition of $\mathsf{Supp}_{\mathfrak{F}^{(k)}}(-u_1)$, we have

$$-u_1 \in \mathsf{C}^\circ\big(\{z^\mathbf{y}_{i_\mathbf{y}} : \mathbf{y} \in \mathcal{A}_\mathbf{x}\}\big) + \mathsf{C}^\circ\big(\downarrow\widetilde{\partial(A_\mathbf{x})}(k)\big) + \mathsf{C}^\circ(Y_k)$$

with $\mathsf{Supp}_{\mathfrak{F}^{(k)}}(-u_1)$ linearly independent. Hence, in view of (6.34) and the definition of the $z^\mathbf{y}_{i_\mathbf{y}}$, we find that

$$-u_1 \in \mathsf{C}^\circ\big(A_\mathbf{x}(k) \cup \downarrow\widetilde{\partial(A_\mathbf{x})}(k) \cup Y_k\big)$$

will follow once we know $A_\mathbf{x}(k) \cup \downarrow\widetilde{\partial(A_\mathbf{x})}(k) \cup Y_k$ is linearly independent, which will then, in turn, also ensure that $A_\mathbf{x}(k) \cup \downarrow\widetilde{\partial(A_\mathbf{x})}(k) \cup Y_k = (A_s \setminus \{x\})(k) \cup$

$\downarrow\widetilde{\partial(A_s)}(k) \cup Y_k$ (cf. (6.29)) minimally encases $-u_1 = \pi_{\mathbf{X}}(\vec{u})$. Thus, to complete the proof, it remains to show the following claim.

Claim D $A_{\mathbf{X}}(k) \cup \downarrow\widetilde{\partial(A_{\mathbf{X}})}(k) \cup Y_k$ is linearly independent for k with all $i_{\mathbf{z}}$ sufficiently large.

Proof The proof is a variation on the argument used to establish the injectivity of $\pi_{\mathbf{X}}$ in Claim B. Since $\downarrow\widetilde{\partial(A_{\mathbf{X}})}(k) \cup Y_k$ is a subset of the linearly independent set $\mathrm{Supp}_{\mathfrak{F}(k)}(-u_1)$, it follows that

$$\downarrow\widetilde{\partial(A_{\mathbf{X}})}(k) \cup Y_k \quad \text{is linearly independent.} \tag{6.37}$$

Since $\mathcal{A}_{\mathbf{X}} = \mathcal{A}_s \setminus \{\mathbf{x}\}$ with $\partial(\{\mathbf{x}\}) = \emptyset$, we have $\mathbb{R}^{\cup}\langle\partial(\mathcal{A}_s)\rangle = \mathbb{R}^{\cup}\langle\partial(\mathcal{A}_{\mathbf{X}})\rangle$. Thus $\pi : \mathbb{R}^d \to \mathbb{R}^{\cup}\langle\partial(\mathcal{A}_{\mathbf{X}})\rangle^{\perp}$ is the orthogonal projection. Since an ordinary Reay system may be considered as an oriented Reay system (for which all half-spaces have trivial boundary) by replacing each element with the ray it defines, Proposition 5.4.1 ensures that $\pi(\mathcal{R}_k)$ is a Reay system. Since $\mathcal{A}_{\mathbf{X}} = \mathcal{A}_s \setminus \{\mathbf{x}\} = \mathcal{A}_{\mathbf{X}}^{\pi_{\mathbf{X}}}$ is a virtual independent set which minimally encases $-u_1 = -\pi_{\mathbf{X}}(u_t)$, it follows from Lemma 5.7 and Proposition 5.3.3 that $\pi(\mathcal{A}_{\mathbf{X}})$ minimally encases $-\pi(u_1)$ with π injective on $\mathcal{A}_{\mathbf{X}}$. However, by definition of π, we have $\mathbb{R}_+\pi(\tilde{\mathbf{y}}(i_{\mathbf{y}})) = \pi(\mathbf{y})$ for all $\mathbf{y} \in \mathcal{A}_{\mathbf{X}}$. Thus $\pi(\tilde{A}_{\mathbf{X}}(k)) \cup \{\pi(u_1)\}$ is a minimal positive basis of size $|\mathcal{A}_{\mathbf{X}}| + 1$ for the subspace $\mathbb{R}^{\cup}\langle\pi(\tilde{A}_{\mathbf{X}}(k)) \cup \pi(u_1)\rangle = \mathbb{R}^{\cup}\langle\pi(\mathcal{A}_{\mathbf{X}}) \cup \pi(\mathbf{x})\rangle = \mathbb{R}^{\cup}\langle\pi(\mathcal{A}_s)\rangle$, and

$$\pi(\mathcal{R}_k) = (\pi(\tilde{A}_{\mathbf{X}}(k)) \cup \{\pi(u_t)\}, \pi(Y_1), \ldots, \pi(Y_\ell))$$

is a Reay system with π injective on $\tilde{A}_{\mathbf{X}}(k) \cup Y \cup \{u_t\}$ (since $\ker\pi \leq \ker\varpi$).

For each $\mathbf{y} \in \mathcal{A}_{\mathbf{X}}$, we see that $\pi(\mathbf{y}(i_{\mathbf{y}}))/\|\pi(\mathbf{y}(i_{\mathbf{y}}))\| \to v_{\mathbf{y}}$, where $v_{\mathbf{y}} = \pi(\tilde{\mathbf{y}}(i_{\mathbf{y}}))/\|\pi(\tilde{\mathbf{y}}(i_{\mathbf{y}}))\|$ is a constant unit vector pointing in the same direction as $\pi(\tilde{\mathbf{y}}(i_{\mathbf{y}}))$ for $\mathbf{y} \in \mathcal{A}_{\mathbf{X}}$ (in view of $\mathbb{R}_+\pi(\tilde{\mathbf{y}}(i_{\mathbf{y}})) = \pi(\mathbf{y})$). Let $\mathfrak{F}_k^{\pi}$ be the associated complete simplicial fan with $V(\mathfrak{F}_k^{\pi}) = \pi(\tilde{A}_{\mathbf{X}}(k)) \cup \{\pi(u_t)\} \cup \pi(Y)$ associated to the Reay system $\pi(\mathcal{R}_k)$. Note $\mathfrak{F}_k^{\pi}$ is fixed (and independent of k) apart from some possible variation in which positive scalar multiples are used for the vertices. Then, by Proposition 4.15.4, for any tuple k with all $i_{\mathbf{z}}$ sufficiently large, there is a simplicial isomorphism φ between $\mathfrak{F}_k^{\pi}$ and a simplicial fan $\mathfrak{F}_k'^{\pi}$ with vertices $\pi(A_{\mathbf{X}}(k)) \cup \{\pi(u_t)\} \cup \pi(Y)$. In particular, support sets map to support sets. Since $\mathfrak{F}_k^{\pi}$ is a simplicial fan associated to the Reay system $\pi(\mathcal{R}_k)$, (6.36) ensures that $\pi(\tilde{A}_{\mathbf{X}}(k)) \cup \pi(Y_k)$ is a support set for $\mathfrak{F}_k^{\pi}$, whence its image $\varphi\big(\pi(\tilde{A}_{\mathbf{X}}(k)) \cup \pi(Y_k)\big) = \pi(A_{\mathbf{X}}(k)) \cup \pi(Y_k)$ is also a support set, and thus linearly independent (as support sets of a simplicial fan are linearly independent by definition). Furthermore, since the simplicial isomorphism φ is injective on vertices, and since π is injective on $\tilde{A}_{\mathbf{X}}(k) \cup Y$, it follows that $|\mathcal{A}_{\mathbf{X}}| + |Y_k| = |\tilde{A}_{\mathbf{X}}(k) \cup Y_k| = |\varphi\big(\pi(\tilde{A}_{\mathbf{X}}(k)) \cup \pi(Y_k)\big)| = |\pi(A_{\mathbf{X}}(k)) \cup \pi(Y_k)|$, forcing π to be injective on $A_{\mathbf{X}}(k) \cup Y_k$. Thus $A_{\mathbf{X}}(k) \cup Y_k$ is linearly independent modulo

$$\ker \pi = \mathbb{R}^{\cup}\langle \partial(\mathcal{A}_{\mathbf{X}})\rangle = \mathbb{R}\langle \downarrow \widetilde{\partial(A_{\mathbf{X}})}(k)\rangle,$$

with the second equality above in view of Proposition 5.3.1. As a result, (6.37) now implies that $A_{\mathbf{X}}(k) \cup \downarrow \widetilde{\partial(A_{\mathbf{X}})}(k) \cup Y_k$ is linearly independent, completing the claim and the proof.

Chapter 7
Finitary Sets

7.1 Core Definitions and Properties

We have now developed the asymptotic framework generalizing the notion of positive basis to the point where we can define our main object of study later in the chapter. We begin by giving three equivalent definitions for the special subset $G_0^\diamond \subseteq G_0 \subseteq \mathbb{R}^d$, which plays a key role in our characterization result for finite elasticities. Note the three equivalent conditions defining $G_0^\diamond$ are all dependent only on notions from Convex Geometry—involving linear combinations over $\mathbb{R}_+$—rather than combinatorial properties dealing with $\mathcal{A}(G_0)$ and linear combinations over $\mathbb{Z}_+$. When we impose additional conditions on G_0, we will later be able to give two further equivalent definitions for $G_0^\diamond$, first in terms of $\mathcal{A}^{\mathsf{elm}}(G_0)$ and linear combinations over $\mathbb{Q}_+$, and then in terms of $\mathcal{A}(G_0)$ and linear combinations over $\mathbb{Z}_+$. Recall that G_0^{lim} was defined in Sect. 3.1. Note the equality $\mathsf{C}(G_0 \cap \mathcal{E}) = \mathcal{E}$ mentioned in the definition below follows by considering $\mathsf{C}(\pi(G_0))$, where $\pi : \mathbb{R}^d \to \mathcal{E}^\perp$ is the orthogonal projection, which must have trivial lineality space as $\mathcal{E}$ is the maximal subspace contained in $\mathsf{C}(G_0)$, thus ensuring that $0 \notin \mathsf{C}^*(\pi(G_0) \setminus \{0\})$, which means any positive linear combination of elements from G_0 that lies in $\mathcal{E}$ must have all it elements lying in $\mathcal{E}$.

Definition 7.1 Let $G_0 \subseteq \mathbb{R}^d$ be a subset with lineality space $\mathcal{E} = \mathsf{C}(G_0) \cap -\mathsf{C}(G_0)$, so $\mathsf{C}(G_0 \cap \mathcal{E}) = \mathcal{E}$. Let $G_0^\diamond \subseteq G_0 \cap \mathcal{E} \subseteq G_0$ be the set of all $g \in G_0 \cap \mathcal{E}$ satisfying the equivalent conditions of Proposition 7.2 (applied to $G_0 \cap \mathcal{E}$ in place of G_0).

Proposition 7.2 *Let $G_0 \subseteq \mathbb{R}^d$ be a subset with $\mathsf{C}(G_0) = \mathcal{E} \subseteq \mathbb{R}^d$ a subspace, where $d \geq 0$, and let $g \in G_0$. The following are equivalent.*

1. *There exists a subset $X \subseteq G_0$ and $\vec{u} \in G_0^{\mathsf{lim}}$ with $g \in X$ and X minimally encasing $-\vec{u}$.*

D. J. Grynkiewicz, *The Characterization of Finite Elasticities*, Lecture Notes in Mathematics 2316, https://doi.org/10.1007/978-3-031-14869-9_7

2. *There exists a linearly independent subset* $X \subseteq G_0$ *with* $g \in X$ *and a sequence* $\{x_i\}_{i=1}^{\infty}$ *of terms* $x_i \in G_0 \cap \mathsf{C}^{\circ}(-X)$ *such that* $-x_i = \sum_{x \in X} \alpha_i^{(x)} x$ *with* $\alpha_i^{(x)} > 0$ *real numbers such that* $\alpha_i^{(g)} \to \infty$.
3. *There exists a linearly independent set* $X \subseteq G_0$ *with* $g \in X$ *and* $\mathsf{C}(X) \cap -G_0$ *not bound to* $\mathsf{C}(X \setminus \{g\})$.

Proof We may w.l.o.g. assume $\mathcal{E} = \mathbb{R}^d$.

1. $\Rightarrow$ 2. Suppose there exists an asymptotically filtered sequence $\{x_i\}_{i=1}^{\infty}$ of terms $x_i \in G_0$ with fully unbounded limit $\vec{u} = (u_1, \ldots, u_t)$ and a subset $X \subseteq G_0$ with $g \in X$ such that X minimally encases $-\vec{u}$. Write each

$$x_i = a_1^{(1)} u_1 + \ldots + a_i^{(t)} u_t + y_i$$

with $a_i^{(j)} > 0$ and $y_i \in \mathbb{R}\langle u_1, \ldots, u_t\rangle^{\perp}$ such that $\|y_i\| \in o(a_i^{(t)})$, and let $x_i^{(t)} = x_i - y_i = a_1^{(1)} u_1 + \ldots + a_i^{(t)} u_t$ be the truncated terms for $i \geq 1$. By hypothesis, $t \geq 1$ and $a_i^{(t)} \to \infty$. Let $\pi : \mathbb{R}^d \to \mathbb{R}\langle X\rangle^{\perp}$ be the orthogonal projection. In view of Proposition 4.20, by removing the first few terms, we can w.l.o.g. assume $X \cup \{x_i^{(t)}\}$ is a minimal positive basis for all i. Since $\mathsf{C}(G_0) = \mathbb{R}^d$, we can find a subset $Y \subseteq G_0$ such that $|\pi(Y)| = |Y|$ and $\pi(Y)$ is a positive basis for $\mathbb{R}\langle X\rangle^{\perp}$ corresponding to a Reay system $\mathcal{R}_{\pi(Y)}$. For each $i \geq 1$, let $Y_i \subseteq Y$ be the subset with $\pi(Y_i) = \mathsf{Supp}_{\mathcal{R}_{\pi}(Y)}(-\pi(x_i))$, so $\pi(Y_i) \cup \{\pi(x_i)\}$ is a minimal positive basis or $Y_i = \emptyset$ with $\pi(x_i) = 0$. By passing to a subsequence, we can assume all Y_i are equal, say $Y_i = Y_1$ for all $i \geq 1$. By Proposition 4.8.3, X and $\pi(Y_1)$ are both linearly independent. As a result, since π is injective on $Y_1 \subseteq Y$ with $\ker \pi = \mathbb{R}\langle X\rangle$, it follows that $X \cup Y_1$ is a linearly independent subset. Since $\pi(Y_1) \cup \{\pi(x_i)\}$ is a minimal positive basis with $\pi(x_i) = \pi(y_i)$, Lemma 4.18 implies

$$\sum_{y \in Y_1} \alpha_i^{(y)} y = -y_i + z_i \tag{7.1}$$

for some $z_i \in \mathbb{R}\langle X\rangle$ and $\alpha_i^{(y)} > 0$ with $\alpha_i^{(y)} \in O(\|\pi(y_i)\|) \subseteq O(\|y_i\|) \subseteq o(a_i^{(t)})$ ($z_i = y_i$ if $Y_1 = \emptyset$ with $\pi(x_i) = 0$). Hence $\|z_i\| \in o(a_i^{(t)})$ and $x_i - y_i + z_i = x_i^{(t)} + z_i \in \mathbb{R}\langle X\rangle$. Thus $x_i - y_i + z_i = x_i^{(t)} + z_i = a_1^{(1)} u_1 + \ldots + a_i^{(t)} u_t + z_i$ is an asymptotically filtered sequence with limit $(u_1, \ldots, u_t)$ (once all i are sufficiently large). Applying Proposition 4.20 to $\{x_i - y_i + z_i\}_{i=1}^{\infty}$, we conclude that

$$\sum_{x \in X} \alpha_i^{(x)} x = -x_i + y_i - z_i \tag{7.2}$$

for some $\alpha_i^{(x)} > 0$ (passing to sufficiently large index terms). Moreover, for each $x \in X$, we have $\alpha_i^{(x)} \in \Theta(a_i^{(j)})$ for some $j \in [1, t]$. Since $g \in X$ and $a_i^{(j)} \to \infty$ for all $j \leq t$, we have $\alpha_i^{(g)} \to \infty$, whence Item 2 follows from (7.1) and (7.2) taking X to be $X \cup Y_1$.

2. $\Rightarrow$ 3. Let $X \subseteq G_0$ and $\{x_i\}_{i=1}^{\infty}$ be as given by Item 2, so $x_i \in G_0$ for all $i \geq 1$. Then each $-x_i = \sum_{x \in X} \alpha_i^{(x)} x \in \mathsf{C}^{\circ}(X) \cap -G_0$ with $\alpha_i^{(x)} > 0$ and $\alpha_i^{(g)} \to \infty$. Assuming by contradiction that $\{-x_i\}_{i=1}^{\infty}$ is bound to $\mathsf{C}(X \setminus \{g\})$, then there is a bound $M > 0$ such that each $-x_i$ has some $z_i \in \mathsf{C}(X \setminus \{g\})$ with $\|-x_i - z_i\| \leq M$. Let $T : \mathbb{R}^d \to \mathbb{R}$ be a linear transformation that sends g to 1 and $\mathbb{R}\langle X \setminus \{g\}\rangle$ to 0, which exists since $g \in X$ with X linearly independent. Then $T(-x_i - z_i) = T(-x_i) = \alpha_i^{(g)}$ with $\alpha_i^{(g)} = \|T(-x_i - z_i)\| \leq C_T \|-x_i - z_i\| \leq C_T M < \infty$, where C_T is the operator norm of T with respect to the Euclidean metric, contradicting that $a_i^{(g)} \to \infty$.

3. $\Rightarrow$ 1. Let $X \subseteq G_0$ be as given by Item 3. Since $\mathsf{C}(X) \cap -G_0$ is not bound to $\mathsf{C}(X \setminus \{g\})$, we can find a sequence $\{-x_i\}_{i=1}^{\infty}$ of terms $-x_i \in \mathsf{C}(X) \cap -G_0$ such that $\mathsf{d}(-x_i, \mathsf{C}(X \setminus \{g\})) \to \infty$. By passing to a subsequence, we can assume $\{x_i\}_{i=1}^{\infty}$ is an asymptotically filtered sequence with complete fully unbounded limit $\vec{u} = (u_1, \ldots, u_t)$, where $t \geq 1$. Thus $\vec{u} \in G_0^{\mathsf{lim}}$ since $\vec{u}$ is fully unbounded and $x_i \in G_0$ for all $i \geq 1$. Since $-x_i \in \mathsf{C}(X)$ for all $i \geq 1$, Proposition 3.12 implies that $\mathsf{C}(X)$ encases $-\vec{u}$, so let $Y \subseteq X$ be a subset for which Y minimally encases $-\vec{u}$. If $g \in Y$, then Item 1 follows. Otherwise, $X \setminus \{g\}$ encases $-\vec{u}$, in which case Proposition 3.11.3 implies that $\{-x_i\}_{i=1}^{\infty}$ is bound to $\mathsf{C}(X \setminus \{g\})$, contradicting that $\mathsf{d}(-x_i, \mathsf{C}(X \setminus \{g\})) \to \infty$. □

We continue with the following basic inclusion for $G_0^{\diamond}$ for subsets G_0 of lattice points.

Proposition 7.3 *Let $\Lambda \subseteq \mathbb{R}^d$ be a full rank lattice, where $d \geq 0$, and let $G_0 \subseteq \Lambda$ be a subset with $\mathsf{C}(G_0) = \mathbb{R}^d$. Then*

$$G_0^{\diamond} \subseteq \Big\{g \in G_0 : \ \sup\{\mathsf{v}_g(U) : \ U \in \mathcal{A}^{\mathsf{elm}}(G_0)\} = \infty\Big\}.$$

Proof Let $g \in G_0^{\diamond}$ be arbitrary. By Proposition 7.2.2, there exists a linearly independent subset $X \subseteq G_0$ with $g \in X$ and a sequence $\{x_i\}_{i=1}^{\infty}$ of terms $x_i \in G_0 \cap -\mathsf{C}^{\circ}(X)$ such that

$$-x_i = \sum_{x \in X} \alpha_i^{(x)} x \tag{7.3}$$

with $\alpha_i^{(x)} > 0$ and $\alpha_i^{(g)} \to \infty$. By Proposition 4.8 (Items 4 and 7), each $X \cup \{x_i\}$ is a minimal positive basis and there exists an elementary atom U_i with $\mathrm{Supp}(U_i) = X \cup \{x_i\}$. Since the vector $(\mathsf{v}_x(U_i))_{x \in X \cup \{x_i\}}$ has $\sum_{x \in X \cup \{x_i\}} \mathsf{v}_x(U_i) x = \sigma(U_i) = 0$, it follows from Proposition 4.8.5 and (7.3) that $\mathsf{v}_x(U_i) = \mathsf{v}_{x_i}(U_i)\alpha_i^{(x)}$ for all $x \in X$. In particular, $\mathsf{v}_g(U_i) = \mathsf{v}_{x_i}(U_i)\alpha_i^{(g)} \geq \alpha_i^{(g)}$ (the inequality follows in view of $x_i \in$

$\mathrm{Supp}(U_i)$) with $\alpha_i^{(g)} \to \infty$, showing that $\sup\{\mathsf{v}_g(U) : U \in \mathcal{A}^{\mathsf{elm}}(G_0)\} = \infty$. This establishes the desired inclusion. □

Lemma 7.4 links the set $\mathcal{A}_j$ from Proposition 6.7 with the diamond subset $G_0^\diamond$.

Lemma 7.4 *Let $\Lambda \subseteq \mathbb{R}^d$ be a full rank lattice, where $d \geq 1$, let $G_0 \subseteq \Lambda$ be a subset of lattice points with $\mathsf{C}(G_0) = \mathbb{R}^d$, let $\mathcal{R} = (\mathcal{X}_1 \cup \{v_1\}, \ldots, \mathcal{X}_s \cup \{v_s\})$ be a virtual Reay system in G_0, let $\mathcal{A}_s$ be the subset given by Proposition 6.12.1 for $j = s$, suppose that the virtual Reay system $\mathcal{R}' = (\mathcal{X}_1 \cup \{v_1\}, \ldots, \mathcal{X}_{s-1} \cup \{v_{s-1}\})$ is anchored, and let $\boldsymbol{x} \in \mathcal{A}_s$.*

1. *If $\vec{u}_{\mathbf{x}}$ is fully unbounded and $\boldsymbol{y} \in \mathcal{A}_s \setminus \{\boldsymbol{x}\}$, then there exists an asymptotically filtered sequence $\{y_i\}_{i=1}^\infty$ of terms $y_i \in \mathsf{C}(G_0^\diamond) \cap \mathbb{R}^\cup\langle\mathcal{A}_s\rangle$ with limit $\vec{u}_{\mathbf{y}}$. Indeed, there is a finite subset $Z \subseteq G_0^\diamond$ such that $y_i \in \mathbf{y}(i) + \mathsf{C}(Z)$ for all sufficiently large i, and $\mathbf{z}(i) \in G_0^\diamond$ for every $\mathbf{z} \in \downarrow\mathcal{A}_s \setminus \{\boldsymbol{x}\}$ and all sufficiently large i.*
2. *Suppose, for each $\mathbf{y} \in \mathcal{A}_s \setminus \{\boldsymbol{x}\}$, that $\{y_{i_{\mathbf{y}}}^{\mathbf{y}}\}_{i_{\mathbf{y}}=1}^\infty$ is an asymptotically filtered sequence of terms $y_{i_{\mathbf{y}}}^{\mathbf{y}} \in \mathbb{R}^\cup\langle\mathcal{A}_s\rangle$ with limit $\vec{u}_{\mathbf{y}}$. Then, for any tuple $k = (i_{\mathbf{z}})_{\mathbf{z} \in \downarrow\mathcal{A}_s \setminus \{\boldsymbol{x}\}}$ with all $i_{\mathbf{z}}$ sufficiently large, the set $\downarrow\partial(A_s)(k) \cup \{y_{i_{\mathbf{y}}}^{\mathbf{y}} : \boldsymbol{y} \in \mathcal{A}_s \setminus \{\boldsymbol{x}\}\}$ minimally encases $-\vec{u}_{\boldsymbol{x}}$ and $\mathbb{R}^\cup\langle\mathcal{A}_s\rangle = \mathbb{R}\langle\downarrow\partial(A_s)(k) \cup \{y_{i_{\mathbf{y}}}^{\mathbf{y}} : \boldsymbol{y} \in \mathcal{A}_s \setminus \{\boldsymbol{x}\}\}\rangle$.*

Proof

1. Let $\vec{u}_{\mathbf{x}} = (u_1^{\mathbf{x}}, \ldots, u_{t_{\mathbf{x}}}^{\mathbf{x}})$ and $\vec{u}_{\mathbf{y}} = (u_1^{\mathbf{y}}, \ldots, u_{t_{\mathbf{y}}}^{\mathbf{y}})$. Let $\pi : \mathbb{R}^d \to \mathbb{R}\langle u_1^{\mathbf{y}}, \ldots, u_{t_{\mathbf{y}}}^{\mathbf{y}}\rangle^\perp$ be the orthogonal projection. Since $\{\mathbf{y}(i)\}_{i=1}^\infty$ is an asymptotically filtered sequence with limit $\vec{u}_{\mathbf{y}} = (u_1^{\mathbf{y}}, \ldots, u_{t_{\mathbf{y}}}^{\mathbf{y}})$, we have

$$\mathbf{y}(i) = a_i^{(1)} u_1^{\mathbf{y}} + \ldots + a_i^{(t_{\mathbf{y}})} u_{t_{\mathbf{y}}}^{\mathbf{y}} + \pi(\mathbf{y}(i)) \quad \text{for all } i,$$

where

$$a_i^{(j)} \in o(a_i^{(j-1)}) \quad \text{for } j \in [2, t_{\mathbf{y}}] \quad \text{and} \quad \|\pi(\mathbf{y}(i))\| \in o(a_i^{(t_{\mathbf{y}})}). \tag{7.4}$$

Our general strategy is as follows. We will partition the terms in $\mathbf{y}(i)$ into a finite number of infinite subsequences, say $I_1 \cup \ldots \cup I_r = \mathbb{Z}_+ \setminus \{0\}$ with this union disjoint and each I_j infinite. We will show Item 1 holds for the sufficiently large index terms in each $\{\mathbf{y}(i)\}_{i \in I_j}$ and, additionally, it does so with each $y_i = \tilde{\mathbf{y}}(i) + \xi_i = \mathbf{y}(i) - \pi(\mathbf{y}(i)) + \xi_i$, for $i \in I_j$, such that $\|\xi_i\| \in o(a_i^{(t_{\mathbf{y}})})$, $\xi_i \in \mathbb{R}^\cup\langle\mathcal{A}_s\rangle$ and $-\pi(\mathbf{y}(i)) + \xi_i \in \mathsf{C}(Z_j)$ for some fixed, finite subset $Z_j \subseteq G_0^\diamond$. Then Item 1 will hold for the sufficiently large index terms in $\{\mathbf{y}(i)\}_{i=1}^\infty$ by setting $Z = \bigcup_{j=1}^r Z_j$ and using the sequence $\{y_i\}_{i=1}^\infty$, where each y_i for $i \in I_j$ was the term defined for the subsequence $\{\mathbf{y}(i)\}_{i \in I_j}$.

Let $Y \subseteq G_0$ be the subset given by Proposition 6.12.3 for $j = s$. Let $k = (i_{\mathbf{z}})_{\mathbf{z}\in\downarrow\mathcal{A}_{\mathbf{x}}}$ be a tuple of indices. In view of Proposition 6.12.3, once all $i_{\mathbf{z}}$ are sufficiently large, then we can assume that $(A_s \setminus \{x\})(k) \cup \downarrow\partial(A_s)(k) \cup Y_k$ minimally encases $-\vec{u}_{\mathbf{x}}$ for some $Y_k \subseteq Y$. Since $\mathcal{R}'$ is anchored, Proposition 6.8 implies that $\tilde{\mathbf{z}}(i_{\mathbf{z}}) = \mathbf{z}(i_{\mathbf{z}})$ for all $\mathbf{z} \in \downarrow\partial(\mathcal{A}_s)$ once $i_{\mathbf{z}}$ is sufficiently large. Thus $(A_s\setminus\{x\})(k)\cup\downarrow\partial(A_j)(k)\cup Y_k$ minimally encases $-\vec{u}_{\mathbf{x}}$ (once all $i_{\mathbf{z}}$ are sufficiently large). As a result, since $\vec{u}_{\mathbf{x}}$ is fully unbounded, it follows that $\mathbf{z}(i_{\mathbf{z}}) \in G_0^\diamond \subseteq G_0$ for all $\mathbf{z} \in (\mathcal{A}_s \setminus \{\mathbf{x}\}) \cup \downarrow\partial(\mathcal{A}_s) = \downarrow\mathcal{A}_s \setminus \{\mathbf{x}\}$, once all $i_{\mathbf{z}}$ are sufficiently large. Now fix all indices $i_{\mathbf{z}}$ with $\mathbf{z} \neq \mathbf{y}$ (chosen sufficiently large that Propositions 6.12.3 and 6.8 are applicable) and consider $i = i_{\mathbf{y}} \to \infty$. Since Y is finite, there are only a finite number of possibilities for the Y_k as $i \to \infty$. Thus (as described at the start of the proof), by passing to a subsequence of $\{\mathbf{y}(i)\}_{i=1}^\infty$, we can w.l.o.g. assume the same set Y_k occurs for every i, in which case $\{\mathbf{y}(i)\} \cup Z$ minimally encases $-\vec{u}_{\mathbf{x}}$ for all i, where $Z := (A_s \setminus \{x, y\})(k) \cup \downarrow\partial(A_j)(k) \cup Y_k \subseteq G_0$ is a fixed subset. Since $\vec{u}_{\mathbf{x}}$ is fully unbounded, we have

$$\{\mathbf{y}(i)\} \cup Z \subseteq G_0^\diamond$$

for all i. By Proposition 4.19, for each i, there are indices $1 = r_1 < \ldots < r_{\ell+1} = t_{\mathbf{x}} + 1$ and a disjoint partition $\{\mathbf{y}(i)\} \cup Z = Z_1 \cup \ldots \cup Z_\ell$ such that $(Z_1 \cup \{u^{\mathbf{x}}_{r_1}\}, \ldots, Z_\ell \cup \{u^{\mathbf{x}}_{r_\ell}\})$ is a Reay system. By Proposition 4.13, for each i, there is some subset $Z' \subseteq Z \cup \{u^{\mathbf{x}}_1, \ldots, u^{\mathbf{x}}_{t_{\mathbf{x}}}\}$ such that $Z' \cup \{\mathbf{y}(i)\}$ is a minimal positive basis. Since there are only a finite number of possibilities for Z', by passing to a subsequence of $\{\mathbf{y}(i)\}_{i=1}^\infty$, we can assume the same set Z' occurs for every i (using the same partitioning argument as before). Let $J \subseteq [1, t_{\mathbf{x}}]$ be the set of indices j for which $u^{\mathbf{x}}_j \in Z'$ and let $U = \{u^{\mathbf{x}}_j : j \in J\}$. Then

$$Z' \setminus U \subseteq Z \subseteq G_0^\diamond.$$

In view of (VR1), we have $u^{\mathbf{y}}_1, \ldots, u^{\mathbf{y}}_{t_{\mathbf{y}}} \in \mathbb{R}\langle \mathbf{y}\rangle \subseteq \mathbb{R}^\cup\langle\mathcal{A}_s\rangle$. Hence, if $\pi(\mathbf{y}(i)) = 0$ for an infinite number of i, then Item 1 follows (on the partition of indices in I having this property) taking $\{y_i\}_{i=1}^\infty$ to be a subsequence of $\{\mathbf{y}(i)\}_{i=1}^\infty$ with all terms having $\xi_i = \pi(\mathbf{y}(i)) = 0$, in view of $\mathbf{y}(i) \in G_0^\diamond$ for all i. Assuming this does not occur (so we restrict to a sub-partition of I where $\pi(\mathbf{y}(i)) = 0$ holds only for a finite number of indices i), by discarding the first few terms in the sequence $\{\mathbf{y}(i)\}_{i=1}^\infty$, we can assume $\pi(\mathbf{y}(i)) \neq 0$ for all i. Thus, since $Z' \cup \{\mathbf{y}(i)\}$ is a minimal positive basis, it follows by Carathéordory's Theorem that $\pi(Z'') \cup \{\pi(\mathbf{y}(i))\}$ is a minimal positive basis of size $|Z''| + 1$ for some subset $Z'' \subseteq Z'$. As there are only a finite number of possibilities for Z'', by once more passing to a subsequence of $\{\mathbf{y}(i)\}_{i=1}^\infty$, we can assume the same set Z'' occurs for every i (by the same partitioning argument as before). As a result, $-\pi(\mathbf{y}(i)) \in \mathsf{C}^\circ(\pi(Z''))$ for all i, in which case Lemma 4.18 implies that

$$\sum_{z\in Z''\setminus U} \alpha_i^z \pi(z) + \sum_{z\in Z''\cap U} \beta_i^z \pi(z) = -\pi(\mathbf{y}(i))$$

for some $\alpha_i^z > 0$ and $\beta_i^z > 0$ with $\alpha_i^z,\ \beta_i^z \in O(\|\pi(\mathbf{y}(i))\|)$ for all $z \in Z''$. Thus

$$\sum_{z\in Z''\setminus U} \alpha_i^z z = -\pi(\mathbf{y}(i)) + \xi_i$$

for some $\xi_i \in \ker\pi + \mathbb{R}\langle U\rangle \subseteq \mathbb{R}\langle u_1^{\mathbf{x}}, \ldots, u_{t_{\mathbf{x}}}^{\mathbf{x}}, u_1^{\mathbf{y}}, \ldots, u_{t_{\mathbf{y}}}^{\mathbf{y}}\rangle$ such that $\|\xi_i\| \in O(\|\pi(\mathbf{y}(i))\|) \subseteq o(a_i^{(t_{\mathbf{y}})})$, with the latter inclusion in view of (7.4). Setting $y_i = \sum_{z\in Z''\setminus U} \alpha_i^z z + \mathbf{y}(i)$, we have

$$y_i = a_i^{(1)} u_1^{\mathbf{y}} + \ldots + a_i^{(t_{\mathbf{y}})} u_{t_{\mathbf{y}}}^{\mathbf{y}} + \xi_i$$

and, by discarding the first few terms in $\{y_i\}_{i=1}^\infty$, we find that $\{y_i\}_{i=1}^\infty$ is an asymptotically filtered sequence with limit $\vec{u}_{\mathbf{y}}$ in view of $\|\xi_i\| \in o(a_i^{(t_{\mathbf{y}})})$. Since $Z'' \setminus U \subseteq Z' \setminus U \subseteq G_0^\diamond$ and $\mathbf{y}(i) \in G_0^\diamond$, it follows that $y_i \in \mathsf{C}(G_0^\diamond)$ for all i. By (VR1), we know $u_1^{\mathbf{y}}, \ldots, u_{t_{\mathbf{y}}}^{\mathbf{y}}, u_1^{\mathbf{x}}, \ldots, u_{t_{\mathbf{x}}}^{\mathbf{x}} \in \mathbb{R}\langle \mathbf{x}\cup\mathbf{y}\rangle \subseteq \mathbb{R}^{\cup}\langle \mathcal{A}_s\rangle$. Consequently, since $\xi_i \in \mathbb{R}\langle u_1^{\mathbf{x}}, \ldots, u_{t_{\mathbf{x}}}^{\mathbf{x}}, u_1^{\mathbf{y}}, \ldots, u_{t_{\mathbf{y}}}^{\mathbf{y}}\rangle$, it follows that $y_i \in \mathbb{R}\langle \mathbf{x} \cup \mathbf{y}\rangle \subseteq \mathbb{R}^{\cup}\langle \mathcal{A}_s\rangle$ for all i, and the sequence $\{y_i\}_{i=1}^\infty$ now has the desired properties (using $Z'' \setminus U$ for Z), completing the proof of Item 1.

2. Define a new virtual Reay system $\widetilde{\mathcal{R}}$ on the set

$$G_0 \cup \{y_{i_{\mathbf{y}}}^{\mathbf{y}} :\ \mathbf{y} \in \mathcal{A}_s \setminus \{\mathbf{x}\} \text{ and } i_{\mathbf{y}} \geq 1\} \subseteq \mathbb{R}^d$$

identical to $\mathcal{R}$ except that we replace each sequence $\{\mathbf{y}(i_{\mathbf{y}})\}_{i_{\mathbf{y}}=1}^\infty$ with the sequence $\{y_{i_{\mathbf{y}}}^{\mathbf{y}}\}_{i_{\mathbf{y}}=1}^\infty$ for $\mathbf{y} \in \mathcal{A}_s \setminus \{\mathbf{x}\}$. Since both these sequences are asymptotically filtered sequences with the same limit for every $\mathbf{y} \in \mathcal{A}_s \setminus \{\mathbf{x}\}$, this leaves all half-spaces and their boundaries unchanged. In particular, $\mathcal{A}_s$ is also the set given by Proposition 6.12.1 for $\widetilde{\mathcal{R}}$ as well as $\mathcal{R}$. Consequently, since $y_{i_{\mathbf{y}}}^{\mathbf{y}} \in \mathbb{R}^{\cup}\langle \mathcal{A}_s\rangle$ for all $i_{\mathbf{y}}$ and $\mathbf{y} \in \mathcal{A}_s \setminus \{\mathbf{x}\}$ by hypothesis, Proposition 6.12.3 applied to $\widetilde{\mathcal{R}}$ implies that

$$\widetilde{\downarrow\partial(A_s)(k)} \cup \{y_{i_{\mathbf{y}}}^{\mathbf{y}} :\ \mathbf{y} \in \mathcal{A}_s \setminus \{\mathbf{x}\}\}$$

minimally encases $-\vec{u}_{\mathbf{x}}$ once all $i_{\mathbf{z}}$ in the tuple $k = (i_{\mathbf{z}})_{\mathbf{z}\in\downarrow\mathcal{A}_s\setminus\{\mathbf{x}\}}$ are sufficiently large. Since $G_0 \subseteq \Lambda$ and $\mathcal{R}'$ is anchored, Proposition 6.8 implies that

$\widetilde{\downarrow\partial(A_s)(k)} = \downarrow\partial(A_s)(k)$ once all $i_{\mathbf{z}}$ are sufficiently large, and thus $\downarrow\partial(A_s)(k) \cup \{y^{\mathbf{y}}_{i_{\mathbf{y}}} : \mathbf{y} \in \mathcal{A}_s \setminus \{\mathbf{x}\}\}$ minimally encases $-\vec{u}_{\mathbf{x}}$ once all $i_{\mathbf{z}}$ are sufficiently large.

By definition, each $y^{\mathbf{y}}_{i_{\mathbf{y}}} = z^{\mathbf{y}}_{i_{\mathbf{y}}} + \epsilon^{\mathbf{y}}_{i_{\mathbf{y}}}$ with $\{z^{\mathbf{y}}_{i_{\mathbf{y}}}\}_{i_{\mathbf{y}}=1}^{\infty}$ a complete asymptotically filtered sequence with limit $\vec{u}_{\mathbf{y}} = (u^{\mathbf{y}}_1, \ldots, u^{\mathbf{y}}_{t_{\mathbf{y}}})$ and $\epsilon^{\mathbf{y}}_{i_{\mathbf{y}}} \in \mathbb{R}^{\cup}\langle\mathcal{A}_s\rangle$ the remainder term, so $z^{\mathbf{y}}_{i_{\mathbf{y}}} = a^{(1)}_{i_{\mathbf{y}},\mathbf{y}} u^{\mathbf{y}}_1 + \ldots + a^{(t_{\mathbf{y}})}_{i_{\mathbf{y}},\mathbf{y}} u^{\mathbf{y}}_{t_{\mathbf{y}}}$ with $\|\epsilon^{\mathbf{y}}_{i_{\mathbf{y}}}\| \in o(a^{(t_{\mathbf{y}})}_{i_{\mathbf{y}},\mathbf{y}})$. Propositions 5.3.1 and 6.12.2(e) imply that

$$\mathbb{R}^{\cup}\langle\partial(\mathcal{A}_s)\rangle = \mathbb{R}^{\cup}\langle\downarrow\partial(A_s)(k)\rangle \quad \text{and}$$
$$\mathbb{R}^{\cup}\langle\mathcal{A}_s\rangle = \mathbb{R}\langle\downarrow\partial(A_s)(k) \cup \{z^{\mathbf{y}}_{i_{\mathbf{y}}} : \mathbf{y} \in \mathcal{A}_s \setminus \{\mathbf{x}\}\}\rangle.$$

Thus, letting $\pi : \mathbb{R}^d \to \mathbb{R}^{\cup}\langle\partial(\mathcal{A}_s)\rangle^{\perp}$ be the orthogonal projection, it follows that

$$\mathbb{R}^{\cup}\langle\pi(\mathcal{A}_s)\rangle = \mathbb{R}\langle\{\pi(z^{\mathbf{y}}_{i_{\mathbf{y}}}) : \mathbf{y} \in \mathcal{A}_s \setminus \{\mathbf{x}\}\}\rangle.$$

Each $z^{\mathbf{y}}_{i_{\mathbf{y}}}$ is a representative for the half-space $\mathbf{y} \in \mathcal{A}_s \setminus \{\mathbf{x}\}$, so Proposition 6.12.2(e) implies that

$$\{\pi(z^{\mathbf{y}}_{i_{\mathbf{y}}}) : \mathbf{y} \in \mathcal{A}_s \setminus \{\mathbf{x}\}\}$$

is linearly independent, and thus a basis of size $|\mathcal{A}_s \setminus \{\mathbf{x}\}|$ for the subspace $\mathbb{R}^{\cup}\langle\pi(\mathcal{A}_s)\rangle$. In view of Proposition 6.12.2(e), we see that the value $\pi(z^{\mathbf{y}}_{i_{\mathbf{y}}})/\|\pi(z^{\mathbf{y}}_{i_{\mathbf{y}}})\| = \pi(u^{\mathbf{y}}_{t_{\mathbf{y}}})/\|\pi(u^{\mathbf{y}}_{t_{\mathbf{y}}})\|$ is constant, where $\vec{u}_{\mathbf{y}} = (u^{\mathbf{y}}_1, \ldots, u^{\mathbf{y}}_{t_{\mathbf{y}}})$, so $\|\pi(z_i)\| = \Theta(a^{(t_{\mathbf{y}})}_{i_{\mathbf{y}},\mathbf{y}})$. In consequence, each $\{\pi(y^{\mathbf{y}}_{i_{\mathbf{y}}})\}_{i_{\mathbf{y}}=1}^{\infty}$ for $\mathbf{y} \in \mathcal{A}_s \setminus \{\mathbf{x}\}$ is a sequence of terms from $\mathbb{R}^{\cup}\langle\pi(\mathcal{A}_s)\rangle$ that radially converges to $\pi(z^{\mathbf{y}}_{i_{\mathbf{y}}})/\|\pi(z^{\mathbf{y}}_{i_{\mathbf{y}}})\|$, and thus Proposition 4.15.5 implies that, once all $i_{\mathbf{y}}$ are sufficiently large, $\{\pi(y^{\mathbf{y}}_{i_{\mathbf{y}}}) : \mathbf{y} \in \mathcal{A}_s \setminus \{\mathbf{x}\}\}$ is also a basis of size $|\mathcal{A}_s \setminus \{\mathbf{x}\}|$ for the subspace $\mathbb{R}^{\cup}\langle\pi(\mathcal{A}_s)\rangle$, which, combined with $\mathbb{R}^{\cup}\langle\partial(\mathcal{A}_s)\rangle = \mathbb{R}^{\cup}\langle\downarrow\partial(A_s)(k)\rangle$, gives the desired conclusion $\mathbb{R}^{\cup}\langle\mathcal{A}_s\rangle = \mathbb{R}\langle\downarrow\partial(A_s)(k) \cup \{y^{\mathbf{y}}_{i_{\mathbf{y}}} : \mathbf{y} \in \mathcal{A}_s \setminus \{\mathbf{x}\}\}\rangle$, completing the proof.

□

We are now in position to give the *key* definition that will be the subject of the remainder of this chapter. As we will later see, finitary sets form an ample class of subsets of lattice points that behave well under inductive arguments and include infinite subsets that nonetheless behave like finite sets, particularly in relation to combinatorial properties related to $\mathcal{A}(G_0)$.

Definition 7.5 Let $\Lambda \leq \mathbb{R}^d$ be a full rank lattice, where $d \geq 0$, and let $G_0 \subseteq \Lambda$. If every purely virtual Reay system over G_0 is anchored, we say that G_0 is **finitary**.

Definition 7.6 Let all setup be as in Definition 7.5. Let $t \in [0, d]$ be a integer. If every purely virtual Reay system $\mathcal{R} = (\mathcal{X}_1 \cup \{\mathbf{v}_1\}, \ldots, \mathcal{X}_s \cup \{\mathbf{v}_s\})$ over G_0 with $\dim \mathbb{R}^{\cup}\langle \mathcal{X}_1 \cup \ldots \cup \mathcal{X}_s \rangle \leq t$ is anchored, then we say that G_0 is **finitary up to dimension** t. (We will need this refined definition for several inductive arguments.)

As a point of clarification, if $\mathcal{R} = (\mathcal{X}_1 \cup \{\mathbf{v}_1\}, \ldots, \mathcal{X}_s \cup \{\mathbf{v}_s\})$ is a virtual Reay system in G_0, then each $\mathbf{x} \in \mathcal{X}_1 \cup \ldots \cup \mathcal{X}_s$ has various possibilities for the asymptotically filtered sequence $\{\mathbf{x}(i)\}_{i=1}^{\infty}$ having limit $\vec{u}_{\mathbf{x}}$. It is possible for some of these sequences to have $\vec{u}_{\mathbf{x}}$ as fully unbounded limit, and some to have $\vec{u}_{\mathbf{x}}$ as an anchored limit. In such case, it is possible to consider $\vec{u}_{\mathbf{x}}$ either as fully unbounded or anchored, meaning the virtual Reay system $\mathcal{R}$ can be considered anchored or not, depending on which sequences $\{\mathbf{x}(i)\}_{i=1}^{\infty}$ are chosen. The definition of finitary requires that all choices of sequences $\{\mathbf{x}(i)\}_{i=1}^{\infty}$ result in $\mathcal{R}$ being anchored. A similar fact holds regarding purely virtual Reay systems, though here we only need each $\mathbf{v}_j$ to have *some* sequence $\{\mathbf{v}_j(i)\}_{i=1}^{\infty}$ having $\vec{u}_{\mathbf{v}_j}$ as fully unbounded limit in order to consider $\mathcal{R}$ as purely virtual.

We now observe that if $\mathcal{R} = (\mathcal{X}_1 \cup \{\mathbf{v}_1\}, \ldots, \mathcal{X}_s \cup \{\mathbf{x}_s\})$ is a virtual Reay system in G_0, then so too is $\mathcal{R}' = (\mathcal{X}_1 \cup \{\mathbf{v}_1\}, \ldots, \mathcal{X}_{s-1} \cup \{\mathbf{v}_{s-1}\}, ((\mathcal{X}_s \cup \{\mathbf{x}_s\}) \setminus \{\mathbf{y}\}) \cup \{\mathbf{y}\})$ for any $\mathbf{y} \in \mathcal{X}_s \cup \{\mathbf{v}_s\}$. This observation will allow us to apply many of the propositions that require $\mathcal{B} \subseteq \mathcal{X}_1 \cup \ldots \cup \mathcal{X}_s$ as a hypothesis when we only have $\mathcal{B} \subseteq (\mathcal{X}_1 \cup \ldots \cup \mathcal{X}_s) \cup \{\mathbf{v}_s\}$ with $\mathcal{X}_s \cup \{\mathbf{v}_s\} \nsubseteq \mathcal{B}$, at least in special circumstances (in theory, many of these propositions could have been stated in this more general form, but the trick above means the added generality is implied by applying the more limited form to a modified oriented Reay system).

We continue by giving some important properties of finitary sets G_0. Recall that $\mathsf{wt}(-g)$ was defined in (5.32).

Proposition 7.7 *Let $\Lambda \subseteq \mathbb{R}^d$ be a full rank lattice, where $d \geq 0$, let $G_0 \subseteq \Lambda$ be a subset with $\mathsf{C}(G_0) = \mathbb{R}^d$, and let $\mathcal{R} = (\mathcal{X}_1 \cup \{\mathbf{v}_1\}, \ldots, \mathcal{X}_s \cup \{\mathbf{v}_s\})$ be a purely virtual Reay system in G_0, where $s \geq 0$. Suppose G_0 is finitary up to dimension $\dim \mathbb{R}^{\cup}\langle \mathcal{X}_1 \cup \ldots \cup \mathcal{X}_s \rangle$.*

1. *$\mathbf{x}(i) \in G_0^{\diamond}$ for every $\mathbf{x} \in \mathcal{X}_1 \cup \ldots \cup \mathcal{X}_s$ and all sufficiently large i.*
2. *If $\mathcal{A}_j$ is the subset given by Proposition 6.12 for $j \in [1, s]$, then $\mathcal{A}_j \setminus \{\mathbf{v}_j\} \subseteq \mathcal{X}_1 \cup \ldots \cup \mathcal{X}_j$.*
3. *If $\vec{u} = (u_1, \ldots, u_t) \in G_0^{\mathsf{lim}}$ with $u_1, \ldots, u_t \in \mathbb{R}^{\cup}\langle \mathcal{X}_1 \cup \ldots \cup \mathcal{X}_s \rangle$, then $-\vec{u}$ is encased by $\mathcal{X}_1 \cup \ldots \cup \mathcal{X}_s$.*
4. *If $\vec{u} = (u_1, \ldots, u_t)$ is an anchored limit of an asymptotically filtered sequence of terms from G_0 with $u_1, \ldots, u_t \in \mathbb{R}^{\cup}\langle \mathcal{X}_1 \cup \ldots \cup \mathcal{X}_s \rangle$, then $-\vec{u}$ is minimally encased urbanely by a support subset $\mathcal{B} \subseteq \mathcal{X}_1 \cup \{\mathbf{v}_1\} \cup \ldots \cup \mathcal{X}_s \cup \{\mathbf{v}_s\}$ with $|\mathcal{B} \cap \{\mathbf{v}_1, \ldots, \mathbf{v}_s\}| \leq 1$. In particular, $\mathsf{wt}(-g) \leq 1$ for all $g \in G_0 \cap \mathbb{R}^{\cup}\langle \mathcal{X}_1 \cup \ldots \cup \mathcal{X}_s \rangle$.*

Proof Let $\mathcal{X} = \mathcal{X}_1 \cup \ldots \cup \mathcal{X}_s$. If $\dim \mathbb{R}^{\cup}\langle \mathcal{X} \rangle = 0$, then $s = 0$ and all items hold vacuously. So we may assume $\dim \mathbb{R}^{\cup}\langle \mathcal{X} \rangle > 0$ (implying $s \geq 1$) and proceed by induction on $\dim \mathbb{R}^{\cup}\langle \mathcal{X} \rangle$. Let $\mathcal{R}'_0 = (\mathcal{X}_1 \cup \{\mathbf{v}_1\}, \ldots, \mathcal{X}_{s-1} \cup \{\mathbf{v}_{s-1}\})$. Applying the induction hypothesis to $\mathcal{R}'_0$, we find that $(\mathcal{X}_1 \cup \ldots \cup \mathcal{X}_{s-1})(k) \subseteq G_0^\diamond$ for any tuple $k = (i_{\mathbf{z}})$ with all $i_{\mathbf{z}}$ sufficiently large, $\mathcal{A}_j \setminus \{\mathbf{v}_j\} \subseteq \mathcal{X}_1 \cup \ldots \cup \mathcal{X}_j$ for all $j \in [1, s-1]$, and Items 3 and 4 both hold for $\mathcal{R}'_0$.

1. Since G_0 is finitary up to dimension $\dim \mathbb{R}^{\cup}\langle \mathcal{X}_1 \cup \ldots \cup \mathcal{X}_s \rangle$, it follows that $\mathcal{R}$ is anchored, meaning $\vec{u}_{\mathbf{x}}$ is anchored for every $\mathbf{x} \in \mathcal{X}_1 \cup \ldots \cup \mathcal{X}_s$. Since $\mathcal{R}$ is purely virtual, $\vec{u}_{\mathbf{v}_s}$ is fully unbounded. Thus Lemma 7.4.1 implies that $\mathbf{x}(i) \in G_0^\diamond$ for every $\mathbf{x} \in \downarrow\mathcal{A}_s \setminus \{\mathbf{v}_s\}$ and all sufficiently large i. Since $\mathcal{X}_s \subseteq \mathcal{A}_s$, Item 1 now follows.
2. Suppose Item 2 fails. Then $\mathcal{A}_s \setminus \{\mathbf{v}_s\}$ contains some $\mathbf{v}_j$ with $j \in [1, s-1]$. Since $\mathcal{R}$ is purely virtual, $\vec{u}_{\mathbf{v}_j}$ is fully unbounded. Let $\mathcal{A}_{\mathbf{v}_s} = (\mathcal{A}_s \setminus \{\mathbf{v}_s\} \cup \partial(\{\mathbf{v}_s\}))^*$. Since $\mathbf{v}_j \in \mathcal{A}_s \setminus \{\mathbf{v}_s\}$ and every $\mathbf{v}_i$ is a maximal element in $\mathcal{X}_1 \cup \ldots \cup \mathcal{X}_s \cup \{\mathbf{v}_1, \ldots, \mathbf{v}_s\}$, it follows that $\mathbf{v}_j \in \mathcal{A}_{\mathbf{v}_s}$. Proposition 6.12.2(b) implies that $\mathcal{A}_{\mathbf{v}_s}$ minimally encases $-\vec{u}_{\mathbf{v}_s}$, which is fully unbounded since $\mathcal{R}$ is purely virtual. Proposition 6.12.2(c) implies that $\mathcal{A}_{\mathbf{v}_s}$ is a support set. Let $\vec{u}_{\mathbf{v}_s} = (u_1, \ldots, u_t)$. Applying Lemma 6.11 to $\mathcal{A}_{\mathbf{v}_s}$ and $\vec{u}_{\mathbf{v}_s}$, we find there is some $\mathcal{A} \preceq \mathcal{A}_{\mathbf{v}_s}$ and $t_0 \in [1, t]$ such that $\mathcal{A}$ minimally encases $-(u_1, \ldots, u_{t_0})$ urbanely and contains some $\mathbf{y} \in \mathcal{A}$ with $\vec{u}_{\mathbf{y}}$ fully unbounded. Since $\mathcal{A} \preceq \mathcal{A}_{\mathbf{v}_s}$ with $\mathcal{A}_{\mathbf{v}_s}$ a support set, it follows that $\mathcal{A} = \mathcal{A}^*$ is also a support set (note $\mathcal{A}^* = \mathcal{A}$ since $\mathcal{A}$ minimally encases $-(u_1, \ldots, u_{t_0})$). Apply Proposition 6.5 to $\mathcal{A}$ and let $\mathcal{R}' = (\mathcal{Y}_1 \cup \{\mathbf{w}_1\}, \ldots, \mathcal{Y}_{s'} \cup \{\mathbf{w}_{s'}\})$ be the resulting virtual Reay system with $\mathcal{Y}_1 \cup \ldots \cup \mathcal{Y}_{s'} = \downarrow\mathcal{A}$. Since $\vec{u}_{\mathbf{v}_s} = (u_1, \ldots, u_t)$ is fully unbounded and $t_0 \geq 1$, it follows that $(u_1, \ldots, u_{t_0})$ is also fully unbounded. Since $G_0 \subseteq \Lambda$, Proposition 6.8 implies that every $\mathbf{x} \in \downarrow\mathcal{A}$ has $\vec{u}_{\mathbf{x}}^{\lhd}$ either trivial or fully unbounded (since each $\vec{u}_{\mathbf{v}_i}$ is fully unbounded, this is trivially true when $\mathbf{x} = \mathbf{v}_i$, while all $\mathbf{x} \in \mathcal{X}_1 \cup \ldots \cup \mathcal{X}_s$ have $\vec{u}_{\mathbf{x}}$ anchored). Thus $\mathcal{R}'$ is purely virtual as noted before Proposition 6.5. As a result, since $\downarrow\mathcal{A} \subseteq \downarrow\mathcal{A}_{\mathbf{v}_s} \subseteq \mathcal{X}_1 \cup \{\mathbf{v}_1\} \cup \ldots \cup \mathcal{X}_s \cup \{\mathbf{v}_s\}$ implies $\dim \mathbb{R}^{\cup}\langle \mathcal{Y}_1 \cup \ldots \cup \mathcal{Y}_{s'} \rangle \leq \dim \mathbb{R}^{\cup}\langle \mathcal{X}_1 \cup \ldots \cup \mathcal{X}_s \rangle$, and since G_0 is finitary up to dimension $\dim \mathbb{R}^{\cup}\langle \mathcal{X}_1 \cup \ldots \cup \mathcal{X}_s \rangle$, it follows that $\mathcal{R}'$ must be anchored. However, this contradicts that $\mathbf{y} \in \mathcal{A} \subseteq \downarrow\mathcal{A} = \mathcal{Y}_1 \cup \ldots \cup \mathcal{Y}_{s'}$ with $\vec{u}_{\mathbf{y}}$ fully unbounded, and Item 2 is now established.
3. To prove Item 3, suppose $\vec{u} = (u_1, \ldots, u_t)$ is a fully unbounded limit of an asymptotically filtered sequence $\{x_i\}_{i=1}^{\infty}$ of terms $x_i \in G_0$ with $u_1, \ldots, u_t \in \mathbb{R}^{\cup}\langle \mathcal{X} \rangle = \mathsf{C}^{\cup}(\mathcal{X} \cup \{\mathbf{v}_1, \ldots, \mathbf{v}_s\})$ (the equality follows by Proposition 5.5) such that $-\vec{u}$ is not encased by $\mathcal{X}$. By choosing such a counter-example with t minimal, we can assume $-\vec{u}^{\lhd}$ is encased by $\mathcal{X}$. Thus there is some $\mathcal{A} \subseteq \mathcal{X}$ which minimally encases $-\vec{u}^{\lhd}$ urbanely. Let $\pi : \mathbb{R}^d \to \mathbb{R}^{\cup}\langle \mathcal{A} \rangle^{\perp}$ be the orthogonal projection. Since $\mathcal{X}$ does not encase $-\vec{u}$, we must have $u_t \notin \mathbb{R}^{\cup}\langle \mathcal{A} \rangle$ (by Proposition 5.9). Let $C \subseteq \mathcal{X}_1 \cup \{\mathbf{v}_1\} \cup \ldots \cup \mathcal{X}_s \cup \{\mathbf{v}_s\}$ be the lift of $\mathsf{Supp}_{\pi(\mathcal{R})}(-\pi(u_t))$, which is a support set by Proposition 5.4.5. Then Propositions 5.9.2 and 5.9.3(a) imply that C minimally encases $-\vec{u}$ urbanely (since $\mathcal{A} \subseteq \mathcal{X}$). Apply Proposition 6.5 to C and let $\mathcal{R}' = (\mathcal{Y}_1 \cup \{\mathbf{w}_1\}, \ldots, \mathcal{Y}_r \cup \{\mathbf{w}_r\})$ be the resulting virtual Reay system in G_0 for the subspace $\mathbb{R}^{\cup}\langle C \rangle \subseteq \mathbb{R}^{\cup}\langle \mathcal{X} \rangle$ with $\mathcal{Y}_r = C$, $\vec{u}_{\mathbf{w}_r} = \vec{u}$ and

$\mathcal{Y}_1 \cup \ldots \cup \mathcal{Y}_r = \downarrow C$. Since $G_0 \subseteq \Lambda$, Proposition 6.8 implies that every $\mathbf{x} \in \downarrow C$ has $\vec{u}^{\triangleleft}_{\mathbf{x}}$ either trivial or fully unbounded (since each $\vec{u}_{\mathbf{v}_i}$ is fully unbounded, this is trivially true when $\mathbf{x} = \mathbf{v}_i$, while all $\mathbf{x} \in \mathcal{X}_1 \cup \ldots \cup \mathcal{X}_s$ have $\vec{u}_{\mathbf{x}}$ anchored). Thus, since $\vec{u} = (u_1, \ldots, u_t)$ is fully unbounded, it follows that $\mathcal{R}'$ is purely virtual as noted before Proposition 6.5. Hence, since G_0 is finitary up to dimension $\dim \mathbb{R}^{\cup}\langle \mathcal{X}_1 \cup \ldots \cup \mathcal{X}_s \rangle$, and since $\mathcal{Y}_1 \cup \ldots \cup \mathcal{Y}_s = \downarrow C \subseteq \mathcal{X}_1 \cup \{\mathbf{v}_1\} \ldots \cup \mathcal{X}_s \cup \{\mathbf{v}_s\}$, it follows that $\mathcal{R}'$ is anchored. In particular, $\vec{u}_{\mathbf{x}}$ is anchored for every $\mathbf{x} \in \mathcal{Y}_r = C \subseteq \mathcal{X}_1 \cup \{\mathbf{v}_1\} \cup \ldots \cup \mathcal{X}_s \cup \{\mathbf{v}_s\}$. However, since $\mathcal{R}$ is purely virtual, this means $\mathbf{x}$ cannot equal any $\mathbf{v}_j$, and thus $C \subseteq \mathcal{X}$, contrary to our assumption that $\vec{u}$ was a counter-example to Item 3. Thus Item 3 is now established.

4. To prove Item 4, suppose $\vec{u} = (u_1, \ldots, u_t)$ is an anchored limit of an asymptotically filtered sequence $\{x_i\}_{i=1}^{\infty}$ of terms $x_i \in G_0$ with $u_1, \ldots, u_t \in \mathbb{R}^{\cup}\langle \mathcal{X} \rangle = \mathsf{C}^{\cup}(\mathcal{X} \cup \{\mathbf{v}_1, \ldots, \mathbf{v}_s\})$. If $t = 0$, then Item 4 follows taking $\mathcal{B} = \emptyset$, so we may assume $t \geq 1$. Let $t'' \in [0, t-1]$ be the maximal index such that $\vec{u}'' = (u_1, \ldots, u_{t''})$ is fully unbounded or trivial. If $t'' = 0$, then $\{x_i\}_{i=1}^{\infty}$ is a bounded sequence of lattice points. Since $\{x_i\}_{i=1}^{\infty}$ is an asymptotically filtered sequence of lattice points, this is only possible if $\{x_i\}_{i=1}^{\infty}$ is eventually constant, in which case $t = 1$ and $\vec{u}^{\triangleleft}$ is trivial. On the other hand, if $t'' > 0$, then Item 3 ensures that there is a subset $\mathcal{A} \subseteq \mathcal{X}$ such that $\mathcal{A}$ minimally encases $-\vec{u}''$ urbanely. This is also true for $t'' = 0$ with $\mathcal{A} = \emptyset$. Let $\pi : \mathbb{R}^d \to \mathbb{R}^{\cup}\langle \mathcal{A} \rangle^{\perp}$ be the orthogonal projection and let $t' \in [t''+1, t]$ be the minimal index such that $\pi(u_{t'}) \neq 0$. Note t' exists else $\mathcal{A}$ minimally encases $-\vec{u}$ as well (by Proposition 5.9), in which case Item 4 follows taking $\mathcal{B} = \mathcal{A}$. Let $C \subseteq \mathcal{X}_1 \cup \{\mathbf{v}_1\} \cup \ldots \cup \mathcal{X}_s \cup \{\mathbf{v}_s\}$ be the lift of $\mathsf{Supp}_{\pi(\mathcal{R})}(-\pi(u_{t'}))$, which is a support set by Proposition 5.4.5. Then Propositions 5.9.2 and 5.9.3(a) imply that C minimally encases $-\vec{u}' := -(u_1, \ldots, u_{t'})$ urbanely (since $\mathcal{A} \subseteq \mathcal{X}$). Apply Proposition 6.5 to C and let $\mathcal{R}' = (\mathcal{Y}_1 \cup \{\mathbf{w}_1\}, \ldots, \mathcal{Y}_r \cup \{\mathbf{w}_r\})$ be the resulting virtual Reay system in G_0 for the subspace $\mathbb{R}^{\cup}\langle C \rangle \subseteq \mathbb{R}^{\cup}\langle \mathcal{X} \rangle$ with $\mathcal{Y}_r = C$, $\vec{u}_{\mathbf{w}_r} = \vec{u}'$ and $\mathcal{Y}_1 \cup \ldots \cup \mathcal{Y}_r = \downarrow C$. Moreover, in the notation of Proposition 6.5, $C_{s_{\ell-1}}^{(\ell-1)} = \partial(\{\mathbf{w}_r\}) = \mathcal{A} \subseteq \downarrow C$. Note $\vec{u}_{\mathbf{w}_r} = \vec{u}' = (u_1, \ldots, u_{t'})$ is anchored in view of the maximality of t''. Since $\mathcal{Y}_1 \cup \ldots \cup \mathcal{Y}_{r-1} \subseteq \mathcal{X}_1 \cup \ldots \cup \mathcal{X}_s$ (by Proposition 6.5(a)), and since $\mathcal{R}$ is anchored, it follows that we can apply Proposition 6.8 to conclude that $\mathbf{w}_r(i) - \tilde{\mathbf{w}}_r^{(t'-1)}(i)$ is constant (and nonzero) for all sufficiently large i. Thus

$$\mathbf{w}_r(i) = x_i = \tilde{\mathbf{w}}_r^{(t'-1)}(i) + a_i^{(t')} u_{t'} + y_i$$

with

$$\tilde{\mathbf{w}}_r^{(t'-1)}(i) \in \mathbb{R}^{\cup}\langle \partial(\{\mathbf{w}_r\}) \rangle = \mathbb{R}^{\cup}\langle \mathcal{A} \rangle \subseteq \mathbb{R}^{\cup}\langle C \rangle, \quad \|y_i\| \in o(a_i^{(t')}),$$

and $\mathbf{w}_r(i) - \tilde{\mathbf{w}}_r^{(t'-1)}(i) = a_i^{(t')} u_{t'} + y_i = \xi \neq 0$ constant, for all sufficiently large i. Since $(u_1, \ldots, u_{t'})$ is anchored, we have $a_i^{(t')} \in O(1)$, so that $\|y_i\| \in o(a_i^{(t)}) \subseteq$

$o(1)$. But now $y_i \to 0$ and $\xi = \lim_{i\to\infty}(a_i^{(t')}u_{t'} + y_i) = (\lim_{i\to\infty} a_i^{(t')})u_{t'} \in \mathbb{R}^{\cup}\langle C\rangle$, with the inclusion since C minimally encases $-(u_1, \ldots, u_{t'})$. Thus $x_i = \mathbf{w}_r(i) = \mathbf{w}_r^{(t'-1)}(i) + \xi \in \mathbb{R}\langle C\rangle$, ensuring that the limit $\vec{u} = (u_1, \ldots, u_t)$ of the asymptotically filtered sequence $\{x_i\}_{i=1}^{\infty}$ has $u_i \in \mathbb{R}^{\cup}\langle C\rangle$ for all $i \in [1, t]$. Proposition 5.9 now ensures that C not only minimally encases $-(u_1, \ldots, u_{t'})$, but also $-\vec{u} = (u_1, \ldots, u_t)$. Consequently, if $\mathrm{wt}(C) \le 1$, then Item 4 follows taking $\mathcal{B} = C$. Assume by contradiction that $\mathrm{wt}(C) \ge 2$. Then there are distinct $\mathbf{v}_{k_1}, \mathbf{v}_{k_2} \in C = \mathcal{Y}_r$, and since $\mathcal{R}$ is purely virtual, both $\vec{u}_{\mathbf{v}_{k_1}}$ and $\vec{u}_{\mathbf{v}_{k_2}}$ are fully unbounded.

As noted after Definition 7.5, $\mathcal{R}'' = (\mathcal{Y}_1 \cup \{\mathbf{w}_1\}, \ldots, ((\mathcal{Y}_r \cup \{\mathbf{w}_r\}) \setminus \{\mathbf{v}_{k_1}\}) \cup \{\mathbf{v}_{k_1}\})$ is also a virtual Reay system in G_0. As noted several times already in the proof, the strict truncation of any limit defining a half-space $\mathbf{x}$ from $\mathcal{R}$ is either trivial or fully unbounded. By construction, $\vec{u}_{\mathbf{v}_{k_1}}$ is fully unbounded. Consider an arbitrary $\mathbf{w}_j$ with $j < r$. Then, in view of Proposition 6.5(b) (recalling that $C_{s_{\ell-1}}^{(\ell-1)} = \mathcal{A}$), it follows that either $\vec{u}_{\mathbf{w}_j} = (u_1, \ldots, u_{r_j})$ for some $r_j \le t''$, in which case $\vec{u}_{\mathbf{w}_j}$ is fully unbounded by definition of t'', or else $\vec{u}_{\mathbf{w}_j}$ is the strict truncation of some defining limit for a half-space from $\downarrow C$, and thus also fully unbounded. Consequently, $\mathcal{R}''$ is purely virtual. Thus, since $\mathbb{R}^{\cup}\langle C\rangle \subseteq \mathbb{R}^{\cup}\langle \mathcal{X}\rangle$ ensures $\dim \mathbb{R}^{\cup}\langle C\rangle \le \mathbb{R}^{\cup}\langle \mathcal{X}\rangle$, and since G_0 is finitary up to dimension $\dim \mathbb{R}^{\cup}\langle \mathcal{X}\rangle$, it follows that $\mathcal{R}''$ is anchored. In particular, $\vec{u}_{\mathbf{x}}$ is anchored for every $\mathbf{x} \in \mathcal{Y}_r \setminus \{\mathbf{v}_{k_1}\} = C \setminus \{\mathbf{v}_{k_1}\}$, contradicting our assumption that $\mathbf{v}_{k_2} \in C$ is a distinct half-space from $\mathbf{v}_{k_1}$ with $\vec{u}_{\mathbf{v}_{k_2}}$ fully unbounded. Thus Item 4 is now also established, completing the proof. □

We will now show that $G_0^{\diamond} \cap -\mathsf{C}(G_0^{\diamond})$ being bounded implies that G_0 is finitary, giving an important example of finitary sets apart from G_0 finite. The converse to this statement is false, as seen by the example

$$G_0 = \{(-1, y) : y \in \mathbb{Z}_+\} \cup \{(x, -1) : x \in \mathbb{Z}_+\} \subseteq \mathbb{Z}^2,$$

which has $G_0^{\diamond} = G_0$ and $G_0^{\mathsf{lim}} = \{(1, 0), (0, 1)\}$. On the other hand, the set

$$G_0 = \{(-1, y) : y \ge 1,\ y \in \mathbb{Z}\} \cup \{(x, 0) : x \ge 1,\ x \in \mathbb{Z}\} \cup \{(0, -1)\}$$

gives a prototypical example of a basic finitary set in $\mathbb{Z}^2$ with $G_0^{\diamond} \cap -\mathsf{C}(G_0^{\diamond}) = \emptyset$, $G_0^{\mathsf{lim}} = \{(1, 0), (0, 1)\}$ and $G_0^{\diamond} = \{(-1, y) : y \ge 1, y \in \mathbb{Z}\} \cup \{(0, -1)\}$.

Theorem 7.8 *Let $\Lambda \subseteq \mathbb{R}^d$ be a full rank lattice, where $d \ge 0$, and let $G_0 \subseteq \Lambda$ be a subset with $\mathsf{C}(G_0) = \mathbb{R}^d$. If $G_0^{\diamond} \cap -\mathsf{C}(G_0^{\diamond})$ is a bounded set, then G_0 is finitary. In particular, if $0 \notin \mathsf{C}^*(G_0^{\diamond})$, then G_0 is finitary.*

Proof Assume by contradiction that the theorem is false. Then, we can let $\mathcal{R} = (\mathcal{X}_1 \cup \{\mathbf{v}_1\}, \ldots, \mathcal{X}_s \cup \{\mathbf{v}_s\})$ be a purely virtual Reay system over G_0 which is not anchored having $\dim \mathbb{R}^{\cup}\langle \mathcal{X}\rangle$ minimal, where $\mathcal{X} = \mathcal{X}_1 \cup \ldots \cup \mathcal{X}_s$. Now $s \ge 1$, and in view of the minimality of $\dim \mathbb{R}^{\cup}\langle \mathcal{X}\rangle$, we conclude that G_0 is finitary up to dimension $\dim \mathbb{R}^{\cup}\langle \mathcal{X}\rangle - 1 \ge 0$. Let $\mathcal{R}'_0 = (\mathcal{X}_1 \cup \{\mathbf{v}_1\}, \ldots, \mathcal{X}_{s-1} \cup \{\mathbf{v}_{s-1}\})$. We can apply Proposition 7.7 to $\mathcal{R}'_0$. In particular, $\mathcal{R}'_0$ is anchored and $(X_1 \cup \ldots \cup X_{s-1})(k) \subseteq G_0^\diamond$ for any tuple $k = (i_\mathbf{z})$ with all $i_\mathbf{z}$ sufficiently large.

Since $\mathcal{R}$ is purely virtual, $\vec{u}_{\mathbf{v}_s}$ is fully unbounded. Thus, since $\mathcal{R}'_0$ is anchored, Lemma 7.4.1 implies that $\mathbf{x}(i) \in G_0^\diamond$ for every $\mathbf{x} \in \downarrow\mathcal{A}_s \setminus \{\mathbf{v}_s\}$ and all sufficiently large i. Since $\mathcal{X}_s \subseteq \mathcal{A}_s$, we now see that $\mathbf{x}(i) \in G_0^\diamond$ for any $\mathbf{x} \in \mathcal{X}$ and sufficiently large i. Moreover, Lemma 7.4.1 also implies that, for each $\mathbf{y} \in \mathcal{X}_s$, there is an asymptotically filtered sequence $\{y_{i_\mathbf{y}}^\mathbf{y}\}_{i_\mathbf{y}=1}^\infty$ of terms $y_{i_\mathbf{y}}^\mathbf{y} \in \mathsf{C}(G_0^\diamond) \cap \mathbb{R}^\cup\langle\mathcal{A}_s\rangle$ with limit $\vec{u}_\mathbf{y}$.

Suppose there were some $\mathbf{z} \in \mathcal{A}_s \setminus \{\mathbf{v}_s\}$ such that $\vec{u}_\mathbf{z}$ is also fully unbounded. Then we could apply Lemma 7.4.1 to conclude that there is also an asymptotically filtered sequence $\{x_i\}_{i=1}^\infty$ of terms $x_i \in \mathsf{C}(G_0^\diamond) \cap \mathbb{R}^\cup\langle\mathcal{A}_s\rangle$ with limit $\vec{u}_{\mathbf{v}_s}$. Indeed, $\mathbf{v}_s(i) \in G_0^\diamond$ for all sufficiently large i and there is a finite set $Z \subseteq G_0^\diamond$ such that every $x_i \in \mathbf{v}_s(i) + \mathsf{C}(Z)$ once i is sufficiently large. By Lemma 7.4.2, the set

$$Y_k := \downarrow\partial(A_s)(k) \cup \{y_{i_\mathbf{y}}^\mathbf{y} \,:\, \mathbf{y} \in \mathcal{A}_s \setminus \{\mathbf{v}_s\}\}$$

minimally encases $-\vec{u}_{\mathbf{v}_s}$ with $\mathbb{R}^\cup\langle\mathcal{A}_s\rangle = \mathbb{R}\langle Y_k\rangle$ for any tuple $k = (i_\mathbf{z})_{\mathbf{z} \in \downarrow\mathcal{A}_s \setminus \{\mathbf{v}_s\}}$ with all $i_\mathbf{z}$ sufficiently large. Fix one such tuple $\kappa = (\iota_\mathbf{z})_{\mathbf{z} \in \downarrow\mathcal{A}_s \setminus \{\mathbf{v}_s\}}$ such that all the above conclusions are true and let $Y = Y_\kappa$. Since $\downarrow\partial(\mathcal{A}_s) \subseteq \mathcal{X}_1 \cup \ldots \cup \mathcal{X}_{s-1}$, we have $\downarrow\partial(A_s)(\kappa) \subseteq G_0^\diamond$, and thus $Y \subseteq \mathsf{C}(G_0^\diamond)$ in view of the definition of the $y_{i_\mathbf{y}}^\mathbf{y}$. Recall that $\mathbb{R}\langle Y\rangle = \mathbb{R}^\cup\langle\mathcal{A}_s\rangle$. Consequently, applying Proposition 4.20 to the sequence $\{x_i\}_{i=1}^\infty$, we find that $Y \cup \{x_i\}$ is a minimal positive basis for all sufficiently large i. Hence, since $x_i \in \mathbf{v}_s(i) + \mathsf{C}(Z)$, we conclude that $\mathbf{v}_s(i) \in -\mathsf{C}(Y \cup Z) \subseteq -\mathsf{C}(G_0^\diamond)$ for all sufficiently large i. However, since $\vec{u}_{\mathbf{v}_s}$ is fully unbounded (in view of $\mathcal{R}$ being purely virtual), we see that $\{\mathbf{v}_s(i)\}_{i=1}^\infty$ is an unbounded sequence of elements contained in $-\mathsf{C}(G_0^\diamond)$ with $\mathbf{v}_s(i) \in G_0^\diamond$ for all sufficiently large i, contradicting the hypothesis that $G_0^\diamond \cap -\mathsf{C}(G_0^\diamond)$ is bounded. So we instead conclude that $\vec{u}_\mathbf{z}$ is anchored for every $\mathbf{z} \in \mathcal{A}_s \setminus \{\mathbf{v}_s\}$. In particular, this ensures that every $\vec{u}_\mathbf{y}$ with $\mathbf{y} \in \mathcal{X}_s \subseteq \mathcal{A}_s \setminus \{\mathbf{v}_s\}$ is anchored, implying that $\mathcal{R}$ is anchored (as $\mathcal{R}'_0$ is anchored), contradicting that $\mathcal{R}$ was a minimal counterexample.

For the in particular statement, suppose $0 \notin \mathsf{C}^*(G_0^\diamond)$. Then, since $0 \notin G_0^\diamond$ (by Proposition 7.2.2), it follows that $G_0^\diamond \cap -\mathsf{C}(G_0^\diamond)$ is bounded (indeed, empty), and so G_0 is finitary by the main part of the theorem. □

Proposition 7.7 has some important consequences beyond Theorem 7.8. Suppose $G_0 \subseteq \Lambda \subseteq \mathbb{R}^d$ is finitary with $\mathsf{C}(G_0) = \mathbb{R}^d$. If $X \subseteq G_0 \cup G_0^\infty$ is a minimal positive basis, then each $x \in X$ defines a one-dimensional half-space $\mathbf{x} = \mathbb{R}_+ x$, and letting $\mathcal{X}$ be the corresponding set of half-spaces, it follows that $((\mathcal{X} \setminus \{\mathbf{x}\}) \cup \{\mathbf{x}\})$ is a virtual Reay system in G_0 for any $\mathbf{x} \in \mathcal{X}$. For a half-space $\mathbf{z} \in \mathcal{X}$, we either have $z \in G_0^\infty$, in which case $\vec{u}_\mathbf{z} \in G_0^\infty$ is unbounded, or else $z \in G_0$, in which case $\vec{u}_\mathbf{z} = z/\|z\|$ is bounded (and in case $z \in G_0$ and $z/\|z\| \in G_0^\infty$, then both are simultaneously possible). Consequently, if $X \cap G_0^\infty \neq \emptyset$, then we can choose $\mathbf{x} \in \mathcal{X}$ so that the resulting virtual Reay system $((\mathcal{X} \setminus \{\mathbf{x}\}) \cup \{\mathbf{x}\})$ is purely virtual. Since G_0 is finitary, the virtual Reay system is anchored, i.e., $|X \cap G_0^\infty| \leq 1$. Moreover, if $|X \cap G_0^\infty| = 1$, then $X \setminus G_0^\infty \subseteq G_0^\diamond$ by Proposition 7.7.1. Thus there are limitations on how minimal positive bases can be formed using elements from $G_0 \cup G_0^\infty$.

In general, if $\{z_i\}_{i=1}^\infty$ is an asymptotically filtered sequence of lattice points with limit u_1, say with corresponding representation $z_i = a_i u_1 + y_i$, then we either have $a_i \to \infty$ or $a_i \to C$ for some $C \geq 0$. If $a_i \to C$, then $\|y_i\| \in o(a_i) \subseteq o(1)$, ensuring that $\|y_i\| \to 0$, which means that $\{z_i\}_{i=1}^\infty$ is a convergent sequence of lattice points with limit $(\lim_{i\to\infty} a_i)u_1 = Cu_1$. Thus, since any convergent sequence of lattice points is eventually constant, it follows that $z_i = (\lim_{i\to\infty} a_i)u_1 = Cu_1$ for all sufficiently large i, forcing $C > 0$ as $z_i = Cu_1 = a_i u_1 + y_i \neq 0$ for all sufficiently large i (as $a_i > 0$ with $y_i \in \mathbb{R}\langle u_1\rangle^\perp$ for all i). Moreover, since $z_i = Cu_1$ for all sufficiently large i, and since u_1 and y_i are orthogonal (and thus linearly independent), it follows that $y_i = 0$ for all sufficiently large i, and if $\{b_i\}_{i=1}^\infty$ is a sequence of positive real numbers with $a_i \in o(b_i)$, then $a_i \to C > 0$ ensures $\{b_i\}_{i=1}^\infty$ is unbounded. Now suppose $\{x_i\}_{i=1}^\infty$ is an asymptotically filtered sequence of terms $x_i \in G_0 \subseteq \Lambda$ having limit $\vec{u} = (u_1, \ldots, u_t)$, say with $x_i = a_i^{(1)} u_1 + \ldots + a_i^{(t)} u_t + y_i$, and that $\pi : \mathbb{R}^d \to \mathcal{E}^\perp$ is the orthogonal projection for some subspace $\mathcal{E} = \mathbb{R}\langle Z\rangle$ generated by a subset of lattice points $Z \subseteq \Lambda$. Then $\pi(G_0) \subseteq \pi(\Lambda)$ with $\pi(\Lambda)$ a full rank lattice in $\mathcal{E}^\perp$ by Proposition 2.1. If $\pi(\vec{u})$ is the empty tuple, then $u_i \in \mathcal{E}$ for all i. Otherwise, there is a minimal index $r_1 \in [1, t]$ with $\pi(u_{r_1}) \neq 0$, we may consider $\{x_i\}_{i=1}^\infty$ as an asymptotically filtered sequence with truncated limit $(u_1, \ldots, u_{r_1})$, and Proposition 3.5.1 implies that the sufficiently large index terms in $\{\pi(x_i)\}_{i=1}^\infty$ form a radially convergent sequence with limit $\overline{u}_{r_1} := \pi(u_{r_1})/\|\pi(u_{r_1})\|$, say with representation $\pi(x_i) = b_i^{(r_1)} \overline{u}_{r_1} + \overline{y}_i$ where $b_i^{(r_1)} \in \Theta(a_i^{(r_1)})$ and $\overline{y}_i$ is a vector orthogonal to $\overline{u}_{r_1}$ with $\|\overline{y}_i\| \in o(b_i^{(r_1)}) = o(a_i^{(r_1)})$. Consequently, if $\{a_i^{(r_1)}\}_{i=1}^\infty$ is bounded, then we can apply the previous observations to conclude that $\pi(x_i)$ is constant with $\overline{y}_i = 0$ for all sufficiently large i, and that $\{a_i^{(j)}\}_{i=1}^\infty$ is unbounded for any $j < r_1$. In such case, $\pi(\vec{u}) = \overline{u}_{r_1}$ is a complete limit for $\{\pi(x_i)\}_{i=1}^\infty$ with $\pi(x_i) = C\overline{u}_{r_1}$ for all sufficiently large i, for some $C > 0$, and $a_i^{(j)} \to \infty$ for any $j < r_1$. This does not imply $r_1 = t$, but does ensure $u_i \in \mathcal{E} + \mathbb{R}u_{r_1}$ for all $i \in [1, t]$. We will need these observations in our discussion below.

Now suppose that

$$\mathcal{R} = (\mathcal{X}_1 \cup \{\mathbf{v}_1\}, \ldots, \mathcal{X}_s \cup \{\mathbf{v}_s\})$$

is a purely virtual Reay system (possibly trivial) in the finitary set G_0 and let

$$\pi : \mathbb{R}^d \to \mathbb{R}^\cup\langle \mathcal{X}_1 \cup \ldots \cup \mathcal{X}_s\rangle^\perp$$

be the orthogonal projection. Since G_0 is finitary, it follows that $\mathcal{R}$ is anchored, ensuring that $\mathbb{R}^\cup\langle \mathcal{X}_1 \cup \ldots \cup \mathcal{X}_s\rangle = \mathbb{R}\langle {\downarrow}(\mathcal{X}_1 \cup \ldots \mathcal{X}_s)(k)\rangle$ for any tuple $k = (i_{\mathbf{z}})_{\mathbf{z} \in {\downarrow}(\mathcal{X}_1 \cup \ldots \cup \mathcal{X}_s)}$ with all $i_{\mathbf{z}}$ sufficiently large (in view of Propositions 5.3.1 and 6.8). Consequently, since ${\downarrow}(\mathcal{X}_1 \cup \ldots \mathcal{X}_s)(k) \subseteq G_0 \subseteq \Lambda$ is a subset of lattice points, it follows from Proposition 2.1 that $\pi(\Lambda)$ is a full rank lattice in $\pi(\mathbb{R}^d) =$

$\mathbb{R}^{\cup}\langle \mathcal{X}_1 \cup \ldots \cup \mathcal{X}_s\rangle^{\perp}$. Let $v \in \pi(G_0) \cup \pi(G_0)^{\infty}$ be nonzero. In view Proposition 3.5.2, there is an asymptotically filtered sequence $\{x_i\}_{i=1}^{\infty}$ of terms $x_i \in G_0$ with limit $\vec{u} = (u_1, \ldots, u_t)$ such that $\{\pi(x_i)\}_{i=1}^{\infty}$ is an asymptotically filtered sequence of terms with limit $\pi(\vec{u}) = \pi(u_t)/\|\pi(u_t)\| = v/\|v\|$. For instance, if $v \in \pi(G_0)$, then $\{x_i\}_{i=1}^{\infty}$ may be taken to be a constant sequence, though there may be other non-constant sequences with $\pi(x_i) = v$ for all i as well which we could choose instead. Moreover, we either have $v \in \pi(G_0)^{\infty}$, in which case $\vec{u}$ is fully unbounded by Proposition 3.5.1 (with $r_{\ell} = r_1 = t$ in the notation of Proposition 3.5.1), or else $v \in \pi(G_0)$, in which case $\vec{u}$ is anchored (and in case $v \in \pi(G_0)$ and $v/\|v\| \in \pi(G_0)^{\infty}$, then both are possible). Let $x_i = a_i^{(1)}u_1 + \ldots + a_i^{(t-1)}u_{t-1} + a_i^{(t)}u_t + y_i$ be the representation of $\{x_i\}_{i=1}^{\infty}$ as an asymptotically filtered sequence with limit $\vec{u}$. If $a_i^{(t)} \to \infty$, then $a_i^{(j)} \to \infty$ for all $j < t$ holds trivially. On the other hand, if $a_i^{(t)}$ is bounded, then the earlier discussion above also ensures that $a_i^{(j)} \to \infty$ for all $j < t$, so $\vec{u}^{\triangleleft}$ is trivial or fully unbounded in all cases. But now Proposition 7.7.3 ensures that there is some $\mathcal{B}_{\mathbf{x}} \subseteq \mathcal{X}_1 \cup \ldots \cup \mathcal{X}_s$ which minimally encases $-\vec{u}^{\triangleleft}$. This allows us to define a half-space $\mathbf{x}$ by setting $\overline{\mathbf{x}} = \mathbb{R}^{\cup}\langle \mathcal{B}_{\mathbf{x}}\rangle + \mathbb{R}_+u_t$, $\partial(\mathbf{x}) = \mathbb{R}^{\cup}\langle \mathcal{B}_{\mathbf{x}}\rangle$ and $\partial(\mathbf{x}) \cap \mathbf{x} = \mathsf{C}(\mathcal{B}_{\mathbf{x}})$. Now suppose $X \subseteq \pi(G_0) \cup \pi(G_0)^{\infty}$ is a minimal positive basis and let $\mathcal{X}$ be the set of half-spaces obtained from the elements $x \in X$ as just described (using some compatible choice of asymptotically filtered sequences for each $x \in X$). Then $\mathcal{R}' = (\mathcal{X}_1 \cup \{\mathbf{v}_1\}, \ldots, \mathcal{X}_s \cup \{\mathbf{v}_s\}, (\mathcal{X} \setminus \{\mathbf{x}\}) \cup \{\mathbf{x}\})$ will be a virtual Reay system over G_0 for any $\mathbf{x} \in \mathcal{X}$. Since $\mathsf{C}(G_0) = \mathbb{R}^d$, we have $\mathsf{C}(\pi(G_0)) = \pi(\mathbb{R}^d) = \mathbb{R}^{\cup}\langle \mathcal{X}_1 \cup \ldots \cup \mathcal{X}_s\rangle^{\perp}$. Consequently, given any nonzero $x \in \pi(G_0) \cup \pi(G_0)^{\infty}$, there exists a minimal positive basis $X \subseteq \pi(G_0) \cup \pi(G_0)^{\infty}$ which contains x. If $X \cap \pi(G_0)^{\infty}$ is nonempty, then we can choose $\mathbf{x} \in \mathcal{X}$ so that $\vec{u}_{\mathbf{x}}$ is fully unbounded, in which case $\mathcal{R}'$ will be purely virtual. Since G_0 is finitary, we conclude that $\mathcal{R}'$ must be anchored, i.e., $|X \cap \pi(G_0)^{\infty}| \leq 1$. Moreover, if $|X \cap \pi(G_0)^{\infty}| = 1$, then $X \setminus \pi(G_0)^{\infty} \subseteq \pi(G_0)^{\diamond}$ by Proposition 7.2.1. Thus we have generalized our initial observation regarding minimal positive bases over $G_0 \cup G_0^{\infty}$. Furthermore, we now see that purely virtual Reay systems over a finitary set G_0 can be constructed greedily by recursively applying the construction just described with our only being prevented from extending the purely virtual Reay system $\mathcal{R}$ to a larger one when $\pi(G_0)^{\infty} = \emptyset$, i.e., when $\pi(G_0)$ is finite (since $\pi(\Lambda)$ is a lattice).

Likewise, if $\mathcal{R}' = (\mathcal{Y}_{s+1} \cup \{\mathbf{w}_{s+1}\}, \ldots, \mathcal{Y}_{s'} \cup \{\mathbf{w}_{s'}\})$ is a purely virtual Reay system over $\pi(G_0)$, then $Y_{s+1} \cup \{w_{s+1}\}$ will be a minimal positive basis, and we can define a virtual set $\mathcal{X}_{s+1} \cup \{\mathbf{v}_{s+1}\}$ (using any compatible choice of asymptotically filtered sequence for each $\mathbf{z} \in \mathcal{Y}_{s+1} \cup \{\mathbf{w}_{s+1}\}$) as above so that

$$\mathcal{R}'' = (\mathcal{X}_1 \cup \{\mathbf{v}_1\}, \ldots, \mathcal{X}_{s+1} \cup \{\mathbf{v}_{s+1}\})$$

is a purely virtual Reay system over G_0 with $\pi(\mathcal{R}'') = (\mathcal{X}_{s+1}^{\pi} \cup \{\pi(\mathbf{v}_{s+1})\}) = (\mathcal{Y}_{s+1} \cup \{\mathbf{w}_{s+1}\})$. It follows that

$$\pi'(\mathcal{R}') = (\pi'(\mathcal{Y}_{s+2}) \cup \{\pi'(\mathbf{w}_{s+2})\}, \ldots, \pi'(\mathcal{Y}_{s'}) \cup \{\pi'(\mathbf{w}_{s'})\})$$

is a purely virtual Reay system over $\pi'(G_0)$ (in view of Proposition 6.3), where $\pi' : \mathbb{R}^d \to \mathbb{R}^{\cup}\langle \mathcal{X}_1 \cup \ldots \cup \mathcal{X}_{s+1}\rangle^{\perp}$ is the orthogonal projection. As before, $\pi'(\Lambda)$ a full rank lattice and, for each $\mathbf{y} \in \mathcal{Y}_{s+2} \cup \{\mathbf{w}_{s+2}\}$, we can choose via Proposition 3.5.2 an asymptotically filtered sequence $\{x_i\}_{i=1}^{\infty}$ of terms $x_i \in G_0$ with limit $\vec{u}_{\mathbf{x}} = (u_1, \ldots, u_t)$ such that $\pi(\vec{u}_{\mathbf{x}}) = \vec{u}_{\mathbf{y}}$ and $\pi(\vec{u}_{\mathbf{x}}^{\triangleleft}) = \pi(\vec{u}_{\mathbf{x}})^{\triangleleft} = \vec{u}_{\mathbf{y}}^{\triangleleft}$, and then $\pi'(\vec{u}_{\mathbf{x}}) = \pi'(\pi(\vec{u}_{\mathbf{x}})) = \pi'(\vec{u}_{\mathbf{y}}) = \vec{u}_{\pi'(\mathbf{y})}$ follows in view of $\ker \pi \leq \ker \pi'$ and Proposition 6.3. Since $\pi'(\mathbf{y})$ lies at depth 1 in the oriented Reay system $\pi'(\mathcal{R}')$, it has trivial boundary, ensuring that $\vec{u}_{\pi'(\mathbf{y})}$ is a single unit vector. Moreover, since $\pi'(\mathbf{y})$ is a relative half-space in $\pi'(\mathcal{R}')$, we must have $\pi'(\pi(u_t)) = \pi'(u_t)$ being a representative for $\pi'(\mathbf{y})$ since $\pi(u_t)$ is a representative for $\mathbf{y}$ (in view of $\pi(\vec{u}_{\mathbf{x}}) = \vec{u}_{\mathbf{y}}$ and $\pi(\vec{u}_{\mathbf{x}}^{\triangleleft}) = \pi(\vec{u}_{\mathbf{x}})^{\triangleleft}$), ensuring that $\vec{u}_{\pi'(\mathbf{y})} = \pi'(u_t)/\|\pi'(u_t)\|$. (Alternatively, since $\partial(\{\mathbf{y}\}) \subseteq \mathcal{Y}_{s+1} = \pi(\mathcal{X}_{s+1})$, $\pi(\vec{u}_{\mathbf{x}}) = \vec{u}_{\mathbf{y}}$ and $\pi(\vec{u}_{\mathbf{x}}^{\triangleleft}) = \pi(\vec{u}_{\mathbf{x}})^{\triangleleft}$, we have $\pi'(u_i) = 0$ for all $i < t$, ensuring the same conclusion.) Thus, per prior discussion, $\vec{u}_{\mathbf{x}}^{\triangleleft}$ will be fully unbounded (or trivial), and thus $-\vec{u}_{\mathbf{x}}^{\triangleleft}$ is minimally encased by some

$$\mathcal{B}_{\mathbf{x}} \subseteq \mathcal{X}_1 \cup \ldots \cup \mathcal{X}_{s+1},$$

allowing us to define a half-space $\mathbf{x}$ with $\partial(\{\mathbf{x}\}) = \mathcal{B}_{\mathbf{x}}$ and $\overline{\mathbf{x}} = \mathbb{R}^{\cup}\langle \mathcal{B}_{\mathbf{x}}\rangle + \mathbb{R}_+ u_t$. Since the support set $\partial(\{\mathbf{x}\}) = \mathcal{B}_{\mathbf{x}}$ minimally encases $-\vec{u}_{\mathbf{x}}^{\triangleleft}$, Proposition 5.9.4 and Proposition 5.4.6 ensure that the support set $\partial(\{\mathbf{x}\})^{\pi} \subseteq \pi(\mathcal{X}_{s+1}) = \mathcal{Y}_{s+1}$ minimally encases $-\pi(\vec{u}_{\mathbf{x}}^{\triangleleft}) = -\pi(\vec{u}_{\mathbf{x}})^{\triangleleft} = -\vec{u}_{\mathbf{y}}^{\triangleleft}$, which means $\partial(\{\mathbf{x}\})^{\pi} = \partial(\{\mathbf{y}\})$. Hence, since the representative u_t for $\mathbf{x}$ maps to the representative $\pi(u_t)$ for $\mathbf{y}$ (in view of $\pi(\vec{u}_{\mathbf{x}}) = \vec{u}_{\mathbf{y}}$ and $\pi(\vec{u}_{\mathbf{x}}^{\triangleleft}) = \pi(\vec{u}_{\mathbf{x}})^{\triangleleft}$), we conclude that $\pi(\mathbf{x}) = \mathbf{y}$. We can iterate this procedure, resulting in a purely virtual Reay system $\mathcal{W} = (\mathcal{X}_1 \cup \{\mathbf{v}_1\}, \ldots, \mathcal{X}_{s'} \cup \{\mathbf{v}_{s'}\})$ over G_0 such that $\pi(\mathcal{W}) = \mathcal{R}'$ and $\mathcal{X}_j^{\pi} \cup \{\pi(\mathbf{v}_j)\} = \mathcal{Y}_j \cup \{\mathbf{w}_j\}$ for all $j \in [s+1, s']$, and $\pi(\vec{u}_{\mathbf{x}}) = \vec{u}_{\pi(\mathbf{x})}$ and $\pi(\vec{u}_{\mathbf{x}}^{\triangleleft}) = \pi(\vec{u}_{\mathbf{x}})^{\triangleleft}$ for all $\mathbf{x} \in \mathcal{X}_{s+1} \cup \{\mathbf{v}_{s+1}\} \cup \ldots \cup \mathcal{X}_{s'} \cup \{\mathbf{v}_{s'}\}$, with each $\vec{u}_{\mathbf{x}}$ the limit of a compatible asymptotically filtered sequence initially chosen for the half-space $\mathbf{y} = \pi(\mathbf{x}) \in \mathcal{Y}_{s+1} \cup \{\mathbf{w}_{s+1}\} \cup \ldots \cup \mathcal{Y}_{s'} \cup \{\mathbf{w}_{s'}\}$. We call $\mathcal{W}$ an **extension** of $\mathcal{R}$.

We call the purely virtual Reay system $\mathcal{R}$ over G_0 with $\pi(G_0)^{\infty} = \emptyset$ a **maximal** purely virtual Reay system over G_0. Since $\pi(G_0)$ is a subset of lattice points, this is equivalent to $\pi(G_0)$ being finite as noted earlier. Now further assume that $\mathcal{R}$ is a maximal purely virtual Reay system over G_0, so $\pi(G_0)$ is a finite set of lattice points. Suppose $\mathcal{R}' = (\mathcal{Y}_{s+1} \cup \{\mathbf{w}_{s+1}\}, \ldots, \mathcal{Y}_{s'} \cup \{\mathbf{w}_{s'}\})$ is a virtual Reay system over $\pi(G_0)$. For instance, we could use Proposition 4.10 to find an ordinary Reay system $(Y_{s+1} \cup \{w_{s+1}\}, \ldots, Y_{s'} \cup \{w_{s'}\})$ (for the entire space $\pi(\mathbb{R}^d)$ if we like), and then replace each element $y \in Y_{s+1} \cup \{w_{s+1}\} \cup \ldots \cup Y_{s'} \cup \{w_{s'}\}$ with the one-dimensional half-space $\mathbf{y} = \mathbb{R}_+ y$ it represents to obtain $\mathcal{R}'$. We can associate to each half-space $\mathbf{y} \in Y_{s+1} \cup \{w_{s+1}\} \cup \ldots \cup Y_{s'} \cup \{w_{s'}\}$ an asymptotically filtered sequence $\{x_i\}_{i=1}^{\infty}$ of terms from G_0 with limit $\vec{u} = (u_1, \ldots, u_t)$ such that $\pi(\vec{u}) = \pi(u_t)/\|\pi(u_t)\| = y/\|y\|$, which ensures $u_i \in \ker \pi = \mathbb{R}^{\cup}\langle \mathcal{X}_1 \cup \ldots \cup \mathcal{X}_s\rangle$ for all $i < t$. For instance, we could take $\{x_i\}_{i=1}^{\infty}$ to be a constant sequence, though there may be other possibilities (we simply need $\pi(x_i)$ to eventually be constant). Per prior discussion, $\vec{u}^{\triangleleft}$ must be trivial or fully unbounded. Thus, since $u_1, \ldots, u_{t-1} \in$

$\mathbb{R}^{\cup}\langle \mathcal{X}_1 \cup \ldots \cup \mathcal{X}_s \rangle$, Proposition 7.7.3 ensures that there is some $\mathcal{B}_{\mathbf{x}} \subseteq \mathcal{X}_1 \cup \ldots \cup \mathcal{X}_s$ which minimally encases $-\vec{u}^{\triangleleft}$. As before, this means we can define a half-space $\mathbf{x}$ by setting $\overline{\mathbf{x}} = \mathbb{R}^{\cup}\langle \mathcal{B}_{\mathbf{x}} \rangle + \mathbb{R}_+ u_t$, $\partial(\mathbf{x}) = \mathbb{R}^{\cup}\langle \mathcal{B}_{\mathbf{x}} \rangle$ and $\partial(\mathbf{x}) \cap \mathbf{x} = \mathsf{C}(\mathcal{B}_{\mathbf{x}})$, and replacing each $\mathbf{y} \in \mathcal{Y}_{s+1} \cup \{\mathbf{w}_{s+1}\} \cup \ldots \cup \mathcal{Y}_{s'} \cup \{\mathbf{w}_{s'}\}$ with the corresponding half-space $\mathbf{x}$ just constructed again results in virtual sets $\mathcal{X}_{s+1} \cup \{\mathbf{v}_{s+1}\}, \ldots, \mathcal{X}_{s'} \cup \{\mathbf{v}_{s'}\}$ such that $\mathcal{W} = (\mathcal{X}_1 \cup \{\mathbf{v}_1\}, \ldots, \mathcal{X}_{s'} \cup \{\mathbf{v}_{s'}\})$ is a virtual Reay system over G_0 with $\pi(\mathcal{R}) = \mathcal{R}'$, $\mathcal{X}_j^{\pi} \cup \{\pi(\mathbf{v}_j)\} = \mathcal{Y}_j \cup \{\mathbf{w}_j\}$ for all $j \in [s+1, s']$, and $\pi(\vec{u}_{\mathbf{x}}) = \vec{u}_{\pi(\mathbf{x})}$ for all $\mathbf{x} \in \mathcal{X}_{s+1} \cup \{\mathbf{v}_{s+1}\} \cup \ldots \cup \mathcal{X}_{s'} \cup \{\mathbf{v}_{s'}\}$, with each $\vec{u}_{\mathbf{x}}$ the limit of the compatible asymptotically filtered sequence initially chosen for the half-space $\mathbf{y} = \pi(\mathbf{x}) \in \mathcal{Y}_{s+1} \cup \{\mathbf{w}_{s+1}\} \cup \ldots \cup \mathcal{Y}_{s'} \cup \{\mathbf{w}_{s'}\}$. We also call $\mathcal{W}$ an **extension** of $\mathcal{R}$. We continue by showing that finitary sets remain finitary modulo the subspace generated by a purely virtual Reay system.

Proposition 7.9 *Let $\Lambda \subseteq \mathbb{R}^d$ be a full rank lattice, where $d \geq 0$, let $G_0 \subseteq \Lambda$ be a finitary subset with $\mathsf{C}(G_0) = \mathbb{R}^d$, let $(\mathcal{X}_1 \cup \{v_1\}, \ldots, \mathcal{X}_s \cup \{v_s\})$ be a purely virtual Reay system over G_0, and let $\pi : \mathbb{R}^d \to \mathbb{R}^{\cup}\langle \mathcal{X}_1 \cup \ldots \cup \mathcal{X}_s \rangle^{\perp}$ be the orthogonal projection. Then $\pi(\Lambda)$ is a full rank lattice in $\mathsf{C}(\pi(G_0)) = \mathbb{R}^{\cup}\langle \mathcal{X}_1 \cup \ldots \cup \mathcal{X}_s \rangle^{\perp}$ and $\pi(G_0)$ is finitary.*

Proof As already remarked after Theorem 7.8, $\pi(\Lambda)$ is a full rank lattice. Since $\mathsf{C}(G_0) = \mathbb{R}^d$, it follows that $\mathsf{C}(\pi(G_0)) = \mathbb{R}^{\cup}\langle \mathcal{X}_1 \cup \ldots \cup \mathcal{X}_r \rangle^{\perp}$. Now let $\mathcal{R}' = (\mathcal{Y}_{s+1} \cup \{\mathbf{w}_{s+1}\}, \ldots, \mathcal{Y}_r \cup \{\mathbf{w}_r\})$ be an arbitrary purely virtual Reay system over $\pi(G_0)$. Let $\mathcal{W} = (\mathcal{X}_1 \cup \{\mathbf{v}_1\}, \ldots, \mathcal{X}_r \cup \{\mathbf{v}_r\})$ be the extension of $\mathcal{R}$ using $\mathcal{R}'$. Then $\mathcal{W}$ will be purely virtual (as both $\mathcal{R}$ and $\mathcal{R}'$ are purely virtual). Thus $\mathcal{W}$ is anchored since G_0 is finitary, implying that $\pi(\mathcal{W}) = \mathcal{R}'$ is also anchored (as $\pi(\vec{u}_{\mathbf{x}}^{\triangleleft}) = \pi(\vec{u}_{\mathbf{x}})^{\triangleleft}$ for all $\mathbf{x} \in \mathcal{X}_1 \cup \ldots \cup \mathcal{X}_s$, so if $\pi(\vec{u}_{\mathbf{x}}) = u_{\pi(\mathbf{x})}$ were fully unbounded, then $\vec{u}_{\mathbf{x}}$ could be fully unbounded too). Since $\mathcal{R}'$ is an arbitrary purely virtual Reay system over $\pi(G_0)$, we conclude that $\pi(G_0)$ is finitary, as desired. □

Proposition 7.10 *Let $\Lambda \subseteq \mathbb{R}^d$ be a full rank lattice, where $d \geq 0$, let $G_0 \subseteq \Lambda$ be a finitary subset with $\mathsf{C}(G_0) = \mathbb{R}^d$, let $\mathcal{R} = (\mathcal{X}_1 \cup \{v_1\}, \ldots, \mathcal{X}_s \cup \{v_s\})$ be a purely virtual Reay system over G_0, and let $\pi : \mathbb{R}^d \to \mathbb{R}^{\cup}\langle \mathcal{X}_1 \cup \ldots \cup \mathcal{X}_r \rangle^{\perp}$ be the orthogonal projection. Then, for any $\pi(g) \in \pi(G_0)^{\diamond}$ with $g \in G_0$, we have $g \in G_0^{\diamond}$. In particular,*

$$\pi(G_0)^{\diamond} \subseteq \pi(G_0^{\diamond}).$$

Proof Let $\mathcal{X} = \mathcal{X}_1 \cup \ldots \cup \mathcal{X}_s$. As G_0 is finitary, $\mathcal{R}$ is anchored with $\pi(\Lambda)$ a full rank lattice by Proposition 7.9. Let $g \in G_0$ be an arbitrary element with $\pi(g) \in \pi(G_0)^{\diamond}$. Then, in view of Proposition 7.2, there exists a subset $Y \subseteq G_0$ such that $|Y| = |\pi(Y)|$, $g \in Y$, and $\pi(Y)$ minimally encases $-\vec{v}$, where $\vec{v}$ is a fully unbounded limit of an asymptotically filtered sequence of terms $\{x_i'\}_{i=1}^{\infty}$ from $\pi(G_0)$. By Proposition 3.5.2, we can w.l.o.g. assume there is an asymptotically filtered sequence $\{x_i\}_{i=1}^{\infty}$ of terms $x_i \in G_0$ with limit $\vec{u} = (u_1, \ldots, u_t)$ such that $\pi(x_i) = x_i'$ for all i and $\pi(\vec{u}) = \vec{v}$, and as $\vec{v}$ is fully unbounded, we can assume by Proposition 3.5.1 that $\vec{u}$ is fully unbounded as well (by choosing such $\vec{u}$ with t minimal, i.e., with $\pi(\vec{u}^{\triangleleft}) = \pi(\vec{u})^{\triangleleft} = \vec{v}^{\triangleleft}$). Thus $\pi(Y)$ minimally encases $-\pi(\vec{u}) = -\vec{v}$, and thus also the equivalent tuple $-(\pi(u_{s_1}), \ldots, \pi(u_{s_r}))$, where $1 \leq s_1 < \ldots < s_r = t$ are the associated indices for $\pi(\vec{u})$. Hence Proposition 4.19 implies there is a partition $Y = Y_1 \cup \ldots \cup Y_\ell$ such that $(\pi(Y_1) \cup \{\pi(u_{r_1})\}, \ldots, \pi(Y_s) \cup \{\pi(u_{r_\ell})\})$ is a Reay system for some $r_1 < \ldots < r_\ell$ with $\{r_1, \ldots, r_\ell\} \subseteq \{s_1, \ldots, s_r\}$ and $r_1 = s_1$. By the same argument described after Theorem 7.8, we can extend the virtual Reay system $\mathcal{R}$ to obtain a virtual Reay system $\mathcal{W} = (\mathcal{X}_1 \cup \{\mathbf{v}_1\}, \ldots, \mathcal{X}_s \cup \{\mathbf{v}_s\}, \mathcal{Y}_1 \cup \{\mathbf{w}_1\}, \ldots, \mathcal{Y}_\ell \cup \{\mathbf{w}_\ell\})$ over G_0 with $\vec{u}_{\mathbf{w}_j} = (u_1, \ldots, u_{r_j})$ and Y_j a set of representatives for the one-dimensional half-spaces in $\mathcal{Y}_j$, for each $j \in [1, \ell]$, where we use the definition of the r_j as given by Proposition 4.19.1, that $u_i \in \mathbb{R}^{\cup}\langle \mathcal{X}_1 \cup \ldots \cup \mathcal{X}_s \cup \mathcal{Y}_1 \cup \ldots \cup \mathcal{Y}_{j-1}\rangle$ for all $i < r_j$ with $u_{r_j} \notin \mathbb{R}^{\cup}\langle \mathcal{X}_1 \cup \ldots \cup \mathcal{X}_s \cup \mathcal{Y}_1 \cup \ldots \cup \mathcal{Y}_{j-1}\rangle$, to ensure each half-space $\mathbf{w}_j$ is well-defined with $\partial(\{\mathbf{w}_j\}) \subseteq \mathcal{X}_1 \cup \ldots \cup \mathcal{X}_s \cup \mathcal{Y}_1 \cup \ldots \cup \mathcal{Y}_{j-1}$ the support set minimally encasing $-\vec{u}^{\triangleleft}$ by Proposition 7.7.3. Note $\mathcal{W}$ is purely virtual as $\mathcal{R}$ is purely virtual and $\vec{u}$ is fully unbounded. Since $g \in Y = Y_1 \cup \ldots \cup Y_\ell$, it follows that g is a representative for some $\mathbf{y} \in \mathcal{Y}_j$ and $j \in [1, \ell]$. Indeed, we may take $g = \mathbf{y}(i) = \tilde{\mathbf{y}}(i)$ for all $i \geq 1$. But now Proposition 7.7.1 implies that $g \in G_0^{\diamond}$, as desired. □

Our next goal is to give a characterization of $G_0^{\diamond}$, for G_0 finitary, in terms of $\mathcal{A}^{\mathsf{elm}}(G_0)$ and linear combinations over $\mathbb{Q}_+$. Towards that end, we continue with the following lemmas.

Lemma 7.11 *Let $\Lambda \subseteq \mathbb{R}^d$ be a full rank lattice, where $d \geq 0$, and let $X \subseteq \Lambda$ be a linearly independent subset. There exists a positive integer $N > 0$ such that, for any $g \in \Lambda \cap \mathbb{R}\langle X\rangle$, we have $Ng \in \mathbb{Z}\langle X\rangle$. Moreover, if we also have $g \in -\mathsf{C}^{\circ}(X)$, then $X \cup \{g\}$ is a minimal positive basis (or $X = \emptyset$ with $g = 0$) and there is an elementary atom U with* $\mathrm{Supp}(U) = X \cup \{g\}$ *and* $1 \leq \mathsf{v}_g(U) \leq N$.

Proof If $X = \emptyset$, then the lemma holds with $N = 1$, so we may assume X is nonempty. Since X is linearly independent, $\mathbb{Z}\langle X\rangle$ is a full rank lattice in $\mathbb{R}\langle X\rangle$. Via the Smith normal form ([101, Theorem III.7.8] and [79, Theorem 2]) applied to the sublattice $\mathbb{Z}\langle X\rangle \leq \Lambda$, we can find a lattice basis $\{e_1, \ldots, e_d\}$ for Λ and positive integers $a_1 \mid \ldots \mid a_s$, where $1 \leq s \leq d$, such that $\{a_1e_1, \ldots, a_se_s\}$ is a lattice basis for $\mathbb{Z}\langle X\rangle$. Note that $\mathbb{R}\langle e_1, \ldots, e_s\rangle = \mathbb{R}\langle a_1e_1, \ldots, a_se_s\rangle = \mathbb{R}\langle X\rangle$. Thus, since $\{e_1, \ldots, e_d\}$ is a lattice basis for Λ, and in particular linearly independent, it follows that any $g \in \Lambda \cap \mathbb{R}\langle X\rangle$ lies in the lattice $\mathbb{Z}\langle e_1, \ldots, e_s\rangle$. Since $\mathbb{Z}\langle X\rangle \leq \mathbb{Z}\langle e_1, \ldots, e_s\rangle$ is a sublattice of full rank, it follows that $\mathbb{Z}\langle e_1, \ldots, e_s\rangle/\mathbb{Z}\langle X\rangle \cong G := \mathbb{Z}/a_1\mathbb{Z} \times \ldots \times \mathbb{Z}/a_s\mathbb{Z}$ is a finite abelian group of exponent $N = a_s > 0$. As a result, any $g \in \Lambda \cap \mathbb{R}\langle X\rangle$ has $Ng \in \mathbb{Z}\langle X\rangle$.

Now suppose $g \in \Lambda \cap -\mathsf{C}^{\circ}(X) \subseteq \Lambda \cap \mathbb{R}\langle X\rangle$. Then Proposition 4.8.4 implies that $X \cup \{g\}$ is a minimal positive basis (since X is linearly independent) and Proposition 4.8.5 implies that there is a strictly positive linear combination $\sum_{x \in X \cup \{g\}} \alpha_x x = 0$, so $\alpha_x > 0$ for all $x \in X \cup \{g\}$, with the property that $\alpha_g = 1$ and, if $\sum_{x \in X \cup \{g\}} \beta_x x = 0$ is another linear combination with $\beta_x \in \mathbb{R}$, then the vector $(\beta_x)_{x \in X \cup \{g\}}$ is a real scalar multiple of the vector $(\alpha_x)_{x \in X \cup \{g\}}$.

Since $Ng \in \mathbb{Z}\langle X\rangle$, it follows that there is a linear combination $\sum_{x\in X\cup\{g\}} \beta_x x = 0$ with $\beta_x \in \mathbb{Z}$ for all $x \in X \cup \{g\}$ and $\beta_g = N > 0$. Thus, by the previous observation, we must have $\beta_x = N\alpha_x$ for all $x \in X \cup \{g\}$. In particular, since $N > 0$ and $a_x > 0$ for all $x \in X \cup \{g\}$, it follows that $\beta_x > 0$ for all $x \in X \cup \{g\}$, implying that $S = \prod^{\bullet}_{x\in X\cup\{g\}} x^{[\beta_x]}$ is a zero-sum sequence. As a result, since $X \cup \{g\}$ is a minimal positive basis, Proposition 4.8.8 implies that there is an elementary atom $U \mid S$ with $\mathsf{Supp}(U) = X \cup \{g\}$ and $S = U^{[m]}$ for some integer $m \geq 1$, ensuring that $1 \leq \mathsf{v}_g(U) \leq \beta_g = N$, which completes the proof. □

Lemma 7.12 *Let $d \geq 1$ and let $X \subseteq \mathbb{R}^d$ be finite. Then $0 \notin \mathsf{C}^*(X)$ if and only if there is a co-dimension one subspace $\mathcal{H} \subseteq \mathbb{R}^d$ such that $X \subseteq \mathcal{H}_+^\circ$.*

Proof Any set $X \subseteq \mathbb{R}^d$ contained in an open half-space clearly has $0 \notin \mathsf{C}^*(X)$. For the reverse inclusion, we proceed by induction on d, with the case $d = 1$ clear. By Proposition 4.3, $X \subseteq \mathcal{H}_+$ for some co-dimension 1 subspace $\mathcal{H} \subseteq \mathbb{R}^d$. Applying the induction hypothesis to $X \cap \mathcal{H}$ yields a subspace $\mathcal{H}' \subseteq \mathcal{H}$ such that $X \cap \mathcal{H}$ lies entirely on one side of the subspace $\mathcal{H}'$. Since X is finite, the space $\mathcal{H}$ can be slightly perturbed by a rotation about $\mathcal{H}'$ (fixing the points in $\mathcal{H}'$) to yield the needed subspace. □

Proposition 7.2.2 means $g \in G_0^\diamond$ when there is a fixed linearly independent set X containing g such that $x_i = -\sum_{x\in X} \alpha_i^{(x)} x \in -\mathsf{C}^\circ(X)$ for some $x_i \in G_0$ with $\alpha_i^{(g)} \to \infty$. By Propositions 4.8.4 and 4.8.5, this means there are elementary atoms U_i with $\mathsf{Supp}(U_i) = X \cup \{x_i\}$ and $\mathsf{v}_x(U_i) = \mathsf{v}_{x_i}(U_i)\alpha_i^{(x)} \geq \alpha_i^{(x)}$ for all $x \in X$. In particular, $\mathsf{v}_g(U_i) \to \infty$ as $\alpha_i^{(g)} \to \infty$. Thus Theorem 7.13 implies that, if $\sup\{\mathsf{v}_g(U) : U \in \mathcal{A}^{\mathsf{elm}}(G_0)\} = \infty$, then this supremum can be obtained by restricting to a subfamily of elementary atoms U_i whose supports are each equal apart from one varying element. Obtaining the $\mathbb{Z}_+$ linear combination analog of Theorem 7.13 using atoms instead of elementary atoms (including the stronger statement regarding the uniform bound N) will be one of the main steps in our characterization result, and one which needs further machinery regarding finitary sets developed in the next sections.

Theorem 7.13 *Let $\Lambda \subseteq \mathbb{R}^d$ be a full rank lattice, where $d \geq 0$, and let $G_0 \subseteq \Lambda$ be a finitary subset with $\mathsf{C}(G_0) = \mathbb{R}^d$. Then*

$$G_0^\diamond = \left\{ g \in G_0 : \sup\{\mathsf{v}_g(U) : U \in \mathcal{A}^{\mathsf{elm}}(G_0)\} = \infty \right\}.$$

Indeed, there is a bound $N > 0$ such that $\mathsf{v}_g(U) \leq N$ for any $U \in \mathcal{A}^{\mathsf{elm}}(G_0)$ and $g \in G_0 \setminus G_0^\diamond$.

Proof By Proposition 7.3, we have the basic inclusion

$$G_0^\diamond \subseteq \left\{ g \in G_0 : \sup\{\mathsf{v}_g(U) : U \in \mathcal{A}^{\mathsf{elm}}(G_0)\} = \infty \right\}.$$

It remains to establish the reverse inclusion, which follows from the stronger conclusion about the existence of the uniform bound N. To this end, assume by contradiction that $\{x_i^{(0)}\}_{i=1}^\infty$ is a sequence of terms $x_i^{(0)} \in G_0 \setminus G_0^\diamond$ and $\{U_i\}_{i=1}^\infty$ is a sequence of elementary atoms, so $U_i \in \mathcal{A}^{\mathsf{elm}}(G_0)$, with $\mathsf{v}_{x_i^{(0)}}(U_i) \to \infty$. Then $|\operatorname{Supp}(U_i)| \leq d+1$ for all i, so by passing to a subsequence, we can w.l.o.g. assume all U_i have the same cardinality support (say) $s+1$ with $s \in [1,d]$ (if $s = 0$, then $U_i = 0$ for all i, contradicting that $\mathsf{v}_{x_i^{(0)}}(U_i) \to \infty$). Let $\operatorname{Supp}(U_i) = \{x_i^{(0)}, x_i^{(1)}, \ldots, x_i^{(s)}\}$ for $i \geq 1$. Since $|\operatorname{Supp}(U_i)| = s+1 \geq 2$, we have $0 \notin \operatorname{Supp}(U_i)$ for all i. If $\{x_i^{(0)}\}_{i=1}^\infty$ is unbounded, then by passing to a subsequence, we can w.l.o.g. assume $\{x_i^{(0)}\}_{i=1}^\infty$ is radially convergent with unbounded limit g_0. On the other hand, if $\{x_i^{(0)}\}_{i=1}^\infty$ is bounded, then, since $x_i^{(0)} \in G_0 \subseteq \Lambda$ and since any bounded subset of lattice points is finite, we can, by passing to a subsequence, w.l.o.g. assume that the sequence $\{x_i^{(0)}\}_{i=1}^\infty$ is constant, say with $x_i^{(0)} = g_0$ for all $i \geq 1$. Repeating this argument for each $\{x_i^{(j)}\}_{i=1}^\infty$ for $j = 1, 2, \ldots, s$, we can likewise assume that, for every $j \in [0,s]$, either $\{x_i^{(j)}\}_{i=1}^\infty$ is constant, say with $x_i^{(j)} = g_j$ for all $i \geq 1$, or else $\{x_i^{(j)}\}_{i=1}^\infty$ is radially convergent with unbounded limit g_j. Partition $[0,s] = I_u \cup I_b$, with I_u consisting of all indices $j \in [0,s]$ such that $\{x_i^{(j)}\}_{i=1}^\infty$ is unbounded, and I_b consisting of all indices $j \in [0,s]$ such that $x_i^{(j)} = g_j$ for all $i \geq 1$. Thus

$$\{g_j : j \in I_b\} \subseteq G_0 \quad \text{and} \quad \{g_j : j \in I_u\} \subseteq G_0^\infty.$$

If $I_u = \emptyset$, then $\operatorname{Supp}(U_i) = \{g_0, \ldots, g_s\}$ is constant for all i, in which case the elementary atoms U_i are all equal to one another (in view of Proposition 4.8.8), contradicting that $\mathsf{v}_{x_i^{(0)}}(U_i) \to \infty$. Therefore we can assume $I_u \neq \emptyset$.

Suppose $0 \notin \mathsf{C}^*(g_0, \ldots, g_s)$. Then Lemma 7.12 gives a co-dimension one subspace $\mathcal{H} \subseteq \mathbb{R}^d$ such that $g_0, \ldots, g_s \in \mathcal{H}_+^\circ$. In consequence, since the sequences $\{x_i^{(j)}\}_{i=1}^\infty$ are radially convergent with limit $g_j/\|g_j\|$, it follows that, for all sufficiently large i, we also have $x_i^{(0)}, \ldots, x_i^{(s)} \in \mathcal{H}_+^\circ$, ensuring that $0 \notin \mathsf{C}^*(x_i^{(0)}, \ldots, x_i^{(s)}) = \mathsf{C}^*(\operatorname{Supp}(U_i))$. However, this contradicts that $U_i \in \mathcal{A}^{\mathsf{elm}}(G_0)$ is an elementary atom. So we instead conclude that $0 \in \mathsf{C}^*(g_0, \ldots, g_s)$. Thus, since $g_j \neq 0$ for all $j \in [0,s]$, there is a minimal positive basis $B \subseteq \{g_0, \ldots, g_s\}$. Let

$$X = B \cap \{g_j : j \in I_b\} \subseteq G_0.$$

Since $I_u \neq \emptyset$, it follows that $X \subset \operatorname{Supp}(U_i)$ is a proper subset of $\operatorname{Supp}(U_i)$ for all $i \geq 1$. Thus, since each $\operatorname{Supp}(U_i)$ is a minimal positive basis, it follows from Proposition 4.8.3 applied to U_i that X is linearly independent. Thus, since B is a minimal positive basis, it follows from Proposition 4.8.2 applied to B that $X \subset B$ is a proper subset, which then implies that $|B \cap \{g_j : j \in I_u\}| \geq 1$. On the other hand, if $|B \cap \{g_j : j \in I_u\}| > 1$, then this contradicts that G_0 is finitary, which ensures (as described after Theorem 7.8) that any minimal positive basis with elements from $G_0 \cup G_0^\infty$ can involve at most one element from G_0^∞. So we conclude that $|B \cap \{g_j : j \in I_u\}| = 1$, and thus $|X| = |B| - 1$. Consequently, if $g_0 \in X$, then Proposition 7.2.1 implies that $g_0 = x_i^{(0)} \in G_0^\diamond$ for all $i \geq 1$, contradicting that $x_i^{(0)} \in G_0 \setminus G_0^\diamond$ by assumption. So we may instead assume $g_0 \notin X$.

Suppose $B = \{g_0, \ldots, g_s\}$. Then $X = \{g_1, \ldots, g_s\}$ as $g_0 \notin X$, ensuring that $-x_i^{(0)} \in \mathsf{C}^\circ(x_i^{(1)}, \ldots, x_i^{(s)}) = \mathsf{C}^\circ(g_1, \ldots, g_s)$ for all $i \geq 1$ with $g_1, \ldots, g_s \in G_0 \subseteq \Lambda$ linearly independent elements (as the sequences $\{x_i^{(j)}\}_{i=1}^\infty$ for $j \in [1, s]$ will all be constant). Applying Lemma 7.11 to the linearly independent set $X = \{g_1, \ldots, g_s\}$, we find that there exists some constant $N > 0$ such that $\mathsf{v}_{x_i^{(0)}}(U_i) \leq N$ for all i (as there is a unique elementary atom with given support by Proposition 4.8.8), contradicting that $\mathsf{v}_{x_i^{(0)}}(U_i) \to \infty$. So we instead conclude that $B \subset \{g_0, \ldots, g_s\}$ is a proper subset.

By re-indexing the g_j with $j \in [1, s]$, we can assume

$$X = \{g_j : j \in [s'+1, s]\} \subseteq G_0 \quad \text{and}$$
$$B = X \cup \{g_t\} \quad \text{for some } t \in [0, s'] \text{ and } s' \in [1, s-1]$$

(we have $s' > 0$ in view of $B \subset \{g_0, \ldots, g_s\}$ being proper). Note $t \in I_u$, and $s' \in [1, s-1]$ is only possible when $d \geq s \geq 2$, which means the proof is now complete for $d = 1$ (and trivial for $d = 0$), allowing us to proceed inductively on d.

Each $g_j \in B$ defines a one-dimensional half-space $\mathbf{g}_j = \mathbb{R}_+ g_j$, and letting $\mathcal{B}$ be the corresponding set of half-spaces, it follows that $\mathcal{R} = ((\mathcal{B} \setminus \{\mathbf{g}_t\}) \cup \{\mathbf{g}_t\})$ is a purely virtual Reay system over G_0 (in view of $t \in I_u$ and since all other $g_j \in B \setminus \{g_t\} = X \subseteq G_0$). Let $\pi : \mathbb{R}^d \to \mathbb{R}^\cup\langle\mathcal{B}\rangle^\perp = \mathbb{R}\langle g_{s'+1}, \ldots, g_s\rangle^\perp = \mathbb{R}\langle X\rangle^\perp$ be the orthogonal projection. Proposition 7.9 ensures that $\pi(\Lambda)$ is a full rank lattice in $\mathsf{C}(\pi(G_0)) = \mathbb{R}\langle X\rangle^\perp$ with $\pi(G_0)$ finitary.

Note that $\pi(x) = 0$ for all $x \in X$. On the other hand, if $J \subset [0, s']$ is a proper subset, then (for any $i \geq 1$) the elements $\pi(x_i^{(j)})$ for $j \in J$ must be distinct and linearly independent, as otherwise $\{x_i^{(j)} : j \in J\} \cup X = \{x_i^{(j)} : j \in J\} \cup \{x_i^{(s'+1)}, \ldots, x_i^{(s)}\}$ would be a linearly dependent, proper subset of $\mathsf{Supp}(U_i)$, contradicting that $\mathsf{Supp}(U_i)$ is a minimal positive basis. In particular, since $s' \geq 1$, it follows that $\pi(x_i^{(j)}) \neq 0$ for all $j \in [0, s']$ and $i \geq 1$. Since there is a strictly positive linear combination of the elements from $\mathsf{Supp}(U_i)$ equal to zero, this is also true for the elements from $\pi(\mathsf{Supp}(U_i)) \setminus \{0\} = \pi(\mathsf{Supp}(U_i) \setminus X)$. Thus it now follows from Proposition 4.8.2 that $\pi(\mathsf{Supp}(U_i)) \setminus \{0\} = \{\pi(x_i^{(0)}), \ldots, \pi(x_i^{(s')})\}$ is a minimal positive basis of size $s' + 1$ for every $i \geq 1$. Consequently, Proposition 4.8.8 now further implies that there is a unique atom $A_i \in \mathcal{A}(\pi(G_0))$ with $\mathsf{Supp}(A_i) \subseteq \pi(\mathsf{Supp}(U_i)) \setminus \{0\}$, and we have $A_i \in \mathcal{A}^{\mathsf{elm}}(\pi(G_0))$ with $\mathsf{Supp}(A_i) =$

$\pi(\mathrm{Supp}(U_i))\setminus\{0\}$ for this atom. Hence each $\pi(U_i) = A_i^{[N_i]} \cdot 0^{[M_i]}$ for some integers $N_i, M_i > 0$, and there is a unique subsequence $V_i \mid U_i$ such that $\pi(V_i) = A_i$ (as π is injective on the elements of $\mathrm{Supp}(U_i) \setminus X$), meaning

$$U_i \cdot W_i^{[-1]} = V_i^{[N_i]},$$

where $W_i \mid U_i$ is the subsequence consisting of all terms from X. Since each U_i is zero-sum, and since $\mathsf{v}_x(W_i) > 0$ for every $x \in X$ with X linearly independent, it follows that

$$0 = \sigma(U_i \cdot W_i^{[-1]}) + \sigma(W_i) = N_i\sigma(V_i) + \sigma(W_i) \in N_i\sigma(V_i) + \mathsf{C}^\circ(X).$$

Thus, since $N_i > 0$ and $\mathrm{Supp}(V_i) \subseteq G_0 \subseteq \Lambda$, it follows that

$$\sigma(V_i) \in \Lambda \cap -\mathsf{C}^\circ(X) \subseteq \Lambda \cap \mathbb{R}\langle X\rangle \quad \text{for all } i \geq 1.$$

Since $X \neq \emptyset$ is linearly independent, we can apply Lemma 7.11 with $g = \sigma(V_i)$ to conclude there is a positive integer $N > 0$ such that $N\sigma(V_i) \in \mathbb{Z}\langle X\rangle$ and, moreover, there is an elementary atom S_i with $\mathrm{Supp}(S_i) = X \cup \{\sigma(V_i)\}$ and $\mathsf{v}_{\sigma(V_i)}(S_i) \leq N$, for every $i \geq 1$. Then

$$T_i = V_i^{[\mathsf{v}_{\sigma(V_i)}(S_i)]} \cdot {\prod}^\bullet_{x \in X} x^{[\mathsf{v}_x(S_i)]} \in \mathcal{F}(G_0)$$

is a zero-sum sequence with $\mathrm{Supp}(T_i) = \mathrm{Supp}(U_i)$. In consequence, since U_i is an elementary atom, it follows from Proposition 4.8.8 that $\mathsf{v}_g(U_i) \leq \mathsf{v}_g(T_i)$ for every $g \in \mathrm{Supp}(U_i)$. In particular, $\mathsf{v}_{x_i^{(0)}}(U_i) \leq \mathsf{v}_{x_i^{(0)}}(T_i) = \mathsf{v}_{\sigma(V_i)}(S_i) \cdot \mathsf{v}_{x_i^{(0)}}(V_i) \leq N \cdot \mathsf{v}_{x_i^{(0)}}(V_i)$. Thus, since $\mathsf{v}_{x_i^{(0)}}(U_i) \to \infty$, we conclude that $\mathsf{v}_{x_i^{(0)}}(V_i) \to \infty$ as well. However, since $\pi(V_i) = A_i \in \mathcal{A}^{\mathsf{elm}}(\pi(G_0))$, this implies that $\mathsf{v}_{\pi(x_i^{(0)})}(A_i) \to \infty$. By the induction hypothesis applied to $\pi(G_0)$, we know there is a bound $N' > 0$ such that $\pi(g) \in \pi(G_0)^\diamond$ whenever $\mathsf{v}_{\pi(g)}(A) > N'$ for some $A \in \mathcal{A}^{\mathsf{elm}}(\pi(G_0))$. Thus, since $\mathsf{v}_{\pi(x_i^{(0)})}(A_i) \to \infty$, it follows that $\pi(x_i^{(0)}) \in \pi(G_0)^\diamond$ for all sufficiently large i. But now Proposition 7.10 implies that $x_i^{(0)} \in G_0^\diamond$ for all sufficiently large i, contradicting our assumption that $x_i^{(0)} \in G_0 \setminus G_0^\diamond$ for all $i \geq 1$, which completes the proof. □

We continue with an important property of purely virtual Reay systems.

Proposition 7.14 *Let $\Lambda \subseteq \mathbb{R}^d$ be a full rank lattice, where $d \geq 0$, let $G_0 \subseteq \Lambda$ be a finitary subset with $\mathsf{C}(G_0) = \mathbb{R}^d$, let $\mathcal{R} = (X_1 \cup \{v_1\}, \ldots, X_s \cup \{v_s\})$ be a purely virtual Reay system over G_0, let $X = X_1 \cup \ldots \cup X_s$, and let $\pi : \mathbb{R}^d \to \mathbb{R}^\cup\langle X\rangle^\perp$ be the orthogonal projection. If $U \in \mathcal{F}(G_0)$ with $\pi(U) \in \mathcal{A}^{\mathsf{elm}}(\pi(G_0))$ and $|\mathrm{Supp}(U)| = |\mathrm{Supp}(\pi(U))|$, then $\mathsf{wt}(-g) \leq 1$, where $g = \sigma(U)$. Moreover, if $\mathsf{wt}(-g) = 1$, then $\mathrm{Supp}(U) \subseteq G_0^\diamond$. In particular, $(G_0 \cap \mathbb{R}^\cup\langle X\rangle) \setminus -\mathsf{C}^\cup(X) \subseteq G_0^\diamond$.*

Proof Since $\pi(U) \in \mathcal{A}^{\mathsf{elm}}(\pi(G_0))$ is an atom, we have $g = \sigma(U) \in \ker \pi = \mathbb{R}^{\cup}\langle \mathcal{X} \rangle$. Furthermore, since $|\operatorname{Supp}(U)| = |\operatorname{Supp}(\pi(U))|$, we conclude that π is injective on $\operatorname{Supp}(U)$ with $\operatorname{Supp}(\pi(U))$ either $\{0\}$ or a minimal positive basis. Note $g = \sigma(U) \in \mathsf{C}_{\mathbb{Z}}(G_0) \subseteq \Lambda$ as $U \in \mathcal{F}(G_0)$. Let $\mathcal{B} = \operatorname{Supp}_{\mathcal{R}}(-g)$, in which case $-g \in \mathsf{C}^{\cup}(\mathcal{B})^{\circ}$. Thus the support set $\mathcal{B}$ minimally encases the limit $-\vec{u} := (-g/\|g\|)$ (cf. the comments after Lemma 5.7), with the encasement trivially urbane as the limit $\vec{u}$ is composed of a single coordinate. We may assume $g \neq 0$ as the proposition is trivial for $g = 0$. By Proposition 6.5 applied to $\mathcal{B}$, there is a virtual Reay system $\mathcal{R}' = (\mathcal{Y}_1 \cup \{\mathbf{w}_1\}, \ldots, \mathcal{Y}_\ell \cup \{\mathbf{w}_\ell\})$ over $G_0 \cup \{g\}$ with $\mathcal{Y}_\ell = \mathcal{B}$ and $\vec{u}_{\mathbf{w}_\ell} = \vec{u}$. Moreover, each $\mathbf{w}_j$ with $j \in [1, \ell - 1]$ is defined by a strict truncation of a limit $\vec{u}_{\mathbf{y}}$ for some $\mathbf{y} \in \downarrow\mathcal{B}$, and is thus fully unbounded by Proposition 6.8 (as $\mathcal{R}$ is purely virtual and anchored in view of G_0 being finitary). Let $\mathcal{U}$ be the set of one-dimensional half-spaces generated by the elements from $\operatorname{Supp}(U)$, in which case $\operatorname{Supp}(U)$ is a set of representatives for $\mathcal{U}$. Let $\pi' : \mathbb{R}^d \to \mathbb{R}^{\cup}\langle \mathcal{Y}_1 \cup \ldots \cup \mathcal{Y}_{\ell-1} \rangle^{\perp}$ be the orthogonal projection. Then

$$\pi'(Y_\ell \cup \{w_\ell\}) = \pi'(B \cup \{g\}) \tag{7.5}$$

is a minimal positive basis of size $|\mathcal{B}| + 1$ by (OR2) with $g = \sigma(U)$. Thus 0 is a strictly positive linear combination of the elements from $\pi'(\operatorname{Supp}(U) \cup B)$. Since $\mathcal{Y}_1 \cup \ldots \cup \mathcal{Y}_{\ell-1} \subseteq \mathcal{X}$ (by Proposition 6.5(a)), we have $\ker \pi' \subseteq \mathbb{R}^{\cup}\langle \mathcal{X} \rangle = \ker \pi$, and since $\mathcal{B} \subseteq \mathcal{X} \cup \{\mathbf{v}_1, \ldots, \mathbf{v}_s\}$, we have $\mathbb{R}\langle B \rangle \subseteq \mathbb{R}^{\cup}\langle \mathcal{B} \rangle \subseteq \mathbb{R}^{\cup}\langle \mathcal{X} \rangle = \ker \pi$.

If $|\operatorname{Supp}(U)| = 1$, then $U = g$ and π' is injective on $\operatorname{Supp}(U) \cup B$ by (OR2) (as it is injective on $B \cup \{g\}$). If $|\operatorname{Supp}(U)| > 1$, then $\pi(y) \neq 0$ for all $y \in \operatorname{Supp}(U)$, while $\pi(x) = 0$ for all $x \in B$, ensuring that $\pi'(x) \neq \pi'(y)$ for $x \in B$ and $y \in \operatorname{Supp}(U)$ in view of $\ker \pi' \leq \ker \pi$. Thus, since π' is injective on $\operatorname{Supp}(U)$ (as π is injective on $\operatorname{Supp}(U)$ with $\ker \pi' \leq \ker \pi$) and since π' is injective on B (by (OR2)), we conclude that π' is injective on $B \cup \operatorname{Supp}(U)$ in both cases.

Let us show that $\pi'(B \cup \operatorname{Supp}(U))$ is a minimal positive basis of size $|\mathcal{B}| + |\mathcal{U}|$ (as π' is injective on $B \cup \operatorname{Supp}(U)$). If $|\operatorname{Supp}(U)| = 1$, then $U = g$ with $\operatorname{Supp}(U) = \{g\}$, in which case this follows from (7.5). Next assume $|\operatorname{Supp}(U)| > 1$ and consider an arbitrary linear combination $\sum_{x \in B} \alpha_x \pi'(x) + \sum_{y \in \operatorname{Supp}(U)} \beta_y \pi'(y) = 0$ for some $\alpha_x, \beta_y \in \mathbb{R}$. Since $\pi'(B \cup \{g\})$ is a minimal positive basis of size $|B| + 1$, it follows that $\pi'(B)$ is linearly independent, and thus $\beta_y \neq 0$ for some $y \in \operatorname{Supp}(U)$, allowing us to w.l.o.g. assume $\beta_y > 0$ for some $y \in \operatorname{Supp}(U)$. Applying π to this linear combination, we find that $\sum_{y \in \operatorname{Supp}(U)} \beta_y \pi(y) = 0$. Thus, since $\operatorname{Supp}(\pi(U))$ is a minimal positive basis of size $|\operatorname{Supp}(U)|$, it now follows from Proposition 4.8.5 that $\beta_y > 0$ for all $y \in \operatorname{Supp}(U)$. Since $\sum_{y \in \operatorname{Supp}(U)} \mathsf{v}_y(U) y = \sigma(U) = g$ with $\pi(g) = 0$ (as $g = \sigma(U) \in \ker \pi$), it follows from Proposition 4.8.5 that $\sum_{y \in \operatorname{Supp}(U)} \beta_y y$ is a positive scalar multiple of g (indeed, each $\beta_y = \alpha \mathsf{v}_y(U)$ for the same $\alpha > 0$). But now $\sum_{x \in B} \alpha_x \pi'(x) = -\sum_{y \in \operatorname{Supp}(U)} \beta_y \pi'(y)$ is positive scalar multiple of $-\pi'(g)$, which combined with $\pi'(B \cup \{g\})$ being a minimal positive basis ensures that $\alpha_x > 0$ for all $x \in B$ (by Proposition 4.8.3). Combined with the fact that $\beta_y > 0$ must also hold for all $y \in \operatorname{Supp}(U)$, shown earlier, we conclude that any proper subset of

$\pi'(B \cup \mathrm{Supp}(U))$ is linearly independent. Thus, since we showed earlier that 0 is a strictly positive linear combination of the elements from $\pi'(B \cup \mathrm{Supp}(U))$, it now follows from Proposition 4.8.2 that $\pi'(B \cup \mathrm{Supp}(U))$ is a minimal positive basis in the case $|\,\mathrm{Supp}(U)| > 1$ as well.

As a result, $\mathcal{R}'' = (\mathcal{Y}_1 \cup \{\mathbf{w}_1\}, \ldots, \mathcal{Y}_{\ell-1} \cup \{\mathbf{w}_{\ell-1}\}, ((\mathcal{B} \cup \mathcal{U}) \setminus \{\mathbf{x}\}) \cup \{\mathbf{x}\})$ will be a virtual Reay system over G_0 for any $\mathbf{x} \in \mathcal{B} \cup \mathcal{U}$ with all half-spaces from $\mathcal{U}$ having trivial boundary with representative sequence a constant term equal to an element from U. Consequently, if $\mathrm{wt}(-g) \geq 1$, then we can choose $\mathbf{x} \in \mathcal{B}$ with $\mathbf{x} = \mathbf{v}_j$ for some $j \in [1, s]$, thus ensuring that $\vec{u}_{\mathbf{x}}$ is fully unbounded (as $\mathcal{R}$ is purely virtual), in which case $\mathcal{R}''$ is purely virtual. In this case, since G_0 is finitary, we must have $\mathcal{R}''$ anchored. In particular, all limits defining the half-spaces from $\mathcal{B} \setminus \{\mathbf{x}\}$ are anchored. Thus $\mathrm{wt}(-g) = 1$ as $\mathcal{R}''$ is purely virtual, and Proposition 7.7.1 applied to $\mathcal{R}''$ yields $\mathrm{Supp}(U) \subseteq G_0^{\diamond}$, as desired. □

Lemma 7.15 *Let $\mathcal{E} \subseteq \mathbb{R}^d$ be a subspace, let $\pi : \mathbb{R}^d \to \mathcal{E}^{\perp}$ be the orthogonal projection, let $G_0 \subseteq \mathbb{R}^d$ be a subset such that $\pi(G_0)$ is finite and let*

$$\widetilde{G}_0 = \{\sigma(U) \colon\ U \in \mathcal{F}(G_0),\ \pi(U) \in \mathcal{A}(\pi(G_0))\}.$$

Suppose $C \subseteq \mathbb{R}^d$ is a polyhedral cone such that C encases every $\vec{u} \in G_0^{\mathsf{lim}}$. Then C also encases every $\vec{u} \in \widetilde{G}_0^{\mathsf{lim}}$.

Proof If C is empty, then G_0^{lim} must also be empty, ensuring that G_0 is bounded. Since $\pi(G_0)$ is finite, it follows by Dickson's Theorem [62, Theorem 1.5.3, Corollary 1.5.4] that $\mathcal{A}(\pi(G_0))$ is finite, and now $\widetilde{G}_0$ is also bounded, ensuring that $\widetilde{G}_0^{\mathsf{lim}}$ is empty. In such case, the lemma holds trivially. Therefore we now assume C is nonempty. Let $\{x_i\}_{i=1}^{\infty}$ be an asymptotically filtered sequence of terms from $\widetilde{G}_0$ with fully unbounded limit $\vec{u} = (u_1, \ldots, u_t)$, say with each $x_i = \sigma(U_i)$ for some $U_i \in \mathcal{F}(G_0)$ with $\pi(U_i) \in \mathcal{A}(\pi(G_0))$. As previously remarked, since $\pi(G_0)$ is finite, it follows that $\mathcal{A}(\pi(G_0))$ is also finite. As a result, by passing to an appropriate subsequence, we can w.l.o.g. assume $\pi(U_i) = V$ is constant and equal to the same atom for all i. Let $V = a_1 \cdot \ldots \cdot a_\ell$ with $a_i \in \pi(G_0)$. Then each $x_i = y_i^{(1)} + \ldots + y_i^{(\ell)}$ for some $y_i^{(j)} \in G_0$ with $\pi(y_i^{(j)}) = a_j$ for all $j \in [1, \ell]$. By passing to a subsequence, we can w.l.o.g. assume each $\{y_i^{(j)}\}_{i=1}^{\infty}$ is either bounded or an asymptotically filtered sequence of terms from G_0 with complete fully unbounded limit $\vec{u}_j$. By hypothesis, C encases every $\vec{u}_j$, whence Proposition 3.11.3 implies that $\{y_i^{(j)}\}_{i=1}^{\infty}$ is bound to C. This is trivially true for any bounded sequence $\{y_i^{(j)}\}_{i=1}^{\infty}$ as C is nonempty, and so $\{y_i^{(j)}\}_{i=1}^{\infty}$ is bound to C for each $j \in [1, \ell]$. Thus, for each $j \in [1, \ell]$, there is a constant $N_j \geq 0$ such that, for every $i \geq 1$, there is some $z_i^{(j)} \in C$ such that $\|y_i^{(j)} - z_i^{(j)}\| \leq N_j$. Since C is convex, it follows that $z_i = z_i^{(1)} + \ldots + z_i^{(\ell)} \in C$ with $\|x_i - z_i\| \leq \sum_{j=1}^{\ell} N_j$ for all $i \geq 1$ (by the triangle inequality). Hence $\{x_i\}_{i=1}^{\infty}$ is bound to C, whence Lemma 3.13 implies that there is an asymptotically filtered sequence $\{x_i'\}_{i=1}^{\infty}$ of terms $x_i' \in C$ having fully unbounded limit $\vec{u}$, in which case Proposition 3.12 ensures that C encases $\vec{u}$, as desired. □

Next, we derive additional properties regarding the geometry of a finitary set G_0 via its maximal purely virtual Reay systems. Theorem 7.16.2 ensures that a finitary set G_0 has a linearly independent subset $X \subseteq G_0^\diamond \subseteq G_0$ such that G_0 is bound to $-\mathsf{C}(X)$, meaning the set G_0 must be concentrated around the simplicial cone $-\mathsf{C}(X)$. Theorem 7.16.3 then further implies that the *finitely* generated convex cone $\mathsf{C}(X)$ perfectly approximates the (possibly) non-finitely generated convex cone $\mathsf{C}^\cup(\mathcal{X})$ in regards to containment of elements from $-\widetilde{G}_0$, giving our first example of finite-like behavior for finitary sets. Combined with Proposition 7.14, we obtain restrictions on the location of the elements from G_0 in relation to the simplicial cone $\mathsf{C}(X)$. Note $G_0 \cap \mathbb{R}\langle X\rangle \subseteq \widetilde{G}_0$ since the single term equal to 0 is always an atom. Since we may not have a maximal Reay system spanning the entire space $\mathbb{R}^d$, the set $\widetilde{G}_0$ as well as Proposition 7.14 will be used for dealing with elements lying outside $\mathbb{R}\langle X\rangle$.

Theorem 7.16 *Let $\Lambda \subseteq \mathbb{R}^d$ be a full rank lattice, where $d \geq 0$, let $G_0 \subseteq \Lambda$ be a finitary subset with $\mathsf{C}(G_0) = \mathbb{R}^d$, let $\mathcal{R} = (\mathcal{X}_1 \cup \{\mathbf{v}_1\}, \ldots, \mathcal{X}_s \cup \{\mathbf{v}_s\})$ be a maximal purely virtual Reay system over G_0, let $\mathcal{X} = \mathcal{X}_1 \cup \ldots \cup \mathcal{X}_s$, and let $\pi : \mathbb{R}^d \to \mathbb{R}^\cup\langle\mathcal{X}\rangle^\perp$ be the orthogonal projection.*

1. *$-\mathcal{X}$ encases every $\vec{u} \in G_0^{\mathsf{lim}}$. In particular, G_0 is bound to $-\mathsf{C}^\cup(\mathcal{X})$.*
2. *For any tuple $k = (i_\mathbf{z})_{\mathbf{z}\in\mathcal{X}}$ with all $i_\mathbf{z}$ sufficiently large, $-\mathsf{C}(X(k))$ encases every $\vec{u} \in G_0^{\mathsf{lim}}$. Moreover, G_0 is bound to $-\mathsf{C}(X(k))$ with $X(k) \subseteq G_0^\diamond$ a linearly independent set.*
3. *Let $\widetilde{G}_0 = \{\sigma(U) : U \in \mathcal{F}(G_0),\ \pi(U) \in \mathcal{A}(\pi(G_0))\}$ and let $\mathcal{X}' \subseteq \mathcal{X}$. Then*

$$\widetilde{G}_0 \cap -\mathsf{C}(\downarrow X'(k)) = \widetilde{G}_0 \cap -\mathsf{C}^\cup(\mathcal{X}')$$

for any tuple $k = (i_\mathbf{z})_{\mathbf{z}\in\downarrow\mathcal{B}}$ with all $i_\mathbf{z}$ sufficiently large.

Proof

1. Let $\{x_i\}_{i=1}^\infty$ be an arbitrary asymptotically filtered sequence of terms $x_i \in G_0$ with fully unbounded limit $\vec{u} = (u_1, \ldots, u_t)$. Since $\mathcal{R}$ is a *maximal* purely virtual Reay system over the finitary set G_0, it follows that $u_1, \ldots, u_t \in \mathbb{R}^\cup\langle\mathcal{X}\rangle$ (otherwise $\{\pi(x_i)\}_{i=1}^\infty$ would be an unbounded sequence of terms $\pi(x_i) \in \pi(G_0)$, contradicting that $\pi(G_0)$ is finite in view of the maximality of $\mathcal{R}$). Thus Proposition 7.7.3 implies that $\mathsf{C}^\cup(\mathcal{X})$ encases $-\vec{u}$. Since $\vec{u} \in G_0^{\mathsf{lim}}$ was arbitrary, and since $\overline{\mathsf{C}^\cup(\mathcal{X})}$ is a polyhedral cone by Proposition 5.3.2, Theorem 3.14.4 implies that G_0 is bound to $-\overline{\mathsf{C}^\cup(\mathcal{X})}$, and thus also to $-\mathsf{C}^\cup(\mathcal{X})$.
2. Since G_0 is finitary and $\mathcal{R}$ is purely virtual, it follows that $\mathcal{R}$ is anchored. For any tuple $k = (i_\mathbf{z})_{\mathbf{z}\in\mathcal{X}}$ with all $i_\mathbf{z}$ sufficiently large, we have $\downarrow\tilde{X}(k) = \downarrow X(k) = X(k)$ by Proposition 6.8, in which case $X(k) = \downarrow X(k)$ is a set of representatives for $\mathcal{X} = \mathcal{X}_1 \cup \ldots \cup \mathcal{X}_s$, ensuring that $X(k)$ is linearly independent by Proposition 4.11.1. Moreover, by Proposition 7.7.1, we have $X(k) \subseteq G_0^\diamond$ once all $i_\mathbf{z}$ are

sufficiently large. It remains to show there is some $N > 0$ so that, so long as all $i_{\mathbf{z}} \geq N$, then G_0 is bound to $-\mathsf{C}(X(k))$. In view of Theorem 3.14, this is equivalent to showing any fully unbounded limit $\vec{u}$ of an asymptotically filtered sequence of terms from G_0 has $-\vec{u}$ encased by $X(k)$. By Item 1, each such $-\vec{u}$ is encased by $\mathcal{X}$, and thus must be minimally encased by some support set $\mathcal{B} \subseteq \mathcal{X}$, with the encasement urbane in view of $\mathcal{B} \subseteq \mathcal{X}$.

Let $\mathfrak{X}$ consist of all subsets $\mathcal{B} \subseteq \mathcal{X}$ for which there is some $\vec{u} \in G_0^{\mathsf{lim}}$ with $-\vec{u}$ minimally encased by $\mathcal{B}$. Note $\mathfrak{X}$ is finite as $\mathcal{X}$ is finite. Let $\mathcal{B} \in \mathfrak{X}$ be arbitrary. Then there is some fully unbounded limit $\vec{u} = (u_1, \ldots, u_t)$ of an asymptotically filtered sequence $\{x_i\}_{i=1}^{\infty}$ of terms from G_0 with $-\vec{u}$ minimally encased by $\mathcal{B}$. We fix the tuple $\vec{u}$ for $\mathcal{B}$ and will show $-\mathsf{C}(\downarrow B(k))$ encases every $\vec{v} = (v_1, \ldots, v_{t'}) \in G_0^{\mathsf{lim}}$ with $v_1, \ldots, v_{t'} \in \mathbb{R}^{\cup}\langle \mathcal{B} \rangle$ for any tuple $k = (i_{\mathbf{z}})_{\mathbf{z} \in \mathcal{X}}$ with all $i_{\mathbf{z}} \geq N_{\mathcal{B}}$, for some $N_{\mathcal{B}}$ that depends only on $\mathcal{B}$ and the fixed tuple $\vec{u}$ (thus not dependent on the potentially infinite number of varying tuples $\vec{v} \in G_0^{\mathsf{lim}}$). Taking $N = \max_{\mathcal{B} \in \mathfrak{X}} N_{\mathcal{B}}$, which exists as $\mathcal{X}$ is finite, Item 2 will follow as every fully unbounded limit of an asymptotically filtered sequence of terms from G_0 is encased by some $\mathcal{B} \in \mathfrak{X}$, as noted above the definition of $\mathfrak{X}$.

Proposition 4.19 ensures that if a subset $X \subseteq \mathbb{R}^d$ minimally encases $-\vec{w} = -(w_1, \ldots, w_r)$ and $\vec{w}' = (w_1, \ldots, w_r, w_{r+1}, \ldots, w_{r+r'})$ has $w_i \in \mathbb{R}\langle X \rangle$ for all $i > r$, then X also minimally encases $-\vec{w}'$. Indeed, if $\mathcal{F} = (\mathcal{E}_1, \ldots, \mathcal{E}_\ell)$ is the filter given by the application of Lemma 4.19 to the minimal encasement of $-\vec{w}$ by X, then the assumption $w_i \in \mathbb{R}\langle X \rangle$ for all $i > r$ ensures that it remains a compatible filter for $-\vec{w}'$ having the same associated set of indices (note X encasing $-\vec{w}$ ensures $w_i \in \mathbb{R}\langle X \rangle$ for all $i \leq r$), and then the oriented Reay system from Lemma 4.19 applied to the minimal encasement of $-\vec{w}$ also shows that X minimally encases $-\vec{w}'$. We use this observation several times below.

For any tuple $k = (i_{\mathbf{z}})_{\mathbf{z} \in \mathcal{B}}$ with all $i_{\mathbf{z}}$ sufficiently large, we have $\downarrow B(k) = \downarrow \tilde{B}(k)$ by Proposition 6.8, in which case $\downarrow B(k)$ is a set of representatives for $\downarrow \mathcal{B} \subseteq \mathcal{X}_1 \cup \ldots \cup \mathcal{X}_s$. Thus Proposition 5.3.1 implies that $\mathbb{R}^{\cup}\langle \mathcal{B} \rangle = \mathbb{R}\langle \downarrow B(k) \rangle$. Since $-\vec{u}$ is minimally encased by $\mathcal{B}$ urbanely, let $1 = r_1 < \ldots < r_\ell < r_{\ell+1} = t + 1$ be the indices and $\emptyset = \mathcal{C}_0 \prec \mathcal{C}_1 \prec \ldots \prec \mathcal{C}_{\ell-1} \prec \mathcal{C}_\ell = \mathcal{B} \subseteq \mathcal{X}$ the support sets given by Proposition 5.9.2. Then $\mathcal{C}_1$ minimally encases $-u_{r_1} = -u_1$, so Lemma 6.9 implies that $\downarrow C_1(k) \cup \{u_{r_1}\}$ is a minimal positive basis for $\mathbb{R}^{\cup}\langle \mathcal{C}_1 \rangle$ with $\downarrow \tilde{C}_1(k) = \downarrow C_1(k)$ minimally encasing $-u_1 = -u_{r_1}$, and thus also minimally encasing $-(u_1, \ldots, u_{r_2-1})$ as $u_i \in \mathbb{R}^{\cup}\langle \mathcal{C}_1 \rangle = \mathbb{R}\langle \downarrow C_1(k) \rangle$ for $i < r_2$, so long as all $i_{\mathbf{z}}$ are sufficiently large. Let $\pi_1 : \mathbb{R}^d \to \mathbb{R}^{\cup}\langle \mathcal{C}_1 \rangle^{\perp}$ be the orthogonal projection. Proposition 5.9 implies that $\mathcal{B}^{\pi_1}$ minimally encases $-\pi_1(\vec{u})$ urbanely with associated support sets $\emptyset \prec \mathcal{C}_2^{\pi_1} \prec \ldots \prec \mathcal{C}_{\ell-1}^{\pi_1} \prec \mathcal{C}_\ell^{\pi_1} = \mathcal{B}^{\pi_1} \subseteq \mathcal{X}^{\pi_1}$ (as in the proof of Proposition 6.5). Note $\downarrow \mathcal{C}_2^{\pi_1} = \pi_1(\downarrow \mathcal{C}_2 \setminus \downarrow \mathcal{C}_1)$ in view of Proposition 5.3.9. Proposition 5.9.2 implies that $\mathcal{C}_2$ minimally encases $-(u_1, \ldots, u_{r_2})$ urbanely, and then Proposition 5.9.4 ensures $\mathcal{C}_2^{\pi_1}$ minimally encases $-\pi_1((u_1, \ldots, u_{r_2})) = -\pi_1(u_{r_2})/\|\pi_1(u_{r_2})\|$ urbanely, allowing us to apply Lemma 6.9 to conclude $(\downarrow \widetilde{C_2^{\pi_1}})(k)$ minimally encases $-\pi_1(u_{r_2})$, once all $i_{\mathbf{z}}$ are sufficiently large.

Since $\mathcal{R}$ is anchored, Proposition 6.3 ensures that $\pi_1(\mathcal{R})$ is also anchored, while $\pi_1(\Lambda)$ is a lattice by Proposition 2.1 since $\ker \pi_1 = \mathbb{R}\langle \downarrow C_1(k)\rangle$ with $\downarrow C_1(k) \subseteq G_0 \subseteq \Lambda$. Thus Proposition 6.8 implies that $(\downarrow \widetilde{C}_2^{\pi_1})(k) = (\downarrow C_2^{\pi_1})(k)$. Combined with the conclusions from the previous paragraph, we find that $(\downarrow \widetilde{C}_2^{\pi_1})(k) = (\downarrow C_2^{\pi_1})(k) = \pi_1(\downarrow C_2 \setminus \downarrow C_1)(k)$ minimally encases $-\pi_1(u_{r_2})$, once all $i_{\mathbf{z}}$ are sufficiently large, so $(\downarrow C_1(k) \cup \{u_{r_1}\}, (\downarrow C_2 \setminus \downarrow C_1)(k) \cup \{u_{r_2}\})$ is a Reay system. Moreover, since $u_1, \ldots, u_{r_3-1} \in \mathbb{R}^{\cup}\langle C_2\rangle = \mathbb{R}^{\cup}\langle \downarrow C_2(k)\rangle$ by Proposition 5.9.2, it follows that $\downarrow C_2(k)$ minimally encases $-(u_1, \ldots, u_{r_3-1})$, once all $i_{\mathbf{z}}$ are sufficiently large. Iterating this argument, we find that

$$(\downarrow C_1(k) \cup \{u_{r_1}\}, (\downarrow C_2 \setminus \downarrow C_1)(k) \cup \{u_{r_2}\}, \ldots, (\downarrow C_\ell \setminus \downarrow C_{\ell-1})(k) \cup \{u_{r_\ell}\})$$

is an ordinary Reay system for any tuple $k = (i_{\mathbf{z}})_{\mathbf{z} \in \mathcal{X}}$ with all $i_{\mathbf{z}} \geq N_{\mathcal{B}}$ sufficiently large. Moreover, for each $j \in [1, \ell]$, we have $-(u_1, \ldots, u_{r_j})^{\triangleleft} = -(u_1, \ldots, u_{r_j-1})$ minimally encased by $\downarrow C_{j-1}(k)$. Thus replacing each element from $\downarrow C_\ell(k) = \downarrow B(k)$ with the one-dimensional half-space it defines and using the asymptotically filtered sequence $\{x_i\}_{i=1}^{\infty}$ as the representative sequence of each $\mathbf{w}_j$ gives rise to a purely virtual Reay system $\mathcal{R}_{\mathcal{B}} = (\mathcal{Y}_1 \cup \{\mathbf{w}_1\}, \ldots, \mathcal{Y}_\ell \cup \{\mathbf{w}_\ell\})$ over G_0 with

$$\mathbb{R}^{\cup}\langle \mathcal{Y}_1 \cup \ldots \cup \mathcal{Y}_\ell\rangle = \mathbb{R}\langle \downarrow B(k)\rangle = \mathbb{R}^{\cup}\langle \mathcal{B}\rangle$$

and $\vec{u}_{\mathbf{w}_j} = (u_1, \ldots, u_{r_j})$ for all $j \in [1, \ell]$ (it is purely virtual as $-\vec{u}$ is fully unbounded). Applying Proposition 7.7.3 to $\mathcal{R}_{\mathcal{B}}$, we conclude that, if $\vec{v} = (v_1, \ldots, v_{t'}) \in G_0^{\mathsf{lim}}$ with all $v_i \in \mathbb{R}^{\cup}\langle \mathcal{B}\rangle$, then $-\vec{v}$ is encased by $\mathcal{Y}_1 \cup \ldots \cup \mathcal{Y}_\ell$, and thus also by $\downarrow C_\ell(k) = \downarrow B(k) \subseteq X(k)$ (as all half-spaces in the sets $\mathcal{Y}_i$ are one-dimensional), completing the proof as remarked earlier.

3. Note $\downarrow \mathcal{X} = \mathcal{X}$ since $\mathcal{R} = (\mathcal{X}_1 \cup \{\mathbf{v}_1\}, \ldots, \mathcal{X}_s \cup \{\mathbf{v}_s\})$ is an oriented Reay system. Each element $g \in \widetilde{G}_0$ has $g = \sigma(U)$ for some $U \in \mathcal{F}(G_0)$, implying $g \in \mathsf{C}_{\mathbb{Z}}(G_0) \subseteq \Lambda$. Thus

$$\widetilde{G}_0 \subseteq \Lambda.$$

Since G_0 is finitary, $\mathcal{R}$ is purely virtual and also anchored. Thus, by Proposition 6.8, we have $X(k) = \widetilde{X}(k)$ for any tuple $k = (i_{\mathbf{z}})_{\mathbf{z} \in \mathcal{X}}$ with all $i_{\mathbf{z}}$ sufficiently large. Since $\mathcal{R}$ is a maximal purely virtual Reay system over G_0, it follows that $\pi(G_0)$ is finite. Thus Item 2 and Lemma 7.15 imply $-\mathsf{C}(X(k))$ encases every $\vec{u} \in \widetilde{G}_0^{\mathsf{lim}}$ for any tuple $k = (i_{\mathbf{z}})_{\mathbf{z} \in \mathcal{X}}$ with all $i_{\mathbf{z}}$ sufficiently large. Thus $\widetilde{G}_0$ is bound to $-\mathsf{C}(X(k))$ by Theorem 3.14. Fix one tuple $\kappa = (\iota_{\mathbf{z}})_{\mathbf{z} \in \mathcal{X}}$ such that $\widetilde{G}_0$ is bound to $-\mathsf{C}(X(\kappa))$ and $X(\kappa) = \widetilde{X}(\kappa)$. By Proposition 6.7.2 (applied to each $\mathbf{x} \in \mathcal{B}$), we can assume $\mathsf{C}(\downarrow B(\kappa)) \subseteq \mathsf{C}(\downarrow B(k))$ for any tuple $k = (i_{\mathbf{z}})_{\mathbf{z} \in \mathcal{X}}$ with all $i_{\mathbf{z}}$ sufficiently large and any $\mathcal{B} \subseteq \downarrow \mathcal{X}' \subseteq \mathcal{X}$. Now $X(\kappa) \subseteq G_0 \subseteq \Lambda$ is a linearly independent set of lattice points. Since $\widetilde{G}_0 \cap \mathbb{R}^{\cup}\langle \mathcal{B}\rangle$ is bound to both $\mathbb{R}^{\cup}\langle \mathcal{B}\rangle$ and $-\mathsf{C}(X(\kappa))$, the former trivially, it follows from Corollary 3.15 that $\widetilde{G}_0 \cap \mathbb{R}^{\cup}\langle \mathcal{B}\rangle$ is bound to $\mathbb{R}^{\cup}\langle \mathcal{B}\rangle \cap -\mathsf{C}(X(\kappa)) = \mathbb{R}\langle \downarrow B(\kappa)\rangle \cap -\mathsf{C}(X(\kappa)) =$

$-\mathsf{C}(\downarrow B(\kappa))$, with former equality by Proposition 5.3.1 and the latter in view of the linear independence of $X(\kappa)$. We can tile $-\mathsf{C}(\downarrow B(\kappa))$ with translates of the fundamental parallelepiped defined using the linearly independent set $\downarrow B(\kappa)$ as the lattice basis for $\Lambda_{\mathcal{B}} := \mathbb{Z}\langle\downarrow B(\kappa)\rangle \leq \Lambda$, and then any point $x \in -\mathsf{C}(\downarrow B(\kappa))$ will be within distance M of some lattice point from $\Lambda_{\mathcal{B}} \cap -\mathsf{C}(\downarrow B(\kappa))$, where M is the maximal distance between a point of the fundamental parallelepiped and the set of vertices for that parallelepiped. Thus, since $\widetilde{G}_0 \cap \mathbb{R}^{\cup}\langle\mathcal{B}\rangle$ is bound to $-\mathsf{C}(\downarrow B(\kappa))$, it follows that there is a bound N such that any $g \in \widetilde{G}_0 \cap \mathbb{R}^{\cup}\langle\mathcal{B}\rangle$ is within distance N of some lattice point $-x_g \in \Lambda_{\mathcal{B}} \cap -\mathsf{C}(\downarrow B(\kappa))$. Moreover, as there are only a finite number of possibilities for $\mathcal{B}$, we can assume the same N suffices for each possible $\mathcal{B} \subseteq \downarrow\mathcal{X}' \subseteq \mathcal{X}$.

Any $g \in \widetilde{G}_0 \cap -\mathsf{C}^{\cup}(\mathcal{X}')$ has $-g$ encased by $\mathcal{X}'$, and thus there is some $\mathcal{B} \preceq \mathcal{X}' \subseteq \mathcal{X}_1 \cup \ldots \cup \ldots \mathcal{X}_s$ which minimally encases $-g$, ensuring that $\mathcal{B} \subseteq \downarrow\mathcal{X}'$ is a support set with $-g \in \mathsf{C}^{\cup}(\mathcal{B})^{\circ} \subseteq \mathbb{R}^{\cup}\langle\mathcal{B}\rangle = \mathbb{R}\langle\downarrow B(\kappa)\rangle$, with the final equality in view of Proposition 5.3.1. Since $\downarrow B(\kappa) \subseteq X(\kappa)$ is linearly independent, it follows that $\downarrow B(\kappa)$ is a linear basis for $\mathbb{R}^{\cup}\langle\mathcal{B}\rangle$. For $\mathbf{x} \in \downarrow\mathcal{B}$, let $x = \tilde{\mathbf{x}}(\iota_{\mathbf{x}}) = \mathbf{x}(\iota_{\mathbf{x}}) \in \downarrow B(\kappa)$, and let

$$\sum_{\mathbf{x}\in\downarrow\mathcal{B}} \alpha_{\mathbf{x}} x = -g \tag{7.6}$$

be the representation of $-g \in \mathbb{R}^{\cup}\langle\mathcal{B}\rangle$ as a linear combination of the basis elements $\downarrow B(\kappa)$, where $\alpha_{\mathbf{x}} \in \mathbb{R}$. Since $\mathcal{B}$ is a support set, and thus virtual independent, Proposition 5.3.4 ensures that any element contained in $\mathsf{C}^{\cup}(\mathcal{B})^{\circ}$ is a *strictly* positive linear combination of some choice of representatives from *all* the half-spaces $\mathbf{x} \in \mathcal{B}$. Moreover, any representative set B for $\mathcal{B}$ is linearly independent modulo $\mathbb{R}^{\cup}\langle\partial(\mathcal{B})\rangle$ by Proposition 5.3.3. Thus, since $-g \in \Lambda \cap \mathsf{C}^{\cup}(\mathcal{B})^{\circ}$, it follows by considering (7.6) modulo $\mathbb{R}^{\cup}\langle\partial(\mathcal{B})\rangle$ that $\alpha_{\mathbf{x}} > 0$ for every $\mathbf{x} \in \mathcal{B}$ (though not necessarily for all $\mathbf{x} \in \downarrow\mathcal{B}$). By the work above, there is some $x_g \in \Lambda_{\mathcal{B}} \cap \mathsf{C}(\downarrow B(\kappa))$ with $\mathsf{d}(-g, x_g) \leq N$. Let

$$\sum_{\mathbf{x}\in\downarrow\mathcal{B}} \beta_{\mathbf{x}} x = x_g$$

be the representation of $x_g \in \mathsf{C}(\downarrow B(\kappa))$ as a positive linear combination of the basis elements $\downarrow B(\kappa)$, so $\beta_{\mathbf{x}} \geq 0$ for all $\mathbf{x} \in \downarrow\mathcal{B}$. Since $x_g \in \Lambda_{\mathcal{B}} \cap \mathsf{C}(\downarrow B(\kappa)) = \mathsf{C}_{\mathbb{Z}}(\downarrow B(\kappa))$, where the latter equality follows in view of the linear independence of $\downarrow B(\kappa)$ and $\Lambda_{\mathcal{B}} = \mathbb{Z}\langle\downarrow B(\kappa)\rangle$, we have $\beta_{\mathbf{x}} \in \mathbb{Z}_+$ for all $\mathbf{x} \in \downarrow\mathcal{B}$.

Observe that $\gamma_{\mathbf{y}}^2 \leq \sum_{\mathbf{x}\in\downarrow\mathcal{B}} \gamma_{\mathbf{x}}^2 = \|\sum_{\mathbf{x}\in\downarrow\mathcal{B}} \gamma_{\mathbf{x}} T(x)\|^2 \leq \|T\|^2 \cdot \|\sum_{\mathbf{x}\in\downarrow\mathcal{B}} \gamma_{\mathbf{x}} x\|^2$ for any $\gamma_{\mathbf{x}} \in \mathbb{R}$ and $\mathbf{y} \in \downarrow\mathcal{B}$, where $T : \mathbb{R}^{\cup}\langle\mathcal{B}\rangle \to \mathbb{R}^{\cup}\langle\mathcal{B}\rangle$ is a linear transformation mapping $\downarrow B(\kappa)$ to an orthonormal basis and $\|T\|$ is the operator norm of T with respect to the Euclidean L_2-norm. Thus, since $\mathsf{d}(-g, x_g) \leq N$, it follows that there is some $N' \geq 0$ such that $|\alpha_{\mathbf{x}} - \beta_{\mathbf{x}}| \leq N'$ for all $\mathbf{x} \in \downarrow\mathcal{B}$ and $-g \in \widetilde{G}_0 \cap \mathsf{C}^{\cup}(\mathcal{B})^{\circ}$ (namely, $N' = \|T\| \cdot N$). For any

$\beta_{\mathbf{x}} \geq \alpha_{\mathbf{x}}$ with $\mathbf{x} \in \mathcal{B}$, let $\beta'_{\mathbf{x}}$ be the greatest integer strictly less than $\alpha_{\mathbf{x}}$. Note $\beta'_{\mathbf{x}} \geq 0$ in view of $\alpha_{\mathbf{x}} > 0$. For all other $\mathbf{x} \in \downarrow\mathcal{B}$, let $\beta'_{\mathbf{x}} = \beta_{\mathbf{x}} \geq 0$. Let

$$\sum_{\mathbf{x} \in \downarrow\mathcal{B}} \beta'_{\mathbf{x}} x = x'_g.$$

Since $|\beta_{\mathbf{x}} - \beta'_{\mathbf{x}}| \leq |\alpha_{\mathbf{x}} - \beta_{\mathbf{x}}| + 1 \leq N' + 1$ for all $\mathbf{x} \in \downarrow\mathcal{B}$, there is some N'' such that $\|x_g - x'_g\| \leq N''$ for all $g \in \widetilde{G}_0 \cap -\mathsf{C}^{\cup}(\mathcal{B})^{\circ}$ (for instance, in view of the triangle inequality, we could take $N'' = (N' + 1)\sum_{\mathbf{x} \in \downarrow\mathcal{B}} \|x\|$). Thus, by replacing N with a larger value and using x'_g in place of x_g, we obtain the additional conclusion that

$$\beta_{\mathbf{x}} \in \mathbb{Z}_+ \quad \text{for all } \mathbf{x} \in \downarrow\mathcal{B} \quad \text{and} \quad \alpha_{\mathbf{x}} > \beta_{\mathbf{x}} \geq 0 \quad \text{for all } \mathbf{x} \in \mathcal{B}, \tag{7.7}$$

ensuring that

$$x_g = x'_g \in \mathsf{C}_{\mathbb{Z}}(\downarrow B(\kappa)) \subseteq \Lambda.$$

Thus, since $g \in \Lambda$ and $\mathsf{d}(-g, x_g) \leq N$, we have $x_g + y = -g$ with $y \in \Lambda$ and $\|y\| \leq N$. Note

$$y = -g - x_g = \sum_{\mathbf{x} \in \downarrow\mathcal{B}} (\alpha_{\mathbf{x}} - \beta_{\mathbf{x}})x$$

with $\alpha_{\mathbf{x}} - \beta_{\mathbf{x}} > 0$ for all $\mathbf{x} \in \mathcal{B}$ in view of (7.7). Thus $y \in \sum_{\mathbf{x} \in \mathcal{B}} (\mathbf{x}^{\circ} + \partial(\mathbf{x})) \subseteq \sum_{\mathbf{x} \in \mathcal{B}} \mathbf{x} \subseteq \mathsf{C}^{\cup}(\mathcal{B})$. In summary, we now have

$$y \in \Lambda \cap \mathsf{C}^{\cup}(\mathcal{B}) \quad \text{and} \quad \|y\| \leq N. \tag{7.8}$$

Since any bounded set of lattice points is finite, there are only a finite number of y satisfying (7.8). In consequence, Proposition 6.7.2 implies that, for any tuple $k = (i_{\mathbf{z}})_{\mathbf{z} \in \mathcal{X}}$ with all $i_{\mathbf{z}}$ sufficiently large, we have $\mathsf{C}(\downarrow B(\kappa)) \subseteq \mathsf{C}(\downarrow B(k)) \subseteq \mathsf{C}(\downarrow X'(k))$ and $y \in \mathsf{C}(\downarrow B(k)) \subseteq \mathsf{C}(\downarrow X'(k))$, for every support set $\mathcal{B} \subseteq \downarrow\mathcal{X}'$ and every possible y satisfying (7.8). But then $y \in \mathsf{C}(\downarrow X'(k))$ and $x_g \in \mathsf{C}(\downarrow B(\kappa)) \subseteq \mathsf{C}(\downarrow B(k)) \subseteq \mathsf{C}(\downarrow X'(k))$, for every $g \in \widetilde{G}_0 \cap -\mathsf{C}^{\cup}(\mathcal{X}')$, ensuring that $-g = y + x_g \in \mathsf{C}(\downarrow X'(k))$ in view of the convexity of $\mathsf{C}(\downarrow X'(k))$. Thus $\widetilde{G}_0 \cap -\mathsf{C}^{\cup}(\mathcal{X}') \subseteq \widetilde{G}_0 \cap -\mathsf{C}(\downarrow X'(k))$. Since $\mathsf{C}^{\cup}(\downarrow X'(k)) = \mathsf{C}^{\cup}(\downarrow \tilde{X}'(k)) \subseteq \mathsf{C}^{\cup}(\mathcal{X}')$ for any tuple k, the reverse inclusion is trivial, and Item 3 follows, completing the proof.

□

7.2 Series Decompositions and Virtualizations

Let $\Lambda \leq \mathbb{R}^d$ be a full rank lattice, let $G_0 \subseteq \Lambda$ be a finitary subset with $\mathsf{C}(G_0) = \mathbb{R}^d$, and let $\mathcal{R} = (\mathcal{X}_1 \cup \{\mathbf{v}_1\}, \ldots, \mathcal{X}_s \cup \{\mathbf{v}_s\})$ be a purely virtual Reay system over G_0. Since G_0 is finitary, any purely virtual Reay system must be anchored. However, if $\mathcal{R}$ is anchored, then $G_0 \subseteq \Lambda$ together with Proposition 6.8 ensures that $\tilde{\mathbf{x}}(i) = \mathbf{x}(i)$ for any $\mathbf{x} \in \bigcup_{i=1}^{s} \mathcal{X}_i$ once i is sufficiently large. Suppose $\mathbf{x} \in \mathcal{X}_j$ with $j \in [1, s]$. Any $\mathbf{x}(i)$ with i sufficiently large is equal to $\tilde{\mathbf{x}}(i)$ and is then an actual lattice point from G_0 which is a representative for $\mathbf{x}$. Let $x' \in G_0$ be any element which is a positive scalar multiple of u_t modulo $\mathbb{R}^{\cup}\langle \bigcup_{i=1}^{j-1} \mathcal{X}_i \rangle$, where $\vec{u}_{\mathbf{x}} = (u_1, \ldots, u_t)$, so $x' \in (\mathbb{R}^{\cup}\langle \bigcup_{i=1}^{j-1} \mathcal{X}_i \rangle + \mathbf{x})^{\circ}$. For instance, $x' \in G_0$ could be any representative for the half-space $\mathbf{x}$, including any $x' = \mathbf{x}(i)$ with i sufficiently large. Define a new half-space $\mathbf{x}' = \mathbb{R}_+ x'$ with $\partial(\mathbf{x}') = \{0\}$. Then $\mathbf{x}$ and $\mathbf{x}'$ are both equal modulo $\mathbb{R}^{\cup}\langle \bigcup_{i=1}^{j-1} \mathcal{X}_i \rangle$, so replacing $\mathbf{x}$ by $\mathbf{x}'$ would preserve (OR2) in the definition of an oriented Reay system. Of course, if $j = 1$, then $\mathbf{x} = \mathbf{x}'$ since $\partial(\mathbf{x}) = \{0\}$ in this case, and we have more or less done nothing apart from changing the representative for $\mathbf{x}$. Suppose, for $j = 1, 2, \ldots, s$, we replace each $\mathbf{x} \in \bigcup_{i=1}^{s} \mathcal{X}_i$ with some half-space $\mathbf{x}'$ with $\partial(\mathbf{x}') = \{0\}$ as just described to result in $(\mathcal{X}'_1 \cup \{\mathbf{v}_1\}, \ldots, \mathcal{X}'_s \cup \{\mathbf{v}_s\})$. Now $(\mathcal{X}'_1 \cup \{\mathbf{v}_1\})$ is still a virtual Reay system over G_0 with $\vec{u}_{\mathbf{x}} = \vec{u}_{\mathbf{x}'}$ for all $\mathbf{x} \in \mathcal{X}_1$, the value of $\vec{u}_{\mathbf{v}_1}$ unchanged and $\mathbb{R}^{\cup}\langle \mathcal{X}_1 \rangle = \mathbb{R}^{\cup}\langle \mathcal{X}'_1 \rangle$. Since $\mathcal{R}$ is purely virtual, $-\vec{u}^{\triangleleft}_{\mathbf{v}_2}$ is fully unbounded or trivial and minimally encased by $\partial(\{\mathbf{v}_2\}) \subseteq \mathbb{R}^{\cup}\langle \mathcal{X}_1 \rangle = \mathbb{R}^{\cup}\langle \mathcal{X}'_1 \rangle$. Thus, since G_0 is finitary and $\mathcal{R}$ is purely virtual, it follows from Proposition 7.7.3 applied to $(\mathcal{X}'_1 \cup \{\mathbf{v}_1\})$ that $-\vec{u}^{\triangleleft}_{\mathbf{v}_2}$ is minimally encased by some subset $\mathcal{B}_2 \subseteq \mathcal{X}'_1$, allowing us to define a new half-space $\mathbf{v}'_2$ with $\vec{u}_{\mathbf{v}'_2} = \vec{u}_{\mathbf{v}_2}$ and $\partial(\{\mathbf{v}'_2\}) = \mathcal{B}_2$. This makes $(\mathcal{X}'_1 \cup \{\mathbf{v}'_1\}, \mathcal{X}'_2 \cup \{\mathbf{v}'_2\})$ into a virtual Reay system, where $\mathbf{v}'_1 := \mathbf{v}_1$, with $\partial(\mathbf{x}') = \{0\}$ and $\mathbf{x}'(i) = x'$ constant for $\mathbf{x}' \in \mathcal{X}'_2$. Moreover, since the value of each $\mathbf{x}' \in \mathcal{X}'_2$ has not changed modulo $\mathbb{R}^{\cup}\langle \mathcal{X}_1 \rangle = \mathbb{R}^{\cup}\langle \mathcal{X}'_1 \rangle$, it follows that $\mathbb{R}^{\cup}\langle \mathcal{X}'_1 \cup \mathcal{X}'_2 \rangle = \mathbb{R}^{\cup}\langle \mathcal{X}_1 \cup \mathcal{X}_2 \rangle$. Iterating this argument, we find that $\mathcal{R}' := (\mathcal{X}'_1 \cup \{\mathbf{v}'_1\}, \ldots, \mathcal{X}'_s \cup \{\mathbf{v}'_s\})$ is a purely Virtual Reay system over G_0 with $\mathbb{R}^{\cup}\langle \bigcup_{i=1}^{j} \mathcal{X}'_i \rangle = \mathbb{R}^{\cup}\langle \bigcup_{i=1}^{j} \mathcal{X}_i \rangle$ and $\vec{u}_{\mathbf{v}'_j} = \vec{u}_{\mathbf{v}_j}$ for all $j \in [1, s]$, and with $\partial(\mathbf{x}') = \{0\}$ and $\mathbf{x}'(i) = x' \in G_0$ constant for all $\mathbf{x}' \in \mathcal{X}'_1 \cup \ldots \cup \mathcal{X}'_s$. In particular, Proposition 7.7.1 applied to $\mathcal{R}'$ ensures that any $x' \in G_0$ which lies in the open half-space $(\mathbb{R}^{\cup}\langle \bigcup_{i=1}^{j-1} \mathcal{X}_i \rangle + \mathbf{x})^{\circ}$, for $\mathbf{x} \in \mathcal{X}_j$, must satisfy $x' \in G_0^{\diamond}$ (assuming $G_0 \subseteq \Lambda \subseteq \mathbb{R}^d$ is finitary with $\mathsf{C}(G_0) = \mathbb{R}^d$ and $\mathcal{R}$ purely virtual).

The above construction of $\mathcal{R}'$ was done using arbitrary elements $x' \in G_0$ lying in the open half-space $(\mathbb{R}^{\cup}\langle \bigcup_{i=1}^{j-1} \mathcal{X}_i \rangle + \mathbf{x})^{\circ}$, for $\mathbf{x} \in \mathcal{X}_j$, and resulting only in the existence of subsets $\partial(\{\mathbf{v}'_j\}) \subseteq \mathcal{X}'_1 \cup \ldots \cup \mathcal{X}'_{j-1}$. Suppose instead we choose each $x' = \mathbf{x}(i_{\mathbf{x}})$ for some fixed but sufficiently large $i_{\mathbf{x}}$. Let $k = (i_{\mathbf{x}})_{\mathbf{x} \in \mathcal{X}}$ be a fixed tuple with all $i_{\mathbf{x}}$ sufficiently large (as determined below). For a half-space $\mathbf{x} \in \mathcal{X}_1 \cup \{\mathbf{v}_1\} \cup \ldots \cup \mathcal{X}_s \cup \{\mathbf{v}_s\}$ from $\mathcal{R}$, let $\mathbf{x}' \in \mathcal{X}'_1 \cup \{\mathbf{v}'_1\} \cup \ldots \cup \mathcal{X}'_s \cup \{\mathbf{v}'_s\}$ denote the corresponding half-space from $\mathcal{R}'$, so each $\mathbf{x}' = \mathbb{R}_+ \mathbf{x}(i_{\mathbf{x}})$ whenever $\mathbf{x} \in \mathcal{X}_1 \cup \ldots \cup \mathcal{X}_s$, and for a subset $\mathcal{X} \subseteq \mathcal{X}_1 \cup \{\mathbf{v}_1\} \cup \ldots \cup \mathcal{X}_s \cup \{\mathbf{v}_s\}$, let $\mathcal{X}' = \{\mathbf{x}' : \mathbf{x} \in$

$\mathcal{X}\} \subseteq \mathcal{X}'_1 \cup \{\mathbf{v}'_1\} \cup \ldots \cup \mathcal{X}'_s \cup \{\mathbf{v}'_s\}$. Assuming the indices $i_{\mathbf{x}}$ are chosen sufficiently large, Proposition 6.7.1 implies that

$$\partial(\{\mathbf{v}'_j\}) = (\downarrow\partial(\{\mathbf{v}_j\}))' = \downarrow\partial(\{v_j\})(k) \quad \text{for each } j \in [1, s]. \tag{7.9}$$

Note, for the equality (7.9) to make sense, we informally identify $\downarrow\partial(\{v_j\})(k)$ with the collection of one-dimensional half-spaces generated by the elements from $\downarrow\partial(\{v_j\})(k)$, with this convention continued at later points of the discussion. If $\mathcal{A}_j$ is the set defined in Proposition 6.12.1 for $\mathcal{R}$, then Proposition 7.7.2 ensures that $\mathcal{A}_j \setminus \{\mathbf{v}_j\} \subseteq \mathcal{X}_1 \cup \ldots \cup \mathcal{X}_j$, in which case Proposition 6.8 implies $(\downarrow A_j \setminus \{v_j\})(k) = (\downarrow \widetilde{A_j \setminus \{v_j\}})(k)$ so long as all indices are sufficiently large. In view of Proposition 6.12.2(e), we have $\mathbb{R}^{\cup}\langle \mathcal{A}_j \rangle = \mathbb{R}^{\cup}\langle \mathcal{A}_j \setminus \{\mathbf{v}_j\} \cup \partial(\{\mathbf{v}_j\})\rangle$, while $(A_j \setminus \{v_j\})(k) \subseteq \mathbb{R}^{\cup}\langle \mathcal{A}_j \setminus \{\mathbf{v}_j\}\rangle$ as the elements of $(A_j \setminus \{v_j\})(k)$ are representatives for the half-spaces from $\mathcal{A}_j \setminus \{\mathbf{v}_j\}$. This ensures that the set Y_k in Proposition 6.12.3 (for the element $\mathbf{v}_j$) is empty, in which case Proposition 6.12.3 implies that $(\downarrow A_j \setminus \{v_j\})(k)$ minimally encases $-\vec{u}_{\mathbf{v}_j}$ so long as all indices are sufficiently large.

In view of (7.9) and Proposition 5.3.1, we have

$$\mathbb{R}^{\cup}\langle \partial(\{\mathbf{v}'_j\})\rangle = \mathbb{R}\langle \downarrow\partial(\{v_j\})(k)\rangle = \mathbb{R}^{\cup}\langle \partial(\{\mathbf{v}_j\})\rangle.$$

Let $\pi_{\mathbf{v}_j} = \mathbb{R}^d \to \mathbb{R}^{\cup}\langle \partial(\{\mathbf{v}_j\})\rangle^{\perp}$ be the orthogonal projection. By Proposition 5.4.1 and Proposition 5.3.9, $\pi_{\mathbf{v}_j}$ is injective on $(\mathcal{X}_1 \cup \ldots \cup \mathcal{X}_s) \setminus \downarrow\partial(\{\mathbf{v}_j\})$ and maps all such elements to non-zero half-spaces. Hence, since $(\downarrow A_j \setminus \{v_j\})(k) \subseteq \mathcal{X}'_1 \cup \ldots \cup \mathcal{X}'_s$ minimally encases $-\vec{u}_{\mathbf{v}_j} = -\vec{u}_{\mathbf{v}'_j}$, it follows from Proposition 5.9.4 that

$$\Big((\downarrow\mathcal{A}_j \setminus \{v_j\})(k)\Big)^{\pi_{\mathbf{v}_j}} = \pi_{\mathbf{v}_j}\big((\downarrow A_j \setminus \downarrow v_j)(k)\big)$$

minimally encases $-\pi_{\mathbf{v}_j}(\vec{u}_{\mathbf{v}'_j})$. Thus, since $\mathcal{A}_j \setminus \{\mathbf{v}_j\} \subseteq \mathcal{X}$, we conclude that $\pi_{\mathbf{v}_j}\big((\downarrow A_j \setminus \downarrow v_j)(k)\big) = \mathsf{Supp}_{\pi_{\mathbf{v}_j}(\mathcal{R}')}(-\pi_{\mathbf{v}_j}(\vec{u}_{\mathbf{v}'_j}))$, meaning $(\downarrow A_j \setminus \downarrow v_j)(k)$ is the pull-back of $\mathsf{Supp}_{\pi_{\mathbf{v}_j}(\mathcal{R}')}(-\pi_{\mathbf{v}_j}(\vec{u}_{\mathbf{v}'_j}))$. As a result, it now follows from Propositions 6.12.1 and 6.12.2(b) that, letting $\mathcal{A}'_j$ and $\mathcal{A}'_{\mathbf{v}'_j}$ denote the sets given by Proposition 6.12 for $\mathcal{R}'$, we have $\mathcal{A}'_j = (\downarrow A_j \setminus \downarrow v_j)(k) \cup \{\mathbf{v}'_j\} = (\downarrow\mathcal{A}_j \setminus \downarrow\partial(\{\mathbf{v}_j\}))'$ and $\mathcal{A}'_{\mathbf{v}'_j} = \big(\mathcal{A}'_j \setminus \{\mathbf{v}'_j\} \cup \partial(\{\mathbf{v}'_j\})\big)^* = ((\downarrow A_j \setminus \{v_j\})(k))^* = (\downarrow A_j \setminus \{v_j\})(k)$, with the final equality holding since all half-spaces from $\mathcal{X}'$ have trivial boundary, and the second in view (7.9). In summary,

$$\partial(\{\mathbf{v}'_j\}) = \big(\downarrow\partial(\{\mathbf{v}_j\})\big)', \quad \mathcal{A}'_{\mathbf{v}'_j} = (\downarrow\mathcal{A}_{\mathbf{v}_j})' \quad \text{and}$$

$$\mathcal{A}'_j = \big(\downarrow\mathcal{A}_j \setminus \downarrow\partial(\{\mathbf{v}_j\})\big)' \text{ for every } j \in [1, s].$$

This ensures the virtual Reay system structure associated to each $\mathbf{v}_j$ is preserved in $\mathbf{v}'_j$ when passing to the virtual Reay system $\mathcal{R}'$ (as much as possible given that $\partial(\mathbf{x}') = \{0\}$ for every $\mathbf{x}' \in \mathcal{X}'_1 \cup \ldots \cup \mathcal{X}'_s$).

In view of the observations just made, when considering a purely virtual Reay system $\mathcal{R} = (\mathcal{X}_1 \cup \{\mathbf{v}_1\}, \ldots, \mathcal{X}_s \cup \{\mathbf{v}_s\})$ over a finitary subset $G_0 \subseteq \Lambda$ with $\mathsf{C}(G_0) = \mathbb{R}^d$, we can often restrict attention to when $\partial(\mathbf{x}) = \{0\}$ for all $\mathbf{x} \in \mathcal{X}_1 \cup \ldots \cup \mathcal{X}_s$, as if this fails for $\mathcal{R}$, then another virtual Reay system $\mathcal{R}'$ over G_0 can be constructed with this property as described above with the values of $\partial(\{\mathbf{v}_j\})$, $\mathcal{A}_j$ and $\mathcal{A}_{\mathbf{v}_j}$ minimally affected.

Let $\Lambda \leq \mathbb{R}^d$ be a full rank lattice and let $G_0 \subseteq \Lambda$ be a finitary subset with $\mathsf{C}(G_0) = \mathbb{R}^d$. We now define the followings sets:

$$\mathfrak{X}(G_0) = \Big\{ \bigcup_{i=1}^{s} \mathcal{X}_i : \text{ there is a purely virtual Reay sytem } (\mathcal{X}_1 \cup \{\mathbf{v}_1\}, \ldots, \mathcal{X}_s \cup \{\mathbf{v}_s\}) \text{ over } G_0 \Big\},$$

$$X(G_0) = \Big\{ X : X \subseteq G_0 \text{ is a set of representatives for the half-spaces from some } \mathcal{X} \in \mathfrak{X}(G_0) \Big\}.$$

Since the empty tuple is by default a purely virtual Reay system, we always have $\emptyset \in X(G_0)$. If $X \in X(G_0)$, then there is a purely virtual Reay System $(\mathcal{X}_1 \cup \{\mathbf{v}_1\}, \ldots, \mathcal{X}_s \cup \{\mathbf{v}_s\})$ over G_0 with $X \subseteq G_0$ being a set of representative for the half-spaces from $\bigcup_{i=1}^{s} \mathcal{X}_i$. Per the discussion above, replacing each $x \in X$ with the half-space $\mathbb{R}_+ x$ results in a purely virtual Reay system $\mathcal{R}' = (\mathcal{X}'_1 \cup \{\mathbf{v}'_1\}, \ldots, \mathcal{X}'_s \cup \{\mathbf{v}'_s\})$ over G_0 such that X is a set of representatives for the half-spaces from $\bigcup_{i=1}^{s} \mathcal{X}'_i$ and $\partial(\mathbf{x}') = \{0\}$ with $\mathbf{x}'(i_{\mathbf{x}}) = x \in G_0$ for all $\mathbf{x}' \in \bigcup_{i=1}^{s} \mathcal{X}'_i$ and $i_{\mathbf{x}} \geq 1$. Thus it can always be assumed that $X \in X(G_0)$ came from a virtual Reay system having these properties. Prior discussion ensures that $X \subseteq G_0^{\diamond}$ for any $X \in X(G_0)$, while any $X \in X(G_0)$ is a linearly independent subset of $G_0 \subseteq \Lambda$ by Proposition 4.11.1, thus generating a sublattice of Λ. Let

$$\begin{aligned}\mathfrak{P}_{\mathbb{Z}}(G_0) &= \{\mathbb{Z}\langle X\rangle : X \in X(G_0)\} \text{ and}\\ \mathfrak{P}_{\mathbb{R}}(G_0) &= \{\mathbb{R}^{\cup}\langle \mathcal{X}\rangle : \mathcal{X} \in \mathfrak{X}(G_0)\}\\ &= \{\mathbb{R}\langle X\rangle : X \in X(G_0)\} = \{\mathbb{R}\langle \Lambda'\rangle : \Lambda' \in \mathfrak{P}_{\mathbb{Z}}(G_0)\},\end{aligned}$$

with the second equality for $\mathfrak{P}_{\mathbb{R}}(G_0)$ by Proposition 5.3.1. Note $\mathfrak{P}_{\mathbb{Z}}(G_0)$ consists of sublattices of Λ as just discussed. The linearly independent set $X \in X(G_0)$ can be recovered from $\mathsf{C}_{\mathbb{Z}}(X)$ by considering each ray defined by a vertex in the polyhedron $\mathsf{C}(X) \cap B_1(0)$ lying on the unit sphere, and then taking the minimal nonzero element of this ray contained in $\mathsf{C}_{\mathbb{Z}}(X)$. Thus $\mathsf{C}_{\mathbb{Z}}(X) = \mathsf{C}_{\mathbb{Z}}(Y)$ implies $X = Y$, allowing us to define a partial order $\preceq_{\mathbb{Z}}$ on $X(G_0)$ by declaring $X \preceq_{\mathbb{Z}} Y$

when $\mathsf{C}_{\mathbb{Z}}(X) \subseteq \mathsf{C}_{\mathbb{Z}}(Y)$. Since each $X \in X(G_0)$ is linearly independent, we have

$$\mathsf{C}_{\mathbb{Z}}(X) = \mathsf{C}(X) \cap \mathbb{Z}\langle X\rangle \quad \text{for } X \in X(G_0). \tag{7.10}$$

Thus

$$X \preceq_{\mathbb{Z}} Y \quad \text{if and only if} \quad \mathsf{C}(X) \subseteq \mathsf{C}(Y) \quad \text{and} \quad \mathbb{Z}\langle X\rangle \leq \mathbb{Z}\langle Y\rangle.$$

Indeed, if $\mathsf{C}_{\mathbb{Z}}(X) \subseteq \mathsf{C}_{\mathbb{Z}}(Y)$, then $x \in \mathsf{C}_{\mathbb{Z}}(Y) \subseteq \mathbb{Z}\langle Y\rangle \cap \mathsf{C}(Y)$ for each $x \in X$, ensuring that $\mathsf{C}(X) \subseteq \mathsf{C}(Y)$ and $\mathbb{Z}\langle X\rangle \leq \mathbb{Z}\langle Y\rangle$, while if $\mathsf{C}(X) \subseteq \mathsf{C}(Y)$ and $\mathbb{Z}\langle X\rangle \leq \mathbb{Z}\langle Y\rangle$, then $\mathsf{C}_{\mathbb{Z}}(X) = \mathsf{C}(X) \cap \mathbb{Z}\langle X\rangle \subseteq \mathsf{C}(Y) \cap \mathbb{Z}\langle Y\rangle = \mathsf{C}_{\mathbb{Z}}(Y)$.

Definition 7.17 For any $X \in X(G_0)$, there is a purely virtual Reay system $\mathcal{R} = (\mathcal{X}_1 \cup \{\mathbf{v}_1\}, \ldots, \mathcal{X}_s \cup \{\mathbf{v}_s\})$ over G_0 and ordered partition $X = X_1 \cup \ldots \cup X_s$ such that each X_j is a set of representatives for the half-spaces from $\mathcal{X}_j$, for $j \in [1, s]$. We call any such $\mathcal{R}$ a **realization** of X and the ordered partition $X = X_1 \cup \ldots \cup X_s$ a **series decomposition** of X. The set $\mathcal{X} = \bigcup_{i=1}^{s} \mathcal{X}_i$ is called a **virtualization** of X. For $\mathcal{X} \in \mathfrak{X}(G_0)$, there is a purely virtual Reay system $\mathcal{R} = (\mathcal{X}_1 \cup \{\mathbf{v}_1\}, \ldots, \mathcal{X}_s \cup \{\mathbf{v}_s\})$ over G_0. We also call $\mathcal{X} = \bigcup_{i=1}^{s} \mathcal{X}_i$ a **series decomposition** of $\mathcal{X}$ and $\mathcal{R}$ a **realization** of $\mathcal{X}$.

To explain the name, it may be helpful to view a series decomposition as an ascending chain $\emptyset \subset X_1 \subset (X_1 \cup X_2) \subset \ldots \subset (X_1 \cup \ldots \cup X_s) = X$ or $\emptyset \subset \mathcal{X}_1 \subset (\mathcal{X}_1 \cup \mathcal{X}_2) \subset \ldots \subset (\mathcal{X}_1 \cup \ldots \cup \mathcal{X}_s) = \mathcal{X}$ If $X = X_1 \cup \ldots \cup X_s$ is a series decomposition of $X \in X(G_0)$, then $Y_j := X_1 \cup \ldots \cup X_j \subseteq X$ is a subset of X with $Y_j \in X(G_0)$ for any $j \in [1, s]$. Likewise, if $\mathcal{X} = \bigcup_{i=1}^{\cup} \mathcal{X}_i$ is a series decomposition of $\mathcal{X} \in \mathfrak{X}(G_0)$, then every $\mathcal{Y}_j := \mathcal{X}_1 \cup \ldots \cup \mathcal{X}_j \subseteq \mathcal{X}$ is a subset of $\mathcal{X}$ with $\mathcal{Y}_j \in \mathfrak{X}(G_0)$ for any $j \in [1, s]$. The converse to both these statements is also true in the following strong sense.

Lemma 7.18 *Let $\Lambda \leq \mathbb{R}^d$ be a full rank lattice and let $G_0 \subseteq \Lambda$ be a finitary subset with $\mathsf{C}(G_0) = \mathbb{R}^d$. Suppose $\mathcal{X}$, $\mathcal{Y} \in \mathfrak{X}(G_0)$ with $\mathcal{Y} \subseteq \mathcal{X}$ and that $\mathcal{Y} = \mathcal{Y}_1 \cup \ldots \cup \mathcal{Y}_t$ is a series decomposition of $\mathcal{Y}$. Let $\pi : \mathbb{R}^d \to \mathbb{R}^{\cup}\langle\mathcal{Y}\rangle^{\perp}$ be the orthogonal projection.*

1. *There is a series decomposition $\mathcal{X} = \mathcal{X}_1 \cup \ldots \cup \mathcal{X}_s$ with $s \geq t$ and $\mathcal{X}_i = \mathcal{Y}_i$ for $i \in [1, t]$.*
2. *$\mathcal{X}^{\pi} = \pi(\mathcal{X} \setminus \mathcal{Y}) \in \mathfrak{X}(\pi(G_0))$.*
3. *If $\mathcal{X} \setminus \mathcal{Y} = \mathcal{X}'_1 \cup \ldots \cup \mathcal{X}'_r$ with $\pi(\mathcal{X} \setminus \mathcal{Y}) = \pi(\mathcal{X}'_1) \cup \ldots \cup \pi(\mathcal{X}'_r)$ a series decomposition, then $\mathcal{X} = \mathcal{Y}_1 \cup \ldots \cup \mathcal{Y}_t \cup \mathcal{X}'_1 \cup \ldots \cup \mathcal{X}'_r$ is a series decomposition.*

Proof

1. Let $\mathcal{R} = (\mathcal{X}_1 \cup \{\mathbf{v}_1\}, \ldots, \mathcal{X}_s \cup \{\mathbf{v}_s\})$ be a realization of $\mathcal{X} \in \mathfrak{X}(G_0)$, so $\mathcal{X} = \mathcal{X}_1 \cup \ldots \cup \mathcal{X}_s$, and let $\mathcal{R}_{\mathcal{Y}} = (\mathcal{Y}_1 \cup \{\mathbf{w}_1\}, \ldots, \mathcal{Y}_t \cup \{\mathbf{w}_t\})$ be a realization of $\mathcal{Y} \subseteq \mathcal{X}$ associated to the series decomposition $\mathcal{Y} = \mathcal{Y}_1 \cup \ldots \cup \mathcal{Y}_t$. Thus $\downarrow\mathcal{Y} = \mathcal{Y}$ when

considered as a subset of half-spaces from $\mathcal{R}_{\mathcal{Y}}$, and thus also (by Proposition 5.11) when considered as a subset of half-spaces from $\mathcal{R}$. In view of Proposition 6.8, we have $\mathbb{R}^{\cup}\langle\mathcal{Y}\rangle = \mathbb{R}\langle\downarrow Y(k)\rangle = \mathbb{R}\langle Y(k)\rangle$ for any tuple $k = (i_{\mathbf{y}})_{\mathbf{y}\in\mathcal{Y}}$ with all $i_{\mathbf{y}}$ sufficiently large. By Proposition 6.3, $\pi(\mathcal{R}) = (\mathcal{X}_j^{\pi} \cup \{\pi(\mathbf{v}_j)\})_{j\in J}$ is a Virtual Reay system over $\pi(G_0)$. Since $\mathcal{R}$ is purely virtual, it follows from Proposition 6.3 that $\pi(\mathcal{R})$ is also purely virtual, and Proposition 6.3 ensures that

$$\vec{u}_{\pi(\mathbf{x})} = \pi(\vec{u}_{\mathbf{x}}), \quad \pi(\vec{u}_{\mathbf{x}})^{\triangleleft} = \pi(\vec{u}_{\mathbf{x}}^{\triangleleft}), \quad \text{and} \quad \partial(\{\pi(\mathbf{x})\}) = \partial(\{\mathbf{x}\})^{\pi} \tag{7.11}$$

for all $\mathbf{x} \in \bigcup_{j\in J}(\mathcal{X}_j \cup \{\mathbf{v}_j\})$ with $\pi(\mathbf{x}) \neq 0$.

Let $j_1 < j_2 < \ldots < j_r$ be the indices from J and let

$$\pi(\mathcal{R}) = (C_1 \cup \{\mathbf{c}_1\}, \ldots, C_r \cup \{\mathbf{c}_r\}),$$

so each $C_i = \mathcal{X}_{j_i}^{\pi}$ and each $\mathbf{c}_i = \pi(\mathbf{v}_{j_i})$ for $i \in [1, r]$. For each $i \in [1, r]$, let $\mathcal{B}_i = \pi^{-1}(C_i) \subseteq \mathcal{X}_{j_i}$. Thus Proposition 5.3.9 and $\downarrow\mathcal{Y} = \mathcal{Y}$ give

$$\mathcal{Y} \cup \bigcup_{i=1}^{r} \mathcal{B}_i = \mathcal{X}, \tag{7.12}$$

with the union disjoint. Each $\mathbf{c}_i = \pi(\mathbf{v}_{j_i})$, for $i \in [1, r]$, is defined by the limit $\vec{u}_{\pi(\mathbf{v}_{j_i})} = \pi(\vec{u}_{\mathbf{v}_{j_i}})$ and satisfies $\pi(\vec{u}_{\mathbf{v}_{j_i}})^{\triangleleft} = \pi(\vec{u}_{\mathbf{v}_{j_i}}^{\triangleleft})$ by (7.11). Moreover, each $-\vec{u}_{\mathbf{v}_{j_i}}^{\triangleleft}$, for $i \in [1, r]$, is minimally encased by $\partial(\{\mathbf{v}_{j_i}\}) \subseteq \bigcup_{k=1}^{j_i-1} \mathcal{X}_k$, and is thus encased by $\mathcal{Y} \cup \mathcal{B}_{\mathbf{c}_i}$, where $\mathcal{B}_{\mathbf{c}_i} := \partial(\{\mathbf{v}_{j_i}\}) \setminus \mathcal{Y}$ with $\mathcal{B}_{\mathbf{c}_i} \subseteq \mathcal{B}_1 \cup \ldots \cup \mathcal{B}_{i-1}$. Also, $\pi(\mathcal{B}_{\mathbf{c}_i}) = \mathcal{B}_{\mathbf{c}_i}^{\pi} = \partial(\{\mathbf{v}_{j_i}\})^{\pi} = \partial(\{\pi(\mathbf{v}_{j_i})\}) = \partial(\{\mathbf{c}_i\})$, where the first equality follows since $\mathcal{B}_{\mathbf{c}_i} \subseteq \mathcal{B}_1 \cup \ldots \cup \mathcal{B}_{i-1}$ with $\pi(\mathbf{x}) \neq \{0\}$ for all $\mathbf{x} \in \mathcal{B}_1 \cup \ldots \cup \mathcal{B}_r$, the third by (7.11), and the other two in view of the definitions of $\mathcal{B}_{\mathbf{c}_i}$ and $\mathbf{c}_i$. But now we have the needed hypotheses to apply Lemma 6.4 (using $\mathcal{R}_{\mathcal{Y}}$ for $\mathcal{R}_{\mathcal{A}}$, and $\pi(\mathcal{R})$ for $\mathcal{R}_C$) to thereby conclude that

$$\mathcal{R}' = (\mathcal{Y}_1 \cup \{\mathbf{w}_1\}, \ldots, \mathcal{Y}_t \cup \{\mathbf{w}_t\}, \mathcal{B}_1 \cup \{\mathbf{b}_1\}, \ldots, \mathcal{B}_r \cup \{\mathbf{b}_r\})$$

is a virtual Reay system in G_0 with each $\mathbf{b}_i$, for $i \in [1, r]$, defined by the limit $\vec{u}_{\mathbf{v}_{j_i}}$. By (7.12), we have $\mathcal{Y}_1 \cup \ldots \mathcal{Y}_t \cup \mathcal{B}_1 \cup \ldots \cup \mathcal{B}_r = \mathcal{X}$. Now $\mathcal{R}_{\mathcal{Y}}$ and $\pi(\mathcal{R})$ are both purely virtual, the former as it is a realization of a set from $\mathfrak{X}(G_0)$, and the latter as observed above (7.11), whence Lemma 6.4 implies that $\mathcal{R}'$ is purely virtual as well, completing Item 1.

2. By Item 1, there exists a realization $\mathcal{R} = (\mathcal{X}_1 \cup \{\mathbf{v}_1\}, \ldots, \mathcal{X}_s \cup \{\mathbf{v}_s\})$ of $\mathcal{X} = \mathcal{X}_1 \cup \ldots \cup \mathcal{X}_s$ with $\mathcal{Y}_i = \mathcal{X}_i$ for $i \in [1, t]$. Proposition 6.3 implies that $\pi(\mathcal{R}) = (\pi(\mathcal{X}_{t+1}) \cup \{\pi(\mathbf{v}_{t+1})\}, \ldots, \pi(\mathcal{X}_s) \cup \{\pi(\mathbf{v}_s)\})$ is a realization of $\mathcal{X}^{\pi} = \pi(\mathcal{X}) \setminus \{\{0\}\} = \pi(\mathcal{X} \setminus \mathcal{Y})$. Thus $\pi(\mathcal{X} \setminus \mathcal{Y}) \in \mathfrak{X}(\pi(G_0))$, establishing Item 2. Note $\mathfrak{X}(\pi(G_0))$ is well-defined by Proposition 7.9.

3. Suppose $\mathcal{X} \setminus \mathcal{Y} = \mathcal{X}'_1 \cup \ldots \cup \mathcal{X}'_r$ with $\pi(\mathcal{X} \setminus \mathcal{Y}) = \pi(\mathcal{X}'_1) \cup \ldots \cup \pi(\mathcal{X}'_r)$ a series decomposition. Let $\mathcal{R} = (\mathcal{X}_1 \cup \{\mathbf{v}_1\}, \ldots, \mathcal{X}_s \cup \{\mathbf{v}_s\})$ be a realization of $\mathcal{X}$. We proceed inductively to show $\mathcal{Y}_1 \cup \ldots \cup \mathcal{Y}_t \cup \mathcal{X}'_1 \cup \ldots \cup \mathcal{X}'_{r'}$ is a series decomposition, for $r' = 1, 2, \ldots, r$. We begin with the case $r' = 1$. Let $\mathcal{R}_C = (\pi(\mathcal{X}'_1) \cup \{\mathbf{c}_1\})$ be a realization of $\pi(\mathcal{X}'_1) \in \mathfrak{X}(G_0)$. Note each $\pi(\mathbf{x})$ with $\mathbf{x} \in \mathcal{X}'_1$ has trivial boundary, so $\vec{u}_{\pi(\mathbf{x})}$ is a tuple consisting of one element, which is a representative for the half-space $\pi(\mathbf{x})$, and $\partial(\mathbf{x}) \subseteq \ker \pi = \mathbb{R}^{\cup}\langle \mathcal{Y} \rangle$, in turn ensuring that $\pi(\vec{u}_{\mathbf{x}})$ is a tuple consisting of one element which is also a representative for the half-space $\pi(\mathbf{x})$ (by (VR1) for $\mathbf{x}$ when considered part of the realization $\mathcal{R}$). Thus $\pi(\vec{u}_{\mathbf{x}}) = \vec{u}_{\pi(\mathbf{x})}$ for $\mathbf{x} \in \mathcal{X}_1$. We aim to use Lemma 6.4 with the realizations $\mathcal{R}$ and $\mathcal{R}_C$ as given above, $\mathcal{A} = \mathcal{Y}$ and $\mathcal{R}_A = (\mathcal{Y}_1 \cup \{\mathbf{w}_1\}, \ldots, \mathcal{Y}_t \cup \{\mathbf{w}_t\})$ a realization of $\mathcal{Y} \in \mathfrak{X}(G_0)$. By Proposition 3.5.2, we have $\vec{u}_{\mathbf{c}_1} = \pi(\vec{u})$ for some limit $\vec{u}$ of an asymptotically filtered sequence of terms from G_0. Moreover, $\pi(\vec{u}^{\triangleleft}) = \pi(\vec{u})^{\triangleleft}$. Let $\vec{u} = (u_1, \ldots, u_\ell)$. Since $\partial(\{\mathbf{c}_1\}) = \emptyset$, we have $u_1, \ldots, u_{\ell-1} \in \ker \pi = \mathbb{R}^{\cup}\langle \mathcal{Y} \rangle$. Since $\vec{u}_{\mathbf{c}_1} = \pi(\vec{u})$ is fully unbounded (as $\mathcal{R}_C$ is purely virtual) with $\pi(\vec{u}^{\triangleleft}) = \pi(\vec{u})^{\triangleleft}$, it follows by Proposition 3.5.1 that $\vec{u}$ is also fully unbounded. Thus $\vec{u}^{\triangleleft}$ is either trivial or fully unbounded, and so $-\vec{u}^{\triangleleft}$ is encased by $\mathcal{Y}$ in view of Proposition 7.7.3 applied to the realization $\mathcal{R}_A$. But now we can apply Lemma 6.4 to conclude $\mathcal{Y}_1 \cup \ldots \cup \mathcal{Y}_t \cup \mathcal{X}'_1$ is a series decomposition (the virtual Reay system given by Lemma 6.4 will be purely virtual since both $\mathcal{R}_C$ and $\mathcal{R}_A$ are purely virtual by assumption). This completes the base case $r' = 1$. However, for $r' > 1$, the induction hypothesis gives that $\mathcal{Y}' := \mathcal{Y}_1 \cup \ldots \cup \mathcal{Y}_t \cup \mathcal{X}'_1 \cup \ldots \cup \mathcal{X}'_{r'-1}$ is a series decomposition of $\mathcal{Y}' \subseteq \mathcal{X}$, showing $\mathcal{Y}' \in \mathfrak{X}(G_0)$. Let $\pi' : \mathbb{R}^d \to \mathbb{R}^{\cup}\langle \mathcal{Y}' \rangle^{\perp}$ be the orthogonal projection. Since $\pi(\mathcal{X} \setminus \mathcal{Y}) = \pi(\mathcal{X}'_1) \cup \ldots \cup \pi(\mathcal{X}'_r)$ is a series decomposition, Proposition 6.3 and Proposition 5.4.1 imply that $\mathcal{X}^{\pi'} = \pi'(\mathcal{X} \setminus \mathcal{Y}') = \pi'(\mathcal{X}_{r'})$ is a series decomposition, and applying the base case with $\mathcal{Y}'$ in place of $\mathcal{Y}$ completes the induction and proof.

□

If $X \in X(G_0)$, then X has a series decomposition $X = X_1 \cup \ldots \cup X_s$ and associated realization $\mathcal{R} = (\mathcal{Y}_1 \cup \{\mathbf{v}_1\}, \ldots, \mathcal{Y}_s \cup \{\mathbf{v}_s\})$. If we replace each $x \in X$ with the one-dimensional half-space $\mathbf{x} = \mathbb{R}_+ x$ and let $\mathcal{X}$ and $\mathcal{X}_j$ for $j \in [1, s]$ be the resulting set of half-spaces represented by X and X_j, respectively, then per the discussion regarding the construction of $\mathcal{R}'$ at the beginning of Sect. 7.2, it follows that $\mathcal{X} \in \mathfrak{X}(G_0)$ with $\mathcal{X} = \mathcal{X}_1 \cup \ldots \cup \mathcal{X}_s$ a series decomposition and $\mathcal{X}$ a virtualization of X. This allows us to translate statements involving $\mathfrak{X}(G_0)$ to ones regarding $X(G_0)$ by replacing any series decomposition $X = X_1 \cup \ldots \cup X_s$ with the series decomposition $\mathcal{X} = \mathcal{X}_1 \cup \ldots \cup \mathcal{X}_s$, applying the appropriate result regarding $\mathfrak{X}(G_0)$ to $\mathcal{X}$, and then returning to $X(G_0)$ by using that each X_i is a set of representatives for the one-dimensional half-spaces from $\mathcal{X}_i$. We do this out concretely as an example in the next lemma.

Lemma 7.19 *Let $\Lambda \leq \mathbb{R}^d$ be a full rank lattice and let $G_0 \subseteq \Lambda$ be a finitary subset with $\mathsf{C}(G_0) = \mathbb{R}^d$. Suppose $X, Y \in X(G_0)$ with $Y \subseteq X$ and that $Y = Y_1 \cup \ldots \cup Y_t$*

is a series decomposition of Y. *Let* $\pi : \mathbb{R}^d \to \mathbb{R}\langle Y\rangle^\perp$ *be the orthogonal projection.*

1. *There is a series decomposition* $X = X_1 \cup \ldots \cup X_s$ *with* $s \geq t$ *and* $X_i = Y_i$ *for* $i \in [1, t]$.
2. $\pi(X \setminus Y) = \pi(X) \setminus \{0\} \in X(\pi(G_0))$.
3. *If* $X \setminus Y = X'_1 \cup \ldots \cup X'_r$ *with* $\pi(X \setminus Y) = \pi(X'_1) \cup \ldots \cup \pi(X'_r)$ *a series decomposition, then* $X = Y_1 \cup \ldots \cup Y_t \cup X'_1 \cup \ldots \cup X'_r$ *is a series decomposition.*

Proof Replace each element $x \in X$ with the half-space $\mathbf{x} = \mathbb{R}_+ x$ to define a new set $\mathcal{X}$. The subset $Y \subseteq X$ then defines a new set $\mathcal{Y} \subseteq \mathcal{X}$. Per the discussion regarding the construction of $\mathcal{R}'$ from the beginning of Sect. 7.2, we have $\mathcal{X}, \mathcal{Y} \in \mathfrak{X}(G_0)$ with X being a set of representatives for $\mathcal{X}$. Indeed, each subset $Y_i \subseteq Y$ defines a new set $\mathcal{Y}_i$ such that $\mathcal{Y} = \mathcal{Y}_1 \cup \ldots \cup \mathcal{Y}_t$ is a series decomposition of $\mathcal{Y}$ with each Y_i a set of representatives for $\mathcal{Y}_i$. Applying Lemma 7.18.1 to $\mathcal{Y} \subseteq \mathcal{X}$ using the series decomposition $\mathcal{Y} = \mathcal{Y}_1 \cup \ldots \cup \mathcal{Y}_t$, we find a series decomposition $\mathcal{X} = \mathcal{X}_1 \cup \ldots \cup \mathcal{X}_s$ with $s \geq t$ and $\mathcal{X}_i = \mathcal{Y}_i$ for $i \in [1, t]$. Since X is a set of representatives for the half-spaces from $\mathcal{X} = \mathcal{X}_1 \cup \ldots \cup \mathcal{X}_s$, we have a partition $X = X_1 \cup \ldots \cup X_s$ with each X_j a set of representatives for the half-spaces from $\mathcal{X}_j$, for $j \in [1, s]$. Thus $X = X_1 \cup \ldots \cup X_s$ is a series decomposition of X. Moreover, since $\mathcal{Y}_j = \mathcal{X}_j$ for $j \in [1, t]$ with $Y_j \subseteq Y \subseteq X$ the set of representatives for $\mathcal{Y}_j$, we have $Y_j = X_j$ for $j \in [1, t]$. This establishes Item 1. Let $\mathcal{R} = (\mathcal{X}_1 \cup \{\mathbf{v}_1\}, \ldots, \mathcal{X}_s \cup \{\mathbf{v}_s\})$ be a realization associated to the series decomposition $\mathcal{X} = \mathcal{X}_1 \cup \ldots \cup \mathcal{X}_s$. Then the realization $\pi(\mathcal{R}) = (\pi(\mathcal{X}_{t+1}) \cup \{\pi(\mathbf{v}_{t+1})\}, \ldots, \pi(\mathcal{X}_s) \cup \{\pi(\mathbf{v}_s)\})$ shows $\pi(\mathcal{X} \setminus \mathcal{Y}) \in \mathfrak{X}(G_0)$ with $\pi(X \setminus Y)$ as a set of representatives, which shows $\pi(X \setminus Y) \in X(G_0)$, giving Item 2. For item 3, each set $X'_j \subseteq X$ defines a set of half-spaces $\mathcal{X}'_j$ by replacing the elements of X'_j with the one-dimensional rays they define. As argued in Item 1, $\pi(\mathcal{X} \setminus Y) = \pi(\mathcal{X}'_1) \cup \ldots \cup \pi(\mathcal{X}'_r)$ is a series decomposition. Applying Lemma 7.18.3 using this decomposition, we find that $\mathcal{X} = \mathcal{Y}_1 \cup \ldots \cup \mathcal{Y}_t \cup \mathcal{X}'_1 \cup \ldots \cup \mathcal{X}'_r$ is a series decomposition, and since each Y_i is a set of representatives for $\mathcal{Y}_i$ and each X'_i is a set of representatives for $\mathcal{X}_i$, Item 3 follows. □

Let $G_0 \subseteq \Lambda \subseteq \mathbb{R}^d$ be a finitary set with $\mathsf{C}(G_0) = \mathbb{R}^d$ and suppose $\mathcal{X} \in \mathfrak{X}(G_0)$. Let $\mathcal{X} = \mathcal{X}_1 \cup \ldots \cup \mathcal{X}_s$ be a series decomposition of $\mathcal{X}$. This corresponds to a chain $\emptyset \subset \mathcal{X}_1 \subset (\mathcal{X}_1 \cup \mathcal{X}_2) \subset \ldots \subset (\mathcal{X}_1 \cup \ldots \cup \mathcal{X}_s) = \mathcal{X}$. A **refinement** of the series decomposition $\mathcal{X} = \mathcal{X}_1 \cup \ldots \cup \mathcal{X}_s$ is a series decomposition $\mathcal{X} = \mathcal{X}'_1 \cup \ldots \cup \mathcal{X}'_r$ such that each $\mathcal{X}_j = \mathcal{X}'_{t_{j-1}+1} \cup \ldots \cup \mathcal{X}'_{t_j}$ for some $0 = t_0 < t_1 < \ldots < t_s = r$. Equivalently, this means that the subsets $(\mathcal{X}_1 \cup \ldots \cup \mathcal{X}_j)$ for $j \in [0, s]$ each occur in the chain $\emptyset \subset \mathcal{X}'_1 \subset (\mathcal{X}'_1 \cup \mathcal{X}'_2) \subset \ldots \subset (\mathcal{X}'_1 \cup \ldots \cup \mathcal{X}'_r) = \mathcal{X}' = \mathcal{X}$. Such a refinement is **proper** if $r > s$, and the series decomposition $\mathcal{X} = \mathcal{X}_1 \cup \ldots \cup \mathcal{X}_s$ is called a **maximal** series decomposition if it has no proper refinements. We say that $\mathcal{X} \in \mathfrak{X}(G_0)$ is **irreducible** if no proper, nonempty subset of $\mathcal{X}$ lies in $\mathfrak{X}(G_0)$. Note this ensures that the only series decomposition of $\mathcal{X}$ is $\mathcal{X} = \mathcal{X}$, since if $\mathcal{X} = \mathcal{X}_1 \cup \mathcal{X}_2$ is a series decomposition, then $\mathcal{X}_1 \in \mathfrak{X}(G_0)$. Moreover, Lemma 7.18.1

ensures the converse also holds, meaning $\mathcal{X} \in \mathfrak{X}(G_0)$ is irreducible if and only if $\mathcal{X} = \mathcal{X}$ is the only series decomposition of $\mathcal{X}$. The above terms are used with the analogous definitions for $X \in X(G_0)$ as well. Using Lemma 7.18, we now characterize maximal series decompositions.

Proposition 7.20 *Let $\Lambda \leq \mathbb{R}^d$ be a full rank lattice, where $d \geq 0$, let $G_0 \subseteq \Lambda$ be a finitary set with $\mathsf{C}(G_0) = \mathbb{R}^d$, and let $\mathcal{X} = \mathcal{X}_1 \cup \ldots \cup \mathcal{X}_s$ be a series decomposition of $\mathcal{X} \in \mathfrak{X}(G_0)$. Then $\mathcal{X} = \mathcal{X}_1 \cup \ldots \cup \mathcal{X}_s$ is maximal if and only if each $\pi_{j-1}(\mathcal{X}_j)$ is irreducible for $j \in [1, s]$, where $\pi_{j-1} : \mathbb{R}^d \to \mathbb{R}^{\cup}\langle \mathcal{X}_1 \cup \ldots \cup \mathcal{X}_{j-1}\rangle^{\perp}$ is the orthogonal projection.*

Proof Suppose the series decomposition $\mathcal{X} = \mathcal{X}_1 \cup \ldots \cup \mathcal{X}_s$ is maximal but by contradiction there is some $j \in [1, s]$ such that $\pi_{j-1}(\mathcal{X}_j)$ is not irreducible. Then there is some series decomposition $\pi_{j-1}(\mathcal{X}_j) = \pi_{j-1}(\mathcal{Y}_1) \cup \ldots \cup \pi_{j-1}(\mathcal{Y}_t)$, where $\mathcal{X}_j = \mathcal{Y}_1 \cup \ldots \cup \mathcal{Y}_t$ with $t \geq 2$. By Proposition 6.3 (applied to a realization $\mathcal{R}$ of $\mathcal{X} = \mathcal{X}_1 \cup \ldots \cup \mathcal{X}_s$), it follows that $\pi_{j-1}(\mathcal{X}_j) \cup \ldots \cup \pi_{j-1}(\mathcal{X}_s)$ and $\pi_j(\mathcal{X}_{j+1}) \cup \ldots \cup \pi_j(\mathcal{X}_s)$ are series decompositions. Using these decomposition and applying Lemma 7.18.3 in $\pi_{j-1}(G_0) \subseteq \pi_{j-1}(\Lambda)$ taking $\pi_{j-1}(\mathcal{X}) = \pi_{j-1}(\mathcal{X}_j) \cup \ldots \cup \pi_{j-1}(\mathcal{X}_s)$ to be $\mathcal{X}$, taking $\pi_{j-1}(\mathcal{X}_j)$ to be $\mathcal{Y}$, and taking π to be π_j, we conclude that

$$\begin{aligned}\pi_{j-1}(\mathcal{X} \setminus (\mathcal{X}_1 \cup \ldots \cup \mathcal{X}_{j-1})) = \pi_{j-1}(\mathcal{Y}_1) \cup \ldots \cup \pi_{j-1}(\mathcal{Y}_t) \\ \cup \pi_{j-1}(\mathcal{X}_{j+1}) \cup \ldots \cup \pi_{j-1}(\mathcal{X}_s)\end{aligned}$$

is a series decomposition, and now a second application of Lemma 7.18.3 yields that $\mathcal{X} = \mathcal{X}_1 \cup \ldots \cup \mathcal{X}_{j-1} \cup \mathcal{Y}_1 \cup \ldots \cup \mathcal{Y}_t \cup \mathcal{X}_{j+1} \cup \ldots \cup \mathcal{X}_s$ is a series decomposition, contradicting that $\mathcal{X} = \mathcal{X}_1 \cup \ldots \cup \mathcal{X}_s$ was maximal in view of $t \geq 2$.

Conversely, now suppose each $\pi_{j-1}(\mathcal{X}_j)$ is irreducible but by contradiction $\mathcal{X} = \mathcal{X}_1 \cup \ldots \cup \mathcal{X}_s$ is not maximal. Then there exists a proper refinement $\mathcal{X} = \mathcal{Y}_1 \cup \ldots \cup \mathcal{Y}_r$. Let $j \in [1, s]$ be the minimal index such that $\mathcal{X}_j \neq \mathcal{Y}_j$, so $\mathcal{X}_i = \mathcal{Y}_i$ for $i \in [1, j-1]$ and $\mathcal{X}_j = \mathcal{Y}_j \cup \ldots \cup \mathcal{Y}_{j+t}$ with $t \geq 1$. Applying Proposition 6.3 to a realization of the series decomposition $\mathcal{Y}_1 \cup \ldots \cup \mathcal{Y}_{j+t}$, we conclude that $\pi_{j-1}(\mathcal{X}_j) = \pi_{j-1}(\mathcal{Y}_j) \cup \ldots \cup \pi_{j-1}(\mathcal{Y}_{j+t})$ is a series decomposition, contradicting that $\pi_{j-1}(\mathcal{X}_j)$ is irreducible and $t \geq 1$. □

Definition 7.21 Let $\Lambda \leq \mathbb{R}^d$ be a full rank lattice, where $d \geq 0$, let $G_0 \subseteq \Lambda$, let $\mathcal{R} = (\mathcal{X}_1 \cup \{\mathbf{v}_1\}, \ldots, \mathcal{X}_s \cup \{\mathbf{v}_s\})$ be an anchored virtual Reay system over G_0, and let $\pi_{\mathbf{x}} : \mathbb{R}^d \to \mathbb{R}^{\cup}\langle \partial(\{\mathbf{x}\})\rangle^{\perp}$ be the orthogonal projection, for $\mathbf{x} \in \mathcal{X} = \mathcal{X}_1 \cup \ldots \cup \mathcal{X}_s$. We say that $\mathcal{R}$ is **Δ-pure** if $\mathbf{x}(i_{\mathbf{x}}) = \tilde{\mathbf{x}}(i_{\mathbf{x}})$ with $\pi_{\mathbf{x}}(\mathbf{x}(i_{\mathbf{x}}))$ constant and $\mathbb{Z}\langle X(k)\rangle = \Delta$, for all tuples $k = (i_{\mathbf{x}})_{\mathbf{x} \in \mathcal{X}}$ and $\mathbf{x} \in \mathcal{X}$.

Let $\Lambda \leq \mathbb{R}^d$ be a full rank lattice, where $d \geq 0$, let $G_0 \subseteq \Lambda$, and let $\mathcal{R} = (\mathcal{X}_1 \cup \{\mathbf{v}_1\}, \ldots, \mathcal{X}_s \cup \{\mathbf{v}_s\})$ be an anchored virtual Reay system over G_0. Set $\mathcal{X} = \mathcal{X}_1 \cup \ldots \cup \mathcal{X}_s$. Then Proposition 6.8 ensures that $\mathbf{x}(i_{\mathbf{x}}) = \tilde{\mathbf{x}}(i_{\mathbf{x}})$ with $\pi_{\mathbf{x}}(\mathbf{x}(i_{\mathbf{x}}))$ constant, for any $\mathbf{x} \in \mathcal{X}$, once $i_{\mathbf{x}}$ is sufficiently large, so the first two conditions in the definition of Δ-pure can always be achieved by discarding the first few terms in each representative sequence $\{\mathbf{x}(i_\mathbf{x})\}_{i=1}^{\infty}$. Assume this is the case. For $j \in [0, s]$, let

$$\mathcal{E}_j = \mathbb{R}\langle \mathcal{X}_1 \cup \ldots \cup \mathcal{X}_j \rangle = \mathbb{R}\langle (X_1 \cup \ldots \cup X_j)(k) \rangle$$

and let $\pi_j : \mathbb{R}^d \to \mathcal{E}_j^\perp$ be the orthogonal projection. Suppose $\mathbf{x} \in \mathcal{X}_j$. Then, since $\pi \mathbf{x}(i_\mathbf{x})$ is constant with $\partial(\{\mathbf{x}\}) \subseteq \mathcal{X}_1 \cup \ldots \cup \mathcal{X}_{j-1}$, it follows that $\pi_{j-1}(\mathbf{x}(i_\mathbf{x}))$ is also constant, say $\pi_{j-1}(\mathbf{x}(i_\mathbf{x})) = \pi_{j-1}(y)$ for all $i_\mathbf{x}$, for some $y \in G_0 \subseteq \Lambda$. This ensures that we have subsets $Y_1, \ldots, Y_s \subseteq G_0 \subseteq \Lambda$ such that each $\pi_{j-1}(X_j(k)) = \pi_{j-1}(Y_j)$ is constant and independent of the tuple $k = (i_\mathbf{x})_{i_\mathbf{x} \in \mathcal{X}}$ with $|Y_j| = |\pi_{j-1}(Y_j)| = |\mathcal{X}_j|$, for $j \in [1, s]$. By definition of an ordinary Reay system, each $\pi_{j-1}(Y_j) = \pi_{j-1}(X_j(k))$, for $j \in [1, s]$, is a linearly independent set of size $|Y_j|$, ensuring that $Y = Y_1 \cup \ldots \cup Y_s$ is linearly independent.

Now consider arbitrary subsets $X_1, \ldots, X_s \subseteq \Lambda$ with $\pi_{j-1}(X_j) = \pi_{j-1}(Y_j)$ and $|X_j| = |\pi_{j-1}(X_j)| = |Y_j|$ for all $j \in [1, s]$, and set $X = X_1 \cup \ldots \cup X_s$. Since π_0 is the identity map, this forces $X_1 = Y_1$, and a short inductive argument then gives

$$\mathbb{R}\langle X_1 \cup \ldots \cup X_j \rangle = \mathbb{R}\langle Y_1 \cup \ldots \cup Y_j \rangle = \mathcal{E}_j \quad \text{for all } j \in [0, s].$$

Thus, since each $\pi_{j-1}(X_j) = \pi_{j-1}(Y_j)$ is a linearly independent set of size $|X_j|$, it follows that $X = X_1 \cup \ldots \cup X_s$ is linearly independent. For $j \in [0, s]$, let

$$\Delta_j = \mathbb{Z}\langle X_1 \cup \ldots \cup X_j \rangle,$$

which is a full rank lattice in $\mathcal{E}_j = \mathbb{R}\langle X_1 \cup \ldots \cup X_j \rangle = \mathbb{R}\langle \Delta_j \rangle$ (as X is linearly independent). Since $\Lambda \cap \mathcal{E}_j$ is also a full rank lattice in $\mathcal{E}_j$ with $\Delta_j = \mathbb{Z}\langle X_1 \cup \ldots \cup X_j \rangle \leq \Lambda \cap \mathcal{E}_j$, it follows that $(\Lambda \cap \mathcal{E}_j)/\Delta_j$ is a finite abelian group, for $j \in [1, s]$.

Let $j \in [1, s]$ be arbitrary, and let $x_1, \ldots, x_r \in X_j$ and $y_1, \ldots, y_r \in Y_j$ be the distinct elements of X_j and Y_j indexed so that $\pi_{j-1}(x_i) = \pi_{j-1}(y_i)$ for all $i \in [1, r]$. For each $i \in [1, r]$, we have $x_i = y_i + \xi_{x_i}$ for some $\xi_i := \xi_{x_i} \in \ker \pi_{j-1} = \mathcal{E}_{j-1}$, and since $x_i, y_i \in \Lambda$, it follows that $\xi_i \in \Lambda \cap \mathcal{E}_{j-1}$. We now have

$$\Delta_j = \Delta_{j-1} + \mathbb{Z}\langle X_j \rangle = \Delta_{j-1} + \mathbb{Z}\langle y_1 + \xi_1, \ldots, y_r + \xi_r \rangle. \tag{7.13}$$

From (7.13), we see that lattice Δ_j is completely determined by the value of the sublattice Δ_{j-1} as well as the values $\xi_i \mod \Delta_{j-1}$ for $i = 1, \ldots, r$. (Recall that the set $Y = Y_1 \ldots \cup Y_s$, and thus also the elements $y_1, \ldots, y_r \in Y_j$, are fixed). Moreover, assuming Δ_{j-1} is fixed, then distinct possibilities for the values of $\xi_1, \ldots, \xi_r$ modulo Δ_{j-1} yield distinct lattices Δ_j, which can be seen by noting that $y_i + (\xi_i + \Delta_{j-1}) \subseteq \Delta_j$ is precisely the subset of all elements $z \in \Delta_j$ with $\pi_{j-1}(z) = \pi_{j-1}(y_i)$ (in view of $\pi_{j-1}(X_j) = \pi_{j-1}(Y_j)$ being a linearly independent set of size $|X_j| = |Y_j|$). Note, if we apply the above setup with X taken to be Y, then $\xi_y = 0$ for all $y \in Y$, meaning $\xi_x \equiv 0 \mod \Delta_{j-1}$ for all $x \in X_j$ whenever $\Delta_j = \mathbb{Z}\langle X_1 \cup \ldots \cup X_j \rangle = \mathbb{Z}\langle Y_1 \cup \ldots \cup Y_j \rangle$ for all $j \in [1, s]$.

Since $(\Lambda \cap \mathcal{E}_{j-1})/\Delta_{j-1}$ is a finite group, there are only a finite number of possibilities for the value of each $\xi_i \in \Lambda \cap \mathcal{E}_{j-1}$ modulo Δ_{j-1}. Thus an inductive argument on $j = 1, \ldots, s$ shows that there are only a finite number of possibilities for each Δ_j. Indeed, since $X_1 = Y_1$, there is only one possibility for Δ_1, completing the base of the induction, while there are only a finite number of possibilities for each Δ_j with Δ_{j-1} fixed, one for each choice of the values of $\xi_1, \ldots, \xi_r$ modulo Δ_{j-1}, with only a finite number of possibilities for Δ_{j-1} by induction hypothesis, leading to only a finite number of possibilities for Δ_j.

All of the above basic observations will be crucial to several of our later arguments, including the following lemma showing how a pure virtual Reay system can be obtained from a given virtual Reay system by passing to appropriate subsequences of the representative sequences.

Lemma 7.22 *Let $\Lambda \leq \mathbb{R}^d$ be a full rank lattice, where $d \geq 0$, let $G_0 \subseteq \Lambda$, let $\mathcal{R} = (\mathcal{X}_1 \cup \{v_1\}, \ldots \mathcal{X}_s \cup \{v_s\})$ be an anchored virtual Reay system over G_0, let $\mathcal{X} = \bigcup_{i=1}^{s} \mathcal{X}_i$. Assume (by Proposition 6.8) that the first few terms in each representative sequence $\{\mathbf{x}(i)\}_{i=1}^{\infty}$ have been discarded so that $\mathbf{x}(i) = \tilde{\mathbf{x}}(i) \in G_0$ is constant modulo $\mathbb{R}\langle\partial(\{\mathbf{x}\})\rangle$ for all $\mathbf{x} \in \mathcal{X} = \mathcal{X}_1 \cup \ldots \cup \mathcal{X}_s$ and $i \geq 1$.*

1. *There are only a finite number of possibilities for $\mathbb{Z}\langle X(k)\rangle$ as we range over all tuples $k = (i_{\mathbf{x}})_{\mathbf{x} \in \mathcal{X}}$.*
2. *If $\Delta = \mathbb{Z}\langle X(\kappa)\rangle$ for some tuple $\kappa = (\iota_{\mathbf{x}})_{\mathbf{x} \in \mathcal{X}}$, then either*
   - *(a) there is a finite set of indices I_Δ such that every tuple $k = (i_{\mathbf{x}})_{\mathbf{x} \in \mathcal{X}}$ with $\Delta = \mathbb{Z}\langle X(k)\rangle$ has $i_{\mathbf{x}} \in I_\Delta$ for some $\mathbf{x} \in \mathcal{X}$, or*
   - *(b) replacing each representative sequence $\{\mathbf{x}(i_{\mathbf{x}})\}_{\mathbf{x} \in \mathcal{X}}$ by an appropriate subsequence, for $\mathbf{x} \in \mathcal{X}$, we can obtain $\mathbb{Z}\langle X(k)\rangle = \Delta$ for all tuples $k = (i_{\mathbf{x}})_{\mathbf{x} \in \mathcal{X}}$,*

   with (b) holding for at least one lattice Δ.

In particular, by discarding the first few terms in each representative sequence $\{\mathbf{x}(i)\}_{i=1}^{\infty}$, for $\mathbf{x} \in \mathcal{X}$, it follows that $\mathcal{R}$ can be made Δ-pure, for any $\Delta = \mathbb{Z}\langle X(k)\rangle$, by replacing each representative sequence $\{\mathbf{x}(i)\}_{i=1}^{\infty}$ by an appropriate subsequence.

Proof Let $Y = Y_1 \cup \ldots \cup Y_s$ with each $Y_j = X_j(\kappa)$, where $\kappa = (\iota_{\mathbf{x}})_{\mathbf{x} \in \mathcal{X}}$ is an arbitrary tuple. Per the discussion above Lemma 7.22 (applied taking X to be $X(k)$ for an arbitrary tuple k), for each $j \in [1, s]$, there are only a finite number of possibilities for $\mathbb{Z}\langle(X_1 \cup \ldots \cup X_j)(k)\rangle$ as we range over all tuples k, yielding Item 1. If $s = 1$, then $\partial(\mathbf{x}) = \{0\}$ for $\mathbf{x} \in \mathcal{X}$, so each $\mathbf{x}(i_{\mathbf{x}})$ with $\mathbf{x} \in \mathcal{X}$ is constant (cf. Proposition 6.8), in which case the lemma holds trivially. Therefore we can assume $s \geq 2$ and proceed by induction on s. Let $\mathbf{x}_1, \ldots, \mathbf{x}_r \in \mathcal{X}_s$ be the distinct half-spaces from $\mathcal{X}_s$, adapt the abbreviations $i_j := i_{\mathbf{x}_j}$ and $\iota_j := \iota_{\mathbf{x}_j}$, and set $y_j = \mathbf{x}_j(\iota_j)$ for $j \in [1, r]$, so $Y_s = \{y_1, \ldots, y_r\}$. Since $\mathbf{x}_j(i_j)$ is constant modulo $\mathbb{R}^{\cup}\langle\partial(\{\mathbf{x}_j\})\rangle \subseteq \mathbb{R}^{\cup}\langle\mathcal{X}_1 \cup \ldots \cup \mathcal{X}_{s-1}\rangle$, we have $\mathbf{x}_j(i_j) - y_j \in \mathbb{R}^{\cup}\langle\mathcal{X}_1 \cup \ldots \cup \mathcal{X}_{s-1}\rangle = \mathbb{R}\langle(X_1 \cup \ldots \cup X_{s-1})(k)\rangle = \mathbb{R}\langle\Delta'(k)\rangle$ for all $j \in [1, r]$ and $i_j \geq 1$.

Let $\Delta = \mathbb{Z}\langle X(\kappa)\rangle$. For a more general tuple $k = (i_{\mathbf{x}})_{i_{\mathbf{x}} \in \mathcal{X}}$, let $\Delta(k) = \mathbb{Z}\langle X(k)\rangle$ and $\Delta'(k) = \mathbb{Z}\langle (X_1 \cup \ldots \cup X_{s-1})(k)\rangle$. For $j \in [1, r]$, let

$$\xi_j(k) = \xi_j(i_j) = \mathbf{x}_j(i_j) - y_j \in \Lambda \cap \mathbb{R}\langle \Delta'(k)\rangle.$$

If there is a finite set of indices I_Δ such that all tuples $k = (i_{\mathbf{x}})_{\mathbf{x} \in \mathcal{X}}$ with $\mathbb{Z}\langle X(k)\rangle = \Delta$ have $i_{\mathbf{x}} \in I_\Delta$ for some $\mathbf{x} \in \mathcal{X}$, then (a) holds and the induction is complete (note this cannot happen for every one of the finite number of possibilities for Δ, by Item 1, as that would mean we could destroy all tuples $k = (i_{\mathbf{x}})_{\mathbf{x} \in \mathcal{X}}$ by only removing a finite number of terms from the $\{\mathbf{x}(i_{\mathbf{x}})\}_{i_{\mathbf{x}}=1}^{\infty}$ with $\mathbf{x} \in \mathcal{X}$, which is absurd). So we can assume otherwise. As we range over all tuples $k = (i_{\mathbf{x}})_{\mathbf{x} \in \mathcal{X}}$ with $\Delta(k) = \Delta$, there are only a finite number of possibilities for the lattices $\Delta'(k)$ (by Item 1). If $k = (i_{\mathbf{x}})_{\mathbf{x} \in \mathcal{X}}$ and $k' = (i'_{\mathbf{x}})_{\mathbf{x} \in \mathcal{X}}$ are two tuples with $\Delta(k) = \Delta(k') = \Delta$ and $\Delta'(k) = \Delta'(k') = \Delta'$, then the comments after (7.13) ensure that

$$\mathbf{x}_j(i_j) - y_j = \xi_j(k) \equiv \xi_j(k') = \mathbf{x}_j(i'_j) - y_j \mod \Delta' \quad \text{for all } j \in [1, r].$$

Indeed, the lattice $\Delta(k)$ is completely determined by $\Delta'(k)$ and the values of $\xi_1(k), \ldots, \xi_r(k)$ modulo $\Delta'(k)$, with different values of $\xi_j(k) \mod \Delta'(k)$ giving rise to different lattices $\Delta(k)$, so there are fixed $\xi_1, \ldots, \xi_r \in \Lambda \cap \mathbb{R}\langle \Delta'\rangle$ such that a tuple $k = (i_{\mathbf{x}})_{\mathbf{x} \in \mathcal{X}}$ has $\Delta(k) = \Delta$ and $\Delta'(k) = \Delta'$ precisely when $\Delta'(k') = \Delta'$ and $\xi_j(k) \equiv \xi_j \mod \Delta'$ for all $j \in [1, r]$, where $k' = (i_{\mathbf{x}})_{\mathbf{x} \in \mathcal{X} \setminus \mathcal{X}_s}$ is the restriction of the tuple k to $\mathcal{X} \setminus \mathcal{X}_s$. Consequently, if we let $\mathcal{I}_{\Delta'}$ consist of all tuples $k' = (i_{\mathbf{x}})_{\mathbf{x} \in \mathcal{X} \setminus \mathcal{X}_s}$ with $\mathbb{Z}\langle (X \setminus X_s)(k')\rangle = \Delta'$ and, for each $j \in [1, r]$, let I_j consist of all indices i_j with $\xi_j(i_j) \equiv \xi_j \mod \Delta'$, then

$$\{k = (i_{\mathbf{x}})_{\mathbf{x} \in \mathcal{X}} : \ (i_{\mathbf{x}})_{\mathcal{X} \setminus \mathcal{X}_s} \in \mathcal{I}_{\Delta'} \text{ and } i_j \in I_j \text{ for all } j \in [1, r]\} = \mathcal{I}_{\Delta'} \times I_1 \times \ldots \times I_r \tag{7.14}$$

is the set of all tuples k with $\Delta(k) = \Delta$ and $\Delta'(k) = \Delta'$.

As we range over all tuples k with $\Delta(k) = \Delta$, there are only a finite number of possibilities for $\Delta'(k)$. If, for some possibility Δ', it is possible to remove a finite number of terms from each $\{\mathbf{x}(i_{\mathbf{x}})\}_{i_{\mathbf{x}}=1}^{\infty}$ with $\mathbf{x} \in \mathcal{X}$ and destroy all occurrences where $\Delta'(k) = \Delta'$ from among tuples the remaining tuples k, then do so (replacing each representative sequence $\{\mathbf{x}(i_{\mathbf{x}})\}_{i_{\mathbf{x}}=1}^{\infty}$ for $\mathbf{x} \in \mathcal{X}$ with an appropriate subsequence of sufficiently large indexed terms). Thus we may w.l.o.g. assume this is not possible for every $\Delta'(k)$ that occurs among tuples k with $\Delta(k) = \Delta$. Note, we cannot have destroyed all such lattices $\Delta'(k)$ by such a procedure, as this would contradict our assumption about there not existing a finite set of indices I_Δ such that every $k = (i_{\mathbf{x}})_{\mathbf{x} \in \mathcal{X}}$ with $\Delta(k) = \Delta$ has $i_{\mathbf{x}} \in I_\Delta$ for some $\mathbf{x} \in \mathcal{X}$. Let Δ' be one possible lattice $\Delta'(k)$ that has survived. Then we can apply the induction hypothesis to $(X_1 \cup \ldots \cup X_{s-1})(k)$ allowing us to replace each $\{\mathbf{x}(i_{\mathbf{x}})\}_{i_{\mathbf{x}}=1}^{\infty}$ for $\mathbf{x} \in \mathcal{X} \setminus \mathcal{X}_s$ with appropriate subsequences resulting in $\Delta'(k) = \Delta'$ for all tuples k. Each I_j is infinite, else removing all indices from the finite set I_j would destroy all tuples k with $\Delta'(k) = \Delta'$ and $\Delta(k) = \Delta$, contrary to assumption. But now, in view of

(7.14), we can replace each $\{\mathbf{x}_j(i_j)\}_{i_j=1}^{\infty}$ with the infinite subsequence $\{\mathbf{x}_j(i_j)\}_{i_j\in I_j}$, for $j\in[1,r]$, and thereby attain (b), which completes the induction and proof. □

Let $\Lambda\leq\mathbb{R}^d$ be a full rank lattice and let $G_0\subseteq\Lambda$ be a finitary set with $\mathsf{C}(G_0)=\mathbb{R}^d$. Let $\mathcal{X}\in\mathfrak{X}(G_0)$ with $\mathcal{R}=(\mathcal{X}_1\cup\{\mathbf{v}_1\},\ldots,\mathcal{X}_s\cup\{\mathbf{v}_s\})$ a purely virtual Reay system over G_0 realizing $\mathcal{X}$. Then Lemma 7.22 ensures that, assuming $\mathcal{X}$ is fixed, there are only a finite number of lattices $\Delta=\mathbb{Z}\langle X(k)\rangle\in\mathfrak{P}_{\mathbb{Z}}(G_0)$ as we range over all tuples $k=(i_{\mathbf{x}})_{\mathbf{x}\in\mathcal{X}}$ with each $i_{\mathbf{x}}$ sufficiently large, and by refining the representative sequences $\{\mathbf{x}(i_{\mathbf{x}})\}_{\mathbf{x}\in\mathcal{X}}$, it can be assumed via Lemma 7.22 that $\mathcal{R}$ is Δ-pure, for any $\Delta=\mathbb{Z}\langle X(k)\rangle$ with $k=(i_{\mathbf{x}})_{\mathbf{x}\in\mathcal{X}}$ so long as all $i_{\mathbf{x}}$ are sufficiently large. We will see below that $\mathfrak{P}_{\mathbb{Z}}(G_0)$ is finite without need to restrict to a fixed $\mathcal{X}\in\mathfrak{X}(G_0)$. If $\{\mathbf{x}(i_{\mathbf{x}})\}_{i_{\mathbf{x}}=1}^{\infty}$ is a representative sequence for $\mathbf{x}\in\mathcal{X}$ in $\mathcal{R}$, then Proposition 5.11 ensures it can also be substituted for a representative sequence for $\mathbf{x}$ in any other realization $\mathcal{R}'$ of $\mathcal{X}$. Thus the possibilities for $\Delta=\mathbb{Z}\langle X(k)\rangle$ depend only on $\mathcal{X}$ and not on the particular series decomposition $\mathcal{X}=\bigcup_{i=1}^{s}\mathcal{X}_i$ nor realization $\mathcal{R}$, and if $\mathcal{X}=\mathcal{X}_1\cup\ldots\cup\mathcal{X}_s$ is a series decomposition having a realization that is Δ-pure, then any series decomposition of $\mathcal{X}$ has a realization that is Δ-pure. Furthermore, if $\Delta=\mathbb{Z}\langle X\rangle\in\mathfrak{P}_{\mathbb{Z}}(G_0)$, where $X\in X(G_0)$, then per earlier discussions, there is a realization $\mathcal{R}=(\mathcal{X}_1\cup\{\mathbf{v}_1\},\ldots,\mathcal{X}_s\cup\{\mathbf{v}_s\})$ with each $\mathbf{x}(i)\in X\subseteq G_0$ constant, for $\mathbf{x}\in\mathcal{X}:=\mathcal{X}_1\cup\ldots\cup\mathcal{X}_s$, ensuring there is a virtualization $\mathcal{X}$ of X having a Δ-pure realization.

Definition 7.23 For $\mathcal{X}\in\mathfrak{X}(G_0)$, let $\mathfrak{P}_{\mathbb{Z}}(\mathcal{X})$ denote all $\Delta\in\mathfrak{P}_{\mathbb{Z}}(G_0)$ for which $\mathcal{X}$ has a Δ-pure realization. This is independent of series decomposition, and every $\Delta\in\mathfrak{P}_{\mathbb{Z}}(G_0)$ has $\Delta\in\mathfrak{P}_{\mathbb{Z}}(\mathcal{X})$ for some $\mathcal{X}\in\mathfrak{X}(G_0)$, as explained above. We let $\mathfrak{X}(G_0,\Delta)\subseteq\mathfrak{X}(G_0)$ consist of all $\mathcal{X}\in\mathfrak{X}(G_0)$ with $\Delta\in\mathfrak{P}_{\mathbb{Z}}(\mathcal{X})$, and we let $X(G_0,\Delta)$ consist of all $X\in X(G_0)$ with $\mathbb{Z}\langle X\rangle=\Delta$, where $\Delta\in\mathfrak{P}_{\mathbb{Z}}(G_0)$.

7.3 Finiteness Properties of Finitary Sets

The sets $X(G_0)$, $\mathfrak{X}(G_0)$ and $\mathfrak{X}(G_0,\Delta)$ may be infinite. Our next goal is to show that, nonetheless, they still exhibit some finite-like behavior. Theorems 7.24, 7.29 and 7.32 contain some of the important finite-like properties possessed by a finitary set, explaining the choice of name. We begin with Theorem 7.24, which contains the essential finitary property of $\mathfrak{X}(G_0)$ and $X(G_0)$ as well as the finiteness of $\mathfrak{P}_{\mathbb{Z}}(G_0)$ and $\mathfrak{P}_{\mathbb{R}}(G_0)$.

Theorem 7.24 *Let $\Lambda\subseteq\mathbb{R}^d$ be a full rank lattice, where $d\geq0$, and let $G_0\subseteq\Lambda$ be a finitary subset with $\mathsf{C}(G_0)=\mathbb{R}^d$.*

1. *There are only a finite number of irreducible sets $\mathcal{X}\in\mathfrak{X}(G_0)$ and $X\in X(G_0)$.*
2. *$\mathfrak{P}_{\mathbb{Z}}(G_0)$ and $\mathfrak{P}_{\mathbb{R}}(G_0)$ are both finite.*

Proof

1. If $\mathcal{X} \in \mathfrak{X}(G_0)$ is irreducible, then any realization $\mathcal{R} = (\mathcal{X}_1 \cup \{\mathbf{v}_1\}, \ldots, \mathcal{X}_s \cup \{\mathbf{v}_s\})$ of $\mathcal{X}$ must have $s = 1$ with $\mathcal{X} = \mathcal{X}_1$, in which case every $\mathbf{x} \in \mathcal{X}$ is a one-dimensional half-space spanned by any representative. Thus to show there are only a finite number of irreducible sets $\mathcal{X} \in \mathfrak{X}(G_0)$, it suffices to show there are only a finite number of irreducible sets $X \in X(G_0)$. Assume by contradiction that this fails and let $\{X_i\}_{i=1}^{\infty}$ be a sequence of distinct irreducible sets $X_i \in X(G_0)$. Now $X \in X(G_0)$ being irreducible implies there exists an unbounded limit $u \in G_0^{\infty}$ of a radially convergent sequence of terms from G_0 such that X minimally encases $-u$, which is equivalent to $X \cup \{u\}$ being a minimal positive basis. Since each X_i is linearly independent, we have $|X_i| \leq d$, and so, by passing to a subsequence of $\{X_i\}_{i=1}^{\infty}$, we can assume all X_i have the same size, say $|X_i| = s$ with $X_i = \{x_i^{(1)}, \ldots, x_i^{(s)}\}$ for all i. For each i, let $x_i^{(s+1)} \in G_0^{\infty}$ be such that $X_i \cup \{x_i^{(s+1)}\}$ is a minimal positive basis. By successively passing to an appropriate subsequence of each $\{x_i^{(j)}\}_{i=1}^{\infty}$, for $j = s+1, s, \ldots, 1$, we can w.l.o.g. assume each $\{x_i^{(j)}\}_{i=1}^{\infty}$, for $j \in [1, s+1]$, is a radially convergent sequence with limit (say) u_j. Since $x_i^{(s+1)} \in G_0^{\infty}$ for all i, it follows from Lemma 3.2 that $u_{s+1} \in G_0^{\infty}$. For $j \in [1, s]$, we have $x_i^{(j)} \in G_0 \subseteq \Lambda$ with Λ a lattice, so each sequence $\{x_i^{(j)}\}_{i=1}^{\infty}$ either eventually stabilizes to a nonzero constant value or else $u_j \in G_0^{\infty}$. Since the X_i are all distinct, the latter cannot occur for all $j \in [1, s]$.

   If $0 \notin \mathsf{C}^*(u_1, \ldots, u_s, u_{s+1})$, then Lemma 7.12 implies there is an open half-space $\mathcal{E}_+^{\circ}$ containing all u_j for $j \in [1, s+1]$. Consequently, since $x_i^{(j)}/\|x_i^{(j)}\| \to u_j$ for each $j \in [1, s+1]$, it follows that each $X_i \cup \{x_i^{(s+1)}\}$, with i sufficiently large, will also be contained in the open half-space $\mathcal{E}_+^{\circ}$, ensuring $0 \notin \mathsf{C}^*(X_i \cup \{x_i^{(s+1)}\})$. However, this contradicts that each $X_i \cup \{x_i^{(s+1)}\}$ is a minimal positive basis. Therefore we instead conclude that $0 \in \mathsf{C}^*(u_1, \ldots, u_s, u_{s+1})$. Thus there is some nonempty subset $J \subseteq [1, s+1]$ such that $U = \{u_j : j \in J\}$ is a minimal positive basis. If $u_j \notin G_0^{\infty}$ for all $j \in J$, then $J \subseteq [1, s]$ and $x_i^{(j)}/\|x_i^{(j)}\| = u_j$ for all sufficiently large i and $j \in J$. In such case, $\{x_i^{(j)} : j \in J\} \subseteq X_i$ will be a minimal positive basis for all sufficiently large i, contradicting that each $X_i \cup \{x_i^{(s+1)}\}$ is a minimal positive basis with $J \subseteq [1, s]$. Therefore we instead conclude that there is some $t \in J$ such that $u_t \in G_0^{\infty}$, and in view of G_0 being finitary, it then follows that $J \cap G_0^{\infty} = \{t\}$ (cf. the comments after Theorem 7.8). In particular, $J \setminus \{t\} \subseteq [1, s]$ as $u_{s+1} \in G_0^{\infty}$. But now $\{x_i^{(j)} : j \in J \setminus \{t\}\} \cup \{u_t\}$ is a minimal positive basis for all sufficiently large i with $u_t \in G_0^{\infty}$, implying $\{x_i^{(j)} : j \in J \setminus \{t\}\} \in X(G_0)$. However, since $\{x_i^{(j)} : j \in J \setminus \{t\}\} \subseteq X_i$ with X_i irreducible, this is only possible if $J \setminus \{t\} = [1, s]$ and $t = s+1$. Hence every $\{x_i^{(j)}\}_{i=1}^{\infty}$ for $j \in [1, s]$ eventually stabilizes, contrary to what we concluded in the previous paragraph. This establishes Item 1.

2. Since every subspace from $\mathfrak{P}_{\mathbb{R}}(G_0)$ is linearly spanned by a lattice from $\mathfrak{P}_{\mathbb{Z}}(G_0)$, it suffices to show $\mathfrak{P}_{\mathbb{Z}}(G_0)$ is finite. For $s \in [1, d]$, let $X_s(G_0) \subseteq X(G_0)$ consist of all $X \in X(G_0)$ having a maximal series decomposition of length s, say $X = X_1 \cup \ldots \cup X_s$. We proceed by induction on $s = 1, 2, \ldots, d$ to show that the number of lattices generated by subsets from $X_1(G_0) \cup \ldots \cup X_s(G_0)$ is finite. Note $X_1(G)$ is precisely the subset of all irreducible subsets $X \in X(G_0)$, so $X_1(G_0)$ is finite by Item 1, ensuring the number of lattices generated by subsets $X \in X_1(G_0)$ is also finite. Thus the base $s = 1$ of the induction is complete, and we assume $s \geq 2$.

   Let $X \in X_s(G_0)$ be arbitrary and let $X = X_1 \cup \ldots \cup X_s$ be a maximal series decomposition of X. Then $X \setminus X_s = X_1 \cup \ldots \cup X_{s-1}$ is a maximal series decomposition by Proposition 7.20. Let $\Delta' = \mathbb{Z}\langle X \setminus X_s \rangle$. By induction hypothesis, there are only a finite number of possibilities for the lattice Δ', so it suffices to show that there are only a finite number of lattices generated by $X' \in X_s(G_0)$ having a maximal series decomposition $X' = X'_1 \cup \ldots \cup X'_s$ with $\mathbb{Z}\langle X' \setminus X'_s \rangle = \Delta'$, for each possible Δ'. To this end, fix an arbitrary possible lattice Δ', let $\mathcal{E} = \mathbb{R}\langle \Delta' \rangle$, let $\pi : \mathbb{R}^d \to \mathcal{E}^\perp$ be the orthogonal projection. Proposition 7.9 implies $\pi(G_0)$ is a finitary subset of the lattice $\pi(\Lambda)$. Proposition 7.20 implies $\pi(X) \setminus \{0\} = \pi(X_s) \in X(\pi(G_0))$ is irreducible. As a result, Item 1 implies that there are only a finite number of possibilities for $\pi(X_s)$. Consequently, it suffices to show, for each possible irreducible set $\pi(Y_s) = \pi(X_s)$, that there are only a finite number of lattices generated by a set $X \in X(G_0)$ having a maximal series decomposition $X = X_1 \cup \ldots \cup X_s$ with $\mathbb{Z}\langle X_1 \cup \ldots \cup X_{s-1} \rangle = \Delta'$ and $\pi(X_s) = \pi(Y)$. However, this follows from the discussion given after (7.13), completing the proof.

□

Lemma 7.25 *Let $\mathcal{R} = (\mathcal{X}_1 \cup \{v_1\}, \ldots, \mathcal{X}_s \cup \{v_s\})$ be an oriented Reay system in $\mathbb{R}^d$, where $d \geq 0$, let $\mathcal{B} \subseteq \mathcal{X}_1 \cup \ldots \cup \mathcal{X}_s$, and let $\pi : \mathbb{R}^d \to \mathbb{R}^{\cup}\langle \mathcal{B} \rangle^\perp$ be the orthogonal projection. For $\mathbf{x} \in \mathcal{B}$, let $\mathcal{B}_{\mathbf{x}} = \mathcal{B} \setminus \{\mathbf{x}\} \cup \partial(\{\mathbf{x}\})$ and let $\pi_{\mathbf{x}} : \mathbb{R}^d \to \mathbb{R}^{\cup}\langle \mathcal{B}_{\mathbf{x}} \rangle^\perp$ be the orthogonal projection. Let $\{x_i\}_{i=1}^\infty$ be an asymptotically filtered sequence of terms $x_i \in \mathbb{R}^d$ with fully unbounded limit $\vec{u} = (u_1, \ldots, u_t)$.*

1. *If $-\vec{u}$ is minimally encased by $\mathcal{B}$, then $\|\pi_{\mathbf{x}}(x_i)\| \to \infty$ for all $\mathbf{x} \in \mathcal{B}$.*
2. *If $\vec{u}$ is a complete fully unbounded limit and $-\vec{u}$ is encased by $\mathcal{B}$ with $\{\pi_{\mathbf{x}}(x_i)\}_{i=1}^\infty$ unbounded for all $\mathbf{x} \in \mathcal{B}$, then $\mathcal{B}$ minimally encases $-\vec{u}$.*
3. *If $-\vec{u}$ is minimally encased by $\mathcal{B}$ and $\{y_i\}_{i=1}^\infty$ is a sequence of terms $y_i \in \mathbb{R}^d$ such that $\|y_i\| \in o(\|\pi_{\mathbf{x}}(x_i)\|)$ for all $\mathbf{x} \in \mathcal{B}$, then the sequence $\{x_i + y_i\}_{i=1}^\infty$ is an asymptotically filtered sequence with fully unbounded limit $(u_1, u_2, \ldots, u_{r_\ell})$ (after discarding the first few terms), where $1 = r_1 < \ldots < r_\ell < r_{\ell+1} = t + 1$ are the indices given by Proposition 5.9 regarding the minimal encasement of $-\vec{u}$ by $\mathcal{B}$.*

Proof Let $x_i = a_i^{(1)}u_1 + \ldots + a_i^{(t)}u_t + \varepsilon_i$ be the representation of $\{x_i\}_{i=1}^{\infty}$ as an asymptotically filtered sequence with fully unbounded limit $\vec{u}$, so $a_i^{(j)} \to \infty$ for all $j \in [1,t]$, $a_i^{(j)} \in o(a_i^{(j-1)})$ for all $j \in [2,t]$, and $\|\varepsilon_i\| \in o(a_i^{(t)})$.

1. Since $\mathcal{B} \subseteq X_1 \cup \ldots \cup X_s$ minimally encases $-\vec{u}$, it must do so urbanely and be a support set. Let $\mathbf{x} \in \mathcal{B}$ be arbitrary. Then $\mathcal{B}_{\mathbf{x}} \prec \mathcal{B} \subseteq X_1 \cup \ldots \cup X_s$ and $\pi_{\mathbf{x}}(x_i) = a_i^{(1)}\pi_{\mathbf{x}}(u_1) + \ldots + a_i^{(t)}\pi_{\mathbf{x}}(u_t) + \pi_{\mathbf{x}}(\varepsilon_i)$. Proposition 5.9.4 implies that $\mathcal{B}^{\pi_{\mathbf{x}}}$ minimally encases $-\pi_{\mathbf{x}}(\vec{u})$. If $\pi_{\mathbf{x}}(u_i) = 0$ for all $i \in [1,t]$, then $\pi_{\mathbf{x}}(\vec{u})$ is the empty tuple, in which case $\mathcal{B}^{\pi_{\mathbf{x}}}$ minimally encasing $-\pi_{\mathbf{x}}(\vec{u})$ is only possible if $\mathcal{B}^{\pi_{\mathbf{x}}} = \emptyset$. Hence $\mathbf{y} \in \mathbb{R}^{\cup}\langle \mathcal{B}_{\mathbf{x}}\rangle$ for all $\mathbf{y} \in \mathcal{B}$, whence Proposition 5.3.9 implies that $\mathcal{B} \subseteq \downarrow\mathcal{B}_{\mathbf{x}}$. However, since $\mathcal{B}$ is a support set, we have $\mathcal{B}^* = \mathcal{B}$, in which case $\mathbf{x} \in \mathcal{B} \setminus \downarrow\mathcal{B}_{\mathbf{x}}$, contradicting that $\mathcal{B} \subseteq \downarrow\mathcal{B}_{\mathbf{x}}$. Therefore we must instead have $\pi_{\mathbf{x}}(u_j) \neq 0$ for some $j \in [1,t]$, and we may assume $r \in [1,t]$ is the minimal index with $\pi_{\mathbf{x}}(u_r) \neq 0$. Since $a_i^{(j)} \in o(a_i^{(j-1)})$ for all $j \in [2,t]$ and $\|\varepsilon_i\| \in o(a_i^{(t)})$, we have $\|\pi_{\mathbf{x}}(x_i)\| \sim a_i^{(r)}\|\pi_{\mathbf{x}}(u_r)\|$, and thus $\|\pi_{\mathbf{x}}(x_i)\| \to \infty$ as $a_i^{(r)} \to \infty$, completing Item 1.
2. Since $\vec{u}$ is a *complete* fully unbounded limit, we have $\|\epsilon_i\|$ bounded. Since $\mathcal{B}$ encases $-\vec{u}$, it follows that there is some $\mathcal{A} \preceq \mathcal{B}$ that minimally encases $-\vec{u}$. Assume be contradiction that $\mathcal{A} \neq \mathcal{B}$. Then $\mathcal{A} \preceq \mathcal{B}_{\mathbf{x}}$ for some $\mathbf{x} \in \mathcal{B}$. Let $\tau : \mathbb{R}^d \to \mathbb{R}^{\cup}\langle \mathcal{A}\rangle^{\perp}$ be the orthogonal projection. Since $\mathcal{A}$ encases $-\vec{u}$, we have $\tau(u_j) = 0$ for all j, and thus $\pi_{\mathbf{x}}(u_j) = 0$ for all j (since $\mathcal{A} \preceq \mathcal{B}_{\mathbf{x}}$ implies $\mathcal{A} \subseteq \downarrow\mathcal{B}_{\mathbf{x}}$). But then $\{\pi_{\mathbf{x}}(x_i)\}_{i=1}^{\infty} = \{\pi_{\mathbf{x}}(\varepsilon_i)\}_{i=1}^{\infty}$ is a bounded sequence (as $\|\epsilon_i\|$ is bounded), contrary to hypothesis.
3. Since $\mathcal{B} \subseteq X_1 \cup \ldots \cup X_s$ minimally encases $-\vec{u}$, it is a support set (so $\mathcal{B}^* = \mathcal{B}$) and must do so urbanely, and since $\vec{u}$ is fully unbounded, this ensures that $\mathcal{B} \neq \emptyset$. Let

$$\emptyset = \mathcal{B}_0 \prec \mathcal{B}_1 \prec \ldots \prec \mathcal{B}_\ell = \mathcal{B}$$

be the support sets and let $1 = r_1 < \ldots < r_\ell < r_{\ell+1} = t+1$ be the indices given by Proposition 5.9. For $j \in [0,\ell]$, let $\pi_j : \mathbb{R}^d \to \mathbb{R}^{\cup}\langle \mathcal{B}_j\rangle^{\perp}$ be the orthogonal projection. For $j \in [1,\ell]$, let $\overline{u}_j = \pi_{j-1}(u_{r_j})/\|\pi_{j-1}(u_{r_j})\|$. Let $\mathcal{F} = (\mathbb{R}^{\cup}\langle \mathcal{B}_1\rangle, \ldots, \mathbb{R}^{\cup}\langle \mathcal{B}_\ell\rangle)$. In view of Propositions 5.9.3 and 4.17, let

$$x_i = (b_i^{(1)}\overline{u}_1 + w_i^{(1)}) + \ldots + (b_i^{(\ell)}\overline{u}_\ell + w_i^{(\ell)}) + \varepsilon_i'$$

be a representation of $\{x_i\}_{i=1}^{\infty}$ as an $\mathcal{F}$-filtered sequence, so $b_i^{(j)} \in \Theta(a_i^{(r_j)})$ for $j \in [1,r]$. Now $\pi_{\ell-1}(u_{r_\ell})/\|\pi_{\ell-1}(u_{r_\ell})\| = \overline{u}_\ell$ and $\mathsf{Supp}_{\pi_{\ell-1}(\mathcal{R})}(-\overline{u}_\ell) = \mathcal{B}^{\pi_{\ell-1}}$ (by Proposition 5.9.2). Thus $B^{\tau} \cup \{\tau(\overline{u}_\ell)\}$ is a minimal positive basis of size $|\mathcal{B}^{\pi_{\ell-1}}|+1$ (by Propositions 5.3.3 and 5.3.4), where $\tau : \mathbb{R}^d \to \mathbb{R}^{\cup}\langle \mathcal{B}_{\ell-1} \cup \partial(\mathcal{B})\rangle^{\perp}$ is the orthogonal projection, whence $\overline{u}_\ell \notin \mathbb{R}\langle \downarrow B \setminus \{x\}\rangle$ for any $\mathbf{x} \in \mathcal{B} \setminus \downarrow\mathcal{B}_{\ell-1}$. (Indeed, if we write $-\overline{u}_\ell$ as a linear combination of elements from $\downarrow B$ and apply τ to this linear combination, then we obtain a linear combination of $-\tau(\overline{u}_\ell)$ using the elements from $\tau(B)$, which, as $B^\tau \cup \{\tau(\overline{u}_\ell)\}$ is a minimal positive basis of size $|\mathcal{B}^{\pi_{\ell-1}}|+1$, is only possible if the coefficient of each element $x \in B$ is strictly positive.) Since $\mathcal{B}_{\ell-1} \prec \mathcal{B}_\ell = \mathcal{B} \neq \emptyset$, we have $\mathcal{B}_{\ell-1} \preceq \mathcal{B}_{\mathbf{x}}$ for some $\mathbf{x} \in \mathcal{B}$, and thus there exists some $\mathbf{x} \in \mathcal{B} \setminus \downarrow\mathcal{B}_{\ell-1}$ (as $\mathcal{B}^* = \mathcal{B}$). For this $\mathbf{x}$, we have $\pi_{\mathbf{x}}(\overline{u}_\ell) \neq 0$, for otherwise $\overline{u}_\ell \in \mathbb{R}^\cup\langle \mathcal{B}_{\mathbf{x}}\rangle = \mathbb{R}\langle \downarrow \mathcal{B}_{\mathbf{x}}\rangle \subseteq \mathbb{R}\langle \downarrow B \setminus \{x\}\rangle$, contrary to what was just noted. Since $\mathcal{B}_{\ell-1} \preceq \mathcal{B}_{\mathbf{x}}$ and $\pi_{\mathbf{x}}(\overline{u}_\ell) \neq 0$, we have

$$\pi_{\mathbf{x}}(x_i) = b_i^{(\ell)} \pi_{\mathbf{x}}(\overline{u}_\ell) + \pi_{\mathbf{x}}(w_i^{(\ell)}) + \pi_{\mathbf{x}}(\varepsilon_i'),$$

whence $\|\pi_{\mathbf{x}}(x_i)\| \in \Theta(b_i^{(\ell)}) = \Theta(a_i^{(r_\ell)})$ (as $\|\varepsilon_i'\|$, $\|w_i^{(\ell)}\| \in o(b_i^{(\ell)})$ in view of the $\mathcal{F}$-filtration representation for $\{x_i\}_{i=1}^\infty$). Thus the hypothesis $\|y_i\| \in o(\|\pi_{\mathbf{x}}(x_i)\|)$ implies that $\|y_i\| \in o(b_i^{(\ell)})$ and $\|y_i\| \in o(a_i^{(r_\ell)})$. It follows that $\{x_i + y_i\}_{i=1}^\infty$ is an asymptotically filtered sequence with fully unbounded limit $(u_1, \ldots, u_{r_\ell})$ (after discarding the first few terms), completing the proof.

□

Suppose $\mathcal{X}, \mathcal{X}' \in \mathfrak{X}(G_0)$. We define a partial order on $\mathfrak{X}(G_0)$ (and thus also on each $\mathfrak{X}(G_0, \Delta)$), by declaring

$$\mathcal{X} \preceq_\cup \mathcal{X}'$$

when the half-spaces from $\mathcal{X}$ are in bijective correspondence with the half-spaces from $\mathcal{X}'$, say with $\mathbf{x} \in \mathcal{X}$ corresponding to $\mathbf{x}' \in \mathcal{X}'$, such that, for every $\mathbf{x} \in \mathcal{X}$, we have $\mathbf{x} \subseteq \mathbf{x}'$ and $\mathbf{x}^\circ \subseteq (\mathbf{x}')^\circ$, the latter meaning any representative for $\mathbf{x}$ is also one for $\mathbf{x}'$. For $\mathcal{A} \subseteq \mathcal{X}$, let $\mathcal{A}' = \{\mathbf{x}' : \mathbf{x} \in \mathcal{A}\} \subseteq \mathcal{X}'$ denote the image of $\mathcal{A}$ under the bijection $\mathbf{x} \mapsto \mathbf{x}'$. The relation $\preceq_\cup$ is clearly transitive and reflexive. We first make some observations regarding the defining condition before giving the argument that $\preceq_\cup$ is anti-symmetric.

Suppose $\mathbf{x}, \mathbf{y} \in \mathcal{X}$ with $\mathbf{y} \subseteq \mathbf{x}$. Then $\mathbf{y} \subseteq \mathbf{x} \subseteq \mathbf{x}'$, so that $\mathbf{x}'$ contains a representative y for $\mathbf{y}$, and thus also one for $\mathbf{y}'$ as a representative for $\mathbf{y}$ is also one for $\mathbf{y}'$. If $\mathbf{y}' \notin \downarrow\mathbf{x}'$, then the representative y for $\mathbf{y}'$ would be linearly independent from the representatives $\downarrow x'$ for $\downarrow\mathbf{x}'$ (as any set of representatives for $\mathcal{X}'$ is linearly independent), contradicting that $y \in \mathbf{x}' \subseteq \mathbb{R}^\cup\langle \downarrow\mathbf{x}'\rangle = \mathbb{R}\langle \downarrow x'\rangle$. Therefore we must have $\mathbf{y}' \in \downarrow\mathbf{x}'$, i.e., $\mathbf{y}' \subseteq \mathbf{x}'$. Moreover, if $\mathbf{y} \subset \mathbf{x}$, then the injectivity of the map $\mathbf{z} \mapsto \mathbf{z}'$ ensures $\mathbf{y}' \neq \mathbf{x}'$. In summary, this shows that

$$\mathbf{y} \subset \mathbf{x} \quad \text{implies} \quad \mathbf{y}' \subset \mathbf{x}' \quad \text{for } \mathbf{x}, \mathbf{y} \in \mathcal{X}. \tag{7.15}$$

In particular, we have

$$(\downarrow\mathcal{A})' \subseteq \downarrow(\mathcal{A}') = \downarrow\mathcal{A}' \quad \text{and} \quad (\mathcal{A}')^* \subseteq (\mathcal{A}^*)' \quad \text{for } \mathcal{A} \subseteq \mathcal{X}. \tag{7.16}$$

To see the second inclusion in (7.16), note that $(\mathcal{A}')^*$ and $(\mathcal{A}^*)'$ are both subsets of $\mathcal{A}'$. Thus, if the inclusion were to fail, then there would be some $\mathbf{y} \in \mathcal{A} \setminus \mathcal{A}^*$ with

$\mathbf{y}' \in (\mathscr{A}')^*$. But then, $\mathbf{y} \in \mathscr{A} \setminus \mathscr{A}^*$ ensures there is some $\mathbf{x} \in \mathscr{A}$ with $\mathbf{y} \subset \mathbf{x}$, in turn implying $\mathbf{y}' \subset \mathbf{x}'$ by (7.15), which contradicts that $\mathbf{y}' \in (\mathscr{A}')^*$. Next, we observe that

$$(\downarrow\partial(\{\mathbf{x}\}))' \subseteq \downarrow\partial(\{\mathbf{x}'\}) \quad \text{for all } \mathbf{x} \in \mathcal{X}. \tag{7.17}$$

Indeed, if $\mathbf{y} \in \downarrow\partial(\{\mathbf{x}\})$, then $\mathbf{y} \subset \mathbf{x}$, which implies $\mathbf{y}' \subset \mathbf{x}'$ by (7.15), yielding (7.17).

The fact that we only have inclusions in (7.16) prompts us to define $\downarrow\overline{\mathscr{A}},\ \mathscr{A}^{\circ *} \subseteq \mathcal{X}$ to be the subsets such that

$$(\downarrow\overline{\mathscr{A}})' = \downarrow\mathscr{A}' \quad \text{and} \quad (\mathscr{A}^{\circ *})' = (\mathscr{A}')^*.$$

In view of (7.16), we have

$$\downarrow\mathscr{A} \subseteq \downarrow\overline{\mathscr{A}} \quad \text{and} \quad \mathscr{A}^{\circ *} \subseteq \mathscr{A}^*.$$

Note $\downarrow(\downarrow\overline{\mathscr{A}}) = \downarrow\overline{\mathscr{A}}$, for if $\mathbf{x} \subseteq \mathbf{y} \in \downarrow\overline{\mathscr{A}}$, then $\mathbf{y}' \in \downarrow\mathscr{A}'$, implying $\mathbf{y}' \subseteq \mathbf{z}'$ for some $\mathbf{z}' \in \mathscr{A}'$, whence (7.15) implies $\mathbf{x}' \subseteq \mathbf{y}' \subseteq \mathbf{z}' \in \mathscr{A}'$, which implies $\mathbf{x}' \in \downarrow\mathscr{A}'$, and thus $\mathbf{x} \in \downarrow\overline{\mathscr{A}}$. We also have $(\mathscr{A}^{\circ *})^* = \mathscr{A}^{\circ *}$. Indeed, $(\mathscr{A}^{\circ *})^* \subseteq \mathscr{A}^{\circ *}$ holds by definition, while if $\mathbf{x} \in \mathscr{A}^{\circ *} \setminus (\mathscr{A}^{\circ *})^*$, then $\mathbf{x} \subset \mathbf{y}$ for some $\mathbf{y} \in \mathscr{A}^{\circ *}$, whence $\mathbf{x}' \subset \mathbf{y}'$ by (7.15) with $\mathbf{x}', \mathbf{y}' \in (\mathscr{A}')^*$ (by definition of $\mathscr{A}^{\circ *}$), which contradicts the definition of $*$. Since $(\downarrow\overline{\mathscr{A}^{\circ *}})' = \downarrow(\mathscr{A}^{\circ *})' = \downarrow((\mathscr{A}')^*) = \downarrow\mathscr{A}' = (\downarrow\overline{\mathscr{A}})'$, the injectivity of the map $\mathbf{z} \mapsto \mathbf{z}'$ implies $\downarrow\overline{\mathscr{A}^{\circ *}} = \downarrow\overline{\mathscr{A}}$. Since $((\downarrow\overline{\mathscr{A}})^{\circ *})' = ((\downarrow\overline{\mathscr{A}})')^* = (\downarrow\mathscr{A}')^* = (\mathscr{A}')^* = (\mathscr{A}^{\circ *})'$, the injectivity of the map $\mathbf{z} \mapsto \mathbf{z}'$ implies $(\downarrow\overline{\mathscr{A}})^{\circ *} = \mathscr{A}^{\circ *}$. In summary,

$$\downarrow(\downarrow\overline{\mathscr{A}}) = \downarrow\overline{\mathscr{A}}, \quad (\mathscr{A}^{\circ *})^* = \mathscr{A}^{\circ *}, \quad \downarrow\overline{\mathscr{A}^{\circ *}} = \downarrow\overline{\mathscr{A}}, \quad \text{and} \quad (\downarrow\overline{\mathscr{A}})^{\circ *} = \mathscr{A}^{\circ *}. \tag{7.18}$$

Thus the arrow closure and star interior defined above behave similar to the closure and interior operations on convex sets.

The condition that any representative for $\mathbf{x}$ also be a representative for $\mathbf{x}'$ may be replaced by the alternate condition that $\partial(\mathbf{x}) \subseteq \partial(\mathbf{x}')$ and $\mathbf{x} \not\subseteq \partial(\mathbf{x}')$, as the following argument shows. Suppose $\mathbf{x} \subseteq \mathbf{x}'$ and every representative for $\mathbf{x}$ is also a representative for $\mathbf{x}'$. Since $x \in \mathbf{x}$ is a representative for $\mathbf{x}'$, we have $x \notin \partial(\mathbf{x}')$, ensuring $\mathbf{x} \not\subseteq \partial(\mathbf{x}')$. If $\partial(\mathbf{x}) \subseteq \partial(\mathbf{x}')$ fails, then $\partial(\mathbf{x}) \subseteq \overline{\mathbf{x}} \subseteq \overline{\mathbf{x}'}$ ensures that there is some representative x' for $\mathbf{x}'$ with $x' \in \partial(\mathbf{x})$. Note that $\{x'\} \cup \downarrow\partial(\{x'\})$ is a basis for $\mathbb{R}^{\cup}\langle\downarrow\mathbf{x}'\rangle$, with the representatives for $\mathbf{x}'$ those elements of $\mathbb{R}^{\cup}\langle\downarrow\mathbf{x}'\rangle$ lying strictly on the positive side of the hyperplane $\mathbb{R}^{\cup}\langle\downarrow\partial(\{x'\})\rangle$. If $x = \alpha x' + \sum_{\mathbf{y}' \in \downarrow\partial(\{\mathbf{x}'\})} \alpha_{\mathbf{y}} y'$ is a representative for $\mathbf{x}$, where $\alpha, \alpha_{\mathbf{y}} \in \mathbb{R}$ (possible as $\mathbf{x} \subseteq \mathbf{x}'$), then so too is $x - (\alpha+1)x'$ in view of $x' \in \partial(\mathbf{x})$. However, the coefficient of x' when expressing $x - (\alpha+1)x'$ as a linear combination of the elements from the basis $\{x'\} \cup \downarrow\partial(\{x'\})$ is negative, meaning $x - (\alpha+1)x'$ lies on the *negative* side of the hyperplane $\mathbb{R}^{\cup}\langle\downarrow\partial(\{x'\})\rangle$, which contradicts that the representative $x - (\alpha+1)x'$ for $\mathbf{x}$ should also be a representative for $\mathbf{x}'$. Therefore we instead conclude that $\partial(\mathbf{x}) \subseteq \partial(\mathbf{x}')$, as desired. Next suppose $\mathbf{x} \subseteq \mathbf{x}'$, $\partial(\mathbf{x}) \subseteq \partial(\mathbf{x}')$ and $\mathbf{x} \nsubseteq \partial(\mathbf{x}')$. Let x be an arbitrary representative for $\mathbf{x}$. Then $x \in \mathbf{x} \subseteq \mathbf{x}'$, so if x is not a representative for $\mathbf{x}'$, then $x \in \partial(\mathbf{x}')$. Thus $\mathbf{x} \subseteq \partial(\mathbf{x}) + \mathbb{R}x \subseteq \partial(\mathbf{x}') + \partial(\mathbf{x}') = \partial(\mathbf{x}')$, contrary to assumption. Therefore we instead conclude that every representative x for $\mathbf{x}$ is also a representative for $\mathbf{x}'$, and the equivalence of the two stated conditions follows.

Finally, to see that $\preceq_{\cup}$ is anti-symmetric, suppose $\mathcal{X} \preceq_{\cup} \mathcal{X}'$ and $\mathcal{X}' \preceq_{\cup} \mathcal{X}$ and let $\mathcal{X} = \mathcal{X}_1 \cup \ldots \cup \mathcal{X}_s$ be a series decomposition of $\mathcal{X}$. Since $\mathcal{X} \preceq_{\cup} \mathcal{X}'$, we have $\partial(\mathbf{x}) \subseteq \partial(\mathbf{x}')$ for $\mathbf{x} \in \mathcal{X}$, implying $\dim \partial(\mathbf{x}) \leq \dim \partial(\mathbf{x}')$. Thus $\sum_{\mathbf{x} \in \mathcal{X}} \dim \partial(\mathbf{x}) \leq \sum_{\mathbf{x} \in \mathcal{X}} \dim \partial(\mathbf{x}')$ when $\mathcal{X} \preceq_{\cup} \mathcal{X}'$. Since we also have $\mathcal{X}' \preceq_{\cup} \mathcal{X}$, the reverse inequality likewise holds, yielding $\sum_{\mathbf{x} \in \mathcal{X}} \dim \partial(\mathbf{x}) = \sum_{\mathbf{x} \in \mathcal{X}} \dim \partial(\mathbf{x}')$. As $\dim \partial(\mathbf{x}) \leq \dim \partial(\mathbf{x}')$, this is only possible if $\dim \partial(\mathbf{x}) = \dim \partial(\mathbf{x}')$ for all $\mathbf{x} \in \mathcal{X}$. Consequently, since $\partial(\mathbf{x}) \subseteq \partial(\mathbf{x}')$ are subspaces, and since $\mathbf{x}$ and $\mathbf{x}'$ share a common representative by definition of $\mathcal{X} \preceq_{\cup} \mathcal{X}'$, it then follows that $\partial(\mathbf{x}) = \partial(\mathbf{x}')$ and $\overline{\mathbf{x}} = \overline{\mathbf{x}}'$ for $\mathbf{x} \in \mathcal{X}$. Now $\partial(\mathbf{x})$ is a subspace of dimension $|{\downarrow}\partial(\{\mathbf{x}\})|$ by Proposition 5.3.1. Likewise, $\partial(\mathbf{x}')$ is a subspace of dimension $|{\downarrow}\partial(\{\mathbf{x}'\})|$. As a result, $\partial(\mathbf{x}) = \partial(\mathbf{x}')$ implies $|{\downarrow}\partial(\{\mathbf{x}\})| = |{\downarrow}\partial(\{\mathbf{x}'\})|$. By (7.17), we know $({\downarrow}\partial(\{\mathbf{x}\}))' \subseteq {\downarrow}\partial(\{\mathbf{x}'\})$ is subset with cardinality $|({\downarrow}\partial(\{\mathbf{x}\}))'| = |{\downarrow}\partial(\{\mathbf{x}\})| = |{\downarrow}\partial(\{\mathbf{x}'\})|$, whence equality must hold in (7.17), i.e.,

$$({\downarrow}\partial(\{\mathbf{x}\}))' = {\downarrow}\partial(\{\mathbf{x}'\}).$$

We can now show $\mathbf{x} = \mathbf{x}'$ by induction on j, where $x \in \mathcal{X}_j$. For $\mathbf{x} \in \mathcal{X}_1$, we have $\partial(\mathbf{x}') = \partial(\mathbf{x}) = \{0\}$, ensuring that $\partial(\mathbf{x}) \cap \mathbf{x} = \{0\} = \partial(\mathbf{x}') \cap \mathbf{x}'$, which forces $\mathbf{x} = \mathbf{x}'$ as $\overline{\mathbf{x}} = \overline{\mathbf{x}}'$. Thus we can assume $j \geq 2$, completing the base of the induction. Applying the induction hypothesis to all half-spaces from ${\downarrow}\partial(\{\mathbf{x}\})$ combined with equality holding in (7.17) yields $\partial(\mathbf{x}) \cap \mathbf{x} = \mathsf{C}^{\cup}\big({\downarrow}\partial(\{\mathbf{x}\})\big) = \mathsf{C}^{\cup}\big(({\downarrow}\partial(\{\mathbf{x}\}))'\big) = \mathsf{C}^{\cup}\big({\downarrow}\partial(\{\mathbf{x}'\})\big) = \partial(\mathbf{x}') \cap \mathbf{x}'$. Combined with $\overline{\mathbf{x}} = \overline{\mathbf{x}}'$, it follows that $\mathbf{x} = \mathbf{x}'$, completing the induction, which shows $\preceq_{\cup}$ is anti-symmetric. Note, it is not initially as evident that the bijection from $\mathcal{X}'$ to $\mathcal{X}$ given by $\mathcal{X}' \preceq_{\cup} \mathcal{X}$ should be the inverse of that from $\mathcal{X}$ to $\mathcal{X}'$, which is why we have argued above without using the assumption $(\mathbf{x}')' = \mathbf{x}$. Of course, the above argument establishes this, since it gives $(\mathbf{x}')' = \mathbf{x}' = \mathbf{x}$. Note this argument shows that $\sum_{\mathbf{x} \in \mathcal{X}} \dim \partial(\mathbf{x}) = \sum_{\mathbf{x} \in \mathcal{X}} \dim \partial(\mathbf{x}')$ implies $\mathbf{x} = \mathbf{x}'$ for all $\mathbf{x} \in \mathcal{X}$.

Next we show that if $\mathcal{B} \subseteq \mathcal{X}$ minimally encases $-\vec{u}$, then $(\mathcal{B}')^*$ minimally encases $-\vec{u}$.

Proposition 7.26 *Let $\Lambda \subseteq \mathbb{R}^d$ be a full rank lattice, where $d \geq 0$, and let $G_0 \subseteq \Lambda$ be a finitary subset with $\mathsf{C}(G_0) = \mathbb{R}^d$. Suppose $\mathcal{X}, \mathcal{X}' \in \mathfrak{X}(G_0)$ with $\mathcal{X} \preceq_{\cup} \mathcal{X}'$ and each $\mathbf{x} \in \mathcal{X}$ in bijective correspondence with the half-space $\mathbf{x}' \in \mathcal{X}'$. Let $\vec{u} = (u_1, \ldots, u_t)$ be a tuple of orthonormal vectors from $\mathbb{R}^d$. If $\mathcal{B} \subseteq \mathcal{X}_1 \cup \ldots \cup \mathcal{X}_s$ minimally encases $\vec{u}$, then $(\mathcal{B}')^*$ minimally encases $\vec{u}$.*

Proof Since $\mathbf{x} \subseteq \mathbf{x}'$ for all $\mathbf{x} \in \mathcal{X}$ (by definition of $\mathcal{X} \preceq_{\cup} \mathcal{X}'$), and since $\mathcal{B}$ minimally encases $\vec{u}$, we conclude that $\mathcal{B}'$ encases $\vec{u} = (u_1, \ldots, u_t)$. Thus there is some $\mathcal{A}' \preceq \mathcal{B}'$ which minimally encases $\vec{u}$, where $\mathcal{A} \subseteq \mathcal{X}$. It remains to show $\mathcal{A}' = (\mathcal{B}')^*$.

Since $\mathcal{A}'$ minimally encases $\vec{u} = (u_1, \ldots, u_t)$, we have $u_1, \ldots, u_t \in \mathbb{R}^{\cup}\langle \mathcal{A}' \rangle = \mathbb{R}\langle \downarrow A' \rangle = \mathbb{R}\langle \downarrow \overline{A} \rangle = \mathbb{R}^{\cup}\langle \downarrow \overline{\mathcal{A}} \rangle$, with the first equality in view of Proposition 5.3.1, the second in view of the definition of $\downarrow \overline{\mathcal{A}}$ and the fact that any representative for a half-space $\mathbf{x} \in \downarrow \overline{\mathcal{A}}$ is a also a representative for the half-space $\mathbf{x}' \in (\downarrow \overline{\mathcal{A}})' = \downarrow \mathcal{A}'$, and the third in view of the first equality in (7.18) and Proposition 5.3.1. Thus, since $\mathcal{B}$ minimally encases $\vec{u}$, Proposition 5.10 and the first equality in (7.18) imply that $\mathcal{B} \subseteq \downarrow \overline{\mathcal{A}}$, in turn implying $\mathcal{B}' \subseteq (\downarrow \overline{\mathcal{A}})' = \downarrow \mathcal{A}'$. Thus $(\mathcal{B}')^* \preceq \mathcal{A}'$ by (5.9). On the other hand, $\mathcal{A}' \preceq \mathcal{B}'$ implies $\mathcal{A}' \subseteq \downarrow \mathcal{B}' = \downarrow (\mathcal{B}')^*$, whence $(\mathcal{A}')^* \preceq (\mathcal{B}')^*$ follows again by (5.9). Since $\mathcal{A}' = (\mathcal{A}')^*$ as $\mathcal{A}'$ minimally encases $\vec{u}$, we see that $(\mathcal{B}')^* \preceq \mathcal{A}' = (\mathcal{A}')^* \preceq (\mathcal{B}')^*$, yielding the desired conclusion $\mathcal{A}' = (\mathcal{B}')^*$. □

Definition 7.27 Let $\mathfrak{X}^*(G_0)$ be the set consisting of all *maximal* series decompositions of the sets $\mathcal{X} \in \mathfrak{X}(G_0)$. We informally write the elements of $\mathfrak{X}^*(G_0)$ in the form $\bigcup_{i=1}^{s} \mathcal{X}_i$, where $\mathcal{X} = \bigcup_{i=1}^{s} \mathcal{X}_i$ is a maximal series decomposition of some $\mathcal{X} \in \mathfrak{X}(G_0)$. For $\Delta \in \mathfrak{P}_{\mathbb{Z}}(G_0)$, let $\mathfrak{X}^*(G_0, \Delta)$ be the set consisting of all maximal series decompositions of the sets $\mathcal{X} \in \mathfrak{X}(G_0, \Delta)$.

If $\bigcup_{i=1}^{s} \mathcal{X}_i \in \mathfrak{X}^*(G_0, \Delta)$ and we set $\mathcal{X} = \bigcup_{i=1}^{s} \mathcal{X}_i \in \mathfrak{X}(G_0, \Delta)$, then there is a realization $\mathcal{R} = (\mathcal{X}_1 \cup \{\mathbf{v}_1\}, \ldots, \mathcal{X}_s \cup \{\mathbf{v}_s\})$ of $\mathcal{X}$ as well as a Δ-pure realization $\mathcal{R}'$ of $\mathcal{X}$, in which case the comments above the definition of $\mathfrak{P}_{\mathbb{Z}}(\mathcal{X})$ allow us to assume (by exchanging the representative sequences in $\mathcal{R}$ for those from $\mathcal{R}'$), that $\mathcal{X}$ has a Δ-pure realization $(\mathcal{X}_1 \cup \{\mathbf{v}_1\}, \ldots, \mathcal{X}_s \cup \{\mathbf{v}_s\})$. We defined a partial order $\preceq_{\cup}$ on $\mathfrak{X}(G_0)$ earlier. We now define a related partial order $\preceq_{\cup}^*$ on $\mathfrak{X}^*(G_0)$ as follows. Let $\bigcup_{i=1}^{s} \mathcal{X}_i, \bigcup_{i=1}^{s} \mathcal{X}_i' \in \mathfrak{X}^*(G_0)$ and set $\mathcal{X} = \bigcup_{i=1}^{s} \mathcal{X}_i$ and $\mathcal{X}' = \bigcup_{i=1}^{s} \mathcal{X}_i'$. Then we declare $\bigcup_{i=1}^{s} \mathcal{X}_i \preceq_{\cup}^* \bigcup_{i=1}^{s} \mathcal{X}_i'$ if there is bijection between $\mathcal{X}$ and $\mathcal{X}'$ showing $\mathcal{X} \preceq_{\cup} \mathcal{X}'$ which restricts to a bijection between $\mathcal{X}_j$ and $\mathcal{X}_j'$ for all $j \in [1, s]$. Thus, if we let $\mathbf{x} \mapsto \mathbf{x}'$ denote the bijection showing $\mathcal{X} \preceq_{\cup} \mathcal{X}'$, then each $\mathcal{X}_i' = \{\mathbf{x}' : \mathbf{x} \in \mathcal{X}_i\}$, so that the notation agrees with our previous definition of $\mathcal{A}'$ for $\mathcal{A} \subseteq \mathcal{X}$. By definition, $\bigcup_{i=1}^{s} \mathcal{X}_i \preceq_{\cup}^* \bigcup_{i=1}^{s} \mathcal{X}_i'$ implies $\mathcal{X} \preceq_{\cup} \mathcal{X}'$, in which case anti-symmetry for $\preceq_{\cup}^*$ follows from that for $\preceq_{\cup}$ (recall that we showed $\mathbf{x}' = \mathbf{x}$ when $\mathcal{X} \preceq_{\cup} \mathcal{X}'$ and $\mathcal{X}' \preceq_{\cup} \mathcal{X}$), while reflexivity and transitivity are immediate. Thus $\preceq_{\cup}^*$ is a partial order on $\mathfrak{X}^*(G_0)$, and so restricts to one on $\mathfrak{X}^*(G_0, \Delta)$ as well. The relation between the partial orders $\preceq_{\cup}$ and $\preceq_{\cup}^*$ is given by the following proposition.

Proposition 7.28 *Let $\Lambda \subseteq \mathbb{R}^d$ be a full rank lattice, where $d \geq 0$, let $G_0 \subseteq \Lambda$ be a finitary subset with $\mathsf{C}(G_0) = \mathbb{R}^d$, and let $\mathcal{X}, \mathcal{X}' \in \mathfrak{X}(G_0)$. Then $\mathcal{X} \preceq_{\cup} \mathcal{X}'$ if and only if there are maximal series decompositions $\mathcal{X} = \bigcup_{i=1}^{s} \mathcal{X}_i$ and $\mathcal{X}' = \bigcup_{i=1}^{s} \mathcal{X}_i'$ such that $\bigcup_{i=1}^{s} \mathcal{X}_i \preceq_{\cup}^* \bigcup_{i=1}^{s} \mathcal{X}_i'$.*

Proof If $\bigcup_{i=1}^{s} \mathcal{X}_i \preceq_{\cup}^* \bigcup_{i=1}^{s} \mathcal{X}_i'$, then $\mathcal{X} \preceq_{\cup} \mathcal{X}'$ follows by definition of $\preceq_{\cup}^*$, so one direction is trivial. Assume $\mathcal{X} \preceq_{\cup} \mathcal{X}'$ and let $\mathcal{X}' = \bigcup_{i=1}^{s} \mathcal{X}_i'$ be any maximal series decomposition of $\mathcal{X}'$. It suffices to show $\mathcal{X} = \bigcup_{i=1}^{s} \mathcal{X}_i$ is also a maximal series decomposition. Since any representative of $\mathbf{x}$ is also a representative for $\mathbf{x}'$, Proposition 5.3.1 implies

$$\mathbb{R}^{\cup}\langle \mathcal{X}_1 \cup \ldots \cup \mathcal{X}_j \rangle = \mathbb{R}\langle X_1 \cup \ldots \cup X_j \rangle = \mathbb{R}\langle X_1' \cup \ldots \cup X_j' \rangle = \mathbb{R}^{\cup}\langle \mathcal{X}_1' \cup \ldots \cup \mathcal{X}_j' \rangle$$

for all $j \in [0, s]$. For $j \in [1, s]$, let $\pi_{j-1} : \mathbb{R}^d \to \mathbb{R}^{\cup}\langle \mathcal{X}_1 \cup \ldots \cup \mathcal{X}_{j-1}\rangle^{\perp}$ be the orthogonal projection.

Let $\mathcal{R}' = (\mathcal{X}'_1 \cup \{\mathbf{v}'_1\}, \ldots, \mathcal{X}'_s \cup \{\mathbf{v}'_s\})$ be a realization of $\mathcal{X}' = \bigcup_{i=1}^{s} \mathcal{X}'_i$. Since each $\mathbf{x} \in \mathcal{X}'_1$ has trivial boundary, it follows that $\mathbf{x} = \mathbf{x}'$ for all $\mathbf{x} \in \mathcal{X}_1$ (as $\mathbf{x} \subseteq \mathbf{x}'$ forces this in view of $\mathbf{x}'$ being one-dimensional). Thus $(\mathcal{X}'_1 \cup \{\mathbf{v}'_1\}) = (\mathcal{X}_1 \cup \{\mathbf{v}'_1\})$ is a purely virtual Reay system with $\mathcal{X}'_1 = \mathcal{X}_1$ irreducible (by Proposition 7.20 applied to the maximal series decomposition $\bigcup_{i=1}^{s} \mathcal{X}'_i$), showing that $\mathcal{X}_1 = \mathcal{X}'_1 \in \mathfrak{X}^*(G_0)$. We proceed by induction on j to show $\bigcup_{i=1}^{j} \mathcal{X}_i \in \mathfrak{X}^*(G_0)$ with $\pi_{j-1}(\mathcal{X}_j) = \pi_{j-1}(\mathcal{X}'_j)$, having just completed the base case $j = 1$. By induction hypothesis, $\bigcup_{i=1}^{j-1} \mathcal{X}_i \in \mathfrak{X}^*(G_0)$, so there is realization $\mathcal{R}_{j-1} = (\mathcal{X}_1 \cup \{\mathbf{v}_1\}, \ldots, \mathcal{X}_{j-1} \cup \{\mathbf{v}_{j-1}\})$. Let $\vec{u}_{\mathbf{v}'_j} = (u_1, \ldots, u_t)$. Then $-\vec{u}^{\triangleleft}_{\mathbf{v}'_j}$ is a fully unbounded (or trivial) limit of an asymptotically filtered sequence of terms from G_0 which is minimally encased by $\partial(\{\mathbf{v}'_j\}) \subseteq \mathcal{X}'_1 \cup \ldots \cup \mathcal{X}'_{j-1}$, thus ensuring that $u_i \in \mathbb{R}^{\cup}\langle \mathcal{X}'_1 \cup \ldots \cup \mathcal{X}'_{j-1}\rangle = \mathbb{R}^{\cup}\langle \mathcal{X}_1 \cup \ldots \cup \mathcal{X}_{j-1}\rangle$ for all $i < t$. Applying Proposition 7.7.3 to $\mathcal{R}_{j-1}$, we conclude that there is a support set $\mathcal{B} \subseteq \mathcal{X}_1 \cup \ldots \cup \mathcal{X}_{j-1}$ which minimally encases $-\vec{u}^{\triangleleft}_{\mathbf{v}'_j}$, allowing us to define a half-space $\mathbf{v}_j$ with $\partial(\{\mathbf{v}_j\}) = \mathcal{B}$ using the limit $\vec{u}_{\mathbf{v}_j} = \vec{u}_{\mathbf{v}'_j}$ with $\mathbf{v}_j(i) = \mathbf{v}'_j(i)$ for all i. By (7.17), we have $\partial(\{\mathbf{x}\})' \subseteq \downarrow\partial(\{\mathbf{x}'\}) \subseteq \mathcal{X}'_1 \cup \ldots \cup \mathcal{X}'_{j-1}$ for all $\mathbf{x} \in \mathcal{X}_j$, ensuring via the bijection $\mathbf{z} \mapsto \mathbf{z}'$ that $\partial(\{\mathbf{x}\}) \subseteq \mathcal{X}_1 \cup \ldots \cup \mathcal{X}_{j-1}$ for all $\mathbf{x} \in \mathcal{X}_j$. Thus, since $\mathbb{R}^{\cup}\langle \mathcal{X}'_1 \cup \ldots \cup \mathcal{X}'_{j-1}\rangle = \mathbb{R}^{\cup}\langle \mathcal{X}_1 \cup \ldots \cup \mathcal{X}_{j-1}\rangle$, it follows that $\pi_{j-1}(\mathcal{X}_j \cup \{\mathbf{v}_j\}) = \pi_{j-1}(\mathcal{X}'_j \cup \{\mathbf{v}'_j\})$ with $(\mathcal{X}_1 \cup \{\mathbf{v}_1\}, \ldots, \mathcal{X}_j \cup \{\mathbf{v}_j\})$ a purely virtual Reay system, showing $\bigcup_{i=1}^{j} \mathcal{X}_i \in \mathfrak{X}^*(G_0)$. This completes the induction. The case $j = s$ shows $\mathcal{X} = \bigcup_{i=1}^{s} \mathcal{X}_i$ is a series decomposition. Since $\mathbb{R}^{\cup}\langle \mathcal{X}'_1 \cup \ldots \cup \mathcal{X}'_{j-1}\rangle = \mathbb{R}^{\cup}\langle \mathcal{X}_1 \cup \ldots \cup \mathcal{X}_{j-1}\rangle$, applying Proposition 7.20 to the maximal decomposition $\bigcup_{i=1}^{s} \mathcal{X}'_i$ gives that $\pi_{j-1}(\mathcal{X}'_j) = \pi_{j-1}(\mathcal{X}_j)$ is irreducible for all $j \in [1, s]$, and then a second application of Proposition 7.20 to $\bigcup_{i=1}^{s} \mathcal{X}_i$ implies that $\mathcal{X} = \bigcup_{i=1}^{s} \mathcal{X}_i$ is maximal. □

Theorem 7.29 and Corollary 7.30 contain the finiteness from above property for the partial orders $\preceq_{\cup}$ and $\preceq^*_{\cup}$ associated to a finitary set G_0.

Theorem 7.29 *Let $\Lambda \subseteq \mathbb{R}^d$ be a full rank lattice, where $d \geq 0$, and let $G_0 \subseteq \Lambda$ be a finitary subset with $\mathsf{C}(G_0) = \mathbb{R}^d$. Then $\mathsf{Max}(\mathfrak{X}^*(G_0, \Delta), \preceq^*_{\cup})$ is finite for every $\Delta \in \mathfrak{P}_{\mathbb{Z}}(G_0)$.*

Proof For $s \in [1, d]$, let $\mathfrak{X}^*_s(G_0, \Delta) \subseteq \mathfrak{X}^*(G_0, \Delta)$ consist of all maximal series decompositions $\mathcal{X} = \bigcup_{i=1}^{s} \mathcal{X}_i$ of some $\mathcal{X} \in \mathfrak{X}(G_0, \Delta)$ having length s. Note $\mathsf{Max}(\mathfrak{X}^*(G_0, \Delta)) = \bigcup_{s=1}^{d} \mathsf{Max}(\mathfrak{X}^*_s(G_0, \Delta))$ since comparable series decompositions under $\preceq^*_{\cup}$ must have the same length. We proceed by induction on $s = 1, 2, \ldots, d$ to show that $\mathsf{Max}(\mathfrak{X}^*_s(G_0, \Delta))$ is finite for every $\Delta \in \mathfrak{P}_{\mathbb{Z}}(G_0)$. Note $\mathfrak{X}^*_1(G_0)$ is in one-to-one correspondence with the subset of all irreducible $\mathcal{X} \in \mathfrak{X}(G_0)$, so $\mathsf{Max}(\mathfrak{X}^*_1(G_0, \Delta))$ is finite by Theorem 7.24.1. Thus the base $s = 1$ of the induction is complete, and we assume $s \geq 2$. Fix $\Delta \in \mathfrak{P}_{\mathbb{Z}}(G_0)$.

Let

$$\bigcup_{i=1}^{s} X_i \in \mathsf{Max}(\mathfrak{X}_s^*(G_0, \Delta))$$

be arbitrary, set $\mathcal{X} = \bigcup_{i=1}^{s} X_i$, and let

$$\mathcal{R} = (X_1 \cup \{\mathbf{v}_1\}, \ldots, X_s \cup \{\mathbf{v}_s\})$$

be a Δ-pure realization. By Lemma 7.22, we can assume (by passing to subsequences of the representative sequences as need be) that $\mathbb{Z}\langle (X \setminus X_s)(k)\rangle = \Delta'$ is constant for all tuples $k = (i_{\mathbf{x}})_{\mathbf{x} \in \mathcal{X}}$, while $\mathcal{X} \setminus X_s = \bigcup_{i=1}^{s-1} X_i$ is a maximal series decomposition by Proposition 7.20. Hence

$$X_1 \cup \ldots \cup X_{s-1} \in \mathfrak{X}_{s-1}^*(G_0, \Delta'). \tag{7.19}$$

□

Claim A $X_1 \cup \ldots \cup X_{s-1} \in \mathsf{Max}(\mathfrak{X}_{s-1}^*(G_0, \Delta'))$.

Proof Suppose by contradiction $X_1 \cup \ldots \cup X_{s-1} \notin \mathsf{Max}(\mathfrak{X}_{s-1}^*(G_0, \Delta'))$. Then, in view of (7.19), there exists some $\bigcup_{i=1}^{s-1} X_i' \in \mathfrak{X}^*(G_0, \Delta')$ with

$$X_1 \cup \ldots \cup X_{s-1} \prec_{\cup}^{*} X_1' \cup \ldots \cup X_{s-1}'.$$

Let $\mathbf{x}' \in (\mathcal{X} \setminus X_s)'$ be the image of $\mathbf{x} \in \mathcal{X} \setminus X_s$ under the bijection showing $\mathcal{X} \setminus X_s \prec_{\cup} (\mathcal{X} \setminus X_s)'$. Let $\mathcal{R}' = (X_1' \cup \{\mathbf{w}_1'\}, \ldots, X_{s-1}' \cup \{\mathbf{w}_{s-1}'\})$ be a Δ'-pure realization of $(\mathcal{X} \setminus X_s)' = \bigcup_{i=1}^{s-1} X_i'$. Since $\mathcal{X} \setminus X_s \prec_{\cup} (\mathcal{X} \setminus X_s)'$ ensures that a set of representatives for $\mathcal{X} \setminus X_s$ is also a set of representatives for $(\mathcal{X} \setminus X_s)'$, we must have $\mathbb{R}^{\cup}\langle \mathcal{X} \setminus X_s\rangle = \mathbb{R}\langle (X \setminus X_s)(k)\rangle = \mathbb{R}\langle (X \setminus X_s)'(k)\rangle = \mathbb{R}^{\cup}\langle (\mathcal{X} \setminus X_s)'\rangle$ by Proposition 5.3.1. For each $\mathbf{y} \in X_s \cup \{\mathbf{v}_s\}$, we have $-\vec{u}_{\mathbf{y}}^{\triangleleft}$ minimally encased by $\partial(\{\mathbf{y}\}) \subseteq X_1 \cup \ldots \cup X_{s-1} = \mathcal{X} \setminus X_s$. Thus Proposition 7.26 (applied to $\mathcal{X} \setminus X_s \prec_{\cup} (\mathcal{X} \setminus X_s)'$) implies that $(\partial(\{\mathbf{y}\})^{\circ *})' = (\partial(\{\mathbf{y}\})')^*$ minimally encases $-\vec{u}_{\mathbf{y}}$. This allows us to define a new half-space $\mathbf{y}'$ such that $\partial(\{\mathbf{y}'\}) = (\partial(\{\mathbf{y}\})')^* \subseteq \partial(\{\mathbf{y}\})' \subseteq (\mathcal{X} \setminus X_s)' = X_1' \cup \ldots \cup X_{s-1}'$ with $\vec{u}_{\mathbf{y}'} = \vec{u}_{\mathbf{y}}$ and $\mathbf{y}(i) = \mathbf{y}'(i)$ for all i, making

$$\mathcal{R}'' = (X_1' \cup \{\mathbf{w}_1'\}, \ldots, X_{s-1}' \cup \{\mathbf{w}_{s-1}'\}, X_s' \cup \{\mathbf{w}_s'\}),$$

where $X_s' = \{\mathbf{y}' : \mathbf{y} \in X_s\}$, $\mathbf{w}_s' = \mathbf{v}_s'$ and $\vec{u}_{\mathbf{w}_s'} = \vec{u}_{\mathbf{v}_s}$, a virtual Reay system over G_0 in view of $\mathbb{R}^{\cup}\langle \mathcal{X} \setminus X_s\rangle = \mathbb{R}^{\cup}\langle (\mathcal{X} \setminus X_s)'\rangle$, which ensures that

$$\pi(X_s' \cup \{\mathbf{w}_s'\}) = \pi(X_s \cup \{\mathbf{v}_s\}) \tag{7.20}$$

with $\pi : \mathbb{R}^d \to \mathbb{R}^{\cup}\langle \mathcal{X} \setminus \mathcal{X}_s \rangle^{\perp}$ the orthogonal projection. For $\mathbf{y} \in \mathcal{X}_s \cup \{\mathbf{v}_s\}$, we have

$$\partial(\mathbf{y}) = \mathbb{R}^{\cup}\langle \downarrow \partial(\{\mathbf{y}\}) \rangle \subseteq \mathbb{R}^{\cup}\langle \downarrow \overline{\partial(\{\mathbf{y}\})} \rangle = \mathbb{R}^{\cup}\langle \downarrow \partial(\{\mathbf{y}\})' \rangle = \mathbb{R}^{\cup}\langle (\partial(\{\mathbf{y}\})')^* \rangle = \partial(\mathbf{y}')$$

and $\partial(\mathbf{y}) \cap \mathbf{y} = \mathsf{C}^{\cup}\big(\partial(\{\mathbf{y}\})\big) \subseteq \mathsf{C}^{\cup}\big(\partial(\{\mathbf{y}\})'\big) \subseteq \mathsf{C}^{\cup}\big(\partial(\{\mathbf{y}'\})\big) = \partial(\mathbf{y}') \cap \mathbf{y}'$ (with the second inclusion by (7.17), and the first as $\mathbf{x} \subseteq \mathbf{x}'$ for all $\mathbf{x}$). Hence, since $\mathbf{y}$ and $\mathbf{y}'$ share a common representative, it follows that $\mathbf{y} \subseteq \mathbf{y}'$ for all $\mathbf{y} \in \mathcal{X}_s$, and that any representative for $\mathbf{y}$ is also a representative for $\mathbf{y}'$. Since $\mathcal{R}'$ is purely virtual, and since $\vec{u}_{\mathbf{w}'_s} = \vec{u}_{\mathbf{v}_s}$ is fully unbounded (as $\mathcal{R}$ is purely virtual), it follows that $\mathcal{R}''$ is purely virtual, showing that $\bigcup_{i=1}^{s} \mathcal{X}'_i$ is a series decomposition. Since $\mathcal{X} = \bigcup_{i=1}^{s} \mathcal{X}_i$ is a maximal series decomposition, Proposition 7.20 ensures that $\pi(\mathcal{X}_s) = \pi(\mathcal{X}'_s)$ is irreducible (the equality follows from (7.20)), and now two further applications of Proposition 7.20, one to the maximal series decomposition $(\mathcal{X} \setminus \mathcal{X}_s)' = \bigcup_{i=1}^{s-1} \mathcal{X}'_i$ and one to the series decomposition $\mathcal{X}' = \bigcup_{i=1}^{s} \mathcal{X}'_i$, implies that $\mathcal{X}' = \bigcup_{i=1}^{s} \mathcal{X}'_i$ is maximal. By construction, each representative of $\mathbf{x} \in \mathcal{X}$ is a representative of $\mathbf{x}' \in \mathcal{X}'_1 \cup \ldots \cup \mathcal{X}'_s$, while $\mathbf{x} \subseteq \mathbf{x}'$ for $\mathbf{x} \in \mathcal{X} \setminus \mathcal{X}_s$ (since $\mathcal{X} \setminus \mathcal{X}_s \prec_{\cup} (\mathcal{X} \setminus \mathcal{X}_s)'$) as well as for $\mathbf{x} \in \mathcal{X}_s$ (as just argued). Thus

$$\mathcal{X}_1 \cup \ldots \cup \mathcal{X}_{s-1} \cup \mathcal{X}_s \prec^*_{\cup} \mathcal{X}'_1 \cup \ldots \cup \mathcal{X}'_{s-1} \cup \mathcal{X}'_s, \tag{7.21}$$

with the strict inequality following as $\mathbf{x} \neq \mathbf{x}'$ for some $\mathbf{x} \in \mathcal{X} \setminus \mathcal{X}_s$ in view of $\mathcal{X} \setminus \mathcal{X}_s \prec_{\cup} (\mathcal{X} \setminus \mathcal{X}_s)'$. Since the representative sequences for $\mathcal{X}'_s$ are the same as those from $\mathcal{X}_s$ with $\mathbb{Z}\langle (X \setminus X_s)(k) \rangle = \mathbb{Z}\langle (X \setminus X_s)'(k) \rangle = \Delta'$, it follows that $\Delta = \mathbb{Z}\langle X(k) \rangle = \mathbb{Z}\langle (X \setminus X_s)(k) \rangle + \mathbb{Z}\langle X_s(k) \rangle = \Delta' + \mathbb{Z}\langle X_s(k) \rangle = \Delta' + \mathbb{Z}\langle X'_s(k) \rangle = \mathbb{Z}\langle X'(k) \rangle$, for any tuple k, ensuring that $\Delta \in \mathfrak{P}_{\mathbb{Z}}(\mathcal{X}')$. Thus $\bigcup_{i=1}^{s} \mathcal{X}'_i \in \mathfrak{X}^*_s(G_0, \Delta)$, which combined with (7.21) contradicts that $\bigcup_{i=1}^{s} \mathcal{X}_i \in \mathsf{Max}(\mathfrak{X}^*_s(G_0, \Delta))$. This completes Claim A. □

By induction hypothesis, $\mathsf{Max}\big(\mathfrak{X}^*_{s-1}(G_0, \Delta')\big)$ is finite for any of the finite choices for $\Delta' \in \mathfrak{P}_{\mathbb{Z}}(G_0)$. Thus, since $\bigcup_{i=1}^{s} \mathcal{X}_i \in \mathsf{Max}(\mathfrak{X}^*_s(G_0, \Delta))$ was arbitrary, it follows in view of (7.19) and Claim A that every $\bigcup_{i=1}^{s} \mathcal{X}_i \in \mathsf{Max}(\mathfrak{X}^*_s(G_0, \Delta))$ has $\bigcup_{i=1}^{s-1} \mathcal{X}_i \in \mathsf{Max}(\mathfrak{X}^*_{s-1}(G_0, \Delta'))$ for some $\Delta' \in \mathfrak{P}_{\mathbb{Z}}(G_0)$. Thus to show $\mathsf{Max}(\mathfrak{X}^*_s(G_0, \Delta))$ is finite, it suffices to show there are only a finite number of $\bigcup_{i=1}^{s} \mathcal{X}_i \in \mathsf{Max}(\mathfrak{X}^*_s(G_0, \Delta))$ extending each possible maximal series decomposition $\bigcup_{i=1}^{s-1} \mathcal{X}_i \in \mathsf{Max}(\mathfrak{X}^*_{s-1}(G_0, \Delta'))$. To this end, fix $\Delta' \in \mathfrak{P}_{\mathbb{Z}}(G_0)$ and $\bigcup_{i=1}^{s-1} \mathcal{X}_i \in \mathsf{Max}(\mathfrak{X}^*_{s-1}(G_0, \Delta'))$ and let $\bigcup_{i=1}^{s} \mathcal{X}_i \in \mathsf{Max}(\mathfrak{X}^*_s(G_0, \Delta))$ be arbitrary having $\mathcal{R} = (\mathcal{X}_1 \cup \{\mathbf{v}_1\}, \ldots, \mathcal{X}_s \cup \{\mathbf{v}_s\})$ as a Δ-pure realization with $(\mathcal{X}_1 \cup \{\mathbf{v}_1\}, \ldots, \mathcal{X}_{s-1} \cup \{\mathbf{v}_{s-1}\})$ a Δ'-pure realization. Set $\mathcal{X} = \bigcup_{i=1}^{s} \mathcal{X}_i$ and $\mathcal{Y} = \bigcup_{i=1}^{s-1} \mathcal{X}_i$, so $\mathcal{Y}$ is fixed and $\mathcal{X}_s$ is arbitrary (subject to the constraints described).

Let $\mathcal{E} = \mathbb{R}^{\cup}\langle \mathcal{X}_1 \cup \ldots \cup \mathcal{X}_{s-1} \rangle = \mathbb{R}^{\cup}\langle \mathcal{Y} \rangle = \mathbb{R}\langle \Delta' \rangle$ and let $\pi : \mathbb{R}^d \to \mathcal{E}^{\perp}$ be the orthogonal projection. Since $\mathcal{R}$ is Δ-pure, we have $\mathbf{x}(i) = \tilde{\mathbf{x}}(i)$ for all $i \geq 1$ and $\mathbf{x} \in \mathcal{X}_s$, with $\pi(\mathbf{x}(i))$ constant. Thus $\pi(\mathcal{X}_s(k))$ is a fixed set for all tuples k. By Proposition 7.20, the set of one-dimensional half-spaces $\pi(\mathcal{X}_s) \in \mathfrak{X}(\pi(G_0))$ is irreducible, and thus $\pi(X_s(k)) \in X(\pi(G_0))$ is also irreducible. As a result, Theorem 7.24.1 implies there are only a finite number of possibilities for $\pi(\mathcal{X}_s)$

and $\pi(X_s(k))$. Thus it suffices to show there are only a finite number of possibilities for $\mathcal{X}$ when $\pi(\mathcal{X}_s) = \mathcal{Y}_s$ and also $\pi(X_s(k)) = Y_s$ are fixed, say for the fixed subsets $Y_s \subseteq \pi(G_0) \subseteq \mathcal{E}^\perp$ and $\mathcal{Y}_s \in \mathfrak{X}(\pi(G_0))$. Since there are only a finite number of possible choices for the sets $\partial(\{\mathbf{x}\}) \subseteq \mathcal{X}_1 \cup \ldots \cup \mathcal{X}_{s-1} = \mathcal{Y}$ for $\mathbf{x} \in \mathcal{X}_s$, we can likewise assume these sets are also fixed for $\mathcal{X}$.

To summarize, we have arbitrary fixed values for Δ, Δ', $\mathcal{Y} = \bigcup_{i=1}^{s-1} \mathcal{X}_i$, $\mathcal{Y}_s$, Y_s and $\mathcal{B}_{\mathbf{y}} \subseteq \mathcal{X}_1 \cup \ldots \cup \mathcal{X}_{s-1}$ for $\mathbf{y} \in \mathcal{Y}_s$ (we will later drop the subscript $\mathbf{y}$ when the corresponding set $\mathcal{B}$ is needed for only one $\mathbf{y} \in \mathcal{Y}_s$). Set $\mathcal{E} = \mathbb{R}^\cup\langle \mathcal{X}_1 \cup \ldots \cup \mathcal{X}_{s-1}\rangle = \mathbb{R}\langle \Delta' \rangle$ and let $\pi : \mathbb{R}^d \to \mathcal{E}^\perp$ be the orthogonal projection. Let Ω consist of all maximal series decompositions

$$\bigcup_{i=1}^{s-1} \mathcal{X}_i \cup \mathcal{X}_s \in \mathsf{Max}(\mathfrak{X}_s^*(G_0, \Delta))$$

having a Δ-pure realization $\mathcal{R} = (\mathcal{X}_1 \cup \{\mathbf{v}_1\}, \ldots, \mathcal{X}_s \cup \{\mathbf{v}_s\})$ of $\mathcal{X} = \bigcup_{i=1}^{s-1} \mathcal{X}_i \cup \mathcal{X}_s$ with $(\mathcal{X}_1 \cup \{\mathbf{v}_1\}, \ldots, \mathcal{X}_{s-1} \cup \{\mathbf{v}_{s-1}\})$ a Δ'-pure realization such that $\pi(\mathcal{X}_s) = \mathcal{Y}_s$, $\pi(X_s(k)) = Y_s$ for all tuples k, and $\partial(\{\mathbf{x}\}) = \mathcal{B}_{\pi(\mathbf{x})}$ for every $\mathbf{x} \in \mathcal{X}_s$. Our goal is to show Ω is finite for the arbitrary choice of fixed values used to define Ω.

Claim B Suppose $\bigcup_{i=1}^{s-1} \mathcal{X}_i \cup \mathcal{X}_s \in \Omega$ and $\bigcup_{i=1}^{s-1} \mathcal{X}_i \cup \mathcal{Z}_s \in \Omega$. If $\mathbf{x} \in \mathcal{X}_s$ and $\mathbf{z} \in \mathcal{Z}_s$ with $\pi(\mathbf{x}) = \pi(\mathbf{x})$, then $\bigcup_{i=1}^{s-1} \mathcal{X}_i \cup (\mathcal{X}_s \setminus \{\mathbf{x}\} \cup \{\mathbf{z}\}) \in \Omega$.

Proof Let $\bigcup_{i=1}^{s-1} \mathcal{X}_i \cup \mathcal{X}_s \in \Omega$ and $\bigcup_{i=1}^{s-1} \mathcal{X}_i \cup \mathcal{Z}_s \in \Omega$, and let

$$\mathcal{R}_{\mathcal{X}} = (\mathcal{X}_1 \cup \{\mathbf{v}_1\}, \ldots, \mathcal{X}_{s-1} \cup \{\mathbf{v}_{s-1}\}, \mathcal{X}_s \cup \{\mathbf{v}_s\})^* \quad \text{and}$$
$$\mathcal{R}_{\mathcal{Z}} = (\mathcal{X}_1 \cup \{\mathbf{w}_1\}, \ldots, \mathcal{X}_{s-1} \cup \{\mathbf{w}_{s-1}\}, \mathcal{Z}_s \cup \{\mathbf{w}_s\})$$

be Δ-pure realizations of $\mathcal{X}$ and $\mathcal{Z}$, respectively, with $(\mathcal{X}_1 \cup \{\mathbf{v}_1\}, \ldots, \mathcal{X}_{s-1} \cup \{\mathbf{v}_{s-1}\})$ and $(\mathcal{X}_1 \cup \{\mathbf{w}_1\}, \ldots, \mathcal{X}_{s-1} \cup \{\mathbf{w}_{s-1}\})$ each Δ'-pure realizations of $\mathcal{Y} = \mathcal{X} \setminus \mathcal{X}_s = \mathcal{Z} \setminus \mathcal{Z}_s$. Since each $\mathbf{w}_j$ is a maximal element in the poset of half-spaces from $\mathcal{R}_{\mathcal{Z}}$, we can swap each $\mathbf{w}_j$ for $\mathbf{v}_j$ (for $j \in [1, s-1]$), allowing us to assume $\mathbf{w}_j = \mathbf{v}_j$ for all $j \in [1, s-1]$, so that $\mathcal{R}_{\mathcal{Z}} = (\mathcal{X}_1 \cup \{\mathbf{v}_1\}, \ldots, \mathcal{X}_{s-1} \cup \{\mathbf{v}_{s-1}\}, \mathcal{Z}_s \cup \{\mathbf{w}_s\})$. For $\mathbf{x} \in \mathcal{X}_s$, there is some nonzero $y \in Y_s \subset \pi(G_0)$ such that $\pi(\mathbf{x}(i)) = y$ for all i. Then $y = \pi(\overline{y})$ for some $\overline{y} \in G_0 \subseteq \Lambda$, meaning each $\mathbf{x}(i) = \overline{y} + \xi_{\mathbf{x}}(i)$ for some $\xi_{\mathbf{x}}(i) \in \Lambda \cap \mathcal{E}$ (as $\ker \pi = \mathcal{E}$ and $\mathbf{x}(i), y \in \Lambda$). As discussed after (7.13), the value of $\mathbb{Z}\langle X(k)\rangle = \mathbb{Z}\langle Y(k)\rangle + \mathbb{Z}\langle X_k\rangle = \Delta' + \mathbb{Z}\langle X_s(k)\rangle$ depends solely on the values of $\xi_{\mathbf{x}}(i) \mod \Delta'$ for $\mathbf{x} \in \mathcal{X}_s$, with distinct possibilities for the $\xi_{\mathbf{x}}(i)$ modulo Δ' giving rise to distinct lattices. Thus, since $\Delta = \mathbb{Z}\langle X(k)\rangle$ and $\Delta' = \mathbb{Z}\langle Y(k)\rangle$ are fixed for all tuples k, it follows that there are $\xi_{\mathbf{y}} \in \Lambda \cap \mathcal{E}$ for $\mathbf{y} \in \mathcal{Y}_s$ such that $\xi_{\mathbf{x}}(i) \equiv \xi_{\mathbf{y}} \mod \Delta'$ for all i and $\mathbf{x} \in \mathcal{X}_s$ with $\pi(\mathbf{x}) = \mathbf{y}$. Applying these same arguments to $\mathcal{Z}_s$ instead of $\mathcal{X}_s$, we likewise conclude that $\xi_{\mathbf{z}}(i) \equiv \xi_{\mathbf{y}} \mod \Delta'$ for all i and $\mathbf{z} \in \mathcal{Z}_s$ with $\pi(\mathbf{z}) = \mathbf{y}$. As a result, if $\pi(\mathbf{x}) = \pi(\mathbf{z}) = \mathbf{y} \in \mathcal{Y}_s$, then $\mathbb{Z}\langle (X \setminus \{x\} \cup \{z\})(k)\rangle = \mathbb{Z}\langle X(k)\rangle = \Delta$ for all tuples k, as the values of Δ' and $\xi_{\mathbf{y}}$ modulo Δ' for $\mathbf{y} \in \mathcal{Y}_s$ completely determine the lattice Δ.

Let $\mathbf{x} \in \mathcal{X}_s$ and $\mathbf{z} \in \mathcal{Z}_s$ with $\pi(\mathbf{x}) = \pi(\mathbf{z})$. Then $\pi(\mathcal{X}_s \setminus \{\mathbf{x}\} \cup \{\mathbf{z}\}) = \pi(\mathcal{X}_s) = \mathcal{Y}_s$ with $\mathcal{R}'_{\mathcal{X}} = (\mathcal{X}_1 \cup \{\mathbf{v}_1\}, \ldots, \mathcal{X}_{s-1} \cup \{\mathbf{v}_{s-1}\}, (\mathcal{X}_s \setminus \{\mathbf{x}\} \cup \{\mathbf{z}\}) \cup \{\mathbf{v}_s\})$ a purely virtual Reay system over G_0, $\partial(\{\mathbf{z}\}) = \partial(\{\mathbf{x}\}) = \mathcal{B}_{\pi(\mathbf{x})} = \mathcal{B}_{\pi(\mathbf{z})}$ and $\pi(X_s \setminus \{x\} \cup \{z\})(k) = Y_s$ for all tuples k. In view of the conclusion of the previous paragraph, we see that $\mathcal{R}'_{\mathcal{X}}$ is Δ-pure, while $(\mathcal{X}_1 \cup \{\mathbf{v}_1\}, \ldots, \mathcal{X}_{s-1} \cup \{\mathbf{v}_{s-1}\})$ is Δ'-pure by assumption. Applying Proposition 7.20 to the maximal decomposition $\bigcup_{i=1}^{s} \mathcal{X}_i$ and using that $\pi(\mathcal{X}_s \setminus \{\mathbf{x}\} \cup \{\mathbf{z}\}) = \mathcal{Y}_s = \pi(\mathcal{X}_s)$ shows that the series decomposition $\bigcup_{i=1}^{s-1} \mathcal{X}_i \cup (\mathcal{X}_s \setminus \{\mathbf{x}\} \cup \{\mathbf{z}\})$ is maximal. It follows that $\bigcup_{i=1}^{s-1} \mathcal{X}_i \cup (\mathcal{X}_s \setminus \{\mathbf{x}\} \cup \{\mathbf{z}\}) \in \mathfrak{X}_s^*(G_0, \Delta)$. It remains to show $\bigcup_{i=1}^{s-1} \mathcal{X}_i \cup (\mathcal{X}_s \setminus \{\mathbf{x}\} \cup \{\mathbf{z}\}) \in \mathsf{Max}(\mathfrak{X}_s^*(G_0, \Delta))$ in order to complete the claim.

Assume by contradiction that $\bigcup_{i=1}^{s-1} \mathcal{X}'_i \cup (\mathcal{X}_s \setminus \{\mathbf{x}\} \cup \{\mathbf{z}\})'$ is a maximal Δ-pure series decomposition with $\bigcup_{i=1}^{s-1} \mathcal{X}_i \cup (\mathcal{X}_s \setminus \{\mathbf{x}\} \cup \{\mathbf{z}\}) \prec_{\cup}^{*} \bigcup_{i=1}^{s-1} \mathcal{X}'_i \cup (\mathcal{X}_s \setminus \{\mathbf{x}\} \cup \{\mathbf{z}\})'$. Let

$$\mathcal{R}' = (\mathcal{X}'_1 \cup \{\mathbf{v}'_1\}, \ldots, \mathcal{X}'_{s-1} \cup \{\mathbf{v}'_{s-1}\}, (\mathcal{X}_s \setminus \{\mathbf{x}\} \cup \{\mathbf{z}\})' \cup \{\mathbf{v}'_s\})$$

be a Δ-pure realization. Since each representative of a half-space $\mathbf{y}$ is also one for $\mathbf{y}'$, it follows from Proposition 5.3.1 that

$$\ker \pi = \mathbb{R}^{\cup}\langle \mathcal{X}_1 \cup \ldots \cup \mathcal{X}_{s-1} \rangle = \mathbb{R}^{\cup}\langle \mathcal{X}'_1 \cup \ldots \cup \mathcal{X}'_{s-1} \rangle. \tag{7.22}$$

For $\mathbf{y} \in \mathcal{X}_s \setminus \{\mathbf{x}\} \cup \{\mathbf{z}\}$, the condition $\mathbf{y} \subseteq \mathbf{y}'$ implies $\pi(\mathbf{y}) \subseteq \pi(\mathbf{y}')$ with both $\pi(\mathbf{y}')$ and $\pi(\mathbf{y})$ one-dimensional half-spaces (which follows by applying the definition of oriented Reay system to $\mathcal{R}_{\mathcal{X}}$, $\mathcal{R}_{\mathcal{Z}}$ and $\mathcal{R}'$), whence

$$\pi(\mathbf{x}) = \pi(\mathbf{z}) = \pi(\mathbf{z}') \quad \text{and} \quad \pi(\mathbf{y}) = \pi(\mathbf{y}') \text{ for all } \mathbf{y} \in \mathcal{X}_s \setminus \{\mathbf{x}\}. \tag{7.23}$$

Since $\mathcal{R}'$ is Δ-pure and $\mathcal{E} = \mathbb{R}^{\cup}\langle \mathcal{X}_1 \cup \ldots \cup \mathcal{X}_{s-1} \rangle = \mathbb{R}^{\cup}\langle \mathcal{X}'_1 \cup \ldots \cup \mathcal{X}'_{s-1} \rangle = \mathbb{R}\langle (X \setminus X_s)'(k) \rangle$ for all tuples k (by Proposition 5.3.1), we have

$$\Delta \cap \mathcal{E} = \Big(\mathbb{Z}\langle (X \setminus X_s)'(k) \rangle + \mathbb{Z}\langle (X_s \setminus \{x\} \cup \{z\})'(k) \rangle\Big) \cap \mathcal{E} = \mathbb{Z}\langle (X \setminus X_s)'(k) \rangle, \tag{7.24}$$

for all tuples k, where the final equality above follows since $\pi((X_s \setminus \{x\} \cup \{z\})'(k))$ is a linearly independent set of size $|X_s|$ by (OR2) for $\mathcal{R}'$. Since $\mathcal{R}_{\mathcal{X}} = (\mathcal{X}_1 \cup \{\mathbf{v}_1\}, \ldots, \mathcal{X}_s \cup \{\mathbf{v}_s\})$ is Δ-pure and $(\mathcal{X}_1 \cup \{\mathbf{v}_1\}, \ldots, \mathcal{X}_{s-1} \cup \{\mathbf{v}_{s-1}\})$ is Δ'-pure, we have

$$\Delta \cap \mathcal{E} = \Big(\mathbb{Z}\langle (X \setminus X_s)(k) \rangle + \mathbb{Z}\langle X_s(k) \rangle\Big) \cap \mathcal{E} = \mathbb{Z}\langle (X \setminus X_s)(k) \rangle = \Delta' \tag{7.25}$$

for all tuples k, with the second equality above following since we have $\ker \pi = \mathcal{E} = \mathbb{R}^{\cup}\langle \mathcal{X}_1 \cup \ldots \cup \mathcal{X}_{s-1} \rangle = \mathbb{R}\langle (X \setminus X_s)(k) \rangle$ (by Proposition 5.3.1) with $\pi(X_s(k))$ a linearly independent set of size $|X_s|$ (by (OR2) for $\mathcal{R}_{\mathcal{X}}$). Combining (7.24) and

(7.25), we find that

$$\Delta' = \Delta \cap \mathcal{E} = \mathbb{Z}\langle (X \setminus X_s)'(k)\rangle$$

for all tuples k, ensuring that $(\mathcal{X}'_1 \cup \{\mathbf{v}'_1\}, \ldots, \mathcal{X}'_{s-1} \cup \{\mathbf{v}'_{s-1}\})$ is Δ'-pure.

We have

$$\begin{aligned}\mathbb{Z}\langle Y_s\rangle &= \mathbb{Z}\langle \pi\big((X_s)(k)\big)\rangle = \pi\Big(\mathbb{Z}\langle (X \setminus X_s)(k)\rangle + \mathbb{Z}\langle X_s(k)\rangle\Big) = \pi(\Delta)\\ &= \pi\Big(\mathbb{Z}\langle (X \setminus X_s)'(k)\rangle + \mathbb{Z}\langle (X_s \setminus \{x\} \cup \{z\})(k)\rangle\Big)\\ &= \mathbb{Z}\langle \pi\Big((X_s \setminus \{x\} \cup \{z\})'(k)\Big)\rangle, \end{aligned} \tag{7.26}$$

for all tuples k, with the first equality since $\bigcup_{i=1}^{s} \mathcal{X}_i \in \Omega$, the third since $\mathcal{R}_{\mathcal{X}}$ is Δ-pure, the fourth since $\mathcal{R}'$ is Δ-pure, and the second and fifth since $\ker \pi = \mathbb{R}^{\cup}\langle \mathcal{X}_1 \cup \ldots \cup \mathcal{X}_{s-1}\rangle = \mathbb{R}\langle (X \setminus X_s)(k)\rangle = \mathbb{R}\langle (X \setminus X_s)'(k)\rangle = \mathbb{R}^{\cup}\langle \mathcal{X}'_1 \cup \ldots \cup \mathcal{X}'_{s-1}\rangle$. In view of (7.23), we have $\pi(\mathbf{z}') = \pi(\mathbf{x})$, which is a one-dimensional ray *positively* spanned by an element from Y_s (per definition of Ω). Likewise, each $\mathbf{y} \in \mathcal{X}_s \setminus \{\mathbf{x}\}$ has $\pi(\mathbf{y}') = \pi(\mathbf{y})$ being a one-dimensional ray *positively* spanned by an element from Y_s. Thus, since Y_s is linearly independent, it follows that only way (7.26) can hold is if $\pi\big((X_s \setminus \{x\} \cup \{z\})'(k)\big) = Y_s$ for all tuples k.

As a result, $\mathcal{R}'$ now satisfies all the same hypotheses as $\mathcal{R}_{\mathcal{X}}$ and $\mathcal{R}_{\mathcal{Z}}$ needed to define analogously the elements $\xi_{\mathbf{y}'}(i)$ for $\mathbf{y} \in \bigcup_{i=1}^{s-1} \mathcal{X}_i \cup (\mathcal{X}_s \setminus \{\mathbf{x}\} \cup \{\mathbf{z}\})$ for $\mathcal{R}'$ as they were defined for $\mathcal{R}_{\mathcal{X}}$ and $\mathcal{R}_{\mathcal{Z}}$. By (7.23) and the discussion after (7.13), it again follows that

$$\xi_{\mathbf{y}'}(i) \equiv \xi_{\pi(\mathbf{y}')} = \xi_{\pi(\mathbf{y})} \mod \Delta' \tag{7.27}$$

for all $\mathbf{y} \in \mathcal{X}_s \setminus \{\mathbf{x}\} \cup \{\mathbf{z}\}$ and $i \geq 1$

Suppose $\mathbf{y} = \mathbf{y}'$ for all half-spaces $\mathbf{y} \in \bigcup_{i=1}^{s-1} \mathcal{X}_i \cup (\mathcal{X}_s \setminus \{\mathbf{x}\})$. Then we must have $\mathbf{z} \subset \mathbf{z}'$ since $\bigcup_{i=1}^{s-1} \mathcal{X}_i \cup (\mathcal{X}_s \setminus \{\mathbf{x}\} \cup \{\mathbf{z}\}) \prec^*_{\cup} \bigcup_{i=1}^{s-1} \mathcal{X}'_i \cup (\mathcal{X}_s \setminus \{\mathbf{x}\} \cup \{\mathbf{z}\})'$. In such case, since $\pi(\mathbf{z}') = \pi(\mathbf{z})$ (as follows by (7.23)) and $\partial(\{\mathbf{z}'\}) \subseteq \bigcup_{i=1}^{s-1} \mathcal{X}'_i = \bigcup_{i=1}^{s-1} \mathcal{X}_i$ (as $\mathbf{y} = \mathbf{y}'$ for all other half-spaces), it follows that $\bigcup_{i=1}^{s-1} \mathcal{X}_i \cup (\mathcal{Z}_s \setminus \{\mathbf{z}\} \cup \{\mathbf{z}'\})$ is a series decomposition. Since $\pi(\mathcal{Z}_s \setminus \{\mathbf{z}\} \cup \{\mathbf{z}'\}) = \pi(\mathcal{Z}_s)$ (by (7.23)), it follows from Proposition 7.20 applied to the maximal series decomposition $\bigcup_{i=1}^{s-1} \mathcal{X}_i \cup \mathcal{Z}_s$, and then to $\bigcup_{i=1}^{s-1} \mathcal{X}_i \cup (\mathcal{Z}_s \setminus \{\mathbf{z}\} \cup \{\mathbf{z}'\})$, that $\bigcup_{i=1}^{s-1} \mathcal{X}_i \cup (\mathcal{Z}_s \setminus \{\mathbf{z}\} \cup \{\mathbf{z}'\})$ is maximal. Moreover, $\bigcup_{i=1}^{s-1} \mathcal{X}_i \cup (\mathcal{Z}_s \setminus \{\mathbf{z}\} \cup \{\mathbf{z}'\})$ is Δ-pure since $\mathcal{R}_{\mathcal{Z}}$ is Δ-pure with $\xi_{\mathbf{z}'}(i) \equiv \xi_{\pi(\mathbf{z}')} = \xi_{\pi(\mathbf{z})} \mod \Delta'$ by (7.27). Thus $\bigcup_{i=1}^{s-1} \mathcal{X}_i \cup \mathcal{Z}_s \prec^*_{\cup} \bigcup_{i=1}^{s-1} \mathcal{X}_i \cup (\mathcal{Z}_s \setminus \{\mathbf{z}\} \cup \{\mathbf{z}'\})$, which contradicts that $\bigcup_{i=1}^{s-1} \mathcal{X}_i \cup \mathcal{Z}_s \in \mathsf{Max}\big(\mathfrak{X}^*_s(G_0, \Delta)\big)$. So we can instead assume $\mathbf{y}_0 \subset \mathbf{y}'_0$ for some $\mathbf{y}_0 \in \bigcup_{i=1}^{s-1} \mathcal{X}_i \cup (\mathcal{X}_s \setminus \{\mathbf{x}\})$.

Let $\vec{u}_{\mathbf{x}} = (u_1, \ldots, u_t)$. Since $\partial(\{\mathbf{x}\}) \subseteq \mathcal{X}_1 \cup \ldots \cup \mathcal{X}_{s-1}$ encases $-\vec{u}^{\lhd}_{\mathbf{x}}$, we have $u_1, \ldots, u_{t-1} \in \mathbb{R}^{\cup}\langle \partial(\{\mathbf{x}\})\rangle \subseteq \mathbb{R}^{\cup}\langle \mathcal{X}_1 \cup \ldots \cup \mathcal{X}_{s-1}\rangle = \mathbb{R}^{\cup}\langle \mathcal{X}'_1 \cup \ldots \cup \mathcal{X}'_{s-1}\rangle$ (by (7.22)). Thus Proposition 7.7.3 applied to a realization of $\bigcup_{i=1}^{s-1} \mathcal{X}'_i$ implies that, there is a subset $C \subseteq \mathcal{X}'_1 \cup \ldots \cup \mathcal{X}'_{s-1}$ that minimally encases $-\vec{u}^{\triangleleft}_{\mathbf{x}}$, allowing us to define a half-space $\mathbf{x}'$ with $\partial(\{\mathbf{x}'\}) = C$, $\vec{u}_{\mathbf{x}'} = \vec{u}_{\mathbf{x}}$ and $\mathbf{x}'(i) = \mathbf{x}(i)$ for all i, so $\overline{\mathbf{x}'} = \mathbb{R}^{\cup}\langle C\rangle + \mathbb{R}_+ u_t$. Then $\pi(\mathbf{x}') = \pi(\mathbf{z}') = \pi(\mathbf{x}) = \mathbb{R}_+\pi(u_t)$ is a one-dimensional half-space (as $\ker \pi = \mathbb{R}^{\cup}\langle \mathcal{X}_1 \cup \ldots \cup \mathcal{X}_{s-1}\rangle$) with

$$\pi((\mathcal{X}_s \setminus \{\mathbf{x}\} \cup \{\mathbf{z}\})') = \pi(\mathcal{X}_s) = \pi(\mathcal{X}'_s) = \mathcal{Y}_s$$

by (7.23). Combined with (7.22) and Proposition 7.20, we find $\bigcup_{i=1}^{s-1} \mathcal{X}'_i \cup \mathcal{X}'_s$ is a maximal series decomposition since $\bigcup_{i=1}^{s-1} \mathcal{X}'_i \cup (\mathcal{X}_s \setminus \{\mathbf{x}\} \cup \{\mathbf{z}\})'$ is a maximal series decomposition. We have $\xi_{\mathbf{z}'}(i) \equiv \xi_{\pi(\mathbf{z}')} = \xi_{\pi(\mathbf{z})} = \xi_{\pi(\mathbf{x})} \equiv \xi_{\mathbf{x}}(i) \mod \Delta'$ for all i by (7.27) and (7.23). Thus, since $\mathbf{x}'(i) = \mathbf{x}(i)$ with $\bigcup_{i=1}^{s-1} \mathcal{X}'_i \cup (\mathcal{X}_s \setminus \{\mathbf{x}\} \cup \{\mathbf{z}\})'$ Δ-pure, it follows that $\bigcup_{i=1}^{s-1} \mathcal{X}'_i \cup \mathcal{X}'_s$ is also Δ-pure. We have now established that $\bigcup_{i=1}^{s} \mathcal{X}'_i \in \mathfrak{X}^*_s(G_0, \Delta)$.

Since $C \subseteq \mathcal{X}'_1 \cup \ldots \cup \mathcal{X}'_{s-1}$ and $\partial(\{\mathbf{x}\}) \subseteq \mathcal{X}_1 \cup \ldots \cup \mathcal{X}_{s-1}$ both minimally encase $-\vec{u}^{\triangleleft}_{\mathbf{x}} = -\vec{u}^{\triangleleft}_{\mathbf{x}'}$, it follows from Proposition 7.26 and Proposition 5.9.1 that $(\partial(\{\mathbf{x}\})')^* = C$, whence $\partial(\{\mathbf{x}\})' \subseteq {\downarrow} C$, ensuring

$$\partial(\mathbf{x}) = \mathbb{R}^{\cup}\langle \partial(\{\mathbf{x}\})\rangle \subseteq \mathbb{R}^{\cup}\langle \partial(\{\mathbf{x}\})'\rangle \subseteq \mathbb{R}^{\cup}\langle {\downarrow} C\rangle = \mathbb{R}^{\cup}\langle {\downarrow}\partial(\{\mathbf{x}'\})\rangle = \partial(\mathbf{x}') \quad \text{and}$$
$$\partial(\{\mathbf{x}\}) \cap \mathbf{x} = \mathsf{C}^{\cup}(\partial(\{\mathbf{x}\})) \subseteq \mathsf{C}^{\cup}(\partial(\{\mathbf{x}\})') \subseteq \mathsf{C}^{\cup}({\downarrow} C) = \mathsf{C}^{\cup}({\downarrow}\partial(\{\mathbf{x}'\})) = \partial(\mathbf{x}') \cap \mathbf{x}',$$

with the first inclusion in both lines above since $\mathbf{y} \subseteq \mathbf{y}'$ for all $\mathbf{y} \in \partial(\{\mathbf{x}\})$. Combined with the fact that u_t is representative for both $\mathbf{x}$ and $\mathbf{x}'$, it follows that $\mathbf{x} \subseteq \mathbf{x}'$ with every representative for $\mathbf{x}$ one for $\mathbf{x}'$. Since $\bigcup_{i=1}^{s-1} \mathcal{X}_i \cup (\mathcal{X}_s \setminus \{\mathbf{x}\} \cup \{\mathbf{z}\}) \prec^*_{\cup} \bigcup_{i=1}^{s-1} \mathcal{X}'_i \cup ((\mathcal{X}_s \setminus \{\mathbf{x}\})' \cup \{\mathbf{z}'\})$, we have $\mathbf{y} \subseteq \mathbf{y}'$ and $\mathbf{y}^\circ \subseteq (\mathbf{y}')^\circ$ for all $\mathbf{y} \in \bigcup_{i=1}^{s-1} \mathcal{X}_i \cup (\mathcal{X}_s \setminus \{\mathbf{x}\})$. It follows that $\bigcup_{i=1}^{s-1} \mathcal{X}_i \cup \mathcal{X}_s \prec^*_{\cup} \bigcup_{i=1}^{s-1} \mathcal{X}'_i \cup \mathcal{X}'_s$, with the relation strict since $\mathbf{y}_0 \subset \mathbf{y}'_0$ for some half-space $\mathbf{y}_0 \in \bigcup_{i=1}^{s-1} \mathcal{X}_i \cup (\mathcal{X}_s \setminus \{\mathbf{x}\})$, which contradicts that $\bigcup_{i=1}^{s-1} \mathcal{X}_i \cup \mathcal{X}_s \in \mathsf{Max}(\mathfrak{X}^*_s(G_0, \Delta))$, completing Claim B.

We continue assuming $\bigcup_{i=1}^{s} \mathcal{X}_i \in \Omega$ with all notation as defined earlier. In particular, $\mathcal{R} = (\mathcal{X}_1 \cup \{\mathbf{v}_1\}, \ldots, \mathcal{X}_s \cup \{\mathbf{v}_s\})$ is Δ-pure realization of $\mathcal{X} = \bigcup_{i=1}^{s} \mathcal{X}_i$. Let $\mathbf{y} \in \mathcal{Y}_s$ be arbitrary. Then there is a support set $\mathcal{B} = \mathcal{B}_{\mathbf{y}} \subseteq \mathcal{X}_1 \cup \ldots \cup \mathcal{X}_{s-1} = \mathcal{Y}$ and some representative $y \in Y_s \subseteq \pi(G_0)$ for $\mathbf{y}$, such that, if $\mathbf{x} \in \mathcal{X}_s$ is any potential half-space with $\pi(\mathbf{x}) = \mathbf{y}$, then $\partial(\{\mathbf{x}\}) = \mathcal{B}$ and $\pi(\mathbf{x}(i)) = y \in \mathcal{E}^{\perp}$ for all $i \geq 1$. Note $y \neq 0$ since it is a representative of a half-space from $\mathcal{Y}_s \in \mathfrak{X}(\pi(G_0))$. Let

$$\pi_{\mathcal{B}} : \mathbb{R}^d \to \mathbb{R}^{\cup}\langle \mathcal{B}\rangle^{\perp}$$

be the orthogonal projection. Letting $\vec{u}_{\mathbf{x}} = (u^{(1)}_{\mathbf{x}}, \ldots, u^{(t_{\mathbf{x}})}_{\mathbf{x}})$, we have (by Proposition 6.8)

$$\mathbf{x}(i) = (a^{(1)}_{i,\mathbf{x}} u^{(1)}_{\mathbf{x}} + \ldots + a^{(t_{\mathbf{x}}-1)}_{i,\mathbf{x}} u^{(t_{\mathbf{x}}-1)}_{\mathbf{x}} + w_{i,\mathbf{x}}) + \xi_{\mathbf{x}} + y \tag{7.28}$$

for some $a^{(j)}_{i,\mathbf{x}} > 0$, $w_{i,\mathbf{x}} \in \mathbb{R}^{\cup}\langle\mathcal{B}\rangle$ and $\xi_{\mathbf{x}} \in \mathcal{E} \cap \mathbb{R}^{\cup}\langle\mathcal{B}\rangle^{\perp}$, with $a^{(j)}_{i,\mathbf{x}} \to \infty$ for $j \in [1, t_{\mathbf{x}} - 1]$, $a^{(j)}_{i,\mathbf{x}} \in o(a^{(j-1)}_{i,\mathbf{x}})$ for $j \in [2, t_{\mathbf{x}} - 1]$, and $\|w_{i,\mathbf{x}}\| \in o(a^{(t_{\mathbf{x}}-1)}_{i,\mathbf{x}})$. Note $\xi_{\mathbf{x}} + y$ is a representative for the half-space $\mathbf{x}$. Since $\xi_{\mathbf{x}} \in \mathcal{E} \cap \mathbb{R}^{\cup}\langle\mathcal{B}\rangle^{\perp}$ and $y \in \mathcal{E}^{\perp}$ is nonzero, it follows that distinct values for $\xi_{\mathbf{x}}$ determine distinct half-spaces $\mathbf{x}$. Also, we have $\pi_{\mathcal{B}}(\mathbf{x}(i)) = \xi_{\mathbf{x}} + y \in \pi_{\mathcal{B}}(G_0) \subseteq \pi_{\mathcal{B}}(\Lambda)$, while $\pi_{\mathcal{B}}(\Lambda)$ is a lattice by Proposition 2.1. The set $\mathcal{B}$ is empty precisely when $t_{\mathbf{x}} = 1$. For each $\mathbf{z} \in \mathcal{B}$, let

$$\mathcal{B}_{\mathbf{z}} = \mathcal{B} \setminus \{\mathbf{z}\} \cup \partial(\{\mathbf{z}\})$$

and let $\pi_{\mathbf{z}} : \mathbb{R}^d \to \mathbb{R}^{\cup}\langle\mathcal{B}_{\mathbf{z}}\rangle^{\perp}$ be the orthogonal projection.

If, for each $\mathbf{y} \in \mathcal{Y}_s$, there are only a finite number of possible half-spaces $\mathbf{x}$ with $\pi(\mathbf{x}) = \mathbf{y}$ that occur for the $\mathcal{X} \in \Omega$, then Ω will be finite, as desired. Therefore we can assume there is some $\mathbf{y} \in \mathcal{Y}_s$ for which this fails, instead having an infinite sequence of distinct half-spaces $\{\mathbf{x}_j\}_{j=1}^{\infty}$ with $\pi(\mathbf{x}_j) = \mathbf{y}$ for all $j \geq 1$, and each $\mathbf{x}_j$ lying in some $\mathcal{X}^{(j)}_s$ with $\bigcup_{i=1}^{s-1} \mathcal{X}_i \cup \mathcal{X}^{(j)}_s \in \Omega$. We use the notation of the previous paragraph with each $\mathbf{x}_j$ except that we replace everywhere the subscript $\mathbf{x} = \mathbf{x}_j$ by j (to lighten notation some). We also write $\mathbf{x}_j(i) := \mathbf{x}(i, j)$ for $i, j \geq 1$, to emphasize the dependence of the representation in (7.28) on both the parameters i and j. Let

$$\mathbf{x}_{\mathcal{B}}(i, j) = a^{(1)}_{i,j} u^{(1)}_j + \ldots + a^{(t_j-1)}_{i,j} u^{(t_j-1)}_j + w_{i,j} \in \mathbb{R}^{\cup}\langle\mathcal{B}\rangle,$$

so $\mathbf{x}(i, j) = \mathbf{x}_{\mathcal{B}}(i, j) + \xi_j + y$.

In view of Claim B, we can assume the other half-spaces in $\mathcal{X}^{(j)}_s$ remain fixed and consider the half-space $\mathbf{x}_j \in \mathcal{X}_s$, corresponding to $\mathbf{y}$, as varying. Then, each choice of a half-space $\mathbf{x}_j$ with $\pi(\mathbf{x}_j) = \mathbf{y}$, gives rise to a set $\mathcal{X}^{(j)}_s$ with $\bigcup_{i=1}^{s-1} \mathcal{X}_i \cup \mathcal{X}^{(j)}_s \in \Omega$ and $\mathcal{X}^{(j)} := \bigcup_{i=1}^{s-1} \mathcal{X}_i \cup \mathcal{X}^{(j)}_s$, where $\mathcal{X}^{(j)} \setminus \{\mathbf{x}_j\}$ is a fixed set. Note, we will for most of the argument drop the super-scripts (j) as this information will be irrelevant to all but the final arguments.

Since the half-spaces $\mathbf{x}_j$ are distinct with $\partial(\{\mathbf{x}_j\}) = \mathcal{B}$ fixed, the representatives $\xi_j + y \in \pi_{\mathcal{B}}(G_0) \subseteq \pi_{\mathcal{B}}(\Lambda)$ are also all distinct, and since these are lattice points, this means the sequence $\{\xi_j\}_{j=1}^{\infty}$ is unbounded. Thus, by passing to a subsequence, we can assume

$$\|\xi_j\| \to \infty \quad \text{with} \quad \|\xi_j\| \geq 1 \quad \text{for all } j,$$

and that $\{\xi_j\}_{j=1}^{\infty}$ is an asymptotically filtered sequence of terms with fully unbounded limit. □

Claim C If $\vec{v} = (v_1, \ldots, v_t) \in G_0^{\lim}$ with $v_1, \ldots, v_t \in \mathbb{R}^{\cup}\langle\mathcal{B}\rangle$, then $\mathcal{B}$ encases $-\vec{v}$.

Proof The set $\mathcal{B} = \partial(\{\mathbf{x}\})$ minimally encases the fully unbounded (or trivial) limit $-\vec{u}^{\triangleleft}_{\mathbf{x}}$ (by Proposition 6.8), and since $\mathcal{B} \subseteq \mathcal{X}_1 \cup \ldots \cup \mathcal{X}_{s-1}$, the encasement is urbane with $\mathcal{B}$ a support set. Proposition 6.5 implies there is a virtual Reay system $\mathcal{R}_{\mathcal{B}} =$

$(\mathcal{Z}_1 \cup \{\mathbf{z}_1\}, \ldots, \mathcal{Z}_{s'} \cup \{\mathbf{z}_{s'}\})$ over G_0 with $\downarrow\mathcal{B} = \bigcup_{i=1}^{s'} \mathcal{Z}_i$. If $\mathcal{B} = \emptyset$, then $\mathcal{R}_\mathcal{B}$ is the trivial (empty) virtual Reay system. Otherwise, $\vec{u}_\mathbf{x}^\triangleleft$ is fully unbounded, and the strict truncation of any $\vec{u}_\mathbf{y}$ with $\mathbf{y} \in \downarrow\mathcal{B} \subseteq \mathcal{X}_1 \cup \ldots \cup \mathcal{X}_{s-1}$ is also either trivial or fully unbounded (by Proposition 6.8). Thus, per the comments above Proposition 6.5, $\mathcal{R}_\mathcal{B}$ is purely virtual. Claim C now follows from Proposition 7.7.3 applied to $\mathcal{R}_\mathcal{B}$. □

We aim to contradict the maximality of $\bigcup_{i=1}^{s} \mathcal{X}_i$ by using the infinite sequence of distinct half-spaces $\{\mathbf{x}_j\}_{j=1}^\infty$ to construct a new, strictly larger half-space $\mathbf{u}$. To define $\mathbf{u}$, we will need to construct an appropriate representation sequence $\{\mathbf{u}(j)\}_{j=1}^\infty$. We define $\mathbf{u}(j) = \mathbf{x}\big(i(j), j\big) \in G_0$, where $i : \mathbb{Z}_+ \setminus \{0\} \to \mathbb{Z}_+$ is a function that associates to each $j \geq 1$ a sufficiently large index i, constructed as follows. If $\mathcal{B} = \emptyset$, then simply take $i(j) = j$ for all $j \geq 1$. Otherwise, for each fixed $j \geq 1$, the sequence $\{\mathbf{x}_\mathcal{B}(i, j)\}_{i=1}^\infty$ is an asymptotically filtered sequence of terms with limit $\vec{u}_j^\triangleleft$ such that $-\vec{u}_j^\triangleleft$ is fully unbounded (by Proposition 6.8) and minimally encased by $\mathcal{B} = \partial(\{\mathbf{x}_j\})$. Thus Lemma 7.25.1 implies that $\|\pi_\mathbf{z}(\mathbf{x}_\mathcal{B}(i, j))\| \to \infty$ for all $\mathbf{z} \in \mathcal{B}$ (as $i \to \infty$). Since ξ_j is a fixed element, it is thus possible to choose $i(j)$ to be sufficiently large so that

$$\|\pi_\mathbf{z}\big(\mathbf{x}_\mathcal{B}\big(i, j\big)\big)\| \geq 2^j \|\xi_j\| \geq 2^j \quad \text{for all } i \geq i(j) \tag{7.29}$$

and all $\mathbf{z} \in \mathcal{B}$. For $j \geq 1$ and $\mathcal{B} \neq \emptyset$, define $i(j)$ to be any sufficiently large index such that (7.29) holds. Note that, even if we pass to a subsequence of $\{\mathbf{x}_j\}_{j=1}^\infty$, then (7.29) remains true under the new re-indexing of the remaining terms from $\{\mathbf{x}_j\}_{j=1}^\infty$. With the function $i : \mathbb{Z}_+ \setminus \{0\} \to \mathbb{Z}_+$ fixed, we set

$$\mathbf{u}_\mathcal{B}(j) = \mathbf{x}_\mathcal{B}\big(i(j), j\big) \in \mathbb{R}^\cup\langle\mathcal{B}\rangle \quad \text{and} \quad \mathbf{u}(j) = \mathbf{x}\big(i(j), j\big) = \mathbf{u}_\mathcal{B}(j) + \xi_j + y \in G_0.$$

If $\mathcal{B} = \emptyset$, then $\mathbf{u}_\mathcal{B}(j) = 0$ and $\mathbf{u}(j) = \xi_j + y$ for all $j \geq 1$. Otherwise, (7.29) ensures that

$$\|\pi_\mathbf{z}(\mathbf{u}_\mathcal{B}(j))\| \to \infty \quad \text{and} \quad \|\xi_j\| \in o(\|\pi_\mathbf{z}(\mathbf{u}_\mathcal{B}(j))\|) \quad \text{for all } \mathbf{z} \in \mathcal{B}. \tag{7.30}$$

In particular, since $\|\mathbf{u}_\mathcal{B}(j)\| \geq \|\pi_\mathbf{z}(\mathbf{u}_\mathcal{B}(j))\|$, we have

$$\|\mathbf{u}_\mathcal{B}(j)\| \to \infty \quad \text{and} \quad \|\xi_j\| \in o(\|\mathbf{u}_\mathcal{B}(j)\|) \quad \text{when } \mathcal{B} \neq \emptyset. \tag{7.31}$$

If $\mathcal{B} \neq \emptyset$, then (7.31) gives $\|\mathbf{u}_\mathcal{B}(j)\| \to \infty$. Thus, by passing to an appropriate subsequence of $\{\mathbf{x}_j\}_{j=1}^\infty$, we can assume $\{\mathbf{u}_\mathcal{B}(j)\}_{j=1}^\infty$ is an asymptotically filtered sequence with complete fully unbounded limit $\vec{v} = (v_1, \ldots, v_t)$. Let

$$\mathbf{u}_\mathcal{B}(j) = \mathbf{x}_\mathcal{B}\big(i(j), j\big) = \alpha_j^{(1)} v_1 + \ldots + \alpha_j^{(t)} v_t + \varepsilon_j$$

be the representation of $\mathbf{u}_\mathcal{B}(j)$ as an asymptotically filtered sequence with limit $\vec{v}$, so $\alpha_j^{(r)} > 0$ and $\alpha_j^{(r)} \to \infty$ for $r \in [1, t]$, $\alpha_j^{(r)} \in o(\alpha_j^{(r-1)})$ for $r \in [2, t]$, and $\|\varepsilon_j\|$

is bounded. Since $\mathbf{u}_{\mathcal{B}}(j) \in \mathbb{R}^{\cup}\langle\mathcal{B}\rangle$ for all j, we must have $v_1, \ldots, v_t \in \mathbb{R}^{\cup}\langle\mathcal{B}\rangle$. When $\mathcal{B} = \emptyset$, we instead set $\vec{v}$ to be the empty tuple, so $t = 0$ and $\varepsilon_j = 0$ for all j.

Claim D $-\vec{v}$ is encased by $\mathcal{B}$.

Proof If $\mathcal{B} = \emptyset$, then the claim is trivial, so we may assume $\mathcal{B} \neq \emptyset$. Note that $\mathbf{u}(j) = \mathbf{u}_{\mathcal{B}}(j) + \xi_j + y \in G_0$ for all j. Since $\alpha_j^{(r)} \to \infty$ for all $r \in [1, t]$, we have the constant sequence $y \in o(\alpha_j^{(r)})$ for all $r \in [1, t]$. Since $\alpha_j^{(1)} \sim \|\mathbf{u}_{\mathcal{B}}(j)\|$, (7.31) implies that $\|\xi_j\| \in o(\alpha_j^{(1)})$. Thus there is a maximal index $t' \in [1, t]$ such that

$$\|\xi_j\| \in o(\alpha_j^{(r)}) \text{ for all } r \in [1, t'].$$

In such case, $\{\mathbf{u}(j)\}_{j=1}^{\infty}$ is an asymptotically filtered sequence of terms from G_0 with fully unbounded limit $\vec{v}' = (v_1, \ldots, v_{t'})$, in which case $-\vec{v}'$ is encased by $\mathcal{B}$ in view of Claim C (note we need $\mathbf{u}(j) \in G_0$ to use Claim C, so we cannot directly apply Claim C to $\{\mathbf{u}_{\mathcal{B}}(j)\}_{j=1}^{\infty}$). Let $\mathcal{B}' \preceq \mathcal{B}$ be such that $\mathcal{B}'$ minimally encases $-\vec{v}'$. Since $\mathcal{B}' \preceq \mathcal{B} \subseteq X_1 \cup \ldots \cup X_{s-1}$, the encasement of $-\vec{v}'$ by $\mathcal{B}'$ is urbane with $\mathcal{B}'$ a support set. Thus we can apply Proposition 5.9 and let

$$\emptyset = \mathcal{B}'_0 \prec \mathcal{B}'_1 \prec \ldots \prec \mathcal{B}'_\ell = \mathcal{B}' \preceq \mathcal{B}$$

be the support sets and $1 = r_1 < \ldots < r_\ell < r_{\ell+1} = t' + 1$ be the indices given by Proposition 5.9 applied to the encasement of $-\vec{v}'$ by $\mathcal{B}'$. If $\mathcal{B}' = \mathcal{B}$, then $\mathcal{B} = \mathcal{B}'$ encases $-\vec{v}$ in view of $v_1, \ldots, v_t \in \mathbb{R}^{\cup}\langle\mathcal{B}\rangle$ and Proposition 5.9.2 (particularly, part (c)), yielding the claim. Therefore we may assume $\mathcal{B}' \prec \mathcal{B}$, whence $\mathcal{B}' \preceq \mathcal{B}_{\mathbf{z}}$ for some $\mathbf{z} \in \mathcal{B}$, implying $\mathcal{B}' \subseteq \downarrow\mathcal{B}_{\mathbf{z}}$. Hence $\pi_{\mathbf{z}}(v_i) = 0$ for all $i \in [1, t']$.

If $\pi_{\mathbf{z}}(v_i) = 0$ for all $i \in [1, t]$, then $\|\pi_{\mathbf{z}}(\mathbf{u}_{\mathcal{B}}(j))\| = \|\pi_{\mathbf{z}}(\varepsilon_j)\|$, which is bounded as $\vec{v}$ is a *complete* fully unbounded limit for $\mathbf{u}_{\mathcal{B}}(j)$, contradicting (7.30). Therefore there must be some minimal $t'' \in [t' + 1, t]$ such that $\pi_{\mathbf{z}}(u_{t''}) \neq 0$. But then $\|\pi_{\mathbf{z}}(\mathbf{u}_{\mathcal{B}}(j))\| \in \Theta(\alpha_j^{(t'')})$, which combined with $\|\xi_j\| \in o(\|\pi_{\mathbf{z}}(\mathbf{u}_{\mathcal{B}}(j))\|)$ from (7.30) yields $\|\xi_j\| \in o(\alpha_j^{(t'')})$. In consequence, since $\alpha_j^{(r)} \in o(\alpha_j^{(r-1)})$ for all $r \in [2, t]$, we have $\|\xi_j\| \in o(\alpha_j^{(r)})$ for all $r \in [1, t'']$, implying by the maximality in the definition of t' that $t' \geq t''$, contradicting that $t'' \in [t' + 1, t]$ as shown above. This establishes Claim D. □

If $\mathcal{B} \neq \emptyset$, then Claim D implies that $\{-\mathbf{u}_{\mathcal{B}}(j)\}_{j=1}^{\infty}$ is an asymptotically filtered sequence of terms with complete fully unbounded limit $-\vec{v}$ encased by $\mathcal{B}$, while $\|\pi_{\mathbf{z}}(\mathbf{u}_{\mathcal{B}}(j))\|$ is unbounded by (7.30) for all $\mathbf{z} \in \mathcal{B}$. Thus Lemma 7.25.2 implies that $\mathcal{B}$ minimally encases $-\vec{v}$, a fact which is trivially true when $\mathcal{B} = \emptyset$ as well. Since $\mathcal{B} \subseteq X_1 \cup \ldots \cup X_{s-1}$, the encasement is urbane. Let

$$\emptyset = \mathcal{B}_0 \prec \mathcal{B}_1 \prec \ldots \prec \mathcal{B}_\ell = \mathcal{B}$$

be the support sets and $1 = r_1 < \ldots < r_\ell < r_{\ell+1} = t+1$ be the indices given by Proposition 5.9 applied to the encasement of $-\vec{v}$ by $\mathcal{B}$. By (7.30), we have $\|\xi_j\| \in o(\|\pi_{\mathbf{z}}(\mathbf{u}_{\mathcal{B}}(j))\|)$ for all $\mathbf{z} \in \mathcal{B}$, and thus also $\|\xi_j + y\| \in o(\|\pi_{\mathbf{z}}(\mathbf{u}_{\mathcal{B}}(j))\|)$ for all $\mathbf{z} \in \mathcal{B}$ as $\|\pi_{\mathbf{z}}(\mathbf{u}_{\mathcal{B}}(j))\| \to \infty$ by (7.30). Thus Lemma 7.25.3 (applied with j replacing i, with x_i taken to be $\mathbf{u}_{\mathcal{B}}(j)$, and with y_i taken to be $\xi_j + y$) implies that $\{\mathbf{u}(j)\}_{j=1}^{\infty}$ is an asymptotically filtered sequence of terms from G_0 with fully unbounded limit $(v_1, \ldots, v_{r_\ell})$ (after discarding the first few terms), when $\mathcal{B} \neq \emptyset$. However, when $\mathcal{B} = \emptyset$, we have $\mathbf{u}(j) = \xi_j + y$, which is also an asymptotically filtered sequence with fully unbounded limit, which we can set equal to $(v_1, \ldots, v_{r_\ell})$. By passing to a subsequence of $\{\mathbf{u}(j)\}_{j=1}^{\infty}$, we may assume $\{\mathbf{u}(j)\}_{j=1}^{\infty}$ is an asymptotically filtered sequence of terms from G_0 with *complete* fully unbounded limit $\vec{w} = (v_1, \ldots, v_{r_\ell}, w_1, \ldots, w_n)$.

Since $\{\mathbf{u}(j)\}_{j=1}^{\infty}$ is asymptotically filtered with complete *fully unbounded* limit $\vec{w}$, so too is the sequence $\{\mathbf{u}(j) - y\}_{j=1}^{\infty}$ obtained by translating all terms by a fixed constant (after discarding the first few terms). As a result, since

$$\mathbf{u}(j) - y = \mathbf{x}_{\mathcal{B}}((i(j), j) + \xi_j \in \mathbb{R}^{\cup}\langle \mathcal{X}_1 \cup \ldots \cup \mathcal{X}_{s-1}\rangle = \mathcal{E},$$

we conclude that $v_1, \ldots, v_{r_\ell}, w_1, \ldots, w_n \in \mathcal{E}$. Consequently, since $\{\mathbf{u}(j)\}_{j=1}^{\infty}$ is an asymptotically filtered sequence of *terms from* G_0 with fully unbounded limit $\vec{w}$, it follows from Proposition 7.7.3 (applied to $(\mathcal{X}_1 \cup \{\mathbf{v}_1\}, \ldots, \mathcal{X}_{s-1} \cup \{\mathbf{v}_{s-1}\})$) that $\mathcal{Y} = \mathcal{X}_1 \cup \ldots \cup \mathcal{X}_{s-1}$ encases $-\vec{w}$. Let $\mathcal{C} \subseteq \mathcal{Y}$ be a subset which minimally encases $-\vec{w}$. Then, since $\mathcal{B} \subseteq \mathcal{Y}$ minimally encases $-(v_1, \ldots, v_{r_\ell})$, it follows in view of Proposition 5.9 that $\mathcal{B} \preceq \mathcal{C}$, with equality only possible if $w_1, \ldots, w_n \in \mathbb{R}^{\cup}\langle \mathcal{B}\rangle$. However, if $w_1, \ldots, w_n \in \mathbb{R}^{\cup}\langle \mathcal{B}\rangle$, then $\{\pi_{\mathcal{B}}(\mathbf{u}(j))\}_{j=1}^{\infty}$ will be a bounded sequence (as $\vec{w}$ is a *complete* fully unbounded limit and $v_1, \ldots, v_{r_\ell} \in \mathbb{R}^{\cup}\langle \mathcal{B}\rangle$), contradicting that $\pi_{\mathcal{B}}(\mathbf{u}(j)) = \pi_{\mathcal{B}}\big(\mathbf{x}((i(j), j))\big) = \xi_j + y$ with $\|\xi_j\| \to \infty$. Therefore we conclude that

$$\mathcal{B} \prec \mathcal{C}.$$

Let $\pi_{\mathcal{C}} : \mathbb{R}^d \to \mathbb{R}^{\cup}\langle \mathcal{C}\rangle^{\perp}$ be the orthogonal projection.

Now $\mathbf{u}(j) = \mathbf{u}_{\mathcal{B}}(j) + \xi_j + y$ with $\mathbf{u}_{\mathcal{B}}(j) \in \mathbb{R}^{\cup}\langle \mathcal{B}\rangle \subseteq \mathbb{R}^{\cup}\langle \mathcal{C}\rangle$ and $y \in \mathcal{E}^{\perp} = \mathbb{R}^{\cup}\langle \mathcal{Y}\rangle^{\perp} \subseteq \mathbb{R}^{\cup}\langle \mathcal{C}\rangle^{\perp}$. Thus $\pi_{\mathcal{C}}(\mathbf{u}(j)) = \pi_{\mathcal{C}}(\xi_j) + y$ with $\{\pi_{\mathcal{C}}(\mathbf{u}(j))\}_{j=1}^{\infty}$ a bounded sequence (as $\vec{w}$ is a *complete* fully unbounded limit) of lattice points from $\pi_{\mathcal{C}}(G_0)$ (in view of Proposition 2.1). By passing once more to a subsequence, we can thus assume $\pi_{\mathcal{C}}(\mathbf{u}(j)) = \pi_{\mathcal{C}}(\xi_j) + y$ is constant (as a bounded set of lattice points is finite), say with $\pi_{\mathcal{C}}(\xi_j) = \xi \in \mathbb{R}^{\cup}\langle \mathcal{C}\rangle^{\perp}$ for all j. As $y \neq 0$ with $y \in \mathcal{E}^{\perp}$ and $\xi_j \in \mathcal{E}$, we have $\xi + y \neq 0$. Let

$$\mathbf{u}(j) = \beta_j^{(1)} v_1 + \ldots + \beta_j^{(r_\ell)} v_{r_\ell} + \gamma_j^{(1)} w_1 + \ldots + \gamma_j^{(n)} w_n + z_j$$

be the representation of $\{\mathbf{u}(j)\}_{j=1}^{\infty}$ as an asymptotically filtered sequence with complete fully unbounded limit $\vec{w}$. Since $\pi_{\mathcal{C}}(\mathbf{u}(j)) = \xi + y \neq 0$, by passing to a subsequence we can assume $\{\mathbf{u}(j)\}_{j=1}^{\infty}$ is an asymptotically filtered sequence with $(v_1, \ldots, v_{r_\ell}, w_1, \ldots, w_n, w_{n+1}, \ldots, w_{n+r})$ as limit, for some $w_{n+1}, \ldots, w_{n+r-1} \in \mathbb{R}^{\cup}\langle C\rangle$ and $w_{n+r} \notin \mathbb{R}^{\cup}\langle C\rangle$, say with representation

$$\mathbf{u}(j) = \beta_j^{(1)} v_1 + \ldots + \beta_j^{(r_\ell)} v_{r_\ell} + \gamma_j^{(1)} w_1 + \ldots + \gamma_j^{(n)} w_n + \gamma_j^{(n+1)} w_{n+1} + \ldots + \gamma_j^{(n+r)} w_{n+r} + z'_j.$$

Then $\xi + y = \pi_C(\mathbf{u}(j)) = \gamma_j^{(n+r)} \pi_C(w_{n+r}) + \pi_C(z'_j)$ is constant. Thus, since $\|z'_j\| \in o(\gamma_j^{(n+r)})$, we cannot have $\gamma_j^{(n+r)} \to 0$, for this would imply $\xi + y = \pi_C(\mathbf{u}(j)) \to 0$, contradicting that $\xi + y \neq 0$. Hence, since $\{\gamma_j^{(n+r)}\}_{j=1}^{\infty}$ is bounded, we must have $\gamma_j^{(n+r)} \to a$ for some $a > 0$, and now $\gamma_j^{(n+r)} \in o(\gamma_j^{(n+r-1)})$ ensures $\gamma_j^{n+r-1} \to \infty$, which is only possible if $r = 1$. Therefore

$$\mathbf{u}(j) = \beta_j^{(1)} v_1 + \ldots + \beta_j^{(r_\ell)} v_{r_\ell} + \gamma_j^{(1)} w_1 + \ldots + \gamma_j^{(n)} w_n + \gamma_j^{(n+1)} w_{n+1} + z'_j$$

with $\gamma_j^{(n+1)} \pi_C(w_{n+1}) + \pi_C(z'_j) = \xi + y$ for all $j \geq 1$. Since $\|z'_j\| \in o(\gamma_j^{(n+1)})$, we have $\|z'_j\| \to 0$, whence $\xi + y = \lim_{j\to\infty}(\gamma_j^{(n+1)} \pi_C(w_{n+1}) + \pi_C(z'_j)) = a\pi_C(w_{n+1})$, ensuring that $y = \pi(\xi + y) = a\pi(w_{n+1})$.

Define a new half-space $\mathbf{u}$ with $\bar{\mathbf{u}} = \mathbb{R}^{\cup}\langle C\rangle + \mathbb{R}_+ w_{n+1}$ and $\partial(\mathbf{u}) \cap \mathbf{u} = \mathsf{C}^{\cup}(C)$. Defining the limit $\vec{u}_{\mathbf{u}} = (v_1, \ldots, v_{r_\ell}, w_1, \ldots, w_n, w_{n+1})$, we see that we then have $C \subseteq \mathcal{X}_1 \cup \ldots \cup \mathcal{X}_{s-1}$ minimally encasing $-\vec{u}_{\mathbf{u}}^{\triangleleft} = -(v_1, \ldots, v_{r_\ell}, w_1, \ldots, w_n)$, allowing us to set $\partial(\{\mathbf{u}\}) = C$. As just shown above, $\xi + y = a\pi_C(w_{n+1})$, ensuring that $\pi_C(\xi_r + y) = \xi + y = a\pi_C(w_{n+1})$ for all r. Thus the representative $\xi_r + y$ for the half-space $\mathbf{x}_r$ will also be a representative for the half-space $\mathbf{u}$. Moreover, $a\pi(w_{n+1}) = \pi(\xi_r + y) = \pi(\xi + y) = y$, ensuring that the representative for the half-space $\mathbf{u}$ modulo $\mathcal{E}$ is equal to a positive multiple of the representative for the half-space $\mathbf{x}_r$ modulo $\mathcal{E}$. Consequently, if we fix some $\mathbf{x}_r \in \mathcal{X}_s$ for defining $\mathcal{X}$, say $\mathbf{x}_r \in \mathcal{X}_s^{(r)}$ with $\mathcal{X} = \mathcal{X}^{(r)}$, and then replace $\mathbf{x}_r \in \mathcal{X}_s^{(r)}$ with $\mathbf{u}$, we obtain a new virtual Reay system

$$\mathcal{R}' = (\mathcal{X}_1 \cup \{\mathbf{v}_1\}, \ldots, \mathcal{X}_{s-1} \cup \{\mathbf{v}_{s-1}\}, (\mathcal{X}_s^{(r)} \setminus \{\mathbf{x}_r\} \cup \{\mathbf{u}\}) \cup \{\mathbf{v}_s\})$$

over G_0 with $\mathcal{X}' := \mathcal{Y} \cup \mathcal{X}_s^{(r)} \setminus \{\mathbf{x}_r\} \cup \{\mathbf{u}\}$. Set $\mathbf{x}'_r = \mathbf{u}$ and $\mathbf{y}' = \mathbf{y}$ for all other half-spaces $\mathbf{y}$ from $\mathcal{R}^{(r)}$. Since $\pi(\mathcal{X}_s^{(r)} \setminus \{\mathbf{x}_r\} \cup \{\mathbf{u}\}) = \pi(\mathcal{X}_s^{(r)}) = \mathcal{Y}_s$, it follows from Proposition 7.20 applied to the maximal series decomposition $\bigcup_{i=1}^{s-1} \mathcal{X}_i \cup \mathcal{X}_i^{(r)}$ that $\bigcup_{i=1}^{s-1} \mathcal{X}_i \cup (\mathcal{X}_s^{(r)} \setminus \{\mathbf{x}_r\} \cup \{\mathbf{u}\})$ is a maximal series decomposition. Since $\mathcal{B} \prec C$, we have $\partial(\mathbf{x}_r) = \mathbb{R}^{\cup}\langle \mathcal{B}\rangle \subseteq \mathbb{R}^{\cup}\langle C\rangle = \partial(\mathbf{u})$. Thus, since we have some representative for $\mathbf{x}_r$ that is also one for $\mathbf{u}$, it follows that any representative for $\mathbf{x}_r$ is also one for $\mathbf{u}$. Since $\mathbb{R}^{\cup}\langle \mathcal{B}\rangle \subseteq \mathbb{R}^{\cup}\langle C\rangle$, Proposition 5.3.9 implies $\mathcal{B} \subseteq {\downarrow} C$, whence $\mathbf{x}_r \cap \partial(\mathbf{x}_r) = \mathsf{C}^{\cup}(\mathcal{B}) \subseteq \mathsf{C}^{\cup}(C) = \mathbf{u} \cap \partial(\mathbf{u})$. If $\mathbb{R}^{\cup}\langle C\rangle = \mathbb{R}^{\cup}\langle \mathcal{B}\rangle$, then Proposition 5.3.9 ensures that $C \subseteq {\downarrow}\mathcal{B}$, implying $C = C^* \preceq \mathcal{B}$, which would contradict that $\mathcal{B} \prec C$ (note $C^* = C$ follows as C minimally encases the limit $-\vec{w}$). In consequence, since any representative for $\mathbf{x}_r$ is one for $\mathbf{u}$, it follows that $\mathbf{x}_r \subset \mathbf{u}$, whence

$$\bigcup_{i=1}^{s-1} \mathcal{X}_i \cup \mathcal{X}_s^{(r)} \prec_{\cup}^{*} \bigcup_{i=1}^{s-1} \mathcal{X}_i \cup (\mathcal{X}_s^{(r)} \setminus \{\mathbf{x}_r\} \cup \{\mathbf{u}\}).$$

Finally, the representative sequence $\{\mathbf{u}(j)\}_{j=1}^{\infty}$ for $\mathbf{u}$ has $\mathbf{u}(j) = \mathbf{x}\big((i(j), j\big)$, which is a term from the representative sequence $\mathbf{x}_j(i)$. By construction, any such term, together with the remaining representatives for the other half-spaces from $\mathcal{X}^{(r)} \setminus \{\mathbf{x}_r\} = \mathcal{X}' \setminus \{\mathbf{u}\}$, generates the lattice Δ (as argued in Claim B). But this ensures $\Delta \in \mathfrak{P}_{\mathbb{Z}}(\mathcal{X}')$, which together with $\bigcup_{i=1}^{s-1} \mathcal{X}_i \cup \mathcal{X}_s^{(r)} \prec_{\cup}^{*} \bigcup_{i=1}^{s-1} \mathcal{X}_i \cup (\mathcal{X}_s^{(r)} \setminus \{\mathbf{x}_r\} \cup \{\mathbf{u}\})$ contradicts that $\bigcup_{i=1}^{s-1} \mathcal{X}_i \cup \mathcal{X}_s^{(r)} \in \mathsf{Max}(\mathfrak{X}_s^{*}(G_0, \Delta))$, completing the proof.

Corollary 7.30 *Let $\Lambda \subseteq \mathbb{R}^d$ be a full rank lattice, where $d \geq 0$, and let $G_0 \subseteq \Lambda$ be a finitary subset with $\mathsf{C}(G_0) = \mathbb{R}^d$. Then $\mathsf{Max}(\mathfrak{X}(G_0), \preceq_{\cup})$, $\mathsf{Max}(\mathfrak{X}^{*}(G_0), \preceq_{\cup}^{*})$, $\mathsf{Max}(\mathfrak{X}(G_0, \Delta), \preceq_{\cup})$ and $\mathsf{Max}(\mathfrak{X}^{*}(G_0, \Delta), \preceq_{\cup}^{*})$ are all finite, with ${\downarrow}\mathsf{Max}(\mathfrak{X}(G_0)) = \mathfrak{X}(G_0)$, ${\downarrow}\mathsf{Max}(\mathfrak{X}^{*}(G_0)) = \mathfrak{X}^{*}(G_0)$, ${\downarrow}\mathsf{Max}(\mathfrak{X}(G_0, \Delta)) = \mathfrak{X}(G_0, \Delta)$ and also ${\downarrow}\mathsf{Max}(\mathfrak{X}^{*}(G_0, \Delta)) = \mathfrak{X}^{*}(G_0, \Delta)$, for any $\Delta \in \mathfrak{P}_{\mathbb{Z}}(G_0)$.*

Proof Let $\mathcal{X} \in \mathfrak{X}(G_0)$ and let $\mathcal{X} = \mathcal{X}_1 \cup \ldots \cup \mathcal{X}_s$ be a maximal series decomposition. If $\mathcal{X}' \in \mathfrak{X}(G_0)$ with $\mathcal{X} \prec_{\cup} \mathcal{X}'$, then $\sum_{\mathbf{x} \in \mathcal{X}} \dim \partial(\mathbf{x}) < \sum_{\mathbf{x}' \in \mathcal{X}'} \dim \partial(\mathbf{x}')$ (as argued when showing anti-symmetry for $\preceq_{\cup}$). Thus, since $0 \leq \sum_{\mathbf{x} \in \mathcal{X}} \dim \partial(\mathbf{x}) \leq d^2$, we see that there can be no infinite ascending (or descending) chains in $(\mathfrak{X}(G_0), \preceq_{\cup})$. Since $\bigcup_{i=1}^{s} \mathcal{X}_i \prec_{\cup}^{*} \bigcup_{i=1}^{s} \mathcal{X}_s'$ implies $\mathcal{X} \prec_{\cup} \mathcal{X}'$, there are also no infinite ascending (or descending) chains in $(\mathfrak{X}^{*}(G_0), \preceq_{\cup}^{*})$. It thus suffices to show $\mathsf{Max}(\mathfrak{X}(G_0), \preceq_{\cup})$, $\mathsf{Max}(\mathfrak{X}^{*}(G_0), \preceq_{\cup}^{*})$, $\mathsf{Max}(\mathfrak{X}(G_0, \Delta), \preceq_{\cup})$ and $\mathsf{Max}(\mathfrak{X}^{*}(G_0, \Delta), \preceq_{\cup}^{*})$ are all finite, for any $\Delta \in \mathfrak{P}_{\mathbb{Z}}(G_0)$. Since each $\mathcal{X} \in \mathsf{Max}(\mathfrak{X}(G_0))$ with $\Delta \in \mathfrak{P}_{\mathbb{Z}}(\mathfrak{X})$ is also an element of $\mathsf{Max}(\mathfrak{X}(G_0, \Delta))$, and since $\mathfrak{P}_{\mathbb{Z}}(G_0)$ is finite by Theorem 7.24.2, it suffices to show $\mathsf{Max}(\mathfrak{X}(G_0, \Delta))$ and $\mathsf{Max}(\mathfrak{X}^{*}(G_0, \Delta))$ are finite for an arbitrary $\Delta \in \mathfrak{P}_{\mathbb{Z}}(G_0)$. If $\mathcal{X} \in \mathsf{Max}(\mathfrak{X}(G_0, \Delta))$ and $\mathcal{X} = \bigcup_{i=1}^{s} \mathcal{X}_i$ is any maximal series decomposition, then we have $\bigcup_{i=1}^{s} \mathcal{X}_i \in \mathsf{Max}(\mathfrak{X}^{*}(G_0, \Delta))$ (since $\bigcup_{i=1}^{s} \mathcal{X}_i \prec_{\cup}^{*} \bigcup_{i=1}^{s} \mathcal{X}_i'$ would imply $\mathcal{X} \prec_{\cup} \mathcal{X}' = \bigcup_{i=1}^{s} \mathcal{X}_i'$). Thus it suffices to prove $\mathsf{Max}(\mathfrak{X}^{*}(G_0, \Delta))$ is finite for an arbitrary $\Delta \in \mathfrak{P}_{\mathbb{Z}}(G_0)$, as all other parts follow routinely from this as just explained, and this was proved in Theorem 7.29. □

We continue with another finite-like property of finitary subsets, showing that, even though a finitary set G_0 can have an infinite number of atoms, every atom must nonetheless contain some term from a fixed *finite* subset of G_0.

Proposition 7.31 *Let $\Lambda \subseteq \mathbb{R}^d$ be a full rank lattice, where $d \geq 0$, and let $G_0 \subseteq \Lambda$ be a finitary subset with $\mathsf{C}(G_0) = \mathbb{R}^d$. Let*

$$\mathcal{A}_{\bullet}(G_0) = \{U \in \mathcal{A}(G_0) : \text{ there exists } X \subseteq \mathrm{Supp}(U) \text{ with } \emptyset \neq X \in X(G_0)\}$$

Then the following hold.

1. $\mathcal{A}(G_0) \setminus \mathcal{A}_{\bullet}(G_0)$ *is finite.*
2. *There are finite sets* $\widetilde{X} \subseteq G_0^{\diamond}$ *and* $\widetilde{Y} \subseteq G_0$ *with* $\mathcal{A}(G_0 \setminus \widetilde{X})$ *finite and* $\mathcal{A}(G_0 \setminus \widetilde{Y}) = \emptyset$.

Proof

1. Let $\mathcal{A}_{\bullet}^{\mathsf{elm}}(G_0) = \mathcal{A}^{\mathsf{elm}}(G_0) \cap \mathcal{A}_{\bullet}(G_0)$. We begin by showing that $\mathcal{A}^{\mathsf{elm}}(G_0) \setminus \mathcal{A}_{\bullet}^{\mathsf{elm}}(G_0)$ is finite. Assume by contradiction that $\{U_i\}_{i=1}^{\infty}$ is an infinite sequence of distinct elementary atoms $U_i \in \mathcal{A}^{\mathsf{elm}}(G_0)$ such that no nonempty subset of $\mathsf{Supp}(U_i)$ lies in $X(G_0)$, for all $i \geq 1$. By passing to a subsequence of $\{U_i\}_{i=1}^{\infty}$, we can w.l.o.g. assume $|U_i| = s \geq 2$ for all i, say with each $U_i = \{x_i^{(1)}, \ldots, x_i^{(s)}\}$. Since the U_i are elementary atoms with $|U_i| \geq 2$, we have $0 \notin \mathsf{Supp}(U_i)$ with $\mathsf{Supp}(U_i)$ a minimal positive basis, for all i. Thus, by again passing to a subsequence of $\{U_i\}_{i=1}^{\infty}$, we can further assume each sequence $\{x_i^{(j)}\}_{i=1}^{\infty}$ is a radially convergent sequence of terms from G_0 with limit u_j, for $j \in [1, s]$. Moreover, since $G_0 \subseteq \Lambda$ is a set of lattice points (so that any bounded subset of G_0 is finite), we either have u_j unbounded or w.l.o.g. $\{x_i^{(j)}\}_{i=1}^{\infty}$ constant, for each $j \in [1, s]$. Since the U_i are all distinct, the latter cannot occur for all $j \in [1, s]$.

   If $0 \notin \mathsf{C}^*(u_1, \ldots, u_s)$, then Lemma 7.12 implies there will be an open half-space $\mathcal{E}_+^{\circ}$ containing all u_j for $j \in [1, s]$. Consequently, since $x_i^{(j)}/\|x_i^{(j)}\| \to u_j$ for each $j \in [1, s]$, it follows that each $\mathsf{Supp}(U_i)$, with i sufficiently large, will also be contained in the open half-space $\mathcal{E}_+^{\circ}$, ensuring $0 \notin \mathsf{C}(\mathsf{Supp}(U_i))$, which contradicts that each U_i is an atom. Therefore we instead conclude that $0 \in \mathsf{C}^*(u_1, \ldots, u_s)$. Thus there is some nonempty subset $J \subseteq [1, s]$ such that $U = \{u_j : j \in J\}$ is a minimal positive basis. If $u_j \notin G_0^{\infty}$ for all $j \in J$, then $J \subset [1, s]$ is a proper, nonempty subset (since at least one u_j with $j \in [1, s]$ is unbounded). In such case, $\{x_i^{(j)} : j \in J\} \subset \mathsf{Supp}(U_i)$ will be a minimal positive basis, contradicting that each $\mathsf{Supp}(U_i)$ is a minimal positive basis, thus ensuring any proper subset is linearly independent. Therefore we instead conclude that there is some $t \in J$ such that $u_t \in G_0^{\infty}$, and in view of G_0 being finitary, it then follows that $J \cap G_0^{\infty} = \{t\}$. But now $\{x_i^{(j)} : j \in J \setminus \{t\}\} \cup \{u_t\}$ is a minimal positive basis. Thus $\{x_i^{(j)} : j \in J \setminus \{t\}\}$ minimally encases $-u_t \in G_0^{\infty}$, ensuring that each $X_i = \{x_i^{(j)} : j \in J \setminus \{t\}\} \subset \mathsf{Supp}(U_i)$ is a subset with $\emptyset \neq X_i \in X(G_0)$, contrary to assumption. So we instead conclude that $\mathcal{A}^{\mathsf{elm}}(G_0) \setminus \mathcal{A}_{\bullet}^{\mathsf{elm}}(G_0)$ is finite.

   By Theorem 4.7, every atom $U \in \mathcal{A}(G_0)$ has a factorization $U = \prod_{i\in[1,\ell]}^{\bullet} U_i^{[\alpha_i]}$ with each $U_i \in \mathcal{A}^{\mathsf{elm}}(G_0)$ and each $\alpha_i \in \mathbb{Q}$ with $0 < \alpha_i \leq 1$. In particular, $\mathsf{Supp}(U) = \bigcup_{i=1}^{\ell} \mathsf{Supp}(U_i)$. Thus, if $U \notin \mathcal{A}_{\bullet}(G_0)$, then $U_i \notin \mathcal{A}_{\bullet}^{\mathsf{elm}}(G_0)$ for every $i \in [1, \ell]$ as well, implying that

$$\mathsf{Supp}(U) \subseteq \bigcup_{V \in \mathcal{A}^{\mathsf{elm}}(G_0) \setminus \mathcal{A}_{\bullet}^{\mathsf{elm}}(G_0)} \mathsf{Supp}(V). \tag{7.32}$$

However, as just shown, $\mathcal{A}^{\mathsf{elm}}(G_0) \setminus \mathcal{A}^{\mathsf{elm}}_{\bullet}(G_0)$ is finite, ensuring that

$$\bigcup_{V \in \mathcal{A}^{\mathsf{elm}}(G_0) \setminus \mathcal{A}^{\mathsf{elm}}_{\bullet}(G_0)} \mathrm{Supp}(V)$$

is finite, implying via Dickson's Lemma [62, Theorem 1.5.3, Corollary 1.5.4] (or see Propositions 2.2 and 2.3) that the set of atoms $\mathcal{A}\big(\bigcup_{V \in \mathcal{A}^{\mathsf{elm}}(G_0) \setminus \mathcal{A}^{\mathsf{elm}}_{\bullet}(G_0)} \mathrm{Supp}(V)\big)$ is also finite. Hence there can only be a finite number of atoms U satisfying (7.32), which implies $\mathcal{A}(G_0) \setminus \mathcal{A}_{\bullet}(G_0)$ is finite, completing the proof of Item 1.

2. Every $U \in \mathcal{A}_{\bullet}(G_0)$ has some nonempty $X \in X(G_0)$ with $X \subseteq \mathrm{Supp}(U)$. Since any $X \in X(G_0)$ contains some $X_1 \subseteq X$ with $X_1 \in X(G_0)$ irreducible, we conclude that every $U \in \mathcal{A}_{\bullet}(G_0)$ has some (nonempty) irreducible $X \in X(G_0)$ with $X \subseteq \mathrm{Supp}(U)$. By Theorem 7.24.1, there are only a finite number of irreducible sets in $X(G_0)$. Let $\widetilde{X}$ be obtained by including one element from each irreducible set from $X(G_0)$. Then $\widetilde{X}$ is finite and $\mathcal{A}(G \setminus \widetilde{X}) \subseteq \mathcal{A}(G_0) \setminus \mathcal{A}_{\bullet}(G_0)$ is finite by Item 1. Moreover, since $X \subseteq G_0^{\diamond}$ for any $X \in X(G_0)$ (as remarked at the start of Sect. 7.2), we have $\widetilde{X} \subseteq G_0^{\diamond}$. Let $\widetilde{Y}$ be obtained by taking $\widetilde{X}$ and including one element from the support of each atom from $\mathcal{A}(G \setminus \widetilde{X})$. Since $\mathcal{A}(G \setminus \widetilde{X})$ and $\widetilde{X}$ are finite, so too is $\widetilde{Y}$, and $\mathcal{A}(G_0 \setminus \widetilde{Y}) = \emptyset$ by construction, completing the proof of Item 2.

□

Next, we give the finiteness from below property for the partial order $\preceq_{\mathbb{Z}}$. We will strengthen the implied consequence that $\mathsf{Min}\big(X(G_0), \preceq_{\mathbb{Z}}\big)$ is finite later in Proposition 7.33.

Theorem 7.32 *Let $\Lambda \subseteq \mathbb{R}^d$ be a full rank lattice, where $d \geq 0$, and let $G_0 \subseteq \Lambda$ be a finitary subset with $\mathsf{C}(G_0) = \mathbb{R}^d$. Then there are neither infinite descending chains nor infinite antichains in $(X(G_0), \preceq_{\mathbb{Z}})$. In particular, $\mathsf{Min}\big(X(G_0), \preceq_{\mathbb{Z}}\big)$ is finite and $\uparrow\mathsf{Min}\big(X(G_0), \preceq_{\mathbb{Z}}\big) = X(G_0)$.*

Proof We need only prove $X(G_0)$ contains no infinite anti-chains nor infinite descending chains, as the remainder of the theorem then follows from Proposition 2.2. In view of Theorem 7.24.2, there are only a finite number of lattices $\Delta \in \mathfrak{P}_{\mathbb{Z}}(G_0)$. Thus it suffices to prove $X(G_0, \Delta)$ contains no infinite anti-chains nor infinite descending chains for an arbitrary $\Delta \in \mathfrak{P}_{\mathbb{Z}}(G_0)$, which we now fix. Recall $X \preceq_{\mathbb{Z}} Y$ for $X, Y \in X(G_0, \Delta)$ is equivalent to $\mathsf{C}(X) \subseteq \mathsf{C}(Y)$. Let $X_s(G_0, \Delta) \subseteq X(G_0, \Delta)$ consist of all $X \in X(G_0, \Delta)$, so $\mathbb{Z}\langle X\rangle = \Delta$, having a maximal series decomposition of length s, say $X = X_1 \cup \ldots \cup X_s$. We proceed by induction on $s = 1, 2 \ldots, d$ to show that $X_s(G_0, \Delta)$ contains no infinite anti-chains nor infinite descending chains, which will complete the proof as there are only a finite number of possible values for $s \in [1, d]$. Note $X_1(G_0, \Delta)$ consists precisely of all irreducible sets $X \in X(G_0, \Delta)$. Thus $X_1(G_0, \Delta)$ is finite by Theorem 7.24.1, in which case it trivially can contain no infinite descending chains nor infinite anti-chains, completing the base $s = 1$ of the induction. So we now assume $s \geq 2$.

Let $X \in X_s(G_0, \Delta)$ be arbitrary having maximal series decomposition $X = X_1 \cup \ldots \cup X_s$. Then $Y = X_1 \cup \ldots \cup X_{s-1} \in X_{s-1}(G_0)$ by Proposition 7.20. Let $\Delta' = \mathbb{Z}\langle Y \rangle \in \mathfrak{P}_\mathbb{Z}(G_0)$. As there are only a finite number of possibilities for Δ' by Theorem 7.24.2, it suffices to show there are no infinite anti-chains nor infinite descending chains involving sets $X \in X_s(G_0, \Delta)$ having a series decomposition $X = X_1 \cup \ldots \cup X_s$ with $\Delta' = \mathbb{Z}\langle X_1 \cup \ldots \cup X_{s-1} \rangle$, for an arbitrary $\Delta' \in \mathfrak{P}_\mathbb{Z}(G_0)$ which we now fix. Let $\mathcal{E} = \mathbb{R}\langle Y \rangle = \mathbb{R}\langle \Delta' \rangle$ and let $\pi : \mathbb{R}^d \to \mathcal{E}^\perp$ be the orthogonal projection. Observe that Δ' and $\Delta \cap \mathcal{E}$ are both full rank lattices in $\mathcal{E}$ with $\Delta' \leq \Delta \cap \mathcal{E}$. Thus $(\Delta \cap \mathcal{E})/\Delta'$ is finite, ensuring there is a finite set $\overline{\Delta} \subseteq \Delta \cap \mathcal{E}$ of coset representatives for Δ'.

Proposition 7.20 implies that $\pi(X_s) \in X(\pi(X_0))$ is irreducible. Proposition 7.9 implies that $\pi(G_0)$ is finitary with $\pi(\Lambda)$ a lattice. Thus Theorem 7.24.1 implies that there are only a finite number of possibilities for $\pi(X_s)$, and so we may add the condition that there is a fixed subset $Y_s \subseteq G_0 \cap \Delta$ with $|\pi(Y_s)| = |Y_s|$ so that $\pi(X_s) = \pi(Y_s)$ is the same fixed irreducible set for all X under consideration. Note $Y \cup Y_s$ is a lattice basis for Δ.

If $x \in X_s$ has $\pi(x) = \pi(y)$ with $y \in Y_s$, then $x - y \in \Delta \cap \ker \pi = \Delta \cap \mathcal{E}$. It follows that $x = y + \xi_x - \varpi_x$ for some $\varpi_x \in \Delta'$ and $\xi_x \in \overline{\Delta}$. Since there are only a finite number of possibilities for $\xi_x \in \overline{\Delta}$, we can add the condition that, for each $y \in Y_s$, there is a fixed $\xi_y \in \overline{\Delta}$, such that, for all X under consideration, the element $x \in X_s$ with $\pi(x) = \pi(y)$ always has $x = y + \xi_y - \varpi_x$ for some $\varpi_x \in \Delta'$. Let $Y_s = \{y_1, \ldots, y_r\}$ be the distinct elements of Y_s and adapt the abbreviation $\xi_j = \xi_{y_j}$ for $j \in [1, r]$. Thus, if $X = \{x_1, \ldots, x_r\}$ are the elements of X indexed so that $\pi(x_j) = \pi(y_j)$ for $j \in [1, r]$, then $x_j = y_j + \xi_j - \varpi_{x_j}$ with $\varpi_{x_j} \in \Delta'$.

By induction hypothesis and Proposition 2.2, we have $\uparrow\mathsf{Min}(X_{s-1}(G_0, \Delta')) = X_{s-1}(G_0, \Delta')$ with $\mathsf{Min}(X_{s-1}(G_0, \Delta'))$ finite. Thus every $Y \in X_{s-1}(G_0, \Delta')$ has some $Y_\emptyset \in \mathsf{Min}(X_{s-1}(G_0, \Delta'))$ with $Y_\emptyset \preceq_\mathbb{Z} Y$. As there are only a finite number of possibilities for $Y_\emptyset$, we may add the condition that there is some fixed subset $Y_\emptyset \in \mathsf{Min}(X_{s-1}(G_0, \Delta'))$ with $Y_\emptyset \preceq_\mathbb{Z} Y = X_1 \cup \ldots \cup X_{s-1}$ for all X under consideration.

In summary, the above work means there are fixed $\Delta, \Delta' \in \mathfrak{P}_\mathbb{Z}(G_0)$, a fixed subset $Y_s = \{y_1, \ldots, y_r\} \subseteq G_0 \cap \Delta$ with $\pi(Y_s) \in X(\pi(G_0))$ irreducible, a fixed set of representatives $\overline{\Delta}$ for the lattice $\Delta \cap \mathcal{E}$ modulo Δ', where $\mathcal{E} = \mathbb{R}\langle \Delta' \rangle$, fixed $\xi_1, \ldots, \xi_r \in \overline{\Delta} \subseteq \Delta \cap \mathcal{E}$, and a fixed $Y_\emptyset \in \mathsf{Min}(X_{s-1}(G_0, \Delta'))$ such that, letting $\Omega \subseteq X(G_0)$ consist of all $X \in X(G_0)$ having a maximal series decomposition $X = X_1 \cup \ldots \cup X_s$ such that $\mathbb{Z}\langle X \rangle = \Delta$, $\mathbb{Z}\langle Y \rangle = \Delta'$, $\mathbb{R}\langle Y \rangle = \mathcal{E}$, $\mathcal{Y}_\emptyset \preceq_\mathbb{Z} Y$, and $X_s = \{x_1, \ldots, x_r\}$ with each

$$x_j = y_j + \xi_j - \varpi_{x_j}$$

for some $\varpi_{x_j} \in \Delta'$, where $j \in [1, r]$ and $Y = X_1 \cup \ldots \cup X_{s-1}$, then it suffices to show there are no infinite anti-chains nor infinite descending chains in Ω, for the arbitrary fixed parameters indicated, to complete the induction and thus also the proof.

For $j \in [1, r]$, let $\widetilde{X}_j = \{\varpi_{x_j} : X \in \Omega\} \subseteq \Delta' \subseteq \mathcal{E}$. Let $\mathcal{R}_\emptyset$ be a realization of $Y_\emptyset \in X(G_0, \Delta)$ in which all half-spaces corresponding to $Y_\emptyset$ are one-dimensional. If $\{z_i\}_{i=1}^\infty$ is an asymptotically filtered sequence of terms from $\widetilde{X}_j$ with fully unbounded limit $\vec{u} = (u_1, \ldots, u_t)$, then $u_1, \ldots, u_t \in \mathcal{E} = \mathbb{R}\langle \Delta' \rangle = \mathbb{R}\langle Y_\emptyset \rangle$ (as $\widetilde{X}_j \subseteq \mathcal{E}$), and then $\{-z_i + \xi_j + y_j\}_{i=1}^\infty$ will be an asymptotically filtered sequence of terms from G_0 with fully unbounded limit $-\vec{u}$ (after discarding the first few terms). Thus Proposition 7.7.3 applied to $\mathcal{R}_\emptyset$ ensures that $\vec{u}$ is encased by $Y_\emptyset$, in which case Theorem 3.14.4 applied to $\mathsf{C}(Y_\emptyset)$ implies that $\widetilde{X}_j$ is bound to $\mathsf{C}(Y_\emptyset)$ for all $j \in [1, r]$, and thus $\widetilde{X}_j$ is also bound to $\mathsf{C}_\mathbb{Z}(Y_\emptyset)$ for all $j \in [1, r]$. This means there is some fixed ball $B \subseteq \mathcal{E}$ such that every $\varpi \in \bigcup_{j=1}^r \widetilde{X}_j$ has some $z \in \mathsf{C}_\mathbb{Z}(Y_\emptyset)$ such that $\varpi \in B + z$. As $\varpi, z \in \Delta'$, we are assured that $\varpi - z \in B \cap \Delta'$, which is a bounded set of lattice points. Thus each $\varpi_j \in \widetilde{X}_j$ has $\varpi_j - \xi'_j \in \mathsf{C}_\mathbb{Z}(Y_\emptyset)$ for some $\xi'_j \in B \cap \Delta'$. As there are only a finite number of choices for ξ'_j, we can make one last restriction by only considering $X \in \Omega$ for which the same fixed value of ξ'_j occurs for all $\varpi_j \in \widetilde{X}_j$, for each $j \in [1, r]$. Let $\Omega' \subseteq \Omega$ be resulting subset. This means it suffices to show there are no infinite anti-chains nor infinite descending chains in Ω' to complete the proof. Moreover, replacing the fixed coset representative ξ_j with the alternative fixed coset representative $\xi_j - \xi'_j$ for each $j \in [1, r]$ (thus replacing ϖ_j by $\varpi_j - \xi'_j$), we obtain

$$\varpi_{x_j} \in \mathsf{C}_\mathbb{Z}(Y_\emptyset)$$

rather than simply $\varpi_{x_j} \in \Delta' = \mathbb{Z}\langle Y_\emptyset \rangle$, for each $j \in [1, r]$.

Each $X \in \Omega'$ corresponds to the tuple

$$\varphi(X) = (Y, \varpi_1, \ldots, \varpi_r) \in X_{s-1}(G_0, \Delta') \times \underbrace{\mathsf{C}_\mathbb{Z}(Y_\emptyset) \times \ldots \times \mathsf{C}_\mathbb{Z}(Y_\emptyset)}_{r}.$$

Indeed, $Y \subseteq X$ consists of all $x \in X$ with $\pi(x) = 0$, while, since $\pi(Y_s)$ consists of $|Y_s| = r$ distinct elements, there is a unique indexing of $X_s = X \setminus Y = \{x_1, \ldots, x_r\}$ such that $\pi(x_j) = \pi(y_j)$ for all $j \in [1, r]$, and then $\varpi_j = \varpi_{x_j} = -x_j + y_j + \xi_j$ as defined above. Moreover, if $X' \in \Omega'$ is a set with corresponding tuple $\varphi(X') = (Y', \varpi'_1, \ldots, \varpi'_r) = (Y, \varpi_1, \ldots, \varpi_r) = \varphi(X)$, then $X = X'$. Thus we see that the elements $X \in \Omega'$ are in bijective correspondence with a subset

$$\overline{\Omega'} \subseteq X_{s-1}(G_0, \Delta') \times \underbrace{\mathsf{C}_\mathbb{Z}(Y_\emptyset) \times \ldots \times \mathsf{C}_\mathbb{Z}(Y_\emptyset)}_{r}.$$

Given a set $Y \in X_{s-1}(G_0, \Delta')$, we define a partial order $\preceq_Y$ on $\mathsf{C}_\mathbb{Z}(Y_\emptyset)$ by declaring

$$\varpi \preceq_Y \varpi' \quad \text{when} \quad \varpi' \in \varpi + \mathsf{C}(Y),$$

for $\varpi, \varpi' \in \mathsf{C}_\mathbb{Z}(Y_\emptyset)$. Note $\preceq_Y$ is transitive as $\mathsf{C}(Y)$ is convex, is reflexive as $0 \in \mathsf{C}(Y)$, and is antisymmetric since Y is linearly independent (which ensures that

$\mathsf{C}(Y) \cap -\mathsf{C}(Y) = \{0\}$). Also, since $\varpi, \varpi' \in \Delta' = \mathbb{Z}\langle Y \rangle$ for all $Y \in X_{s-1}(G_0, \Delta')$ with Y linearly independent, we have

$$\varpi' \in \varpi + \mathsf{C}(Y) \quad \text{if and only if} \quad \varpi' \in \varpi + \mathsf{C}_{\mathbb{Z}}(Y).$$

□

Claim A For $X, X' \in \Omega'$ with $\varphi(X) = (Y, \varpi_1, \ldots, \varpi_r)$ and $\varphi(X') = (Y', \varpi_1', \ldots, \varpi_r')$, we have $X \preceq_{\mathbb{Z}} X'$ if and only if $Y \preceq_{\mathbb{Z}} Y'$ and $\varpi_j \preceq_{Y'} \varpi_j'$ for all $j \in [1, r]$.

Proof Suppose $X \preceq_{\mathbb{Z}} X'$. Then $\mathsf{C}_{\mathbb{Z}}(X) \subseteq \mathsf{C}_{\mathbb{Z}}(X')$. Let $\sum_{y \in Y} \alpha_y y \in \mathsf{C}_{\mathbb{Z}}(Y)$ be an arbitrary element, where $\alpha_y \in \mathbb{Z}_+$. Then $\sum_{y \in Y} \alpha_y y \in \mathsf{C}_{\mathbb{Z}}(Y) \subseteq \mathsf{C}_{\mathbb{Z}}(X) \subseteq \mathsf{C}_{\mathbb{Z}}(X')$, whence $\sum_{y \in Y} \alpha_y y = \sum_{x' \in X'} \beta_{x'} x'$ for some $\beta_{x'} \in \mathbb{Z}_+$. Thus, since $\mathbb{R}\langle Y \rangle = \mathbb{R}\langle Y' \rangle = \mathcal{E} = \ker \pi$ with $\pi(X' \setminus Y')$ a set of $|X' \setminus Y'|$ linearly independent elements, it follows that $\beta_{x'} = 0$ for all $x' \in X' \setminus Y'$, whence $\sum_{y \in Y} \alpha_y y = \sum_{x' \in Y'} \beta_{x'} x' \in \mathsf{C}_{\mathbb{Z}}(Y')$, which shows $\mathsf{C}_{\mathbb{Z}}(Y) \subseteq \mathsf{C}_{\mathbb{Z}}(Y')$, and thus that $Y \preceq_{\mathbb{Z}} Y'$. Let $X \setminus Y = X_s = \{x_1, \ldots, x_r\}$ and $X' \setminus Y' = X_s' = \{x_1', \ldots, x_r'\}$ with

$$x_j = y_j + \xi_j - \varpi_j \quad \text{and} \quad x_j' = y_j + \xi_j - \varpi_j' \quad \text{for } j \in [1, r].$$

Let $j \in [1, r]$ be arbitrary. Then $x_j \in \mathsf{C}_{\mathbb{Z}}(X) \subseteq \mathsf{C}_{\mathbb{Z}}(X')$, implying that $x_j = \sum_{x' \in X'} \beta_{x'} x'$ for some $\beta_{x'} \in \mathbb{Z}_+$. Since $\pi(x_j) = \pi(y_j)$ with $\pi(X' \setminus Y') = \{\pi(y_1), \ldots, \pi(y_r)\}$ linearly independent, it follows that $\beta_{x'} = 0$ for all $x' \in X_s' \setminus \{x_j'\}$ and $\beta_{x_j'} = 1$. Hence

$$y_j + \xi_j - \varpi_j = x_j = x_j' + \sum_{y' \in Y'} \beta_{y'} y' = y_j + \xi_j - \varpi_j' + \sum_{y' \in Y'} \beta_{y'} y' \in y_j + \xi_j - \varpi_j' + \mathsf{C}_{\mathbb{Z}}(Y'),$$

implying that $\varpi_j' \in \varpi_j + \mathsf{C}_{\mathbb{Z}}(Y')$, and thus that $\varpi_j \preceq_{Y'} \varpi_j'$. This establishes one direction of the claim.

Next suppose that $Y \preceq_{\mathbb{Z}} Y'$ and $\varpi_j \preceq_{Y'} \varpi_j'$ for all $j \in [1, r]$. Let $\sum_{x \in X} \alpha_x x \in \mathsf{C}_{\mathbb{Z}}(X)$ be arbitrary, where $\alpha_x \in \mathbb{Z}_+$. Since $Y \preceq_{\mathbb{Z}} Y'$, we have $\mathsf{C}_{\mathbb{Z}}(Y) \subseteq \mathsf{C}_{\mathbb{Z}}(Y')$, ensuring that $\sum_{y \in Y} \alpha_y y \in \mathsf{C}_{\mathbb{Z}}(Y')$. Now $X_s = X \setminus Y = \{x_1, \ldots, x_r\}$ with $x_j = y_j + \xi_j - \varpi_j$ and $X_s' = X' \setminus Y' = \{x_1', \ldots, x_r'\}$ with $x_j' = y_j + \xi_j - \varpi_j'$. Thus, since $\varpi_j \preceq_{Y'} \varpi_j'$, which ensures $-\varpi_j \in -\varpi_j' + \mathsf{C}_{\mathbb{Z}}(Y')$, we have $x_j \in x_j' + \mathsf{C}_{\mathbb{Z}}(Y')$, for $j \in [1, r]$. As a result, since $\mathsf{C}_{\mathbb{Z}}(Y')$ is a convex lattice cone (closed under addition and positive scalar multiplication by integers), we have

$$\sum_{x \in X} \alpha_x x = \sum_{y \in Y} \alpha_y y + \sum_{j=1}^{r} \alpha_{x_j} x_j \in \sum_{j=1}^{r} \alpha_{x_j} x_j' + \mathsf{C}_{\mathbb{Z}}(Y') \subseteq \mathsf{C}_{\mathbb{Z}}(X_s') + \mathsf{C}_{\mathbb{Z}}(Y') = \mathsf{C}_{\mathbb{Z}}(X').$$

Since $\sum_{x\in X}\alpha_x x \in \mathsf{C}_{\mathbb{Z}}(X)$ was arbitrary, this shows $\mathsf{C}_{\mathbb{Z}}(X) \subseteq \mathsf{C}_{\mathbb{Z}}(X')$, implying $X \preceq_{\mathbb{Z}} X'$, which completes the other direction in Claim A. □

If $\mathsf{C}_{\mathbb{Z}}(Y_1) \subseteq \mathsf{C}_{\mathbb{Z}}(Y_2)$, where $Y_1, Y_2 \in X_{s-1}(G_0, \Delta')$, then $\varpi \preceq_{Y_1} \varpi'$ implies $\varpi \preceq_{Y_2} \varpi'$. In particular, $\varpi \preceq_{Y_\emptyset} \varpi'$ implies $\varpi \preceq_{Y} \varpi'$ for any Y occurring as the first coordinate of an element from $\overline{\Omega'}$ (recall that $\mathsf{C}_{\mathbb{Z}}(Y_\emptyset) \subseteq \mathsf{C}_{\mathbb{Z}}(Y)$ since $Y_\emptyset \preceq_{\mathbb{Z}} Y$ by construction of Ω). We make $\overline{\Omega'}$ into a poset by declaring $(Y, \varpi_1, \ldots, \varpi_r) \preceq (Y', \varpi_1', \ldots, \varpi_r')$ when $Y \preceq_{\mathbb{Z}} Y'$ and $\varpi_j \preceq_{Y'} \varpi_j'$ for all $j \in [1, r]$, with transitivity guaranteed by the observation just noted. Claim A then ensures that φ gives an isomorphism between the posets Ω' and $\overline{\Omega'}$. Moreover, it is readily seen (since $Y_\emptyset$ is linearly independent) that there is an isomorphism of posets $(\mathsf{C}_{\mathbb{Z}}(Y_\emptyset), \preceq_{Y_\emptyset}) \cong \mathbb{Z}_+^{|Y_\emptyset|}$ using the product order on $\mathbb{Z}_+^{|Y_\emptyset|}$. Consequently, in view of Claim A, we see that we can remove relations from the poset $\Omega' \cong \overline{\Omega'}$ to result in a poset isomorphic to a subset of $X_{s-1}(G_0, \Delta') \times \underbrace{\mathbb{Z}_+^{|Y_\emptyset|} \times \ldots \times \mathbb{Z}_+^{|Y_\emptyset|}}_{r}$. By induction hypothesis, there are no infinite anti-chains nor infinite descending chains in $X_{s-1}(G_0, \Delta')$, and there are also no infinite anti-chains nor infinite descending chains in $\mathbb{Z}_+^{|Y_\emptyset|}$ (recall the discussion in Sect. 2.2). Therefore iterated application of Proposition 2.3 implies there are no infinite anti-chains in $X_{s-1}(G_0, \Delta') \times \underbrace{\mathbb{Z}_+^{|Y_\emptyset|} \times \ldots \times \mathbb{Z}_+^{|Y_\emptyset|}}_{r}$, and thus also no infinite anti-chains in Ω', as this lattice can be obtained from $X_{s-1}(G_0, \Delta') \times \underbrace{\mathbb{Z}_+^{|Y_\emptyset|} \times \ldots \times \mathbb{Z}_+^{|Y_\emptyset|}}_{r}$ by adding additional relations and removing elements (which both can only reduce the size of an anti-chain). It remains to show that Ω' also has no infinite descending chains.

Suppose by contradiction that $\{X_i\}_{i=1}^{\infty}$ is a strictly descending sequence of $X_i \in \Omega'$ and let $\varphi(X_i) = (Y_i, \varpi_i^{(1)}, \ldots, \varpi_i^{(r)})$ for $i \geq 1$. Then, in view of Claim A, it follows that $\{Y_i\}_{i=1}^{\infty}$ is a descending sequence of $Y_i \in X_{s-1}(G_0, \Delta')$. As a result, since $X_{s-1}(G_0, \Delta')$ contains no infinite descending chain by induction hypothesis, it follows that, by discarding the first few terms in $\{X_i\}_{i=1}^{\infty}$, we obtain $Y_i = Y$ for all i. Since $\{X_i\}_{i=1}^{\infty}$ is a descending sequence, Claim A now ensures that $\{\varpi_i^{(j)}\}_{i=1}^{\infty}$ is a descending sequence in $(\mathsf{C}_{\mathbb{Z}}(Y_\emptyset), \preceq_Y)$ for each $j \in [1, r]$. Since $\mathsf{C}_{\mathbb{Z}}(Y_\emptyset) \subseteq \mathsf{C}_{\mathbb{Z}}(Y)$ in view of $Y_\emptyset \preceq_{\mathbb{Z}} Y$, it follows that $(\mathsf{C}_{\mathbb{Z}}(Y_\emptyset), \preceq_Y)$ is a sub-poset of $(\mathsf{C}_{\mathbb{Z}}(Y), \preceq_Y)$, which is isomorphic to $\mathbb{Z}_+^{|Y|}$ as Y is linearly independent. However, the poset $\mathbb{Z}_+^{|Y|}$ contains no infinite descending chains (as discussed in Sect. 2.2), meaning neither the isomorphic poset $(\mathsf{C}_{\mathbb{Z}}(Y), \preceq_Y)$ nor the sub-poset $(\mathsf{C}_{\mathbb{Z}}(Y_\emptyset), \preceq_Y)$ contain infinite descending chains. It follows that each chain $\{\varpi_i^{(j)}\}_{i=1}^{\infty}$ for $j \in [1, r]$ eventually stabilizers, contradicting that $\{X_i\}_{i=1}^{\infty}$ is a strictly descending chain. With this contradiction, we instead conclude that Ω' contains no infinite descending chains, completing the induction and proof.

7.4 Interchangeability and the Structure of $X(G_0)$

The goal of this section is study the structure of $X(G_0)$. One of our main aims is to partition $X(G_0)$ into a *finite* number of subsets, each consisting of various $X \in X(G_0)$ having the same type, such that sets of the same type posses common regularity properties. We begin with the basic definitions.

Let $\Lambda \leq \mathbb{R}^d$ be a full rank lattice and let $G_0 \subseteq \Lambda$ be a finitary subset with $\mathsf{C}(G_0) = \mathbb{R}^d$. Let $X \in X(G_0)$, let $X = X_1 \cup \ldots \cup X_s$ be a maximal series decomposition of X. For $j \in [0, s]$, let $\Delta_j = \mathbb{Z}\langle X_1 \cup \ldots \cup X_j\rangle$ and let $\pi_j : \mathbb{R}^d \to \mathbb{R}\langle X_1 \cup \ldots \cup X_j\rangle^\perp$ be the orthogonal projection. Note $\Delta_j = \mathbb{R}\langle X_1 \cup \ldots \cup X_j\rangle \cap \Delta$, for $j \in [0, s]$, as X is linearly independent, where $\Delta := \Delta_s = \mathbb{Z}\langle X\rangle$. Also, if $x \in X_j$, then j is the minimal index such that $\pi_j(x) = 0$. Suppose $X' \in X(G_0)$ is another set with maximal series decomposition $X' = X'_1 \cup \ldots \cup X'_s$ such that $\pi_{j-1}(X_j) = \pi_{j-1}(X'_j)$ for $j = 1, 2, \ldots, s$. Then $\mathbb{R}\langle X_1 \cup \ldots \cup X_j\rangle = \mathbb{R}\langle X'_1 \cup \ldots \cup X'_j\rangle$ for $j = 0, 1, 2 \ldots, s$. Moreover, each $x \in X$ identifies uniquely with some $x' \in X'$. Namely, $x \in X_j$ with $j \in [1, s]$ the minimal index such that $\pi_j(x) = 0$, and then $x' \in X'_j$ is the unique element with $\pi_{j-1}(x) = \pi_{j-1}(x')$. This allows us to define a bijection between and X and X' by sending $x \mapsto x'$ as just defined. In such case, we say that X and X' have the same **linear type**. If we additionally have

$$\mathbb{Z}\langle X'_1 \cup \ldots \cup X'_j\rangle = \mathbb{Z}\langle X_1 \cup \ldots \cup X_j\rangle \quad \text{for every } j \in [1, s],$$

then we say that X and X' have the same **lattice type**. For each possible lattice type, we can fix a representative set $Z \in X(G_0)$ with series decomposition $Z = Z_1 \cup \ldots \cup Z_s$ having this lattice type. Then, given any other $X \in X(G_0)$ having the same lattice type, say with associated maximal series decomposition $X = X_1 \cup \ldots \cup X_s$, there is a bijection $\phi : X \to Z$ such that $\mathbb{Z}\langle X_1 \cup \ldots \cup X_j\rangle = \mathbb{Z}\langle \phi(X_1) \cup \ldots \cup \phi(X_j)\rangle$ with $\phi(X_j) = Z_j$ for $j \in [1, s]$. If $X' \in X(G_0)$ is another set with the same lattice type, we likewise have a bijection $\phi' : X' \to Z$. Moreover, if $x \in X \cap X'$, then $\phi(x) = \phi'(x)$ by how the bijections ϕ and ϕ' are defined above. Thus we can extend the domain of ϕ to obtain a map $\phi : \bigcup_X X \to Z$, where the union runs over all $X \in X(G_0)$ having the same lattice type as Z, and we identify ϕ as the lattice type of Z itself, allowing us to say a set $X \in X(G_0)$ has lattice type ϕ. Note a given $X \in X(G_0)$ may have multiple types. However, each maximal series decomposition $X = X_1 \cup \ldots \cup X_s$ corresponds to precisely one. The type ϕ associated to the empty set is called the trivial type, which corresponds to when $s = 0$. Only the empty set has trivial type. We remark that the lattice type ϕ depends on the associated linear type and the lattices Δ_j for $j = 1, \ldots, s$. However, once the linear type is fixed, in turn fixing the values of the subspaces $\mathbb{R}\langle X_1 \cup \ldots \cup X_j\rangle$ for $j \in [0, s]$, the value of $\Delta_s = \mathbb{Z}\langle X\rangle$ then completely determines the other values Δ_j with $j < s$ since $\mathbb{Z}\langle Z_1 \cup \ldots \cup Z_j\rangle = \mathbb{Z}\langle Z_1 \cup \ldots \cup Z_s\rangle \cap \mathbb{R}\langle Z_1 \cup \ldots \cup Z_j\rangle = \Delta_s \cap \mathbb{R}\langle X_1 \cup \ldots \cup X_j\rangle = \mathbb{Z}\langle X_1 \cup \ldots \cup X_s\rangle \cap \mathbb{R}\langle X_1 \cup \ldots \cup X_j\rangle = \mathbb{Z}\langle X_1 \cup \ldots \cup X_j\rangle$ as Z and X are linearly independent.

The linear type of $X \in X(G_0)$ associated to the maximal series decomposition $X = X_1 \cup \ldots \cup X_s$ is determined by the values of $\pi_{j-1}(X_j)$ for $j \in [1, s]$, where $\pi_{j-1} : \mathbb{R}^d \to \mathbb{R}\langle X_1 \cup \ldots \cup X_{j-1}\rangle^\perp$ is the orthogonal projection. The sets $\phi_{j-1}(X_j)$ are each irreducible by Proposition 7.20, while Theorem 7.24.1 ensures that each finitary set $\pi_{j-1}(G_0)$ (by Proposition 7.9) only has a finite number of irreducible sets as possibilities for $\phi_{j-1}(X_j)$. It follows that there are only a finite number of linear types for G_0. The discussion after (7.13) ensures that each linear type only has a finite number of associated lattice types. Thus the number of lattice types for G_0 is also finite. Indeed, per the discussion after (7.13), the value of the lattice $\Delta_j = \mathbb{Z}\langle X_1 \cup \ldots \cup X_j\rangle$ is uniquely determined by Δ_{j-1} and the values of ξ_x mod Δ_{j-1} for $x \in X_j$, where

$$\xi_x := x - \phi(x) \in \Lambda \cap \mathbb{R}\langle \Delta_{j-1}\rangle.$$

Note $\xi_y = 0$ for all $y \in Y$, ensuring $\xi_{\phi(x)} = 0$ for all $x \in X$. Thus, given a lattice type ϕ having $Z = Z_1 \cup \ldots \cup Z_s$ as the associated maximal series decomposition of its codomain, there are elements $\xi_z \in \Lambda \cap \mathbb{R}\langle \Delta_{j-1}\rangle$ for $z \in Z$ such that, if $X \in X(G_0)$ has the same linear type as ϕ, then X also has the same lattice type as ϕ precisely when

$$\xi_x \equiv \xi_{\phi(x)} = 0 \mod \Delta_{j-1} \quad \text{for every } x \in X_j \text{ and } j \in [1, s], \tag{7.33}$$

where $X = X_1 \cup \ldots \cup X_s$ is the associated maximal series decomposition and $\Delta_{j-1} = \mathbb{Z}\langle Z_1 \cup \ldots \cup Z_{j-1}\rangle$.

We let $\mathfrak{T}(G_0)$ consist of the set of possible lattice types ϕ for the sets $X \in X(G_0)$. Given $\phi \in \mathfrak{T}(G_0)$, let

$$\mathfrak{X}(\phi) \subseteq X(G_0)$$

consist of all $X \in X(G_0)$ that have lattice type ϕ and set

$$\mathfrak{X}^\cup(\phi) = \bigcup_{X \in \mathfrak{X}(\phi)} X,$$

so each $\phi \in \mathfrak{T}(G_0)$ is a map $\phi : \mathfrak{X}^\cup(\phi) \to Z$, where $Z \in \mathfrak{X}(\phi)$ is a distinguished set with type ϕ having associated series decomposition $Z = Z_1 \cup \ldots \cup Z_s$. Note the lattice types associated to $X \in X(G_0)$ are in bijective correspondence with the maximal series decompositions of X.

Suppose $X, X' \in \mathfrak{X}(\phi)$ and $x \in X$ and $x' \in X'$ are elements with $\phi(x) = \phi(x')$. Define a new set $X'' = X \setminus \{x\} \cup \{x'\} \subseteq G_0$. Then $X'' \in \mathfrak{X}(\phi)$ as well, as the following short argument shows. We must have $x \in X_j$ and $x' \in X'_j$ for some $j \in [1, s]$, where $X = X_1 \cup \ldots \cup X_s$ and $X' = X'_1 \cup \ldots \cup X'_s$ are the associated maximal series decompositions. Since X and X' both have the same type, we have $\mathbb{Z}\langle X_1 \cup \ldots \cup X_{j-1}\rangle = \Delta_{j-1} = \mathbb{Z}\langle X'_1 \cup \ldots \cup X'_{j-1}\rangle$, and since $\phi(x) = \phi(x')$, it follows that $\xi_x \equiv \xi_{x'} \equiv 0 \mod \Delta_{j-1}$, for $j \in [1, s]$, ensuring that

$$\Delta_j = \mathbb{Z}\langle X_1 \cup \ldots \cup X_j \rangle = \mathbb{Z}\langle X_1 \cup \ldots \cup X_{j-1} \cup (X_j \setminus \{x\} \cup \{x'\}) \rangle \quad \text{for } j \in [1, s], \tag{7.34}$$

since the values of Δ_{j-1} and each $\xi_x \mod \Delta_{j-1}$ for $x \in X_j$ determine the lattice Δ_j. Let $\mathcal{R} = (\mathcal{X}_1 \cup \{\mathbf{v}_1\}, \ldots, \mathcal{X}_s \cup \{\mathbf{v}_s\})$ be a realization of $X \in X(G_0)$ associated to the maximal series decomposition $X = X_1 \cup \ldots \cup X_s$ with all $\mathbf{x} \in \mathcal{X} = \mathcal{X}_1 \cup \ldots \cup \mathcal{X}_s$ having dimension one. Note $\mathcal{R}$ exists per the discussion at the beginning of Chap. 7.2, and by this same discussion, replacing the half-space $\mathbf{x} = \mathbb{R}_+ x$ with $\mathbf{x}' = \mathbb{R}_+ x'$ results in another virtual Reay system (after possibly modifying the $\mathbf{v}_i$ for $i \in [j+1, s]$), showing that $X_1 \cup \ldots \cup X_{j-1} \cup (X_j \setminus \{x\} \cup \{x'\}) \cup X_{j+1} \cup \ldots \cup X_s$ is a series decomposition of X'', which is maximal in view of Proposition 7.20 since $X = X_1 \cup \ldots \cup X_s$ was maximal and $\pi_{j-1}(x) = \pi_{j-1}(x')$. In view of (7.34), it now follows that X'' also has type ϕ, as claimed.

What the above important observation means is, if $X,\ X' \in \mathfrak{X}(\phi)$ have the same type ϕ, then it is possible to exchange elements between X and X' equal under the mapping ϕ and have the resulting set remain in the class $\mathfrak{X}(\phi)$. Indeed, if $Z_\phi \in X(\phi)$ is the representative for the type ϕ and $X \subseteq \mathfrak{X}^{\cup}(\phi) = \bigcup_{Z \in \mathfrak{X}(\phi)} Z$ is *any* subset consisting of precisely one element $x \in X$ with $\phi(x) = z$ for each $z \in Z_\phi$, then $X \in \mathfrak{X}(\phi)$ (as X can be constructed by a process of at most $|Z_\phi| - 1$ exchanges just described). We call this the **interchangeability property** of $\mathfrak{X}(\phi)$, which essentially amounts to saying each $\mathfrak{X}^{\cup}(\phi)$ has the structure of a direct product. Thus

$$X(G_0) = \bigcup_{\phi \in \mathfrak{T}(G_0)} \mathfrak{X}(\phi)$$

is a decomposition of $X(G_0)$ into a finite number of exchange closed subsets (with the union not necessarily disjoint).

A lattice type corresponds to a map $\phi : \mathfrak{X}^{\cup}(\phi) = \bigcup_{X \in \mathfrak{X}(\phi)} X \to Z_\phi$, where $Z_\phi \in \mathfrak{X}(\phi)$ is a distinguished representative. The choice of representative Z_ϕ is rather arbitrary. A map

$$\varphi : \bigcup_{X \in \mathfrak{X}_m(\varphi)} X \to Z_\varphi$$

obtained from ϕ by first changing the representative set used for the codomain from Z_ϕ to some $Z_\varphi \in \mathfrak{X}(\phi)$ and then restricting the domain to a union of sets from some subset $\mathfrak{X}_m(\varphi) \subseteq \mathfrak{X}(\phi)$ which contains the new codomain, so $Z_\varphi \in \mathfrak{X}_m(\varphi)$, will be called a) **refinement** of ϕ. In such case, we set

$$\mathfrak{X}^{\cup}(\varphi) = \bigcup_{X \in \mathfrak{X}_m(\varphi)} X$$

and note $\varphi : \mathfrak{X}^{\cup}(\varphi) \to Z_\varphi$.

In view of Proposition 7.33 below, we refer to the sets $X \in \mathfrak{X}_m(\varphi)$ as having the same **minimal type** φ. Note, if X and X' both have the same minimal type φ, then they must also have the same lattice type, as the notion of minimal type refines that of lattice type, and by a small abuse of notation, we will often use φ for this lattice type as well. In what follows, we say an element $x \in X \in \mathfrak{X}_m(\varphi)$ is at **depth** $j \in [1, s]$ when $\varphi(x) \in Z_j$, where $Z_\varphi = Z_1 \cup \ldots \cup Z_s$ is the maximal series decomposition associated to the codomain Z_φ of the minimal type φ, and we call

$$\mathsf{dep}(\varphi) = s$$

the **depth** of the minimal type φ. We likewise define all such terms, as well as $\mathsf{dep}(\phi)$, for a lattice type $\phi \in \mathfrak{T}(G_0)$. The argument for the existence of the finite set $\mathfrak{T}_m(G_0)$ is similar to that used for Theorem 7.32.

Proposition 7.33 *Let $\Lambda \subseteq \mathbb{R}^d$ be a full rank lattice, where $d \geq 0$, and let $G_0 \subseteq \Lambda$ be a finitary subset with $\mathsf{C}(G_0) = \mathbb{R}^d$. Then there is a finite collection $\mathfrak{T}_m(G_0)$ with each $\varphi \in \mathfrak{T}_m(G_0)$ being a map $\varphi : \mathfrak{X}_m^{\cup}(\varphi) = \bigcup_{X \in \mathfrak{X}_m(\varphi)} X \to Z_\varphi$ satisfying the following properties:*

(a) $\mathfrak{X}(G_0) = \bigcup_{\varphi \in \mathfrak{T}_m(G_0)} \mathfrak{X}_m(\varphi)$.
(b) φ is a refinement of some lattice type for G_0.
(c) $Z_\varphi \preceq_{\mathbb{Z}} X$ for every $X \in \mathfrak{X}_m(\varphi)$, i.e., $\mathsf{C}_{\mathbb{Z}}(Z_\varphi) \subseteq \mathsf{C}_{\mathbb{Z}}(X)$.
(d) For any $X \in \mathfrak{X}_m(\varphi)$ and $x \in X$ at depth $j \in [1, s]$, we have

$$\varphi(x) \in x + \mathsf{C}_{\mathbb{Z}}(Z_1 \cup \ldots \cup Z_{j-1}),$$

where $Z_\varphi = Z_1 \cup \ldots \cup Z_s$ is the maximal series decomposition associated to Z_φ.
(e) The interchangeability property holds for $\mathfrak{X}_m(\varphi)$, meaning, given any $X, X' \in \mathfrak{X}_m(\varphi)$ and $x \in X$ and $x' \in X'$ with $\varphi(x) = \varphi(x')$, we have

$$X \setminus \{x\} \cup \{x'\} \in \mathfrak{X}_m(\varphi).$$

Proof Since $\mathfrak{T}(G_0)$ is finite and $\mathfrak{X}(G_0) = \bigcup_{\phi \in \mathfrak{T}(G_0)} \mathfrak{X}(\phi)$, it suffices to show each $\mathfrak{X}(\phi)$, for $\phi \in \mathfrak{T}(G_0)$, can be written as a finite union of sets satisfying (b)–(e). Let $\phi \in \mathfrak{T}(G_0)$ be an arbitrary lattice type with codomain $Y \in \mathfrak{X}(\phi)$ and associated maximal series decomposition $Y = Y_1 \cup \ldots \cup Y_s$, so

$$\phi : \mathfrak{X}^{\cup}(\phi) \to Y_1 \cup \ldots \cup Y_{s-1} \cup Y_s.$$

We proceed by induction on $s = 0, 1, \ldots, d$. If $s = 0$, then (b)–(e) hold with $\varphi = \phi$ the trivial lattice type and $\mathfrak{X}_m(\varphi) = \mathfrak{X}(\phi) = \{\emptyset\}$. If $s = 1$, then (b)–(e) hold with $\varphi = \phi$, $\mathfrak{X}_m(\varphi) = \mathfrak{X}(\phi) = \{Y\} = \{Y_1\}$ and $Z_\varphi = Y$. Therefore we may assume

$s \geq 2$. Let $X \in \mathfrak{X}(\phi)$ be arbitrary with associated maximal series decomposition $X = X_1 \cup \ldots \cup X_s$. By definition of a lattice type,

$$\Delta := \mathbb{Z}\langle Y_1 \cup \ldots \cup Y_s\rangle = \mathbb{Z}\langle X_1 \cup \ldots \cup X_s\rangle \quad \text{and}$$
$$\Delta' := \mathbb{Z}\langle Y_1 \cup \ldots \cup Y_{s-1}\rangle = \mathbb{Z}\langle X_1 \cup \ldots \cup X_{s-1}\rangle$$

are fixed (with $\Delta = \Delta_s$ and $\Delta' = \Delta_{s-1}$ in the notation from the start of Sect. 7.4).

By induction hypothesis, there are only a finite number of possible minimal types φ' for the set $X_1 \cup \ldots \cup X_{s-1} \in X(G_0)$ that are refinements of the lattice type defined for $Y \setminus Y_s$ by the maximal series decomposition $Y \setminus Y_s = Y_1 \cup \ldots \cup Y_{s-1}$ (maximality follows by Proposition 7.20 since $Y = Y_1 \cup \ldots \cup Y_s$ is maximal). Note any $X \in \mathfrak{X}(\varphi)$ with associated maximal series decomposition $X = X_1 \cup \ldots \cup X_s$ has $X \setminus X_s = X_1 \cup \ldots \cup X_{s-1}$ of the same lattice type as $Y \setminus Y_s = Y_1 \cup \ldots \cup Y_{s-1}$, as X and Y both have lattice type ϕ. For any such minimal type φ' (refining the lattice type defined by $Y_1 \cup \ldots \cup Y_{s-1}$), let $\mathfrak{X}_{\varphi'} \subseteq \mathfrak{X}(\phi)$ consist of all $X \in \mathfrak{X}(\phi)$ having associated maximal series decomposition $X = X_1 \cup \ldots \cup X_s$ such that $X \setminus X_s \in \mathfrak{X}_m(\varphi')$. Note the condition that φ' refine the lattice type defined by $Y_1 \cup \ldots \cup Y_{s-1}$, and thus also that defined by $X_1 \cup \ldots \cup X_{s-1}$, guarantees the subtle condition that $\varphi'(x) = \varphi'(z)$ if and only if $\phi(x) = \phi(z)$, for any x and z in the common domain of both φ' and ϕ. Now $\mathfrak{X}(\phi) = \bigcup_{\varphi'} \mathfrak{X}_{\varphi'}$ is a finite union of sets $\mathfrak{X}_{\varphi'}$ with all $X \in \mathfrak{X}_{\varphi'}$ having $X \setminus X_s$ of the same minimal type φ', so it suffices to show each $\mathfrak{X}_{\varphi'}$ can be written as a finite union of sets satisfying (b)–(e). To this end, let φ' be arbitrary and suppose $Z_{\varphi'} = Z_1 \cup \ldots \cup Z_{s-1}$ is the codomain of φ', so

$$\varphi' : \mathfrak{X}_m^{\cup}(\varphi') \to Z_1 \cup \ldots \cup Z_{s-1}.$$

In view of the interchangeability properties for $\mathfrak{X}(\phi)$ and $\mathfrak{X}(\varphi')$ and that guaranteed in (e) by induction for $\mathfrak{X}_m(\varphi')$, it follows that $\mathfrak{X}_{\varphi'}$ also has the interchangeability property. Note that

$$\mathbb{Z}\langle Z_{\varphi'}\rangle = \mathbb{Z}\langle Z_1 \cup \ldots \cup Z_{s-1}\rangle = \Delta'$$

as φ' refined the lattice type associated to $Y_1 \cup \ldots \cup Y_{s-1}$.

If $Z_1' \cup \ldots \cup Z_{s-1}' \in \mathfrak{X}_m(\varphi') \subseteq \mathfrak{X}(\varphi')$ and $X_1' \cup \ldots \cup X_s' \in \mathfrak{X}(\phi)$, then we have $\mathbb{Z}\langle Z_1' \cup \ldots \cup Z_{s-1}'\rangle = \Delta' = \mathbb{Z}\langle Y_1 \cup \ldots \cup Y_{s-1}\rangle = \mathbb{Z}\langle X_1' \cup \ldots \cup X_{s-1}'\rangle$, with the first equality as φ' refines the lattice type associated to $Y_1 \cup \ldots \cup Y_{s-1}$. Thus

$$\Delta = \mathbb{Z}\langle X_1' \cup \ldots \cup X_{s-1}'\rangle + \mathbb{Z}\langle X_s'\rangle = \mathbb{Z}\langle Z_1' \cup \ldots \cup Z_{s-1}'\rangle + \mathbb{Z}\langle X_s'\rangle$$
$$= \mathbb{Z}\langle Z_1' \cup \ldots \cup Z_{s-1}' \cup X_s'\rangle. \tag{7.35}$$

Since φ' refines the lattice type associated to $X_1 \cup \ldots \cup X_{s-1}$ and $X_1' \cup \ldots \cup X_s' \in \mathfrak{X}(\phi)$, it follows that $Z_1' \cup \ldots \cup Z_{s-1}' \cup X_s'$ has the same linear type as ϕ, which combined with (7.35) ensures that $Z_1' \cup \ldots \cup Z_{s-1}' \cup X_s'$ has the same lattice type as ϕ. Hence $Z'_1 \cup \ldots \cup Z'_{s-1} \cup X'_s \in \mathfrak{X}(\phi)$ with $Z'_1 \cup \ldots \cup Z'_{s-1} \in \mathfrak{X}_m(\varphi')$, ensuring by definition of $\mathfrak{X}_{\varphi'}$ that $Z'_1 \cup \ldots \cup Z'_{s-1} \cup X'_s \in \mathfrak{X}_{\varphi'}$. Applying this with $Z'_1 \cup \ldots \cup Z'_{s-1} = Z_1 \cup \ldots \cup Z_{s-1}$ and $X'_1 \cup \ldots \cup X'_s = Y_1 \cup \ldots \cup Y_s$, we conclude that

$$Y_{\varphi'} := Z_1 \cup \ldots \cup Z_{s-1} \cup Y_s \in \mathfrak{X}_{\varphi'}. \tag{7.36}$$

Since all $X \in \mathfrak{X}_{\varphi'} \subseteq \mathfrak{X}(\phi)$ have the same lattice type, it follows by (7.33) that any $x \in \mathfrak{X}^{\cup}(\phi)$ with $\mathsf{dep}(x) = s$ has

$$\xi_x := x - \phi(x) \equiv 0 \mod \Delta'. \tag{7.37}$$

For $y \in Y_s$, define

$$\widetilde{X}_y = \{\xi_x : \text{ there is some } X \in \mathfrak{X}_{\varphi'} \text{ and } x \in X \text{ with } \phi(x) = y\} \subseteq \Delta'.$$

By definition of ξ_x and $\widetilde{X}_y$, we find that

$$\xi \in \widetilde{X}_y \quad \text{implies} \quad \xi + y = \xi_x + \phi(x) = x \in \mathfrak{X}^{\cup}(\varphi) \subseteq G_0. \tag{7.38}$$

Let $\mathcal{R}_{\varphi'}$ be a realization of $Z_{\varphi'} = Z_1 \cup \ldots \cup Z_{s-1}$ in which all anchored half-spaces have dimension one. If $\{-\xi_i\}_{i=1}^{\infty}$ is an asymptotically filtered sequence of terms from $-\widetilde{X}_y$ with fully unbounded limit $\vec{u} = (u_1, \ldots, u_t)$, then $u_1, \ldots, u_t \in \mathbb{R}\langle \Delta' \rangle = \mathbb{R}\langle Z_{\varphi'} \rangle$, and then (7.38) ensures $\{\xi_i + y\}_{i=1}^{\infty}$ is an asymptotically filtered sequence of terms from G_0 with fully unbounded limit $-\vec{u}$. Thus Proposition 7.7.3 applied to $\mathcal{R}_{\varphi'}$ ensures that $\vec{u}$ is encased by $Z_{\varphi'}$, in which case Theorem 3.14.4 applied to $\mathsf{C}(Z_{\varphi'})$ implies that $-\widetilde{X}_y$ is bound to $\mathsf{C}(Z_{\varphi'})$, and thus $-\widetilde{X}_y$ is also bound to $\mathsf{C}_{\mathbb{Z}}(Z_{\varphi'})$. This means there is some fixed ball $B_y \subseteq \mathbb{R}\langle Z_{\varphi'} \rangle$ such that every $\xi \in \widetilde{X}_y$ has some $z \in \mathsf{C}_{\mathbb{Z}}(Z_{\varphi'})$ with $-\xi \in B_y + z$. As $\xi,\ z \in \Delta'$, we are assured that $-\xi - z \in \Delta' \cap B_y$, which is a bounded set of lattice points, and thus finite. Consequently,

$$\text{each } \xi \in \widetilde{X}_y \text{ has } \varpi + \xi \in \mathsf{C}_{\mathbb{Z}}(-Z_{\varphi'}) \text{ for some } \varpi \in \Delta' \cap B_y. \tag{7.39}$$

For each $y \in Y_s$, there are only a finite number of choices for an element $\varpi_y \in \Delta' \cap B_y$. For each fixed choice of elements $\varpi_y \in \Delta' \cap B_y$ for each $y \in Y_s$, we can define a subset $\Omega \subseteq \mathfrak{X}_{\varphi'}$ consisting of all $X \in \mathfrak{X}_{\varphi'}$ such that, whenever $x \in X$ with $\phi(x) = y \in Y_s$, then $\varpi_y + \xi_x \in \mathsf{C}_{\mathbb{Z}}(-Z_{\varphi'})$. In view of (7.39), we see that $\mathfrak{X}_{\varphi'}$ is the union of all possible Ω, as we range over the finite number of possible choices for the ϖ_y. It thus suffices to show (b)–(e) hold for each possible $\Omega \subseteq \mathfrak{X}_{\varphi'}$. Fix one such arbitrary possibility for Ω. For $y \in Y_s$, define

$$\widetilde{X}_y^{\Omega} = \{\xi_x : \text{ there is some } X \in \Omega \text{ and } x \in X \text{ with } \phi(x) = y\} \subseteq \widetilde{X}_y.$$

Then, by definition of Ω, we are assured that, for all $y \in Y_s$, we have

$$\varpi_y + \widetilde{X}_y^{\Omega} \subseteq \mathsf{C}_{\mathbb{Z}}(-Z_{\varphi'}) \text{ and } \tag{7.40}$$

$$y + \widetilde{X}_y^{\Omega} \subseteq \{x \in G_0 : x \in X \text{ for some } X \in \Omega \text{ and } \phi(x) = y\}, \tag{7.41}$$

with the latter inclusion above following by noting that $\xi \in \widetilde{X}_y^{\Omega}$ implies $\xi = x - \phi(x) = x - y$ for some $x \in X$ with $X \in \Omega$ (by (7.37) and definition of $\widetilde{X}_y^{\Omega}$).

The restriction added to a set $X = X_1 \cup \ldots \cup X_s$ when passing from $\mathfrak{X}_{\varphi'}$ to $\Omega \subseteq \mathfrak{X}_{\varphi'}$ only applies to conditions involving X_s, and the restrictions imposed for each $x \in X_s$ are independent of each other. Thus, since the interchangeability property holds for $\mathfrak{X}_{\varphi'}$, it follows that it also holds for Ω, i.e., if $X, X' \in \Omega$ and $x \in X$ and $x' \in X'$ with $\phi(x) = \phi(x')$, then $X \setminus \{x\} \cup \{x'\} \in \Omega$. Moreover, if $X' = X'_1 \cup \ldots \cup X'_s \in \mathfrak{X}_{\varphi'}$ and $X = X_1 \cup \ldots \cup X_s \in \Omega$, then $X'_1 \cup \ldots \cup X'_{s-1} \cup X_s \in \Omega$. In particular, the case $X' = Y_{\varphi'} \in \mathfrak{X}_{\varphi'}$ (by (7.36)) tell us that, if $X = X_1 \cup \ldots \cup X_{s-1} \cup X_s \in \Omega$, then $Z_1 \cup \ldots \cup Z_{s-1} \cup X_s \in \Omega$. If $\xi \in \widetilde{X}_y^{\Omega}$, then (7.41) implies there is some $X \in \Omega$ with

$$y + \xi = \phi(x) + \xi = x \in X = X_1 \cup \ldots \cup X_{s-1} \cup X_s,$$

and by the previous conclusion, we can w.l.o.g. assume $X = Z_1 \cup \ldots \cup Z_{s-1} \cup X_s$ (note $x \in X_s$ as $\phi(x) = y \in Y_s$). As a result, if, for each $y \in Y_s$, we have some $\xi_y \in \widetilde{X}_y^{\Omega}$, then swapping the elements $y+\xi_y$ into the series decomposition $X = Z_1 \cup \ldots \cup Z_{s-1} \cup X_s$, one by one, yields a series decomposition $Z_1 \cup \ldots \cup Z_{s-1} \cup Z_s \in \Omega$ with $Z_s = \{y + \xi_y : y \in Y_s\}$.

The natural partial order on $\mathsf{C}_{\mathbb{Z}}(-Z_{\varphi'})$, by declaring $a \preceq b$ for $a, b \in \mathsf{C}_{\mathbb{Z}}(-Z_{\varphi'})$ when $b \in a + \mathsf{C}_{\mathbb{Z}}(-Z_{\varphi'})$, makes $\mathsf{C}_{\mathbb{Z}}(-Z_{\varphi'})$ into a partially ordered set isomorphic to $\mathbb{Z}_+^{|Z_{\varphi'}|}$ (since $Z_{\varphi'}$ is a linearly independent set). Since $\varpi_y + \widetilde{X}_y^{\Omega} \subseteq \mathsf{C}_{\mathbb{Z}}(-Z_{\varphi'})$ by (7.40), Hilbert's Basis Theorem (see Sect. 2.2) ensures that $\mathsf{Min}(\varpi_y + \widetilde{X}_y^{\Omega})$ is finite with $\uparrow\mathsf{Min}(\varpi_y + \widetilde{X}_y^{\Omega}) = \varpi_y + \widetilde{X}_y^{\Omega}$, for each $y \in Y_s$. For each $y \in Y_s$, we can choose some minimal element $\varpi_y + \xi_y \in \mathsf{Min}(\varpi_y + \widetilde{X}_y^{\Omega})$ and then set

$$z_y = y + \xi_y.$$

Letting $Z_s = \{z_y : y \in Y_s\}$, the discussion of the previous paragraph ensures that

$$Z_\varphi := Z_1 \cup \ldots \cup Z_{s-1} \cup Z_s \in \Omega.$$

As $\mathsf{Min}(\varpi_y + \widetilde{X}_y^{\Omega})$ is finite, there are only a finite number of possibilities for how to construct the set Z_s. For each possible Z_s, we can define a subset $\mathfrak{X}_m(\varphi) \subseteq \Omega$ consisting of all $X \in \Omega$ such that, for every $y \in Y_s$ and $x \in X$ with $\phi(x) = y$, we have $\varpi_y + \xi_y \preceq \varpi_y + \xi_x$. Changing the codomain of ϕ to Z_φ and restricting to the domain $\bigcup_{X \in \mathfrak{X}_m(\varphi)} X$ gives rise to a refinement $\varphi : \bigcup_{X \in \mathfrak{X}_m(\varphi)} X \to Z_\varphi$ of ϕ. We now have Ω written as a finite union of sets $\mathfrak{X}_m(\varphi)$ such that (b) holds.

Let φ be an arbitrary possible restriction defined above, let $X \in \mathfrak{X}_m(\varphi)$ be arbitrary and let $x \in X$. If x is at depth $j < s$, then, since $X \in \mathfrak{X}_m(\varphi) \subseteq \Omega \subseteq X_{\varphi'}$, we have $X_1 \cup \ldots \cup X_{s-1} \in \mathfrak{X}_m(\varphi')$ satisfying (d), meaning

$$\varphi(x) = \varphi'(x) \in x + \mathsf{C}_{\mathbb{Z}}(Z_1 \cup \ldots \cup Z_{j-1}).$$

If x is at depth s, then $\varpi_y + \xi_y \preceq \varpi_y + \xi_x$, where $y = \phi(x)$. In consequence, $\xi_x \in \xi_y - \mathsf{C}_{\mathbb{Z}}(Z_1 \cup \ldots \cup Z_{s-1})$, in turn implying

$$\varphi(x) = z_y = \xi_y + y \in \xi_x + y + \mathsf{C}_{\mathbb{Z}}(Z_1 \cup \ldots \cup Z_{s-1}) = x + \mathsf{C}_{\mathbb{Z}}(Z_1 \cup \ldots \cup Z_{s-1}),$$

with the final equality by (7.37). In particular, (d) holds, and (as $\varphi(X_s) = Z_s$)

$$z \in \mathsf{C}_{\mathbb{Z}}(Z_1 \cup \ldots \cup Z_{s-1}) + \mathsf{C}_{\mathbb{Z}}(X_s) \quad \text{for every } z \in Z_\varphi.$$

Since $X \in \mathfrak{X}_m(\varphi) \subseteq \Omega \subseteq X_{\varphi'}$, we have $X_1 \cup \ldots \cup X_{s-1} \in \mathfrak{X}_m(\varphi')$ satisfying (c), meaning $\mathsf{C}_{\mathbb{Z}}(Z_1 \cup \ldots \cup Z_{s-1}) \subseteq \mathsf{C}_{\mathbb{Z}}(X_1 \cup \ldots \cup X_{s-1})$, whence

$$z \in \mathsf{C}_{\mathbb{Z}}(Z_1 \cup \ldots \cup Z_{s-1}) + \mathsf{C}_{\mathbb{Z}}(X_s) \subseteq \mathsf{C}_{\mathbb{Z}}(X) \quad \text{for every } z \in Z_\varphi.$$

Thus $Z_\varphi \subseteq \mathsf{C}_{\mathbb{Z}}(X)$, implying $\mathsf{C}_{\mathbb{Z}}(Z_\varphi) \subseteq \mathsf{C}_{\mathbb{Z}}(X)$, showing that (c) holds for $\mathfrak{X}_m(\varphi)$. It remains to show that the interchangeability property holds for $\mathfrak{X}_m(\varphi)$. However, the restriction added to a set $X = X_1 \cup \ldots \cup X_s$ when passing from Ω to $\mathfrak{X}_m(\varphi) \subseteq \Omega$ only applies to conditions involving X_s, and the restrictions imposed for each $x \in X_s$ are independent of each other. Thus, since the interchangeability property holds for Ω, it follows that it also holds for $\mathfrak{X}_m(\varphi)$, i.e., if $X,\ X' \in \mathfrak{X}_m(\varphi)$ and $x \in X$ and $x' \in X'$ with $\varphi(x) = \varphi(x')$, and hence $\phi(x) = \phi(x')$ too, then $X \setminus \{x\} \cup \{x'\} \in \mathfrak{X}_m(\varphi)$. Thus (e) holds, which completes the induction and proof. □

Chapter 8
Factorization Theory

8.1 Lambert Subsets and Elasticity

In this final chapter, we apply the machinery regarding Convex Geometry and finitary sets developed in prior chapters to derive some striking consequences regarding the behavior of factorizations. Our first goal, which is one of the most difficult and crucial steps in the characterization of finite elasticities, is to obtain a multi-dimensional generalization of a result of Lambert for subsets of $\mathbb{Z}$. After we have achieved this, we then characterize when $\rho_{d+1}(G_0)$ is finite. Recall that $\mathsf{v}_X(S) = \sum_{x \in X} \mathsf{v}_x(S)$ for $S \in \mathcal{F}(G_0)$ and $X \subseteq G_0$.

Definition 8.1 Let G be an abelian group and let $G_0 \subseteq G$ be a subset. We say that $X \subseteq G_0$ is a **Lambert** subset if there exists a bound $N \geq 0$ such that $\mathsf{v}_X(U) = \sum_{x \in X} \mathsf{v}_x(U) \leq N$ for all $U \in \mathcal{A}(G_0)$.

Lambert [99] (see also [71, Lemma 4.3] and [16, Theorem 3.2]) showed that, if $G_0 \subseteq \mathbb{Z}$ is a subset with $G_0 \cap -\mathbb{Z}_+$ finite, then $\sum_{x \in G_0 \cap \mathbb{Z}_+} \mathsf{v}_x(U) \leq N$ for every $U \in \mathcal{A}(G_0)$, where $N = |\min(G_0 \cap -\mathbb{Z}_+)|$, which, in terms of the notation just introduced, means $G_0 \cap \mathbb{Z}_+ \subseteq G_0$ is a Lambert subset. His proof was a clever adaptation of the well-known argument for obtaining the basic upper bound $\mathsf{D}(G) \leq |G|$ for the Davenport Constant [82, Theorem 10.2]. The hypothesis $|G_0 \cap -\mathbb{Z}_+| < \infty$ readily implies that G_0 is finitary. Indeed, it is essentially a characterization of being a finitary set in $\mathbb{Z}$ (either $G_0 \cap -\mathbb{Z}_+$ or $G_0 \cap \mathbb{Z}_+$ must be finite for a subset of $\mathbb{Z}$ to be finitary). Moreover, if G_0 is infinite, then $G_0^\diamond = (G_0 \cap -\mathbb{Z}_+) \setminus \{0\}$, so Lambert's conclusion can be reworded as saying $G_0 \setminus G_0^\diamond \subseteq G_0 \subseteq \mathbb{Z}$ is a Lambert subset.

Consider a more general subset $G_0 \subseteq \Lambda$, where $\Lambda \leq \mathbb{R}^d$ is a lattice and $d \geq 0$. If $\mathsf{C}(G_0) \neq \mathbb{R}^d$, then terms from $G_0 \setminus \mathcal{E}$, where $\mathcal{E} = \mathsf{C}(G_0) \cap -\mathsf{C}(G_0)$ is the lineality space of $\mathsf{C}(G_0)$, are contained in no atom, and can essentially be discarded. Assuming $\mathsf{C}(G_0) = \mathbb{R}^d$, then Proposition 7.3 implies $X \subseteq G_0 \setminus G_0^\diamond$ for any Lambert subset $X \subseteq G_0$. We will later in this chapter see that, when G_0 is finitary, then

D. J. Grynkiewicz, *The Characterization of Finite Elasticities*, Lecture Notes in Mathematics 2316, https://doi.org/10.1007/978-3-031-14869-9_8

$G_0 \setminus G_0^\diamond$ is the unique maximal Lambert subset of G_0, giving a multi-dimensional analog of Lambert's result. However, we begin first with two propositions entailing some of the basic relationships between finitary sets and finite elasticities.

Proposition 8.2 *Let G be an abelian group and let $G_0 \subseteq G$ be a subset. Suppose $X \subseteq G_0$ is a subset with $\mathcal{A}(X) = \emptyset$ and $G_0 \setminus X \subseteq G_0$ is a Lambert subset with bound N. Then $\rho(G_0) \leq N < \infty$. In particular, $\rho_k(G) \leq \rho(G_0)k \leq Nk < \infty$ for any $k \geq 1$.*

Proof Let $U_1, \ldots, U_k \in \mathcal{A}(G_0)$ be atoms. Suppose $U_1 \cdot \ldots \cdot U_k = V_1 \cdot \ldots \cdot V_\ell$ for some $V_1, \ldots, V_\ell \in \mathcal{A}(G)$. Since $\mathcal{A}(X) = \emptyset$, it follows that each V_i must contain a term from $G_0 \setminus X$. However, since $G_0 \setminus X \subseteq G_0$ is a Lambert subset with bound N, there are at most kN terms in $U_1 \cdot \ldots \cdot U_k$ from $G_0 \setminus X$. Thus $\ell \leq kN$, implying $\ell/k \leq N$, and the proposition follows by definition of $\rho(G_0)$. □

Proposition 8.3 *Let G be an abelian group with torsion-free rank $d \geq 0$ and let $G_0 \subseteq G$ be a subset. Then, regarding the statements below, we have the implications 1. ⇒ 2. ⇒ 3. ⇒ 4.*

1. *There exists a subset $X \subseteq G_0$ such that $\mathcal{A}(X) = \emptyset$ and $G_0 \setminus X \subseteq G_0$ is a Lambert subset.*
2. *$\rho(G_0) < \infty$.*
3. *$\rho_k(G_0) < \infty$ for all $k \geq 1$.*
4. *$\rho_{d+1}(G_0) < \infty$.*

Proof The implication 1. ⇒ 2. follows by Proposition 8.2, while the implications 2. ⇒ 3. and 3. ⇒ 4. follow from the definition of $\rho(G_0)$. □

Proposition 8.4 *Let $\Lambda \leq \mathbb{R}^d$ be a full rank lattice in $\mathbb{R}^d$, where $d \geq 0$, and let $G_0 \subseteq \Lambda$ be a subset with $\mathsf{C}(G_0) = \mathbb{R}^d$. If $\rho_{d+1}(G_0) < \infty$, then $0 \notin \mathsf{C}^*(G_0^\diamond)$. In particular, G_0 is finitary.*

Proof Note $0 \notin G_0^\diamond$ by Proposition 7.2.2. Assume by contradiction that $0 \in \mathsf{C}^*(G_0^\diamond)$. Then Carathéordory's Theorem implies that there is a minimal positive basis $X \subseteq G_0^\diamond$. Let $X = \{x_1, \ldots, x_s\}$ be the distinct elements of X, where $2 \leq s \leq d+1$, let $V \in \mathcal{A}^{\mathsf{elm}}(G_0)$ be the unique elementary atom with $\mathsf{Supp}(V) = X$ (by Proposition 4.8), and let $N = \max\{\mathsf{v}_{x_j}(V) \;:\; j \in [1, s]\}$. Since each $x_j \in G_0^\diamond$, Proposition 7.3 implies that, for each $j \in [1, s]$, there is a sequence $\{U_i^{(j)}\}_{i=1}^\infty$ of atoms $U_i^{(j)} \in \mathcal{A}^{\mathsf{elm}}(G_0)$ with $\mathsf{v}_{x_j}(U_i^{(j)}) \to \infty$. Let $M_i = \min\{\mathsf{v}_{x_j}(U_i^{(j)}) \;:\; j \in [1, s]\}$ and observe that $M_i \to \infty$ since each $\mathsf{v}_{x_j}(U_i^{(j)}) \to \infty$. Consider the product $W_i = U_i^{(1)} \cdot \ldots \cdot U_i^{(s)}$ for $i \geq 1$ and let $m_i = \lfloor \frac{M_i}{N} \rfloor$. Then $W_i = V^{[m_i]} \cdot W_i'$ is a factorization of W_i, where $W_i' = W_i \cdot V^{[-m_i]} \in \mathcal{B}(G_0)$ and $V \in \mathcal{A}(G_0)$. Moreover, since N is fixed and $M_i \to \infty$, it follows that $m_i \to \infty$, showing that $\rho_s(G_0) = \infty$. However, since $s \leq d+1$, we have $\rho_s(G_0) \leq \rho_{d+1}(G_0)$ by (2.9), forcing $\rho_{d+1}(G_0) = \infty$ as well, contrary to hypothesis. Note $0 \notin \mathsf{C}^*(G_0^\diamond)$ implies G_0 is finitary by Theorem 7.8. □

We will need the following lemma for our generalization of Lambert's result.

Lemma 8.5 *Let $\Lambda \subseteq \mathbb{R}^d$ be a full rank lattice, where $d \geq 0$, let $G_0 \subseteq \Lambda$ be a finitary subset with $\mathsf{C}(G_0) = \mathbb{R}^d$, let $\mathcal{R} = (\mathcal{X}_1 \cup \{v_1\}, \ldots, \mathcal{X}_s \cup \{v_s\})$ be a purely virtual Reay system over G_0, let $\mathcal{X} = \mathcal{X}_1 \cup \ldots \cup \mathcal{X}_s$, and let $\pi : \mathbb{R}^d \to \mathbb{R}^{\cup}\langle\mathcal{X}\rangle^{\perp}$ be the orthogonal projection. If $U \in \mathcal{F}_{\mathsf{rat}}(G_0)$ with $\pi(U) = W^{[\alpha]}$ for some rational number $\alpha > 0$ and $W \in \mathcal{A}^{\mathsf{elm}}(\pi(G_0))$, and there exists some $w \in \mathsf{Supp}(W)$ such that $\pi^{-1}(w) \cap \mathsf{Supp}(U) \subseteq G_0 \setminus G_0^{\diamond}$, then $-\sigma(U) \in \mathsf{C}^{\cup}(\mathcal{X})$.*

Proof Suppose U is a counter-example with $|\mathsf{Supp}(U)|$ minimal. By replacing U with $U^{[1/\alpha]}$, we can w.l.o.g. assume $\alpha = 1$. Let $w_1, \ldots, w_t \in \mathsf{Supp}(W)$ be the distinct elements of $\mathsf{Supp}(W)$ and let $\alpha_i = \mathsf{v}_{w_i}(W) \in \mathbb{Z}_+$ for $i \in [1, t]$. Since $W \in \mathcal{A}^{\mathsf{elm}}(\pi(G_0))$, it follows that $\{w_1, \ldots, w_t\}$ is either $\{0\}$ or a minimal positive basis with $\alpha_1 w_1 + \ldots + \alpha_t w_t = 0$. By hypothesis, we have some w_i, say w_1, such that $\pi^{-1}(w_1) \cap \mathsf{Supp}(U) \subseteq G_0 \setminus G_0^{\diamond}$, i.e., every $u \in \mathsf{Supp}(U)$ with $\pi(u) = w_1$ has $u \in G_0 \setminus G_0^{\diamond}$. Partition $\mathsf{Supp}(U) = U_1 \cup \ldots \cup U_t$ such that each U_i, for $i \in [1, t]$, consists of those $u \in \mathsf{Supp}(U)$ with $\pi(u) = w_i$. Note $U_1 \subseteq G_0 \setminus G_0^{\diamond}$.

Let $X \subseteq \mathsf{Supp}(U)$ be an arbitrary subset such that $|X \cap U_i| = 1$ for every $i \in [1, t]$, say with $X \cap U_i = \{x_i\}$ for $i \in [1, t]$. Then $U_X := \prod^{\bullet}_{i \in [1,t]} x_i^{[\alpha_i]} \in \mathcal{F}(G_0)$ with $\pi(U_X) = W$ and $|\mathsf{Supp}(U_X)| = |\mathsf{Supp}(W)|$. As a result, since $x_1 \in U_1 \subseteq G_0 \setminus G_0^{\diamond}$, it follow from Proposition 7.14 applied to U_X that $\mathsf{wt}(-\sigma(U_X)) = 0$, i.e.,

$$-\sigma(U_X) \in \mathsf{C}^{\cup}(\mathcal{X}). \tag{8.1}$$

Let

$$\beta = \min\left\{\frac{\mathsf{v}_{x_i}(U)}{\mathsf{v}_{x_i}(U_X)} : i \in [1, t]\right\} = \min\left\{\frac{\mathsf{v}_{x_i}(U)}{\alpha_i} : i \in [1, t]\right\} > 0.$$

Since $\mathsf{v}_{x_i}(U_X) = \alpha_i = \sum_{u \in U_i} \mathsf{v}_u(U) \geq \mathsf{v}_{x_i}(U)$ for each $i \in [1, t]$ (as $\pi(U) = W$), we have $\beta \leq 1$ with equality only possible if $U_X^{[\beta]} = U_X = U$.

In view of the definition of β, we have $U_X^{[\beta]} \mid U$ with $\mathsf{v}_{x_i}(U_X^{[\beta]}) = \mathsf{v}_{x_i}(U)$ for every $i \in [1, t]$ attaining the minimum in the definition of β. Thus $U_X^{[\beta]} \in \mathcal{F}_{\mathsf{rat}}(G_0)$ with

$$|\mathsf{Supp}(U \cdot U_X^{[-\beta]})| < |\mathsf{Supp}(U)| \quad \text{and} \quad \pi(U_X^{[\beta]}) = W^{[\beta]}.$$

By (8.1), we have $-\sigma(U_X^{[\beta]}) = -\beta\sigma(U_X) \in \mathsf{C}^{\cup}(\mathcal{X})$ (since $\mathsf{C}^{\cup}(\mathcal{X})$ is a convex cone by Proposition 5.3.2). Consequently, if $\beta = 1$, so that $U_X^{[\beta]} = U$, then $-\sigma(U) = -\sigma(U_X^{[\beta]}) \in \mathsf{C}^{\cup}(\mathcal{X})$, as desired. On the other hand, if $\beta < 1$, then we have $U \cdot U_X^{[-\beta]} \in \mathcal{F}_{\mathsf{rat}}(G_0)$ with $\pi(U \cdot U_X^{[-\beta]}) = W^{[1-\beta]}$. Thus the hypotheses hold using $U \cdot U_X^{[-\beta]}$, in which case the minimality of $|\mathsf{Supp}(U)|$ for the counter-example U ensures that $-\sigma(U \cdot U_X^{[-\beta]}) \in \mathsf{C}^{\cup}(\mathcal{X})$. Combined with (8.1) and the convexity of $\mathsf{C}^{\cup}(\mathcal{X})$ (Proposition 5.3.2), we conclude that $-\sigma(U) = -\sigma(U_X^{[\beta]}) - \sigma(U \cdot U_X^{[-\beta]}) \in \mathsf{C}^{\cup}(\mathcal{X})$, as desired. □

We now come to the key generalization of Lambert's result. Theorem 8.6 requires the hypothesis $0 \notin \mathsf{C}^*(G_0^\diamond)$, which is slightly stronger than being finitary (in view of Theorem 7.8). In exchange, we actually attain a decomposition of an arbitrary atom $U \in \mathcal{A}(G_0)$, reminiscent of Theorem 4.7 (Carathéordory's Theorem), which implies that $G_0 \setminus G_0^\diamond \subseteq G_0$ is a Lambert subset. A simplification of the argument used to prove Theorem 8.6 works when we only have G_0 being finitary, though it only yields that $G \setminus G_0^\diamond \subseteq G_0$ is a Lambert subset and not the additional decomposition result for atoms. We deal with this in more detail afterwards in Corollary 8.7. The proof of Theorem 8.6 is algorithmic and yields recursively defined values for the constants N_S and N_T, which are quite large and dependent on the structure of the individual set G_0. We have not optimized the estimates for N_S and N_T, instead opting for arguments that simplify the recursive definitions and presentation. We remark that the proof of Theorem 8.6 does not require minimal types, that is:

Theorem 8.6 remains valid when $\varphi_1, \ldots, \varphi_\mathfrak{s} \in \mathfrak{T}(G_0)$ are the distinct nontrivial lattice types, rather than the distinct nontrivial minimal types.

In such case $\mathfrak{s}+1 = |\mathfrak{T}(G_0)|$ rather than $|\mathfrak{T}_m(G_0)|$, which is in general smaller, thus requiring less iterations of the algorithm, and so yielding improved bounds for N_T and N_S. However, we will need the added refinements provided by minimal types later for Theorem 8.13.

Theorem 8.6 *Let $\Lambda \leq \mathbb{R}^d$ be a full rank lattice, where $d \geq 0$, and let $G_0 \subseteq \Lambda$ be a subset with $\mathsf{C}(G_0) = \mathbb{R}^d$. Suppose $0 \notin \mathsf{C}^*(G_0^\diamond)$ and let $\varphi_1, \ldots, \varphi_\mathfrak{s} \in \mathfrak{T}_m(G_0)$ be the distinct nontrivial minimal types $\varphi_j : \mathfrak{X}_m^{\cup}(\varphi_j) \to Z_{\varphi_j}$ indexed so that $\mathsf{dep}(\varphi_i) \leq \mathsf{dep}(\varphi_j)$ whenever $i \leq j$. Then there are bounds $N_S \geq 0$ and $N_T \geq 0$ such that any atom $U \in \mathcal{A}(G_0)$ has a factorization*

$$U = A_0 \cdot A_1 \cdot \ldots \cdot A_\mathfrak{s}, \quad \textit{where} \quad A_i \in \mathcal{B}_{\mathsf{rat}}(G_0),$$

such that, for every $j \in [1, \mathfrak{s}]$,

(a) no subset $X \subseteq \mathrm{Supp}(A_0 \cdot \ldots \cdot A_{j-1})$ with $X \in \mathcal{X}(G_0)$ has minimal type φ_j;
(b) $|\mathrm{Supp}(\{A_0\})| \leq N_S$ and $|\mathrm{Supp}(\{A_j\})| \leq N_S$;
(c) $|A_0| \leq N_T$ and $\mathsf{v}_{G_0 \setminus \mathfrak{X}_m^{\cup}(\varphi_j)}(A_j) = \sum_{x \in G_0 \setminus \mathfrak{X}_m^{\cup}(\varphi_j)} \mathsf{v}_x(A_j) \leq N_T$; and
(d) $\sum_{x \in G_0 \setminus \mathfrak{X}_m^{\cup}(\varphi_j)} \mathsf{v}_x(A_j) x \in -\mathsf{C}(Z_{\varphi_j})$.

Proof If $d = 0$, then there are no nontrivial types and $U = 0$ is the only possible atom, in which case the theorem holds with $\mathfrak{s} = 0$, $N_S = 0$ and $N_T = 1$. Therefore we may assume $d \geq 1$. Since $0 \notin \mathsf{C}^*(G_0^\diamond)$, it follows from Theorem 7.8 that G_0 is finitary. We construct the bounds N_S and N_T as well as the subsequences A_j inductively for $j = \mathfrak{s}, \ldots, 1, 0$. Assume, for some $\mathfrak{t} \in [1, \mathfrak{s}+1]$, we have already constructed bounds $N_S \geq 0$ and $N_T \geq 0$ so that, for any atom $U \in \mathcal{A}(G_0)$, we can find a factorization

$$U = V \cdot A_\mathfrak{t} \cdot \ldots \cdot A_\mathfrak{s}, \quad \text{where} \quad V, A_i \in \mathcal{B}_{\mathsf{rat}}(G_0),$$

such that, for every $j \in [\mathfrak{t}, \mathfrak{s}]$, the following hold:

(a) no subset $X \subseteq \mathrm{Supp}(V \cdot A_{\mathfrak{t}} \cdot \ldots \cdot A_{j-1})$ with $X \in X(G_0)$ has minimal type φ_j;
(b) $|\mathrm{Supp}(\{V\})| \leq N_S$ and $|\mathrm{Supp}(\{A_j\})| \leq N_S$;
(c) $\mathsf{v}_{G_0 \setminus \mathfrak{X}_m^{\cup}(\varphi_j)}(A_j) \leq N_T$; and
(d) $\sum_{x \in G_0 \setminus \mathfrak{X}_m^{\cup}(\varphi_j)} \mathsf{v}_x(A_j)x \in -\mathsf{C}(Z_{\varphi_j})$.

For instance, (a)–(d) hold trivially when $\mathfrak{t} = \mathfrak{s}+1$ taking $V = U$ and $N_S = N_T = 0$, completely the base of the inductive argument.

Suppose the inductive process is finished, that is, (a)–(d) hold with $\mathfrak{t} = 1$. Set $A_0 = V$. Then the theorem holds apart from showing $|V| = |A_0| \leq N_T$. From (a) we conclude that there is no nonempty subset $X \subseteq \mathrm{Supp}(V)$ with $X \in X(G_0)$ (as any such X has some minimal type φ_j). In view of Proposition 7.31, there are only a finite number of atoms $W \in \mathcal{A}(G_0)$ such that there is no subset $X \subseteq \mathrm{Supp}(W)$ with $\emptyset \neq X \in X(G_0)$. In particular, there are only a finite number of such elementary atoms, say $W_1, \ldots, W_\ell \in \mathcal{A}^{\mathsf{elm}}(G_0)$, such that there is no nonempty subset $X \subseteq \mathrm{Supp}(W_j)$ with $X \in X(G_0)$, for $j \in [1, \ell]$. By Theorem 4.7, any zero-sum rational sequence has a factorization as a product of rational powers of elementary atoms. For the rational zero-sum $V \in \mathcal{B}_{\mathsf{rat}}(G_0)$, only the elementary atoms $W_1, \ldots, W_\ell$ can occur in such a factorization (in view of our prior work). Thus $V = \prod^\bullet_{j \in [1,\ell]} W_j^{[\alpha_j]}$ for some rational numbers $\alpha_j \geq 0$. As $V \mid U$ with $U \in \mathcal{A}(G_0)$, we must have $\alpha_j \leq 1$ for all $j \in [1, \ell]$. Hence $|V| = \sum_{j=1}^{\ell} \alpha_j |W_j| \leq \sum_{j=1}^{\ell} |W_j|$. Observing that $\sum_{j=1}^{\ell} |W_j|$ is a fixed constant independent of U, we can now replace N_T by $\max\{N_T, \sum_{j=1}^{\ell} |W_j|\}$ and thereby obtain the final remaining conclusion $|V| \leq N_T$, which would complete the proof. So we may instead assume $\mathfrak{t} > 1$ and proceed with the construction of $A_{\mathfrak{t}-1}$.

Let $\varphi = \varphi_{\mathfrak{t}-1}$, let $Z_\varphi = Z_1 \cup \ldots \cup Z_s$ be the codomain of φ, let

$$\mathbb{Z}_\varphi = \{z_1^{(\varphi)}, \ldots, z_n^{(\varphi)}\},$$

with $z_1^{(\varphi)}, \ldots, z_n^{(\varphi)} \in \mathbb{Z}_\varphi$ the $n \leq d$ distinct element in Z_φ, and let

$$\pi : \mathbb{R}^d \to \mathbb{R}\langle Z_\varphi \rangle^\perp$$

be the orthogonal projection. Since φ is nontrivial, we have $Z_\varphi \neq \emptyset$, whence $\dim \mathbb{R}\langle Z_\varphi \rangle^\perp < d$. Let $\mathfrak{X} = \mathfrak{X}_m^{\cup}(\varphi) = \bigcup_{X \in \mathfrak{X}_m(\varphi)} X \subseteq G_0^\diamond$ (the inclusion follows by the remarks from the beginning of Sect. 7.2) and, for each $i \in [1, n]$, let $\mathfrak{X}_i \subseteq \mathfrak{X}$ consist of all $x \in \mathfrak{X}$ with $\varphi(x) = z_i^{(\varphi)}$. Thus

$$\mathfrak{X} = \bigcup_{i=1}^{n} \mathfrak{X}_i \subseteq G_0^\diamond.$$

In view of Proposition 7.9, $\pi(G_0)$ is also finitary. Let $U \in \mathcal{A}(G_0)$ be an arbitrary atom. By induction hypothesis, we have bounds $N_S \geq 0$ and $N_T \geq 0$ and a factorization $U = V \cdot A_\mathfrak{t} \cdot \ldots \cdot A_\mathfrak{s}$ such that (a)–(d) hold. Let $V_\mathfrak{X} \mid V$ be the rational subsequence consisting of all terms from $\mathfrak{X}$ and let $V_{G_0 \setminus \mathfrak{X}} \mid V$ be the rational subsequence consisting of all terms from $G_0 \setminus \mathfrak{X}$. Thus

$$V = V_\mathfrak{X} \cdot V_{G_0\setminus\mathfrak{X}} \quad \text{with } V_\mathfrak{X} \in \mathcal{F}_{\mathsf{rat}}(\mathfrak{X}) \text{ and } V_{G_0\setminus\mathfrak{X}} \in \mathcal{F}_{\mathsf{rat}}(G_0 \setminus \mathfrak{X}).$$

Note, since $\mathrm{Supp}(V_\mathfrak{X}) \cap \mathrm{Supp}(V_{G_0\setminus\mathfrak{X}}) = \emptyset$, that

$$|\{V\}| = |\{V_\mathfrak{X}\}| + |\{V_{G_0\setminus\mathfrak{X}}\}|.$$

Since any $X \in \mathfrak{X}_m(\varphi)$ has $\mathbb{R}\langle X\rangle = \mathbb{R}\langle Z_\varphi\rangle = \ker \pi$, we have $\pi(x) = 0$ for all $x \in \mathfrak{X}$ by definition of $\mathfrak{X}$. Thus $V \in \mathcal{B}_{\mathsf{rat}}(G_0)$ implies

$$\pi(V_{G_0\setminus\mathfrak{X}}) \in \mathcal{B}_{\mathsf{rat}}(\pi(G_0)).$$

If there is no subset $X \subseteq \mathrm{Supp}(V)$ with $X \in X(G_0)$ having minimal type $\varphi = \varphi_{\mathfrak{t}-1}$, then we may take $A_{\mathfrak{t}-1}$ to be the trivial sequence and find that (a)–(d) hold for the factorization $U = V \cdot A_{\mathfrak{t}-1} \cdot \ldots \cdot A_\mathfrak{s}$, completing the induction and proof. Therefore we may assume otherwise that there is some $Y \in \mathfrak{X}_m(\varphi)$ with $Y \subseteq \mathrm{Supp}(V)$, say with corresponding maximal series decomposition $Y = Y_1 \cup \ldots \cup Y_s$. In particular, $Y \subseteq \mathrm{Supp}(V_\mathfrak{X})$ with $V_\mathfrak{X}$ nontrivial (as $\varphi = \varphi_{\mathfrak{t}-1}$ is nontrivial).

If there were some nonempty subset $Y_{s+1} \subseteq \mathrm{Supp}(V)$ such that $|\pi(Y_{s+1})| = |Y_{s+1}|$ and $\pi(Y_{s+1}) \in X(\pi(G_0))$ is irreducible, then we could greedily extend the realization associated to $Y = Y_1 \cup \ldots \cup Y_s$ by that associated to $\pi(Y_{s+1})$ to conclude that $Y' = Y_1 \cup \ldots \cup Y_s \cup Y_{s+1} \in X(G_0)$ with $Y' \subseteq \mathrm{Supp}(V)$. By Proposition 7.20, $Y' = Y_1 \cup \ldots \cup Y_s \cup Y_{s+1}$ is a maximal series decomposition. Letting φ' be a minimal type associated to $Y' = Y_1 \cup \ldots \cup Y_s \cup Y_{s+1}$, it follows that $\mathsf{dep}(\varphi') = s + 1 > s = \mathsf{dep}(\varphi) = \mathsf{dep}(\varphi_{\mathfrak{t}-1})$, which in view of the choice of indexing for the φ_i forces $\varphi' = \varphi_j$ for some $j \geq \mathfrak{t}$, contrary to the conclusion of (a). Therefore we can instead assume no nonempty subset $X \subseteq \mathrm{Supp}(\pi(V))$ has $X \in X(\pi(G_0))$, as any such set must contain an irreducible subset. In consequence, since $\mathrm{Supp}(\pi(V)) = \mathrm{Supp}(\pi(V_{G_0\setminus\mathfrak{X}})) \cup \{0\}$, we also find that no nonempty subset $X \subseteq \mathrm{Supp}(\pi(V_{G_0\setminus\mathfrak{X}}))$ has $X \in X(\pi(G_0))$.

Definition and Properties of the T'_k Since $\pi(G_0)$ is finitary, it follows from Proposition 7.31 that there are only a finite number of atoms $W \in \mathcal{A}(\pi(G_0))$ such that there is no nonempty subset $X \subseteq \mathrm{Supp}(W)$ with $X \in X(\pi(G_0))$. In particular, there are only a finite number of such elementary atoms, say $W_0, W_1 \ldots, W_{\ell_\varphi} \in \mathcal{A}^{\mathsf{elm}}(\pi(G_0))$, where W_0 is the zero-sum consisting of a single term equal to 0, such that there is no nonempty subset $X \subseteq \mathrm{Supp}(W_j)$ with $X \in X(\pi(G_0))$, for $j \in [0, \ell_\varphi]$. By Theorem 4.7, any zero-sum rational sequence has a factorization as a product of rational powers of elementary atoms. For the rational zero-sum $\pi(V_{G_0\setminus\mathfrak{X}}) \in \mathcal{B}_{\mathsf{rat}}(\pi(G_0))$, only the elementary atoms

$W_0, \ldots, W_{\ell_\varphi}$ can occur in such a factorization (in view of our prior comments). Thus $\pi(V_{G_0 \setminus \mathfrak{X}}) = \prod^\bullet_{j \in [0,\ell_\varphi]} W_j^{[w_i]}$ for some rational numbers $w_j \geq 0$. Since $\pi(U)$ is not an atom, we cannot conclude $w_j \leq 1$. However, letting $m_j = \lfloor w_j \rfloor$ and $\varepsilon_j = w_j - m_j \in [0,1) \cap \mathbb{Q}$, we have

$$\pi(V_{G_0 \setminus \mathfrak{X}}) = \Big(\prod\nolimits^\bullet_{j \in [0,\ell_\varphi]} W_j^{[m_j]}\Big) \cdot \Big(\prod\nolimits^\bullet_{j \in [0,\ell_\varphi]} W_j^{[\varepsilon_j]}\Big).$$

Let $\ell'_U = \sum_{i=0}^{\ell_\varphi} m_i + \sum_{i=0}^{\ell_\varphi} \lceil \varepsilon_i \rceil$. Then there must be a factorization

$$V_{G_0 \setminus \mathfrak{X}} = T'_1 \cdot \ldots \cdot T'_{\ell'_U}, \qquad \text{where each } T'_k \in \mathcal{F}_{\mathsf{rat}}(G_0),$$

such that

- for each $k \in [1, \sum_{i=0}^{\ell_\varphi} m_i]$, we have $\pi(T'_k) = W_j$ for some $j \in [0, \ell_\varphi]$, and
- for each $k \in [\sum_{i=0}^{\ell_\varphi} m_i + 1, \ell'_U]$, say with $\alpha \in [1, \sum_{i=0}^{\ell_\varphi} \lceil \varepsilon_j \rceil] \subseteq [1, \ell_\varphi + 1]$ and $k = \sum_{i=0}^{\ell_\varphi} m_i + \alpha$, we have $\pi(T'_k) = W_j^{[\varepsilon_j]}$ for the α-th largest $j \in [0, \ell_\varphi]$ with $\varepsilon_j > 0$.

Indeed, the T'_k can be sequentially constructed for $k = 1, 2, \ldots, \ell'_U$ by always first attempting to only include terms in T'_k from $V_{G_0 \setminus \mathfrak{X}} \cdot (T'_1 \cdot \ldots \cdot T'_{k-1})^{[-1]}$ with integer multiplicities until this is no longer possible, after which we attempt to only include terms in T'_k from $V_{G_0 \setminus \mathfrak{X}} \cdot (T'_1 \cdot \ldots \cdot T'_{k-1})^{[-1]}$ with their full remaining multiplicity in $V_{G_0 \setminus \mathfrak{X}} \cdot (T'_1 \cdot \ldots \cdot T'_{k-1})^{[-1]}$ until this is no longer possible, and then finally (potentially) including one last term from $V_{G_0 \setminus \mathfrak{X}} \cdot (T'_1 \cdot \ldots \cdot T'_{k-1})^{[-1]}$ with multiplicity less than one, which we need only do at most once for each element of $\mathsf{Supp}(\pi(T'_k))$. Additionally, when choosing a term y to include in T'_k according to the previous guidelines, always choose one with $y \in G_0 \setminus G_0^\diamond$ whenever possible. Assume the T'_k have been constructed according to such restrictions.

Let us analyse how the fractional subsequence $\{V_{G_0 \setminus \mathfrak{X}}\}$ compares to the modification $\{V_{G_0 \setminus \mathfrak{X}} \cdot (\prod^\bullet_{k \in I} T'_k)^{[-1]}\}$, for a subset $I \subseteq [1, \ell'_\varphi]$. For each T'_k, we have $\pi(T'_k) = W_j$ or $W_j^{[\varepsilon_j]}$ for some $j \in [0, \ell_\varphi]$. The process of constructing T'_k first takes available terms with integer multiplicities in $V_{G_0 \setminus \mathfrak{X}} \cdot (T'_1 \cdot \ldots \cdot T'_{k-1})^{[-1]}$ to construct $\lfloor T'_k \rfloor$. Removing these terms cannot create a new fractional term. The process then includes terms with their full available fractional multiplicity. Including these terms actually removes a term with fractional multiplicity. Finally, only the last term y included into T'_k, for each $w \in \mathsf{Supp}(W_j)$ (so $\pi(y) = w$), can actually create a new fractional term that was not already present in $V_{G_0 \setminus \mathfrak{X}} \cdot (T'_1 \cdot \ldots \cdot T'_{k-1})^{[-1]}$, and this can only occur if y has multiplicity at least one in $V_{G_0 \setminus \mathfrak{X}} \cdot (T'_1 \cdot \ldots \cdot T'_{k-1} \cdot \lfloor T'_k \rfloor)^{[-1]}$. Let $L_k \subseteq G$ consist of all y included as a last term into T'_k, potentially one for each $w \in \mathsf{Supp}(W_j)$, such that the term y is included with non-integer multiplicity in T'_k. Note $|\mathsf{Supp}(W_j)| \leq d$, since $W_j \in \mathcal{A}^{\mathsf{elm}}(\pi(G_0))$ is an elementary atom with $\dim(\operatorname{im}\pi) = \dim \mathbb{R}\langle Z_\varphi\rangle^\perp < d$, so $|L_k| \le |\operatorname{Supp}(W_j)| \le d$. Note that $L_{\ell'_U} \subseteq \operatorname{Supp}(\{V_{G_0\backslash\mathfrak{X}}\}) \cup L_1 \cup \ldots \cup L_{\ell'_U-1}$, since every term is included into $T'_{\ell'_U}$ with its full remaining multiplicity. Moreover, if $j = 0$, then we obtain the improved estimate $|\operatorname{Supp}(W_0)| = |W_0| = 1$ as an upper bound rather than d.

For $k \in [1, \sum_{i=0}^{\ell_\varphi} m_i]$, we have $\pi(T'_k) = W_j$, with the multiplicity of each term in W_j being an integer (since W_j is an elementary atom). If $T'_k \in \mathcal{F}(G_0)$, then $T'_k \mid \lfloor V_{G_0\backslash\mathfrak{X}}\rfloor$, and removing T'_k creates no new fractional term. If $T'_k \notin \mathcal{F}(G_0)$, then there is some $w \in \operatorname{Supp}(W_j)$ such that $\mathsf{v}_w(W_j) > \mathsf{v}_w(\pi(\lfloor T'_k\rfloor))$. However, since both $\mathsf{v}_w(W_j)$ and $\mathsf{v}_w(\pi(\lfloor T'_k\rfloor))$ are integers, we have $\mathsf{v}_w(W_j) - \mathsf{v}_w(\pi(\lfloor T'_k\rfloor)) \ge 1$. Thus all remaining terms y to be included into T'_k, each with $\pi(y) = w$ for some $w \in \operatorname{Supp}(W_j)$ satisfying $\mathsf{v}_w(W_j) > \mathsf{v}_w(\pi(\lfloor T'_k\rfloor))$, must come from $\{V_{G_0\backslash\mathfrak{X}}\}$ and have their full remaining multiplicity in $V_{G_0\backslash\mathfrak{X}} \cdot (T'_1 \cdot \ldots \cdot T'_{k-1} \cdot \lfloor T'_k\rfloor)^{[-1]}$ strictly less than one (else y could be included with integer multiplicity in view of $\mathsf{v}_w(W_j) - \mathsf{v}_w(\pi(\lfloor T'_k\rfloor)) \ge 1$). In particular, including the last term into T'_k, for each $w \in \operatorname{Supp}(W_j)$, does not create a new fractional term, meaning $L_k \subseteq \operatorname{Supp}(\{V_{G_0\backslash\mathfrak{X}}\})$ for $k \in [1, \sum_{i=1}^{\ell_\varphi} m_i]$. Moreover, for each $w \in \operatorname{Supp}(W_j)$ satisfying $\mathsf{v}_w(W_j) > \mathsf{v}_w(\pi(\lfloor T'_k\rfloor))$, of which there is at least one in view of $T'_k \notin \mathcal{F}(G_0)$, the next term y included into T'_k with $\pi(y) = w$ is from $\{V_{G_0\backslash\mathfrak{X}}\}$ and included with its full remaining multiplicity as $\mathsf{v}_y(V_{G_0\backslash\mathfrak{X}} \cdot (T'_1 \cdot \ldots \cdot T'_{k-1} \cdot \lfloor T'_k\rfloor)^{[-1]}) < 1 \le \mathsf{v}_w(W_j) - \mathsf{v}_w(\pi(\lfloor T'_k\rfloor))$. Thus we lose one fractional term from $\{V_{G_0\backslash\mathfrak{X}}\}$ for each $k \in [1, \sum_{i=1}^{\ell_\varphi} m_i]$ with $T'_k \notin \mathcal{F}(G_0)$.

The results of the previous two paragraphs can now be combined to derive the following summary statements. For each T'_k with $k \in [1, \sum_{i=0}^{\ell_\varphi} m_i]$, we either have $T'_k \in \mathcal{F}(G_0)$ or else

$$|\operatorname{Supp}(\left\{V_{G_0\backslash\mathfrak{X}} \cdot (T'_1 \cdot \ldots \cdot T'_k)^{[-1]}\right\})| < |\operatorname{Supp}(\left\{V_{G_0\backslash\mathfrak{X}} \cdot (T'_1 \cdot \ldots \cdot T'_{k-1})^{[-1]}\right\})|. \tag{8.2}$$

Consequently, (b) ensures there are at most $N_S + \sum_{i=0}^{\ell_\varphi} \lceil \varepsilon_i \rceil \le N_S + \ell_\varphi + 1$ indices $k \in [1, \ell'_U]$ with $T'_k \notin \mathcal{F}(G_0)$. Moreover, for any $I \subseteq [1, \ell'_U]$, we have

$$\operatorname{Supp}(\left\{V_{G_0\backslash\mathfrak{X}} \cdot (\textstyle\prod^\bullet_{k\in I} T'_k)^{[-1]}\right\}) \subseteq \operatorname{Supp}(\{V_{G_0\backslash\mathfrak{X}}\}) \cup \bigcup_{k=1}^{\max I} L_k. \tag{8.3}$$

We have $|L_k \setminus \operatorname{Supp}(\{V_{G_0\backslash\mathfrak{X}}\})| = 0$ for $k \in [1, \sum_{i=0}^{\ell_\varphi} m_i]$; we have $|L_k| \le d$ for each of the $\sum_{i=0}^{\ell_\varphi} \lceil \varepsilon_i \rceil \le \ell_\varphi + 1$ values $k \in [\sum_{i=0}^{\ell_\varphi} m_i + 1, \ell'_U]$; and we have $L_{\ell'_U} \subseteq \operatorname{Supp}(\{V_{G_0\backslash\mathfrak{X}}\}) \cup L_1 \cup \ldots \cup L_{\ell'_U-1}$. Moreover, if equality holds in the estimate $\sum_{i=0}^{\ell_\varphi} \lceil \varepsilon_i \rceil \le \ell_\varphi + 1$, then $\pi(T'_k) = W_0^{[\varepsilon_0]}$ for $k = \sum_{i=0}^{\ell_\varphi} m_i + 1$ with corresponding improved estimate $|L_k| \le 1$. Thus (8.3) applied to $\prod^\bullet_{k\in I} T'_k =$

$V_{G_0 \setminus \mathfrak{X}} \cdot \left(\prod^\bullet_{k \in [1,\ell'_U] \setminus I} T'_k\right)^{[-1]}$, combined with (b), implies (since $d \geq 1$) that

$$|\operatorname{Supp}(\{\prod^\bullet_{k\in I} T'_k\})| \leq N_S + \max\{0, (\ell_\varphi - 1)d + 1\} \leq N_S + \ell_\varphi d, \quad \text{for any } I \subseteq [1, \ell'_U]. \tag{8.4}$$

Each T'_k with $k \in [1, \ell'_U]$ has $\pi(T'_k) = W_j$ or $W_j^{[\varepsilon_j]}$ for some $j \in [0, \ell_\varphi]$, with $\varepsilon_j \leq 1$. Thus $|\lfloor T'_k \rfloor| \leq |T'_k| \leq |W_j|$. Let

$$M_W = \max_{j \in [0, \ell_\varphi]} |W_j|.$$

Then, for every $k \in [1, \ell'_U]$, (8.4) applied with $I = \{k\}$ yields

$$|\operatorname{Supp}(\{T'_k\})| \leq N_S + \max\{0, (\ell_\varphi - 1)d + 1\} \quad \text{and} \quad |\lfloor T'_k \rfloor| \leq |T'_k| \leq M_W. \tag{8.5}$$

Definition and Properties of the T_k We next proceed to define a new factorization

$$V_{G_0 \setminus \mathfrak{X}} = T_1 \cdot \ldots \cdot T_{\ell_U}, \qquad \text{where each } T_k \in \mathcal{F}_{\mathsf{rat}}(G_0),$$

by modifying the sequences T'_k with $k \in [1, \ell'_U]$ as follows. Let

$$\Delta = \mathbb{Z}\langle Z_\varphi \rangle \quad \text{and} \quad \Lambda' = \Lambda \cap \mathbb{R}\langle Z_\varphi \rangle.$$

Then Δ and Λ' are both full rank lattices in $\mathbb{R}\langle Z_\varphi \rangle$ with $\Delta \leq \Lambda'$. It follows that Λ'/Δ is a finite abelian group with Davenport Constant $D_\Delta := \mathsf{D}(\Lambda'/\Delta) \leq |\Lambda'/\Delta|$ (see Sect. 2.3).

Consider an arbitrary sequence T'_k with $k \in [1, \ell'_U]$. Then $\pi(T'_k) = W_j$ or $W_j^{[\varepsilon_j]}$ for some $j \in [0, \ell_\varphi]$. Let us call T'_k *pure* if there is some $w \in \operatorname{Supp}(W_j)$ such that $\pi^{-1}(w) \cap \operatorname{Supp}(T'_k) \subseteq G_0 \setminus G_0^\diamond$.

By definition of the T'_k, we have $\sigma(T'_k) \in \ker \pi = \mathbb{R}\langle Z_\varphi \rangle$ for every $k \in [1, \ell'_U]$. Thus, if $T'_k \in \mathcal{F}(G_0)$, then we have $\sigma(T'_k) \in \mathbb{R}\langle Z_\varphi \rangle \cap \Lambda = \Lambda'$. If $I \subseteq [1, \ell'_U]$ is any subset of indices $k \in [1, \ell'_U]$ with $T'_k \in \mathcal{F}(G_0)$ and $|I| \geq D_\Delta$, then the definition of D_Δ ensures that there is some minimal nonempty subset $I' \subseteq I$ such that $|I'| \leq D_\Delta$ and $\sigma(\prod^\bullet_{k \in I'} T'_k) \in \Delta$.

The rational sequences T_k and sets $I_k \subseteq [1, \ell'_U]$ such that $T_1 \cdot \ldots \cdot T_k = \prod^\bullet_{i \in I_k} T'_i$ will be constructed sequentially for $k = 1, 2, \ldots, \ell_U$ (though we will sometimes need to define multiple T_k at the same time). Set $I_0 = \emptyset$ and take T_0 to be the trivial sequence. Assume we have already constructed $T_1 \cdot \ldots \cdot T_k = \prod^\bullet_{i \in I_k} T'_i$. Then we construct T_{k+1} (or possibly T_{k+1} and T_{k+2} simultaneously, in which case only I_{k+2} is defined and not I_{k+1}) according to the following three possibilities.

(i) Suppose there is a nonempty subset $I \subseteq [1, \ell'_U] \setminus I_k$ of indices $i \in [1, \ell'_U] \setminus I_k$ such that $T'_i \in \mathcal{F}(G_0)$ is pure for every $i \in I$ and $\sigma(\prod^\bullet_{i \in I} T'_i) \in \Delta$. Then choose a minimal such subset I and set $T_{k+1} = \prod^\bullet_{i\in I} T_i'$ and $I_{k+1} = I_k \cup I$. Note $|I| \leq D_\Delta$ in such case, as remarked previously. Since each $T_i' \in \mathcal{F}(G_0)$, for $i \in I$, it follows that $\pi(T_i') \in \mathcal{B}(\pi(G_0))$ is a zero-sum *sequence* (not just rational sequence) with $\mathsf{Supp}(\pi(T_i')) = \mathsf{Supp}(W)$ for some elementary atom $W \in \mathcal{A}^{\mathsf{elm}}(\pi(G_0))$, which forces $\pi(T_i')$ to be an *integer* power of W (by definition of an elementary atom), and thus $\pi(T_i') = W$ by construction of the T_i'.

(ii) Suppose the hypotheses of (i) fail but there is some $i \in [1, \ell_U'] \setminus I_k$ with $\mathsf{Supp}(T_i') \nsubseteq G_0^\diamond$. We have $\pi(T_i') = W^{[\varepsilon]}$ for some elementary atom $W \in \mathcal{A}^{\mathsf{elm}}(\pi(G_0))$ and rational number $\varepsilon \in (0, 1]$. For each $w \in \mathsf{Supp}(W)$, write $\alpha_w := \mathsf{v}_w(W^{[\varepsilon]}) = \alpha_w^{\bar\diamond} + \alpha_w^\diamond$, where

$$\alpha_w^{\bar\diamond} = \sum_{\substack{\pi(x)=w \\ x\in\mathsf{Supp}(T_i')\setminus G_0^\diamond}} \mathsf{v}_x(T_i') \quad \text{and} \quad \alpha_w^{\diamond} = \sum_{\substack{\pi(x)=w \\ x\in\mathsf{Supp}(T_i')\cap G_0^\diamond}} \mathsf{v}_x(T_i').$$

Let

$$\varepsilon' = \varepsilon \cdot \max\{\alpha_w^{\bar\diamond}/\alpha_w \,:\, w \in \mathsf{Supp}(W)\} \in (0, \epsilon] \cap \mathbb{Q} \subseteq (0, 1] \cap \mathbb{Q}.$$

Note that $\varepsilon' > 0$ in view of $\mathsf{Supp}(T_i') \nsubseteq G_0^\diamond$ and that $\varepsilon' = \varepsilon$ precisely when T_i' is pure. Now $W^{[\varepsilon']}$ has $W^{[\varepsilon']} \mid W^{[\varepsilon]}$ and $\mathsf{v}_w(W^{[\varepsilon']}) \geq \alpha_w^{\bar\diamond}$ for every $w \in \mathsf{Supp}(W)$, with equality holding for each w obtaining the maximum in the definition of ε'. Define a new rational subsequence $T_{k+1} \mid T_i'$ with $\pi(T_{k+1}) = W^{[\varepsilon']}$ as follows. Include in T_{k+1} all terms $x \in \mathsf{Supp}(T_i') \setminus G_0^\diamond$ with their full multiplicity from T_i', which is possible in view of $\mathsf{v}_w(W^{[\varepsilon']}) \geq \alpha_w^{\bar\diamond}$. Continue to include terms in T_{k+1} from $x \in \mathsf{Supp}(T_i') \cap G_0^\diamond$ with their full multiplicity from T_i', so long as this is possible, and finally (potentially) add one last term to T_{k+1}, for each $w \in \mathsf{Supp}(W)$, with only part of its multiplicity from T_i'. Note $T_{k+1} = T_i'$ when T_i' pure. In such case, we set $I_{k+1} = I_k \cup \{i\}$. If T_i' is not pure, then $T_{k+2} = T_i' \cdot T_{k+1}^{[-1]}$ is a nonempty sequence, and in such case, we set $I_{k+2} = I_k \cup \{i\}$, defining T_{k+1} and T_{k+2} simultaneously under these circumstances. For any $w \in \mathsf{Supp}(W)$ attaining the maximum in the definition of ε', we have

$$\pi^{-1}(w) \cap \mathsf{Supp}(T_{k+1}) \subseteq G_0 \setminus G_0^\diamond. \tag{8.6}$$

Also, $\mathsf{Supp}(T_{k+2}) \subseteq G_0^\diamond$ since all terms from $\mathsf{Supp}(T_i') \setminus G_0^\diamond$ were included in T_{k+1} with their full multiplicity from T_i'.

(iii) Suppose the hypotheses of (i) and (ii) both fail, i.e., $\mathsf{Supp}(T_i') \subseteq G_0^\diamond$ for every $i \in [1, \ell_U'] \setminus I_k$. Then take any remaining $i \in [1, \ell_U'] \setminus I_k$ and set $T_{k+1} = T_i'$ and $I_{k+1} = I_k \cup \{i\}$. If no such i exists, i.e., if $I_k = [1, \ell_U']$, then set $\ell_U = k$.

We partition the interval

$$[1, \ell_U] = I_{\mathbb{Z}} \cup I_{\mathbb{Q}} \cup I_{\diamond},$$

where $I_{\mathbb{Z}} \subseteq [1, \ell_U]$ consists of all $k+1 \in [1, \ell_U]$ for which T_{k+1} was constructed under condition (i), where $I_{\mathbb{Q}}$ consists of all $k+1 \in [1, \ell_U]$ for which T_{k+1} was constructed under condition (ii), and where $I_{\diamond}$ consists of all $k+1 \in [1, \ell_U]$ for which T_{k+1} was constructed under condition (iii) as well as all $k+2 \in [1, \ell_U]$ for which T_{k+2} was constructed under condition (ii). By construction, $I_{\mathbb{Z}}$ consists of the first $|I_{\mathbb{Z}}|$ elements from $[1, \ell_U]$. We have the following observations.

If $k \in I_{\mathbb{Z}}$, then there is a nonempty subset $I \subseteq [1, \ell'_U]$ with $|I| \leq D_\Delta$ and $T_k = \prod^\bullet_{i \in I} T'_i$ such that, for every $i \in I$, we have

$$T'_i \in \mathcal{F}(G_0), \quad \pi(T'_i) \in \mathcal{A}^{\mathsf{elm}}(\pi(G_0)), \quad \text{and} \tag{8.7}$$

$$\pi^{-1}(w) \cap \mathsf{Supp}(T'_i) \subseteq G_0 \setminus G_0^\diamond \quad \text{for some } w \in \mathsf{Supp}(\pi(T'_i)). \tag{8.8}$$

Moreover,

$$|T_k| \leq D_\Delta M_W, \quad T_k \in \mathcal{F}(G_0) \quad \text{and} \quad \sigma(T_k) \in \Delta, \quad \text{for every } k \in I_{\mathbb{Z}}, \tag{8.9}$$

where the first inequality follows in view of (8.5) and $|I| \leq D_\Delta$.

If $k \in I_{\mathbb{Q}} \cup I_\diamond$, then $T_k \mid T'_i$ for some $i \in [1, \ell'_U]$, and $\pi(T_k^{[\epsilon_k]}) \in \mathcal{A}^{\mathsf{elm}}(\pi(G_0))$ for some rational $\epsilon_k \geq 1$. Thus (8.5) implies

$$|T_k| \leq |T'_i| \leq M_W \quad \text{for all } k \in I_{\mathbb{Q}} \cup I_\diamond. \tag{8.10}$$

If $k \in I_{\mathbb{Q}}$, then (8.6) implies that

$$\pi^{-1}(w) \cap \mathsf{Supp}(T_k) \subseteq G_0 \setminus G_0^\diamond \quad \text{for some } w \in \mathsf{Supp}(\pi(T_k)). \tag{8.11}$$

If $k \in I_\diamond$, then (per the hypotheses of Condition (iii) and the remarks regarding T_{k+2} in Condition (ii))

$$\mathsf{Supp}(T_k) \subseteq G_0^\diamond. \tag{8.12}$$

□

Claim A $|I_{\mathbb{Q}}| \leq D_\Delta + N_S + 2\ell_\varphi$.

Proof If $k \in I_{\mathbb{Q}}$, then there was some $i \in [1, \ell'_U]$ such that $T_k \mid T'_i$ with $\mathsf{Supp}(T'_i) \nsubseteq G_0^\diamond$. By construction, the same i does not occur for two distinct $k \in I_{\mathbb{Q}}$. We can assume there are at most $D_\Delta - 1$ such T'_i with $T'_i \in \mathcal{F}(G_0)$ and T'_i pure, else the algorithm for producing T_k would have used condition (i) rather than (ii). As noted after (8.2), there are at most $N_S + \ell_\varphi + 1$ indices $i \in [1, \ell'_U]$ with $T'_i \notin \mathcal{F}(G_0)$, and this estimate includes all indices from $[\sum_{\alpha=0}^{\ell_\varphi} m_\alpha + 1, \ell'_U]$. It remains to find an upper bound for the number of $i \in [1, \sum_{\alpha=1}^{\ell_\varphi} m_\alpha]$ such that $\mathsf{Supp}(T'_i) \nsubseteq G_0^\diamond$, $T'_i \in \mathcal{F}(G_0)$ but T'_i is not pure. For any such T'_i, we have $\pi(T'_i) = W_j$ for some $j \in [0, \ell_\varphi]$ (since $i \in [1, \sum_{\alpha=1}^{\ell_\varphi} m_\alpha]$). Since $T'_i \in \mathcal{F}(G_0)$, it follows that T'_i was filled entirely with terms from $\lfloor V \rfloor$ (cf. (2.6)). Since $\mathsf{Supp}(T'_i) \nsubseteq G_0^\diamond$, there is at least one $x \in \mathsf{Supp}(T'_i)$ with $x \in G_0 \setminus G_0^\diamond$. Thus we must have $j \neq 0$, else $\mathsf{Supp}(T'_i) = \{x\} \subseteq G_0 \setminus G_0^\diamond$ (since $|W_0| = 1$), contradicting that T'_i is not pure. Since T'_i is not pure, we have $\pi^{-1}(w) \cap \mathsf{Supp}(T_i) \cap G_0^\diamond \neq \emptyset$ for every $w \in \mathsf{Supp}(W_j) = \mathsf{Supp}(\pi(T'_i))$.

Suppose $T'_{i'}$ is another sequence with $i' \in [1, \sum_{\alpha=0}^{\ell_\varphi} m_\alpha]$ such that

$$T'_i \in \mathcal{F}(G_0) \quad \text{and} \quad \pi(T'_{i'}) = W_j = \pi(T'_i).$$

Since $T'_i, T'_{i'} \in \mathcal{F}(G_0)$, it follows that both T'_i and $T'_{i'}$ were constructed by only including terms with integer multiplicities. Moreover, when there was a choice between a term from $G \setminus G_0^\diamond$ and another from $G_0^\diamond$ (with available integer multiplicity), those from $G \setminus G^\diamond$ always had preference. If $i' < i$, then, since there is some $x \in \mathsf{Supp}(T'_i)$ with $x \in G_0 \setminus G_0^\diamond$, it follows that, when choosing terms y to place into $T'_{i'}$ with $\pi(y) = \pi(x)$, we did not exhaust all possible choices from $G_0 \setminus G_0^\diamond$. Thus, due to our preference to include terms from $G_0 \setminus G_0^\diamond$ when possible when constructing the T'_i, we conclude that every term $y \in \mathsf{Supp}(T'_{i'})$ with $\pi(y) = \pi(x)$ satisfies $y \in G_0 \setminus G_0^\diamond$, ensuring that $T'_{i'}$ is pure. On the other hand, if $i' > i$, then, since every $w \in \mathsf{Supp}(W_j)$ has $\pi^{-1}(w) \cap \mathsf{Supp}(T'_i) \cap G_0^\diamond \neq \emptyset$, we must have exhausted all possible terms from $G_0 \setminus G_0^\diamond$ that could every be included in $T'_{i'}$ (again, in view of our preference to choose terms from $G_0 \setminus G_0^\diamond$). Hence $\mathsf{Supp}(T'_{i'}) \subseteq G_0^\diamond$. In conclusion, we see that, for each $j \in [1, \ell_\varphi]$, there is at most one $i \in [1, \sum_{\alpha=0}^{\ell_\varphi} m_\alpha]$ such that $\mathsf{Supp}(T'_i) \nsubseteq G_0^\diamond$, $T'_i \in \mathcal{F}(G_0)$ but T'_i is not pure. Combined with our previous estimates, we now conclude that $|I_\mathbb{Q}| \le (D_\Delta - 1) + (N_S + \ell_\varphi + 1) + \ell_\varphi = D_\Delta + N_S + 2\ell_\varphi$, completing Claim A.

□

Claim B For each $k \in I_\mathbb{Z} \cup I_\mathbb{Q}$, we have $-\sigma(T_k) \in \mathsf{C}(X)$ for any $X \in \mathfrak{X}_m(\varphi)$. For each $k \in I_\mathbb{Z}$, we have $-\sigma(T_k) \in \mathsf{C}_\mathbb{Z}(X)$ for any $X \in \mathfrak{X}_m(\varphi)$. In particular, both these statements hold with $X = Z_\varphi$.

Proof Let $\mathcal{R}_X = (\mathcal{X}_1 \cup \{\mathbf{v}_1\}, \ldots, \mathcal{X}_s \cup \{\mathbf{v}_s\})$ be a realization of X (associated to the minimal type φ) with all half-spaces from $\bigcup_{j=1}^{s} \mathcal{X}_j$ having dimension one, in which case $\mathbb{R}^\cup\langle \mathcal{X} \rangle = \mathbb{R}\langle Z_\varphi \rangle = \ker \pi$ and $\mathbb{Z}\langle X \rangle = \mathbb{Z}\langle Z_\varphi \rangle = \Delta$. Since $\varphi \in \mathfrak{X}_m(G_0)$, $\mathcal{R}_X$ is purely virtual. Since all half-spaces in $\mathcal{R}_X$ have dimension one with the elements from X being representatives for the half-spaces from $\mathcal{X}$, we have $\mathsf{C}^\cup(\mathcal{X}) = \mathsf{C}(X)$. If $k \in I_\mathbb{Q}$, then (8.11) allows us to apply Lemma 8.5 to T_k and thereby conclude $-\sigma(T_k) \in \mathsf{C}^\cup(\mathcal{X}) = \mathsf{C}(X)$, as desired. If $k \in I_\mathbb{Z}$, then $T_k = \prod_{i \in I} T'_i$ for some $I \subseteq [1, \ell'_U]$. Then (8.7) implies that each $T'_i \in \mathcal{F}(G_0)$ with $\pi(T'_i) = W$ for some

$W \in \mathcal{A}^{\mathsf{elm}}(\pi(G_0))$. Hence, in view of (8.8), we can apply Lemma 8.5 to each T_i' to conclude $-\sigma(T_i') \in \mathsf{C}^{\cup}(X) = \mathsf{C}(X)$, for $i \in I$. Thus, in view of the convexity of $\mathsf{C}(X)$, we have $-\sigma(T_k) = -\sum_{i \in I} \sigma(T_i') \in \mathsf{C}(X)$. In view of (8.9), we have $\sigma(T_k) \in \Delta$. Thus $-\sigma(T_k) \in \mathsf{C}(X) \cap \Delta = \mathsf{C}_{\mathbb{Z}}(X)$, with the latter equality in view of the linear independence of X and $\mathbb{Z}\langle X \rangle = \Delta$, completing Claim B. □

Each $k \in I_{\mathbb{Q}}$ has $T_k \mid T'_{i(k)}$ for some $i(k) \in [1, \ell_U']$, so $\mathsf{Supp}(\prod^\bullet_{k \in I_{\mathbb{Q}}} T_k) \subseteq \mathsf{Supp}(\prod^\bullet_{k \in I_{\mathbb{Q}}} T'_{i(k)})$. By (8.4), we have $|\,\mathsf{Supp}(\left\{\prod^\bullet_{k \in I_{\mathbb{Q}}} T'_{i(k)}\right\})| \le N_S + \ell_\varphi d$. Claim A and (8.5) imply $|\lfloor \prod^\bullet_{k \in I_{\mathbb{Q}}} T'_{i(k)} \rfloor| \le \sum_{k \in I_{\mathbb{Q}}} |T'_{i(k)}| \le |I_{\mathbb{Q}}| M_W \le (D_\Delta + N_S + 2\ell_\varphi) M_W$. As a result of all these calculations, we now find that (via (2.2))

$$\begin{aligned} |\,\mathsf{Supp}(\prod\nolimits^\bullet_{k \in I_{\mathbb{Q}}} T_k)| &\le |\,\mathsf{Supp}(\prod\nolimits^\bullet_{k \in I_{\mathbb{Q}}} T'_{i(k)})| \\ &\le |\lfloor \prod\nolimits^\bullet_{k \in I_{\mathbb{Q}}} T'_{i(k)} \rfloor| + |\,\mathsf{Supp}(\left\{\prod\nolimits^\bullet_{k \in I_{\mathbb{Q}}} T'_{i(k)}\right\})| \\ &\le (D_\Delta + N_S + 2\ell_\varphi) M_W + N_S + \ell_\varphi d. \end{aligned} \tag{8.13}$$

Recall the definition of $\mathfrak{X} = \bigcup_{i=1}^n \mathfrak{X}_i$. In view of the interchangeability property of $\mathfrak{X}_m(\varphi)$ (Proposition 7.33), the sets $X \in \mathfrak{X}_m(\varphi)$ with $X \subseteq \mathsf{Supp}(V_{\mathfrak{X}})$ are precisely those obtained by choosing one element $x_i \in \mathfrak{X}_i \cap \mathsf{Supp}(V_{\mathfrak{X}})$ for each $i \in [1, n]$. Moreover, since $Y \subseteq \mathsf{Supp}(V_{\mathfrak{X}})$ with $Y \in \mathfrak{X}_m(\varphi)$, the sets $\mathfrak{X}_i \cap \mathsf{Supp}(V_{\mathfrak{X}})$ are all nonempty, for $i \in [1, n]$.

Let $X \in \mathfrak{X}_m(\varphi)$ be arbitrary, say with $X = \{x_1, \ldots, x_n\}$ the distinct elements $x_i \in X$, and let $g \in \Delta \cap -\mathsf{C}(X)$. Then $X \subseteq G_0^\diamond$ is a linearly independent set with $\mathbb{Z}\langle X \rangle = \Delta$ (cf. the remarks at the beginning of Sect. 7.2). It follows that

$$\mathsf{C}_{\mathbb{Z}}(X) = \mathsf{C}(X) \cap \Delta. \tag{8.14}$$

As a result, if $-g = \alpha_1 x_1 + \ldots + \alpha_n x_n$ is a linear combination with $\alpha_j \in \mathbb{R}_+$ expressing that $-g \in \mathsf{C}(X)$, then it follows by the linear independence of X (and recalling that $g \in \Delta = \mathbb{Z}\langle X \rangle$) that

$$\alpha_j \in \mathbb{Z}_+ \quad \text{for all } j \in [1, n]. \tag{8.15}$$

The Outer Loop We proceed to recursively construct fractional sequences $B_k \in \mathcal{F}_{\mathsf{rat}}(G_0)$, for $k = 0, 1, \ldots$, with

$$B_k \mid V_{\mathfrak{X}} \cdot (B_0 \cdot \ldots \cdot B_{k-1})^{[-1]}.$$

We refer to the process which constructs the B_k as the *Outer Loop*, and initially set B_0 to be the trivial sequence. We view

$$V_k := V_{\mathfrak{X}} \cdot (B_0 \cdot \ldots \cdot B_k)^{[-1]} = V_{\mathfrak{X}} \cdot (B_1 \cdot \ldots \cdot B_k)^{[-1]}$$

as the current state resulting after k iterations of the outer loop. We view each $j \in [1, n]$ (indexing $\mathfrak{X}_j$) as a box which contains some of the damaged elements

$$D_k := \mathsf{Supp}(\{V_k\}) \subseteq G_0^\diamond.$$

A box $j \in [1, n]$ is *depleted* if $\mathfrak{X}_j \cap \mathsf{Supp}(\lfloor V_k \rfloor) = \emptyset$, which means every $x \in \mathsf{Supp}(V_k) \cap \mathfrak{X}_j$ has $\mathsf{v}_x(V_k) < 1$. All elements contained in a depleted box are damaged. A box $j \in [1, n]$ with $\mathfrak{X}_j \cap \mathsf{Supp}(\lfloor V_k \rfloor) \neq \emptyset$, meaning there is some $x \in \mathsf{Supp}(V_k) \cap \mathfrak{X}_j$ with $\mathsf{v}_x(V_k) \geq 1$, is called *undepleted*. An undepleted box contains at least one undamaged element. Let $J_k^d \subseteq [1, n]$ consist of all depleted boxes $j \in [1, n]$ after k iterations, and let $J_k^u \subseteq [1, n]$ consist of the remaining undepleted boxes $j \in [1, n]$ after k iterations. Note, once a box becomes depleted, it remains depleted in all further iterations of the outer loop. If there is some box $j \in [1, n]$ with $\mathfrak{X}_j \cap \mathsf{Supp}(V_{k-1}) = \emptyset$, that is, an empty box, then the outer loop process halts after $(k - 1)$ iterations. Assuming this is not the case, then the next sequence B_k, for $k \geq 1$, must satisfy the following properties.

1. $|D_k \cap \mathfrak{X}_j| \leq |D_{k-1} \cap \mathfrak{X}_j| + 1$ for all $j \in J_{k-1}^u$.
2. $|D_k \cap \mathfrak{X}_j| \leq |D_{k-1} \cap \mathfrak{X}_j|$ for all $j \in J_{k-1}^d$.
3. Either $|J_k^d| > |J_{k-1}^d|$ or else $|D_k \cap \mathfrak{X}_j| < |D_{k-1} \cap \mathfrak{X}_j|$ for some $j \in J_k^d = J_{k-1}^d$.
4. $\sigma(B_k) = -\beta_k g_k$ for some $\beta_k \in \mathbb{Q}_+$ with $0 < \beta_k \leq 1$, where

$$g_k = \begin{cases} \sigma(T_k), & \text{for } k \leq |I_\mathbb{Z}| \\ \sum_{i=1}^{|I_\mathbb{Z}|}(1 - \beta_i)g_i + \sum_{i \in I_\mathbb{Q}} \sigma(T_i), & \text{for } k = |I_\mathbb{Z}| + 1 \\ g_k = (1 - \beta_{k-1})g_{k-1}, & \text{for } k \geq |I_\mathbb{Z}| + 2. \end{cases}$$

Moreover, $\beta_k = 1$ is only possible if either $k \geq |I_\mathbb{Z}| + 1$ or else $k = 1 = |I_\mathbb{Z}|$, $|I_\mathbb{Q}| = |I_\diamond| = 0$ and $V = U$, and in either case, the Outer Loop process then halts after k iterations.

Recall that $I_\mathbb{Z} = [1, |I_\mathbb{Z}|]$ by construction.

Suppose it is possible to construct such sequences B_k as described above (the algorithm for their construction will be given afterwards). We proceed to give some basic properties which must then hold, as well as an estimate for how many iterations the outer loop process can run.

First observe that Claim B combined with $\mathsf{C}(X)$ being a convex cone ensures that

$$-g_k \in \mathsf{C}(X) \quad \text{for all } X \in \mathfrak{X}_m(\varphi) \text{ and } k \geq 1, \qquad \text{and} \tag{8.16}$$
$$-g_k \in \mathsf{C}_\mathbb{Z}(X) \quad \text{for all } X \in \mathfrak{X}_m(\varphi) \text{ and } k \leq |I_\mathbb{Z}|.$$

If $g_k = 0$ with $k \leq |I_\mathbb{Z}|$, then $\sigma(T_k) = 0$ with $T_k \mid U$ and $T_k \in \mathcal{F}(G_0)$ (by (8.9)), in which case $U \in \mathcal{A}(G_0)$ being an atom forces $U = T_k \mid V_{G_0 \setminus \mathfrak{X}}$, contradicting that $V_\mathfrak{X} \mid U$ is nontrivial. Therefore we instead conclude that $g_k \neq 0$ for all $k \leq |I_\mathbb{Z}|$.

We next show $g_k \neq 0$ for $k = |I_{\mathbb{Z}}| + 1$. In this case, $g_k = \sum_{i=1}^{|I_{\mathbb{Z}}|}(1-\beta_i)\sigma(T_i) + \sum_{i\in I_{\mathbb{Q}}}\sigma(T_i)$ with $\sigma(T_i) \in -\mathsf{C}(X)$ for all $i \in I_{\mathbb{Z}} \cup I_{\mathbb{Q}}$ (in view of Claim B). Since $0 \notin \mathsf{C}^*(X)$ (as each $X \in \mathfrak{X}_m(\varphi)$ is linearly independent), a sum of elements from $\mathsf{C}(X)$ can only equal zero if all elements in the sum are themselves zero. Thus, assuming by contradiction that $g_k = 0$, it follows that $(1-\beta_i)\sigma(T_i) = 0$ for all $i \in I_{\mathbb{Z}}$ and $\sigma(T_i) = 0$ for all $i \in I_{\mathbb{Q}}$. As we have already established that $\sigma(T_i) = g_i \neq 0$ for each $i \in I_{\mathbb{Z}}$, this forces $\beta_i = 1$ for all $i \leq |I_{\mathbb{Z}}|$. However, per Item 4, we know the Outer Loop halts immediately at step i if $\beta_i = 1$. Thus, if $I_{\mathbb{Z}} \neq \emptyset$, then the Outer Loop must have halted at step $i = 1$, meaning $k = |I_{\mathbb{Z}}| + 1 \geq 2$ does not exist. We are left to conclude that $g_k = 0$ for $k = |I_{\mathbb{Z}}| + 1$ is only possible if $I_{\mathbb{Z}} = \emptyset$ and $\sigma(T_i) = 0$ for all $i \in I_{\mathbb{Q}}$. As a result, since $V \in \mathcal{B}_{\mathsf{rat}}(G_0)$ has sum zero, it follows that $\sigma(V_{\mathfrak{X}} \cdot \prod^\bullet_{i\in I_\diamond} T_i) = \sigma(V \cdot (\prod^\bullet_{i\in I_{\mathbb{Q}}\cup I_{\mathbb{Z}}} T_i)^{[-1]}) = 0$. However, $\mathsf{Supp}(V_{\mathfrak{X}} \cdot \prod^\bullet_{i\in I_\diamond} T_i) \subseteq G_0^\diamond$ by (8.12) and $\mathfrak{X} \subseteq G_0^\diamond$. Thus our hypothesis $0 \notin \mathsf{C}^*(G_0^\diamond)$ (note this is the first time we have used the hypothesis $0 \notin \mathsf{C}^*(G_0^\diamond)$ rather than the weaker consequence that G_0 is finitary) forces $V_{\mathfrak{X}} \cdot \prod^\bullet_{i\in I_\diamond} T_i$ to be the trivial sequence, contradicting that $V_{\mathfrak{X}}$ is nontrivial. So we instead conclude that $g_k \neq 0$ also for $k = |I_{\mathbb{Z}}| + 1$. As a consequence, a simple inductive argument now shows $g_k \neq 0$ for all $k \geq |I_{\mathbb{Z}}| + 1$ (since Item 4 ensures that $\beta_{k-1} \neq 1$, else the Outer Loop halts immediately at step $k-1$). Summarizing:

$$g_k \neq 0 \quad \text{for all } k. \tag{8.17}$$

Next observe that $|D_0 \cap \mathfrak{X}_j| \leq N_S$ for all $j \in [1, n]$ (in view of (b)), which combined with Items 1 and 2 ensures that

$$|D_k \cap \mathfrak{X}_j| \leq k + N_S \quad \text{for all } j \in [1, n]. \tag{8.18}$$

The value $|J_k^d| \in [0, n]$ is nondecreasing with k. There are therefore at most $n' \leq n$ 'jump' values of k where $|J_k^d| > |J_{k-1}^d|$, say occurring for $1 \leq \kappa_1 < \ldots < \kappa_{n'}$. Let $\kappa_0 = 0$ and $\kappa_{n'+1} = \infty$. Note that Item 3 ensures that $|J_1^d| \geq 1$.

We wish to estimate how long the outer loop process can run without halting. Items 2 and 3 ensure that $|D_k \cap \bigcup_{j\in J_k^d} \mathfrak{X}_j|$ decreases by at least one after each iteration except when $|J_k^d| > |J_{k-1}^d|$, in which case it instead increases by at most $(|J_k^d| - |J_{k-1}^d|)(k + N_S)$ in view of (8.18). Thus

$$|D_k \cap \bigcup_{j\in J_k^d} \mathfrak{X}_j| - |D_{k-1} \cap \bigcup_{j\in J_{k-1}^d} \mathfrak{X}_j| \leq (|J_k^d| - |J_{k-1}^d|)(k+1+N_S) - 1 \quad \text{for all } k \geq 1. \tag{8.19}$$

Consequently, since $|D_k \cap \bigcup_{j\in J_k^d} \mathfrak{X}_j| \geq 0$ must hold for all $k \geq 0$, it follows that every

$$k \leq \kappa_{n'} + |D_{\kappa_{n'}} \cap \bigcup_{j\in J_{\kappa_{n'}}^d} \mathfrak{X}_j|.$$

Indeed, apart from (possibly) $k = 0$ and the final value of k, we find we must have $|D_k \cap \bigcup_{j \in J_k^d} \mathfrak{X}_j| \geq |J_k^d| \geq |J_1^d| \geq 1$, as otherwise the mechanism for halting the outer loop process is triggered. In particular, either $\kappa_{n'} = 1$ (in which case $n' = 1$) or $|D_{\kappa_{n'}-1} \cap \bigcup_{j \in J^d_{\kappa_{n'}-1}} \mathfrak{X}_j| \geq |J^d_{\kappa_{n'}-1}| \geq |J_1^d| \geq 1$, which is slightly better then the estimate that this quantity be non-negative. Thus we likewise obtain estimates

$$\kappa_t \leq \kappa_{t-1} + |D_{\kappa_{t-1}} \cap \bigcup_{j \in J^d_{\kappa_{t-1}}} \mathfrak{X}_j| \quad \text{for all } t \geq 2 \tag{8.20}$$

and $\kappa_1 \leq 1 + |D_0 \cap \bigcup_{j \in J_0^d} \mathfrak{X}_j|$ (with equality only possible if $\kappa_1 = 1$ and also $|D_0 \cap \bigcup_{j \in J_0^d} \mathfrak{X}_j| = 0$). For $k < \kappa_{t+1}$, we have passed $t' \leq t$ jump values, which add at most

$$\sum_{\alpha=1}^{t'} (|J^d_{\kappa_\alpha}| - |J^d_{\kappa_{\alpha-1}}|)(\kappa_\alpha + 1 + N_S) - t' \leq \sum_{\alpha=1}^{t} (|J^d_{\kappa_\alpha}| - |J^d_{\kappa_{\alpha-1}}|)(\kappa_\alpha + 1 + N_S) - t'$$

damaged elements into the depleted boxes (in view of (8.19)), on top of the initial number $|D_0 \cap \bigcup_{j \in J_0^d} \mathfrak{X}_j|$ of damaged elements. At the same time, for each of the $k - t'$ non-jump steps, the number of damaged elements decreases by at least one (as remarked above (8.19)). Thus

$$|D_k \cap \bigcup_{j \in J_k^d} \mathfrak{X}_j| \leq |D_0 \cap \bigcup_{j \in J_0^d} \mathfrak{X}_j| - k + \sum_{\alpha=1}^{t} (|J^d_{\kappa_\alpha}| - |J^d_{\kappa_{\alpha-1}}|)(\kappa_\alpha + 1 + N_S) \quad \text{for } k < \kappa_{t+1}. \tag{8.21}$$

The above estimates gets larger when the values of κ_α are each as large as possible, that is, delaying when the jumps $|J_k^d| > |J_{k-1}^d|$ occur can only increase the estimates. Also, increasing the number of times that we have jumps $|J_k^d| > |J_{k-1}^d|$ only increases these estimates, since this breaks multiple simultaneous jumps into several jumps spaced out, with the later jumps having larger contributing factor. Thus we can obtain an upper bound for how long the outer loop process will run by taking $n' = n$ and delaying each jump as long as possible per the estimates above. In particular, (8.21) now yields

$$|D_{\kappa_t} \cap \bigcup_{j \in J^d_{\kappa_t}} \mathfrak{X}_j| \leq |D_0 \cap \bigcup_{j \in J_0^d} \mathfrak{X}_j| - \kappa_t + \sum_{\alpha=1}^{t} (\kappa_\alpha + 1 + N_S). \tag{8.22}$$

Note $|D_0 \cap \bigcup_{j \in J_0^d} \mathfrak{X}_j| \leq N_S$ (in view of (b)). Thus $\kappa_1 \leq 1 + N_S$ and

$$|D_{\kappa_1} \cap \bigcup_{j \in J^d_{\kappa_1}} \mathfrak{X}_j| \leq |D_0 \cap \bigcup_{j \in J_0^d} \mathfrak{X}_j| - \kappa_1 + (\kappa_1 + 1 + N_S) \leq 2(1 + N_S).$$

A simple inductive argument, using (8.20) and (8.22) and the estimate $|D_0 \cap \bigcup_{j\in J_0^d} \mathfrak{X}_j| \leq 1+N_S$, now gives $\kappa_t \leq (2^t-1)(N_S+1)$ and $|D_{\kappa_t} \cap \bigcup_{j\in J_{\kappa_t}^d} \mathfrak{X}_j| \leq 2^t(N_S+1)$ for all $t \in [1,n]$. We thus find every $k \leq \kappa_n + |D_{\kappa_n} \cap \bigcup_{j\in J_{\kappa_n}^d} \mathfrak{X}_j| \leq (2^n-1)(1+N_S)+2^n(1+N_S) = (2^{n+1}-1)(1+N_S)$, meaning the outer loop process must halt after at most

$$(2^{n+1}-1)(1+N_S)$$

steps, which is independent of U.

Suppose we run the above outer loop process and it stops at step $k \leq (2^{n+1}-1)(1+N_S)$. For each $i \in [1,k]$ with $i \leq |I_{\mathbb{Z}}|$, let $C_i = B_i \cdot T_i^{[\beta_i]}$. If $\beta_k = 1 = k = |I_{\mathbb{Z}}|$ and $|I_{\mathbb{Q}}| = |I_\diamond| = 0$, then $V_{G_0\setminus\mathfrak{X}} = T_1 \mid C_1$, ensuring $\mathsf{Supp}(V \cdot C_1^{[-1]}) \subseteq \mathfrak{X} \subseteq G_0^\diamond$. If $k \geq |I_{\mathbb{Z}}|+1$, then we have

$$\begin{aligned} & V \cdot (C_1 \cdot \ldots \cdot C_{|I_{\mathbb{Z}}|})^{[-1]} \\ & = \left(V_{\mathfrak{X}} \cdot (B_1 \cdot \ldots \cdot B_{|I_{\mathbb{Z}}|})^{[-1]}\right) \cdot \left(V_{G_0\setminus\mathfrak{X}} \cdot T_1^{[-\beta_1]} \cdot \ldots \cdot T_{|I_{\mathbb{Z}}|}^{[-\beta_{|I_{\mathbb{Z}}|}]}\right) \\ & = \left(V_{\mathfrak{X}} \cdot (B_1 \cdot \ldots \cdot B_{|I_{\mathbb{Z}}|})^{[-1]}\right) \cdot \left(\prod\nolimits^\bullet_{i\in I_{\mathbb{Z}}} T_i^{[1-\beta_i]} \cdot \prod\nolimits^\bullet_{i\in I_{\mathbb{Q}}} T_i \cdot \prod\nolimits^\bullet_{i\in I_\diamond} T_i\right). \end{aligned}$$

Let $S_{|I_{\mathbb{Z}}|+1} = \prod^\bullet_{i\in I_{\mathbb{Z}}} T_i^{[1-\beta_i]} \cdot \prod^\bullet_{i\in I_{\mathbb{Q}}} T_i$ and $C_{|I_{\mathbb{Z}}|+1} = B_{|I_{\mathbb{Z}}|+1} \cdot S_{|I_{\mathbb{Z}}|+1}^{[\beta_{|I_{\mathbb{Z}}|+1}]}$. For an index $i \in [|I_{\mathbb{Z}}|+2, k+1]$, let $S_i = S_{i-1}^{[1-\beta_{i-1}]}$. For $i \in [|I_{\mathbb{Z}}|+2, k]$, set $C_i = B_i \cdot S_i^{[\beta_i]}$. For $i \geq |I_{\mathbb{Z}}|+1$, we have $g_i = \sigma(S_i)$, while S_j is the subsequence obtained from $\prod^\bullet_{i\in I_{\mathbb{Z}}} T_i \cdot \prod^\bullet_{i\in I_{\mathbb{Q}}} T_i$ by removing the terms from the subsequence $\prod^\bullet_{i\in I_{\mathbb{Z}}} T_i^{[\beta_i]} \cdot \prod^\bullet_{i\in[|I_{\mathbb{Z}}|+1,j-1]} S_i^{[\beta_i]}$, as can be seen by a short inductive argument on $j = |I_{\mathbb{Z}}|+1,\ldots,k+1$. Thus $C_1 \cdot \ldots \cdot C_k \mid V$ and $\prod^\bullet_{i\in I_{\mathbb{Z}}} T_i^{[\beta_i]} \cdot \prod^\bullet_{i\in[|I_{\mathbb{Z}}|+1,k]} S_i^{[\beta_i]} \mid \prod^\bullet_{i\in I_{\mathbb{Z}}} T_i \cdot \prod^\bullet_{i\in I_{\mathbb{Q}}} T_i$ with equality holding precisely when $\beta_k = 1$. In particular, combined with the previous observation for what happens when $\beta_k = 1 = k = |I_{\mathbb{Z}}|$ with $|I_{\mathbb{Q}}| = |I_\diamond| = 0$, we conclude that

$$\mathsf{Supp}(V \cdot (C_1 \cdot \ldots \cdot C_k)^{[-1]}) \subseteq G_0^\diamond \quad \text{when } \beta_k = 1. \tag{8.23}$$

In view of Item 4, we have $C_i \in \mathcal{B}_{\mathsf{rat}}(G_0)$ for all $i \in [1,k]$. Define

$$A_{\mathfrak{t}-1} = C_1 \cdot \ldots \cdot C_k \quad \text{and} \quad V' = V \cdot (C_1 \cdot \ldots \cdot C_k)^{[-1]} = V \cdot A_{\mathfrak{t}-1}^{[-1]}$$

Since the $V, C_1, \ldots, C_k \in \mathcal{B}_{\mathsf{rat}}(G_0)$, it follows that $V' \in \mathcal{B}_{\mathsf{rat}}(G_0)$. Let us show that there are bounds $N'_S \geq N_S$ and $N'_T \geq N_T$ such that (a)–(d) hold for $U = V' \cdot A_{\mathfrak{t}-1} \cdot A_{\mathfrak{t}} \cdot \ldots \cdot A_{\mathfrak{s}}$. Note, since (a) held originally for $U = V \cdot A_{\mathfrak{t}} \cdot \ldots \cdot A_{\mathfrak{s}}$, it suffices to show $\mathsf{Supp}(V')$ contains no subset $X \in \mathfrak{X}_m(\varphi) = \mathfrak{X}_m(\varphi_{\mathfrak{t}-1})$ in order to show (a) holds for the new factorization.

If $k \le |I_{\mathbb{Z}}|$, then $\prod^\bullet_{i\in[1,k]} T_i^{[\beta_i]}$ is the subsequence of $A_{\mathfrak{t}-1} = C_1 \cdot \ldots \cdot C_k$ consisting of all terms from $G_0 \setminus \mathfrak{X}$, and $\sum_{i=1}^k \sigma(T_i^{[\beta_i]}) = \sum_{i=1}^k \beta_i \sigma(T_k) \in -\mathsf{C}(Z_\varphi)$ in view of Step B and the convexity of $\mathsf{C}(Z_\varphi)$. If $k \ge |I_{\mathbb{Z}}|+1$, then

$$\sigma(S_{|I_{\mathbb{Z}}|+1}) = \sigma(\prod^\bullet_{i\in I_{\mathbb{Z}}} T_i^{[1-\beta_i]} \cdot \prod^\bullet_{i\in I_{\mathbb{Q}}} T_i) \in -\mathsf{C}(Z_\varphi)$$

in view of Claim B and the convexity of $\mathsf{C}(Z_\varphi)$. A short inductive argument now shows $\sigma(S_i) \in -\mathsf{C}(Z_\varphi)$ for all $i \in [|I_{\mathbb{Z}}|+1, k]$, as $\mathsf{C}(Z_\varphi)$ is a convex cone. Now $\prod^\bullet_{i\in I_{\mathbb{Z}}} T_i^{[\beta_i]} \cdot \prod^\bullet_{i\in[|I_{\mathbb{Z}}|+1,k]} S_i^{[\beta_i]}$ is the subsequence of $A_{\mathfrak{t}-1} = C_1 \cdot \ldots \cdot C_k$ consisting of all terms from $G_0 \setminus \mathfrak{X}$. Thus, since $\sigma(S_i) \in -\mathsf{C}(Z_\varphi)$ for all $i \in [|I_{\mathbb{Z}}|+1, k]$, the convexity of the convex cone $\mathsf{C}(Z_\varphi)$ ensures that

$$\sigma(\prod^\bullet_{i\in I_{\mathbb{Z}}} T_i^{[\beta_i]} \cdot \prod^\bullet_{i\in[|I_{\mathbb{Z}}|+1,k]} S_i^{[\beta_i]}) \in -\mathsf{C}(Z_\varphi).$$

We conclude that (d) holds for the factorization $U = V' \cdot A_{\mathfrak{t}-1} \cdot A_{\mathfrak{t}} \cdot \ldots \cdot A_{\mathfrak{s}}$.

There are two ways the outer loop process can halt. First, we may have $\beta_k = 1$. In this case, (8.23) implies $\mathrm{Supp}(V') \subseteq G_0^\diamond$, while $V' \in \mathcal{B}_{\mathrm{rat}}(G_0)$ as already remarked above. In consequence, since $0 \notin \mathsf{C}^*(G_0^\diamond)$ by hypothesis, we conclude that V' is the trivial sequence. (Here, we have again used the hypothesis $0 \notin \mathsf{C}^*(G_0^\diamond)$ rather than weaker consequence that G_0 is finitary.) In such case, the remaining part of (a) holds trivially. Moreover, V' being trivial forces V_k to be trivial, as $V_k = V_{\mathfrak{X}} \cdot (B_1 \cdot \ldots \cdot B_k)^{[-1]}$ consists of all terms of V' from $\mathfrak{X}$. In particular, $D_k = \emptyset$ and $J_k^d = [1, n]$, so that

$$\text{Item 4 holding with } \beta_k = 1 \text{ implies Items 1–3 trivially hold in the Outer Loop.} \tag{8.24}$$

The second way the outer loop process can halt is if there is an empty box $j \in [1, n]$ with $\mathfrak{X}_j \cap \mathrm{Supp}\, V_k = \emptyset$. In view of the definition of V', we have $V'_{\mathfrak{X}} = V_{\mathfrak{X}} \cdot (B_1 \cdot \ldots \cdot B_k)^{[-1]} = V_k$, where $V'_{\mathfrak{X}} \mid V'$ is the subsequence of terms from $\mathfrak{X}$. Thus the remaining part of (a) holds in this case as well.

If $k \le |I_{\mathbb{Z}}|$, then (8.9) implies

$$\mathsf{v}_{G_0 \setminus \mathfrak{X}}(A_{\mathfrak{t}-1}) = \sum_{i=1}^k \beta_i |T_i| \le \sum_{i=1}^k |T_i| \le k D_\Delta M_W \le (2^{n+1}-1)(1+N_S) D_\Delta M_W.$$

By Claim A and (8.10), we have $\sum_{k\in I_{\mathbb{Q}}} |T_k| \le |I_{\mathbb{Q}}| M_W \le (D_\Delta + N_S + 2\ell_\varphi) M_W$. As a result, if $k \ge |I_{\mathbb{Z}}|+1$, then $|I_{\mathbb{Z}}| \le k-1 \le (2^{n+1}-1)(1+N_S)-1$, and combined with (8.9) we now obtain

$$\mathsf{v}_{G_0\setminus\mathfrak{X}}(A_{t-1}) \le \sum_{i\in I_{\mathbb{Z}}} |T_i| + \sum_{i\in I_{\mathbb{Q}}} |T_i| \le ((2^{n+1}-1)(1+N_S)-1)D_\Delta M_W$$
$$+ (D_\Delta + N_S + 2\ell_\varphi)M_W = ((2^{n+1}-1)(1+N_S)D_\Delta + N_S + 2\ell_\varphi)M_W.$$

Thus, setting

$$N_T' = \max\{N_T,\ ((2^{n+1}-1)(1+N_S)D_\Delta + N_S + 2\ell_\varphi)M_W\}, \tag{8.25}$$

which is independent of U, we see that (c) holds.

If $k \le |I_{\mathbb{Z}}|$, then

$$V' = V\boldsymbol{\cdot}(C_1\boldsymbol{\cdot}\ldots\boldsymbol{\cdot}C_k)^{[-1]} = \Big(V_{\mathfrak{X}}\boldsymbol{\cdot}(B_1\boldsymbol{\cdot}\ldots\boldsymbol{\cdot}B_k)^{[-1]}\Big)\boldsymbol{\cdot}\Big(V_{G_0\setminus\mathfrak{X}}\boldsymbol{\cdot}(T_1^{[\beta_1]}\boldsymbol{\cdot}\ldots\boldsymbol{\cdot}T_k^{[\beta_k]})^{[-1]}\Big)$$

with $V_k = V_{\mathfrak{X}} \boldsymbol{\cdot} (B_1 \boldsymbol{\cdot} \ldots \boldsymbol{\cdot} B_k)^{[-1]}$ and $T_i \in \mathcal{F}(G_0)$ for all $i \le k$ (by (8.9)). In view of (8.18), we have $|D_k \cap \mathfrak{X}_j| \le k + N_S$ for all $j \in [1, n]$; in view of (b) and $\mathsf{Supp}(V_{\mathfrak{X}}) \cap \mathsf{Supp}(V_{G_0\setminus\mathfrak{X}}) = \emptyset$, we have $|\mathsf{Supp}(\{V_{G_0\setminus\mathfrak{X}}\})| = |\mathsf{Supp}(\{V\})| - |\mathsf{Supp}(\{V_{\mathfrak{X}}\})| \le |\mathsf{Supp}(\{V\})| \le N_S$; in view of $T_i \in \mathcal{F}(G_0)$ and (8.9), we have $|\mathsf{Supp}(\{T_i^{[\beta_i]}\})| \le |\mathsf{Supp}(T_i)| \le |T_i| \le D_\Delta M_W$. Thus (2.4) and (2.5) yield

$$\begin{aligned}
|\mathsf{Supp}(\{V'\})| &\le |\mathsf{Supp}(\{V_k\})| + |\mathsf{Supp}(\{V_{G_0\setminus\mathfrak{X}}\})| + \sum_{i=1}^{k} |\mathsf{Supp}(T_i)| \\
&\le \sum_{j=1}^{n} |D_k \cap \mathfrak{X}_j| + |\mathsf{Supp}(\{V_{G_0\setminus\mathfrak{X}}\})| + \sum_{i=1}^{k} |\mathsf{Supp}(T_i)| \\
&\le n(k+N_S) + N_S + kD_\Delta M_W \\
&\le (n + D_\Delta M_W)(2^{n+1}-1)(1+N_S) + (n+1)N_S.
\end{aligned}$$

If $k \ge |I_{\mathbb{Z}}| + 1$, then

$$V' = V \boldsymbol{\cdot} (C_1 \boldsymbol{\cdot} \ldots \boldsymbol{\cdot} C_k)^{[-1]} = \Big(V_{\mathfrak{X}} \boldsymbol{\cdot} (B_1 \boldsymbol{\cdot} \ldots \boldsymbol{\cdot} B_k)^{[-1]}\Big) \boldsymbol{\cdot} \Big(V_{G_0\setminus\mathfrak{X}} \boldsymbol{\cdot} \prod^{\bullet}_{i\in I_{\mathbb{Z}}\cup I_{\mathbb{Q}}} T_i^{[-\gamma_i']}\Big)$$

for some $\gamma_i' \le 1$ with $V_k = V_{\mathfrak{X}} \boldsymbol{\cdot} (B_1 \boldsymbol{\cdot} \ldots \boldsymbol{\cdot} B_k)^{[-1]}$. In view of (8.18), we have $|D_k \cap \mathfrak{X}_j| \le k + N_S$ for all $j \in [1, n]$; in view of (a) and $\mathsf{Supp}(V_{\mathfrak{X}}) \cap \mathsf{Supp}(V_{G_0\setminus\mathfrak{X}}) = \emptyset$, we have $|\mathsf{Supp}(\{V_{G_0\setminus\mathfrak{X}}\})| = |\mathsf{Supp}(\{V\})| - |\mathsf{Supp}(\{V_{\mathfrak{X}}\})| \le |\mathsf{Supp}(\{V\})| \le N_S$; in view of $T_i \in \mathcal{F}(G_0)$ for $i \in I_{\mathbb{Z}}$ and (8.9), we have $|\mathsf{Supp}(T_i)| \le |T_i| \le D_\Delta M_W$

for $i \in I_{\mathbb{Z}}$; in view of $k \geq |I_{\mathbb{Z}}| + 1$, we have $|I_{\mathbb{Z}}| < k \leq (2^{n+1} - 1)(1 + N_s)$. Combining these estimates with (8.13), we obtain (via (2.4) and (2.5))

$$\begin{aligned}
&|\operatorname{Supp}(\{V'\})| \\
&\quad \leq |\operatorname{Supp}(\{V_k\})| + |\operatorname{Supp}(\{V_{G_0 \setminus \mathfrak{x}}\})| + \sum_{i \in I_{\mathbb{Z}}} |\operatorname{Supp}(T_i)| + |\operatorname{Supp}(\textstyle\prod^{\bullet}_{i \in I_{\mathbb{Q}}} T_i)| \\
&\quad \leq \sum_{j=1}^{n} |D_k \cap \mathfrak{X}_j| + |\operatorname{Supp}(\{V_{G_0 \setminus \mathfrak{x}}\})| + \sum_{i \in I_{\mathbb{Z}}} |\operatorname{Supp}(T_i)| + |\operatorname{Supp}(\textstyle\prod^{\bullet}_{i \in I_{\mathbb{Q}}} T_i)| \\
&\quad \leq n(k + N_S) + N_S + |I_{\mathbb{Z}}| D_{\Delta} M_W + |\operatorname{Supp}(\textstyle\prod^{\bullet}_{i \in I_{\mathbb{Q}}} T_i)| \\
&\quad \leq n(k + N_S) + N_S + k D_{\Delta} M_W + (D_{\Delta} + N_S + 2\ell_{\varphi}) M_W + N_S + \ell_{\varphi} d \\
&\quad \leq (n + D_{\Delta} M_W)(2^{n+1} - 1)(1 + N_S) + (n+2) N_S + \ell_{\varphi} d + (D_{\Delta} + N_S + 2\ell_{\varphi}) M_W.
\end{aligned}$$

Setting $N''_S = (n + D_{\Delta} M_W)(2^{n+1} - 1)(1 + N_S) + (n+2) N_S + \ell_{\varphi} d + (D_{\Delta} + N_S + 2\ell_{\varphi}) M_W \geq N_S$, which is independent of U, we see that the first bound in (b) holds.

Finally, (2.5) implies $\operatorname{Supp}(\{A_{\mathfrak{t}-1}\}) \subseteq \operatorname{Supp}(\{V'\}) \cup \operatorname{Supp}(\{V' \cdot A_{\mathfrak{t}-1}\}) = \operatorname{Supp}(\{V'\}) \cup \operatorname{Supp}(\{V\})$. Thus, in view of the above work and (b) for the original factorization $U = V \cdot V_{\mathfrak{t}} \cdot \ldots \cdot V_{\mathfrak{s}}$, we have $|\operatorname{Supp}(\{A_{\mathfrak{t}-1}\})| \leq |\operatorname{Supp}(\{V'\})| + |\operatorname{Supp}(\{V\})| \leq N''_S + N_S$. Thus, letting

$$\begin{aligned}
N'_S := (n + D_{\Delta} M_W)(2^{n+1} - 1)(1 + N_S) + (n+3) N_S + \ell_{\varphi} d \\
+ (D_{\Delta} + N_S + 2\ell_{\varphi}) M_W \geq N_S,
\end{aligned} \tag{8.26}$$

we see that the second bound in (b) also holds, which completes the induction. It remains only to show that it is indeed possible to construct the sequences B_k of the outer loop with the desired list of properties, and then the proof will be complete.

The Inner Loop Assume that the rational sequences $B_1, \ldots, B_{k-1} \in \mathcal{F}_{\mathrm{rat}}(G_0)$ of the Outer Loop have already been constructed, for $k \geq 1$, and that $\mathcal{X}_j \cap \operatorname{Supp}(V_{k-1}) \neq \emptyset$ for all $j \in [1, n]$, and $0 < \beta_i < 1$ for all $i < k$, so that the Outer Loop process has not terminated. Let

$$V' = V_{k-1} = V_{\mathfrak{X}} \cdot (B_1 \cdot \ldots \cdot B_{k-1})^{[-1]}.$$

We then construct the sequence $B_k \mid V'$ by a separate recursive process which we refer to as the Inner Loop. Since $\mathcal{X}_j \cap \operatorname{Supp}(V') \neq \emptyset$ for all $j \in [1, n]$, we can select a fixed element $z_j \in \mathcal{X}_j \cap \operatorname{Supp}(V')$ from each depleted box $j \in J^d_{k-1}$. Let $X = \{z_j : j \in J^d_{k-1}\}$. Observe that

$$X \subseteq D_{k-1} \cap \bigcup_{j \in J^d_{k-1}} \mathfrak{X}_j \quad \text{and} \quad X \subseteq \operatorname{Supp}(\{V'\}) \tag{8.27}$$

by its definition and that of a depleted box. Let

$$W_X = \prod^{\bullet}_{x \in X} x^{\lfloor \mathsf{v}_x(V') \rfloor} \mid \{V'\}.$$

Since all the terms in W_X are depleted in $V_{k-1} = V'$, it follows that

$$\mathrm{Supp}(\lfloor V' \rfloor) \cap \mathrm{Supp}(W_X) = \emptyset.$$

Assume we have already constructed a sequence $C \in \mathcal{F}_{\mathrm{rat}}(G_0)$ which satisfies

$$C \mid \lfloor V' \rfloor \cdot W_X$$

such that $\sigma(C) = -\beta g_k$ for some $\beta \in \mathbb{Q}_+$ with $0 \le \beta \le 1$, and such that

$$Z' = \mathrm{Supp}(C) \cap \mathrm{Supp}(\lfloor V' \rfloor \cdot W_X \cdot C^{[-1]}) \tag{8.28}$$

is a subset $Z' \subseteq \mathfrak{X}$ satisfying

(a) $|Z' \cap \mathfrak{X}_j| \le 1$ for all $j \in [1, n]$;
(b) $Z' \cap \bigcup_{j \in J^d_{k-1}} \mathfrak{X}_j \subseteq X$;
(c) If $\beta = 0$, then C is trivial; and
(d) If $\beta = 1$, then either $k \ge |I_{\mathbb{Z}}| + 1$ or else $k = 1 = |I_{\mathbb{Z}}|$, $|I_{\mathbb{Q}}| = |I_{\diamond}| = 0$ and $U = V$.

For instance, we could initially start with C taken to be the trivial sequence, in which case $Z' = \emptyset$ and $\beta = 0$. In view of (8.28), we see that the terms $z \in \mathrm{Supp}(C) \setminus Z'$ are those which are completely removed from $\lfloor V' \rfloor \cdot W_X$ when we remove the sequence C. Thus,

$$\text{if } z \in \mathrm{Supp}(C) \setminus Z' \text{ and } z \notin X, \text{ then } \mathsf{v}_z(V' \cdot C^{[-1]}) = \mathsf{v}_z(\{V'\}) < 1, \tag{8.29}$$

and if $z \in \mathrm{Supp}(C) \setminus Z'$ and $z \in X$, then $\mathsf{v}_z(V' \cdot C^{[-1]}) = 0 < \mathsf{v}_z(V') < 1$ (as the elements from X come from depleted boxes). As a result, if we were to halt the Inner Loop process and set $B_k = C$ and $V_k = V' \cdot C^{[-1]}$, then D_k would be obtained from D_{k-1} by removing all elements from $(X \cap \mathrm{Supp}(C)) \setminus Z'$ and including (possibly) some of the elements from Z', meaning

$$D_k \subseteq \big(D_{k-1} \setminus (X \cap \mathrm{Supp}(C))\big) \cup Z'. \tag{8.30}$$

If $X \not\subseteq \mathrm{Supp}(V' \cdot C^{[-1]})$, which is equivalent to $X \not\subseteq \mathrm{Supp}(\lfloor V' \rfloor \cdot W_X \cdot C^{[-1]})$ in view of the definition of W_X and the fact that all elements of X are from depleted boxes, then the Inner loop halts and we set $B_k = C$ and $V_k = V' \cdot C^{[-1]}$. Note, since $X \subseteq \mathrm{Supp}(V')$, that this is only possible if C is nontrivial. Thus Item 4 of the Outer

Loop holds with $\beta_k = \beta$ by definition of C, (c) and (d). It follows in view of (8.30) and (a) that Item 1 of the Outer Loop holds. It follows in view of (8.30), (b), and (8.27) that Item 2 from the Outer Loop holds. Since $X \nsubseteq \mathsf{Supp}(\lfloor V' \rfloor \cdot W_X \cdot C^{[-1]})$ and $X = \mathsf{Supp}(W_X)$, it follows that there must be some term from $X \cap \mathsf{Supp}(C)$ completely removed from $\lfloor V' \rfloor \cdot W_X$, whence $X \cap \mathsf{Supp}(C) \nsubseteq Z'$. If there is some $j \in J_k^d \setminus J_{k-1}^d$, then Item 3 of the Outer Loop holds. Otherwise, we have $J_k^d = J_{k-1}^d$. In this case, any element $x \in (X \cap \mathsf{Supp}(C)) \setminus Z'$, which exists as we just observed $(X \cap \mathsf{Supp}(C)) \nsubseteq Z'$, is an element of D_{k-1} (by (8.27)) not in D_k (by (8.30)), which combined with the already established Item 2 of the Outer Loop yields Item 3 of the Outer Loop. Thus $B_k = C$ satisfies all conditions for the Outer Loop, as required. Therefore instead assume $X \subseteq \mathsf{Supp}(V' \cdot C^{[-1]})$, equivalent to

$$X \subseteq \mathsf{Supp}(\lfloor V' \rfloor \cdot W_X \cdot C^{[-1]}). \tag{8.31}$$

If $\beta = 1$, then (d) ensures either $k \geq |I_{\mathbb{Z}}| + 1$ or else $k = 1 = |I_{\mathbb{Z}}|$, $|I_{\mathbb{Q}}| = |I_\diamond| = 0$ and $U = V$. In this case, the Inner loop requirement that $\sigma(C) = -\beta g_k = -g_k$ ensures that Item 4 of the Outer Loop holds with $\beta_k = \beta = 1$ taking $B_k = C$, and then Items 1–3 do as well by (8.24), meaning we can set $B_k = C$ for the Outer Loop, as required. Therefore we may instead assume

$$\beta < 1. \tag{8.32}$$

If $x \in \mathfrak{X}_j$ with $j \in J_{k-1}^u$, then $x \notin X = \mathsf{Supp}(W_X)$, as all terms from X are from depleted boxes. Thus $C \mid \lfloor V' \rfloor \cdot W_X$ ensures that $\mathsf{v}_x(C) \leq \mathsf{v}_x(\lfloor V' \rfloor)$. Hence either $\mathsf{v}_x(C) = \mathsf{v}_x(\lfloor V' \rfloor)$ or $x \in \mathfrak{X}_j \cap \mathsf{Supp}(\lfloor V' \rfloor \cdot W_X \cdot C^{[-1]}\rfloor)$. Now suppose there were some $x \in \mathfrak{X}_j \cap \mathsf{Supp}(\lfloor V' \cdot C^{[-1]} \rfloor)$ with $x \notin \mathfrak{X}_j \cap \mathsf{Supp}(\lfloor V' \rfloor \cdot W_X \cdot C^{[-1]})$, where $j \in J_{k-1}^u$. Then $\mathsf{v}_x(V') - \mathsf{v}_x(C) \geq 1$ and $\mathsf{v}_x(C) = \mathsf{v}_x(\lfloor V' \rfloor)$, ensuring that $\mathsf{v}_x(\{V'\}) = \mathsf{v}_x(V') - \mathsf{v}_x(\lfloor V' \rfloor) = \mathsf{v}_x(V') - \mathsf{v}_x(C) \geq 1$, contrary to the definition of $\{V'\}$. So we instead conclude that

$$\mathfrak{X}_j \cap \mathsf{Supp}(\lfloor V' \cdot C^{[-1]} \rfloor) \subseteq \mathfrak{X}_j \cap \mathsf{Supp}(\lfloor V' \rfloor \cdot W_X \cdot C^{[-1]}) \quad \text{for every } j \in J_{k-1}^u. \tag{8.33}$$

If there is some $j \in J_{k-1}^u$ with $\mathfrak{X}_j \cap \mathsf{Supp}(\lfloor V' \cdot C^{[-1]} \rfloor) = \emptyset$, then the Inner loop halts and we set $B_k = C$ and $V_k = V' \cdot C^{[-1]}$. Note this ensures that $|J_k^d| > |J_{k-1}^d|$ (as $j \in J_k^d \setminus J_{k-1}^d$), so Item 3 of the Outer Loop holds. Since $\mathfrak{X}_j \cap \mathsf{Supp}(\lfloor V' \rfloor) \neq \emptyset$ for every $j \in J_{k-1}^u$ by definition of an undepleted box, we must have C nontrivial. Thus Item 4 of the Outer Loop holds with $\beta_k = \beta$ by definition of C, (c) and (8.32). By the same arguments used to establish (8.31), it follows from (a), (b), (8.27) and (8.30) that Items 1 and 2 of the Outer loop hold. Thus $B_k = C$ satisfies all conditions for the Outer Loop, as required. Therefore instead assume every $j \in J_{k-1}^u$ has some $z_j \in \mathfrak{X}_j \cap \mathsf{Supp}(\lfloor V' \cdot C^{[-1]} \rfloor) \subseteq \mathfrak{X}_j \cap \mathsf{Supp}(\lfloor V' \rfloor \cdot W_X \cdot C^{[-1]})$, with the inclusion in view of (8.33).

In view (b), for every $j \in J^d_{k-1}$, the set Z' either contains $z_j \in X \cap \mathfrak{X}_j$ or no element from $\mathfrak{X}_j$ at all. In view of (8.31), every $j \in J^d_{k-1}$ has $z_j \in \mathsf{Supp}(\lfloor V' \rfloor \cdot W_X \cdot C^{[-1]})$. In view of (a), for every $j \in J^u_{k-1}$, the set Z' contains at most one element from $\mathfrak{X}_j$. In view of the conclusion of the previous paragraph, for every $j \in J^u_{k-1}$ for which $Z' \cap \mathfrak{X}_j = \emptyset$, there is some

$$z_j \in \mathfrak{X}_j \cap \mathsf{Supp}(\lfloor V' \cdot C^{[-1]} \rfloor) \subseteq \mathfrak{X}_j \cap \mathsf{Supp}(\lfloor V' \rfloor \cdot W_X \cdot C^{[-1]}).$$

As a result of all these observations, (8.28) and the Interchangeability Property of $\mathfrak{X}_m(\varphi)$ (Proposition 7.33), we can find some subset (say)

$$Z = \{z_1, \ldots, z_n\}$$

with $Z \in \mathfrak{X}_m(\varphi)$ and

$$X \cup Z' \subseteq Z \subseteq \mathsf{Supp}(\lfloor V' \rfloor \cdot W_X \cdot C^{[-1]}) \subseteq \mathsf{Supp}(V' \cdot C^{[-1]}), \tag{8.34}$$

where the latter inclusion follows by recalling that $W_X \mid \{V'\}$. In particular, since $X = \mathsf{Supp}(W_X)$, we have

$$Z \setminus X \subseteq \mathsf{Supp}(\lfloor V' \rfloor). \tag{8.35}$$

In view of the definition of the g_i, it follows that g_k (indeed, every g_i) is a positive rational linear combination of the $\sigma(T_i)$ with $i \in I_\mathbb{Z} \cup I_\mathbb{Q}$. Let $\beta < 1$ be the rational number $\beta \in \mathbb{Q}_+$ from the definition of C. Since $\beta < 1$, it follows that $(1-\beta)g_k$ is a positive scalar multiple of g_k, and thus $-(1-\beta)g_k \in \mathsf{C}(Z)$ in view of Claim B, $Z \in \mathfrak{X}_m(\varphi)$, and the convexity of $\mathsf{C}(Z)$. Since $Z \in \mathfrak{X}_m(\varphi)$ is linearly independent, let $\alpha_1, \ldots, \alpha_n \in \mathbb{R}_+$ be the unique real numbers such that

$$\alpha_1 z_1 + \ldots + \alpha_n z_n = -(1-\beta)g_k.$$

Since g_k is a positive rational linear combination of the $\sigma(T_i)$ with $i \in I_\mathbb{Z} \cup I_\mathbb{Q}$, with each $\sigma(T_i)$ a positive rational linear combination of terms from $G_0 \subseteq \Lambda$ (since $T_i \in \mathcal{F}_{\mathsf{rat}}(G_0)$), and since $\beta \in \mathbb{Q}_+$, it follows that $-m'(1-\beta)g_k \in \Lambda \cap \mathsf{C}(Z) \subseteq \Lambda \cap \mathbb{R}\langle\Delta\rangle = \Lambda'$, for some integer $m' \geq 1$ (recall that $\mathbb{Z}\langle Z\rangle = \mathbb{Z}\langle Z_\varphi\rangle = \Delta$ since $Z, Z_\varphi \in \mathfrak{X}_m(\varphi)$). Thus, since Λ'/Δ is a finite abelian group, it follows that there is some integer $m \geq 1$ such that $-m(1-\beta)g_k \in \Delta \cap \mathsf{C}(Z) = \mathsf{C}_\mathbb{Z}(Z)$, with the equality in view of (8.14). Consequently, (8.15) ensures that $m\alpha_i \in \mathbb{Z}_+$ for all $i \in [1, n]$, showing that $\alpha_1, \ldots, \alpha_n \in \mathbb{Q}_+$.

Claim C Either $\mathsf{v}_{z_i}(\lfloor V' \rfloor \cdot W_X \cdot C^{[-1]}) < \alpha_i$ for some $i \in [1, n]$, or $k \geq |I_\mathbb{Z}| + 1$, or $k = 1 = |I_\mathbb{Z}|$, $|I_\mathbb{Q}| = |I_\diamond| = 0$ and $U = V$.

Proof Suppose $k \le |I_{\mathbb{Z}}|$ and $\mathsf{v}_{z_i}(\lfloor V' \rfloor \cdot W_X \cdot C^{[-1]}) \ge \alpha_i$ for all $i \in [1, n]$. We will show that $k = 1 = |I_{\mathbb{Z}}|$, $|I_{\mathbb{Q}}| = |I_{\diamond}| = 0$ and $U = V$. Since $k \le |I_{\mathbb{Z}}|$, we have

$$-g_k = -\sigma(T_k) \in \mathsf{C}_{\mathbb{Z}}(Z) \subseteq \mathbb{Z}\langle Z \rangle = \Delta$$

from (8.16) and $Z \in \mathfrak{X}_m(\varphi)$. Also, $W_X \mid \{V'\}$ ensures that $\lfloor V' \rfloor \cdot W_X \mid V' = V_{k-1}$, which is a subsequence of $V_{\mathfrak{X}}$, thus disjoint from $T_k \mid V_{G_0 \setminus \mathfrak{X}}$. Hence, since $\sigma(C) = -\beta g_k = -\beta\sigma(T_k)$ by hypothesis of the Inner Loop process, we conclude that

$$T := T_k \cdot C \cdot {\prod}^{\bullet}_{i \in [1,n]} z_i^{[\alpha_i]} \in \mathcal{B}_{\mathsf{rat}}(G_0) \quad \text{and} \quad T \mid U.$$

In particular, T is zero-sum. By (8.9), we have

$$T_k \in \mathcal{F}(G_0).$$

By (8.29), we have $\mathsf{v}_x(V' \cdot C^{[-1]}) = \mathsf{v}_x(\{V'\})$ for all $x \in \mathsf{Supp}(C) \setminus (Z' \cup X)$. Thus

$$\begin{aligned} \mathsf{v}_x(C) &= \mathsf{v}_x(V') - \mathsf{v}_x(V' \cdot C^{[-1]}) \\ &= \mathsf{v}_x(V') - \mathsf{v}_x(\{V'\}) = \mathsf{v}_x(\lfloor V' \rfloor) \in \mathbb{Z}_+ \quad \text{for all } x \in \mathsf{Supp}(C) \setminus (Z' \cup X). \end{aligned} \tag{8.36}$$

Since T is zero-sum and $\mathsf{Supp}(T_k) \cap Z \subseteq \mathsf{Supp}(V_{G_0 \setminus \mathfrak{X}}) \cap \mathfrak{X} = \emptyset$, we have

$$\sum_{i=1}^{n} \mathsf{v}_{z_i}(T) z_i = -\sum_{x \in \mathsf{Supp}(C) \setminus Z} \mathsf{v}_x(C)x - \sigma(T_k) = -\sum_{x \in \mathsf{Supp}(C) \setminus Z} \mathsf{v}_x(C)x - g_k.$$

As the elements from any $\mathfrak{X}_j$ are contained in some set from $\mathfrak{X}_m(\varphi)$, with every such set being a lattice basis for Δ, it follows that $\mathsf{Supp}(C) \subseteq \mathfrak{X} \subseteq \Delta$. Thus, since $g_k \in \Delta$ and $\mathsf{v}_x(C) \in \mathbb{Z}_+$ for all $x \in \mathsf{Supp}(C) \setminus Z \subseteq \mathsf{Supp}(C) \setminus (Z' \cup X)$ (in view of (8.36) and (8.34)), it follows that $\sum_{i=1}^{n} \mathsf{v}_{z_i}(T) z_i \in \Delta$. Since $\mathsf{v}_{z_i}(T) = \mathsf{v}_{z_i}(C) + \alpha_i \ge 0$, we have $\sum_{i=1}^{n} \mathsf{v}_{z_i}(T) z_i \in \mathsf{C}(Z)$. But now $\sum_{i=1}^{n} \mathsf{v}_{z_i}(T) z_i \in \Delta \cap \mathsf{C}(Z) = \mathsf{C}_{\mathbb{Z}}(Z)$ by (8.14), whence $\mathsf{v}_{z_i}(T) \in \mathbb{Z}_+$ for all $i \in [1, n]$ by (8.15). Combined with (8.36), $X \cup Z' \subseteq Z$ (in view of (8.34)) and $T_k \in \mathcal{F}(G_0)$, we conclude that $T \in \mathcal{F}(G_0)$ is an ordinary zero-sum sequence. However, since $T \mid U$ is nontrivial (as each T_k is nontrivial) with $U \in \mathcal{A}(G_0)$ an atom, this is only possible if $T = U$, which, in turn, is only possible if $|I_{\mathbb{Z}}| = 1 = k$, $|I_{\mathbb{Q}}| = |I_{\diamond}| = 0$, $V_{G_0 \setminus \mathfrak{X}} = T_1$, and $U = V$. This establishes Claim C. □

Let

$$\gamma = \min(\{1\} \cup \{\mathsf{v}_{z_i}(\lfloor V' \rfloor \cdot W_X \cdot C^{[-1]})/\alpha_i \,:\, i \in [1, n],\ \alpha_i \ne 0\}) \le 1.$$

We cannot have $\alpha_i = 0$ for all $i \in [1, n]$ in view of $\beta < 1$ and (8.17). Step C ensures that $\gamma = 1$ is only possible if $k \geq |I_{\mathbb{Z}}|+1$ or else $k = 1 = |I_{\mathbb{Z}}|$, $|I_{\mathbb{Q}}| = |I_\diamond| = 0$ and $U = V$. In view of (8.34), $\mathsf{v}_{z_i}(\lfloor V' \rfloor \cdot W_X \cdot C^{[-1]}) > 0$ for all $i \in [1, n]$, implying $\gamma > 0$.

By definition, we have $\mathsf{v}_{z_i}(\lfloor V' \rfloor \cdot W_X \cdot C^{[-1]}) \geq \gamma\alpha_i$ for all $i \in [1, n]$, with equality holding for any $i \in [1, n]$ attaining the minimum in the definition of γ. Moreover, since V', $C \in \mathcal{F}_{\mathrm{rat}}(G_0)$, we have α_i, $\mathsf{v}_{z_i}(\lfloor V' \rfloor \cdot W_X \cdot C^{[-1]}) \in \mathbb{Q}_+$, implying $\gamma \in \mathbb{Q}_+$ is a positive rational number, and thus $\gamma\alpha_i \in \mathbb{Q}_+$ for every $i \in [1, n]$. Define

$$C' = C \cdot \prod\nolimits^{\bullet}_{i \in [1,n]} z_i^{[\gamma\alpha_i]}.$$

Note C' is nontrivial in view of $\gamma > 0$ and not all $\alpha_i = 0$. By construction, we have $C' \mid \lfloor V' \rfloor \cdot W_X$ and $C' \in \mathcal{F}_{\mathrm{rat}}(G_0)$ (since $C \in \mathcal{F}_{\mathrm{rat}}(G_0)$ with $\gamma\alpha_i \in \mathbb{Q}_+$ for all $i \in [1, n]$). Note

$$\sigma(C') = \sigma(C) - \gamma(1-\beta)g_k = -\beta g_k - \gamma(1-\beta)g_k = -(\beta + \gamma(1-\beta))g_k.$$

Since β, $\gamma \in \mathbb{Q}_+$ with $0 < \gamma \leq 1$ and $0 \leq \beta < 1$, we have $\beta' := \beta+\gamma(1-\beta) \in \mathbb{Q}_+$ with $0 < \beta' \leq 1$. Furthermore, $\beta' = 1$ if and only if $\gamma = 1$. Since $Z' \subseteq Z = \{z_1, \ldots, z_n\}$ and $C \mid C'$, it follows from (8.28) that

$$Z'' = \mathrm{Supp}(C') \cap \mathrm{Supp}(\lfloor V' \rfloor \cdot W_X \cdot C'^{[-1]}) \subseteq Z.$$

Since $Z \subseteq \mathfrak{X}$ with $|Z \cap \mathfrak{X}_j| = 1$ for all $j \in [1, n]$ (as $Z \in \mathfrak{X}_m(\varphi)$), it follows that the subset $Z'' \subseteq Z$ satisfies $Z'' \subseteq \mathfrak{X}$ with $|Z'' \cap \mathfrak{X}_j| \leq 1$ for all $j \in [1, n]$. Note $Z'' \cap \bigcup_{j \in J^d_{k-1}} \mathfrak{X}_j \subseteq Z \cap \bigcup_{j \in J^d_{k-1}} \mathfrak{X}_j = X$, with the equality in view of $X \subseteq Z \in \mathfrak{X}_m(\varphi)$ (from (8.34)). Finally, if $\beta' = 1$, then $\gamma = 1$, whence either $k \geq |I_{\mathbb{Z}}| + 1$ or else $k = 1 = |I_{\mathbb{Z}}|$, $|I_{\mathbb{Q}}| = |I_\diamond| = 0$ and $U = V$ (as remarked after the definition of γ). Thus the rational sequence C' satisfies all the requirements for our recursive construction of the Inner Loop. Hence we may replace C by C' and repeat the Inner Loop process just described once more.

If at any point while iterating the Inner Loop Process we obtain $\gamma = 1$, then $\beta' = 1$ follows, and the Inner Loop immediately halts and outputs $C' = B_k$ as described earlier (when arguing that we could assume $\beta < 1$). Otherwise, there will always be an index $i \in [1, n]$ attaining the minimum in the definition of γ, in which case $z_i \in Z \subseteq \mathrm{Supp}(\lfloor V' \rfloor \cdot W_X \cdot C^{[-1]})$ but $z_i \notin \mathrm{Supp}(\lfloor V' \rfloor \cdot W_X \cdot C'^{[-1]})$, thus ensuring $|\mathrm{Supp}(\lfloor V' \rfloor \cdot W_X \cdot C^{[-1]})| > |\mathrm{Supp}(\lfloor V' \rfloor \cdot W_X \cdot C'^{[-1]})|$ has strictly decreased. As a result, we cannot iterate the Inner Loop process indefinitely, as the non-negative integer $|\mathrm{Supp}(\lfloor V' \rfloor \cdot W_X \cdot C^{[-1]})|$ strictly decreases after each iteration, ensuring that the process must halt for one of the three possibilities described earlier, all of which lead to the construction of the next sequence B_k in the outer loop process. As this shows it is always possible to construct the sequence B_k with the needed properties for the Outer Loop process, the proof is now complete.

Corollary 8.7 *Let $\Lambda \leq \mathbb{R}^d$ be a full rank lattice, where $d \geq 0$, and let $G_0 \subseteq \Lambda$ be a finitary subset with $\mathsf{C}(G_0) = \mathbb{R}^d$. Then $G_0 \setminus G_0^\diamond \subseteq G_0$ is a Lambert subset.*

Proof Suppose $0 \notin \mathsf{C}^*(G_0)$ (which implies G_0 is finitary by Theorem 7.8). Let $U \in \mathcal{A}(G_0)$ be arbitrary. Per the remarks above Theorem 8.6, apply the lattice type version of Theorem 8.6 to the atom U, so $\mathfrak{s}$ equals the number of nontrivial lattice types rather than minimal types. Letting $\mathfrak{X}^{\cup}(\varphi_j) = \bigcup_{X \in \mathfrak{X}(\varphi_j)} X$, observe that $\mathfrak{X}^{\cup}(\varphi_j) \subseteq G_0^\diamond$, for all $j \in [1, \mathfrak{s}]$, since $\mathfrak{X}^{\cup}(\varphi_j)$ is a union of sets $X \in X(G_0)$ with each $X \subseteq G_0^\diamond$, as discussed at the beginning of Sect. 7.2. Thus $G_0 \setminus G_0^\diamond \subseteq G_0 \setminus \mathfrak{X}^{\cup}(\varphi_j)$ for all $j \in [1, \mathfrak{s}]$, whence Theorem 8.6(c) ensures $\mathsf{v}_{G_0 \setminus G_0^\diamond}(U) = \sum_{i=0}^{\mathfrak{s}} \mathsf{v}_{G_0 \setminus G_0^\diamond}(A_i) \leq (\mathfrak{s}+1)N_T$, showing that $G_0 \setminus G_0^\diamond \subseteq G_0$ is a Lambert subset with bound $(\mathfrak{s}+1)N_T$, where N_T is bound given in Theorem 8.6.

If G_0 is only finitary, with $0 \in \mathsf{C}^*(G_0)$, then the corollary does not follow directly from Theorem 8.6. However, most parts of the proof of Theorem 8.6 only require the hypothesis that G_0 is finitary, not the stronger hypothesis that $0 \in \mathsf{C}^*(G_0)$. Indeed, this stronger hypothesis is only used twice in the proof: first, in the paragraph after (8.16), where it is used to show $g_k \neq 0$ for $k > |I_{\mathbb{Z}}|$, which is needed for the Inner Loop to construct B_k, and second, in the paragraph above (8.24), where it is used in the case when $\beta_k = 1$, which is needed to show (a) holds for the next iteration $U = V' \cdot A_{\mathfrak{t}-1} \cdot A_{\mathfrak{t}} \cdot \ldots \cdot A_{\mathfrak{s}}$. Note the latter only occurs when $k \geq |I_{\mathbb{Z}}|+1$ or when $k = 1 = |I_{\mathbb{Z}}|, |I_{\mathbb{Q}}| = |I_\diamond| = 0$ and $V = U$.

One can modify the Outer Loop by requiring it to prematurely halt during the construction of $A_{\mathfrak{t}-1}$ if ever $k = |I_{\mathbb{Z}}|$ is reached. In this way, both cases where the hypothesis $0 \notin \mathsf{C}^*(G_0)$ is used are never encountered. To avoid confusion, let $N_T^{(\mathfrak{t})}$ and $N_S^{(\mathfrak{t})}$ denote the recursively defined constants for which (a)–(d) holds for $U = V \cdot A_t \cdot \ldots \cdot A_{\mathfrak{s}}$ (assuming the process did not prematurely halt before A_t could be constructed), and let N_T and N_S denote the final values, so $N_S = N_S^{(1)}$ and $N_T = \max\{N_T^{(1)}, \sum_{j=1}^{\ell} |W_j|\}$, where $W_1, \ldots, W_\ell \in \mathcal{A}(G_0)$ are the distinct atoms having no nonempty $X \subseteq \mathrm{Supp}(W_j)$ with $X \in X(G_0)$, of which there are a finite number as noted at the beginning of the proof of Theorem 8.6 (which only requires the assumption that $G_0^\diamond$ is finitary, not that $0 \notin \mathsf{C}^*(G_0^\diamond)$). Note N_T and N_S still exist, with their values independent of U, even if we cannot complete the construction of all $A_{\mathfrak{t}-1}$ for some atom U. Indeed, $N_T^{(\mathfrak{t})}$ and $N_S^{(\mathfrak{t})}$ are simply defined by the dual recursion given by (8.25) and (8.26), so $N_T^{(\mathfrak{t}-1)} = N'_T$ and $N_S^{(\mathfrak{t}-1)} = N'_S$ with N'_T and N'_S defined by (8.25) and (8.26) using $N_S^{(\mathfrak{t})}$ and $N_T^{(\mathfrak{t})}$ in place of N_S and N_T in the formulas (8.25) and (8.26) (and initial values $N_T^{(\mathfrak{s}+1)} = N_S^{(\mathfrak{s}+1)} = 0$). Since all constants used in the recursive formulas (8.25) and (8.26) depend only on the individual lattice types φ for the finitary subset G_0, the resulting values of N_T and N_S are well-defined.

If, for the atom U, the Outer Loop process never prematurely halts during the construction of any A_{t-1}, then we obtain $\mathsf{v}_{G_0 \setminus G_0^\diamond}(U) \leq (\mathfrak{s}+1)N_T$, as we did when $0 \notin \mathsf{C}^*(G_0^\diamond)$. Now instead suppose, for the atom U, that the process prematurely halts during the construction of $A_{\mathfrak{t}-1}$ for some $\mathfrak{t} \in [2, \mathfrak{s}+1]$, so we have $U =$

$V \cdot A_{\mathfrak{t}} \cdot \ldots \cdot A_{\mathfrak{s}}$ satisfying (a)–(d) but fail to construct $A_{\mathfrak{t}-1}$. There are two ways this failure can arise. First, we may have completed the Outer Loop process with it ending when $\beta_k = 1$, $k = 1 = |I_{\mathbb{Z}}|$, $|I_{\mathbb{Q}}| = |I_\diamond| = 0$ and $V = U$, leaving us unable to conclude (a) holds for $A_{\mathfrak{t}-1}$. In such case, $V_{G_0\setminus\mathfrak{X}} = T_1$ with $\mathsf{v}_{G_0\setminus G_0^\diamond}(U) = \mathsf{v}_{G_0\setminus G_0^\diamond}(V) \leq |V_{G_0\setminus\mathfrak{X}}| \leq |T_1| \leq D_\Delta M_W \leq N_T^{(\mathfrak{t}-1)} \leq N_T$ by (8.9) and (8.25), as desired. Second, the Outer Loop process has constructed $B_1, \ldots, B_{|I_{\mathbb{Z}}|}$ and has not finished, so the terminal value $k > |I_{\mathbb{Z}}|$, yet $g_{|I_{\mathbb{Z}}|+1} = 0$, leaving us unable to construct the next sequence $B_{|I_{\mathbb{Z}}|+1}$ using the Inner Loop. In this case, since $\mathrm{Supp}(T_i) \subseteq G_0^\diamond$ for $i \in I_\diamond$ by (8.12), and since $\mathrm{Supp}(V_{\mathfrak{X}}) \subseteq \mathfrak{X} \subseteq G_0^\diamond$, we obtain

$$\begin{aligned}
\mathsf{v}_{G_0\setminus G_0^\diamond}(V) &\leq \sum_{i\in I_{\mathbb{Z}}\cup I_{\mathbb{Q}}} |T_i| \leq |I_{\mathbb{Z}}| D_\Delta M_W + |I_{\mathbb{Q}}| M_W \\
&\leq (k-1) D_\Delta M_W + (D_\Delta + N_S^{(t)} + 2\ell_\varphi) M_W \\
&\leq ((2^{n+1}-1)(1+N_S^{(t)}) D_\Delta + N_S^{(t)} + 2\ell_\varphi) M_W \leq N_T^{(\mathfrak{t}-1)} \leq N_T,
\end{aligned} \tag{8.37}$$

with the second inequality in view of (8.9) and (8.10), with the third inequality in view of $k > |I_{\mathbb{Z}}|$ and Claim A, with fourth in view of the upper bound $k \leq (2^{n+1}-1)(1+N_S^{(\mathfrak{t})})$, estimating how long the Outer Loop Process can run, and with the fifth in view of (8.25). Thus $\mathsf{v}_{G\setminus G_0^\diamond}(U) = \mathsf{v}_{G\setminus G_0^\diamond}(V) + \sum_{i=\mathfrak{t}}^{\mathfrak{s}} \mathsf{v}_{G\setminus G_0^\diamond}(A_i) \leq N_T^{(\mathfrak{t}-1)} + (\mathfrak{s}-\mathfrak{t}+1) N_T^{(\mathfrak{t})} \leq (\mathfrak{s}+1) N_T$ now follows from (c) holding for $U = V \cdot A_{\mathfrak{t}} \cdot \ldots \cdot A_{\mathfrak{s}}$. In all cases, $G_0 \setminus G_0^\diamond \subseteq G_0$ is a Lambert subset with bound $(\mathfrak{s}+1) N_T$, completing the proof. □

We can now extend Theorem 7.13 to characterize the set $G_0^\diamond$ in terms of $\mathbb{Z}_+$-linear combinations and atoms, rather than $\mathbb{Q}_+$-linear combinations and elementary atoms.

Corollary 8.8 *Let $\Lambda \leq \mathbb{R}^d$ be a full rank lattice, where $d \geq 0$, and let $G_0 \subseteq \Lambda$ be a finitary subset with $\mathsf{C}(G_0) = \mathbb{R}^d$. Then*

$$\begin{aligned}
G_0^\diamond &= \Big\{ g \in G_0 : \ \sup\{\mathsf{v}_g(U) : U \in \mathcal{A}^{\mathsf{elm}}(G_0)\} = \infty \Big\} \\
&= \Big\{ g \in G_0 : \ \sup\{\mathsf{v}_g(U) : U \in \mathcal{A}(G_0)\} = \infty \Big\}.
\end{aligned}$$

Proof The equality $G_0^\diamond = \{g \in G_0 : \ \sup\{\mathsf{v}_g(U) : U \in \mathcal{A}^{\mathsf{elm}}(G_0)\} = \infty\}$ holds by Theorem 7.13. Since $\mathcal{A}^{\mathsf{elm}}(G_0) \subseteq \mathcal{A}(G_0)$, the inclusion

$$\begin{aligned}
&\Big\{ g \in G_0 : \ \sup\{\mathsf{v}_g(U) : U \in \mathcal{A}^{\mathsf{elm}}(G_0)\} = \infty \Big\} \\
&\subseteq \Big\{ g \in G_0 : \ \sup\{\mathsf{v}_g(U) : U \in \mathcal{A}(G_0)\} = \infty \Big\}
\end{aligned}$$

is trivial. To see the reverse inclusion, let $g \in G_0$ with $\sup\{\mathsf{v}_g(U) : U \in \mathcal{A}(G_0)\} = \infty$. Then Corollary 8.7 ensures

$$g_0 \in G_0^\diamond = \Big\{g \in G_0 : \sup\{\mathsf{v}_g(U) : U \in \mathcal{A}^{\mathsf{elm}}(G_0)\} = \infty\Big\},$$

as needed. □

We can now achieve one of our mains goals: the characterization of finite elasticities for subsets $G_0 \subseteq \mathbb{Z}^d$. Note that Theorem 8.9.1 characterizes $\rho_{d+1}(G_0) < \infty$ in terms of a basic, combinatorial property of G_0, one which trivially implies $\rho_{d+1}(G_0) < \infty$ (cf. Proposition 8.3), while Theorem 8.9.5 characterizes $\rho_{d+1}(G_0) < \infty$ in terms of the subset $G_0^\diamond \subseteq G_0$, which can be defined purely in terms of Convex Geometry (cf. Proposition 7.2).

Theorem 8.9 *Let $\Lambda \le \mathbb{R}^d$ be a full rank lattice in $\mathbb{R}^d$, where $d \ge 0$, and let $G_0 \subseteq \Lambda$ be a subset. Then the following are equivalent.*

1. *There exists a subset $X \subseteq G_0$ such that $\mathcal{A}(X) = \emptyset$ and $G_0 \setminus X \subseteq G_0$ is a Lambert subset.*
2. $\rho(G_0) < \infty$
3. $\rho_k(G_0) < \infty$ *for all* $k \ge 1$
4. $\rho_{d+1}(G_0) < \infty$.
5. $0 \notin \mathsf{C}^*(G_0^\diamond)$.

Proof The implications 1. ⇒ 2. ⇒ 3. ⇒ 4. follow by Proposition 8.3. Let $\mathcal{E} = \mathsf{C}(G_0) \cap -\mathsf{C}(G_0)$ be the lineality space of $\mathsf{C}(G_0)$. Let

$$\widetilde{G}_0 = \{g \in G_0 : g \in \mathsf{Supp}(U) \text{ for some } U \in \mathcal{A}(G_0)\}.$$

Corollary 4.5 implies

$$G_0 \cap \mathcal{E} = \widetilde{G}_0 \quad \text{and} \quad \mathsf{C}(\widetilde{G}_0) = \mathsf{C}(G_0) \cap -\mathsf{C}(G_0) = \mathcal{E}. \tag{8.38}$$

Thus, by definition of $G_0^\diamond$, we have $G_0^\diamond = \widetilde{G}_0^\diamond$. Note that $\widetilde{\Lambda} = \Lambda \cap \mathcal{E} \le \Lambda$ is a sublattice. Since $\widetilde{G}_0 \subseteq \widetilde{\Lambda}$ is a subset which linearly generates $\mathcal{E}$, it follows that $\widetilde{\Lambda} \le \mathcal{E}$ is a full rank lattice in $\mathcal{E}$. Let $\tilde{d} = \dim \mathcal{E}$. Now suppose we knew Theorem 8.9 held for $\widetilde{G}_0 \subseteq \widetilde{\Lambda} \le \mathcal{E}$. Then $\rho_{d+1}(G_0) < \infty$ would imply $\rho_{\tilde{d}+1}(\widetilde{G}_0) \le \rho_{d+1}(\widetilde{G}_0) = \rho_{d+1}(G_0) < \infty$, with the first inequality in view of $\tilde{d} \le d$ and (2.9), in turn implying $0 \notin \mathsf{C}^*(\widetilde{G}_0^\diamond) = \mathsf{C}^*(G_0^\diamond)$, in turn implying there is a subset $\widetilde{X} \subseteq \widetilde{G}_0$ with $\mathcal{A}(\widetilde{X}) = \emptyset$ and $\widetilde{G}_0 \setminus \widetilde{X} \subseteq \widetilde{G}_0$ a Lambert subset. But then clearly $X = \widetilde{X} \cup (G_0 \setminus \widetilde{G}_0) \subseteq G_0$ is a subset with $\mathcal{A}(X) = \emptyset$ and $G_0 \setminus X = \widetilde{G}_0 \setminus \widetilde{X} \subseteq G_0$ a Lambert subset. Thus the remaining implications follow from the case when $G_0 = \widetilde{G}_0$, which we now assume. In this case, the implication 4. ⇒ 5. follows by Proposition 8.4 applied to $G_0 = \widetilde{G}_0$, which we can apply in view of (8.38) and $\rho_{\tilde{d}+1}(G_0) \le \rho_{d+1}(G_0) < \infty$ (by (2.9)). It remains to prove 5. ⇒ 1. To this end, suppose $0 \notin \mathsf{C}^*(G_0)$. Then G_0 is finitary by Theorem 7.8 and (8.38), whence Corollary 8.7 and (8.38) imply

$G_0 \setminus G_0^\diamond \subseteq G_0$ is a Lambert subset. Since $0 \notin \mathsf{C}^*(G_0^\diamond)$, Proposition 4.2 implies $\mathcal{A}(G_0^\diamond) = \emptyset$. Thus Item 1 holds with $X = G_0^\diamond$, completing the proof. □

Let $G_0 \subseteq G$ be a subset of an abelian group G. For $k \geq 1$, we define the elementary elasticity $\rho_k^{\mathsf{elm}}(G_0)$ to be the minimal integer N such that, if

$$U_1 \cdot \ldots \cdot U_k = V_1 \cdot \ldots \cdot V_\ell$$

with $U_1, \ldots, U_k \in \mathcal{A}^{\mathsf{elm}}(G_0)$ and $V_1, \ldots, V_\ell \in \mathcal{A}(G_0)$, then $\ell \leq N$. If no such N exists, then we set $\rho_k^{\mathsf{elm}}(G_0) = \infty$. It is readily noted (by the same simple argument establishing (2.9)) that $\rho_1^{\mathsf{elm}}(G_0) \leq \rho_2^{\mathsf{elm}}(G_0) \leq \ldots$. Indeed, if $U_1 \cdot \ldots \cdot U_{k-1} = V_1 \cdot \ldots \cdot V_\ell$ with $U_1, \ldots, U_{k-1} \in \mathcal{A}^{\mathsf{elm}}(G_0)$ and $V_1, \ldots, V_\ell \in \mathcal{A}(G_0)$, then the factorization $U_1 \cdot \ldots \cdot U_{k-1} \cdot U_1 = V_1 \cdot \ldots \cdot V_\ell \cdot U_1$ shows $\rho_k^{\mathsf{elm}}(G_0) \geq \ell + 1$. If $\rho_{k-1}^{\mathsf{elm}}(G_0)$ is finite, then we may take $\ell = \rho_{k-1}^{\mathsf{elm}}(G_0)$, and if $\rho_{k-1}^{\mathsf{elm}}(G_0)$ is infinite, then we may take ℓ to be arbitrarily large. In either case, $\rho_k^{\mathsf{elm}}(G_0) \geq \rho_{k-1}^{\mathsf{elm}}(G_0)$ follows. Trivially, we also have

$$\rho_k^{\mathsf{elm}}(G_0) \leq \rho_k(G_0).$$

Thus if $\rho_k(G_0)$ is finite, then so is $\rho_k^{\mathsf{elm}}(G_0)$. As a consequence of Theorem 8.9, we have the following converse.

Corollary 8.10 *Let $\Lambda \subseteq \mathbb{R}^d$ be a full rank lattice, where $d \geq 0$, and let $G_0 \subseteq \Lambda$ be a subset. Then $\rho_{d+1}(G_0) < \infty$ if and only if $\rho_{d+1}^{\mathsf{elm}}(G_0) < \infty$.*

Proof We may w.l.o.g. suppose every $g \in G_0$ is contained in an atom, whence $\mathsf{C}(G_0) = \mathcal{E} := \mathsf{C}(G_0) \cap -\mathsf{C}(G_0)$ is the lineality space of $\mathsf{C}(G_0)$ by Corollary 4.5. Let $\tilde{d} = \dim \mathcal{E} \leq d$. As already remarked, one direction of the corollary is trivial. Suppose $\rho_{d+1}^{\mathsf{elm}}(G_0) < \infty$. Then $\rho_{\tilde{d}+1}^{\mathsf{elm}}(G_0) \leq \rho_{d+1}^{\mathsf{elm}}(G_0) < \infty$. If $0 \notin \mathsf{C}^*(G_0^\diamond)$, then $\rho_{d+1}(G_0) < \infty$ follows by Theorem 8.9.3, as desired. Therefore we may assume $0 \in \mathsf{C}^*(G_0^\diamond)$, in which case Carthédory's Theorem implies that there is a minimal positive basis $X \subseteq G_0^\diamond$ with $s = |X| \leq d + 1$. Let $V \in \mathcal{A}^{\mathsf{elm}}(X)$ be the unique elementary atom with support $X = \{g_1, \ldots, g_s\} \subseteq G_0^\diamond$ (by Proposition 4.8) and let $m = \max\{\mathsf{v}_{g_j}(V) : j \in [1, s]\}$. By Proposition 7.3, every $g \in G_0^\diamond$ has $\mathsf{Supp}\{\mathsf{v}_g(U) : U \in \mathcal{A}^{\mathsf{elm}}(G_0)\} = \infty$. Thus, for every $i \geq 0$ and $j \in [1, s]$, we can find an elementary atom $U_i^{(j)} \in \mathcal{A}^{\mathsf{elm}}(G_0)$ with $\mathsf{v}_{g_j}(U_i^{(j)}) \geq i$. Then $V^{[\lfloor i/m \rfloor]} \mid U_i^{(1)} \cdot \ldots U_i^{(s)}$ for every $i \geq 1$. Hence, for every $i \geq 1$, we have a factorization

$$U_i^{(1)} \cdot \ldots \cdot U_i^{(s)} = \Big(\prod_{j=1}^{\lfloor i/m \rfloor} V\Big) \cdot B_i$$

for some $B_i \in \mathcal{B}(G_0)$. Since $i \to \infty$ and m is a fixed constant, it follows that $\rho_s^{\mathsf{elm}}(G_0) = \infty$. However, since $s \leq d + 1$, we have $\rho_s^{\mathsf{elm}}(G_0) \leq \rho_{d+1}^{\mathsf{elm}}(G_0)$, so that $\rho_{d+1}^{\mathsf{elm}}(G_0) = \infty$ as well, contrary to assumption. □

8.2 The Structure of Atoms and Arithmetic Invariants

In this section, we give a weak structural description of the atoms $U \in \mathcal{A}(G_0)$ when $0 \notin \mathsf{C}^*(G_0)$, equivalently, when $\rho_{d+1}(G_0) < \infty$. We then use this result, along with our characterization of finite elasticities and prior machinery, to show that having finite elasticities implies nearly all standard invariants of Factorization Theory are also finite, meaning the hypothesis $\rho_{d+1}(G_0) < \infty$ is sufficient to guarantee factorizations are as well-behaved as could be hoped.

We begin by describing what a set X with minimal type φ looks like when expressed using the lattice basis Z_φ.

Lemma 8.11 *Let $\Lambda \leq \mathbb{R}^d$ be a full rank lattice, where $d \geq 0$, let $G_0 \subseteq \Lambda$ be a finitary subset with $\mathsf{C}(G_0) = \mathbb{R}^d$, let $\varphi \in \mathfrak{X}_m(G_0)$ be a minimal type with codomain $Z_\varphi = Z_1 \cup \ldots \cup Z_s$, and let $X \in \mathfrak{X}_m(\varphi)$ be a subset with minimal type φ and associated maximal series decomposition $X = X_1 \cup \ldots \cup X_s$. For $k \in [1, s]$, let $z_1^{(k)}, \ldots z_{t_k}^{(k)} \in Z_k$ and $x_1^{(k)}, \ldots, x_{t_k}^{(k)} \in X_k$ be the distinct elements of Z and X, respectively, indexed so that $\varphi(x_i^{(k)}) = z_i^{(k)}$ for all $k \in [1, s]$ and $i \in [1, t_k]$. Then, for every $k \in [1, s]$ and $i \in [1, t_k]$, we have*

$$x_i^{(k)} = z_i^{(k)} - \sum_{\kappa=1}^{k-1} \sum_{\iota=1}^{t_\kappa} \xi_\iota^{(\kappa)} z_\iota^{(\kappa)} \quad \textit{for some} \quad \xi_\iota^{(\kappa)} \in \mathbb{Z}_+. \tag{8.39}$$

Proof For $k \in [0, s]$, let $\Delta_k = \mathbb{Z}\langle Z_1 \cup \ldots \cup Z_k \rangle$ and let $\pi_k : \mathbb{R}^d \to \mathbb{R}\langle \Delta_k \rangle^\perp$ be the orthogonal projection. Set $\Delta = \Delta_s$. Since X has minimal type φ, we have $\mathbb{Z}\langle X_1 \cup \ldots \cup X_j \rangle = \Delta_j$ for all $j \in [1, s]$. Let $k \in [1, s]$ and $i \in [1, t_k]$ be arbitrary. Let $Y = Z_\varphi \setminus \{z_i^{(k)}\} \cup \{x_i^{(k)}\}$. By the interchangeability property of $\mathfrak{X}_m(\varphi)$ (Proposition 7.33), we have $Y \in \mathfrak{X}_m(\varphi)$. Thus Y also has minimal type φ, and so $Z_\varphi \subseteq \mathsf{C}_\mathbb{Z}(Z_\varphi) \subseteq \mathsf{C}_\mathbb{Z}(Y)$. In particular, we can write $z_i^{(k)}$ as a positive integer combination of the elements from Y, say

$$z_i^{(k)} = \sum_{\kappa=1}^{k-1} \sum_{\iota=1}^{t_\kappa} \xi_\iota^{(\kappa)} z_\iota^{(\kappa)} + \xi_i^{(k)} x_i^{(k)} + \sum_{\substack{\iota=1 \\ \iota \neq i}}^{t_k} \xi_\iota^{(k)} z_i^{(k)} + \sum_{\kappa=k+1}^{s} \sum_{\iota=1}^{t_\kappa} \xi_\iota^{(\kappa)} z_\iota^{(\kappa)}$$

for some $\xi_\iota^{(\kappa)} \in \mathbb{Z}_+$. The coefficients of $z_\iota^{(\kappa)}$ with $\kappa \geq k+1$ are all zero in view of the linear independence of $Z_{k+1} \cup \ldots \cup Z_s$ modulo $\mathbb{R}\langle \Delta_k \rangle$ (cf. Proposition 4.11). Since X and Z_φ share the same minimal type φ with $\varphi(x_i^{(k)}) = z_i^{(k)}$, we have $z_i^{(k)} - x_i^{(k)} \in \mathbb{R}\langle \Delta_{k-1} \rangle \cap \Delta = \Delta_{k-1}$. In particular, $x_i^{(k)}$ and $z_i^{(k)}$ are equal modulo $\mathbb{R}\langle \Delta_{k-1} \rangle$, in which case the linear independence of Z_k modulo $\mathbb{R}\langle \Delta_{k-1} \rangle$ ensures that the coefficients of $z_\iota^{(k)}$ with $\iota \neq i$ are also zero, and that the coefficient of $x_i^{(k)}$ is one. □

For the next proposition, we need to view our sequences as indexed, so that two distinct terms of a sequence $S \in \mathcal{F}(G_0)$ that are equal as elements can still be viewed as distinct terms in the sequence S. We follow the notation introduced in Sect. 2.3.

Proposition 8.12 *Let $\Lambda \leq \mathbb{R}^d$ be a full rank lattice, where $d \geq 0$, let $G_0 \subseteq \Lambda$ be a finitary subset with $\mathsf{C}(G_0) = \mathbb{R}^d$, let $\varphi \in \mathfrak{X}_m(G_0)$ be a minimal type with codomain $Z_\varphi = Z_1 \cup \ldots \cup Z_s$, let $\mathfrak{X} = \mathfrak{X}^{\cup}_m(\varphi) = \bigcup_{X \in \mathfrak{X}_m(\varphi)} X$, let $S = g_1 \cdot \ldots \cdot g_\ell \in \mathcal{F}(\mathfrak{X})$ be a sequence, and for every $k \in [1, s]$, let $I_k \subseteq [1, \ell]$ be the subset of all $a \in [1, \ell]$ with g_a at depth k. Suppose $\sigma(S) \in \mathsf{C}(Z_\varphi)$. Then there exists a system of subsets $T_a \subseteq [1, \ell]$ for $a \in [1, \ell]$ such that the following hold.*

1. *For every $a \in [1, \ell]$, say $a \in I_k$, there is some $\partial(a) \subseteq I_1 \cup \ldots \cup I_{k-1}$ such that $T_a = \{a\} \cup \bigcup_{b \in \partial(a)} T_b$ is a disjoint union.*
2. *If $b \notin T_a$ and $a \notin T_b$, then $T_a \cap T_b = \emptyset$, for every $a, b \in [1, \ell]$.*
3. *$\sigma(S(T_a)) = \varphi(g_a)$, for every $a \in [1, \ell]$.*

   In particular, for any system of sets $T_a \subseteq [1, \ell]$ satisfying Items 1–3, the following hold.
4. *If $a \in I_k$, then $T_a \setminus \{a\} \subseteq I_1 \cup \ldots \cup I_{k-1}$.*
5. *If $b \in T_a$, then $T_b \subseteq T_a$.*
6. *For every $k \in [1, s]$, there exists a subset $J_k \subseteq I_1 \cup \ldots \cup I_k$ such that $\bigcup_{i \in J_k} T_i = I_1 \cup \ldots \cup I_k$ is a disjoint union, ensuring that*

$$\prod\nolimits^{\bullet}_{i \in [1,k]} S(I_i) = \prod\nolimits^{\bullet}_{i \in J_k} S(T_i)$$

   is a factorization of the subsequence of all terms in S with depth at most k into a product of subsequences whose sums each lie in Z_φ.

Proof Item 4 follows from Item 1 and a quick inductive argument on $k = 1, 2, \ldots, s$. Likewise, Item 5 follows from Item 1 and a quick inductive argument on k, where $a \in I_k$. To show Item 6 holds, let $k \in [1, s]$. The set J_k is then constructed recursively in k steps. Set $J_k^{(k)} = I_k$. Assuming $J_k^{(j)}$ has been constructed, with $j \in [2, k]$, define $J_k^{(j-1)} = J_k^{(j)} \cup \left(I_{j-1} \setminus \bigcup_{c \in J_k^{(j)}} T_c\right) \subseteq I_{j-1} \cup I_j \cup \ldots \cup I_k$. Set $J_k = J_k^{(1)}$. In view of Item 4, any $a \in I_j$ with $j \leq k$ has $T_a \subseteq I_1 \cup \ldots \cup I_k$. Thus $J_k = J_k^{(1)} \subseteq I_1 \cup \ldots \cup I_k$ and $\bigcup_{i \in J_k} T_i = I_1 \cup \ldots \cup I_k$ by construction. By Item 4, we have $b \notin T_a$ and $a \notin T_b$ for any distinct $a, b \in I_j$ and $j \in [1, s]$. Hence Item 2 ensures that $\bigcup_{i \in I_j} T_i$ is a disjoint union for any $j \in [1, s]$. In particular, $\bigcup_{i \in J_k^{(k)}} T_i$ is a disjoint union. Also, if $a \in J_k^{(j)}$ and $b \in I_{j-1} \setminus \bigcup_{c \in J_k^{(j)}} T_c$, then $b \notin T_a$. Since $b \in I_{j-1}$ ensures that $T_b \subseteq I_1 \cup \ldots \cup I_{j-1}$ (by Item 4), while $a \in J_k^{(j)} \subseteq I_j \cup \ldots \cup I_k$, we also have $a \notin T_b$. Thus Item 2 ensures that T_a is disjoint from T_b, for every $a \in J_k^{(j)}$. We also have T_b disjoint from any other $T_{b'}$ with $b' \in I_{j-1}$ as already mentioned. An inductive argument now shows that $\bigcup_{i \in J_k^{(j)}} T_i$ is a disjoint union for $j = k, k-1, \ldots, 1$. The case $j = 1$ implies $\bigcup_{i \in J_k^{(1)}} T_i = \bigcup_{i \in J_k} T_i$ is a disjoint union, and we now see that Item 6 follows. It remains to establish Items 1–3.

For $k \in [1, s]$, let $z_1^{(k)}, \ldots z_{t_k}^{(k)} \in Z_k$ be the distinct elements of Z_k. Let $\mathfrak{X} = \bigcup_{k=1}^{s} \bigcup_{i=1}^{t_k} \mathfrak{X}_{k,i}$, with $\mathfrak{X}_{k,i} \subseteq \mathfrak{X}$ consisting of all $x \in \mathfrak{X}$ with $\varphi(x) = z_i^{(k)}$. Given any $x \in \mathbb{R}\langle Z_\varphi \rangle$, we have

$$x = \sum_{k=1}^{s} \sum_{i=1}^{t_k} \xi_{k,i}(x) z_i^{(k)} \quad \text{for some uniquely defined } \xi_{k,i}(x) \in \mathbb{R},$$

as $Z_\varphi \in \mathfrak{X}_m(\varphi)$ is linearly independent. If $x \in \mathfrak{X}$, then Lemma 8.11 ensures that $\xi_{k,i}(x) \in \mathbb{Z}$ for all k and i. Moreover, if $\varphi(x) = z_\iota^{(\kappa)}$, then $\xi_{k,i}(x) = 0$ whenever $k \geq \kappa$, apart from the value $\xi_{\kappa,\iota}(x) = 1$, and $\xi_{k,i}(x) \leq 0$ whenever $k < \kappa$.

We construct the subsets $\partial(a)$ and T_a recursively, first for all $a \in I_1$, then for all $a \in I_2$, and so forth. For $a \in I_1$, define $\partial(a) = \emptyset$ and $T_a = \{a\}$. Then $\partial(a) \subseteq I_1 \cup \ldots \cup I_0 = \emptyset$, and Items 1 and 2 trivially hold. Moreover, since $g_a \in X$ for some $X \in \mathfrak{X}_m(\varphi)$ with $\varphi(g_a) \in Z_1$ (since $a \in I_1$), it follows from the definition of lattice type (each minimal type is a refinement of a lattice type) that $g_a = \varphi(g_a)$. Thus Item 3 holds. Assume the T_a and $\partial(a)$ have been constructed for all $a \in I_1 \cup \ldots \cup I_{\kappa-1}$, where $\kappa \geq 2$, such that Items 1–3 hold. Let $J_{\kappa-1} \subseteq I_1 \cup \ldots \cup I_{\kappa-1}$ be a subset such that $\bigcup_{i \in J_{\kappa-1}} T_i = I_1 \cup \ldots \cup I_{\kappa-1}$ is a disjoint union with

$$\prod\nolimits^{\bullet}_{i \in [1,\kappa-1]} S(I_i) = \prod\nolimits^{\bullet}_{i \in J_{\kappa-1}} S(T_i)$$

a factorization of the subsequence of all terms in S with depth at most $\kappa - 1$ into a product of subsequences whose sums each lie in Z_φ, which exists by the argument used to derive Item 6 applied to the subsequence of S consisting of all terms having depth at most $\kappa - 1$. For $k \in [1, \kappa - 1]$ and $j \in [1, t_k]$, let $J_{\kappa-1}^{(k,j)} \subseteq J_{\kappa-1}$ be all those $b \in J_{\kappa-1}$ with $\sigma(S(T_b)) = z_j^{(k)}$. Then $J_{\kappa-1} = \bigcup_{k=1}^{\kappa-1} \bigcup_{j=1}^{t_k} J_{\kappa-1}^{(k,j)}$ is a disjoint union. We proceed to construct $\partial(a)$ and T_a for $a \in I_\kappa$ as follows.

For each $k \in [1, \kappa - 1]$ and $j \in [1, t_k]$, we have $\xi_{k,j}(g_a) \leq 0$, for $a \in I_\kappa$, in view of Lemma 8.11 as remarked earlier. Since $I_k \subseteq \bigcup_{i \in J_{\kappa-1}} T_i$ (as $k \leq \kappa - 1$), it follows that $\prod^{\bullet}_{i \in J_{\kappa-1}} S(T_i)$ contains all terms x from S with $\xi_{k,j}(x) > 0$. Thus, since $\sigma(S) \in \mathsf{C}(Z_\varphi)$ by hypothesis with Z_φ linearly independent, we must have $\sum_{a \in I_\kappa} |\xi_{k,j}(g_a)| \leq |J_{\kappa-1}^{(k,j)}|$. Consequently, it is possible to find disjoint subsets $D_a^{(k,j)} \subseteq J_{\kappa-1}^{(k,j)}$, for $a \in I_\kappa$, such that $|D_a^{(k,j)}| = |\xi_{k,j}(g_a)|$ for each $a \in I_\kappa$. Note $D_a^{(k,j)} \subseteq J_{\kappa-1}^{(k,j)} \subseteq J_{\kappa-1} \subseteq I_1 \cup \ldots \cup I_{\kappa-1}$. Define $\partial(a) = \bigcup_{k=1}^{\kappa-1} \bigcup_{j=1}^{t_k} D_a^{(k,j)} \subseteq J_{\kappa-1} \subseteq I_1 \cup \ldots \cup I_{\kappa-1}$ and $T_a = \{a\} \cup \bigcup_{c \in \partial(a)} T_c$. By construction, Items 1 and 3 both hold for all $a \in I_\kappa$, with disjointness in Item 1 following from the disjointness of the union $\bigcup_{i \in J_{\kappa-1}} T_i$. Since

$$\bigcup_{i \in J_{\kappa-1}} T_i = I_1 \cup \ldots \cup I_{\kappa-1} \tag{8.40}$$

is a disjoint union, and since $\bigcup_{a\in I_\kappa} \partial(a) \subseteq J_{\kappa-1}$ is also a disjoint union (as the $D_a^{(k,j)}$ are disjoint), it follows that $T_a \cap T_b = \emptyset$ for all distinct $a,\ b \in I_\kappa$. If $a \in I_\kappa$ and $b \in I_1 \cup \ldots \cup I_{\kappa-1}$ with $b \notin T_a$, then $b \in T_c$ for some $c \in J_{\kappa-1} \setminus \partial(a)$ (in view of (8.40)), in which case the disjointness of the union in (8.40) ensures that $T_a \cap T_c = \emptyset$. However, since Item 4 holds for $I_1 \cup \ldots \cup I_{\kappa-1}$, it follows that $b \in T_c$ implies $T_b \subseteq T_c$. Hence $T_a \cap T_b = \emptyset$ follows in view of $T_a \cap T_c = \emptyset$. Finally, if $a,\ b \in I_1 \cup \ldots \cup I_{\kappa-1}$ with $a \notin T_b$ and $b \notin T_a$, then Item 2 holding for $I_1 \cup \ldots \cup I_{\kappa-1}$ ensures that $T_a \cap T_b = \emptyset$. Hence Item 2 holds for $I_1 \cup \ldots \cup I_\kappa$, and iterating the construction for $\kappa = 1, 2, \ldots, s$ completes the proof. □

Item 2 in Proposition 8.12 implies that either $b \in T_a$ (implying $T_b \subseteq T_a$ by Item 5) or $a \in T_b$ (implying $T_1 \subseteq T_a$ by Item 5) or $T_a \cap T_b = \emptyset$. Thus any system of sets $T_a \subseteq [1, \ell]$ satisfying Items 1–3 in Proposition 8.12 also satisfies

1′. $a \in T_a$ and $T_a \setminus \{a\} \subseteq I_1 \cup \ldots \cup I_{k-1}$, where $a \in I_k$,
2′. either $T_a \subseteq T_b$ or $T_b \subseteq T_a$ or $T_a \cap T_b = \emptyset$, and
3′. $\sigma(S(T_a)) = \varphi(g_a)$,

for any $a,\ b \in [1, \ell]$. However, a system of sets $T_a \subseteq [1, \ell]$ satisfying Items 1′–3′ must also satisfy Items 1–3, meaning these are equivalent defining conditions for the set system $T_a \subseteq [1, \ell]$—as the following short argument shows. If $a \notin T_b$, then Item 1′ ensures $T_a \nsubseteq T_b$. Likewise, if $b \notin T_a$, then Item 1′ ensures $T_b \nsubseteq T_a$. As a result, if $a \notin T_b$ and $b \notin T_b$, then Item 2′ implies $T_a \cap T_b = \emptyset$, yielding Item 2. Item 3 is the same as Item 3′. Let $a \in [1, \ell]$. If $b \in T_a \setminus \{a\}$, then Item 1′ ensures $\mathsf{dep}(g_c) \le \mathsf{dep}(g_b) < \mathsf{dep}(g_a)$ for all $c \in T_b$, implying $a \notin T_b$. Thus $T_a \nsubseteq T_b$ (as $a \in T_a$ but $a \notin T_b$) and $T_a \cap T_b \neq \emptyset$ (as $b \in T_a \cap T_b$), in which case Item 2′ yields $T_b \subseteq T_a$, showing Item 5 holds. We conclude (by Items 1′ and 5) that $b \in T_b \subseteq T_a$ for all $b \in T_a$, implying $T_a = \{a\} \cup \bigcup_{b\in T_a\setminus\{a\}} T_b$. Let $\partial(a) \subseteq T_a \setminus \{a\}$ consist of all $b \in T_a \setminus \{a\}$ such that there does not exist any $c \in T_a \setminus \{a, b\}$ with $T_b \subseteq T_c$. By Item 1′, we have $\partial(a) \subseteq I_1 \cup \ldots \cup I_{k-1}$, where $a \in I_k$. By definition of $\partial(a)$, we have $T_a = \{a\} \cup \bigcup_{b\in T_a\setminus\{a\}} T_b = \{a\} \cup \bigcup_{b\in\partial(a)} T_b$. If $b,\ c \in \partial(a)$ are distinct, then the definition of $\partial(a) \subseteq T_a \setminus \{a\}$ ensures that $T_b \nsubseteq T_c$ and $T_c \nsubseteq T_b$, so that Item 2′ implies T_b and T_c are disjoint. By Item 1′, any $b \in \partial(a) \subseteq I_1 \cup \ldots \cup I_{k-1}$ has $T_b \subseteq I_1 \cup \ldots \cup I_{k-1}$, ensuring that $a \notin T_b$ (as $a \in I_k$). It follows that the union $T_a = \{a\} \cup \bigcup_{b\in\partial(a)} T_b$ is disjoint, yielding Item 1, which completes the equivalence of the conditions.

It is entirely possible for the size of $|\operatorname{Supp}(U)|$, for an atom $U \in \mathcal{A}(G_0)$, to be arbitrarily large even under the assumption $\rho_{d+1}(G_0) < \infty$. This means that, if we partition the terms of the sequence U according to whether they are equal as elements from G_0, then we cannot hope to achieve a global bound on the number of partition classes we must use. However, the following theorem shows that there is a less restrictive notion of support that *does* have the property that any atom $U \in \mathcal{A}(G_0)$ can have its terms partitioned into at most $N + \sum_{\varphi\in\mathfrak{X}_m(G_0)} |Z_\varphi| \le N + d|\mathfrak{X}_m(G_0)|$ types of elements, with elements of the same type behaving in the same essential manner as described by Proposition 8.12. Theorem 8.13 can be viewed as a weak structural description of the atoms $U \in \mathcal{A}(G_0)$, allowing us to effectively simulate globally bounded finite support. We will later give an example of how this can be accomplished in the proof of Theorem 8.14.

Theorem 8.13 *Let $\Lambda \leq \mathbb{R}^d$ be a full rank lattice, where $d \geq 0$, and let $G_0 \subseteq \Lambda$ be a subset with $\mathsf{C}(G_0) = \mathbb{R}^d$. Suppose $0 \notin \mathsf{C}^*(G_0^\diamond)$. Then there exists a bound $N \geq 0$ such that any atom $U \in \mathcal{A}(G_0)$ has a factorization*

$$U = R \cdot {\prod}^{\bullet}_{\varphi \in \mathfrak{T}_m(G_0)} S_\varphi, \quad \text{with} \quad R \in \mathcal{F}(G_0) \quad \text{and} \quad S_\varphi \in \mathcal{F}(\mathfrak{X}_m^{\cup}(\varphi)),$$

such that $|R| \leq N$ and $\sigma(S_\varphi) \in \mathsf{C}_{\mathbb{Z}}(Z_\varphi)$ for all $\varphi \in \mathfrak{T}_m(G_0)$, where Z_φ is the codomain of φ. In particular, all conclusions of Proposition 8.12 hold for each S_φ.

Proof Since $0 \notin \mathsf{C}^*(G_0)$, Theorem 7.8 ensures that G_0 is finitary. Let $U \in \mathcal{A}(G_0)$ be an arbitrary atom and let $U = A_0 \cdot \prod_{\varphi \in \mathfrak{T}_m(G_0)} A_\varphi$ be the factorization of U given by Theorem 8.6, with bounds $N_S \geq 0$ and $N_T \geq 0$ and $A_0,\ A_\varphi \in \mathcal{B}_{\text{rat}}(G_0)$ (with A_φ the trivial sequence when $\varphi \in \mathfrak{T}_m(G_0)$ is the trivial type). For each $\varphi \in \mathfrak{T}_m(G_0)$, we define the subsequence $S_\varphi \mid \lfloor A_\varphi \rfloor$ as follows.

For the trivial type φ, we set S_φ to be the trivial sequence. Let $\varphi \in \mathfrak{T}_m(G_0)$ be an arbitrary, nontrivial type, let $Z_\varphi = Z_1 \cup \ldots \cup Z_s$ be the codomain of φ and associated maximal series decomposition, and let $\mathfrak{X} = \mathfrak{X}_m^{\cup}(\varphi) = \bigcup_{X \in \mathfrak{X}_m(\varphi)} X$. For $k \in [1, s]$, let $z_1^{(k)}, \ldots z_{t_k}^{(k)} \in Z_k$ be the distinct elements of Z_k. Let $\mathfrak{X} = \bigcup_{k=1}^{s} \bigcup_{i=1}^{t_k} \mathfrak{X}_{k,i}$, with $\mathfrak{X}_{k,i} \subseteq \mathfrak{X}$ consisting of all $x \in \mathfrak{X}$ with $\varphi(x) = z_i^{(k)}$. Let $\mathfrak{X}_k = \bigcup_{i \in [1, t_k]} \mathfrak{X}_{k,i}$,

$$A = A_\varphi \quad \text{and} \quad A = A_{\mathfrak{X}} \cdot A_{G_0 \setminus \mathfrak{X}},$$

where $A_{\mathfrak{X}} \mid A$ is the rational subsequence consisting of all terms from $\mathfrak{X}$, and where $A_{G_0 \setminus \mathfrak{X}} \mid A$ is the rational subsequence consisting of all terms from $G_0 \setminus \mathfrak{X}$. For $k \in [1, s]$ and $i \in [1, t_k]$, let $A_{\mathfrak{X}_{k,i}} \mid A_{\mathfrak{X}}$ be the rational subsequence consisting of all terms x with $\varphi(x) = z_i^{(k)}$, i.e., all terms $x \in \mathfrak{X}_{k,i}$, and let $A_{\mathfrak{X}_k}$ be the rational subsequence consisting of all terms from $\mathfrak{X}_k$. By Theorem 8.6(d), we have $\sigma(A_{G_0 \setminus \mathfrak{X}}) \in -\mathsf{C}(Z_\varphi)$, whence

$$\sigma(A_{\mathfrak{X}}) \in \mathsf{C}(Z_\varphi) = \mathsf{C}\big(\{z_i^{(k)} : k \in [1, s],\ i \in [1, t_k]\}\big) \tag{8.41}$$

follows in view of $A = A_\varphi \in \mathcal{B}_{\text{rat}}(G_0)$ being zero-sum.

Given any $x \in \mathbb{R}\langle Z_\varphi \rangle$, which includes any $x \in \mathfrak{X}$, we have

$$x = \sum_{k=1}^{s} \sum_{i=1}^{t_k} \xi_{k,i}(x) z_i^{(k)} \quad \text{for some uniquely defined } \xi_{k,i}(x) \in \mathbb{R}, \tag{8.42}$$

as Z_φ is linearly independent. If $x \in \mathfrak{X}$, then Lemma 8.11 ensures that $\xi_{k,i} \in \mathbb{Z}$ for all k and i. Moreover, if $x \in \mathfrak{X}_{\kappa,\iota}$, then $\xi_{k,i}(x) = 0$ whenever $k \geq \kappa$, apart from the value $\xi_{\kappa,\iota}(x) = 1$, and $\xi_{k,i}(x) \leq 0$ whenever $k < \kappa$. Consequently, $\xi_{k,i}(x) \leq 0$

for all $x \in \mathrm{Supp}(A_{\mathfrak{X}})$, apart from those terms $x \in \mathfrak{X}_{k,i}$, for which we instead have $\xi_{k,i}(x) = 1$.

We proceed to iteratively define rational subsequences $D_r \mid A_{\mathfrak{X}}$, for $r = 0, 1, \ldots, s-1$, such that $D_0 \mid D_1 \mid \ldots \mid D_{s-1}$ and

(a) $A_{\mathfrak{X}} \cdot D_r^{[-1]} \in \mathcal{F}(\mathfrak{X})$,
(b) $\xi_{k,i}(\sigma(A_{\mathfrak{X}} \cdot D_r^{[-1]})) \in \mathbb{Z}_+$ for all $k \in [1, r]$ and $i \in [1, t_k]$,
(c) $D_r^{(j)} = D_k^{(j)}$ for all $k \in [0, r-1]$ and $j \in [1, k+1]$,
(d) $|D_r| \le N_S + (|D_0^{(1)}| + |D_1^{(2)}| + \ldots + |D_{r-1}^{(r)}|)$, and
(e) $|D_{k-1}^{(k)}| \le N_S$ for $k \in [1, r]$,

where $D_k^{(j)} \mid D_k$ denotes the subsequence of all terms from $\mathfrak{X}_j$, for $k \in [0, s-1]$ and $j \in [1, s]$. Set $D_0 = \{A_{\mathfrak{X}}\}$. Then (a)–(e) hold with $|D_0| = |\{A_{\mathfrak{X}}\}| \le |\mathrm{Supp}(\{A_{\mathfrak{X}}\})| \le N_S$ in view of Theorem 8.6(b) and (2.1). Assume we have already constructed the sequences $D_0, \ldots, D_{r-1}$, with $r \ge 1$. Then we construct D_r as follows.

For $k \in [1, s]$ and $i \in [1, t_k]$, let $D_{r-1}^{(k,i)} \mid D_{r-1}$ denote the rational subsequence of all terms from $\mathfrak{X}_{k,i}$. The following makes implicit use of the comments after (8.42). All terms $x \in \mathrm{Supp}(A_{\mathfrak{X}})$ with depth less than r have $\xi_{r,i}(x) = 0$, while all terms $x \in \mathrm{Supp}(A_{\mathfrak{X}})$ with depth greater than r either have $\xi_{r,i}(x) = 0$ or $\xi_{r,i}(x) \le -1$ (as they must be integer values), for $i \in [1, t_r]$. Thus, for each $i \in [1, t_r]$, we have

$$|D_{r-1}^{(r,i)}| = \xi_{r,i}(\sigma(D_{r-1}^{(r,i)})) \ge \xi_{r,i}(\sigma(D_{r-1})).$$

For $i \in [1, t_r]$, let $B_i \mid A_{\mathfrak{X}} \cdot D_{r-1}^{[-1]}$ be a minimal length sequence $B_i \in \mathcal{F}(G_0)$ such that either $\xi_{r,i}(\sigma(B_i)) \le -\xi_{r,i}(\sigma(D_{r-1}))$ (if such B_i exists) or else let $B_i \mid A_{\mathfrak{X}} \cdot D_{r-1}^{[-1]}$ consist of all terms x with $\xi_{r,i}(x) < 0$. Note, if $\xi_{r,i}(\sigma(D_{r-1})) \le 0$, then B_i is the trivial sequence.

Regardless of which case holds in the definition of B_i, we have $\xi_{r,i}(x) \in \mathbb{Z}$ with $\xi_{r,i}(x) \le -1$ for all $x \in \mathrm{Supp}(B_i)$, ensuring B_i only contains terms with depth at least $r+1$ and that $|B_i| \le |\xi_{r,i}(\sigma(B_i))| = -\xi_{r,i}(\sigma(B_i))$. If the latter case holds in the definition of B_i (but not the former), then $\xi_{r,i}(\sigma(B_i)) > -\xi_{r,i}(\sigma(D_{r-1}))$, implying $|B_i| \le -\xi_{r,i}(\sigma(B_i)) \le \xi_{r,i}(\sigma(D_{r-1})) \le |D_{r-1}^{(r,i)}|$. If the former case holds in the definition of B_i and $\xi_{r,i}(\sigma(D_{r-1})) \ge 0$, then, we have $|B_i| \le |\xi_{r,i}(\sigma(D_{r-1}))| = \xi_{r,i}(\sigma(D_{r-1})) \le |D_{r-1}^{(r,i)}|$ in view of $\xi_{r,i}(x) \le -1$ for all $x \in \mathrm{Supp}(B_i)$ and the minimality of $|B_i|$. Finally, if the former case holds in the definition of B_i and $\xi_{r,i}(\sigma(D_{r-1})) \le 0$, then B_i is trivial, whence $|B_i| \le |D_{r-1}^{(r,i)}|$ holds trivially. Consequently, in all cases, $|B_i| \le |D_{r-1}^{(r,i)}|$ for $i \in [1, t_r]$. Thus, letting $B = \mathrm{lcm}(B_1, \ldots, B_{t_r}) \in \mathcal{F}(G_0)$ (which is the smallest subsequence of $A_{\mathfrak{X}} \cdot D_{r-1}^{[-1]}$ containing each B_i as a subsequence), we have

$$|B| \le |B_1| + \ldots + |B_{t_r}| \le \sum_{i=1}^{t_r} |D_{r-1}^{(r,i)}| = |D_{r-1}^{(r)}|. \tag{8.43}$$

Set $D_r = D_{r-1} \cdot B$. Since $B \in \mathcal{F}(G_0)$ and $A_{\mathfrak{X}} \cdot D_{r-1}^{[-1]} \in \mathcal{F}(\mathfrak{X})$ (in view of (a) for D_{r-1}), it follows that $A_{\mathfrak{X}} \cdot D_r^{[-1]} \in \mathcal{F}(\mathfrak{X})$, whence (a) holds for D_r. For $i \in [1, t_r]$, we either have $\xi_{r,i}(\sigma(D_r)) = \xi_{r,i}(\sigma(D_{r-1}))+\xi_{r,i}(\sigma(B)) \le \xi_{r,i}(\sigma(D_{r-1}))+\xi_{r,i}(\sigma(B_i)) \le 0$ (with the first inequality in view of all terms $x \in \mathrm{Supp}(B)$ having depth at least $r+1$, which ensures $\xi_{r,i}(x) \le 0$), or else B_i contains all terms x with $\xi_{r,i}(x) < 0$, in which case $\xi_{r,i}(\sigma(A_{\mathfrak{X}} \cdot D_r^{[-1]})) \ge 0$. However, we can also obtain this latter inequality in the former case by noting that (8.41) combined with $\xi_{r,i}(\sigma(D_r)) \le 0$ implies

$$\xi_{r,i}(\sigma(A_{\mathfrak{X}} \cdot D_r^{[-1]})) = \xi_{r,i}(\sigma(A_{\mathfrak{X}})) - \xi_{r,i}(\sigma(D_r)) \ge \xi_{r,i}(\sigma(A_{\mathfrak{X}})) \ge 0.$$

By the comments after (8.42), $\xi_{r,i}(x) \in \mathbb{Z}$ for each $x \in \mathrm{Supp}(A_{\mathfrak{X}} \cdot D_r^{[-1]})$, whence

$$\xi_{r,i}(\sigma(A_{\mathfrak{X}} \cdot D_r^{[-1]})) = \sum_{x \in \mathrm{Supp}(A_{\mathfrak{X}} \cdot D_r^{[-1]})} \mathsf{v}_x(A_{\mathfrak{X}} \cdot D_r^{[-1]})\xi_{r,i}(x) \in \mathbb{Z},$$

with the final inclusion since (a) for D_r holds. Thus $\xi_{r,i}(\sigma(A_{\mathfrak{X}} \cdot D_r^{[-1]})) \in \mathbb{Z}_+$ for $i \in [1, t_r]$. For $k < r$ and $i \in [1, t_k]$, since $B \in \mathcal{F}(\mathfrak{X})$ only contains terms with depth at least $r+1 > k$, we have $-\xi_{k,i}(\sigma(B)) \in \mathbb{Z}_+$, which combined with (b) holding for D_{r-1} implies $\xi_{k,i}(\sigma(A_{\mathfrak{X}} \cdot D_r^{[-1]})) = \xi_{k,i}(\sigma(A_{\mathfrak{X}} \cdot D_{r-1}^{[-1]})) - \xi_{k,i}(\sigma(B)) \in \mathbb{Z}_+$. Thus (b) holds for D_r. As B contains only terms with depth at least $r+1$, it follows that $D_r^{(j)} = D_{r-1}^{(j)}$ for all $j \in [1, r]$, whence (c) holds for D_r (as (c) held for D_{r-1}). We have $|D_r| = |D_{r-1}| + |B| \le N_S + (|D_0^{(1)}| + |D_1^{(2)}| + \ldots + |D_{r-2}^{(r-1)}|) + |D_{r-1}^{(r)}|$ in view of (8.43) and (d) for D_{r-1}, whence (d) holds for D_r. Finally,

$$\begin{aligned}
|D_{r-1}^{(r)}| &\le |D_{r-1}| - (|D_{r-1}^{(1)}| + |D_{r-1}^{(2)}| + \ldots + |D_{r-1}^{(r-1)}|) \\
&= |D_{r-1}| - (|D_0^{(1)}| + |D_1^{(2)}| + \ldots + |D_{r-2}^{(r-1)}|) \\
&\le (N_S + |D_0^{(1)}| + \ldots + |D_{r-2}^{(r-1)}|) - (|D_0^{(1)}| + |D_1^{(2)}| + \ldots + |D_{r-2}^{(r-1)}|) = N_S,
\end{aligned}$$

with the first equality in view of (c) and the second inequality in view of (d), both for D_{r-1}. Thus (e) holds for D_r as well, completing the construction, which shows that the D_r exist.

Set $S_\varphi = A_{\mathfrak{X}} \cdot D_{s-1}^{[-1]}$. Since $D_{s-1} \mid A_{\mathfrak{X}}$, it follows in view of (a) that $S_\varphi \in \mathcal{F}(\mathfrak{X})$. In view of (b), we have $\xi_{k,i}(\sigma(S_\varphi)) \in \mathbb{Z}_+$ for all $k \in [1, s-1]$ and $i \in [1, t_k]$. However, since $\xi_{s,i}(x) \in \mathbb{Z}_+$ for all $x \in \mathfrak{X}$ (per the comments after (8.42)), we trivially have $\xi_{s,i}(\sigma(S_\varphi)) \in \mathbb{Z}_+$ for all $i \in [1, t_s]$. Thus $\sigma(S_\varphi) \in \mathsf{C}_{\mathbb{Z}}(Z_\varphi)$. Set $R = U \cdot \left(\prod_{\varphi \in \mathfrak{X}_m(G_0)} S_\varphi\right)^{[-1]}$. Since $U \in \mathcal{F}(G_0)$ and every $S_\varphi \in \mathcal{F}(G_0)$, it follows that $R \in \mathcal{F}(G_0)$. It remains only to bound $|R|$ independent of U. To this end, we can combine (d) and (e) (applied with $r = s-1$) to obtain the estimate $|D_{s-1}| \le sN_S \le dN_S$. We also have $|A_0| \le N_T$ and $|A_{G_0 \setminus \mathfrak{X}}| \le N_T$ by Theorem 8.6(c), while $|\mathfrak{X}_m(G_0)| < \infty$ in view of Proposition 7.33. Thus $|R| \le |A_0| + \sum_\varphi(|A_{G_0 \setminus \mathfrak{X}}| +$

$|D_{s-1}|) \leq N$ with $N = N_T + (N_T + dN_S)(|\mathfrak{T}_m(G_0)| - 1) < \infty$ (with the summation over all nontrivial minimal types φ), which is a finite bound independent of the atom U, completing the proof. □

The tame degree (see [62]) is an invariant of factorization theory whose finiteness implies the finiteness of numerous other factorization invariants. The following theorem shows that having finite elasticities ensures a weaker tameness property holds in $\mathcal{B}(G_0)$, though one which is just sufficiently strong to still deduce the desired finiteness for the other invariants. We let $\mathsf{t}_w(G_0)$ be the minimal integer $N \geq 1$ for which Theorem 8.14 holds, which we call the **weak tame degree** of G_0. The proof of Theorem 8.14 illustrates how Theorem 8.13 can be used to simulate finite support in an argument.

Theorem 8.14 *Let $\Lambda \leq \mathbb{R}^d$ be a full rank lattice, where $d \geq 0$, and let $G_0 \subseteq \Lambda$ be a subset with $\mathsf{C}(G_0) = \mathbb{R}^d$ and $0 \notin \mathsf{C}^*(G_0^\diamond)$. Then there exists an integer $N \geq 1$ such that, given any $U_1, \ldots, U_k, V_1, \ldots, V_\ell \in \mathcal{A}(G_0)$ with*

$$U_1 \cdot \ldots \cdot U_k = V_1 \cdot \ldots \cdot V_\ell,$$

where $k, \ \ell \geq 1$, there exist atoms $W_1, \ldots, W_{\ell'} \in \mathcal{A}(G_0)$, $r \in [1, k]$ and $I \subseteq [1, \ell']$ such that

$$U_1 \cdot \ldots \cdot U_k = W_1 \cdot \ldots \cdot W_{\ell'},$$

$\ell' \geq \ell$ and $U_r \mid \prod^\bullet_{x \in I} W_x$ with $|I| \leq N$.

Proof Since $0 \notin \mathsf{C}^*(G_0^\diamond)$, Theorem 8.9 and Proposition 8.2 imply there is an integer $N_\rho \geq 1$ such that

$$\rho_k(G_0) \leq N_\rho k \quad \text{for all } k \geq 1.$$

Since $0 \notin \mathsf{C}^*(G_0^\diamond)$, it follows from Theorem 7.8 that G_0 is finitary. In view of Proposition 7.33, there are only a finite number of minimal types. Let $\varphi_1, \ldots, \varphi_m \in \mathfrak{T}_m(G_0)$ be the distinct nontrivial minimal types for G_0. For each $j \in [1, m]$, let

$$\mathcal{Z}_{\varphi_j} = \mathcal{Z}_1^{(j)} \cup \ldots \cup \mathcal{Z}_{s_j}^{(j)}$$

be the codomain of φ_j with $s_j \leq d$ and $|\mathcal{Z}_{\varphi_j}| \leq d$.

Consider $U_1, \ldots, U_k, V_1, \ldots, V_\ell \in \mathcal{A}(G_0)$ with

$$S := U_1 \cdot \ldots \cdot U_k = V_1 \cdot \ldots \cdot V_\ell,$$

where $k, \ \ell \geq 1$. By definition of $\rho_k(G_0)$, we have

$$\ell \leq \rho_k(G_0) \leq N_\rho k. \tag{8.44}$$

Let $S = g_1 \cdot \ldots \cdot g_{|S|}$ be an indexing of the terms of S. Let $I_1 \cup \ldots \cup I_k = [1, |S|] = J_1 \cup \ldots \cup J_\ell$ be disjoint partitions such that

$$S(I_i) = U_i \quad \text{and} \quad S(J_j) = V_j \quad \text{for all } i \in [1,k] \text{ and } j \in [1,\ell].$$

Since $0 \notin \mathsf{C}^*(G_0^\diamond)$ and $\mathsf{C}(G_0) = \mathbb{R}^d$, we can apply Theorem 8.13 to each atom V_i for $i \in [1, \ell]$. Let $N_R \geq 0$ be the global bound from Theorem 8.13 (which we can assume is an integer) and let $V_i = R_i \cdot \prod_{j=1}^m S_i^{(j)}$, for $i \in [1, \ell]$, be the resulting factorization given by Theorem 8.13, with $S_i^{(j)} \in \mathcal{F}(G_0)$ corresponding to the minimal type φ_j. Then $\sigma(S_i^{(j)}) \in \mathsf{C}_\mathbb{Z}(\mathbb{Z}_{\varphi_j})$, meaning we can apply Proposition 8.12 to each $S_i^{(j)}$. Let $J_i^{(j)} \subseteq [1, |S|]$ be disjoint subsets such that

$$S(J_i^{(0)}) = R_i \quad \text{and} \quad S(J_i^{(j)}) = S_i^{(j)}, \quad \text{for } i \in [1,\ell] \text{ and } j \in [1,m].$$

Moreover, for each $i \in [1, \ell]$, $j \in [1, m]$ and $n \in [1, s_j]$, let $J_i^{(j,n)} \subseteq J_i^{(j)}$ be the subset of all $x \in J_i^{(j)}$ with $\varphi_j(x) \in Z_n^{(j)}$ at depth n. Then

$$J_i = \bigcup_{j=0}^{m} J_i^{(j)} = J_i^{(0)} \cup \bigcup_{j=1}^{m} \bigcup_{n=1}^{s_j} J_i^{(j,n)} \quad \text{for every } i \in [1,\ell].$$

Let

$$X_0 = \bigcup_{i=1}^{\ell} J_i^{(0)}, \quad \Omega_0 = \{(0, g_x) : x \in X_0\} \quad \text{and}$$

$$\Omega_\diamond = \{(j, z) : j \in [1,m],\ z \in Z_{\varphi_j}\}.$$

Moreover, partition

$$\Omega_\diamond = \Omega_1 \cup \ldots \cup \Omega_d$$

such that Ω_n consists of all $(j, z) \in \Omega_\diamond$ with $z \in Z_n^{(j)}$ at depth n.

Since Theorem 8.13 implies $|J_i^{(0)}| = |R_i| \leq N_R$ for all $i \in [1, \ell]$, and since $|Z_{\varphi_j}| \leq d$ for all $j \in [1, m]$, we have

$$|X_0| \leq \ell N_R \quad \text{and} \quad |\Omega_\diamond| = \sum_{j=1}^{m} |Z_{\varphi_j}| \leq md. \tag{8.45}$$

We view $\Omega := \Omega_0 \cup \Omega_\diamond = \Omega_0 \cup \Omega_1 \cup \ldots \cup \Omega_d$ as the set of *support types* for $S = V_1 \cdot \ldots \cdot V_\ell$. A support type $\tau \in \Omega_n$ is said to be at depth n. Note, if $\tau = (j, z)$ with $j \geq 1$, then the depth of τ equals the depth of $z \in Z_{\varphi_j}$. For each $x \in [1, |S|]$, we have $x \in J_i^{(j)}$ for some unique $i \in [1, \ell]$ and $j \in [0, m]$, allowing us to define $\mathsf{s}(x) = (0, g_x) \in \Omega_0$ when $j = 0$ and $\mathsf{s}(x) = (j, \varphi_j(g_x)) \in \Omega_\diamond$ when $j \geq 1$. For $I \subseteq [1, |S|]$, $\mathsf{s}(I) \in \mathcal{F}(\Omega)$ is a sequence of support types from Ω. We associate the depth of $\mathsf{s}(x)$ (defined above) as the depth of $x \in [1, |S|]$.

Let

$$\alpha = \min\left\{ |X_0 \cap I_i| + \sum_{\tau \in \Omega_\diamond} \left(\frac{\ell \mathsf{v}_\tau(\mathsf{s}(I_i))}{\mathsf{v}_\tau\big(\mathsf{s}([1, |S|])\big)} + 1 \right) : \, i \in [1, k] \right\}.$$

Technically, we exclude terms $\tau \in \Omega_\diamond$ in the sum defining α with $\mathsf{v}_\tau\big(\mathsf{s}([1, |S|])\big) = 0$. Then

$$\begin{aligned} k\alpha &\leq \sum_{i=1}^{k} |X_0 \cap I_i| + \sum_{i=1}^{k} \sum_{\tau \in \Omega_\diamond} \frac{\ell \mathsf{v}_\tau(\mathsf{s}(I_i))}{\mathsf{v}_\tau\big(\mathsf{s}([1, |S|])\big)} + k|\Omega_\diamond| \\ &= |X_0| + \sum_{\tau \in \Omega_\diamond} \sum_{i=1}^{k} \frac{\ell \mathsf{v}_\tau(\mathsf{s}(I_i))}{\mathsf{v}_\tau\big(\mathsf{s}([1, |S|])\big)} + k|\Omega_\diamond| \leq |X_0| + (\ell + k)|\Omega_\diamond| \\ &\leq \ell(N_R + md) + kmd \leq kN_\rho(N_R + md) + kmd, \end{aligned} \tag{8.46}$$

with the first inequality in (8.46) in view of (8.45), and the second in view of (8.44). Thus

$$\alpha \leq N := N_\rho(N_R + md) + md,$$

which is a global bound independent of the U_i and V_j.

Let $r \in [1, k]$ be an index attaining the minimum in the definition of α. Then $|X_0 \cap I_r| \leq \alpha \leq N$, ensuring that there is some subset $\mathcal{I}_0 \subseteq [1, \ell]$ with

$$X_0 \cap I_r \subseteq \bigcup_{i \in \mathcal{I}_0} J_i^{(0)} \quad \text{and} \quad |\mathcal{I}_0| \leq |X_0 \cap I_r| \leq N. \tag{8.47}$$

Likewise, letting

$$n_\tau = \left\lceil \frac{\mathsf{v}_\tau(\mathsf{s}(I_r))}{\mathsf{v}_\tau\big(\mathsf{s}([1, |S|])\big)/\ell} \right\rceil < \frac{\ell \mathsf{v}_\tau(\mathsf{s}(I_r))}{\mathsf{v}_\tau\big(\mathsf{s}([1, |S|])\big)} + 1 \leq \ell + 1 \quad \text{for } \tau \in \Omega_\diamond,$$

we have

$$\sum_{\tau \in \Omega_\diamond} n_\tau \leq \alpha - |X_0 \cap I_r| \leq N - |\mathcal{I}_0|. \tag{8.48}$$

We interpret $n_\tau = 0$ when $\mathsf{v}_\tau\big(\mathsf{s}([1, |S|])\big) = 0$.

We now describe how the $W_1, \ldots, W_{\ell'} \in \mathcal{B}(G_0)$ can be constructed. The idea is as follows. An index set $I \subseteq [1, |S|]$ indexes a sequence $S(I) \mid S$, but it also indexes a sequence $\mathsf{s}(I) \in \mathcal{F}(\Omega)$, obtained by replacing each indexed term in the sequence $S(I)$ with its corresponding support type from Ω, so $\mathsf{s}(I) = \prod^\bullet_{x \in I} \mathsf{s}(x)$. When $I \subseteq [1, |S|] \setminus X_0$, we have $\mathsf{s}(I) \in \mathcal{F}(\Omega_\diamond)$ with $\Omega_\diamond$ a fixed, finite set independent of S. Let $\tau \in \Omega_\diamond$. If we select a subset $\mathcal{I}_\tau \subseteq [1, \ell]$ with $|\mathcal{I}_\tau| = n_\tau$ such that the $\mathsf{v}_\tau(\mathsf{s}(J_x))$, for $x \in \mathcal{I}_\tau$, are the n_τ largest values occurring over all $\mathsf{v}_\tau(\mathsf{s}(J_i))$ with $i \in [1, \ell]$, then the definition of n_τ ensures that

$$\mathsf{v}_\tau\Big(\mathsf{s}\Big(\bigcup_{z \in \mathcal{I}_\tau} J_z\Big)\Big) \geq n_\tau\Big(\mathsf{v}_\tau\big(\mathsf{s}([1, |S|])\big)/\ell\Big) \geq \mathsf{v}_\tau(\mathsf{s}(I_r)),$$

with the first inequality holding since the sum of the n_τ largest terms in a sum of ℓ non-negative terms is always at least n_τ times the average value of all terms being summed. As a result, $\mathsf{s}(I_r) \mid \prod^\bullet_{z \in \mathcal{I}} \mathsf{s}(J_z)$, where $\mathcal{I} = \mathcal{I}_0 \cup \bigcup_{\tau \in \Omega_\diamond} \mathcal{I}_\tau$, with $|\mathcal{I}| \leq N$ in view of (8.48). However, since the map s is not injective, this does not guarantee that the associated sequence $U_r = S(I_r)$ is a subsequence of the associated sequence $\prod^\bullet_{z \in \mathcal{I}} V_z = \prod^\bullet_{z \in \mathcal{I}} S(J_z)$. We do, however, have $S(X_0 \cap I_r) \mid \prod^\bullet_{z \in \mathcal{I}_0} S(J_z) = \prod^\bullet_{z \in \mathcal{I}_0} V_z$ in view of (8.47). To deal with the terms from $\Omega_\diamond$, we must use the sequences $S(T_x)$ given by Proposition 8.12 to exchange terms between the V_i.

If there are terms $x \in J_i^{(j)}$ and $y \in J_{i'}^{(j)}$ with $\mathsf{s}(x) = \mathsf{s}(y)$, $i \neq i'$ and $j \geq 1$, say with g_x and g_y at depth n, then Proposition 8.12 implies that there are subsets $T_x \subseteq J_i^{(j)}$ and $T_y \subseteq J_{i'}^{(j)}$ such that $\sigma(S(T_x)) = \varphi_j(g_x) = \varphi_j(g_y) = \sigma(S(T_y))$. If we exchange these sets, defining

$$K_i^{(j)} = (J_i^{(j)} \setminus T_x) \cup T_y \quad \text{and} \quad K_{i'}^{(j)} = (J_{i'}^{(j)} \setminus T_y) \cup T_x,$$

and correspondingly define

$$K_i = (J_i \setminus T_x) \cup T_y \quad \text{and} \quad K_{i'} = (J_{i'} \setminus T_y) \cup T_x,$$

then the new sequences $W_i = S(K_i)$ and $W_{i'} = S(K_{i'})$ (replacing V_i and $V_{i'}$) will still be zero-sum, though we do not guarantee that they remain atoms. However, since $y \in K_i$ and $x \in K_{i'}$, they are non-empty. Consequently, if either W_i or $W_{i'}$ is not an atom, then we can re-factor them to write $V_i \cdot V_{i'} = W_i \cdot W_{i'} = V_1' \cdot \ldots \cdot V_\omega'$ as a product of $\omega \geq 3$ atoms. This leads to a factorization $U_1 \cdot \ldots \cdot U_k = V_1 \cdot \ldots \cdot V_\ell \cdot V_i^{[-1]} \cdot V_{i'}^{[-1]} \cdot V_1' \cdot \ldots \cdot V_\omega'$ into $\ell' = \ell - 2 + \omega > \ell$ atoms. In this case, we begin from scratch using this factorization in place of the original one $U_1 \cdot \ldots \cdot U_k = V_1 \cdot \ldots \cdot V_\ell$. As $\ell' \leq |S| < \infty$, we cannot start from scratch endlessly, meaning eventually we will never encounter this problem, allowing us to w.l.o.g. assume $W_i = S(K_i)$ and $W_{i'} = S(K_{i'})$ are always atoms (where $\ell' = \ell$ may have increased in size from the original ℓ given in the hypotheses). Furthermore, we still have $\sigma(S(K_i^{(j)})) = \sigma(S(J_i^{(j)})) \in \mathsf{C}(Z_{\varphi_j})$ and $\sigma(S(K_{i'}^{(j)})) = \sigma(S(J_{i'}^{(j)})) \in \mathsf{C}(Z_{\varphi_j})$, with the inclusions originating from our application of Theorem 8.13 at the start of the proof. Thus Proposition 8.12 can still be applied to $K_i^{(j)}$ and $K_{i'}^{(j)}$ if we later wish to continue with further such swaps between these sets. Proposition 8.12 guarantees that the set T_x contains *no* terms with depth greater than n (the depth of x), and has x as the *unique* $a \in T_x$ having depth equal to n. Likewise for T_y. Thus when swapping the sets T_x and T_y, we leave unaffected all terms in $J_i^{(j)}$ and $J_{i'}^{(j)}$ with depth at least n, apart from the exchanging of x for y. This ensures that terms previously swapped but at a higher or equal depth will remain unaffected by exchanging T_x and T_y. The value of $\mathsf{s}(a)$ does not change whether regarding $a \in [1, |S|]$ with respect to the original factorization $S = V_1 \cdot \ldots \cdot V_\ell$ or to the one obtained after swapping T_x and T_y and replacing V_i and $V_{i'}$ by W_i and $W_{i'}$. In particular, the value of $\mathsf{v}_\tau\big(\mathsf{s}([1, |S|])\big)$ remains unchanged, ensuring that the value of α is unaffected when replacing J_i and $J_{i'}$ by K_i and $K_{i'}$, and that $r \in [1, k]$ remains an index attaining the minimum in the definition of α (note, the numerators in the definition of α depend on the U_i, not the V_i). Swapping the elements x and y in this fashion leaves all elements from X_0, as well as any $J_c^{(b)}$ with $b \neq j$, unaltered, and the sequences W_i and $W_{i'}$ remain nontrivial, as W_i contains g_y, and $W_{i'}$ contains g_x. Since, apart from x and y, only terms with depth less than n are affected by the swap, it follows that the sets $I_{\tau'}$, corresponding to any type $\tau' \in \Omega_\diamond$ with depth at least n, still have the property that they index the $\mathsf{v}_\tau(\mathsf{s}(J_z))$, for $z \in I_{\tau'}$, with the $n_{\tau'}$ largest values occurring over all $\mathsf{v}_{\tau'}(\mathsf{s}(J_i))$ with $i \in [1, \ell]$.

With these observations in mind, we can now describe how the zero-sums V_i must be modified. Begin with any type $\tau \in \Omega_\diamond$ having maximal available depth. Construct the subset I_τ for the current factorization $S = V_1 \cdot \ldots \cdot V_\ell$ as described above. Then $\mathsf{v}_\tau\big(\mathsf{s}(\bigcup_{z\in I_\tau} J_z)\big) \geq \mathsf{v}_\tau(\mathsf{s}(I_r))$. If $\bigcup_{z\in I_\tau} J_z$ contains all elements from I_r having type τ, then nothing need be done, we discard τ from the list of available types from $\Omega_\diamond$, we select the next available type from $\Omega_\diamond$ with maximal depth, and continue once more. On the other hand, if there is some $x \in I_r$ with type τ not contained in $\bigcup_{z\in I_\tau} J_z$, then $\mathsf{v}_\tau\big(\mathsf{s}(\bigcup_{z\in I_\tau} J_z)\big) \geq \mathsf{v}_\tau(\mathsf{s}(I_r))$ ensures that there must be some $y \in \bigcup_{z\in I_\tau} J_z$ having type τ with $y \notin I_r$. In this case, use Proposition 8.12 to define the sequences T_x and T_y, perform the swap of T_x and T_y described above, and redefine our factorization $V_1 \cdot \ldots \cdot V_\ell$ by replacing $J_i^{(j)}$ and $J_{i'}^{(j)}$ by $K_i^{(j)}$ and $K_{i'}^{(j)}$, where $x \in J_i^{(j)}$ and $y \in J_{i'}^{(j)}$, and correspondingly replacing V_i and $V_{i'}$ by W_i and $W_{i'}$. To simplify notation, redefine V_i, $V_{i'}$, $J_i^{(j)}$ and $J_{i'}^{(j)}$ accordingly so as to reflect the new current state that now has $x \in \bigcup_{z\in I_\tau} J_z$ and $y \notin \bigcup_{z\in I_\tau} J_z$. If we now have $\bigcup_{z\in I_\tau} J_z$ containing all elements from I_r having type τ, then nothing need be done, we discard τ from the list of available types in $\Omega_\diamond$ and carry on as before. If this is not the case, we again find a new term $x' \in I_r$ with type τ not contained in $\bigcup_{z\in I_\tau} J_z$, find a new term $y' \notin \bigcup_{z\in I_\tau} J_z$ with type τ and swap x' and y' as before by use of Proposition 8.12. Since the depth of x and x' are the same, we will not swap x back out of $\bigcup_{z\in I_\tau} J_z$ when doing so, nor indeed any other element from $\bigcup_{z\in I_\tau} J_z$ having type τ. Thus, iterating such a procedure, we will eventually obtain that $\bigcup_{z\in I_\tau} J_z$ contains every element of I_r having type τ, in which case we move on to the next available type τ' with maximal available depth. We repeat the same for procedure for τ' as we did for τ (and, later, as we did for any support types previously discarded before selecting τ'). We first construct the subset $\mathcal{I}_{\tau'}$ for the current state for $S = V_1 \cdot \ldots \cdot V_\ell$, and then swap elements into $\bigcup_{z\in\mathcal{I}_{\tau'}} J_z$ until it contains all elements from I_r having type τ'. While doing so, since we always first choose support types with maximal available depth, we are assured that any type τ'' that has already been discarded had depth at least that of τ', and thus no element of type τ'' will be moved when swapping at the later stage for τ', ensuring that prior work cannot be undone. Continue until all support types from $\Omega_\diamond$ have been exhausted. Once the process ends, we now have a new factorization $U_1 \cdot \ldots \cdot U_k = S = W_1 \cdot \ldots \cdot W_\ell$, where W_i reflects the final state of V_i after running the above process, such that $U_r = S(I_r) \mid \prod^\bullet_{z\in\mathcal{I}} W_i$, where $\mathcal{I} = \mathcal{I}_0 \cup \bigcup_{\tau\in\Omega_\diamond} \mathcal{I}_\tau$, with $|\mathcal{I}| \leq |X_0 \cap I_r| + \sum_{\tau\in\Omega_\diamond} n_\tau \leq N$, completing the proof. □

With the aid of Theorem 8.14, it is now possible to establish that both the set of distances $\Delta(G_0)$ and Catenary degree $\mathsf{c}(G_0)$ are finite, and that there can be no arbitrarily large jumps in the elasticities $\rho_k(G_0)$, which in turn implies that the Structure Theorem for Unions holds in $\mathcal{B}(G_0)$.

Theorem 8.15 *Let $\Lambda \leq \mathbb{R}^d$ be a full rank lattice, where $d \geq 0$, and let $G_0 \subseteq \Lambda$ be a subset with $\rho_{d+1}(G_0) < \infty$. Let $N = \max\{2, \mathsf{t}_w(G_0)\}$.*

1. *$\rho_k(G_0) - \rho_{k-1}(G_0) \leq N < \infty$ for all $k \geq 2$.*
2. *$\max \Delta(G_0) \leq \rho_N(G_0) - N < \infty$. In particular, $\Delta(G_0)$ is finite.*
3. *The Structure Theorem for Unions holds in $\mathcal{B}(G_0)$.*
4. *The catenary degree $\mathsf{c}(G_0) \leq \rho_N(G_0) < \infty$ is finite.*

Proof Removing elements from G_0 that are contained in no atom does not affect the quantities $\rho_k(G_0)$ nor $\Delta(G_0)$, $\mathsf{c}(G_0)$, $\mathsf{t}_w(G_0)$ or $\mathcal{B}(G_0)$. Thus we may w.l.o.g. assume every $g \in G_0$ is contained in some atom, in which case Proposition 4.4 implies $\mathsf{C}(G_0) = \mathbb{R}\langle G_0\rangle$ is a subspace, and $\Lambda \cap \mathbb{R}\langle G_0\rangle$ is a full rank lattice in $\mathbb{R}\langle G_0\rangle$. Thus we may w.l.o.g. assume $\mathsf{C}(G_0) = \mathbb{R}^d$. Then, in view of $\rho_{d+1}(G_0) < \infty$ and Theorem 8.9, we have $0 \notin \mathsf{C}^*(G_0^\diamond)$ and can apply Proposition 8.2 to conclude that there is a constant $N_\rho \geq 1$ such that

$$\rho_k(G_0) \leq N_\rho k < \infty \quad \text{for all } k \geq 1.$$

Since $0 \notin \mathsf{C}^*(G_0^\diamond)$ and $\mathsf{C}(G_0) = \mathbb{R}^d$, it follows from Theorem 7.8 that G_0 is finitary.

1. Apply Theorem 8.14 to G_0 and let $N = \mathsf{t}_w(G_0) \in \mathbb{Z}_+$ be the resulting constant. Let $k \geq 2$ and let $U_1, \ldots, U_k, V_1, \ldots, V_\ell \in \mathcal{A}(G_0)$ be atoms with

$$U_1 \cdot \ldots \cdot U_k = V_1 \cdot \ldots \cdot V_\ell \quad \text{and} \quad \ell = \rho_k(G_0).$$

Then Theorem 8.14 implies that there are atoms $W_1, \ldots, W_{\ell'} \in \mathcal{A}(G_0)$, where $\ell' \geq \ell$, and $r \in [1, k]$ and $\mathcal{I} \subseteq [1, \ell']$ such that

$$U_1 \cdot \ldots \cdot U_k = W_1 \cdot \ldots \cdot W_{\ell'}$$

and $U_r \mid \prod^\bullet_{x\in I} W_x$ with $|I| \le N$. Since $\ell = \rho_k(G_0)$, the definition of $\rho_k(G_0)$ ensures $\ell' \le \ell$, whence $\ell' = \ell'$. By re-indexing, we may w.l.o.g assume $r = k$ and $I = [1, m]$ with

$$m \le N.$$

Since $U_k \mid \prod^\bullet_{i\in[1,m]} W_i$, there is a factorization $U_k \cdot W_2' \cdot \ldots \cdot W_{m'} = W_1 \cdot \ldots \cdot W_m$ with $W_i' \in \mathcal{A}(G_0)$ for all $i \in [2, m']$. Note $m' \ge 1$. But now

$$U_1 \cdot \ldots \cdot U_{k-1} = W_2' \cdot \ldots \cdot W_{m'}' \cdot W_{m+1} \cdot \ldots \cdot W_\ell$$

with $W_i', W_j \in \mathcal{A}(G_0)$ for all i and j. Thus

$$\rho_{k-1}(G_0) \ge \ell - m + m' - 1 \ge \ell - m = \rho_k(G_0) - m \ge \rho_k(G_0) - N,$$

implying $\rho_k(G_0) - \rho_{k-1}(G_0) \le N = \mathsf{t}_w(G_0)$, as desired.

2. To show $\Delta(G_0)$ is finite, it suffices to show $\max \Delta(G_0)$ is finite. Apply Theorem 8.14 to G_0 and let $N \ge 2$ be the resulting constant (if $N = 1$, replace it with $N = 2$). We will show that

$$\max \Delta(G_0) \le \rho_N(G_0) - N. \tag{8.49}$$

For $S \in \mathcal{B}(G_0)$, let $\mathsf{L}(S) \subseteq [1, |S|]$ be the set of lengths for S, which consists of all $k \in [1, |S|]$ for which there exists a factorization $S = U_1 \cdot \ldots \cdot U_k$ with $U_i \in \mathcal{A}(G_0)$ for all i. Then $\delta \in \Delta(G_0)$ means there is some $S \in \mathcal{B}(G_0)$ with $\delta = \ell - k$ for two consecutive elements $k, \ell \in \mathsf{L}(S)$ with $k < \ell$. In other words, there must exist some $S \in \mathcal{B}(G_0)$ and atoms $U_1, \ldots, U_k, V_1, \ldots, V_\ell \in \mathcal{A}(G_0)$ such that

$$U_1 \cdot \ldots \cdot U_k = S = V_1 \cdot \ldots \cdot V_\ell$$

with $k < \ell$ and $\ell - k = \delta$, and there cannot exist a factorization $W_1 \cdot \ldots \cdot W_r = S$ with $W_i \in \mathcal{A}(G_0)$ and $k < r < \ell$. Note $\Delta(G_0) = \bigcup_{S\in\mathcal{B}(G_0)} \Delta(\mathsf{L}(S))$, where $\Delta(\mathsf{L}(S))$ consists of all δ for which there exist consecutive elements $k, \ell \in \mathsf{L}(S)$ with $k < \ell$ and $\ell - k = \delta$.

If (8.49) fails, then there must be some $S \in \mathcal{B}(G_0)$ and $k, \ell \in \mathsf{L}(S)$ with $k < \ell$ consecutive elements of $\mathsf{L}(S)$ and

$$\ell - k \ge \rho_N(G_0) - (N-1) \ge \rho_n(G_0) - (n-1) \quad \text{for all } n \in [1, N], \tag{8.50}$$

where the second inequality in (8.50) follows in view of iterated application of the basic inequality $\rho_{k+1}(G_0) > \rho_k(G_0)$ (see (2.9) in Sect. 2.4). Choose such a counter example with ℓ minimal. Note $k \ge 2$, else $\ell = k$, contradicting that $\ell - k \ge \rho_N(G_0) - N + 1 \ge 1$ by (8.50). Thus (8.50) ensures $\ell \ge \rho_N(G_0) -$

$N + 1 + k \geq k + 1 \geq 3$. Let

$$U_1 \cdot \ldots \cdot U_k = S = V_1 \cdot \ldots \cdot V_\ell \tag{8.51}$$

be factorizations exhibiting that $k, \ell \in \mathsf{L}(S)$, where $U_1, \ldots, U_k, V_1, \ldots, V_\ell \in \mathcal{A}(G_0)$. Apply Theorem 8.14 to the factorization given in (8.51) with the roles of the U_i and the V_j swapped. Then there are atoms $W_1, \ldots, W_{k'} \in \mathcal{A}(G_0)$, where $k' \geq k$, and $r \in [1, \ell]$ and $I \subseteq [1, k']$ such that

$$W_1 \cdot \ldots \cdot W_{k'} = V_1 \cdot \ldots \cdot V_\ell$$

with $V_r \mid \prod^\bullet_{x \in I} W_x$ and

$$m := |I| \leq N. \tag{8.52}$$

By re-indexing, we may w.l.o.g. assume $r = \ell$ and $I = [1, m]$.

The algorithm which constructs the sequences W_i given in Theorem 8.14 proceeds by successively taking two zero-sums U_i and $U_{i'}$ in the factorization $U_1 \cdot \ldots \cdot U_k$ and replacing them with a re-factorization of $U_i \cdot U_{i'} = W_i \cdot W_{i'}$ into $\omega \geq 2$ atoms. We begin will all $U_j \in \mathcal{A}(G_0)$ being atoms. If, at some point during the process of constructing the W_i, we take two atoms and find that their replacement zero-sum sequences W_i and $W_{i'}$ are not both themselves atoms, that is, $\omega \geq 3$, then the first time that this occurs, we can re-factor S by replacing these two zero-sums with a factorization of length $\omega \in [3, \rho_2(G_0)]$, to thereby find that S has a factorization of length $k - 2 + \omega$ with $k < k - 2 + \omega \leq k - 2 + \rho_2(G_0) < \ell$, where the last inequality follows from (8.50) and $N \geq 2$. However, this contradicts that $k < \ell$ are *consecutive* elements of $\mathsf{L}(S)$. Therefore, we instead conclude that we never need to re-factor the sequences $W_i \cdot W_{i'}$ when applying the algorithm for Theorem 8.14, whence $k' = k$ follows.

Since $V_\ell \mid W_1 \cdot \ldots \cdot W_m$, we have a factorization $V_\ell \cdot W'_2 \cdot \ldots \cdot W'_{m'} = W_1 \cdot \ldots \cdot W_m$ with $W'_i \in \mathcal{A}(G_0)$ for all $i \in [2, m']$. Note

$$1 \leq m' \leq \rho_m(G_0). \tag{8.53}$$

But now

$$W'_2 \cdot \ldots \cdot W'_{m'} \cdot W_{m+1} \cdot \ldots \cdot W_k = S \cdot V_\ell^{[-1]} = V_1 \cdot \ldots \cdot V_{\ell-1},$$

showing that

$$k' - 1,\ \ell - 1 \in \mathsf{L}(S'),$$

where

$$S' = S \cdot V_\ell^{[-1]} \in \mathcal{B}(G_0) \quad \text{and} \quad k' = k - m + m'.$$

Observe that any factorization of length t for S' gives a factorization of S of length $t+1$ by concatenating the atom V_ℓ onto the end of the factorization of S'. Consequently, since $k < \ell$ are consecutive elements of $\mathsf{L}(S)$, it follows that

$$[k, \ell-2] \cap \mathsf{L}(S') = \emptyset. \tag{8.54}$$

In view of (8.53), (8.52), and (8.50), we have $k' = k-m+m' \leq k+\rho_m(G_0)-m \leq \ell-1$. Thus, since $k'-1 \in \mathsf{L}(S')$, we conclude from (8.54) that $k'-1 \leq k-1$. Hence, in view of (8.54) again, let $r \in [k'-1, k-1]$ be the maximal element of $\mathsf{L}(S')$ less than $\ell-1 \in \mathsf{L}(S')$. Since r, $\ell-1 \in \mathsf{L}(S')$ with $r \leq k-1 < \ell-1$, the minimality of ℓ ensures that

$$\rho_N(G_0) - N \geq (\ell-1) - r \geq (\ell-1) - (k-1) = \ell - k \geq \rho_N(G_0) - N + 1,$$

with the final inequality in view of (8.50), which is a contradiction. Thus (8.49) is now established, and since $\rho_N(G_0) < \infty$, we conclude that $\max \Delta(G_0)$, and thus also $\Delta(G_0)$ itself, are both finite, completing Item 2.

3. This follows from Items 1 and 2 and [54, Theorem 4.2].
4. Let $N \geq 2$ be a constant given by Theorem 8.14 applied to G_0. We will show that

$$\mathsf{c}(G_0) \leq \rho_N(G_0) < \infty.$$

Let $U_1, \ldots, U_k, V_1, \ldots, V_\ell \in \mathcal{A}(G_0)$ be arbitrary atoms with

$$U_1 \boldsymbol{\cdot} \ldots \boldsymbol{\cdot} U_k = V_1 \boldsymbol{\cdot} \ldots \boldsymbol{\cdot} V_\ell.$$

By re-indexing the U_i and V_i, we may collect together all the atoms which occur in both factorizations, say let $I \subseteq [1,k] \cap [1,\ell]$ consist of all i such that $U_i = V_i$. Apply Theorem 8.14 to the factorization $\prod^\bullet_{i\in[1,k]\setminus I} U_i = \prod^\bullet_{i\in[1,\ell]\setminus I} V_i$. Then there exists a factorization $\prod^\bullet_{i\in[1,\ell']\setminus I} W_i = \prod^\bullet_{i\in[1,k]\setminus I} U_i$ with $\ell' \geq \ell$ and $W_i \in \mathcal{A}(G_0)$ satisfying the conclusion of Theorem 8.14. Moreover the algorithm from the proof of Theorem 8.14 constructs the W_i by sequentially modifying pairs of atoms W_i and $W_{i'}$ by replacing such pairs with a re-factorization into $\omega \in [2, \rho_2(G_0)]$ new atoms. Thus each successive factorization in the algorithm constructing the W_i differs from the prior one by at most $\rho_2(G_0)$ factors. Theorem 8.14 guarantees that there is some $r \in [1,k] \setminus I$ with U_r a subsequence of a product of at most N of the atoms W_i, say $U_r \mid \prod^\bullet_{i\in J} W_i$ with $|J| \leq N$ and $J \subseteq [1,\ell'] \setminus I$. We may re-factor $\prod^\bullet_{i\in J} W_i$ into a product of at most $\rho_N(G_0)$ atoms that includes the atom U_r. We thus obtain a factorization $U_1 \boldsymbol{\cdot} \ldots \boldsymbol{\cdot} U_k = V'_1 \boldsymbol{\cdot} \ldots \boldsymbol{\cdot} V'_{\ell''}$ that now has one more shared factor among the U_i and V'_i than the original one, and which can be constructed sequentially with each new factorization differing from the prior one by at most $\rho_N(G_0) \geq \rho_2(G_0) \geq 2$ factors. Iterating this process at most k times thus transforms the factorization

$V_1 \cdot \ldots \cdot V_\ell$ into the factorization $U_1 \cdot \ldots \cdot U_k$ with each successive factorization differing by at most $\rho_N(G_0)$ factors from the previous one, yielding the desired bound for the catenary degree.

□

Summary

We can now summarize our results regarding what we have shown regarding $\mathcal{B}(G_0)$ under an assumption of finite elasticities. Let $\Lambda \leq \mathbb{R}^d$ be a full rank lattice, where $d \geq 0$, and let $G_0 \subseteq \Lambda$ be a subset. There is little loss of generality to assume every element $g \in G_0$ occurs in some atom (else we can pass to the subset of G_0 having this property). Then Proposition 4.4 implies that $\mathsf{C}(G_0) = \mathbb{R}\langle G_0\rangle$ with $\Lambda \cap \mathbb{R}\langle G_0\rangle \leq \mathbb{R}\langle G_0\rangle$ a full rank lattice. Hence, replacing $\mathbb{R}^d$ with $\mathbb{R}\langle G_0\rangle$, we can w.l.o.g. assume $\mathsf{C}(G_0) = \mathbb{R}^d$, which is simply a normalization hypothesis to avoid trivial degeneracies. Under these assumptions, we now summarize some of the key results.

1. There exists a minimal $s \in [1, d+1]$ such that $\rho_s(G_0) < \infty$ implies $\rho_k(G_0) < \infty$ for all $k \geq 1$.

Item 1 follows in view of Theorem 8.9, which also shows that

$$\rho_{d+1}(G_0) < \infty \text{ is equivalent to } \rho(G_0) < \infty.$$

We remark that it would be interesting to know whether the estimate $s \leq d+1$ is tight or can be improved; we have focussed primarily on its existence.

Corollary 8.10 ensures that $\rho_{d+1}^{\mathsf{elm}}(G_0) < \infty$ implies $\rho_{d+1}(G_0) < \infty$, meaning it is sufficient to know no product of $d+1$ elementary atoms can be re-factored into an arbitrarily large number of atoms. Theorem 8.9 characterizes when Item 1 occurs either in terms of a basic combinatorial property of the atoms $\mathcal{A}(G_0)$, or the geometric property $0 \notin \mathsf{C}^*(G_0^\diamond)$, which involves $\mathbb{R}_+$-linear combinations of elements of G_0 rather than $\mathbb{Z}_+$-linear combinations. Assuming additionally that $\rho_{d+1}(G_0) < \infty$, so that the conclusion of Item 1 holds, we obtain the following properties for $\mathcal{B}(G_0)$ and subset $G_0 \subseteq \Lambda$.

2. $G_0 \subseteq \mathbb{R}^d$ is finitary (by Theorem 7.8). In particular, all the results of Sect. 7 are available for studying the set G_0, including Theorems 7.16, 7.24, 7.29 and 7.32.
3. Besides the three equivalent defining definitions of the subset $G_0^\diamond \subseteq G_0$ (given in Proposition 7.2), we also have (by Corollary 8.8)

$$\begin{aligned} G_0^\diamond &= \left\{g \in G_0 : \ \sup\{\mathsf{v}_g(U) : \ U \in \mathcal{A}^{\mathsf{elm}}(G_0)\right\} = \infty\} \\ &= \left\{g \in G_0 : \ \sup\{\mathsf{v}_g(U) : \ U \in \mathcal{A}(G_0)\} = \infty\right\}. \end{aligned}$$

4. There is a finite subset $X \subseteq G_0^\diamond$ such that $\mathcal{A}(G_0 \setminus X)$ is finite, and thus also a finite subset $Y \subseteq G_0$ such that $\mathcal{A}(G_0 \setminus Y) = \emptyset$ (by Proposition 7.31).
5. $G_0 \setminus G_0^\diamond \subseteq G_0$ is a Lambert subset, indeed, the unique maximal Lambert subset. (Corollaries 8.7 and 8.8)
6. The Weak Tame Degree (as defined in Theorem 8.14) is finite: $\mathsf{t}_w(G_0) < \infty$, and we set $N := \max\{2, \mathsf{t}_w(G_0)\}$.
7. The elasticities of G_0 do not contain arbitrarily large gaps: $\rho_k(G_0) - \rho_{k-1}(G_0) \le N < \infty$ for all $k \ge 2$. In particular, $\rho_k(G_0) \le Nk$ grows linearly. (Theorem 8.15.1)
8. $\max \Delta(G_0) \le \rho_N(G_0) - N < \infty$ is finite. (Theorem 8.15.2)
9. The Set of Distances is finite: $|\Delta(G_0)| < \infty$. (Theorem 8.15.2)
10. The Catenary Degree is finite: $\mathsf{c}(G_0) \le \rho_N(G_0) < \infty$. (Theorem 8.15.4)
11. The Structure Theorem for Unions holds for $\mathcal{B}(G_0)$. (Theorem 8.15.3)
12. A weak structure theorem holds for the atoms, effectively allowing simulation of globally bounded finite support for the atoms $U \in \mathcal{A}(G_0)$. (Theorem 8.13)

8.3 Transfer Krull Monoids Over Subsets of Finitely Generated Abelian Groups

In this final section, we show how our prior results regarding subsets $G_0 \subseteq \Lambda \le \mathbb{R}^d$ can be extended to cover (Transfer) Krull Monoids H over a subset $G_0 \subseteq G$, where $G = \mathbb{Z}^d \oplus G_T$ is a finitely generated abelian group with torsion-free rank $d \ge 0$ and torsion subgroup $G_T \le G$.

When $G = G_T$ is a finite group, all factorization invariants we have encountered are trivially finite. It is natural to suppose their finiteness, for subset of a more general finitely generated abelian group G, is thus principally affected by the torsion-free portion of G, which naturally embeds as a lattice $\mathbb{Z}^d \le \mathbb{R}^d$. To this end, we have the following basic relation between $\mathcal{A}(G_0)$ and $\mathcal{A}(\pi(G_0))$, where $\pi : G \to \mathbb{Z}^d$ is the natural projection homomorphism with kernel G_T.

Proposition 8.16 *Let $G = \mathbb{Z}^d \oplus G_T$ be a finitely generated abelian group with torsion subgroup $G_T \le G$, where $d \ge 0$, let $\pi : G \to \mathbb{Z}^d$ be the projection homomorphism with $\ker \pi = G_T$, let $\mathsf{D}(G_T)$ be the Davenport constant for G_T, let $m = \exp(G_T)$, and let $G_0 \subseteq G$.*

1. *If $U \in \mathcal{A}(G_0)$ is an atom and $\pi(U) = W_1 \cdot \ldots \cdot W_\ell$ is a factorization of $\pi(U)$ with $W_i \in \mathcal{A}(\pi(G_0))$ for all $i \in [1, \ell]$, then $\ell \le \mathsf{D}(G_T)$.*
2. *If $U \in \mathcal{F}(G_0)$ with $\pi(U) \in \mathcal{A}^{\mathsf{elm}}(\pi(G_0))$, then $U^{[m]} \in \mathcal{B}(G_0)$. Moreover, if $U^{[m]} = V_1 \cdot \ldots \cdot V_\ell$ is a factorization with $V_i \in \mathcal{A}(G_0)$ for all $i \in [1, \ell]$, then $\ell \le m = \exp(G_T)$.*

Proof

1. Factor $U = V_1 \cdot \ldots \cdot V_\ell$, with the $V_i \in \mathcal{F}(G_0)$ such that $\pi(V_i) = W_i$ for all $i \in [1, \ell]$. Since $U \in \mathcal{A}(G_0)$ is an atom, $\pi(U) \in \mathcal{B}(\pi(G_0))$ is a zero-sum sequence. Since $\pi(V_i) = W_i \in \mathcal{A}(\pi(G_0))$, we have $\sigma(V_i) \in \ker \pi = G_T$ for each $i \in [1, \ell]$. If $\ell > \mathsf{D}(G_T)$, then applying the definition of $\mathsf{D}(G_T)$ to the sequence $\prod^\bullet_{i\in[1,\ell]} \sigma(V_i) \in \mathcal{F}(G_T)$, we can find some nontrivial, proper subset $I \subseteq [1, \ell]$ such that $\prod^\bullet_{i\in I} V_i \in \mathcal{F}(G_0)$ is a nontrivial, proper zero-sum subsequence of U, contradicting that $U \in \mathcal{A}(G_0)$ is an atom, which completes Item 1.
2. Since $\pi(U) \in \mathcal{A}^{\mathsf{elm}}(G_0)$ is an elementary atom, we have $\sigma(U) \in \ker \pi = G_T$, whence $\sigma(U^{[m]}) = m\sigma(U) = \exp(G_T)\sigma(U) = 0$, ensuring $U^{[m]} \in \mathcal{B}(G_0)$. If $\pi(U)$ is the subsequence consisting a single term equal to 0, then $|U^{[m]}| = m$, in which case $\ell \le m$ is trivial. Therefore we may instead assume $0 \notin \mathsf{Supp}(\pi(U))$. As a result, since $\pi(U) \in \mathcal{A}^{\mathsf{elm}}(G_0)$ is an elementary atom, it follows by Proposition 4.8 that $X = \mathsf{Supp}(\pi(U))$ is a minimal positive basis, and $\pi(U)$ is the unique atom whose support is contained in X. Since each $V_i \in \mathcal{A}(G_0)$ with $V_i \mid U^{[m]}$, it follows that $\pi(V_i) \in \mathcal{B}(X)$. However, as already remarked, the unique atom with support contained in X is $\pi(U)$, ensuring that each $\pi(V_i) = \pi(U)^{[m_i]}$ for some $m_i \ge 1$. But now

$$\pi(U)^{[m]} = \pi(U^{[m]}) = \pi(V_1) \cdot \ldots \cdot \pi(V_\ell) = \pi(U)^{[m_1]} \cdot \ldots \cdot \pi(U)^{[m_\ell]},$$

implying $\ell \le \sum_{i=1}^{\ell} m_i = m$, with the inequality since $m_i \ge 1$ for all i, completing the proof.

□

The next proposition gives a correspondence between finite elasticities in $G_0 \subseteq G$ and finite elasticities in $\pi(G_0) \subseteq \mathbb{Z}^d$. In particular, Theorem 8.9 also characterizes when $\rho_{(d+1)m}(G_0) < \infty$ is finite by applying it to the set $\pi(G_0) \subseteq \mathbb{Z}^d$.

Proposition 8.17 *Let $G = \mathbb{Z}^d \oplus G_T$ be a finitely generated abelian group with torsion subgroup $G_T \le G$, where $d \ge 0$, let $\pi : G \to \mathbb{Z}^d$ be the projection homomorphism with $\ker \pi = G_T$, let $\mathsf{D}(G_T)$ be the Davenport constant for G_T, let $m = \exp(G_T)$, and let $G_0 \subseteq G$.*

1. *If $\rho_{d+1}(\pi(G_0)) < \infty$, then $\rho_k(G_0) \le \rho_{k\mathsf{D}(G_T)}(\pi(G_0)) < \infty$ for all $k \ge 1$.*
2. *If $\rho_{(d+1)m}(G_0) < \infty$, then $\rho_k(\pi(G_0)) < \infty$ for all $k \ge 1$.*

In particular, $\rho_k(G_0) < \infty$ for all $k \ge 1$ if and only if $\rho_k(\pi(G_0)) < \infty$ for all $k \ge 1$. Moreover, if $\rho_{(d+1)m}(G_0) < \infty$, then $\rho_k(G_0) < \infty$ for all $k \ge 1$.

Proof We may w.l.o.g. discard terms from G_0 contained in no atom, as such terms have no bearing on $\rho_k(G_0)$ nor $\rho_k(\pi(G_0))$ (as G_T has finite exponent), and thereby assume every $g \in G_0$ is contained in some atom. We may also w.l.o.g. assume $\langle G_0 \rangle = G$ in view of (2.9), and may embed $\mathbb{Z}^d \le \mathbb{R}^d$, which is a full rank lattice in $\mathbb{R}^d$. Since $\langle G_0 \rangle = G$, we have $\mathbb{R}\langle \pi(G_0) \rangle = \mathbb{R}^d$. Since every $g \in G_0$ is contained in some atom, it follows that every $\pi(g) \in \pi(G_0)$ is also contained in some atom, whence Proposition 4.4 implies $\mathsf{C}(\pi(G_0)) = \mathbb{R}\langle\pi(G_0)\rangle = \mathbb{R}^d$ with $\pi(G_0) \subseteq \mathbb{Z}^d$.

1. Suppose $\rho_{d+1}(\pi(G_0)) < \infty$. Then Theorem 8.9 implies that

$$\rho_k(\pi(G_0)) < \infty \quad \text{for all } k \geq 1.$$

Consider an arbitrary factorization

$$U_1 \cdot \ldots \cdot U_k = V_1 \cdot \ldots \cdot V_\ell$$

with $U_j,\ V_i \in \mathcal{A}(G_0)$ for all i and j. Re-factor each $\pi(U_j) = \prod^\bullet_{x \in I_j} W_x^{(j)}$ into a product of $|I_j| \leq \mathsf{D}(G_T)$ atoms $W_x^{(j)} \in \mathcal{A}(\pi(G_0))$ via Proposition 8.16.1. Then $\prod^\bullet_{j\in[1,k]} \prod^\bullet_{x\in I_j} W_x^{(j)}$ is a product of at most $k\mathsf{D}(G_T)$ atoms. As a result, since

$$\prod\nolimits^\bullet_{j\in[1,k]} \prod\nolimits^\bullet_{x\in I_j} W_x^{(j)} = \pi(U_1) \cdot \ldots \cdot \pi(U_k) = \pi(V_1) \cdot \ldots \cdot \pi(V_\ell)$$

with each $\pi(V_j) \in \mathcal{B}(\pi(G_0))$ nontrivial, it follows in view of (2.9) that $\ell \leq \rho_{k\mathsf{D}(G_T)}(\pi(G_0)) < \infty$, as desired.

2. Suppose $\rho_{(d+1)m}(G_0) < \infty$ but, by way of contradiction, that there is some $k \geq 1$ such that $\rho_k(\pi(G_0)) = \infty$. If $\rho_{d+1}(\pi(G_0)) < \infty$, then Theorem 8.9 ensures that $\rho_k(\pi(G_0)) < \infty$, contrary to assumption. Therefore we can assume $\rho_{d+1}(\pi(G_0)) = \infty$. Thus Corollary 8.10 implies that $\rho^{\mathsf{elm}}_{d+1}(\pi(G_0)) = \infty$. Consequently, there is a sequence of factorizations

$$\pi(U_{i,1}) \cdot \ldots \cdot \pi(U_{i,d+1}) = \pi(V_{i,1}) \cdot \ldots \cdot \pi(V_{i,\ell_i}) \quad \text{for } i = 1, 2, \ldots$$

such that

$$U_{i,j},\ V_{i,j} \in \mathcal{F}(G_0), \quad \pi(U_{i,j}) \in \mathcal{A}^{\mathsf{elm}}(\pi(G_0)), \quad \text{and}$$
$$\pi(V_{i,j}) \in \mathcal{A}(\pi(G_0)), \quad \text{for all } i \text{ and } j,$$

with

$$U_{i,1} \cdot \ldots \cdot U_{i,d+1} = V_{i,1} \cdot \ldots \cdot V_{i,\ell_i} \quad \text{for every } i \geq 1, \quad \text{and} \quad \ell_i \to \infty. \tag{8.55}$$

Since the $U_{i,j}$ and $V_{i,j}$ are zero-sum modulo G_T with $m = \exp(G_T)$, we have $U_{i,j}^{[m]},\ V_{i,j}^{[m]} \in \mathcal{B}(G_0)$ for all i and j. Re-factor each $U_{i,j}^{[m]} = \prod^\bullet_{x\in I_{i,j}} W_x^{(i,j)}$ into a product of $|I_{i,j}| \leq m$ atoms $W_x^{(i,j)} \in \mathcal{A}(G_0)$ via Proposition 8.16.2. Then each $\prod^\bullet_{j\in[1,d+1]} \prod^\bullet_{x\in I_{i,j}} W_x^{(i,j)}$ is a product of at most $(d+1)m$ atoms from $\mathcal{A}(G_0)$, for every $i \geq 1$. In view of (8.55), we have

$$\prod^{\bullet}_{j\in[1,d+1]}\prod^{\bullet}_{x\in I_{i,j}} W_x^{(i,j)} = U_{i,1}^{[m]}\cdot\ldots\cdot U_{i,d+1}^{[m]} = V_{i,1}^{[m]}\cdot\ldots\cdot V_{i,\ell_i}^{[m]}$$

with $V_{i,j}^{[m]} \in \mathcal{B}(G_0)$ a nontrivial zero-sum sequence for every $i \geq 1$ (since $\pi(V_{i,j}) \in \mathcal{A}(\pi(G_0))$ with $m = \exp(G_T)$). In consequence, since $\ell_i \to \infty$, it follows via (2.9) that $\rho_{(d+1)m}(G_0) = \infty$, contrary to assumption, which completes Item 2.

□

We next extend Proposition 8.2.

Corollary 8.18 *Let H be a Transfer Krull Monoid over a subset G_0 of a finitely generated abelian group $G = \mathbb{Z}^d \oplus G_T$ with torsion subgroup $G_T \leq G$, where $d \geq 0$. If $\rho_{(d+1)\exp(G_T)}(H) < \infty$, then there is a constant $N_\rho \geq 1$ such that $\rho_k(H) \leq N_\rho k < \infty$ for all $k \geq 1$.*

Proof Since $\rho_k(H) = \rho_k(G_0)$ for all $k \geq 1$ by definition of a Transfer Krull Monoid, it suffices to prove the corollary when $H = \mathcal{B}(G_0)$, which we now assume. Let $\pi : G \to \mathbb{Z}^d$ be the projection homomorphism with kernel G_T. In view of (2.9), we may embed $\mathbb{Z}^d \leq \mathbb{R}^d$ and w.l.o.g. assume $\langle G_0\rangle = G$ and that every $g \in G_0$ is contained in some atom, whence $\mathsf{C}(\pi(G_0)) = \mathbb{R}^d$ (as in the proof of Proposition 8.17). Since $\rho_{(d+1)\exp(G_T)}(G_0) < \infty$, Proposition 8.17 implies that $\rho_k(\pi(G_0)) < \infty$ and $\rho_k(G_0) \leq \rho_{k\mathsf{D}(G_T)}(\pi(G_0)) < \infty$ for all $k \geq 1$. As a result, Theorem 8.9 implies that we can apply Proposition 8.2 to conclude there is a constant $N_\rho \geq 1$ (namely $\mathsf{D}(G_T)$ times the constant given by Proposition 8.2) such that $\rho_k(G_0) \leq \rho_{k\mathsf{D}(G_T)}(\pi(G_0)) \leq N_\rho k < \infty$ for all $k \geq 1$, as desired. □

We now extend Theorem 8.14 to show the **weak tame degree** $\mathsf{t}_w(H)$ (the minimal integer $N \geq 1$ such that the conclusion of Main Theorem 8.19 holds) is finite when $\rho_{(d+1)\exp(G_T)}(H) < \infty$. The definition is a variation on the ω constant often used in conjunction with the tame degree [58, 66]. The proof is a variation on that used to prove Theorem 8.14. We have opted to first give a proof of the simpler Theorem 8.14, to first present a version of the algorithm that avoids the delicate (and distracting) technicalities needed to get the algorithm to work in the general setting as stated in Theorem 8.19.

Main Theorem 8.19 *Let H be a Krull Monoid over a subset G_0 of a finitely generated abelian group $G = \mathbb{Z}^d \oplus G_T$ with torsion subgroup $G_T \leq G$, where $d \geq 0$. Suppose $\rho_{(d+1)\exp(G_T)}(H) < \infty$. Then there exists an integer $N \geq 1$ such that, given any atoms $U_1, \ldots, U_k, V_1, \ldots, V_\ell \in \mathcal{A}(H)$ with*

$$U_1\cdot\ldots\cdot U_k = V_1\cdot\ldots\cdot V_\ell,$$

where $k, \ell \geq 1$, then there exist atoms $W_1, \ldots, W_{\ell'} \in \mathcal{A}(H)$, $r \in [1, k]$ and $I \subseteq [1, \ell']$ such that

$$U_1 \cdot \ldots \cdot U_k = W_1 \cdot \ldots \cdot W_{\ell'},$$

$\ell' \geq \ell$ and $U_r \mid \prod_{x \in I} W_x$ with $|I| \leq N$.

Proof Let us first show that the theorem reduces to the case when $H = \mathcal{B}(G_0)$. To this end, assume the theorem holds for $\mathcal{B}(G_0)$ with bound $N \geq 1$. We can w.l.o.g. assume H is reduced, so we replace H by $H_{\text{red}} = H/H^\times$, and let $\varphi : H \to \mathcal{F}(P)$ be a divisor homomorphism and $\theta : H \to \mathcal{B}(G_0)$ the associated transfer homomorphism, with $G_0 \subseteq G$ and $[\cdot]$ as defined in Sect. 2 (above (2.7)), so $\theta(S) = [\varphi(S)]$ for $S \in H$. Then $\theta(U_1) \boldsymbol{\cdot} \ldots \boldsymbol{\cdot} \theta(U_k) = \theta(U_1 \cdot \ldots \cdot U_k) = \theta(V_1 \cdot \ldots \cdot V_\ell) = \theta(V_1) \boldsymbol{\cdot} \ldots \boldsymbol{\cdot} \theta(V_\ell)$ with all $\theta(U_i), \theta(V_j) \in \mathcal{A}(G_0)$. Applying the theorem to this factorization, we find atoms $W''_1, \ldots, W''_{\ell'} \in \mathcal{A}(G_0)$ with

$$\theta(U_1 \cdot \ldots \cdot U_k) = \theta(U_1) \boldsymbol{\cdot} \ldots \boldsymbol{\cdot} \theta(U_k) = W''_1 \boldsymbol{\cdot} \ldots \boldsymbol{\cdot} W''_{\ell'} \tag{8.56}$$

and $I \subseteq [1, \ell']$ such that $\ell' \geq \ell$ and $\theta(U_r) \mid \prod^\bullet_{x \in I} W''_x$ with $|I| \leq N$, for some $r \in [1, k]$. For $i \in [1, \ell']$, let $W'_i \in \mathcal{F}(P)$ be a sequence with $[W'_i] = W''_i$. Then using the definition of the composition map θ in $\theta(U_r) \mid \prod^\bullet_{x \in I} W''_x$ and (8.56), we find

$$[\varphi(U_r)] \mid \prod\nolimits^\bullet_{x \in I} [W'_x] \quad \text{and} \quad [\varphi(U_1)] \boldsymbol{\cdot} \ldots \boldsymbol{\cdot} [\varphi(U_k)] = [W'_1] \boldsymbol{\cdot} \ldots \boldsymbol{\cdot} [W'_{\ell'}]. \tag{8.57}$$

Consequently, we can choose the pre-image sequences $W'_i \in \mathcal{F}(P)$ such that

$$\varphi(U_r) \mid \prod\nolimits^\bullet_{x \in I} W'_x \quad \text{and} \quad \varphi(U_1) \boldsymbol{\cdot} \ldots \boldsymbol{\cdot} \varphi(U_k) = W'_1 \boldsymbol{\cdot} \ldots \boldsymbol{\cdot} W'_{\ell'}.$$

Since each $W'_i \in \mathcal{F}(P)$ with $[W'_i] = W''_i \in \mathcal{A}(G_0) \subseteq \mathcal{B}(G_0)$, for $i \in [1, \ell']$, it follows from (2.7) that $W'_i \in \varphi(H)$ for all $i \in [1, \ell']$. Thus there are $W_i \in H$ with $\varphi(W_i) = W'_i$ for $i \in [1, \ell']$. We now have $\varphi(U_r) \mid \prod^\bullet_{x \in I} \varphi(W_x)$ and

$$\varphi(U_1 \cdot \ldots \cdot U_k) = \varphi(U_1) \boldsymbol{\cdot} \ldots \boldsymbol{\cdot} \varphi(U_k) = \varphi(W_1) \boldsymbol{\cdot} \ldots \boldsymbol{\cdot} \varphi(W_{\ell'}) = \varphi(W_1 \cdot \ldots \cdot W_{\ell'}).$$

As a result, since $\varphi : H \to \varphi(H)$ is an isomorphism (by (2.7)), it follows that $U_1 \cdot \ldots \cdot U_k = W_1 \cdot \ldots \cdot W_{\ell'}$ with $W_i \in H$ for all i, and since φ is a divisor homomorphism, it follows from $\varphi(U_r) \mid \prod^\bullet_{x \in I} \varphi(W_x)$ that $U_r \mid \prod^\bullet_{x \in I} W_x$. Finally, since θ is a transfer homomorphism with $\theta(W_i) = [\varphi(W_i)] = [W'_i] = W''_i \in \mathcal{A}(G_0)$, it follows that $W_i \in \mathcal{A}(H)$ for all i, showing the theorem holds for H. It remains to prove the theorem when $H = \mathcal{B}(G_0)$, which we now assume.

Let $\pi : G \to \mathbb{Z}^d$ and $\pi_T : G \to G_T$ be the projection homomorphisms with respective kernels G_T and $\mathbb{Z}^d$. In view of (2.9), we may embed $\mathbb{Z}^d \leq \mathbb{R}^d$ and w.l.o.g. assume $\langle G_0 \rangle = G$ and that every $g \in G_0$ is contained in some atom, whence $\mathsf{C}(\pi(G_0)) = \mathbb{R}^d$ (by Proposition 4.4). Since $\rho_{(d+1)\exp(G_T)}(G_0) < \infty$, Proposition 8.17 implies that $\rho_k(\pi(G_0)) < \infty$ and $\rho_k(G_0) \leq \rho_{k\mathsf{D}(G_T)}(\pi(G_0)) < \infty$ for all $k \geq 1$. As a result, Theorem 8.9 implies that $0 \notin \mathsf{C}^*(\pi(G_0)^\diamond)$, while Corollary 8.18 implies there is an integer $N_\rho \geq 1$ such that

$$\rho_k(G_0) \leq N_\rho k \quad \text{for all } k \geq 1.$$

Since $0 \notin \mathsf{C}^*(\pi(G_0)^\diamond)$, it follows from Theorem 7.8 that $\pi(G_0)$ is finitary. In view of Proposition 7.33, there are only a finite number of minimal types. Let $\varphi_1, \ldots, \varphi_m \in \mathfrak{T}_m(\pi(G_0))$ be the distinct nontrivial minimal types for $\pi(G_0)$. For each $j \in [1, m]$, let

$$\mathsf{Z}_{\varphi_j} = \mathsf{Z}_1^{(j)} \cup \ldots \cup \mathsf{Z}_{s_j}^{(j)}$$

be the codomain of φ_j with $s_j \leq d$ and $|\mathsf{Z}_{\varphi_j}| \leq d$.

Consider $U_1, \ldots, U_k, V_1, \ldots, V_\ell \in \mathcal{A}(G_0)$ with

$$S := U_1 \cdot \ldots \cdot U_k = V_1 \cdot \ldots \cdot V_\ell,$$

where $k, \ell \geq 1$. Now

$$\ell \leq \rho_k(G_0) \leq N_\rho k. \tag{8.58}$$

Let $S = g_1 \cdot \ldots \cdot g_{|S|}$ be an indexing of the terms of S. Let $I_1 \cup \ldots \cup I_k = [1, |S|] = J_1 \cup \ldots \cup J_\ell$ be disjoint partitions such that

$$S(I_i) = U_i \quad \text{and} \quad S(J_j) = V_j \quad \text{for all } i \in [1, k] \text{ and } j \in [1, \ell].$$

By Proposition 8.16.1, each $\pi(V_i)$ for $i \in [1, \ell]$ factors into a product of

$$d_i \leq \mathsf{D}(G_T)$$

atoms modulo G_T, say

$$V_i = {\prod}^\bullet_{t \in [1, d_i]} V_{i,t} \quad \text{with every } \pi(V_{i,t}) \in \mathcal{A}(\pi(G_0)).$$

We thus have partitions $J_i = J_{i,1} \cup \ldots \cup J_{i,d_i}$ such that

$$S(J_{i,t}) = V_{i,t} \quad \text{for every } i \in [1, \ell] \text{ and } t \in [1, d_i].$$

Since $0 \notin \mathsf{C}^*(\pi(G_0)^\diamond)$, we can apply Theorem 8.13 to each atom $\pi(V_{i,t})$ for $i \in [1, \ell]$ and $t \in [1, d_i]$. Let $N_R \geq 0$ be the global bound from Theorem 8.13 (which we can assume is an integer) and let $V_{i,t} = R_{i,t} \cdot \prod_{j=1}^m S_{i,t}^{(j)}$, for $i \in [1, \ell]$ and $t \in [1, d_i]$, be the resulting factorization given by Theorem 8.13, with $\pi(S_{i,t}^{(j)}) \in \mathcal{F}(\pi(G_0))$ corresponding to the minimal type φ_j, so $\sigma(\pi(S_{i,t}^{(j)})) \in \mathsf{C}_{\mathbb{Z}}(Z_{\varphi_j})$. We can then apply Proposition 8.12 to each $\pi(S_{i,t}^{(j)})$. Let $J_{i,t}^{(j)} \subseteq [1, |S|]$ be disjoint subsets with

$$S(J_{i,t}^{(0)}) = R_{i,t} \quad \text{and} \quad S(J_{i,t}^{(j)}) = S_{i,t}^{(j)}, \qquad \text{for } i \in [1, \ell], t \in [1, d_i] \text{ and } j \in [1, m].$$

Moreover, for each $i \in [1, \ell]$, $t \in [1, d_i]$, $j \in [1, m]$ and $n \in [1, s_j]$, let $J_{i,t}^{(j,n)} \subseteq J_{i,t}^{(j)}$ be the subset of all $x \in J_{i,t}^{(j)}$ with $\varphi_j(x) \in Z_n^{(j)}$ at depth n. Then

$$J_{i,t} := \bigcup_{j=0}^{m} J_{i,t}^{(j)} = J_{i,t}^{(0)} \cup \bigcup_{j=1}^{m} \bigcup_{n=1}^{s_j} J_{i,t}^{(j,n)} \quad \text{for every } i \in [1, \ell] \text{ and } t \in [1, d_i].$$

Let

$$X_0 = \bigcup_{i=1}^{\ell} \bigcup_{t=1}^{d_i} J_{i,t}^{(0)}, \quad \Omega_0 = \{(0, \pi(g_x), \pi_T(g_x)) : x \in X_0\} \quad \text{and}$$

$$\Omega_\diamond = \{(j, z, a) : j \in [1, m], z \in Z_{\varphi_j}, a \in G_T\}.$$

Moreover, partition

$$\Omega_\diamond = \Omega_1 \cup \ldots \cup \Omega_d$$

such that Ω_n consists of all $(j, z, a) \in \Omega_\diamond$ with $z \in Z_n^{(j)}$ at depth n. Apply Proposition 8.12 to each $\pi(S_{i,t}^{(j)})$ (for $i \in [1, \ell]$, $t \in [1, d_i]$ and $j \in [1, m]$) and fix a system of subsets $T_x \subseteq J_{i,t}^{(j)}$, for each $x \in J_{i,t}^{(j)}$, such that the conclusions of Proposition 8.12 hold.

Since Theorem 8.13 implies $|J_{i,t}^{(0)}| = |R_{i,t}| \le N_R$ for all $i \in [1, \ell]$ and $t \in [1, d_i]$, since $d_i \le \mathsf{D}(G_T)$ for all $i \in [1, \ell]$, and since $|Z_{\varphi_j}| \le d$ for all $j \in [1, m]$, we have

$$|X_0| \le \ell \mathsf{D}(G_T) N_R \quad \text{and} \quad |\Omega_\diamond| = |G_T| \cdot \sum_{j=1}^{m} |Z_{\varphi_j}| \le md|G_T|. \tag{8.59}$$

We view $\Omega = \Omega_0 \cup \Omega_\diamond = \Omega_0 \cup \Omega_1 \cup \ldots \cup \Omega_d$ as the set of *support types* for $S = V_1 \cdot \ldots \cdot V_\ell$. A support type $\tau \in \Omega_n$ is said to be at depth n. Note, if $\tau = (j, z, a)$ with $j \ge 1$, then the depth of τ equals the depth of $z \in Z_{\varphi_j}$. For each $x \in [1, |S|]$, we have $x \in J_{i,t}^{(j)}$ for some unique $i \in [1, \ell]$, $t \in [1, d_i]$ and $j \in [0, m]$, allowing us to define

$$\mathsf{s}(x) = (0, \pi(g_x), \pi_T(g_x)) \in \Omega_0 \quad \text{when } j = 0, \qquad \text{and}$$
$$\mathsf{s}(x) = (j, \varphi_j(\pi(g_x)), \pi_T\Big(\sigma(S(T_x))\Big)) \in \Omega_\diamond \quad \text{when } j \geq 1.$$

Note Proposition 8.12 implies $\varphi_j(\pi(g_x)) = \sigma(\pi(S)(T_x)) = \pi\Big(\sigma(S(T_x))\Big)$. For $I \subseteq [1, |S|]$, $\mathsf{s}(I) \in \mathcal{F}(\Omega)$ is a sequence of support types from Ω. We associate the depth of $\mathsf{s}(x)$ (defined above) as the depth of $x \in [1, |S|]$.

Let

$$\alpha = \min\left\{ |X_0 \cap I_i| + \sum_{\tau \in \Omega_\diamond} \Big(\frac{\ell \mathsf{v}_\tau(\mathsf{s}(I_i))}{\mathsf{v}_\tau(\mathsf{s}([1, |S|]))} + 1\Big) : \ i \in [1, k] \right\}.$$

Technically, we exclude terms $\tau \in \Omega_\diamond$ in the sum defining α with $\mathsf{v}_\tau\big(\mathsf{s}([1, |S|])\big) = 0$. Then

$$\begin{aligned} k\alpha &\leq \sum_{i=1}^{k} |X_0 \cap I_i| + \sum_{i=1}^{k} \sum_{\tau \in \Omega_\diamond} \frac{\ell \mathsf{v}_\tau(\mathsf{s}(I_i))}{\mathsf{v}_\tau\big(\mathsf{s}([1, |S|])\big)} + k|\Omega_\diamond| \qquad (8.60) \\ &= |X_0| + \sum_{\tau \in \Omega_\diamond} \sum_{i=1}^{k} \frac{\ell \mathsf{v}_\tau(\mathsf{s}(I_i))}{\mathsf{v}_\tau\big(\mathsf{s}([1, |S|])\big)} + k|\Omega_\diamond| \\ &\leq |X_0| + (\ell + k)|\Omega_\diamond| \\ &\leq \ell(N_R \mathsf{D}(G_T) + md|G_T|) + kmd|G_T| \leq kN_\rho(N_R \mathsf{D}(G_T) + md|G_T|) + kmd|G_T|, \end{aligned}$$

with the first inequality in (8.60) in view of (8.59), and the second in view of (8.58). Thus

$$\alpha \leq N := N_\rho(N_R \mathsf{D}(G_T) + md|G_T|) + md|G_T|,$$

which is a global bound independent of the U_i and V_j.

Let $r \in [1, k]$ be an index attaining the minimum in the definition of α. Then $|X_0 \cap I_r| \leq \alpha \leq N$, ensuring that there is some subset $\mathcal{I}_0 \subseteq [1, \ell]$ with

$$X_0 \cap I_r \subseteq \bigcup_{i \in \mathcal{I}_0} J_i^{(0)} \quad \text{and} \quad |\mathcal{I}_0| \leq |X_0 \cap I_r| \leq N. \qquad (8.61)$$

Likewise, letting

$$n_\tau = \left\lceil \frac{\mathsf{v}_\tau(\mathsf{s}(I_r))}{\mathsf{v}_\tau\big(\mathsf{s}([1, |S|])\big)/\ell} \right\rceil < \frac{\ell \mathsf{v}_\tau(\mathsf{s}(I_r))}{\mathsf{v}_\tau\big(\mathsf{s}([1, |S|])\big)} + 1 \leq \ell + 1 \quad \text{for } \tau \in \Omega_\diamond,$$

we have

$$\sum_{\tau\in\Omega_\diamond} n_\tau \le \alpha - |X_0 \cap I_r| \le N - |\mathcal{I}_0|. \tag{8.62}$$

We interpret $n_\tau = 0$ when $\mathsf{v}_\tau\big(\mathsf{s}([1,|S|])\big) = 0$.

We now describe how the $W_1, \ldots, W_{\ell'} \in \mathcal{B}(G_0)$ can be constructed. The idea is as follows. An index set $I \subseteq [1,|S|]$ indexes a sequence $S(I) \mid S$, but it also indexes a sequence $\mathsf{s}(I) \in \mathcal{F}(\Omega)$, obtained by replacing each indexed term in the sequence $S(I)$ with its corresponding support type from Ω, so $\mathsf{s}(I) = \prod^\bullet_{x\in I} \mathsf{s}(x)$. When $I \subseteq [1,|S|] \setminus X_0$, we have $\mathsf{s}(I) \in \mathcal{F}(\Omega_\diamond)$ with $\Omega_\diamond$ a fixed, finite set independent of S. Let $\tau \in \Omega_\diamond$. If we select a subset $\mathcal{I}_\tau \subseteq [1,\ell]$ with $|\mathcal{I}_\tau| = n_\tau$ such that the $\mathsf{v}_\tau(\mathsf{s}(J_x))$, for $x \in \mathcal{I}_\tau$, are the n_τ largest values occurring over all $\mathsf{v}_\tau(\mathsf{s}(J_i))$ with $i \in [1,\ell]$, then the definition of n_τ ensures that

$$\mathsf{v}_\tau\Big(\mathsf{s}\Big(\bigcup\nolimits_{z\in\mathcal{I}_\tau} J_z\Big)\Big) \ge n_\tau\Big(\mathsf{v}_\tau\big(\mathsf{s}([1,|S|])\big)/\ell\Big) \ge \mathsf{v}_\tau(\mathsf{s}(I_r)),$$

with the first inequality holding since the sum of the n_τ largest terms in a sum of ℓ non-negative terms is always at least n_τ times the average value of all terms being summed. As a result, $\mathsf{s}(I_r) \mid \prod^\bullet_{z\in\mathcal{I}} \mathsf{s}(J_z)$, where $\mathcal{I} = \mathcal{I}_0 \cup \bigcup_{\tau\in\Omega_\diamond} \mathcal{I}_\tau$, with $|\mathcal{I}| \le N$ in view of (8.62). However, since the map s is not injective, this does not guarantee that the associated sequence $U_r = S(I_r)$ is a subsequence of the associated sequence $\prod^\bullet_{z\in\mathcal{I}} V_z = \prod^\bullet_{z\in\mathcal{I}} S(J_z)$. We do, however, have $S(X_0 \cap I_r) \mid \prod^\bullet_{z\in\mathcal{I}_0} S(J_z) = \prod^\bullet_{z\in\mathcal{I}_0} V_z$ in view of (8.61). To deal with the terms from $\Omega_\diamond$, we must use the sequences $S(T_x)$ given by Proposition 8.12 to exchange terms between the V_i.

If there are terms $x \in J^{(j)}_{i,t}$ and $y \in J^{(j)}_{i',t'}$ with $\mathsf{s}(x) = \mathsf{s}(y)$, $i \ne i'$ and $j \ge 1$, say with g_x and g_y at depth n, then $\pi_T\Big(\sigma(S(T_x))\Big) = \pi_T\Big(\sigma(S(T_y))\Big)$, while Proposition 8.12 implies that

$$\begin{aligned}\pi\Big(\sigma(S(T_x))\Big) &= \sigma(\pi(S)(T_x)) = \varphi_j(\pi(g_x))\\ &= \varphi_j(\pi(g_y)) = \sigma(\pi(S)(T_y)) = \pi\Big(\sigma(S(T_y))\Big).\end{aligned}$$

Combined with $\pi_T\Big(\sigma(S(T_x))\Big) = \pi_T\Big(\sigma(S(T_y))\Big)$, we conclude that

$$\sigma(S(T_x)) = \sigma(S(T_y)). \tag{8.63}$$

If we exchange these sets, defining

$$K^{(j)}_{i,t} = (J^{(j)}_{i,t} \setminus T_x) \cup T_y \quad \text{and} \quad K^{(j)}_{i',t'} = (J^{(j)}_{i',t'} \setminus T_y) \cup T_x,$$

and correspondingly define

$$K_i = (J_i \setminus T_x) \cup T_y \quad \text{and} \quad K_{i'} = (J_{i'} \setminus T_y) \cup T_x,$$

then the new sequences $W_i = S(K_i)$ and $W_{i'} = S(K_{i'})$ will still be zero-sum, though we do not guarantee that they remain atoms. However, since $y \in K_i$ and $x \in K_{i'}$, they are non-empty. Consequently, if either W_i or $W_{i'}$ is not an atom, then we can re-factor them to write $V_i \cdot V_{i'} = W_i \cdot W_{i'} = V_1' \cdot \ldots \cdot V_\omega'$ as a product of $\omega \geq 3$ atoms. This leads to a factorization $U_1 \cdot \ldots \cdot U_k = V_1 \cdot \ldots \cdot V_\ell \cdot V_i^{[-1]} \cdot V_{i'}^{[-1]} \cdot V_1' \cdot \ldots \cdot V_\omega'$ into $\ell' = \ell - 2 + \omega > \ell$ atoms. In this case, we begin from scratch using this factorization in place of the original one $U_1 \cdot \ldots \cdot U_k = V_1 \cdot \ldots \cdot V_\ell$. As $\ell' \leq |S| < \infty$, we cannot start from scratch endlessly, meaning eventually we will never encounter this problem, allowing us to w.l.o.g. assume $W_i = S(K_i)$ and $W_{i'} = S(K_{i'})$ are always atoms (where $\ell' = \ell$ may have increased in size from the original ℓ given in the hypotheses). Furthermore, we still have $\sigma(\pi(S)(K_{i,t}^{(j)})) = \sigma(\pi(S)(J_{i,t}^{(j)})) \in \mathsf{C}(Z_{\varphi_j})$ and $\sigma(\pi(S)(K_{i',t'}^{(j)})) = \sigma(\pi(S)(J_{i',t'}^{(j)})) \in \mathsf{C}(Z_{\varphi_j})$ by (8.63), with the inclusions originating from our application of Theorem 8.13 at the start of the proof. Thus Proposition 8.12 can still be applied to $K_{i,t}^{(j)}$ and $K_{i',t'}^{(j)}$ if we later wish to continue with further such swaps between these sets (though we do not guarantee nor need that $\pi(S)(K_{i,t}^{(j)})$ and $\pi(S)(K_{i',t'}^{(j)})$ remain atoms). Proposition 8.12 guarantees that the set T_x contains *no* terms with depth greater than g_x, and has x as the *unique* $a \in T_x$ with g_a having depth equal to that of x. Likewise for T_y. Thus when swapping the sets T_x and T_y, we leave unaffected all terms in $J_{i,t}^{(j)}$ and $J_{i',t'}^{(j)}$ with depth at least n, apart from the exchanging of x for y. This ensures that terms previously swapped but at a higher or equal depth will remain unaffected by exchanging T_x and T_y.

As in the proof of Theorem 8.14, the sequences $\mathsf{s}(K_{i,t}^{(j)}) = \mathsf{s}(J_{i,t}^{(j)})$ and $\mathsf{s}(K_{i',t'}^{(j)}) = \mathsf{s}(J_{i',t'}^{(j)})$ remain unchanged, though this now requires a short argument. The value of $\mathsf{s}(z)$ for $z \in [1, |S|]$ depends upon the sequence T_z. After swapping T_x and T_y, the sets T_z need to be redefined via Proposition 8.12. We must show that this can be done in such a way that $\sigma(S(T_z))) = \sigma(S(T_z'))$ for all z, where T_z' is the new sequence associated to z after performing the swap. Unless $z \in J_{i,t}^{(j)}$ or $z \in J_{i',t'}^{(j)}$, the values for T_z can be left unaffected, so $T_z = T_z'$. Let us consider the case when $z \in J_{i,t}^{(j)}$. The case $z \in J_{i',t'}^{(j)}$ will then follow by an analogous argument. If $T_z \cap T_x = \emptyset$, then the value of T_z can also be left unaffected. Otherwise, Proposition 8.12.2 implies that either $z \in T_x$ or $x \in T_z$. If $z \in T_x$, the Proposition 8.12.1 implies that $T_z \subseteq T_x$, and we can again leave the value of T_z unchanged, though note that $T_z' \subseteq K_{i',t'}^{(j)}$ while $T_z \subseteq J_{i,t}^{(j)}$. In the remaining case $x \in T_z$ but $z \notin T_x$ (so $x \neq z$), we have $T_x \subseteq T_z$ by Proposition 8.12.1 with $\sigma(S(T_x)) = \sigma(S(T_y))$ by (8.63). This allows us to define $T_z' = T_z \setminus T_x \cup T_y \subseteq K_{i,t}^{(j)}$, which then satisfies $\sigma(S(T_z')) = \sigma(S(T_z)) - \sigma(S(T_x)) + \sigma(S(T_y)) = \sigma(S(T_z))$. The equivalent defining conditions $1'$ and $2'$ for Proposition 8.12 also hold for the newly defined set system T_z'. This shows it is possible to adjust the sequences T_z after swapping T_x for T_y in such a way that the value $\mathsf{s}(z)$ remains unaffected. In particular, the value of $\mathsf{v}_\tau\big(\mathsf{s}([1,|S|])\big)$ remains unchanged, ensuring that the value of α is unaffected when replacing J_i and $J_{i'}$ by K_i and $K_{i'}$, and that $r \in [1,k]$ remains an index attaining the minimum in the definition of α (note, the numerators in the definition of α depend on the U_i, not the V_j). Swapping the elements x and y in this fashion leaves all elements from X_0, as well as any $J_{c,d}^{(b)}$ with $b \neq j$, unaltered, and the sequences W_i and $W_{i'}$ remain nontrivial, as W_i contains g_y, and $W_{i'}$ contains g_x. Since, apart from x and y, only terms with depth less than n are affected by the swap, it follows that the sets $\mathcal{I}_{\tau'}$, corresponding to any type $\tau' \in \Omega_\diamond$ with depth at least n, still have the property that they index the $\mathsf{v}_\tau(\mathsf{s}(J_z))$, for $z \in \mathcal{I}_{\tau'}$, with the $n_{\tau'}$ largest values occurring over all $\mathsf{v}_{\tau'}(\mathsf{s}(J_i))$ with $i \in [1,\ell]$.

With these observations in mind, we can now describe how the zero-sums V_i must be modified. Begin with any type $\tau \in \Omega_\diamond$ having maximal available depth. Construct the subset $\mathcal{I}_\tau$ for the current factorization $S = V_1 \cdot \ldots \cdot V_\ell$ as described above. Then $\mathsf{v}_\tau\big(\mathsf{s}(\bigcup_{z\in\mathcal{I}_\tau} J_z)\big) \geq \mathsf{v}_\tau(\mathsf{s}(I_r))$. If $\bigcup_{z\in\mathcal{I}_\tau} J_z$ contains all elements from I_r having type τ, then nothing need be done, we discard τ from the list of available types from $\Omega_\diamond$, we select the next available type from $\Omega_\diamond$ with maximal depth, and continue once more. On the other hand, if there is some $x \in I_r$ with type τ not contained in $\bigcup_{z\in\mathcal{I}_\tau} J_z$, then $\mathsf{v}_\tau\big(\mathsf{s}(\bigcup_{z\in\mathcal{I}_\tau} J_z)\big) \geq \mathsf{v}_\tau(\mathsf{s}(I_r))$ ensures that there must be some $y \in \bigcup_{z\in\mathcal{I}_\tau} J_z$ having type τ with $y \notin I_r$. In this case, perform the swap of T_x and T_y described above, and redefine our factorization $V_1 \cdot \ldots \cdot V_\ell$ by replacing $J_{i,t}^{(j)}$ and $J_{i',t'}^{(j)}$ by $K_{i,t}^{(j)}$ and $K_{i',t'}^{(j)}$, where $x \in J_{i,t}^{(j)}$ and $y \in J_{i',t'}^{(j)}$, and correspondingly replacing V_i and $V_{i'}$ by W_i and $W_{i'}$. Also, adjust the values of the auxiliary sets T_z as described above. To simplify notation, redefine V_i, $V_{i'}$, $J_{i,t}^{(j)}$, $J_{i',t'}^{(j)}$ and T_z accordingly so as to reflect the new current state that now has $x \in \bigcup_{z\in\mathcal{I}_\tau} J_z$ and $y \notin \bigcup_{z\in\mathcal{I}_\tau} J_z$. If we now have $\bigcup_{z\in\mathcal{I}_\tau} J_z$ containing all elements from I_r having type τ, then nothing need be done, we discard τ from the list of available types in $\Omega_\diamond$ and carry on as before. If this is not the case, we again find a new term $x' \in I_r$ with type τ not contained in $\bigcup_{z\in\mathcal{I}_\tau} J_z$, find a new term $y' \notin \bigcup_{z\in\mathcal{I}_\tau} J_z$ with type τ and swap x' and y' as before by use of Proposition 8.12. Since the depth of x and x' are the same, we will not swap x back out of $\bigcup_{z\in\mathcal{I}_\tau} J_z$ when doing so, nor indeed any other element from $\bigcup_{z\in\mathcal{I}_\tau} J_z$ having type τ. Thus, iterating such a procedure, we will eventually obtain that $\bigcup_{z\in\mathcal{I}_\tau} J_z$ contains every element of I_r having type τ, in which case we move on to the next available type τ' with maximal available depth. We repeat the same for procedure for τ' as we did for τ (and, later, as we did for any support types previously discarded before selecting τ'). We first construct the subset $\mathcal{I}_{\tau'}$ for the current state for $S = V_1 \cdot \ldots \cdot V_\ell$, and then swap elements into $\bigcup_{z\in\mathcal{I}_{\tau'}} J_z$ until it contains all elements from I_r having type τ'. While doing so, since we always first choose support types with maximal available depth, we are assured that any type τ'' that has already been discarded had depth at least that of τ', and thus no element of type τ'' will be moved when swapping at the later stage for τ', ensuring that prior work cannot be undone. Continue until all support types from $\Omega_\diamond$ have been exhausted. Once the process ends, we now have a new factorization

$U_1 \cdot \ldots \cdot U_k = S = W_1 \cdot \ldots \cdot W_\ell$, where W_i reflects the final state of V_i after running the above process, such that $U_r \mid \prod^\bullet_{z \in \mathcal{I}} W_i$, where $\mathcal{I} = I_0 \cup \bigcup_{\tau \in \Omega_\diamond} I_\tau$, with $|\mathcal{I}| \le |X_0 \cap I_r| + \sum_{\tau \in \Omega_\diamond} n_\tau \le N$, completing the proof. □

The following basic proposition shows that having finite elasticities implies a Krull Monoid is always locally tame.

Proposition 8.20 *Let H be a Krull Monoid, let $\varphi : H \to \mathcal{F}(P)$ be a divisor homomorphism for H, and let $U \in \mathcal{A}(H)$. Then*

$$\mathsf{t}(H, U) \le \rho_{|\varphi(U)|}(H).$$

Proof Let $U_1, \ldots, U_k \in \mathcal{A}(H)$ with $U \mid U_1 \cdot \ldots \cdot U_k$ and let $I \subseteq [1, k]$ be a minimal subset with $U \mid \prod_{i \in I} U_i$. Then $\varphi(U) \mid \varphi(\prod_{i \in I} U_i) = \prod_{i \in I} \varphi(U_i)$. Each $\varphi(U_i)$ is a nontrivial sequence as each $U_i \in \mathcal{A}(H)$. Thus we trivially have $\varphi(U) \mid \prod_{i \in J} \varphi(U_i)$ for some $J \subseteq I$ with $|J| \le |\varphi(U)|$. By definition of a divisor homomorphism, it follows that $U \mid \prod_{i \in J} U_i$, and now the minimality of I ensures $I = J$ with $|I| = |J| \le |\varphi(U)|$. Since $U \mid \prod_{i \in I} U_i$, we have $\prod_{i \in I} U_i = U \cdot V_2 \cdot \ldots \cdot V_r$ for some $V_2, \ldots, V_r \in \mathcal{A}(H)$, and by definition of the elasticities and (2.9), we must have $r \le \rho_{|I|}(H) \le \rho_{|\varphi(U)|}(H)$. The result now follows. □

Having now established Theorem 8.19, we can immediately extend Theorem 8.15 to the more general finitely generated group setting, showing that finite elasticity implies most other arithmetic factorization invariants are also well-behaved.

Main Theorem 8.21 *Let H be a Transfer Krull Monoid over a subset G_0 of a finitely generated abelian group $G = \mathbb{Z}^d \oplus G_T$ with torsion subgroup $G_T \le G$, where $d \ge 0$. Suppose*

$$\rho_{(d+1)\exp(G_T)}(H) < \infty$$

and let $N = \max\{2, \mathsf{t}_w(G_0)\}$.

1. *$\rho_k(H) - \rho_{k-1}(H) \le N < \infty$ for all $k \ge 2$.*
2. *$\max \Delta(H) \le \rho_N(H) - N < \infty$. In particular, $\Delta(H)$ is finite.*
3. *The Structure Theorem for Unions holds in H.*
4. *If H is also a Krull Monoid, the catenary degree $\mathsf{c}(H) \le \rho_N(H) < \infty$ is finite.*
5. *If H is also a Krull Monoid, then H is locally tame.*

Proof Since H is a Transfer Krull Monoid over G_0, we have $\rho_k(H) = \rho_K(G_0)$ and $\mathcal{U}_k(H) = \mathcal{U}_k(G_0)$ for all $k \ge 1$, and $\Delta(H) = \Delta(G_0)$. Moreover, when H is a Krull Monoid, we have $\mathsf{c}(H) \le \max\{\mathsf{c}(G_0), 2\}$ (see Sect. 2). It thus suffices to prove the theorem when $H = \mathcal{B}(G_0)$, which we now assume. By Corollary 8.18, there is a constant $N_\rho \ge 1$ such that $\rho_k(G_0) \le N_\rho k < \infty$ for all $k \ge 1$. Items 1–4 now follows by the identical arguments given in Theorem 8.15, simply replacing the use of Theorem 8.14 with Theorem 8.19. Item 5 follows from Proposition 8.20 and Item 1 (which inductively shows $\rho_k(H) = \rho_k(G_0) < \infty$ for all k). □

Consider an infinite abelian group G and let $G_0 \subseteq G$ be a subset such that there is some $m \geq 0$ so that every element $g \in G$ is the sum of at most m elements from G_0 (e.g., this assumption holds for integrally closed finitely generated algebras over perfect fields). Then $\Delta(G_0)$ is infinite by [88, Theorem 1.1], so Main Theorem 8.21 further implies that $\rho_{(d+1)\exp(G_T)}(G_0) = \infty$ must also be infinite.

We now extend the definition of $G_0^\diamond \subseteq G_0$.

Definition 8.22 Let $G = \mathbb{Z}^d \oplus G_T$ be a finitely generated abelian group with torsion subgroup $G_T \leq G$, where $d \geq 0$, and let $\pi : G \to \mathbb{Z}^d \leq \mathbb{R}^d$ be the projection homomorphism with kernel G_T. For $G_0 \subseteq G$, we define

$$G_0^\diamond = \{g \in G_0 : \pi(g) \in \pi(G_0)^\diamond\}.$$

The following extends Corollary 8.8.

Proposition 8.23 *Let $G = \mathbb{Z}^d \oplus G_T$ be a finitely generated abelian group with torsion subgroup $G_T \leq G$, where $d \geq 0$, and let $G_0 \subseteq G$ be a subset. Suppose $\rho_{(d+1)\exp(G_T)}(G_0) < \infty$. Then*

$$\begin{aligned} G_0^\diamond &= \Big\{g \in G_0 : \sup\{\mathsf{v}_g(U) : U \in \mathcal{A}^{\mathsf{elm}}(G_0)\} = \infty\Big\} \\ &= \Big\{g \in G_0 : \sup\{\mathsf{v}_g(U) : U \in \mathcal{A}(G_0)\} = \infty\Big\}. \end{aligned}$$

Proof Since G_T has finite exponent, an element $g \in G_0$ is contained in an atom if and only if $\pi(g) \in \pi(G_0)$ is contained in an atom. Thus Corollary 4.5 and the definition of $G_0^\diamond$ ensure that removing elements from G_0 contained in no atom does not affect $G_0^\diamond$, allowing us to w.l.o.g. assume every $g \in G_0$ is contained in an atom. We may embed $\mathbb{Z}^d \leq \mathbb{R}^d$. Let $\pi : G \to \mathbb{Z}^d$ be the projection homomorphism with kernel G_T, and let $m = \exp(G_T)$. In view of (2.9), we may w.l.o.g. assume $\langle G_0 \rangle = G$. Since $\langle G_0 \rangle = G$ and every $g \in G_0$ is contained in an atom, Corollary 4.5 yields $\mathsf{C}(\pi(G_0)) = \mathbb{R}^d$. Since $\rho_{(d+1)\exp(G_T)}(G_0) < \infty$, Proposition 8.17 implies $\rho_{d+1}(\pi(G_0)) < \infty$, whence Theorem 8.9 implies that $0 \notin \mathsf{C}^*(\pi(G_0)^\diamond)$, in turn implying that $\pi(G_0)$ is finitary (by Theorem 7.8).

Let $g \in G_0^\diamond$ be arbitrary. Then $\pi(g) \in \pi(G_0)^\diamond$, whence Corollary 8.8 implies that there is a sequence of elementary atoms $V_i \in \mathcal{A}^{\mathsf{elm}}(\pi(G_0))$, for $i = 1, 2, \ldots$, with $\mathsf{v}_{\pi(g)}(V_i) \to \infty$. Moreover, since $0 \notin \pi(G_0)^\diamond$ (by Proposition 7.2.2), we have $\pi(g) \neq 0$. For $i \geq 1$, let $U_i \in \mathcal{F}(G_0)$ be a sequence with $\pi(U_i) = V_i$, $\mathsf{v}_g(U_i) = \mathsf{v}_{\pi(g)}(V_i)$ and $|\operatorname{Supp}(U_i)| = |\operatorname{Supp}(V_i)|$. Thus $\mathsf{v}_g(U_i^{[m]}) \to \infty$. Since V_i is an elementary atom containing the nonzero element $\pi(g)$, Proposition 4.8 implies that $\operatorname{Supp}(V_i)$ is a minimal positive basis, in which case $\mathcal{A}(Y) = \emptyset$ for any proper subset $Y \subset \operatorname{Supp}(V_i)$. Thus, since $|\operatorname{Supp}(U_i)| = |\operatorname{Supp}(V_i)|$, it follows that $\mathcal{A}(\pi(Y)) = \emptyset$ for any proper subset $Y \subset \operatorname{Supp}(U_i)$, in turn implying $\mathcal{A}(Y) = \emptyset$ as well. In view of Proposition 8.16.2, each $U_i^{[m]}$ factors into a product of at most m atoms. Thus, letting $W_i \mid U_i^{[m]}$ be a subsequence which is an atom with $\mathsf{v}_g(W_i)$ maximal, it follows that $\mathsf{v}_g(W_i) \to \infty$ in view of $\mathsf{v}_g(U_i^{[m]}) \to \infty$. Moreover, since $\mathcal{A}(Y) = \emptyset$ for any proper subset $Y \subset \mathrm{Supp}(U_i)$, it follows that each $W_i \in \mathcal{A}^{\mathsf{elm}}(G_0)$ is an elementary atom with $\mathsf{v}_g(W_i) \to \infty$, which establishes the inclusion

$$G_0^\diamond \subseteq \Big\{g \in G_0 : \sup\{\mathsf{v}_g(U) : U \in \mathcal{A}^{\mathsf{elm}}(G_0)\} = \infty\Big\}.$$

The inclusion

$$\Big\{g \in G_0 : \sup\{\mathsf{v}_g(U) : U \in \mathcal{A}^{\mathsf{elm}}(G_0)\} = \infty\Big\}$$
$$\subseteq \Big\{g \in G_0 : \sup\{\mathsf{v}_g(U) : U \in \mathcal{A}(G_0)\} = \infty\Big\}$$

holds trivially in view of $\mathcal{A}^{\mathsf{elm}}(G_0) \subseteq \mathcal{A}(G_0)$.

To establish the final reverse inclusion, let $g \in G_0$ be an element with

$$\sup\{\mathsf{v}_g(U) : U \in \mathcal{A}(G_0)\} = \infty.$$

Let $U_i \in \mathcal{A}(G_0)$, for $i = 1, 2, \ldots$, be a sequence of atoms with $\mathsf{v}_g(U_i) \to \infty$. By Proposition 8.16.1, each $\pi(U_i)$ factors as a product of at most $\mathsf{D}(G_T)$ atoms. Thus, letting $V_i \mid \pi(U_i)$ be an atom $V_i \in \mathcal{A}(\pi(G_0))$ with $\mathsf{v}_{\pi(g)}(V_i)$ maximal, it follows that $\mathsf{v}_{\pi(g)}(V_i) \to \infty$ in view of $\mathsf{v}_g(U_i) \to \infty$. Hence

$$\pi(g) \in \{x \in \pi(G_0) : \sup\{\mathsf{v}_x(V) : V \in \mathcal{A}(\pi(G_0))\} = \infty\} = \pi(G_0)^\diamond,$$

with the equality in view of Corollary 8.8, which implies $g \in G_0^\diamond$ by definition of $G_0^\diamond$. This establishes the reverse inclusion

$$\Big\{g \in G_0 : \sup\{\mathsf{v}_g(U) : U \in \mathcal{A}(G_0)\} = \infty\Big\} \subseteq G_0^\diamond,$$

completing the proof. □

Next, we extend Corollary 8.7.

Proposition 8.24 *Let $G = \mathbb{Z}^d \oplus G_T$ be a finitely generated abelian group with torsion subgroup $G_T \leq G$, where $d \geq 0$, and let $G_0 \subseteq G$ be a subset. Suppose $\rho_{(d+1)\exp(G_T)}(G_0) < \infty$. Then $G_0 \setminus G_0^\diamond \subseteq G_0$ is a Lambert subset with $\mathcal{A}(G_0^\diamond) = \emptyset$.*

Proof In view of (2.9) and Corollary 4.5, we may w.l.o.g. assume $\langle G_0 \rangle = G$ and that every $g \in G_0$ is contained in an atom. We may embed $\mathbb{Z}^d \leq \mathbb{R}^d$. Let $\pi : G \to \mathbb{Z}^d$ be the projection homomorphism with kernel G_T, and let $m = \exp(G_T)$. As argued in Proposition 8.23, we have $\mathsf{C}(\pi(G_0)) = \mathbb{R}^d$ and $0 \notin \mathsf{C}^*(\pi(G_0)^\diamond)$ with $\pi(G_0)$ finitary. Note $\pi(G_0^\diamond) = \pi(G_0)^\diamond$ by definition of $G_0^\diamond$. Thus Proposition 4.2 and $0 \notin \mathsf{C}^*(\pi(G_0)^\diamond)$ together imply $\mathcal{A}(\pi(G_0^\diamond)) = \mathcal{A}(\pi(G_0)^\diamond) = \emptyset$, in turn implying $\mathcal{A}(G_0^\diamond) = \emptyset$. By Corollary 8.7, $\pi(G_0) \setminus \pi(G_0)^\diamond \subseteq \pi(G_0)$ is a Lambert subset, say with bound $N \geq 1$. By definition of $G_0^\diamond$, we have $\pi(G_0 \setminus G_0^\diamond) = \pi(G_0) \setminus \pi(G_0)^\diamond$. Let $U \in \mathcal{A}(G_0)$ be an arbitrary atom. Proposition 8.16.1 implies that $\pi(U) = W_1 \cdot \ldots \cdot W_\ell$ for some atoms $W_1, \ldots, W_\ell \in \mathcal{A}(\pi(G_0))$ with $\ell \leq \mathsf{D}(G_T)$. Thus

$$\mathsf{v}_{G_0\setminus G_0^\diamond}(U) \leq \sum_{i=1}^{\ell} \mathsf{v}_{\pi(G_0\setminus G_0^\diamond)}(W_i) = \sum_{i=1}^{\ell} \mathsf{v}_{\pi(G_0)\setminus\pi(G_0)^\diamond}(W_i) \leq \ell N \leq \mathsf{D}(G_T)N,$$

with the second inequality in view of $\pi(G_0) \setminus \pi(G_0)^\diamond \subseteq \pi(G_0)$ being a Lambert subset with bound $N \geq 1$, which shows that $G_0 \setminus G_0^\diamond \subseteq G_0$ is a Lambert subset with bound $\mathsf{D}(G_T)N < \infty$. □

The following extends Proposition 7.31 to our current setting.

Proposition 8.25 *Let $G = \mathbb{Z}^d \oplus G_T$ be a finitely generated abelian group with torsion subgroup $G_T \leq G$, where $d \geq 0$, and let $G_0 \subseteq G$ be a subset. Suppose $\rho_{(d+1)\exp(G_T)}(G_0) < \infty$. Then there are finite subsets $X \subseteq G_0^\diamond$ and $Y \subseteq G_0$ such that $\mathcal{A}(G_0 \setminus X)$ is finite and $\mathcal{A}(G_0 \setminus Y) = \emptyset$.*

Proof In view of (2.9) and Corollary 4.5, we may w.l.o.g. assume $\langle G_0 \rangle = G$ and that every $g \in G_0$ is contained in an atom. We may embed $\mathbb{Z}^d \leq \mathbb{R}^d$. Let $\pi : G \to \mathbb{Z}^d$ be the projection homomorphism with kernel G_T. As argued in Proposition 8.23, we have $\mathsf{C}(\pi(G_0)) = \mathbb{R}^d$ and $0 \notin \mathsf{C}^*(\pi(G_0)^\diamond)$ with $\pi(G_0)$ finitary. Thus Proposition 7.31 implies that there is a finite subset $\widetilde{X} \subseteq \pi(G_0)^\diamond$ such that $\mathcal{A}(\pi(G_0) \setminus \widetilde{X})$ is finite. Hence, since $\ker \pi = G_T$ is finite, it follows that $X := \pi^{-1}(\widetilde{X}) \cap G_0 \subseteq G_0$ is finite with $\pi(X) = \widetilde{X} \subseteq \pi(G_0)^\diamond$, ensuring $X \subseteq G_0^\diamond$ by definition of $G_0^\diamond$. It follows that $\mathcal{A}(\pi(G_0) \setminus \widetilde{X}) = \mathcal{A}(\pi(G_0 \setminus X))$ is finite.

Consider an arbitrary atom $U \in \mathcal{A}(G_0 \setminus X)$. Then Proposition 8.16.1 ensures that $\pi(U) = W_1 \cdot \ldots \cdot W_\ell$ for some atoms $W_1, \ldots, W_\ell \in \mathcal{A}(\pi(G_0 \setminus X))$ with $\ell \leq \mathsf{D}(G_T)$. Since $\mathcal{A}(\pi(G_0 \setminus X))$ is finite, there are at most $|\mathcal{A}(\pi(G_0 \setminus X))| \cdot \ell \leq |\mathcal{A}(\pi(G_0 \setminus X))| \cdot \mathsf{D}(G_T) < \infty$ possibilities for $\pi(U)$, for our arbitrary atom $U \in \mathcal{A}(G_0 \setminus X)$. Thus, since $\ker \pi = G_T$ is finite, it follows that there are only finitely many possibilities for $U \in \mathcal{A}(G_0 \setminus X)$, meaning $\mathcal{A}(G_0 \setminus X)$ is finite. Including in X one element from each of the finite number of atoms from $\mathcal{A}(G_0 \setminus X)$ then yields a subset $Y \subseteq G_0$ with $\mathcal{A}(G_0 \setminus Y) = \emptyset$, completing the proof. □

We conclude with the extension of Theorem 8.9.

Main Theorem 8.26 *Let H be a Transfer Krull Monoid over a subset G_0 of a finitely generated abelian group $G = \mathbb{Z}^d \oplus G_T$ with torsion subgroup $G_T \leq G$, where $d \geq 0$. Then the following are equivalent.*

1. *$\rho(H) < \infty$.*
2. *$\rho_k(H) < \infty$ for all $k \geq 1$.*
3. *$\rho_{(d+1)\exp(G_T)}(H) < \infty$.*
4. *There exists a subset $X \subseteq G_0$ such that $\mathcal{A}(X) = \emptyset$ and $G_0 \setminus X \subseteq G_0$ is a Lambert subset.*

5. $\mathcal{A}(G_0^\diamond) = \emptyset$.
6. $0 \notin \mathsf{C}^*(\pi(G_0)^\diamond)$, *where* $\pi : G \to \mathbb{Z}^d \le \mathbb{R}^d$ *is the projection with kernel* G_T.

Proof Since H is a Transfer Krull Monoid, we have $\rho(H) = \rho(G_0)$ and $\rho_k(H) = \rho_k(G_0)$ for all $k \ge 1$. It thus suffices to prove the theorem when $H = \mathcal{B}(G_0)$, which we now assume. The implications 4. $\Rightarrow$ 1. $\Rightarrow$ 2. follows from proposition 8.3, while the implication 2. $\Rightarrow$ 3. is trivial. The implication 3. $\Rightarrow$ 5. follows by Proposition 8.24. In view of Proposition 4.2, Item 6 is equivalent to $\mathcal{A}(\pi(G_0)^\diamond) = \emptyset$. By definition of $G_0^\diamond$, we have $\pi(G_0^\diamond) = \pi(G_0)^\diamond$. Since G_T has finite exponent, it follows that $\mathcal{A}(G_0^\diamond) = \emptyset$ if and only if $\mathcal{A}(\pi(G_0)^\diamond) = \mathcal{A}(\pi(G_0^\diamond)) = \emptyset$. Thus Items 5 and 6 are equivalent. It remains to establish the implication 6. $\Rightarrow$ 4. Suppose $0 \notin \mathsf{C}^*(\pi(G_0)^\diamond)$. Then Theorem 8.9 implies that $\rho_{d+1}(\pi(G_0)) < \infty$, whence Proposition 8.17.1 implies that $\rho_{(d+1)\exp(G_T)}(G_0) < \infty$, allowing us to apply Proposition 8.24, which shows that Item 4 holds with $X = G_0^\diamond$. □

Summary

We can now summarize our results regarding what we have shown under an assumption of finite elasticities. Let H be a Transfer Krull Monoid over a subset G_0 of a finitely generated abelian group $G = \mathbb{Z}^d \oplus G_T$ with torsion subgroup $G_T \le G$, where $d \ge 0$. We may embed $\mathbb{Z}^d \le \mathbb{R}^d$. For example, $H = \mathcal{B}(G_0)$. Let $\pi : G \to \mathbb{Z}^d$ be the projection homomorphism with kernel G_T. There is little loss of generality to assume every element $g \in G_0$ occurs in some atom (else we can pass to the subset of G_0 having this property) and that $\langle G_0 \rangle = G$ (else we can replace G with $\langle G_0 \rangle$), which we now do. Note, since G_T has finite exponent, that $g \in G_0$ is in an atom from $\mathcal{A}(G_0)$ if and only if $\pi(g) \in \pi(G_0)$ is contained in an atom from $\mathcal{A}(\pi(G_0))$. Thus Proposition 4.4 implies that $\mathsf{C}(\pi(G_0)) = \mathbb{R}\langle G_0 \rangle = \mathbb{R}^d$. Under these assumptions, we now summarize some of the key results.

1. There exists a minimal $s \in [1, (d+1)\exp(G_T)]$ such that $\rho_s(H) < \infty$ implies $\rho_k(H) < \infty$ for all $k \ge 1$. (Corollary 8.18)

As for the torsion-free case, it would be interesting to know if the estimate $s \le (d+1)\exp(G_T)$ is tight or can be improved. Main Theorem 8.26 characterizes when Item 1 occurs either in terms of a basic combinatorial property of the atoms $\mathcal{A}(G_0)$, or the geometric property $0 \notin \mathsf{C}^*(\pi(G_0)^\diamond)$. It also shows that

$$\rho_{(d+1)\exp(G_T)}(H) < \infty \text{ is equivalent to } \rho(H) < \infty.$$

Assuming additionally that $\rho_{(d+1)\exp(G_T)}(H) < \infty$, so that the conclusion of Item 1 holds, we obtain the following properties.

2. $0 \notin \mathsf{C}(\pi(G_0)^\diamond)$. In particular, $\pi(G_0) \subseteq \mathbb{R}^d$ is finitary and all the results of Sect. 7 are available for studying the set $\pi(G_0)$, including Theorems 7.16, 7.24, 7.29 and 7.32. (Main Theorem 8.26 and Theorem 7.8)
3. Besides the defining definition of the subset $G_0^\diamond \subseteq G_0$ (stated earlier in Sect. 8.3 combined with Proposition 7.2), we also have (by Proposition 8.23)

$$G_0^\diamond = \left\{g \in G_0 : \sup\{\mathsf{v}_g(U) : U \in \mathcal{A}^{\mathsf{elm}}(G_0)\} = \infty\right\}$$
$$= \left\{g \in G_0 : \sup\{\mathsf{v}_g(U) : U \in \mathcal{A}(G_0)\} = \infty\right\}.$$

4. There is a finite subset $X \subseteq G_0^\diamond$ such that $\mathcal{A}(G_0 \setminus X)$ is finite, and thus also a finite subset $Y \subseteq G_0$ such that $\mathcal{A}(G_0 \setminus Y) = \emptyset$. (Proposition 8.25).
5. $G_0 \setminus G_0^\diamond \subseteq G_0$ is a Lambert subset with $\mathcal{A}(G_0^\diamond) = \emptyset$, indeed, $G_0 \setminus G_0^\diamond \subseteq G_0$ is the unique maximal Lambert subset. (Propositions 8.24 and 8.23)
6. If H is a Krull Monoid, the Weak Tame Degree (as defined in Main Theorem 8.19) is finite: $\mathsf{t}_w(H) < \infty$. In particular, $N := \max\{2, \mathsf{t}_w(G_0)\} < \infty$ is finite.
7. The elasticities of H do not contain arbitrarily large gaps: $\rho_k(H) - \rho_{k-1}(H) \le N < \infty$ for all $k \ge 2$. In particular, $\rho_k(H) \le Nk$ grows linearly. (Main Theorem 8.21.1)
8. $\max \Delta(H) \le \rho_N(H) - N < \infty$ is finite. (Main Theorem 8.21.2)
9. The Set of Distances is finite: $|\Delta(H)| < \infty$. (Main Theorem 8.21.2)
10. If H is a Krull Monoid, the catenary degree is finite: $\mathsf{c}(H) \le \rho_N(H) < \infty$. (Main Theorem 8.21.4)
11. If H is a Krull Monoid, then H is locally tame. (Main Theorem 8.21.5)
12. The Structure Theorem for Unions holds for H. (Main Theorem 8.21.3)

We remark that there are a few even stronger regularity properties of factorization that are *not* implied by the finiteness of the elasticities, including the finiteness of the monotone catenary degree [71, Section 7], the finiteness of the (global) tame degree [63, Theorem 4.2], and that the Structure Theorem for Sets of Lengths holds [71, Section 6].

References

1. D.F. Anderson, Elasticity of factorizations in integral domains: a survey, in *Factorization in Integral Domains*. Lecture Notes in Pure and Applied Mathematics, vol. 189 (Marcel Dekker, New York, 1997), pp. 1–29
2. D.D. Anderson, D.F. Anderson, Elasticity of factorizations in integral domains. J. Pure Appl. Algebra **80**, 217–235 (1992)
3. D.D. Anderson, D.F. Anderson, Elasticity of factorizations in integral domains II. Houston J. Math. **20**, 1–15 (1994)
4. D.D. Anderson, D.F. Anderson, S.T. Chapman, W.M. Smith, Rational elasticity of factorizations in Krull Domains. Proc. Amer. Math. Soc. **117**(1), 37–43 (1993)
5. D. Bachman, N. Baeth, J. Gossell, Factorizations of upper triangular matrices. Linear Algebra Appl. **450**, 138–157 (2014)
6. N.R. Baeth, A Krull-Schmidt theorem for one-dimensional rings of finite Cohen-Macaulay type. J. Pure Appl. Algebra **208**(3), 923–940 (2007)
7. N.R. Baeth, Direct sum decompositions over two-dimensional local domains. Commun. Algebra **37**(5), 1469–1480 (2009)
8. N.R. Baeth, A. Geroldinger, Monoids of modules and arithmetic of direct-sum decompositions. Pac. J. Math. **271**, 257–319 (2014)
9. N.R. Baeth, M.R. Luckas, Monoids of torsion-free modules over rings with finite representation type. J. Commut. Algebra **3**(4), 439–458 (2011)
10. N.R. Baeth, S. Saccon, Monoids of modules over rings of infinite Cohen-Macaulay type. J. Commut. Algebra **4**(3), 297–326 (2012)
11. N.R. Baeth, D. Smertnig, Factorization theory: from commutative to noncommutative settings. J. Algebra **441**, 475–551 (2015)
12. N.R. Baeth, D. Smertnig, Lattices over Bass rings and graph agglomerations (submitted). https://arxiv.org/abs/2006.10002
13. N.R. Baeth, R. Wiegand, Factorization theory and decompositions of modules. Amer. Math. Monthly **120**(1), 3–34 (2013)
14. N. Baeth, A. Geroldinger, D.J. Grynkiewicz, D. Smertnig, A semigroup-theoretical view of direct-sum decompositions and associated combinatorial problems. J. Algebra Appl. 14(2), 1550016, 60 pp. (2015)
15. P. Baginski, S.T. Chapman, M.T. Holden, T.A. Moore, Asymptotic elasticity in atomic monoids. Semigroup Forum **72**(1), 134–142 (2006)
16. P. Baginski, S.T. Chapman, R. Rodriguez, G. Schaeffer, Y. She, On the delta set and catenary degree of Krull Monoids with infinite cyclic divisor class group. J. Pure Appl. Algebra **214**, 1334–1339 (2010)

D. J. Grynkiewicz, *The Characterization of Finite Elasticities*, Lecture Notes in Mathematics 2316, https://doi.org/10.1007/978-3-031-14869-9

17. M. Banister, J. Chaika, S.T. Chapman, W. Meyerson, A theorem on accepted elasticity in certain local arithmetical congruence monoids. Abh. Math. Semin. Univ. Hambg. **79**(1), 79–86 (2009)
18. T. Barron, C. O'Neill, R. Pelayo, On the set of elasticities in numerical monoids. Semigroup Forum **94**(1), 37–50 (2017)
19. M. Batell, J. Coykendall, Elasticity in polynomial-type extensions. Proc. Edin. Math. Soc. **59**(3), 581–590 (2016)
20. W. Bonnice, V.L. Klee, The generation of convex hulls. Math. Ann. **152**, 1–29 (1963)
21. W. Bonnice, J.R. Reay, Relative interiors of convex hulls. Proc. Amer. Math. Soc. **20**, 246–250 (1969)
22. N. Bourbaki, Éléments de mathématique. Fasc. XXXI. Algèbre commutative, Actualités Scientifiques et Industrielles, No. 1314 Hermann, Paris (1965)
23. N. Bourbaki, *Commutative Algebra*, Chapters 1–7, Translated from the French, Reprint of the 1989 English translation, Elements of Mathematics (Berlin) (Springer, Berlin, 1998)
24. W. Bruns, J. Gubeladze, *Polytopes, Rings, and K-Theory* (Springer, Berlin, 2009)
25. V. Buchstaber, T. Panov, *Toric Topology*. Mathematical Surveys and Monographs, vol. 204 (American Mathematical Society, Providence, 2015)
26. P.J. Cahen, J.L. Chabert, Elasticity for integer-valued polynomials. J. Pure Appl. Algebra **103**, 303–311 (1995)
27. J.W.S. Cassels, *An Introduction to the Geometry of Numbers*. Die Grundlehren der Mathematischen Wissenschaften, Band 99 (Springer, Berlin, 1959)
28. S.T. Chapman, B. McClain, Irreducible polynomials and full elasticity in rings of integer-valued polynomials. J. Algebra **293**, 595–610 (2005)
29. S.T. Chapman, W.W. Smith, On factorization in block monoids formed by $\{\overline{1}, \overline{a}\}$ in $\mathbb{Z}_n$. Proc. Edinb. Math. Soc. **46**(2), 257–267 (2003)
30. S.T. Chapman, J.I. García-García, P.A. García-Sánchez, J.C. Rosales, Computing the elasticity of a Krull Monoid. Linear Algebra Appl. **336**, 191–200 (2001)
31. S.T. Chapman, F. Halter-Koch, U. Krause, Inside factorial monoids and integral domains. J. Algebra **252**, 350–375 (2002)
32. S.T. Chapman, U. Krause, E. Oeljeklaus, On Diophantine monoids and their class groups. Pac. J. Math. **207**, 125–147 (2002)
33. S.T. Chapman, M.T. Holden, T.A. Moore, Full elasticity in atomic monoids and integral domains. Rocky Mountain J. Math. **36**(5), 1437–1455 (2006)
34. M. Conforti, G. Cornuéjols, G. Zambelli, Polyhedral approaches to mixed integer programming, in *50 Years of Integer Programming: From the Early Years to the State-of-the-Art*, ed. by M. Jünger, T.M. Liebling, D. Naddef, G.L. Nemhauser, W.R. Pulleyblank, G. Reinelt, G. Rinaldi, L.A. Wolsey (Springer, Berlin, 2008)
35. A.R. Conn, K. Scheinberg, L.N. Vicente, *Introduction to Derivative-Free Optimization*. MPS-SIAM Book Series on Optimization (SIAM, Philadelphia, 2009)
36. C. Davis, Theory of positive linear dependence. American J. Math. **76**(4), 733–746 (1954)
37. J.A. De Loera, X. Goaoc, F. Meunier, N.H. Mustafa, The discrete yet ubiquitous theorems of Carathéodory, Helly, Sperner, Tucker, and Tverberg. Bull. Amer. Math. Soc. **56**(3), 415–511 (2019)
38. G. Denge, X. Zeng, Elasticities of Krull monoids with infinite cyclic class groups. J. Commut. Alg. **13**(3), 449–459 (2021)
39. P. Diaconis, R.L. Graham, B. Sturmfels, Primitive partition identities, in *Combinatorics, Paul Erdős is Eighty*, vol. 1 (János Bolyai Mathematical Society, Budapest, 1993), pp. 173–192
40. P. Eakin, W. Heinzer, More noneuclidean PID's and Dedekind domains with prescribed class groups. Proc. Am. Math. Soc. **40**, 66–68 (1973)
41. J. Eckhoff, Helly, radon, and carathéordory type theorems, in *Handbook of Convex Geometry* (Elsevier, Amsterdam, 1993)
42. A. Facchini, Direct sum decomposition of modules, semilocal endomorphism rings, and Krull Monoids. J. Algebra **256**, 280–307 (2002)

43. A. Facchini, Geometric regularity of direct-sum decompositions in some classes of modules. J. Math. Sci. **139**, 6814–6822 (2006)
44. A. Facchini, *Semilocal Categories and Modules with Semilocal Endomorphism Rings*. Progress in Mathematics, vol. 331 (Birkhäuser/Springer, Cham, 2019)
45. A. Facchini, R. Wiegand, Direct-sum decomposition of modules with semilocal endomorphism rings. J. Algebra **274**, 689–707 (2004)
46. W. Fenchel, *Convex Cones, Sets, and Functions* (Princeton University, Princeton, 1953)
47. R.M. Fossum, *The Divisor Class Group of a Krull Domain*. Ergebnisse der Mathematik und ihrer Grenzgebiete, vol. 74 (Springer, New York, 1973)
48. M. Freeze, A. Geroldinger, Unions of sets of lengths. Funct. Approx. Comment. Math. **39**, part 1, 149–162 (2008)
49. M. Freeze, W.A. Schmid, Remarks on a generalization of the Davenport constant. Discrete Math. **310**, 3373–3389 (2010)
50. C. Frei, S. Frisch, Non-unique factorization of polynomials over residue class rings of the integers. Commut. Algebra **39**(4), 1482–1490 (2011)
51. S. Frisch, Relative polynomial closure and monadically Krull Monoids of integer-valued polynomials, in *Multiplicative Ideal Theory and Factorization Theory*, ed. by S.T. Chapman, M. Fontana, A. Geroldinger, B. Olberding (Springer, Berlin, 2016), pp. 145–157
52. A. Fujiwara, J. Gibson, M.O. Jenssen, D. Montealegre, V. Ponomarenko, A. Tenzer, Arithmetic of congruence monoids. Commut. Algebra **44**(8), 3407–3421 (2016)
53. W. Gao, A. Geroldinger, Zero-sum problems in finite abelian groups: a survey. Expositiones Mathematicae **24**(4), 337–369 (2006)
54. W. Gao, A. Geroldinger, On products of k atoms. Monatsh. Math. **156**, 141–157 (2009)
55. W. Gao, A. Geroldinger, W.A. Schmid, Local and global tameness in Krull Monoids. Commut. Algebra **43**(1), 262–296 (2015)
56. P.A. García-Sánchez, D. Llena, J.C. Rosales, Strongly taut finitely generated monoids. Monatsh. Math. **155**(2), 119–124 (2008)
57. P.A. García Sánchez, I. Ojeda, J.C. Rosales, Affine semigroups having a unique Betti element. J. Algebra Appl. **12**(3), 1250177 (2013)
58. A. Geroldinger, Chains of factorizations in weakly krull domains. Colloq. Math. **72**, 53–81 (1997)
59. A. Geroldinger, Non-commutative Krull Monoids: a divisor theoretic approach and their arithmetic. Osaka J. Math. **50**, 503–539 (2013)
60. A. Geroldinger, Sets of lengths. Amer. Math. Month. **123**, 960–988 (2016)
61. A. Geroldinger, D.J. Grynkiewicz, On the arithmetic of Krull Monoids with finite Davenport constant. J. Algebra **321**, 1256–1284 (2009)
62. A. Geroldinger, F. Halter-Koch, *Non-Unique Factorizations: Algebraic, Combinatorial and Analytic Theory*, Pure and Applied Mathematics: A Series of Monographs and Textbooks, vol. 278 (Chapman & Hall, Boca Raton, 2006), an imprint of Taylor & Francis Group, Boca Raton, FL
63. A. Geroldinger, F. Kainrath, On the arithmetic of tame monoids with applications to Krull Monoids and Mori domains. J. Pure Appl. Algebra **214**, 2199–2218 (2010)
64. A. Geroldinger, F. Kainrath, On transfer homomorphisms of Krull monoids. Boll. Unione Mat. Ital. **14**, 629–646 (2021)
65. A. Geroldinger, G. Lettl, Factorization problems in semigroups. Semigroup Forum **40**, 23–38 (1990)
66. A. Geroldinger, I. Ruzsa, *Combinatorial Number Theory and Additive Group Theory* (Birkhâuser, Basel, 2009)
67. A. Geroldinger, P. Yuan, The monotone catenary degree of Krull Monoids. Result. Math. **63**, 999–1031 (2013)
68. A. Geroldinger, Q. Zhong, Long sets of lengths with maximal elasticity. Can. J. Math. **70**, 1284–1318 (2018)
69. A. Geroldinger, Q. Zhong, Sets of arithmetical invariants in transfer Krull Monoids. J. Pure Appl. Algebra **223**, 3889–3918 (2019)

70. A. Geroldinger, Q. Zhong, Factorization theory in commutative monoids. Semigroup Forum **100**, 22–51 (2020)
71. A. Geroldinger, D.J. Grynkiewicz, G.J. Schaeffer, W.A. Schmid, On the arithmetic of Krull Monoids with infinite cyclic class group. J. Pure and Appl. Alg. **214**(12), 2219–2250 (2010)
72. A. Geroldinger, F. Kainrath, A. Reinhart, Arithmetic of seminormal weakly Krull Monoids and Domains. J. Algebra **444**, 201–245 (2015)
73. A. Geroldinger, W. Schmid, Q. Zhong, Systems of sets of lengths: transfer krull monoids versus weakly krull monoids, in *Rings, Polynomials, and Modules* (Springer, Cham, 2017), pp. 191–235
74. M. Giaquinta, G. Modica, *Mathematical Analysis, Foundations and Advanced Techniques for Functions of Several Variables* (Birkhäuser, Basel, 2012)
75. R. Gilmer, W. Heinzer, W.W. Smith, On the distribution of prime ideals within the ideal class group. Houston J. Math. **22**, 51–59 (1996)
76. K.R. Goodearl, *von Neumann Regular Rings*. Monographs and Studies in Mathematics, vol. 4 (Pitman Advanced Publishing Program, Boston, 1979)
77. F. Gotti, C. O'Neill, The elasticity of Puiseux monoids. J. Commut. Alg. **12**(3), 319–331 (2020)
78. P. Gritzmann, J.M. Wills, Lattice points, in *Handbook of Convex Geometry* (Elsevier, Amsterdam, 1993)
79. P.M. Gruber, Geometry of numbers, in *Handbook of Convex Geometr* (Elsevier, Amsterdam, 1993)
80. P.M. Gruber, *Convex and Discrete Geometry*, Grundlehren der Mathematischen Wissenschaften [Fundamental Principles of Mathematical Sciences], vol. 336 (Springer, Berlin, 2007)
81. P.M. Gruber, J.M. Wills (Eds.), *Handbook of Convex Geometry* (Elsevier, Amsterdam, 1993)
82. D.J. Grynkiewicz, *Structural Additive Theory*. Developments in Mathematics, vol. 30 (Springer, Berlin, 2013)
83. J. Haarmann, A. Kalauli, A. Moran, C. O'Neill, R. Pelayo, Factorization properties of Leamer monoids. Semigroup Forum **89**(2), 409–421 (2014)
84. F. Halter-Koch, Elasticity of factorizations in atomic monoids and integral domains. J. Théor. Nombres Bordeaux **7**(2), 367–385 (1995)
85. F. Halter-Koch, Finitely generated monoids, finitely primary monoids, and factorization properties of integral domains, in *Factorization in Integral Domains*. Lecture Notes in Pure and Applied Mathematics (Iowa City, IA, 1996), vol. 189 (Dekker, New York, 1997), pp. 31–72
86. F. Halter-Koch, *Ideal Systems: An Introduction to Multiplicative Ideal Theory* (Marcel Dekker, New York, 1998)
87. W. Hansen, V. Klee, Intersection theorems for positive sets. Proc. Amer. Math. Soc. **22**, 450–457 (1969)
88. W. Hassler, Factorization properties of Krull Monoids with infinite class group. Colloq. Math. **92**, 229–242 (2002)
89. W. Hassler, Elasticity of factorizations in $R_0 + XR_1 + \ldots + X^l R_l[X]$. Result. Math. **41**, 316–319 (2002)
90. M. Hazewinkel, N. Gubareni, V.V. Kirichenko, *Algebras, rings and Modules. Vol. 1*. Mathematics and its Applications, vol. 575 (Kluwer Academic Publishers, Dordrecht, 2004)
91. T. Hungerford, *Algebra*. Graduate Textbooks in Mathematics, vol. 73 (Springer, New York, 1974)
92. E. Jespers, J. Okniński, *Noetherian Semigroup Algebras*. Algebra and Applications, vol. 7 (Springer, Berlin, 2007)
93. F. Kainrath, Elasticity of finitely generated domains. Houston J. Math. **31**, 43–64 (2005)
94. F. Kainrath, Arithmetic of Mori domains and monoids: the global case, in *Multiplicative Ideal Theory and Factorization Theory*, Springer Proceedings in Mathematics Statistics, vol. 170 (Springer, Berlin, 2016), pp. 183–218

95. K.M. Kattchee, On factorization in Krull domains with divisor class group $\mathbb{Z}_{2^k}$, in *Arithmetical Properties of Commutative Rings And Monoids*. Lecture Notes in Pure and Applied Mathematics, vol. 241 (Chapman & Hall/CRC, Boca Raton, 2005), pp. 325–336
96. H. Kim, The distribution of prime divisors in Krull monoid domains. J. Pure Appl. Algebra **155**(2–3), 203–210 (2001)
97. H. Kim, Elasticity of factorization in graded integral domains, in *Commutative Rings* (Nova Science Publishers, Hauppauge, 2002), pp. 183–191
98. J.B. Kruskal, The theory of well-quasi-ordering: a frequently discovered concept. J. Combin. Theor. Ser. A **13**, 297–305 (1972)
99. J.L. Lambert, Une borne pour les générateurs des solutions entiéres positives d'une équation diophantienne linéaire. C. R. Acad. Sci. Paris Ser. I **305**, 39–40 (1987)
100. S. Lang, *Fundamentals of Diophantine Geometry* (Springer, Berlin, 1983)
101. S. Lang, *Algebra*. Graduate Textbooks in Mathematics, vol. 211, rev. 3rd edn. (Springer, Berlin, 2000)
102. G. Lettl, Subsemigroups of finitely generated groups with divisor-theory. Monatsh. Math. **106**, 205–210 (1988)
103. R. McKinney, Positive bases for linear spaces. Trans. Amer. Math. Soc. **103**, 131–148 (1962)
104. C. O'Neill, R. Pelayo, Factorization invariants in numerical monoids, in *Algebraic and Geometric Methods in Discrete Mathematics*. Contemporary Mathematics, vol. 685 (American Mathematical Society, Providence, 2017), pp. 231–249
105. J.R. Reay, *Generalizations of a Theorem of Carathéodory*. Memoirs of the American Mathematical Society, vol. 54 (American Mathematical Society, Providence, 1965)
106. J.R. Reay, A new proof of the Bonnice-Klee theorem. Proc. Amer. Math. Soc. **16**(4), 585–587 (1965)
107. J.R. Reay, Unique minimal representations with positive bases. Amer. Math. Month. **73**, 253–261 (1966)
108. A. Reinhart, On monoids and domains whose monadic submonoids are Krull, in *Commutative Algebra: Recent Advances in Commutative Rings, Integer-Valued Polynomials, and Polynomial Functions* (Springer, Berlin, 2014), pp. 307–330
109. A. Reinhart, On the divisor-class group of monadic submonoids of rings of integer-valued polynomials. Commun. Korean Math. Soc. **32**, 233–260 (2017)
110. R. Regis, On the properties of positive spanning sets and positive bases. Optim. Eng. **17**(1), 229–262 (2016)
111. R.T. Rockafellar, The elementary vectors of a subspace of $\mathbb{R}^N$, in *Combinatorial Mathematics and its Applications* (University of North Carolina Press, Chapel Hill, 1969), pp. 104–127
112. S. Roman, *Lattices and Ordered Sets* (Springer, New York, 2008)
113. J.C. Rosales, P.A. García-Sánchez, J.I. García-García, Atomic commutative monoids and their elasticity. Semigroup Forum **68**(1), 64–86 (2004)
114. W.A. Schmid, Some recent results and open problems on sets of lengths of Krull Monoids with finite class group, in *Multiplicative Ideal Theory and Factorization Theory* (Springer, Berlin, 2016), pp. 323–352
115. G.C. Shephard, Diagrams for positive bases. J. Lond. Math. Soc. **4**(2), 165–175 (1971)
116. D. Smertnig, Sets of lengths in maximal orders in central simple algebras. J. Algebra **390**, 1–43 (2013)
117. D. Smertnig, Factorizations of elements in noncommutative rings: a survey, in *Multiplicative Ideal Theory and Factorization Theory* (Springer, Berlin, 2016), pp. 353–402
118. D. Smertnig, Factorizations in bounded hereditary noetherian prime rings. Proc. Edinburgh Math. Soc. **62**, 395–442 (2019)
119. V. Soltan, *Lectures on Convex Sets* (World Scientific, Hackensack, 2015)
120. B. Sturmfels, *Gröbner Bases and Convex Polytopes*. University Lecture Series, vol. 8 (American Mathematical Society, Providence, 1996)

121. R.J. Valenza, Elasticity of factorization in number fields. J. Number Theory **36**(2), 212–218 (1990)
122. Q. Zhong, On elasticities of locally finitely generated monoids. J. Algebra **534**, 145–167 (2019)
123. Q. Zhong, On the arithmetic of Mori monoids and domains. Glasgow Math. J. **62**, 313–322 (2020)

Index

D. J. Grynkiewicz, *The Characterization of Finite Elasticities*, Lecture Notes in Mathematics 2316, https://doi.org/10.1007/978-3-031-14869-9

LECTURE NOTES IN MATHEMATICS

Springer

Editors in Chief: J.-M. Morel, B. Teissier;

Editorial Policy

1. Lecture Notes aim to report new developments in all areas of mathematics and their applications – quickly, informally and at a high level. Mathematical texts analysing new developments in modelling and numerical simulation are welcome.

   Manuscripts should be reasonably self-contained and rounded off. Thus they may, and often will, present not only results of the author but also related work by other people. They may be based on specialised lecture courses. Furthermore, the manuscripts should provide sufficient motivation, examples and applications. This clearly distinguishes Lecture Notes from journal articles or technical reports which normally are very concise. Articles intended for a journal but too long to be accepted by most journals, usually do not have this "lecture notes" character. For similar reasons it is unusual for doctoral theses to be accepted for the Lecture Notes series, though habilitation theses may be appropriate.

2. Besides monographs, multi-author manuscripts resulting from SUMMER SCHOOLS or similar INTENSIVE COURSES are welcome, provided their objective was held to present an active mathematical topic to an audience at the beginning or intermediate graduate level (a list of participants should be provided).

   The resulting manuscript should not be just a collection of course notes, but should require advance planning and coordination among the main lecturers. The subject matter should dictate the structure of the book. This structure should be motivated and explained in a scientific introduction, and the notation, references, index and formulation of results should be, if possible, unified by the editors. Each contribution should have an abstract and an introduction referring to the other contributions. In other words, more preparatory work must go into a multi-authored volume than simply assembling a disparate collection of papers, communicated at the event.

3. Manuscripts should be submitted either online at www.editorialmanager.com/lnm to Springer's mathematics editorial in Heidelberg, or electronically to one of the series editors. Authors should be aware that incomplete or insufficiently close-to-final manuscripts almost always result in longer refereeing times and nevertheless unclear referees' recommendations, making further refereeing of a final draft necessary. The strict minimum amount of material that will be considered should include a detailed outline describing the planned contents of each chapter, a bibliography and several sample chapters. Parallel submission of a manuscript to another publisher while under consideration for LNM is not acceptable and can lead to rejection.

4. In general, **monographs** will be sent out to at least 2 external referees for evaluation.

   A final decision to publish can be made only on the basis of the complete manuscript, however a refereeing process leading to a preliminary decision can be based on a pre-final or incomplete manuscript.

   Volume Editors of **multi-author works** are expected to arrange for the refereeing, to the usual scientific standards, of the individual contributions. If the resulting reports can be forwarded to the LNM Editorial Board, this is very helpful. If no reports are forwarded or if other questions remain unclear in respect of homogeneity etc, the series editors may wish to consult external referees for an overall evaluation of the volume.

5. Manuscripts should in general be submitted in English. Final manuscripts should contain at least 100 pages of mathematical text and should always include

   – a table of contents;
   – an informative introduction, with adequate motivation and perhaps some historical remarks: it should be accessible to a reader not intimately familiar with the topic treated;
   – a subject index: as a rule this is genuinely helpful for the reader.
   – For evaluation purposes, manuscripts should be submitted as pdf files.

6. Careful preparation of the manuscripts will help keep production time short besides ensuring satisfactory appearance of the finished book in print and online. After acceptance of the manuscript authors will be asked to prepare the final LaTeX source files (see LaTeX templates online: https://www.springer.com/gb/authors-editors/book-authors-editors/manuscriptpreparation/5636) plus the corresponding pdf- or zipped ps-file. The LaTeX source files are essential for producing the full-text online version of the book, see http://link.springer.com/bookseries/304 for the existing online volumes of LNM). The technical production of a Lecture Notes volume takes approximately 12 weeks. Additional instructions, if necessary, are available on request from lnm@springer.com.

7. Authors receive a total of 30 free copies of their volume and free access to their book on SpringerLink, but no royalties. They are entitled to a discount of 33.3 % on the price of Springer books purchased for their personal use, if ordering directly from Springer.

8. Commitment to publish is made by a *Publishing Agreement*; contributing authors of multiauthor books are requested to sign a *Consent to Publish form*. Springer-Verlag registers the copyright for each volume. Authors are free to reuse material contained in their LNM volumes in later publications: a brief written (or e-mail) request for formal permission is sufficient.

Addresses:

Professor Jean-Michel Morel, CMLA, École Normale Supérieure de Cachan, France
E-mail: moreljeanmichel@gmail.com

Professor Bernard Teissier, Equipe Géométrie et Dynamique,
Institut de Mathématiques de Jussieu – Paris Rive Gauche, Paris, France
E-mail: bernard.teissier@imj-prg.fr

Springer: Ute McCrory, Mathematics, Heidelberg, Germany,
E-mail: lnm@springer.com

Printed in the United States
by Baker & Taylor Publisher Services